MW00604392

SIGNAL PROCESSING FOR WIRELESS COMMUNICATIONS

ABOUT THE AUTHOR

Joseph Boccuzzi, Ph.D., is a Principal Systems Design Scientist with Broadcom Corp. He previously worked as an architect and a technical director at Infineon/Morphics Technology, a senior technical manager at Cadence Design Systems, a member of the technical staff at AT&T Bell Labs, a design engineer at Motorola, and an engineer at Eaton Corp. Dr. Boccuzzi also teaches graduate-level courses in electrical engineering; has written numerous articles and papers for the IEEE, AT&T, and international conferences; holds over 15 domestic and international patents; and has made public presentations to various organizations worldwide.

SIGNAL PROCESSING FOR WIRELESS COMMUNICATIONS

Joseph Boccuzzi, Ph.D.

New York Chicago San Francisco Lisbon London Madrid
Mexico City Milan New Delhi San Juan Seoul
Singapore Sydney Toronto

The McGraw-Hill Companies

Library of Congress Cataloging-in-Publication Data

Boccuzzi, Joseph.
Signal processing for wireless communications / Joseph Boccuzzi.
p. cm.
ISBN 978-0-07-148905-8 (alk. paper)
1. Signal processing. 2. Wireless communication systems. I. Title.
TK5102.9.B575 2008
621.382′2—dc22 2007038835

1 2 3 4 5 6 7 8 9 0 FGR/FGR 0 1 3 2 1 0 9 8 7

ISBN 978-0-07-148905-8
MHID 0-07-148905-3

Printed and bound by Quebecor/Fairfield.

This book is printed on acid-free paper.

McGraw-Hill books are available at special quantity discounts to use as premiums and sales promotions, or for use in corporate training programs. For more information, please write to the Director of Special Sales, McGraw-Hill Professional, Two Penn Plaza, New York, NY 10121-2298. Or contact your local bookstore.

Sponsoring Editor
Stephen S. Chapman

Production Supervisor
Pamela A. Pelton

Editing Supervisor
Stephen M. Smith

Project Manager
Gita Raman

Copy Editor
Yumnam Ojen

Proofreader
Barbara Danziger

Indexer
Valerie Perry

Art Director, Cover
Jeff Weeks

Composition
International Typesetting and Composition

To my wonderful loving family Ninamarie, Giovanni, and Giacomo

CONTENTS

PREFACE

The proliferation of wireless digital communications has created a foundation of what will become an unprecedented demand for capacity while simultaneously supporting global mobility. This growth of ubiquitous systems will force designers to make both challenging and difficult technology decisions. As in any field, decisions made with a sound engineering foundation will prevail. It is for this reason that I have written this book.

The topics covered in this book are intended to provide the reader with a good understanding of the issues designers are faced with every day. The book's philosophy is to start with a comparison of modulation schemes and continue to the multipath fading channel, where computer simulation models are contrasted. Various coherent and noncoherent demodulation techniques used in the receiver are presented and explained in detail. Next, performance improvement techniques are provided followed by select digital signal processing functions. An overview of the WCDMA cellular system is provided prior to presenting its evolutionary path. A comparison of computer simulation techniques used in error estimation is discussed. The intention is that the book be used for graduate level courses in engineering as well as a reference for engineers and scientists designing wireless digital communications systems.

Chapter 1 provides an overview of some of the past and present wireless communications systems used. This introduction is used to compare certain design choices and identify system parameters that are essential in reliable communications. The systems covered include personal area networks, wireless local area networks, cellular communications, and paging systems. A discussion on the trend of mobile devices is provided and, in this context, reconfigurable receivers are presented to aid support of the multiple wireless standards required for acceptable communications.

Chapter 2 presents various modulation schemes from the perspective of spectral efficiency, receiver complexity, and link performance. In addition, an evolutionary path is provided highlighting the roadmap for the modulation scheme migration. Treatment of this chapter from the perspective of the transmitter lends itself to discussions on spectral regrowth issues. This is made possible with a model of the nonlinear transmit power amplifier.

Chapter 3 emphasizes the wireless communication channel model. An introduction of frequency flat and frequency selective fading is presented. A concise discussion on propagation path loss of both micro- and macro-cell deployments in various countries worldwide is provided. Moreover, both indoor and outdoor delay spread measurements are summarized and presented in an easy to understand and coherent fashion. A discussion of frequency dependency on channel model phenomenon (i.e., path loss, shadowing, and so forth) is provided. Lastly, computer simulation models for the multipath fading channel are given to support the wireless system designer in computer simulations.

Chapter 4 turns our attention to the receiver by presenting various demodulation techniques for both coherent and noncoherent detection. This comparison will be made on the grounds of bit error rate (BER) performance. II/4-DQPSK and GMSK were the two chosen exemplary modulation schemes; however, with slight modifications, the detection techniques can be used for other modulation schemes. An effort to unify the BER performance for each of the detectors chosen is given.

Chapter 5 presents topics on performance improvement. In many applications the resulting system performance based on the above material is insufficient to provide a reliable communication link. Hence performance improvement techniques are discussed that allow the system designer to be able to create a reliable communication system. Interleaving and de-interleaving operations are provided to emphasize their ability to redistribute bursts of errors. Forward error correction techniques are discussed in depth to compare block codes, convolutional codes, Reed-Solomon codes, turbo codes, and puncturing methods. An attempt to compare their performances to concatenated coding is undertaken to emphasize the proximity of the operating point to the Shannon capacity bound. Next, a treatment on spatial antenna diversity-combining techniques including switching, maximal ratio combining, and optimal combining is provided. The BER performance of all these techniques is compared and theoretical equations are provided where necessary. Lastly, two methods of space-time transmit diversity are discussed. At this point a link budget example can be leveraged into the discussions and used as a demonstration vehicle as to where performance improvements can have significant impact on the overall system.

Chapter 6 discusses various digital signal-processing algorithms typically used in the wireless digital communications systems. Equalization is presented as a means to combat the inter-symbol interference of the frequency-selective fading channel. An introduction to LMS, RLS, and DMI is given in the context of equalization training and tracking. The performance of space-time equalization in various channel environments is compared. The eigenspectral decomposition tool is used to emphasize the relationship between performance improvement and eigenvalues. The equivalence of the maximum likelihood sequence estimator and various antenna combing methods is provided. As the wireless systems evolve toward orthogonal frequency-division multiplexing, performing the signal processing functions in the frequency domain may be more feasible than performing them in the time domain. It is for this reason that an introduction to frequency-domain equalization is presented. Also, various symbol (and bit) timing recovery algorithms are provided and performance degradation due to timing offsets is discussed. Next, various channel quality estimation techniques are compared and their potential usage scenarios are provided. Lastly, a few frequency correction algorithms are provided and their performance is discussed.

Chapter 7 provides an in-depth overview of the wideband CDMA (WCDMA) standard covering topics such as high-speed downlink and uplink packet access. This contains the network (UTRAN), UE, and physical layer introduction. DS-CDMA overview is provided with the North American spread-spectrum cellular system used as an example. PN code (Maximal Length, Gold, Walsh, Kasami, and so forth) generation methods and statistical properties are discussed. RAKE receiver and Generalized RAKE operations such as channel estimation and time tracking are presented. There is a discussion on the impact of inter-chip interference on the RAKE output, where a new derivation is presented. Several essential signal processing functions required in spread-spectrum systems are discussed such as multipath searching, RAKE finger combining, signal-to-interference ratio based power control (i.e., open, closed inner, and closed outer loop), transport format combination decoding, and access channel procedures. Simulation results for the link performance of advanced receivers are provided, as discussed by the 3GPP standard.

Chapter 8 provides a concise treatment of computer simulation techniques that are available to the system designer. As communication systems evolve and become increasingly complex, deriving a single, exact, mathematically tractable equation accurately describing system performance is difficult. Hence computer simulations have gained global acceptance as a method to evaluate system performance. The intention of this chapter is to make the designer aware of the multiple computer simulation techniques available. The chapter starts with the well-known Monte Carlo technique making no assumptions about the noise statistics, but having the caveat of long runtime. Various methods are introduced to reduce the simulation runtime, such as importance sampling, improved importance sampling, and tail extrapolation. Lastly, a semi-analytic approach is provided to offer the fastest computer simulation runtime benefit.

Chapter 9 provides insight into numerous topics such as orthogonal frequency-division multiplexing (OFDM) and the usage of such for third-generation long-term evolution (3G LTE) and Mobile TV. A discussion of delivering broadcasting services through the broadcast and multicast channels (MBMS) is provided. An overview of the DVB-H standard is presented for the sake of discussing delivery of Mobile TV services to the handheld terminal. A short comparison of competing delivery mechanisms is also presented (i.e., MediaFLO). As the networks and user equipment (UE) continue to evolve toward a packet-based system, it is imperative to address certain system related issues that arise from those techniques. So an overview of the 3GPP continuous packet connectivity (CPC) feature is provided to address certain UE dynamic behavior. A discussion of a canonical UE architecture is compared and performance bottlenecks are highlighted to reveal certain strains placed within the UE. The HSPA evolution (64QAM, 16QAM, MIMO, VoIP related, and so forth) is discussed along with the additional requirements they bring.

Last and definitely not least, the appendices are provided to contain, as completely as possible, various mathematical equations characterizing the BER performance for the modulation techniques discussed herein. This is intended to provide the reader with a single source of reference when considering the ideal performance.

Throughout the book I have used both theoretical and computer simulation results to explain the relevant concepts. I hope you enjoy the material and find this book interesting.

Joseph Boccuzzi, Ph.D.

ACKNOWLEDGMENTS

This book is a result of my 20 years of professional experience in the field of digital communications, having had the honor of working with reputable companies such as Eaton Corporation, Motorola, Inc., AT&T Bell Labs, Cadence Design Systems, Morphics Technology, Infineon Technologies, and Broadcom Corporation. Moreover, the material has been effectively taught over the years while I was an instructor for Besser Associates, and adjunct for Polytechnic University of New York. I have had tremendous feedback during these sessions, which have improved the content and delivery of this book.

I cannot conclude without offering my appreciation to my editor Steve Chapman of McGraw-Hill and to Gita Raman, project manager at ITC, for providing corrections, suggestions, and a continued sense of urgency to keep this book on schedule.

SIGNAL PROCESSING FOR WIRELESS COMMUNICATIONS

CHAPTER 1
WIRELESS TOPICS

1.1 INTRODUCTION

Wireless communications has enabled a variety of services starting from Voice continuing to Data and now to Multimedia. We are about to enter an era that has never been seen before as cell phones, laptops, cameras, personal digital assistants (PDAs), and televisions all converge into a potentially single consumer electronic device.

Since the conception of cellular communications, we have learned to enjoy the freedom that comes with mobility. The first generation cellular system was AMPS, Analog Mobile Phone System. It used the traditional frequency modulation (FM) scheme to enable communication among voice users. A user's speech was directly converted by an FM modulator to be sent over the wireless medium. This system had voice quality issues, in particular at the cell fringe, and suffered from low user capacity [1–3].

Here comes Digital! The second generation (2G) cellular systems used digital modulation schemes such as $\pi/4$-Shifted Differential Quaternary Phase Shift Keying ($\pi/4$-DQPSK) and Gaussian Filtered Minimum Shift Keying (GMSK). The speech was digitized, error protected, and transmitted in a specific time, frequency, or code domain. These advances helped improve voice quality and user capacity. These 2G systems are North American TDMA System (IS-136), Group Special Mobile (GSM), North American Spread Spectrum System (IS-95), and Japan Digital Cellular (JDC), to name a few, with GSM being the de facto Global Standard, in other words, the more widely used system throughout the world.

These 2G cellular systems suffered not only from low user capacity, but also from low user data rates, since applications were being created that really wanted higher data rates in order to fully exploit their benefits. With advances in signal processing and advancements in technology used in communication devices, the 2G systems had their lifetime extended into what has been commonly accepted as 2.5G, the second and one-half generation. Here, GSM evolved into General Packet Radio Service (GPRS) and Enhanced Data Rates for Global Evolution (EDGE) and still continues its evolution. Similarly, IS-95 evolved into CDMA2000 and it also continues to evolve.

During this time, various studies were undertaken in order to compare time division multiple access (TDMA) against code division multiple access (CDMA) for the next-generation cellular system. For reasons that will become apparent in latter sections, CDMA prevailed and we now have third generation (3G) systems based on wideband CDMA (WCDMA). WCDMA continued to increase user capacity and user data rates, thus further opening doors for wireless applications in both the circuit switch (CS) and packet switch (PS) domains.

As we progress and gain global adaptation of WCDMA/CDMA2000 systems, we continue to identify the fourth generation (4G) system which would not only continue on the road to increase data rates and user capacity, but also support a variety of mixed services. These accommodations will also be taken care of by improvements on the network side.

However, with history repeating itself, we see evolutions in the form of releases in the 3G systems. These releases particularly address the Internet Protocol (IP) services, along with increased user data rates to support such applications. In particular, we are referring to evolving WCDMA to HSDPA/HSUPA and continuing to Long Term Evolution (LTE) and CDMA2000 to 1xEVDO/DV (Evolution Data Optimized/Data Voice) systems.

1.2 WIRELESS STANDARDS OVERVIEW

This section will provide an overview of various wireless standards. It is our intention to present a few wireless standards that encompass the full spectrum of service capabilities. On one side of the spectrum, we have the Personal Area Network (PAN), which accommodates low-mobility users operating in a small coverage area. We then discuss the Local Area Network (LAN) and continue to the other side with the Wide Area Network (WAN), which accommodates high-mobility users operating in large coverage areas.

As we progress through our book, we will learn why certain system design decisions were made for particular wireless standards. It is also our intention to present 2G cellular systems as well as their respective 3G cellular systems. These overviews will serve as a reference as well as for the sake of comparison as we latter discuss the evolutionary paths of these standards.

Since the cellular standards have classically been devoted to voice communications, we would like to present a block diagram on a transmitter for such an application (see Fig. 1.1).

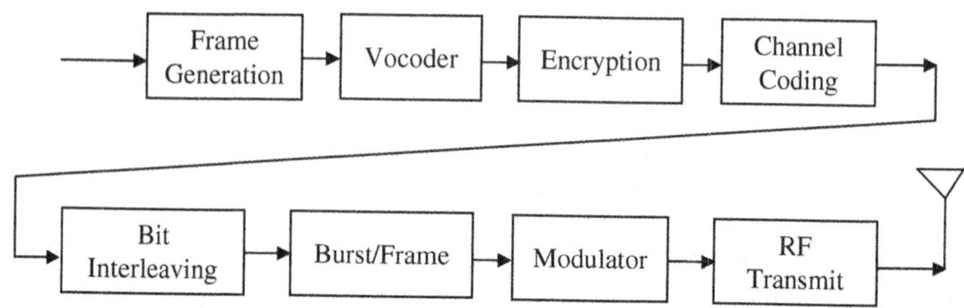

FIGURE 1.1 A typical transmit functional block diagram.

1.2.1 IS-136 North American TDMA

In this section, we will present the physical layer details for the North American Digital Cellular (NADC) TDMA system [4]. This system uses Frequency Division Duplex (FDD) for uplink and downlink communications. Below we list some high-level features of this digital cellular system:

Bit rate = 48.6 kbps (aggregate channel bit rate)

Modulation = $\pi/4$-DQPSK w/SRC filtering with a roll-off of $\alpha = 0.35$

VSELP 7.95-kbps voice coding

3 users per 30-kHz channel for full rate

6 users per 30-kHz channel for half rate

Beginning of Mobile Assisted Handoff (MAHO)

TDMA frame length = 40 msec

Forward error correction (FEC) on voice

In-band signaling (FACCH)

Out-of-band signaling (SACCH)

The time-slot relationship between the downlink and uplink is given in Fig. 1.2, but first understand that the downlink can be continuously transmitted while the uplink is only transmitted in bursts. The users will share the channel frequency bandwidth by being time division multiplexed onto the frame structure. A user will typically transmit twice (pair of time slots) per 40-msec frame, for example, the user data bits could be transmitted in slots 1 and 4, or slots 2 and 5, or slots 3 and 6. This mode of operation is called full rate. The half-rate mode is achieved when users occupy a single time slot during the 40-msec frame.

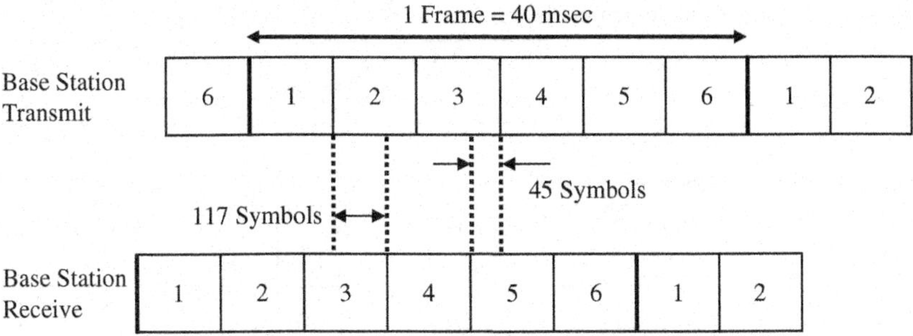

FIGURE 1.2 NADC uplink and downlink timing relationship.

We will use this standard to convey some important observations that were made when we evolved the AMPS to Digital [5]. In Fig. 1.3, we plot the voice quality as a function of carrier to interference ratio (C/I). There are quite a few ways that can be used in viewing the figure, we will present one. The x-axis can be viewed as we move closer to the serving base station, the C/I increases. Continuing along these lines, then moving away from the base station will produce a smaller C/I ratio, such as at

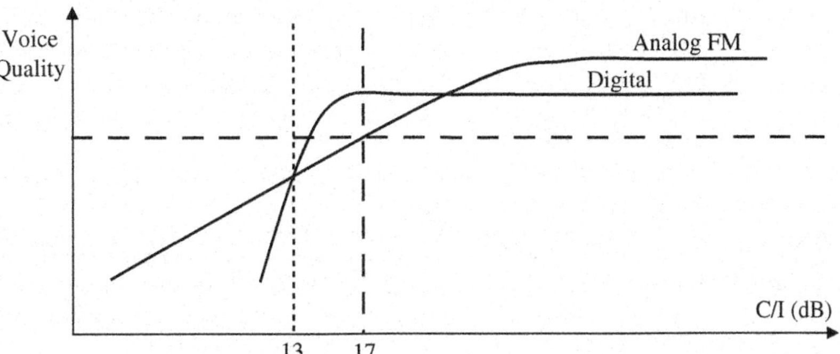

FIGURE 1.3 Voice quality performance.

the cell boundary. A method used to quantify the voice quality is to subject listeners to various samples of voice recordings under various channel conditions and then rate the voice quality on a scale from 1 to 5. This is commonly called the Mean Opinion Score (MOS[1]) for that particular vocoder.

When we are close to the base station, the analog system produces slightly better voice quality since the digital vocoder is a lossy compression algorithm. However, as we travel away from the cell site, the performance of the digital system is better because the system characteristics (i.e., error correction, interleaving) are able to compensate for multipath fading errors. Lastly, as we travel to the edge of the cell, depending on the system planning, the digital system has a sharper roll-off than the analog system. A reason for this is that the error correction technique used begins to break down. In other words, the amount of errors present exceeded the error correction capability of the code.

The time-slot structures for both uplink and downlink are given in Fig. 1.4. Each time slot contains 324 bits which is equivalent to 6.66 msec in time duration. The bit fields in the time-slot structures are defined as:

Guard = 6 bits are allocated to the guard field to prevent uplink transmitted time-slot bursts from overlapping.

[1] MOS voting has a 5-point rating with 5 = excellent, 4 = good, 3 = fair, 2 = poor, and 1 = unacceptable.

Base Station Transmit

Sync	SACCH	Data	CDVCC	Data	Rsvd
28	12	130	12	130	12

Base Station Receive

Guard	Ramp	Data	Sync	Data	SACCH	CDVCC	Data
6	6	16	28	122	12	12	122

FIGURE 1.4 Time-slot structure.

Ramp = 6 bits are allocated to the ramp field to allow the mobile transmitter to gradually reach its desired power level while maintaining emission mask requirements.

Sync = 28 bits are allocated to the sync field to allow the mobile and base station to obtain bit-slot synchronization in addition to a variety of signal processing functions such as channel estimation, equalizer training, frequency offset estimation, and so forth.

CDVCC = 12 coded bits are allocated to promote reliable control between the mobile and base station, Coded Digital Verification Color Code (CDVCC). The uncoded bits, DVCC, are retransmitted back to the base station in order to assist the base station in separating the desired traffic channel from a co-channel signal.

Rsvd = 12 bits were allocated as reversed for future usage. In the meantime, all zeros were transmitted in their places.

SACCH = 12 bits were allocated for the slow associated control channel (SACCH).

The modulation scheme used is $\pi/4$-DQPSK with the signal constellation diagram in Fig 1.5, where we have purposely drawn the axis states as circles and the nonaxis states as boxes; the reason will become apparent in the next chapter, where we discuss modulation theory [6, 7].

This type of differential encoding allows for four phase changes per symbol time interval. The four possible phase trajectories shown below will be controlled by the following differential encoding rule

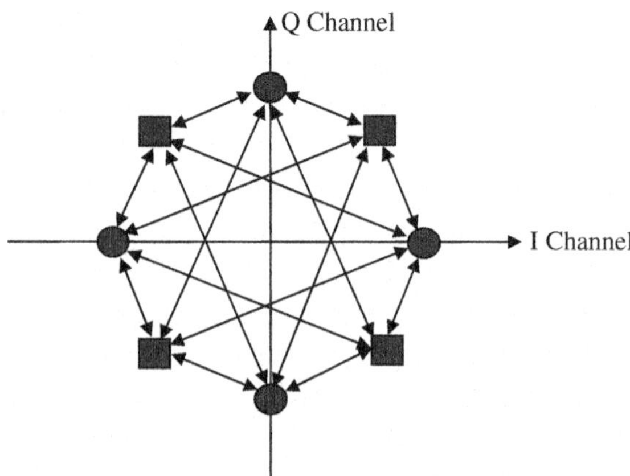

FIGURE 1.5 $\pi/4$-DQPSK signal constellation diagram.

given in the form of a phase state table (see Table 1.1). The input bit stream is converted to symbols or di-bit pairs using the mapping rule in the phase state table.

TABLE 1.1 $\pi/4$-DQPSK Phase State Table

Di-bit pair	Phase change ($\Delta\phi$)
0–0	$+\pi/4$
1–0	$-\pi/4$
0–1	$+3\pi/4$
1–1	$-3\pi/4$

1.2.2 Group Special Mobile (GSM)

In this section, we will provide the physical layer details of the European-born TDMA system, GSM, which has been adapted practically worldwide. This system uses FDD for uplink and downlink communications [8–11].

The time-slot structures for both uplink and downlink are given in Fig. 1.6. There are a total of 8 time slots per radio frame. The aggregate data rate is 270.8 kbps and the per user data rate is 33.85 kbps. The uplink and downlink frames have an offset of 3 time slots.

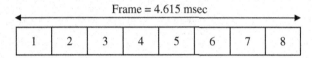

Frame = 4.615 msec

1	2	3	4	5	6	7	8

FIGURE 1.6 Time-slot structure.

The modulation scheme used is GMSK with the signal constellation diagram shown in Fig. 1.7. Some of the benefits of this modulation scheme are constant envelope, spectrally efficient, and excellent Bit Error Rate (BER) performance. We will discuss these properties and more in the next chapter.

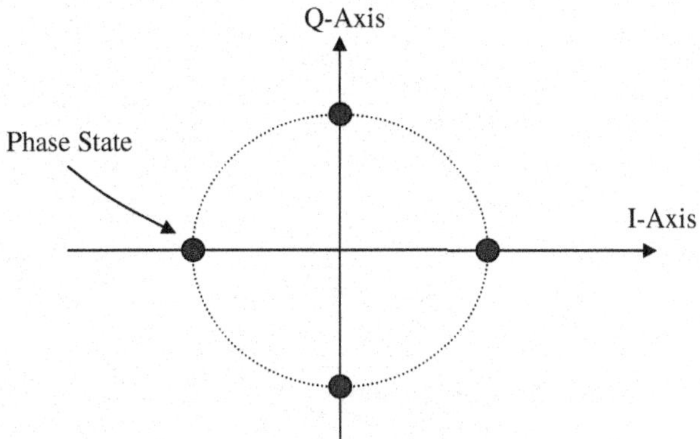

FIGURE 1.7 MSK signal constellation diagram.

The Gaussian premodulation filter response is given below

$$h(t) = \frac{\exp\left(\frac{-t^2}{2\partial^2 T^2}\right)}{\partial \cdot T \cdot \sqrt{2\pi}} \tag{1.1}$$

with the following defined variable

$$\partial = \frac{\sqrt{\ln(2)}}{2\pi \cdot BT} \tag{1.2}$$

Recall T is the bit time (3.69 μsec = 1/270.833 kbps), and B is the 3-dB bandwidth of the filter $h(t)$. This is usually defined together with the bit-time interval and given as $BT = 0.3$.

TABLE 1.2 An Overview of Some GSM System Parameters

System parameters	Values
Channel BW	200 kHz
Modulation scheme	GMSK
Channel bit rate	270.833 kbps
LPF	Gaussian $BT = 0.3$
Forward error correction	Convolutional coding
Raw data rate	13 kbps
Voice frame length	4.615 msec
Interleaving depth	40 msec
Users/Channel	8
Spectral efficiency	1.35 bps/Hz

The supported user data rates for full-rate mode are 13 and 12.2 kbps, and 5.6 kbps for the half rate. We provide some GSM system design parameters in Table 1.2. A simplified GSM architecture is given in Fig. 1.8. Handheld devices called Mobile Stations (MS) communicate with Base Station Transceiver Station (BTS). The BTS performs the modulation/demodulation (modem) functionality, provides synchronization signals (time and frequency), frequency hopping, and performs measurements.

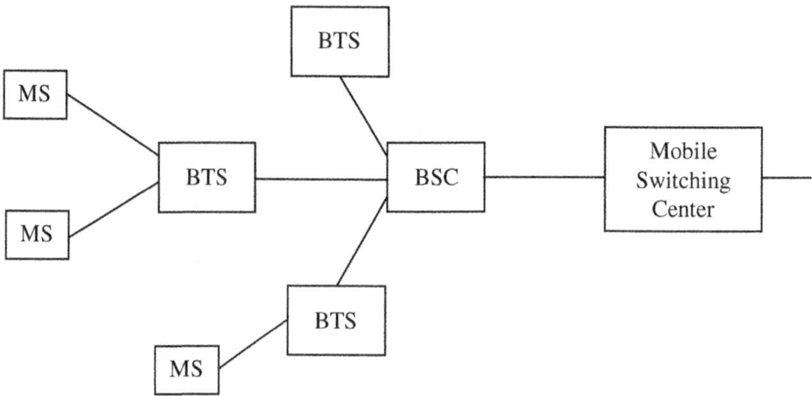

FIGURE 1.8 Simplified GSM architecture block diagram.

The demodulated user signals communicate to a Base Station Controller (BSC). The BSC is responsible for frequency channel assignment, time slots, and controls frequency hopping [12]. Notice a BSC can support multiple BTS to allow for handoffs to occur.

Next we aim to provide an overview of the logical and physical channels. The logical channels are grouped into either traffic channels (TCH) or control channels (CCH). The TCH will carry either speech or data information samples in the uplink and downlink. The CCH will carry control information, either signaling or synchronization samples. We provide a pictorial presentation in Fig. 1.9.

The **broadcast CCH (BCCH)** are used for the following purposes:

- Downlink only
- Broadcast system information to the MS (such as frequency hopping sequence and RF channels)
- Configurations of the CCH
- Always transmitted (not frequency hopped)
- Sends FCCH and SCH for frequency correction and frame synchronization info.

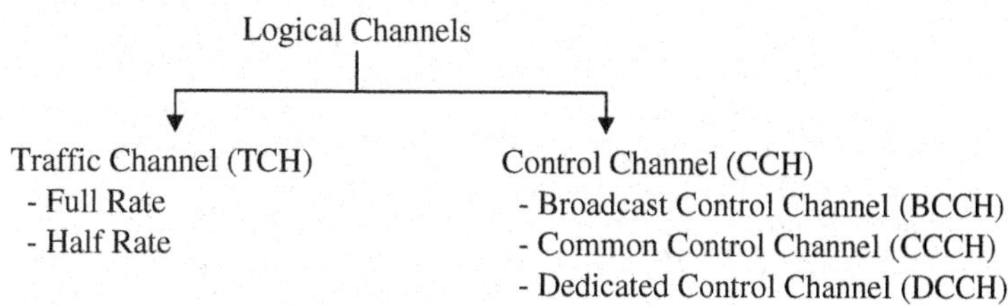

FIGURE 1.9 Logical channel partitioning.

The **common CCH (CCCH)** are used for the following purposes:

- Uplink and downlink
- MS access management information
- Sends PCH (paging channel) used to page the MS
- Sends AGCH (access grant channel) used to allocate SDCCH (stand-alone dedicated control channel) or TCH (traffic channel) to MS
- Receives RACH (random access channel) used for access request; can either be a response to a page or a call origination request. This is based on the Slotted Aloha Protocol.

The **dedicated CCH (DCCH)** are used in the following manner:

- Used for both uplink and downlink
- Sends/receives SDCCH for system signaling and call setup
- Sends/receives SACCH for measurement reporting and cell info
- Sends/receives FACCH for exchange info for rates higher than SACCH. The 20-msec speech is deleted and FACCH is inserted into the frame.

For us to define the time-slot structures, we will present the normal burst and synchronization burst time-slot structures (see Fig. 1.10).

Normal Time Slot. The training sequence is a known 26-bit pattern that can be used for a variety of signal processing reasons, for example, training of the equalizer and timing recovery. The guard

Normal Time Slot

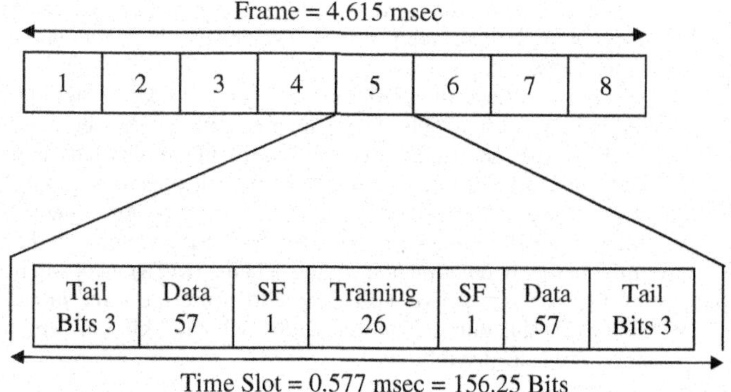

FIGURE 1.10 Normal burst time-slot structure.

period is equal to 156.25 − 148 = 8.25 bits, or 4.125 bits on either side of the burst, and is an empty space. These are inserted to avoid overlapping between adjacent time slots due to the MSs geographically located in different parts of the cell. A scenario will occur when two mobiles are using adjacent time slots and one is located near the BTS and the other is located far away. Due to the propagation time differences, the time slots would overlap in the absence of guard bits. The total guard time is 30.46 μsec in duration. The tail bits are always transmitted with values of 0. Lastly, there are two stealing flags (SF) associated with each on the interleaved data bits.

Synchronization Time Slot. This burst is used to aid time synchronization at the mobile, a 64-bit sequence is used. A block diagram of this time slot is shown in Fig. 1.11. The TDMA frame number and BTS Identification Code (BSIC) are transmitted. The tail bits and guard bits are the same. The training sequence is extended to 64 bits in duration.

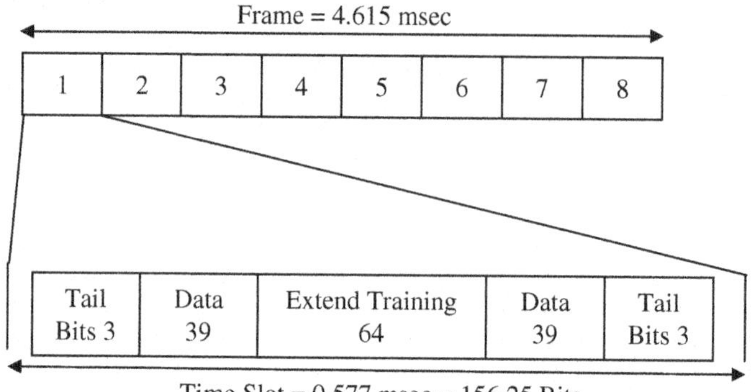

FIGURE 1.11 Synchronization time-slot structure.

EDGE Evolution. The original intention of GSM was to support CS-based services. In particular, they refer to voice and fax of up to 9.6 kbps. PS-based services will be better handled in the evolution of GSM, such as GPRS [13] and EDGE [14–16]. The benefits will take the form of increased user capacity, efficient usage of radio resources, higher throughput rate, and so forth. Multimedia-based services and Internet access are among the many usage scenarios. Obviously, all these targeted improvements must be able to coexist with the traditional voice network/service.

Due to the bursty nature of the PS services, a system that can dynamically allocate resources (i.e., time slots) depending on the user's demands is essential. This concept is called bandwidth on demand. GPRS keeps the modulation scheme to GMSK, but adds additional coding methods for varying input requirements. EDGE introduces 8-PSK modulation scheme as well as further coding methods for varying throughput requirements. The channel baud rate and time-slot structures remain unchanged at the physical layer.

Let us consider the EDGE system. For our sake, we will focus on the coding schemes and modulation schemes available. There are nine coding schemes shown in Table 1.3.

From the above table, we can see the theoretical maximum data rate is 59.2 kbps times 8 time slots = 473.6 kbps using 8-PSK. These schemes allow the network and MS to trade off data rate, depending on the channel conditions. The data rate changes occur by changing the modulation scheme and the amount of parity bits inserted (code rate) in the transmitted bit stream. These are captured in the modulation and coding scheme (MCS) values in the above table.

As discussed above, GSM and GPRS use the same modulation scheme, which is GMSK. The EDGE modulation scheme is 8-PSK where 3 bits are grouped together to form a symbol. The complex-valued symbol is given as

$$s(k) = \exp\left(j\frac{2\pi}{8}d(k)\right) \tag{1.3}$$

TABLE 1.3 Coding Parameters for EDGE

Scheme	Code rate	Modulation	Data rate per time slot (kbps)
MCS-9	1	8-PSK	59.2
MCS-8	0.92	8-PSK	54.4
MCS-7	0.76	8-PSK	44.8
MCS-6	0.49	8-PSK	29.6
MCS-5	0.37	8-PSK	22.4
MCS-4	1	GMSK	17.6
MCS-3	0.85	GMSK	14.8
MCS-2	0.66	GMSK	11.2
MCS-1	0.53	GMSK	8.8

The mapping of the bits to symbols are governed by the Gray coding rule, which is shown in the signal constellation diagram in Fig. 1.12.

Prior to performing the pulse-shaping operation, the symbols are rotated by $3\pi/8$ radians every symbol time. The pulse shaping for EDGE is based on a linearized GMSK pulse and is described as:

$$h(t) = \begin{cases} \prod_{i=0}^{3} S(t + iT) \ , & 0 \leq t \leq 5T \\ 0 \ , & \text{otherwise} \end{cases} \tag{1.4}$$

$$S(t) = \begin{cases} \sin\left(\pi \int_0^t g(x)\,dx\right) \ , & 0 \leq t \leq 4T \\ \sin\left(\frac{\pi}{2} - \pi \int_0^{t-4T} g(x)\,dx\right) \ , & 4T \leq t \leq 8T \\ 0 \ , & \text{otherwise} \end{cases} \tag{1.5}$$

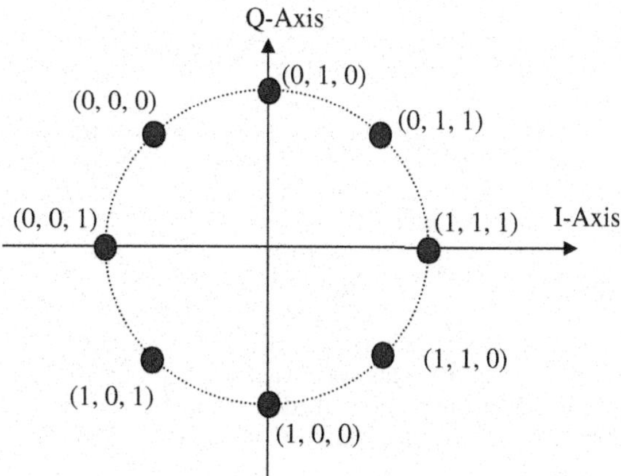

FIGURE 1.12 EDGE signal constellation diagram mapping for 8-PSK.

$$g(t) = \frac{1}{2T} \cdot \left[Q\left(2\pi \cdot BT \cdot \frac{t - 5T/2}{T\sqrt{\ln(2)}} \right) - Q\left(2\pi \cdot BT \cdot \frac{t - 3T/2}{T\sqrt{\ln(2)}} \right) \right] \quad (1.6)$$

$$Q(t) = \frac{1}{\sqrt{2\pi}} \int_{t}^{\infty} e^{-\frac{x^2}{2}} dx \quad (1.7)$$

A technique used in PS transmissions is called Incremental Redundancy (IR). This technique estimates the link quality and selects a strategic MCS combination in order to maximize the user bit rate. With IR, the packet is first sent with little coding overhead in order to produce a high bit rate, if decoding is successful. If packet decoding fails, additional coding (or redundancy) is sent until the packet is received successfully or a time-out occurs. Obviously, the more retransmission and coding performed, the lower is the user bit rate and higher is the overall system delay. Differing code rates are accomplished by puncturing different bits each time a packet retransmission occurs.

Some of the technical hurdles to overcome with EDGE (and GPRS) deployment are:

1. Linear power amplifiers (PAs): GSM terminals have enjoyed a quasi- constant envelope modulation scheme which in term allows us to design a transmitter with nonlinear PAs. With 8-PSK, the modulation schemes became linear, increasing the modulation scheme peak to average ratios, forcing us to use more linear and inefficient PAs.
2. Network upgrade: The network must now be upgraded not only to accommodate an increase in the throughput, but also to provide improved quality of service (QoS) for overall operation. The Serving GPRS Support Node (SGSN) and Gateway GPRS Support Node (GGSN) can handle CS/PS now.
3. BTS/BSC upgrades: The BTS will need to be upgraded to support additional MCS. There may be a need for cell size reduction on the system design due to effects of the interference, modulation scheme, and vehicle speed on link budget. With the move toward a PS domain, increased attention to buffering, latency, and scheduling will be required. The aggregate cell throughput will be dependent on the type and quality of the scheduling algorithm used.
4. Equalizer: A commonly used technique to mitigate and combat frequency selective fading (FSF) is to use an equalizer. The equalizer can take on the form of a decision feedback equalizer (DFE) or maximal likelihood sequential estimation (MLSE). The MLSE maintains a trellis or state machine that is dependent on not only the delay spread but also the number of modulation levels. Going from GMSK to 8-PSK dramatically increases implementation and computational complexity.
5. Cell planning: To offer coverage with sufficient C/I ratio in order to make use of the available bit rates. The higher-order modulation schemes will be more effective closer to the BTS transmitter than at the cell edge, due to their higher SNR requirements.
6. Terminal complexity: The need to maintain GSM voice + GPRS + EDGE in the MS will require a prudent hardware/software (HW/SW) partitioning design to handle higher data rates and multislot operations. Lastly, now that PS services are available, the MS must have the applications that make use of the increased data rates. Typical communications between the BTS and MS are asymmetric in nature, meaning the downlink (DL) has a higher data rate than the uplink (UL). When mentioning throughput, we must also address the IR memory required to store subsequent transmissions until a correct payload/packet is received.

1.2.3 IS-95/CDMA2000

In this section, we will present the physical layer details of the North American Spread Spectrum evolution of IS-95 CDMA. This system uses FDD for uplink and downlink communications. First, let us describe IS-95, and then, we will address CDMA2000 cellular system [17].

The high-level technical details of the IS-95 CDMA system are

- 1.5-MHz channel spacing
- Base stations that are time-synchronous with the aid of Global Positioning System (GPS)
- Common pilot channel for coherent detection (code-multiplexed)

- The chip rate = 1.2288 Mcps
- Modulation scheme = QPSK/OQPSK
- 20-msec frame duration
- Variable rate voice encoding (9600, 4800, 2400, and 1200 bps)

A typical IS-95 system will have up to 7 paging channels, 55 TCH, a sync channel, and a pilot channel. The sum of all the channels cannot exceed 64, due to the length of the Walsh channelization codes used. An example of the downlink (or forward link) operations are shown in Fig. 1.13, where we have chosen to show the pilot and traffic channels only.

The pilot channel consists of a stream of zeros, which is then channelized with a Walsh code (W_0); this channelized signal then becomes quadrature spread by I- and Q-channel PN sequences. The I- and Q-channel PN sequences are maximal length sequences of order 15. As discussed above, the base stations are time-synchronized through GPS; so, the timing phase offset between base stations allows the mobile to distinguish between base stations. Each base station will have an offset that is a multiple of 64 PN chips relative to another. This extended PN sequence repeats 75 times every 2 sec. In practice, this time shift or offset is performed using a PN code mask. This function will be discussed later in the spread spectrum chapter. Suffice it to say that with this mask approach, the PN output sequence is applied to another set of combinatorial logic (i.e., mask) to produce the desired offset in the PN sequence.

Also shown is 1 of 55 TCH, whose data rate can vary, depending on the voice activity, from 1200 to 9600 bps. Regardless of voice activity, the generated coded bit stream is 19.2 kbps. The TCH gets scrambled to provide voice privacy by using a decimated version of a long PN code sequence of order 42. The downlink uses a rate $R = {}^1/_2$ convolutional code. A multiplexed 800-bps power control subchannel is also shown to aid in the power control mechanism. The base station will measure the received signal power and compare this value to a desired value. Based on this comparison, the mobile is instructed to either reduce or increase its transmission power. On the downlink, each user is distinguished from a Walsh channelization code sequence.

Although not shown, the paging channel is generated in a similar fashion as the TCH, with the exception that no power control subchannel is multiplexed into the signal. As its name implies, this channel is used to instruct the mobile of incoming calls or other information. Lastly, the sync channel is a constant 1.2-kbps stream that is Walsh channelized by W_{32} consistently. The sync channel provides

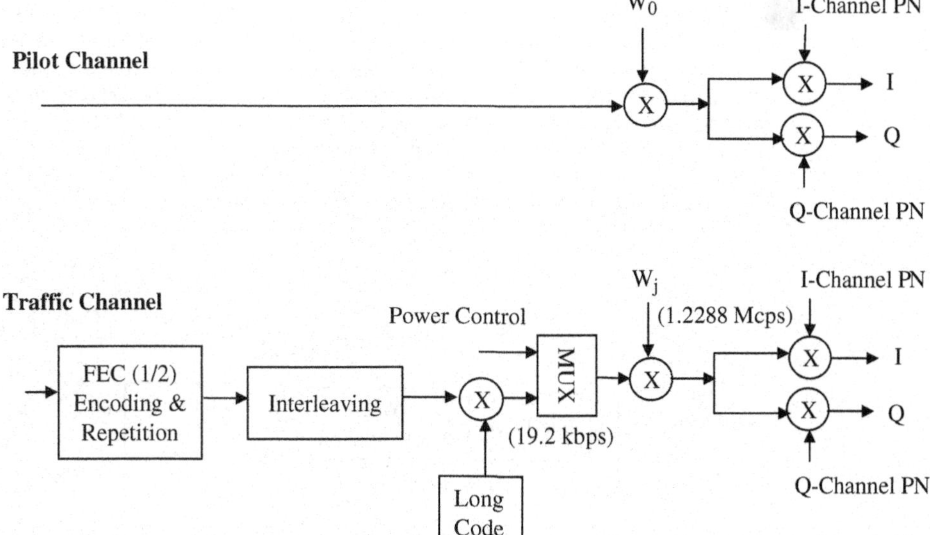

FIGURE 1.13 Generation of forward link pilot and traffic channels.

information used by the mobile to obtain system timing. Once timing has been established, the mobile can transmit on the uplink access channel, when desired.

All the active downlink channels are summed to give a single I- and Q-baseband signal. These baseband signals are then filtered and sent to a linear transmitter. Each channel uses QPSK modulation, and no attempt was made to reduce the transitions through the origin because once the relevant channels are summed, the resulting signal will have a large envelope variation (i.e., peak to average value) as the number of users in the cell increase.

The transmission data rates are classified into two rate sets (RS). First, RS1 has a maximum bit rate of 9.6 kbps while RS2 has 14.4 kbps. Both RS allow for variable rate data transmission as well.

The reverse link (RL) channels are discussed next. Here, the mobile can transmit either an access channel or a TCH. The access channel is used to initiate a call or respond to a call from either a page or other message. The TCH is used to transmit user data, using a rate $R = {}^1/_3$ convolutional code. The uplink also uses 64-ary orthogonal modulation (see Fig. 1.14). On the uplink, each user is distinguished by a specific time offset in the long PN sequence of order 42.

This spread signal enters the I and Q PN sequences for quadrature spreading, where the phase or time offsets are set to zero for each user. The quadrature spread signal then enters an OQPSK modulation scheme. To accomplish this task, the Q-channel is delayed by Tc/2 chips prior to entering the pulse shaping filter and quadrature modulator. This was chosen in order to avoid the constellation origin and use more nonlinear, efficient PAs. An additional block inserted in the uplink is the data burst randomizer. Here, repeated bits are not transmitted in order to reduce the uplink interference for the cell. After coding, the data rate is set to 28.8 ksps, and for the low vocoder data rates, this coded rate is achieved by repetition. Hence, these repeated bits are not transmitted by a pseudorandom selection.

Since this is a CDMA system, each user will continuously transmit in the same frequency band, and as such, the system will be interference limited. A mechanism commonly used to improve user capacity, as well as system performance (i.e., near-far problem, multipath fading), is called power control. Here, the 20-msec frame is divided into 16 time segments called Power Control Groups (PCG), each having a time duration of 1.25 msec. The forward link (FL) has an 800-bps subchannel, where power control commands are sent to the mobile instructing it to either increase or decrease its transmission power. On the RL, the base station receives messages from the mobile representing the power received by the mobile. The base station then uses this information to adjust its transmit power.

CDMA2000 Discussion. Next we move our attention to the CDMA2000 air interface [18]. The CDMA2000 system was created in an effort to evolve IS-95 to provide 3G functionality in the form of higher data rates, increased user capacity, and improved QoS control. The evolved standard must perform this and provide backward compatibility to IS-95, which is also known as cdmaOne.

The RL has been improved in the following sense: a reverse pilot channel is now used to establish a phase reference (among other signal processing functions) for coherent detection. This contrasts to the noncoherent detection of the 64-ary orthogonal modulation used in IS-95. Hence, the E_b/N_o is now lower, and thus, capacity and system performance have been improved. Also a power control subchannel is multiplexed on the pilot channel to instruct the base station to adjust its transmit power level. Moreover, a DCCH, fundamental channel, and one or more supplemental channels were added.

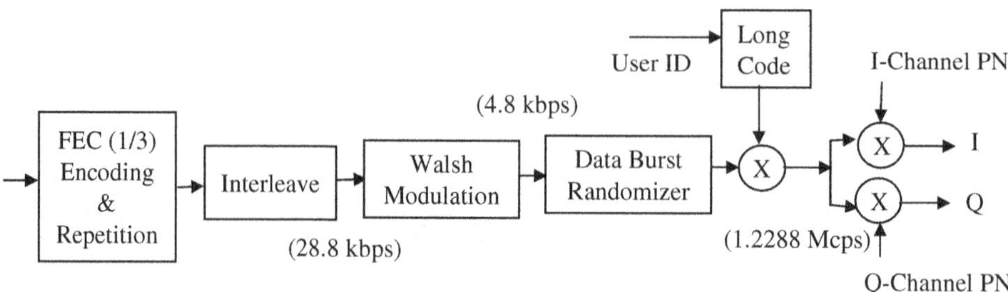

FIGURE 1.14 Uplink transmission traffic channel.

Any upper-layer signaling can be sent over the DCCH now. This means that the service on the fundamental channel is not compromised anymore when the need to transmit upper-layer signaling arises. The fundamental channel supports the existing IS-95 data rates and can be used to transmit information, voice, and part of data signal. The supplemental channels are used to transmit the data signals. Also, the real-valued PN spreading operations were changed to complex multiplications to better make use of nonlinear PAs.

CDMA200 supports multirate users through the usage of various modulation schemes, multiple code channels, and FEC techniques. These are best described through the multiple radio configurations available on the reverse TCH. We list some characteristics in Table 1.4.

In the table below, we have a column named number of carriers; the CDMA2000 air interface specification calls this spreading rate. The PN chip rate for spreading rate 1 is 1.2288 Mcps, while for a spreading rate of 3, the chip rate is 3.6864 Mcps. With this in mind, the reverse fundamental channel and supplemental channel structures were actually shown in Fig. 1.14 for the radio configuration numbers 1 and 2.

A block diagram of the RL physical channel multiplexing and QPSK spreading is shown in Fig. 1.15. This corresponds to radio configurations 3 and 4 for the reverse TCH and using spreading rate = 1.

For radio configurations 5 and 6, the RL channel multiplexing and spreading block diagram is similar to the one shown in Fig. 1.15. The exception being using spreading rate = 3, which translates to a chip rate of 3.6864 Mcps.

Convolutional coding is used on the fundamental and supplemental channels. Their coding rates depend on the radio configuration used. The supplemental channel also supports turbo coding, provided the amount of bits per frame exceeds a certain threshold value. All convolutional codes have a constraint length of 9. The octal representations of the generator functions are given below:

Code rate	G_0	G_1	G_2	G_3
$R = 1/2$	753	561		
$R = 1/3$	557	663	711	
$R = 1/4$	765	671	513	473

TABLE 1.4 Reverse Link Radio Configuration Characteristics

Radio configuration	Number of carriers	Data rates supported	FEC technique	Modulation scheme
1	1	1.2K, 2.4K, 4.8K, 9.6K	$R = 1/3$ Convolutional	64-ary orthogonal
2	1	1.8K, 3.6K, 7.2K, 14.4K	$R = 1/2$ Convolutional	64-ary orthogonal
3	1	1.2K, 1.35K, 1.5K, 2.4K, 2.7K, 4.8K, 9.6K, 19.2K, 38.4K, 76.8K, 153.6K, 307.2K	$R = 1/4$ and $R = 1/2$ Convolutional	Coherent BPSK
4	1	1.8K, 3.6K, 7.2K, 14.4K, 28.8K, 57.6K, 115.2K, 230.4K	$R = 1/4$ Convolutional	Coherent BPSK
5	3	1.2K, 1.35K, 1.5K, 2.4K, 2.7K, 4.8K, 9.6K, 19.2K, 38.4K, 76.8K, 153.6K, 307.2K, 614.4K	$R = 1/3$ Convolutional	Coherent BPSK
6	3	1.8K, 3.6K, 7.2K, 14.4K, 28.8K, 57.6K, 115.2K, 230.4K, 460.8K, 1036.8K	$R = 1/4$ and $R = 1/2$ Convolutional	Coherent BPSK
7	1	19.2K, 40.8K, 79.2K,156K, 309.6K, 463.2K, 616.8K, 924K, 1231.2K, 1538.4K, 1845.6K	$R = 1/5$ Convolutional	Coherent BPSK, QPSK and 8-PSK

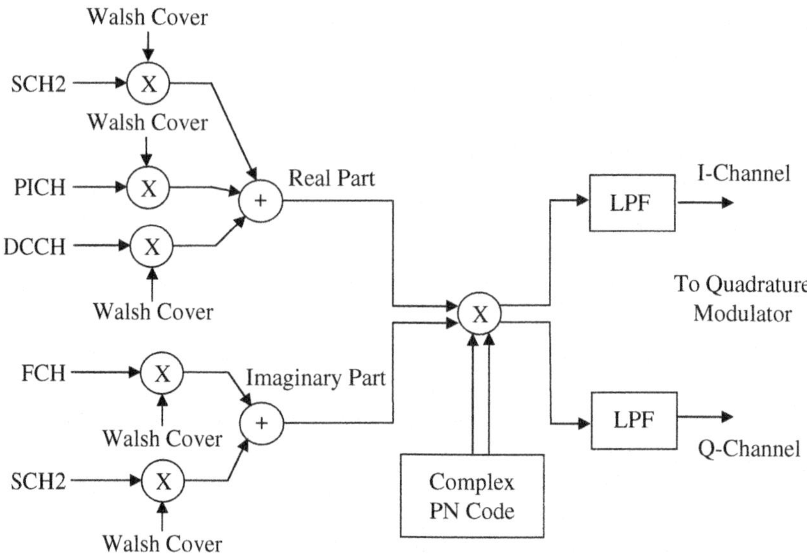

FIGURE 1.15 Reverse link channel multiplexing and spreading.

The turbo encoders have a common coder whose outputs are punctured according to a specific pattern, to achieve the desired code rates of $^1/_2$, $^1/_3$, $^1/_4$, and $^1/_5$.

The RL complex PN code is generated as shown in Fig. 1.16, where we have used the spread rate = 3 chip rate, as an example.

The FL improvements come in the form of adding forward supplemental channels for data services. A quick paging channel was introduced for increased mobile stand by times. Transmit diversity is supported to reduce the required E_b/N_o and thus increase user capacity. QPSK modulation is used for higher data rates and to support true complex-valued spreading. The FEC was enhanced to include turbo codes for performance improvement at the higher data rates. In Table 1.5, we list the radio configurations for the forward traffic channel.

A block diagram of the PN-code spreading and FL physical channel mapping is shown in Fig. 1.17 for radio configurations 3, 4, and 5, using spreading rate = 1.

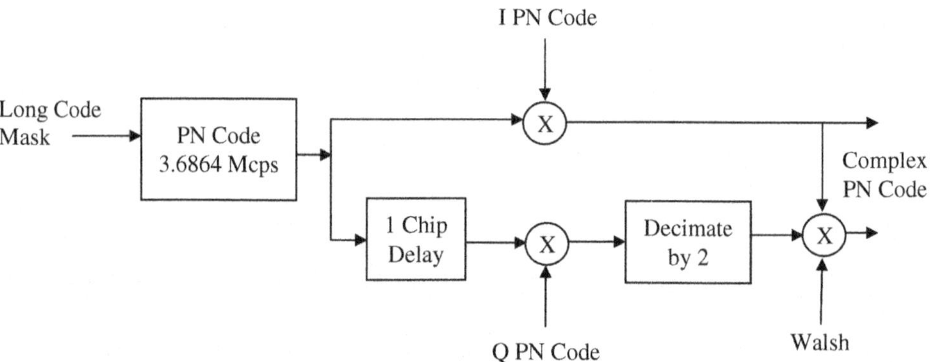

FIGURE 1.16 Reverse link complex PN-coded generation.

TABLE 1.5 Forward Link Radio Configuration Characteristics

Radio configuration	Number of carriers	Data rates supported	FEC technique	Modulation scheme
1	1	1.2K, 2.4K, 4.8K, 9.6K	$R = 1/2$	BPSK
2	1	1.8K, 3.6K, 7.2K, 14.4K	$R = 1/2$	BPSK
3	1	1.2K, 1.35K, 1.5K, 2.4K, 2.7K, 4.8K, 9.6K, 19.2K, 38.4K, 76.8K, 153.6K	$R = 1/4$	QPSK
4	1	1.2K, 1.35K, 1.5K, 2.4K, 2.7K, 4.8K, 9.6K, 19.2K, 38.4K, 76.8K, 153.6K, 307.2K	$R = 1/2$	QPSK
5	1	1.8K, 3.6K, 7.2K, 14.4K, 28.8K, 57.6K, 115.2K, 230.4K	$R = 1/4$	QPSK
6	3	1.2K, 1.35K, 1.5K, 2.4K, 2.7K, 4.8K, 9.6K, 19.2K, 38.4K, 76.8K, 153.6K, 307.2K	$R = 1/6$	QPSK
7	3	1.2K, 1.35K, 1.5K, 2.4K, 2.7K, 4.8K, 9.6K, 19.2K, 38.4K, 76.8K, 153.6K, 307.2K, 614.4K	$R = 1/3$	QPSK
8	3	1.8K, 3.6K, 7.2K, 14.4K, 28.8K, 57.6K, 115.2K, 230.4K, 460.8K	$R = 1/3$ and $R = 1/4$	QPSK
9	3	1.8K, 3.6K, 7.2K, 14.4K, 28.8K, 57.6K, 115.2K, 230.4K, 259.2K, 460.8K, 518.4K, 1036.8K	$R = 1/2$ and $R = 1/3$	QPSK
10	1	43.2K, 81.6K, 86.4K, 158.4K, 163.2K, 172.8K, 312K, 316.8K, 326.4K, 465.6, 619.2K, 624K, 633.6K, 772.8K, 931.2K, 1238.4K, 1248K, 1545.6K, 1862.4K, 2476.8K, 3091.2K	$R = 1/5$	QPSK, 8-PSK, 16-QAM

The signal is now mapped to a modulation format and then quadrature spread (see Fig. 1.18). For radio configurations 6, 7, 8, and 9, using spreading rate = 3, we have the forward-spreading operations shown in Fig. 1.19.

Convolutional coding is used on the fundamental and supplemental channels. Their coding rates depend on the radio configuration used. The supplemental channel also supports turbo coding, provided

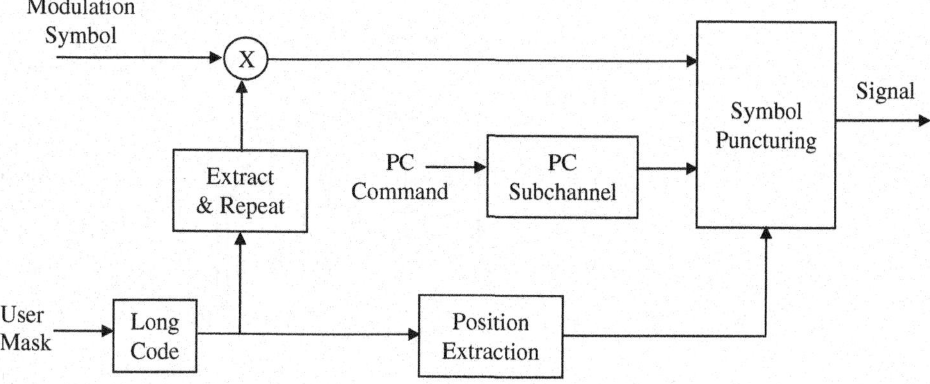

FIGURE 1.17 Forward link traffic channel multiplexing and spreading.

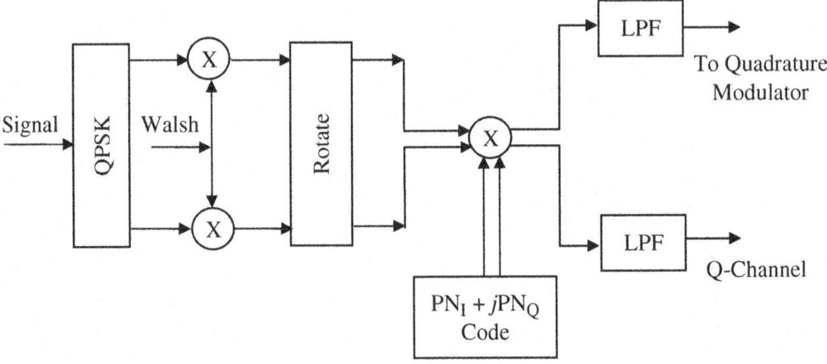

FIGURE 1.18 Forward link traffic channel for spreading rate = 1.

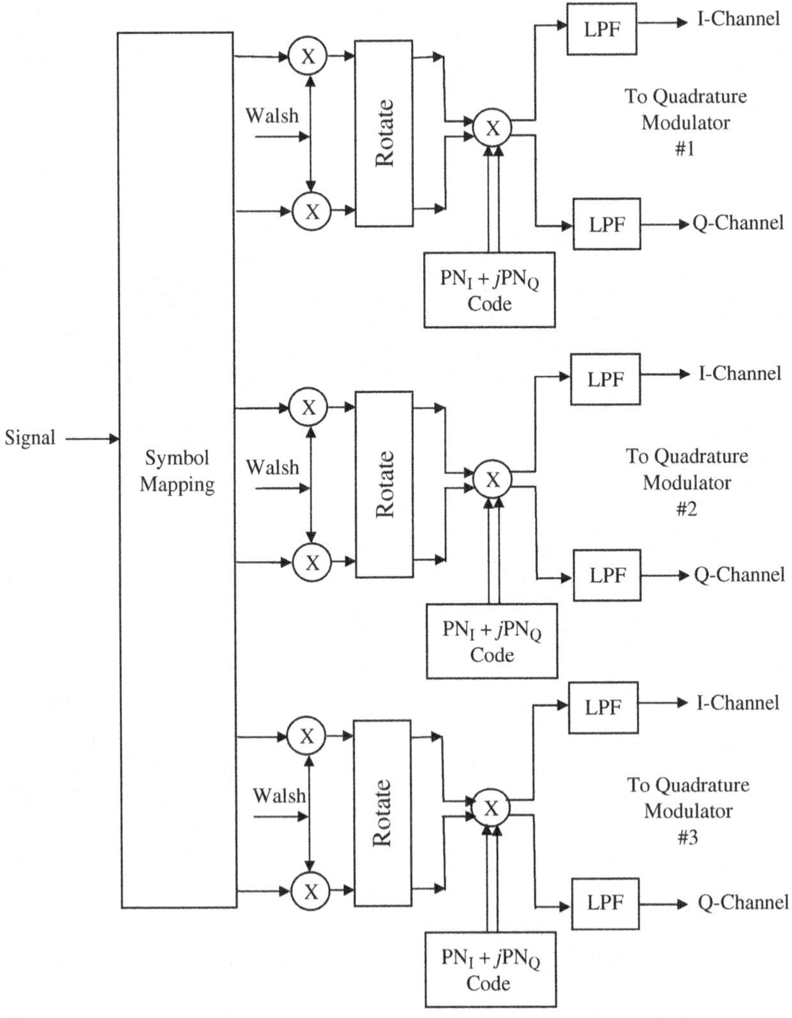

FIGURE 1.19 Forward link traffic channel for spreading rate = 3.

the amount of bits per frame exceeds a certain threshold value. All convolutional codes have a constraint length of 9. The octal representations of the generator functions are given below.

Code rate	G_0	G_1	G_2	G_3	G_4	G_5
$R = {}^1/_2$	753	561				
$R = {}^1/_3$	557	663	711			
$R = {}^1/_4$	765	671	513	473		
$R = {}^1/_5$	457	755	551	637	625	727

The turbo encoders have a common coder whose outputs are punctured according to a specific pattern, to achieve the desired code rates of ${}^1/_2$, ${}^1/_3$, ${}^1/_4$, and ${}^1/_5$. The reader is encouraged to consult [18] for further details.

1.2.4 Wideband CDMA

In this section, we will present the physical layer details of the 3G cellular system called wideband CDMA (WCDMA). This system uses FDD for uplink and downlink communications [19–23].

We will begin our discussion of WCDMA by reviewing the uplink. In the uplink, the data channel and control channel are quadrature multiplexed. By this, we mean that the dedicated physical data channel (DPDCH) is transmitted on the I-axis and the dedicated physical control channel (DPCCH) is transmitted on the Q-axis. The uplink time-slot structure is given in Fig. 1.20.

In the figure, we see that the radio frame is 10 msec in time duration which consists of 15 time slots. The time slots and frames will contain varying data rates, and thus, we have varying spreading factors. However, each time slot consists of 2560 chips in duration. The bit field lengths for the data and control channels are shown in Table 1.6.

Above we notice that as the source data rate changes, the spreading factor changes, since we have a fixed chip rate. This leads to transmitting more bits per frame time, using a smaller spreading factor.

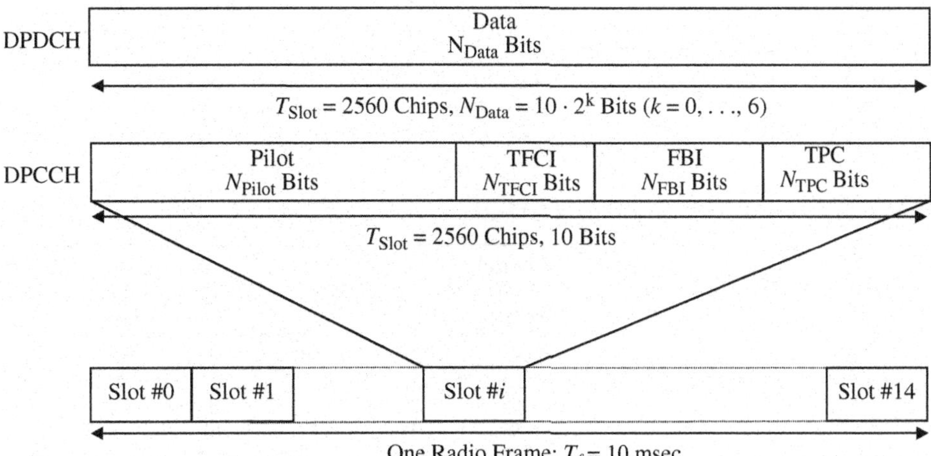

FIGURE 1.20 Frame structure for uplink DPDCH/DPCCH.

TABLE 1.6 Uplink DPDCH Fields

Slot format #*i*	Channel bit rate (kbps)	Channel symbol rate (ksps)	SF	Bits/Frame	Bits/Slot	N_{Data}
0	15	15	256	150	10	10
1	30	30	128	300	20	20
2	60	60	64	600	40	40
3	120	120	32	1200	80	80
4	240	240	16	2400	160	160
5	480	480	8	4800	320	320
6	960	960	4	9600	640	640

On the contrary, the CCH is transmitted with a fixed data rate of spreading factor = 256 all the time (see Table 1.7).

The uplink CCH contains pilot bits that are used by the NodeB receiver to estimate the channel phase rotation to aid in coherent detection (see Table 1.7). Although not specifically stated by the 3G Partnership Project (3GPP) standard, these pilot bits can also be used for the purposes of timing synchronization, frequency offset estimation, signal to interference ratio (SIR) estimation, and so forth [24–25].

A mechanism used to combat fading and the near-far problem is closed inner loop power control. For this uplink discussion, the user equipment (UE) will calculate the received SIR and compare it to a reference or SIR target value; it will then derive a power control command to send to the NodeB, instructing it to either increase or decrease its downlink transmit power. This closed-loop mechanism will be discussed in more detail in the latter chapters.

The standard is flexible enough to allow for multiple transport channels to be multiplexed onto the same physical channel; when this occurs, we should reveal this combination to the receiver so that it can properly perform the transport channel demultiplexing. The transport format combination indicator (TFCI) bits are used to identify the transport channel combination. We should also mention that support also exists for blind transport format detection.

Two operating scenarios arise when there exists a need for the UE to feedback information to the NodeB with low latency. These are fast cell selection and closed-loop transmit diversity. Let us discuss the latter of the two. Here, the NodeB transmits the dedicated waveform via two transmit antennas. The UE estimates the channel impulse response (CIR) seen by each transmit antenna and then derives an antenna weight to be fed back to the NodeB, in order to maximize the UE-received power. This information is conveyed in the feedback indicator (FBI) bit field. In the absence of these cases, this bit field is not required.

Next we move our discussion to the downlink time-slot structure (see Fig. 1.21). Here, the DPCCH and DPDCH are time-multiplexed. The same description given earlier about the bit fields applies here as well.

TABLE 1.7 Uplink DPCCH Fields

Slot format #*i*	Channel bit rate (kbps)	Channel symbol rate (ksps)	SF	Bits/ Frame	Bits/ Slot	N_{pilot}	N_{TPC}	N_{TFCI}	N_{FBI}
0	15	15	256	150	10	6	2	2	0
1	15	15	256	150	10	8	2	0	0
2	15	15	256	150	10	5	2	2	1
3	15	15	256	150	10	7	2	0	1
4	15	15	256	150	10	6	2	0	2
5	15	15	256	150	10	5	1	2	2

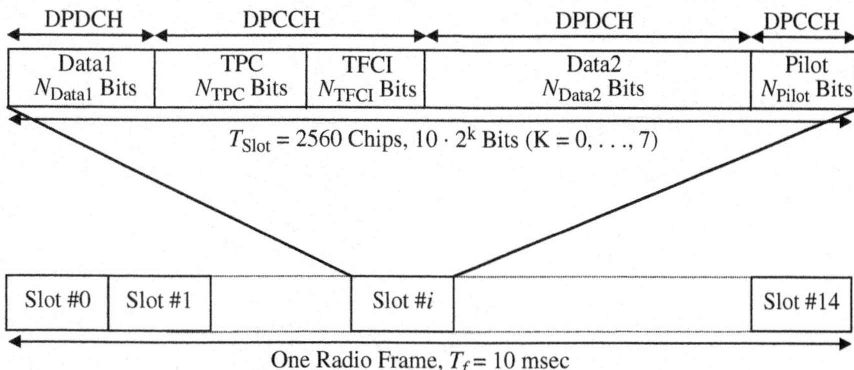

FIGURE 1.21 Frame structure for downlink DPCCH.

In Table 1.8, we can see that the spreading factor can vary from SF = 4 to SF = 512, depending on the data rate that is being transmitted.

Let us now turn our attention to the spreading operations performed. The uplink spreading functionality is shown in Fig. 1.22. Here, the separate dedicated data and control channels are channelized (and spread) by an Orthogonal Variable Spreading Factor (OVSF) sequence. We have labeled the OVSF codes as Walsh codes, due to their similarity. The resulting sequence is then made to be complex valued, and then, QPSK spreading is applied prior to entering the pulse-shaping filter. Note that, in order to acknowledge the terminology used by 3GPP, this QPSK spreading operation is also called scrambling within the standard. On one hand, this is clearly understandable since the OVSF multiplication did indeed spread the information frequency bandwidth. The second PN multiplication is acting more like a scrambling function than a frequency-spreading function.

The QPSK spreading (or scrambling) is actually called hybrid PSK (HPSK). This was introduced to the UE transmitter since this technique will dramatically reduce the number of phase trajectories that come dangerously close to the origin.

TABLE 1.8 Downlink DPDCH and DPCCH Fields

Slot format #i	Channel bit rate (kbps)	Channel symbol rate (ksps)	SF	Bits/Slot	DPDCH bits/slot		DPCCH bits/slot		
					N_{Data1}	N_{Data2}	N_{TPC}	N_{TFCI}	N_{pilot}
0	15	7.5	512	10	0	4	2	0	4
1	15	7.5	512	10	0	2	2	2	4
2	30	15	256	20	2	14	2	0	2
3	30	15	256	20	2	12	2	2	2
4	30	15	256	20	2	12	2	0	4
5	30	15	256	20	2	10	2	2	4
6	30	15	256	20	2	8	2	0	8
7	30	15	256	20	2	6	2	2	8
8	60	30	128	40	6	28	2	0	4
9	60	30	128	40	6	26	2	2	4
10	60	30	128	40	6	24	2	0	8
11	60	30	128	40	6	22	2	2	8
12	120	60	64	80	12	48	4	8	8
13	240	120	32	160	28	112	4	8	8
14	480	240	16	320	56	232	8	8	16
15	960	480	8	640	120	488	8	8	16
16	1920	960	4	1280	248	1000	8	8	16

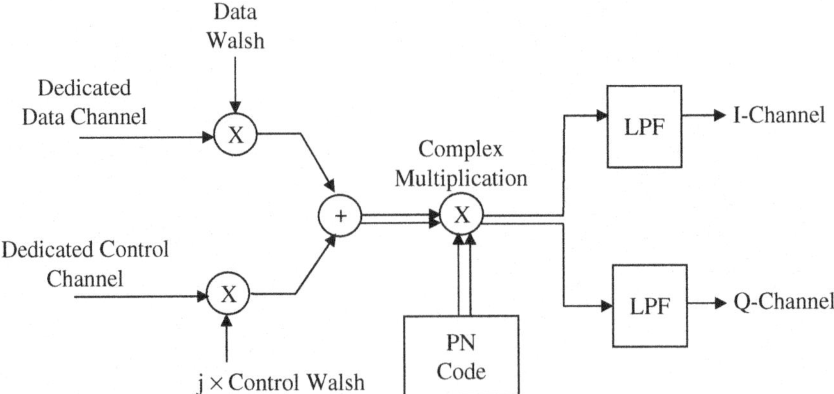

FIGURE 1.22 Uplink channelization and spreading operations.

Next we will discuss the downlink operations. The downlink spreading functionality is shown in Fig. 1.23, where the coded composite transport channel is mapped to a QPSK symbol. These symbols are channelized by an OVSF (Walsh) code assigned to a UE. The channelized symbol is then spread by a complex-valued gold PN code.

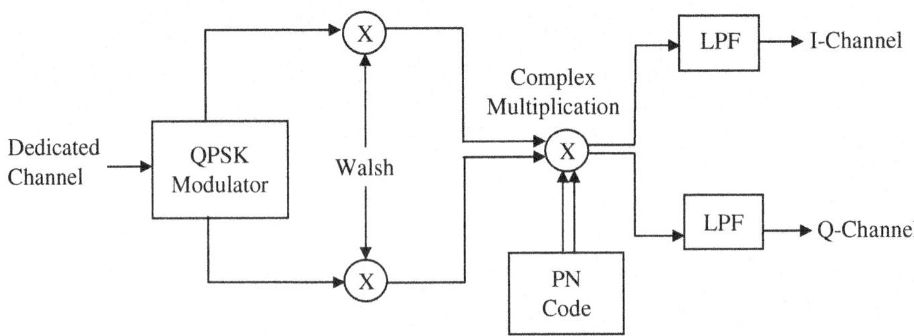

FIGURE 1.23 Downlink channelization and spreading operations.

The baseband, low-pass filter is a square root–raised cosine (SRC) with a roll-off factor of $\alpha = 0.22$. This filter is used at both the transmitter and receiver sections. The power control update rate is given as 1500 Hz. As indicated above, the maximum number of power control bits per time slot is 8. These will all be transmitted with the same sign to aid the receiver in detecting the transmit power control (TPC) command reliably.

An example of transport channel multiplexing is given in Fig. 1.24. We show an example of how a data-based transport channel (DTCH) and low-rate control transport channel (DCCH) are each encoded, interleaved, and segmented into frames. Then, they are multiplexed into a physical channel to be interleaved and time slot segmented. The downlink multiplexing operations are shown in Fig. 1.24.

As we can see in the figure, the downlink spreading factor for the data-based service is 32, and the associated uplink spreading factor is 16. This was obtained from the fact that there are 160 bits per time slot (140 data plus 20 control) that become 80 QPSK symbols. There are 2560 chips per time slot, so each symbol is spread by 32 chips. Here, we have chosen to use a 64-kbps data service application.

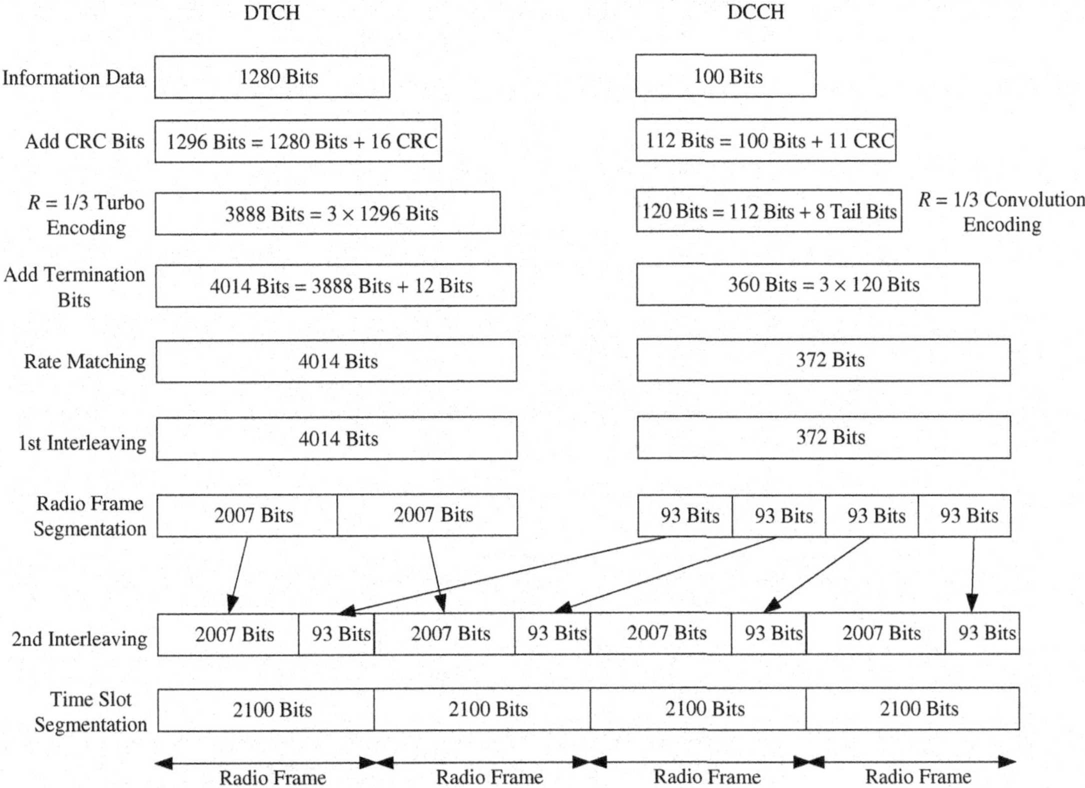

FIGURE 1.24 Downlink transport channel multiplexing example.

1.2.5 IEEE802.11 Wireless LAN

In this section, we will present the physical layer details of the IEEE802.11 (a, b, g) Wireless LAN (WLAN) systems [26–29]. This standard has gained wide acceptance with such services offered in airports, coffee shops, public areas, the workplace, and the home. The WLAN devices can be configured to operate either in an ad hoc mode where all the devices are allowed to communicate to each other and act as access points, or in an infrastructure mode where a single WLAN device behaves as the access point and others connect to that access point.

IEEE802.11b. This system allows computers, printers, and mobile phones to communicate to each other in the 2.4-GHz frequency band. The medium access protocol allows for stations to support different sets of data rates. The basic set of data rates is 1 and 2 Mbps operating in a Direct Sequence Spread Spectrum (DSSS) system. A high-rate extension exists that pushes the data rate to 5.5 and 11 Mbps. In order to provide the higher data rates, the complementary code keying (CCK) modulation scheme is used. The chip rate of the DSSS system is 11 Mcps. The preamble and header fields are transmitted at 1-Mbps DBPSK. The physical layer frequency range is

United States, Canada, and Europe	2.4–2.4835 GHz
Japan	2.471–2.497 GHz
France	2.4465–2.4835 GHz
Spain	2.445–2.475 GHz

The higher layer service data units (SDU) are appended to the preamble and header fields to assist demodulation signal processing functions. Two formats are discussed next (see Figs. 1.25 and 1.26). The PSDU (PLCP SDU) is transmitted at 1-Mbps DBPSK, 2-Mbps DQPSK, and 5.5/11-Mbps CCK.

Below we provide some definitions to the fields shown in Figs. 1.25 and 1.26:

Sync Field = consists of 128 bits for the long format and 56 bits for the short format; these bits are used for synchronization functions.

Start Frame Delimiter (SFD) = consists of 16 bits; these bits are used to indicate the start of the PHY (physical layer) dependent parameters within the PLCP preamble.

Signal = consists of 8 bits, these bits are used to indicate the modulation that shall be used during the PSDU.

Service = consists of 8 bits; these bits are used to indicate the modulation choice of CCK or other and to indicate if certain clocks are locked.

Length = consists of 16 bits; these bits are used to indicate the number of microseconds for PSDU.

Long Physical Layer Convergence Procedure (PLCP)

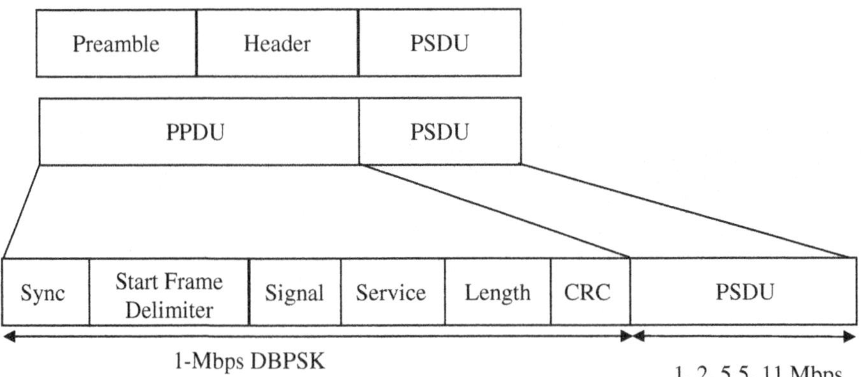

FIGURE 1.25 Long PLCP PSDU format.

Short Physical Layer Convergence Procedure (PLCP)

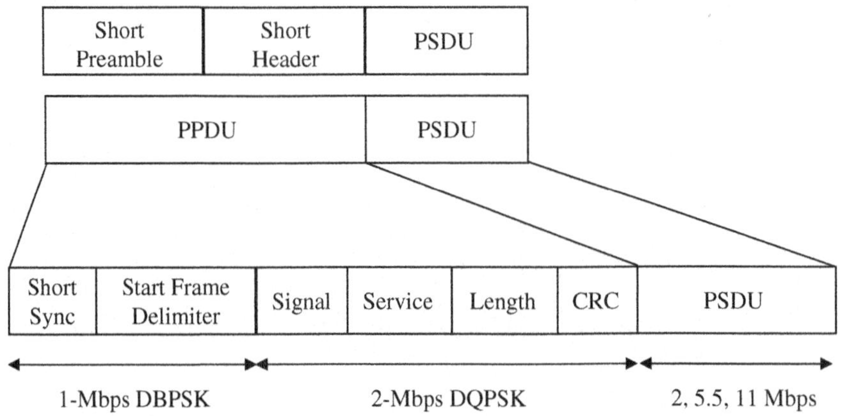

FIGURE 1.26 Short PLCP PPDU format.

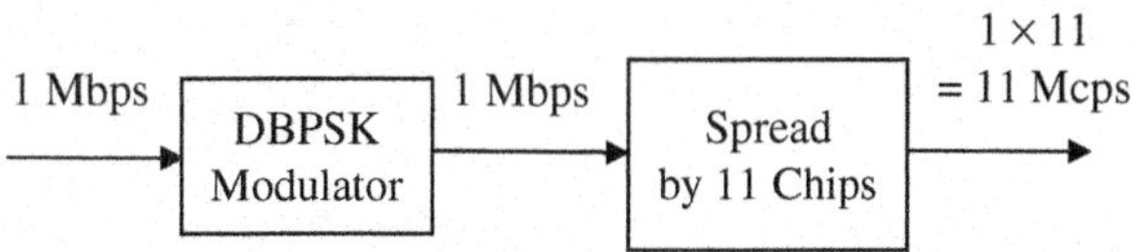

FIGURE 1.27 A 1-Mbps DBPSK transmission block diagram.

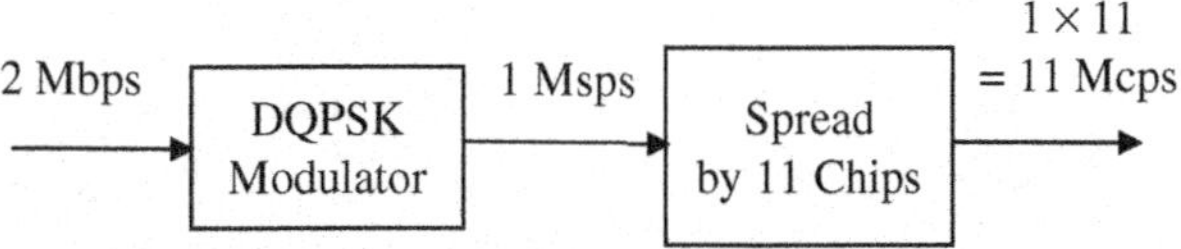

FIGURE 1.28 A 2-Mbps DQPSK transmission block diagram.

Modulation Scheme Discussion. The 1-Mbps data rate service will use DBPSK modulation, the 2-Mbps data rate service will use DQPSK, and the 5.5/11-Mbps data rate services will use CCK modulation. The 1- and 2-Mbps data rate services will be discussed first (see Figs. 1.27 and 1.28). Since this is a DSSS system, with a chip rate of 11 Mcps, we need to discuss the PN code sequence, which in this case is a Barker sequence and is given as

$$\{+1,-1,+1,+1,-1,+1,+1,+1,-1,-1,-1\} \tag{1.8}$$

Here, each symbol (or bit) of DQPSK (or DBPSK) shall be spread by this 11-chip sequence given in Eq. (1.8). The DBPSK encoding phase state table is given as

Input	Output
0	0
1	π

The DQPSK encoding phase state table is given as

Input	Output
00	0
01	$\pi/2$
11	π
10	$-\pi/2$

For the higher–data rate services, CCK modulation is used. Each symbol shall be spread by 8 chips that are complex valued. The 5.5-Mbps data rate groups 4 bits to transmit per symbol. The CCK spreading code is derived as follows (8 complex-valued chips):

$$\{e^{j(\phi_1+\phi_2+\phi_3+\phi_4)},\ e^{j(\phi_1+\phi_3+\phi_4)},\ e^{j(\phi_1+\phi_2+\phi_4)},\ -e^{j(\phi_1+\phi_4)},\ e^{j(\phi_1+\phi_2+\phi_3)},\ e^{j(\phi_1+\phi_3)},\ -e^{j(\phi_1+\phi_2)},\ e^{j\phi_1}\} \tag{1.9}$$

Here we take the input 5.5-Mbps data stream and group incoming bits into groups of 4 bits. These 4 bits are used to determine the phase values used in the PN codes described above. Thus every group of 4 bits leads to different complex values of the PN shown above. The block diagram is shown in Fig. 1.29.

Lastly, the 11-Mbps data rate is accomplished by grouping 8 bits together to transmit a symbol (see Fig. 1.30). Using Eq. (1.9) for the generation of the complex-valued PN code sequence, but with different encoding rules, we simply group the incoming bits into groups of 8 bits and use those 8 bits to calculate a CCK chip sequence of length 8.

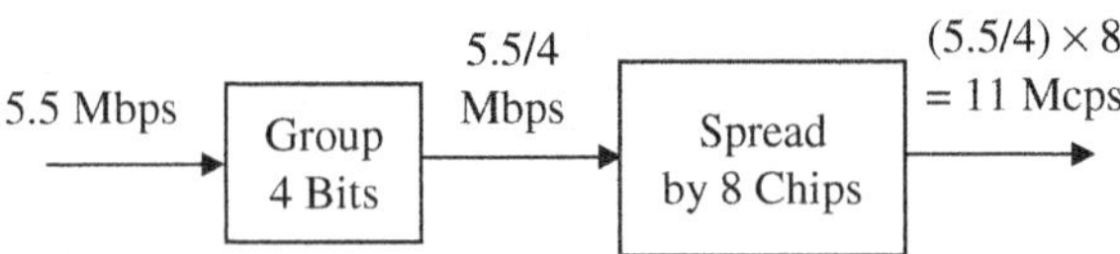

FIGURE 1.29 A 5.5-Mbps CCK transmission block diagram.

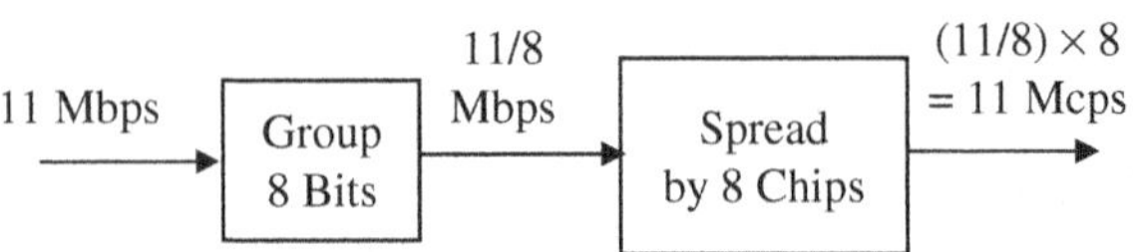

FIGURE 1.30 An 11-Mbps CCK transmission block diagram.

Some general system characteristics are a transmit frequency tolerance of less than or equal to 25 ppm, and an Error Vector Magnitude (EVM) requirement of less than 35%, utilizing a 25-MHz channel spacing, output power less than 1 W, and so forth.

In addition to the DSSS physical layer mentioned above, there is a Frequency Hopping Spread Spectrum (FHSS) physical layer as well. The data rates supported are from 1 to 4.5 Mbps in increments of 0.5 Mbps. The FHSS PPDU structure is shown in Fig. 1.31.

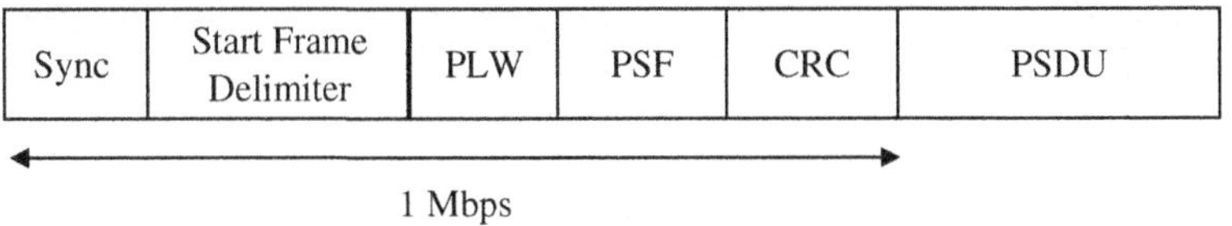

FIGURE 1.31 FHSS PLCP PPDU format.

The modulation scheme is GFSK 2 level and 4 level. A Gaussian filter with the BT product = 0.5 is used. The frequency deviation is greater than 110 kHz, thus having a modulation index of $\beta \geq \frac{2 \times 110\,\text{kHz}}{1\,\text{Mbps}} \geq 0.22$.

A simplified block diagram of the transmit section is shown in Fig. 1.32.

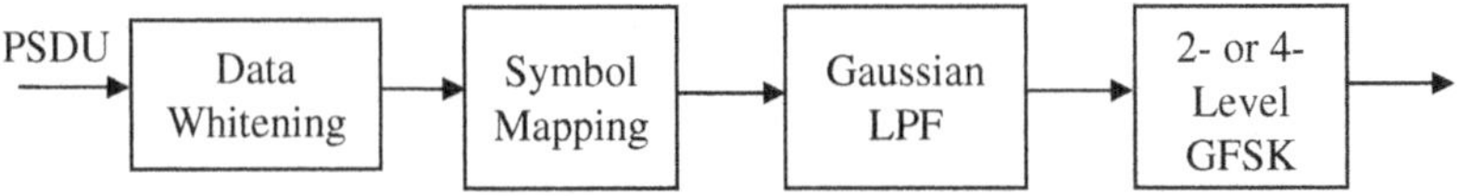

FIGURE 1.32 An FHSS transmission block diagram.

IEEE802.11a. This WLAN system was targeted for use in the 5.15–5.35 GHz and 5.725–5.825 GHz unlicensed bands. The standard is based on the orthogonal frequency division multiplexing (OFDM) physical layer. The data rates supported are 6, 9, 12, 18, 24, 36, 48, and 54 Mbps, where 6, 12, and 24 Mbps are mandatory.

The OFDM physical layer uses 52 subcarriers using BPSK/QPSK, 16-QAM (quadrature amplitude modulation), and 64-QAM, with the FEC technique being convolutional coding with code rates of $^1/_2$, $^2/_3$, and $^3/_4$. These coded bits are mapped to 48 subcarriers plus 4 pilot subcarriers, providing a total of 52 subcarriers. The pilot bits are BPSK modulated.

TABLE 1.9 An Overview of Modulation and Coding Parameters

Data rate (Mbps)	Modulation	Coding rate	Coded bits per subcarrier
6	BPSK	$^1/_2$	1
9	BPSK	$^3/_4$	1
12	QPSK	$^1/_2$	2
18	QPSK	$^3/_4$	2
24	16-QAM	$^1/_2$	4
36	16-QAM	$^3/_4$	4
48	64-QAM	$^2/_3$	6
54	64-QAM	$^3/_4$	6

The possible system data rates along with their modulation and coding parameters are given in Table 1.9. The PPDU frame format is given in Fig. 1.33.

The bit fields have the following description:

PLCP preamble = can be used for the following signal processing functions: AGC convergence, diversity selection, time and frequency acquisition, channel estimation, and so forth. The preamble consists of 10 repetitions of a short training sequence followed by 2 repetitions of a long training sequence, where a short symbol consists of 12 subcarriers and a long symbol consists of 53 subcarriers. The PLCP preamble is BPSK OFDM modulated at 6 Mbps and of length 16 μsec.

Rate = 4 bits used to indicate the data rate shown in Table 1.9.

Length = 12 bits indicating the number of octets in the PSDU that the MAC (medium access control) is currently requesting to transmit.

This system operates in a 5-MHz channel spacing, with a transmit frequency tolerance of less than ±20 ppm.

IEEE802.11g. Thus, WLAN system provides further data rate extensions into the 2.4-GHz frequency band. The data rates supported are 1, 2, 5.5, 11, 6, 9, 12, 18, 24, 36, 48, and 54 Mbps, with 1, 2, 5.5, 11, 6, 9, 12, and 24 Mbps being mandatory to support.
Here, a few operating modes are supported:

a. DSSS/CCK based

b. OFDM based

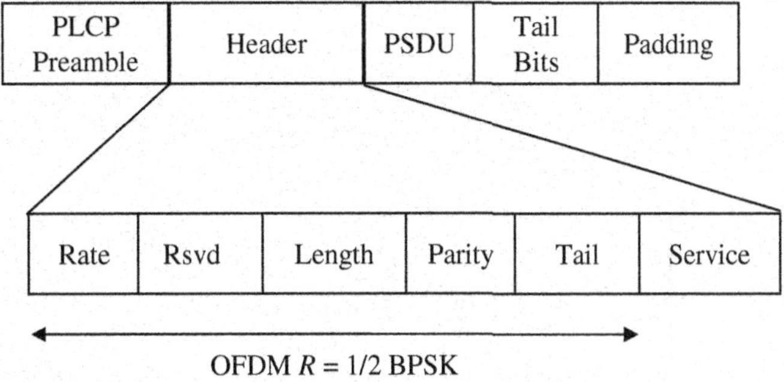

FIGURE 1.33 An OFDM PPDU frame format.

c. DSSS-OFDM (optional)—this is a hybrid modulation scheme combining a DSSS preamble and header with an OFDM payload.

The discussion on OFDM will be delayed for now and discussed in later chapters.

1.2.6 Integrated Digital Enhanced Network (iDEN)

Motorola introduced a proprietary Land Mobile Radio air interface protocol called Motorola Integrated Radio System (MIRS) [30, 31]. This system is based on TDMA using the TDD (time division duplex) method for uplink and downlink communications, operating with a channel spacing of 25 kHz. This system was proposed around approximately the same time the cellular TDMA standards NADC and GSM were being proposed and used. It is well known how the combination of increasing the transmission bit rate and frequency selectivity of the channel leads to a limitation on the transmitted bit rate, while keeping implementation complexity low. Here we mean that as the frequency selective fading (FSF) increases, the amount of intersymbol interference (ISI) present increases (keeping the symbol time constant), thus producing a BER floor.

With this in mind, iDEN introduced a multichannel (carrier) 16-QAM modulation scheme to combat the fast FSF channel commonly encountered in practice. This was latter called M16-QAM, for Motorola 16-QAM.

Some of the physical aspects of the air interface protocol are time slot length = 15 msec, creating a frame length of 90 msec, assuming 6 users per carrier. M16-QAM was chosen to produce 64 kbps in a 25-kHz channel. There are 4 subcarriers, each carrying 16 kbps of information. A block diagram of the M16-QAM transmitter is shown in Fig. 1.34 with 4 subcarriers.

The data rate on each subcarrier has been reduced by 4, compared to the more conventional single carrier 16-QAM modulation scheme. Hence assuming the channel conditions (i.e., RMS delay spread) stay the same, then relative to each subcarrier, the channel behaves more frequency flat, thus reducing the need of an equalizer (or at least its complexity). In fact, with the above block diagram of 4 subcarriers, a max delay spread of approximately 10 μsec can be accommodated without the need of a complex equalizer.

Now let's turn our attention to the receiver. Each subchannel has sync words and pilot symbols inserted into the 16-QAM symbol stream. Assuming that each subchannel will encounter a flat frequency fading channel, then the pilot symbols can be used to estimate the channel to be used for coherent detection. The channel estimate between pilot groups can be estimated through interpolation techniques.

There exist a few options on the receiver side. First, we can simply demodulate each subcarrier independently and combine all subchannels to create the aggregate symbol rate to be passed to either the vocoder or multimedia device. Second, we can jointly demodulate all subchannels through the use of a variety of array signal processing techniques available.

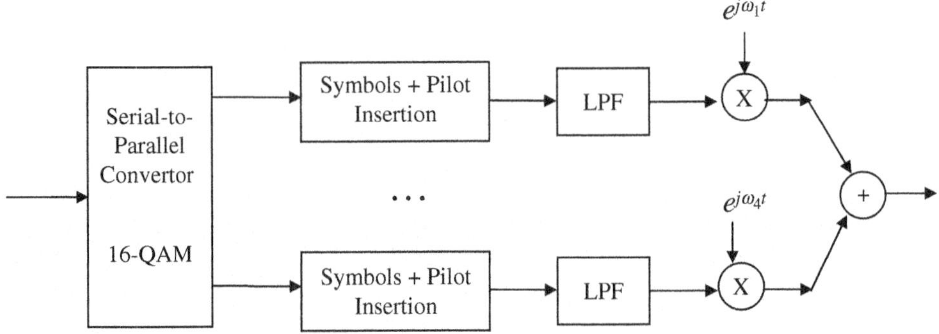

FIGURE 1.34 M16-QAM transmitter block diagram using 4 subcarriers.

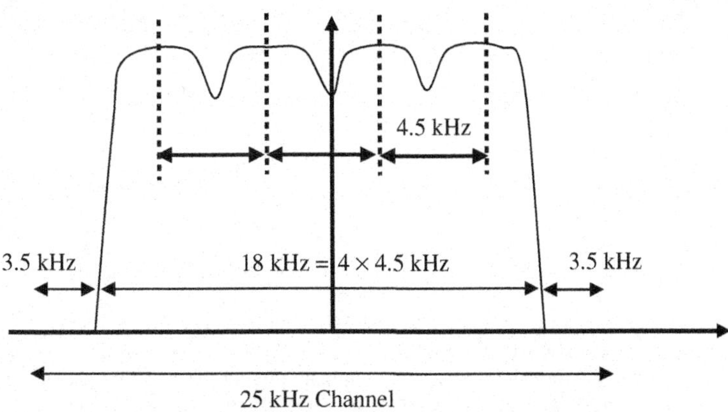

FIGURE 1.35 Transmission RF bandwidth showing subcarrier spacing.

The wireless link bit rate is 64 kbps, where the frequency channel has been split up into 6 time slots, where each time slot is a radio link. The iDEN RF signal consisted of 4 subcarriers, whose center frequencies are separated by 4.5 kHz from each other. A pictorial representation of the modulation riding on each subcarrier is shown in Fig. 1.35.

Each subcarrier has 16 kbps, and the resulting signal produced an aggregate data rate of 64 kbps.

1.2.7 Bluetooth (BT) Personal Area Network

Bluetooth is a low-power, short-range radio technology that was originally intended to be used in device connection applications such as mobile phone headsets, computer peripherals, and cable replacement. The short-range aspect of this system makes this a Personal Area Network (PAN) [32, 33].

Bluetooth operates in the 2.4-GHz industrial, scientific, and medical (ISM) band. The channel spacing is 1 MHz, with each channel having a data rate of at least 1 Mbps. The system uses a TDMA with TDD communication.

Packets of voice and data are transmitted between Bluetooth devices. After each transmission, the device hops to another frequency channel; this is called FHSS. This hopping pattern is across 79 channels. Since the frequency band will have varying degrees of interference, certain channels can be excluded from the hopping sequence to provide further robustness to interference. This allows Bluetooth device manufacturers to design creative frequency hopping criteria in order to have product differentiation. The hop rate is 1600 hops/sec.

A Bluetooth device can essentially operate in two modes: master and slave. The master device controls the frequency hopping pattern, controls which other devices are allowed to transmit, and provides synchronization information. A group of devices operating in this fashion comprise what is called a piconet. For data applications, slave devices are only allowed to transmit in response to a transmission from the master device. In voice applications, slave devices transmit periodically in their designated time slots.

Bluetooth was initially deployed with version 1.0 specification transmitting with a basic data rate of 1 Mbps. It was later enhanced to provide 2 and 3 Mbps using the Enhanced Data Rate (EDR) modifications [34].

Physical layer links are formed only between master and slave devices in the piconet. Each Bluetooth device has an address which is used for identification and consists of 48 bits in length.

We have already discussed the piconet scenario above; however, there are occasions where a Bluetooth device is participating in two or more piconets; this is now called a scatter net. Each piconet has its unique pseudorandom hopping pattern. The frequency channel is divided in time by what are called time slots. Each time slot corresponds to an RF hop frequency.

The basic rate feature is accomplished by the use of 2-level GFSK with $BT = 0.5$. The modulation index is between 0.28 and 0.35. The data rate is 1 Mbps, which allows us to calculate the peak-to-peak frequency deviation ($2f_d$) as follows:

$$0.28 \leq \frac{2f_d}{1\ \text{Mbps}} \leq 0.35 \tag{1.10}$$

$$140\ \text{kHz} \leq f_d \leq 175\ \text{kHz} \tag{1.11}$$

As discussed above, in order to increase the system capacity, the EDR was introduced. The concept here is that the modulation scheme is changed within the packet. That is the access code and packet header are transmitted in 1 Mbps with the GFSK basic rate, while the payload is transmitted using the EDR modulation schemes.

The 2-Mbps data rate is accomplished using the $\pi/4$-DQPSK modulation scheme. The 3 Mbps data rate is accomplished using the 8-DPSK modulation scheme. For the EDR case, the low pass filter is now an SRC filter with roll off factors $\alpha = 0.4$ and given as follows

$$h(t) = \begin{cases} 1 & 0 \leq |f| \leq \dfrac{1-\alpha}{2T} \\ \sqrt{\dfrac{1}{2}\left[1 - \sin\left(\dfrac{\pi(2fT - 1)}{2\alpha}\right)\right]} & \dfrac{1-\alpha}{2T} \leq f \leq \dfrac{1+\alpha}{2T} \\ 0 & \text{elsewhere} \end{cases} \tag{1.12}$$

where T = symbol period = 1 μsec.

Differential phase encoding is accomplished and defined as (using k as the time index)

$$S_k = S_{k-1}e^{j\phi_k} \tag{1.13}$$

where S_k is the symbol to be presently transmitted, S_{k-1} is the previously transmitted symbol, and $e^{j\phi_k}$ is the present phase change. Also, the initial phase being arbitrary

$$S_0 = e^{j\phi} \qquad \phi \in (0, 2\pi) \tag{1.14}$$

a $\pi/4$-DQPSK modulation scheme is generated as shown in Fig. 1.36.

Initially, the allowable frequency offset is ±75 kHz, which is then reduced to ±10 kHz prior to receiving the packet header. Bluetooth provides point-to-point connections and point-to-multipoint connections. Up to seven slaves can be active in the piconet.

The basic rate packet format is shown in Fig. 1.37. The access code duration is 68–72 bits. The header consists of 54 bits, and the payload can range from 0 to 2745 bits in duration.

The EDR packet structure is shown in Fig. 1.38. The CRC (cyclic redundancy check) is used to establish integrity of the received packet, payload is used to transmit the user data, payload header is used as a logical channel identifier and disclose the length of the packet, the packet header is used to determine if the packet is address to the Bluetooth device, and the channel access code is used to identify communications on a particular channel.

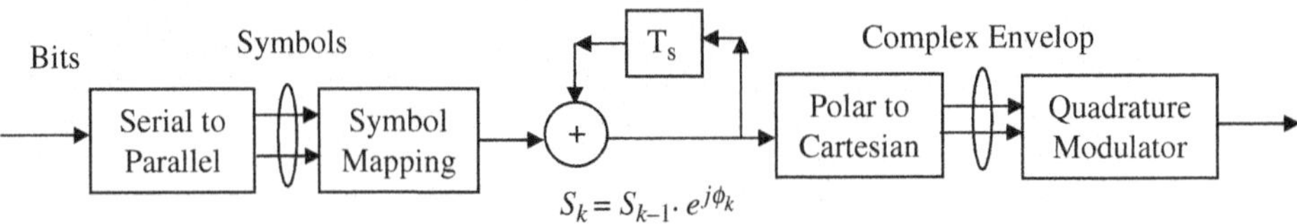

FIGURE 1.36 $\pi/4$-DQPSK modulation transmitter block diagram.

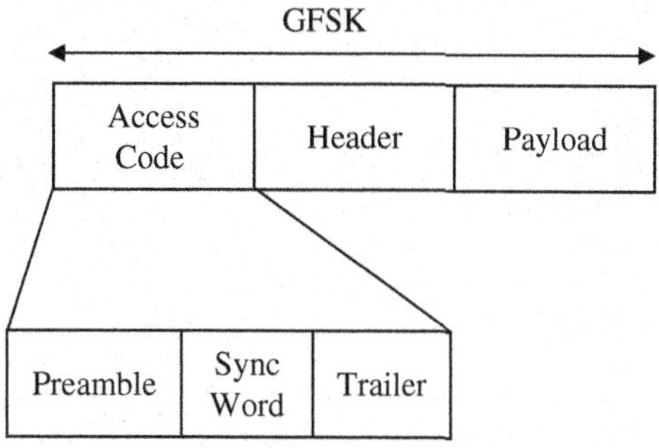

FIGURE 1.37 Basic rate packet format.

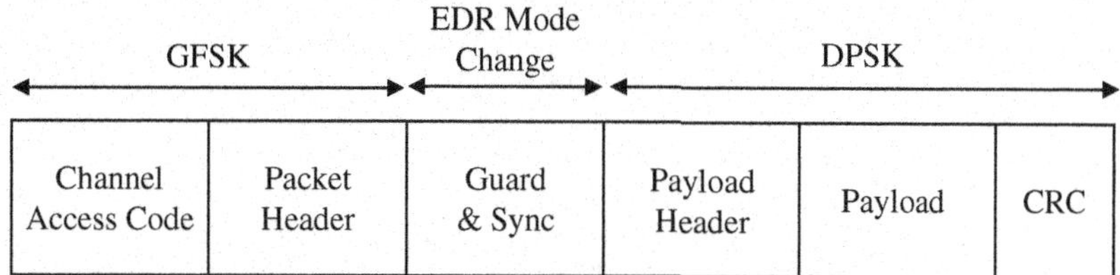

FIGURE 1.38 Enhanced Data Rate (EDR) packet format.

A time slot has a duration = 625 μsec. The time slots are numbered and packets shall be aligned with the start of the time slot. Packets may extend up to 1, 3, or 5 time slots. The master transmits on the even-numbered time slots and the slave transmits on the odd-numbered time slot. The packet payload allows for both data and voice to be sent. As discussed above, the payload size can vary, where the maximum length is determined by the minimum switching time between the Tx and Rx, which is specified at 200 μsec. The devices would therefore demodulate (625 − 200) = 425 μsec of the time slot before switching to another frequency for the next time slot.

As far as packet protection or error correction is concerned, there are three mechanisms defined in Bluetooth for this purpose:

a. Rate $^1/_3$ code, where each bit is repeated 3 times
b. Rate $^2/_3$ code, where a (15, 10) Hamming block code is used
c. Automatic repeat request (ARQ) scheme, where the data packets with CRC protection are used for retransmission if the CRC fails

1.2.8 Flex Paging

Flex is a Motorola trademark which stands for Flexible, Wide Area, Synchronous Paging Protocol [35]. This paging protocol was meant to significantly increase the capacity and messaging reliability over other paging protocols, such as Post Office Code Standardization Advisory Group (POCSAG). Capacity increase is achieved by an increase in data rate from 1200/2400 to 6400 bps. The improved messaging reliability is achieved by the improved synchronous protocol as well as error correction techniques against multipath fading and transmitter simulcasting. Moreover, battery life improvement is achieved through protocol methods, specifically due to the synchronous mechanism found in TDM. Flex is a one-way paging protocol, and Re-Flex is the two-way paging protocol.

The Flex paging protocol has base station transmitters that are synchronized by GPS. The physical layer Flex frame structure is given in Fig. 1.39. In the figure, we see a frame has a time duration of 1.875 sec. The frames are numbered modulo 128, that is, from 0 to 127, where 128 frames together is called a cycle and has a time duration of 4 minutes. There are 15 Flex cycles that repeat each hour and they are synched to the start of the GPS hour. Cycles are numbered from 0 through 14.

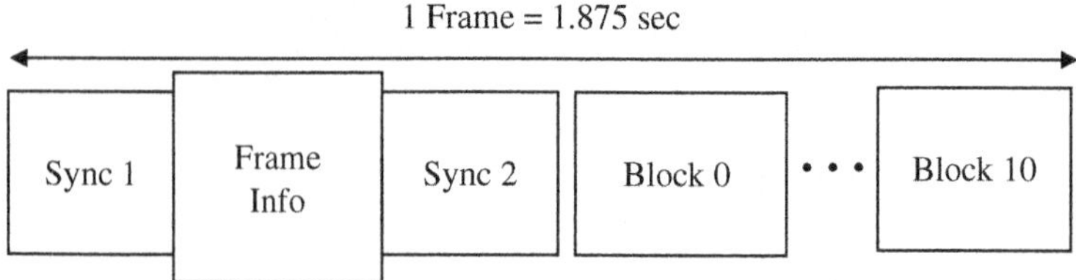

FIGURE 1.39 Flex protocol frame structure.

Sync 1 is used for time and frequency synchronization. Frame info is used to identify the frame number of the Flex cycle number. Sync 2 indicates the data rate used in the messaging part (1600, 3200, or 6400 bps). The modulation scheme used is FSK (2 level and 4 level).

The bursty errors caused by multipath fading are compensated or minimized by the use of a BCH[2] (32, 21) error correcting block code, along with an interleaver that can resist a certain length of a burst of errors. Also a few paging base stations will normally transmit the page simultaneously to a wide area; these pages or waveforms arrive at the pager with different propagation time delays. This is a phenomenon similar to delay spread or FSF. As the symbol time increases, the differential propagation time delay will consist of a smaller portion of the entire symbol and thus cause less BER degradation. Proper base station deployment will reduce the differential propagation time delays and improve paging coverage and system performance. This is especially noticed in going from 2-level to 4-level FSK. The four possible frequency deviations for the FSK modulation schemes are given as

$$\begin{aligned} f_{d1} &= \pm 1600 \text{ Hz} \\ f_{d2} &= \pm 4800 \text{ Hz} \end{aligned} \tag{1.15}$$

The 2-level FSK modulator is simplified and can be drawn as shown in Fig. 1.40. Here, the input bits are mapped to a bipolar signal, then low pass filtered to reduce the out-of-band emissions and then sent to a Voltage Control Oscillator (VCO) to deviate the frequency around the nominal carrier frequency. The signal entering the VCO is a binary signal.

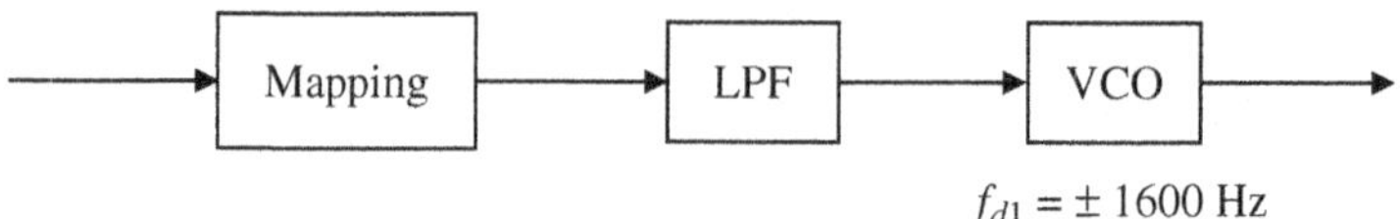

FIGURE 1.40 Two-level FSK modulation transmit block diagram.

Similarly, the 4-level FSK modulator is simplified and can be drawn, as shown in Fig. 1.41. Here we added the grouping of 2 bits to create a symbol. This function is spread across the serial to parallel and mapping blocks. This time, the signal entering the VCO is a 4-level signal.

In Table 1.10, we provide a summary of some of the above-discussed wireless systems.

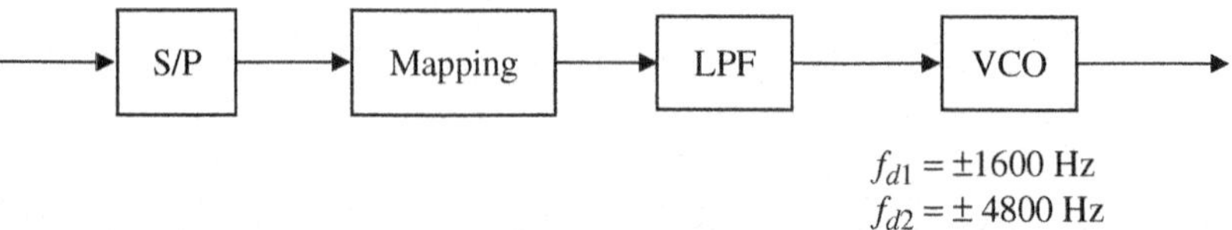

FIGURE 1.41 Four-level FSK modulation transmit block diagram.

[2] BCH = Bose-Chadhuri-Hocquenghem.

TABLE 1.10 Parameter Summary of Some Digital Cellular Systems

Design parameter	IS-136	GSM	IS-95	CDMA2000	WCDMA
Duplex	FDD	FDD	FDD	FDD	FDD
Modulation scheme	π/4-DQPSK	GMSK	QPSK OQPSK	QPSK HPSK	QPSK HPSK
Frame size	20 msec	4.615 msec	10 msec	10 msec	10 msec
FEC	Convolution	Convolution	Convolution	Convolution + turbo	Convolution+ turbo
Data rates	48.6 kbps	270.8 kbps	14.4 kbps	3 Mbps	2 Mbps
Detection technique	CD/DD	CD	CD/NC	CD	CD
Multipath mitigation	EQ	EQ	RAKE + PC	RAKE + PC	RAKE + PC
Scrambling codes	NA	NA	m-Sequence	m-Sequence	Gold
Channelization codes	NA	NA	Walsh	Walsh	OVSF
Spectral efficiency	48.6K/30K = 1.62	270.8K/200K= 1.354	14.4/1.25K = 0.0116	3M/2.5M = 1.2	2M/5M = 0.4

1.3 REASONS BEHIND CONVERGENCE OF WIRELESS SERVICES

We must first mention that there has been and continues to be a tremendous amount of standards activity that can basically be classified into two regions. The first region aims to keep alive certain standards that are already successfully deployed but are or have been coming up short in delivering the increasing demand for capacity and services, along with the user data rates needed to support these services/applications.

The second region aims to create new standards (albeit some built from existing ones) that are to satisfy the expected worldwide growth in capacity, as well as the increase in data rate required for multimedia services. This can also be referred to as evolution.

As a result of these regions, we see a proliferation of standards available to terminal and infrastructure manufacturers. In Fig 1.42, we show an evolutionary path for certain standards that support our observations. Similar observations can be made for both the WLAN and WPAN areas as well. The migration paths are shown as a function of time as we slide from the left to the right side of the page. As time and technology progress, we can take notes on the state of the consumer electronic devices [36].

The end user would much rather prefer carrying a single device compared to many devices. The amount of effort to design not only the devices but also the network to support the convergence of all these functions is tremendous. A design manufacturer may slice the problem into smaller problems, integrating functions as time and market demands (see Fig. 1.43). This also poses several hurdles to the service providers, since they must have content and applications to deliver the customer satisfaction.

1.3.1 Marketing Overview

In the previous section, we have presented a path of cellular standards. Similar paths exist for PAN as well as WLAN. They all increase data rates and throughput, not to mention improvement in QoS, as well as the addition of mobility to some degree. If we simply focus on data rate/throughput, we can then present Fig. 1.44 for comparison.

The observations that can be made from above is that the PAN and WLAN systems are growing upward and to the right (into the larger coverage area or increasing mobility), while cellular systems are growing straight up. It is worth mentioning that since the cellular systems need to handle indoor environments as well as high-speed vehicular environments, the actual slope of the cellular throughput data rates will tend to be negative. The slope is with respect to support of mobility.

A few scenarios can occur here; let's discuss two of them. The first being that WLAN + PAN continue to grow at a greater rate than that of WAN. In this case, the WLAN + PAN provide solutions, cellular provides another, and convergence continues, as long as there are distinctive advantages of one technique over another.

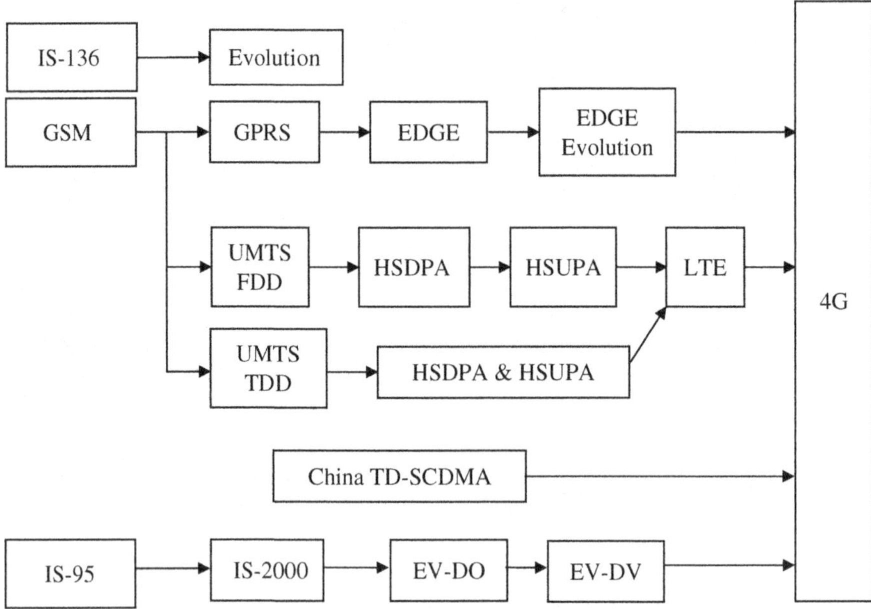

FIGURE 1.42 Evolutionary path for certain wireless standards.

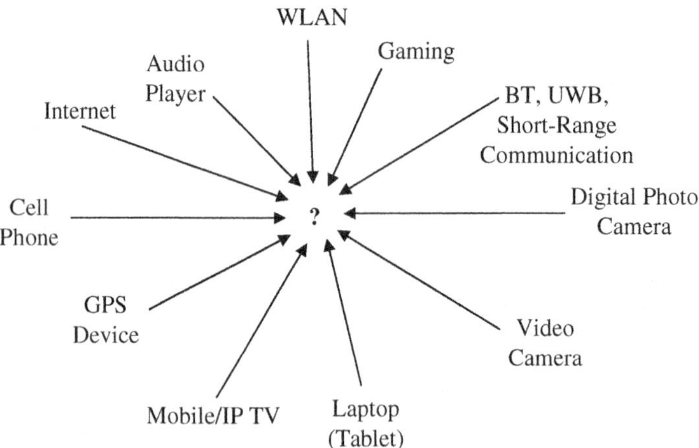

FIGURE 1.43 Convergence of services.

The second case is that WAN dramatically approaches the throughput of WLAN + PAN. In this case, the WLAN + PAN must win on the cost curve, to say the least. WAN will encounter growing pains in keeping mobility, while dramatically increasing user data rates. We can continue along these lines and realize that there is a need for PAN, WLAN, and WAN services, and thus, creating devices that bundle these converged services into a single platform is highly desirable. The integration of these technologies into a terminal is a challenging task. Let us take a look into a simplified handset block diagram in Fig. 1.45, where we emphasize the actual locations of the above-mentioned services inside the handset (see Fig. 1.46).

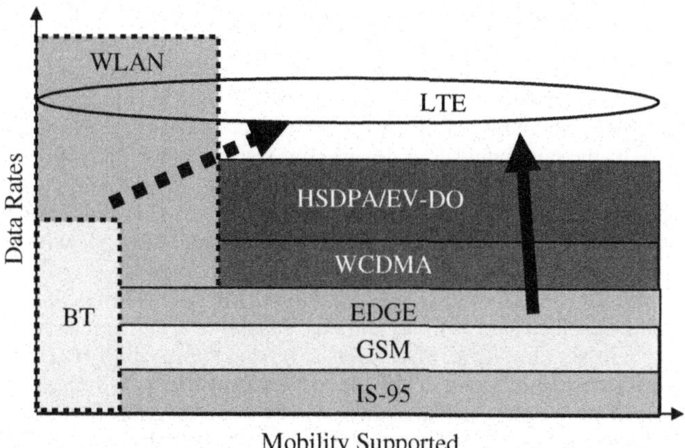

FIGURE 1.44 General overview of the trends in wireless systems.

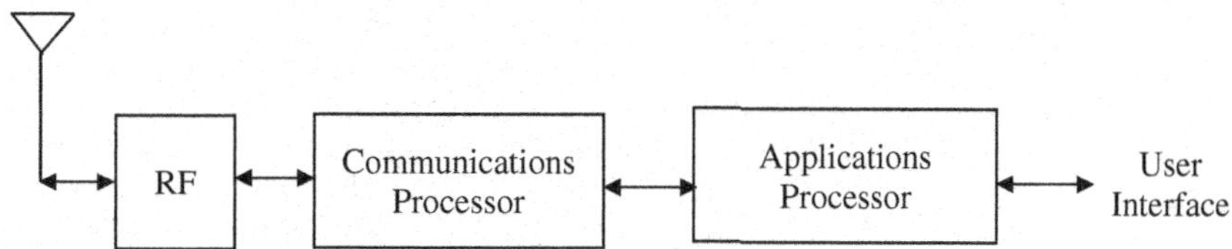

FIGURE 1.45 Simplified handset functional overview block diagram.

As you can see in Fig. 1.46, the WLAN + PAN services have been on the applications side of the handset with a possible connection to the optional or "add-on" device. A critical design issue is the simultaneous operation of all/some of these technologies. Now as IP communication technology enters, specifically VoIP, the mainstream of this partitioning may change.

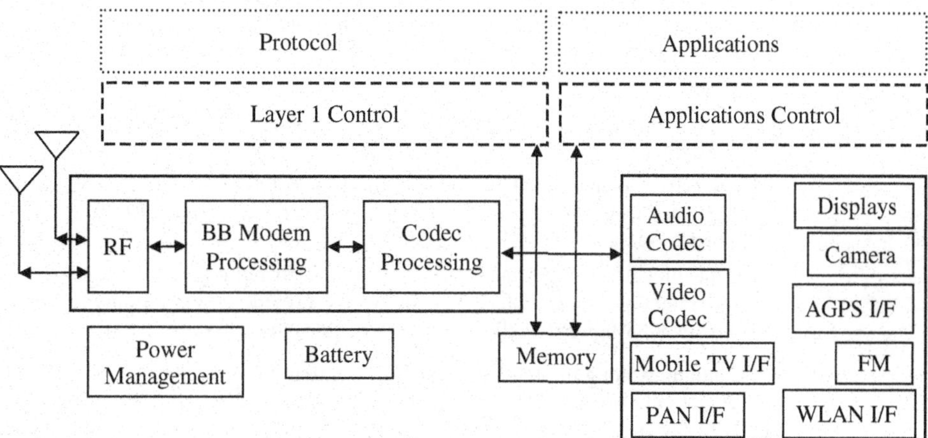

FIGURE 1.46 Functional partitioning between communications and applications.

1.3.2 Software Defined Radio (SDR) Introduction

There are many reasons and benefits in achieving an SDR platform [37]. We will use this as a demonstration vehicle to support the multimode receiver for the purposes of convergence.

There are a few reasons why the WLAN and WPAN components have been on the applications side of the communication device. For starters, these technologies (WLAN, BT, GPS, UWB, etc.) exist, in the most part, as a single-chip solution—thus making its intellectual property separated from the WAN, but at the same time, allowing these solutions to be considered as an "add-on" to the PDA or laptop community. Another reason is, historically speaking, the IC manufacturers of WAN, WLAN, and PAN devices have been different.There are very few successful cases of one manufacturer providing all of the above-mentioned solutions. However, the landscape has been changing and continues to change as time, services, capabilities, and demands rise.

In its simplest form, SDR refers to having the capability to reconfigure the terminal (i.e., handset) by software. More specifically the software must also encompass the protocol stack plus applications. Moreover, we must state that there will be some functionality (at least initially) in dedicated HW with a roadmap to be reconfigurable at some later date. In the simplest form, SDR entails being able to change the personality of the terminal (or device) by downloading or changing the software that resides on the device. This requires control not only over the RF, but also over the baseband signal processing, of which includes the layer 1 as well as the higher layers.

TABLE 1.11 Signal Processing Complexity Estimates

Cellular system	Processing complexity
GSM	100 MIPS
IS-95	500 MIPS
WCDMA	5000 MIPS

A most likely path to achieve an SDR terminal would be to start at the top of the protocol stack and work your way down. Once you get close to the physical layer, you are essentially operating at or below the RTOS so care must be taken when considering various implementations. A few public estimates are available for the required signal processing operations of MIPS and are given in Table 1.11 [38].

As these complexity estimates show, having a programmable device by itself may not be the best solution when considering the implications of device cost, size, as well as power consumption [39–41].

This may initially mean that, as far as the physical layer is concerned, we can either use a multimode receiver approach or a reconfigurable approach (this can also take on the form of an accelerator/ coprocessor, etc.). This reconfigurability can be further broken down into two parts: first, being fully reconfigurable in the same sense as an FPGA or, second, being partially reconfigurable. Here we choose a space or a vector subspace if you will, and create functions in that space that are reconfigurable. For example, you may create a spread spectrum engine that can be manipulated and configured to perform a wide variety of spread spectrum standards only. Hence, if you are looking for a TDMA standard, then you would either move to a point solution or create another vector space engine targeted toward mathematical operations that are typically performed in the TDMA systems.

The benefits of SDR are given below:

- For the end user, he/she will experience more personal and flexible services. This would also allow for coordinated effort to simplify international roaming.
- For service providers, SDR gives them the opportunity to provide personal services, more differentiated services, and performance. All of these services will be offered with the hopes of increasing their revenue.
- For manufacturers, they would trade in their multiple point solution devices to get a single platform (ideally) that can scale with the economy and provide longer product lifetimes. Moreover, they allow for some concerted effort in the development process.

Continuing along these lines, we should mention the concept of Cognitive Radio [42]. Here, the device not only has the ability to change its personality via a software change, but it can also sense the available spectrum and then decide the best band to operate in and with what technology to use. Presently, handsets are designed considering dual as well as quad band RF sections. As more frequency bands become available, these numbers will increase. If the intended area is not included in the presently designed handset, then the handset RF section will be redesigned to accommodate the change in frequency band. At first sight, we see an immediate need for such a radio in the military communications domain. When troops are deployed, more often than not, the frequency bands are different as well as the possible interference present. This Cognitive Radio will basically scan the frequency bands to identify safe and reliable frequency bands and then use them for communication. The potential for various standardized air specifications can be considered since this discussion was started with a software-defined radio in mind. Or the cognitive radio may be able to use a new and nonintrusive air interface. There are many questions and possibilities that will require the attention of the research community to help answer. Another application is in the first response domain (i.e., emergency medical, security, etc.). In this case, the terminals will scan other frequencies and decide on another RAT (radio access technology) to use that will maximize communication reliability.

1.3.3 Reconfigurable Modem Architecture

In this section, we will address reconfigurable architectures with the intention of being used instead of, or in conjunction with, the SDR approach discussed above. If we can revisit the simplified terminal block diagram shown earlier, we can see the "Velcro" approach is shown in Fig. 1.47.

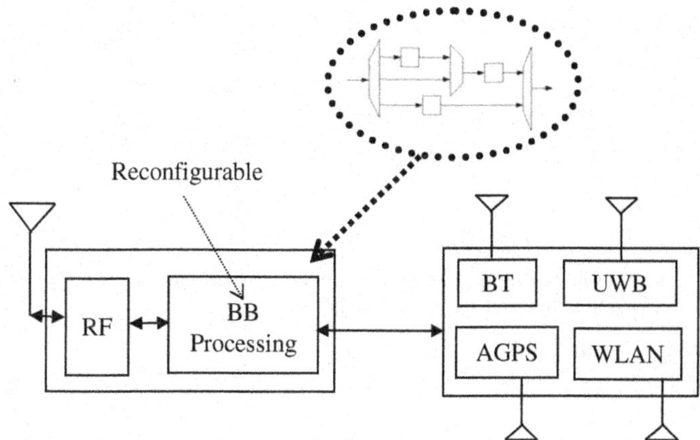

FIGURE 1.47 Multifunction terminal.

Here point solutions are integrated to reduce board space, component listing, phone volume, and of course bill of material (BOM). But, as time progresses, it may make more sense to observe the mathematics involved in each of the standards and discuss the similarities of the regions of overlap.

For example, a WLAN and cellular system using OFDM can conceivably reuse or share a common frequency domain–based receiver, with the appropriate parameters changed. Similarly, DSSS systems may be able to accommodate a single reconfigurable chip rate engine.

During this initial era of convergence, time, money, and common sense dictate that the point solutions are a viable solution. However, as time progresses and the technology landscape changes, an alternative solution may prevail.

1.4 STOCHASTIC AND SIGNAL PROCESSING REVIEW

In this section, we will present a brief review of some relevant probability and statistical signal processing. We will use these equations to describe and analyze various signal properties and signal processing functions [43, 44].

1.4.1 Statistical Foundation

A random variable is described by a cumulative distribution function (CDF), which is defined as

$$F(x) = P(X \leq x) \qquad (-\infty < x < \infty) \tag{1.16}$$

It is confined in the interval $0 \leq F(x) \leq 1$, with the additional constrains of $F(-\infty) = 0$ and $F(\infty) = 1$.

The derivative of the CDF is described by the probability density function (PDF), which is defined as

$$p(x) = \frac{d}{dx}F(x) \qquad (-\infty < x < \infty) \tag{1.17}$$

The PDF is always a positive function with a total area equal to 1:

$$\int_{-\infty}^{\infty} p(x)\,dx = 1 \tag{1.18}$$

Now, using the earlier definitions, we have rewritten the relationship between the CDF and the PDF.

$$F(x) = P(X \leq x) = \int_{-\infty}^{x} p(v)\,dv \tag{1.19}$$

Next we present a few methods used to analyze a random variable. The mean or expected value of a random variable X is defined as

$$E\{X\} = m_X = \int_{-\infty}^{\infty} xP(x)\,dx \tag{1.20}$$

$E\{g\}$ is the expected value of the random variable g. The expected value is a linear operator with the following properties (assuming K is a constant):

$$E\{Kg\} = K \cdot E\{g\} \tag{1.21}$$

$$E\{g + y\} = E\{g\} + E\{y\} \tag{1.22}$$

The variance of a random variable is defined as

$$\mathrm{Var}\{g\} = E\{[g - E\{g\}]^2\} \tag{1.23}$$

which is also represented in the following manner (given the above definitions).

$$\sigma_g^2 = \int_{-\infty}^{\infty} (g - m_g)^2 p(g)\,dg \tag{1.24}$$

Assuming the mean value is a constant, we then obtain

$$\mathrm{Var}\{g\} = E\{g^2\} - [E\{g\}]^2 \tag{1.25}$$

$\mathrm{Var}\{g\}$ is the variance of the random variable g. The variance is used to measure the concentration (or variability) of g around its expected value. The square root of the variance is called the standard deviation of the random variable being analyzed.

A commonly used random variable is the Gaussian or normal random variable. The PDF is defined as

$$p(x) = \frac{1}{\sigma\sqrt{2\pi}} e^{-\frac{(x-m_x)^2}{2\sigma^2}} \tag{1.26}$$

The CDF is defined as

$$F(x) = \frac{1}{2}\left(1 + \text{erf}\left[\frac{x - m_x}{\sigma\sqrt{2}}\right]\right) \tag{1.27}$$

where erf(x) is the error function and is defined as

$$\text{erf}(x) = \frac{2}{\sqrt{\pi}}\int_0^x e^{-t^2}dt \tag{1.28}$$

The CDF can also be defined in terms of the complementary error function given below:

$$F(x) = 1 - \frac{1}{2}\text{erfc}\left[\frac{x - m_x}{\sigma\sqrt{2}}\right] \tag{1.29}$$

where the erfc(x) function is defined below along with its relationship to the earlier defined error function.

$$\text{erfc}(x) = \frac{2}{\sqrt{\pi}}\int_x^{\infty} e^{-t^2}dt = 1 - \text{erf}(x) \tag{1.30}$$

1.4.2 Autocorrelation and Power Spectral Density (PSD)

Let us define the autocorrelation of a random process $X(t)$ to be a function of two time instances, t_1 and t_2.

$$R_X(t_1, t_2) = E\{x(t_1)x(t_2)\} \tag{1.31}$$

which can also be expressed, assuming $\tau = t_2 - t_1$.

$$R_X(\tau) = E\{x(t + \tau)x(t)\} \tag{1.32}$$

Let us consider the effects of a linear filter on a random process. In Fig. 1.48, the random variable $X(t)$ is input to the filter with impulse response $h(t)$.

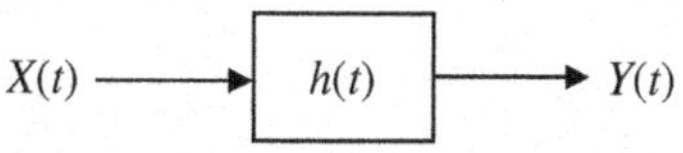

$X(t)$ → $h(t)$ → $Y(t)$

FIGURE 1.48 Filter convolution example.

The filter output is easily written in the time domain by convolving the impulse response of the filter with the input signal.

$$Y(t) = \int_{-\infty}^{\infty} h(\tau)X(t - \tau)\,d\tau \tag{1.33}$$

Using the frequency domain, we can simply write down the power spectral density (PSD) of the input as (units are Watts/Hz)

$$S_X(f) = \int_{-\infty}^{\infty} R_X(\tau)e^{-j2\pi f\tau}\,d\tau \tag{1.34}$$

We can write down the PSD of the linear filter output as

$$S_Y(f) = |H(f)|^2 S_X(f) \tag{1.35}$$

Now let us apply the above observations to the case when the linear time-invariant filter is excited with noise $N(t)$ that has a Gaussian distribution and whose spectral density is white. The PSD is expressed as

$$S_N(f) = \frac{N_o}{2} \tag{1.36}$$

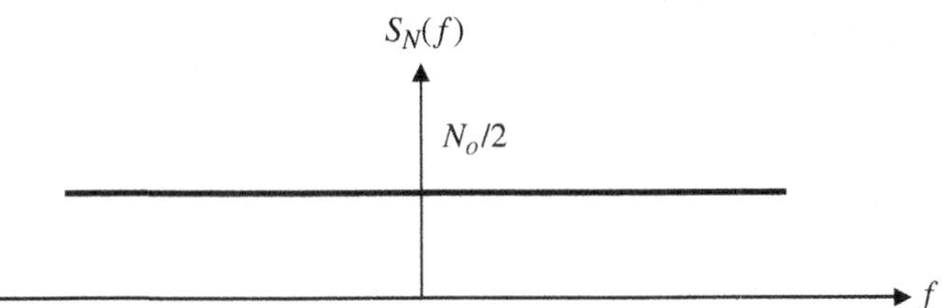

FIGURE 1.49 Power spectral density of white noise.

and shown graphically in Fig. 1.49.

To obtain the autocorrelation function, we simply use the inverse Fourier transform to get

$$R_N(\tau) = \frac{N_o}{2}\partial(\tau) \tag{1.37}$$

where $\partial(t)$ is the Dirac delta function whose value equals 1 only when $t = 0$. The autocorrelation function is graphically expressed in Fig. 1.50.

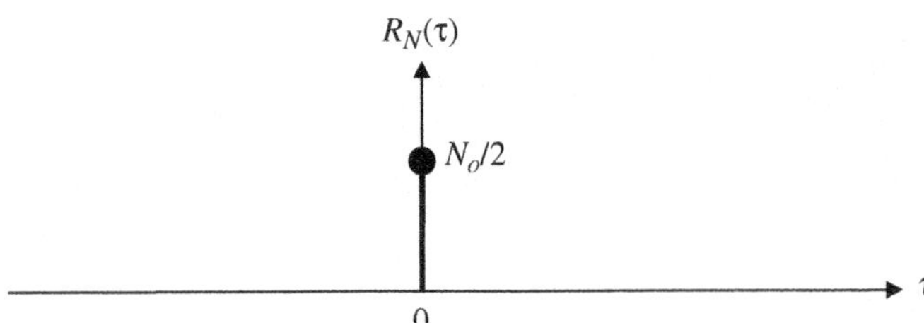

FIGURE 1.50 Autocorrelation function of white noise.

What this tells us is that if we take two different samples of this noise signal, they are uncorrelated with each other. Now let us insert an ideal low pass filter of bandwidth BW; then the PSD at the filter output is

$$S_Y(f) = \begin{cases} \frac{N_o}{2} & -\text{BW} \leq f \leq \text{BW} \\ 0 & |f| > \text{BW} \end{cases} \tag{1.38}$$

The inverse Fourier transform provides the autocorrelation function as

$$R_Y(\tau) = \int_{-\text{BW}}^{\text{BW}} \frac{N_o}{2} \cdot e^{j2\pi f\tau}\, d\tau \tag{1.39}$$

$$R_Y(\tau) = \frac{N_o}{2} \cdot \frac{\sin(2\pi\tau\text{BW})}{\pi\tau} \tag{1.40}$$

1.5 REPRESENTING BAND PASS SIGNALS AND SUBSYSTEMS

In this section, we aim to provide tools the reader can use to analyze band pass signals and systems. This tool will take the form of complex envelope theory and will be used extensively throughout this book.

1.5.1 Complex Envelope Theory

We will use the complex envelope theory to model band pass signals and systems (see [45] and [46]). We have decided to present the complex envelope theory with the hope of accomplishing the following three goals:

1. Provide mathematical insight into the signal/system being considered.
2. Draw relationships to the actual HW being designed.
3. Provide manageable means of computer simulation.

In order to derive the complex envelope of a signal, we approach this in two directions: First, we provide the mathematical derivation shown in a step-by-step procedure. Second, we simply use the brute force approach and write down the complex envelope directly from the signal being observed.[3]

Let us first start with the canonical form of a band pass signal.

$$g(t) = A \cdot \cos[\omega_c t + \theta(t)] \tag{1.41}$$

or

$$g(t) = \mathrm{Re}\{A \cdot e^{j\omega_c t} \cdot e^{j\theta(t)}\} \tag{1.42}$$

which can be written as

$$g(t) = A \cdot \cos[\theta(t)]\cos[\omega_c t] - A \cdot \sin[\theta(t)]\sin[\omega_c t] \tag{1.43}$$

We will define the following

$$g_c(t) = A \cdot \cos[\theta(t)] \tag{1.44}$$

$$g_s(t) = A \cdot \sin[\theta(t)] \tag{1.45}$$

where $g_c(t)$ is the in-phase component of the band pass signal and $g_s(t)$ is the quadrature component of the band pass signal. Note that the results equally apply when there is a time-varying amplitude as well, $A(t)$.

The complex envelope of $g(t)$ is now written as

$$\tilde{g}(t) = g_c(t) + jg_s(t) \tag{1.46}$$

Note that $\tilde{g}(t)$ is called the "complex envelope" or sometimes referred to as the "analytic signal." In order to provide some insight into the analytic signal, we see that, in the definition of $g(t)$, we can simply obtain its complex envelope by writing down whatever gets multiplied by the carrier phasor within the real part extraction bracket.

Our next direction requires the introduction of pre-envelope, Hilbert transform, and then culminating to the complex envelope. So the concept of the pre-envelope is as follows:

$$g_p(t) = g(t) + j\hat{g}(t) \tag{1.47}$$

where the pre-envelope consists of the real-valued signal plus a quadrature component that is the Hilbert transform of the real-valued signal.

[3] Various trigonometric identities are supplied in the Appendix.

The Hilbert transform of $g(t)$ is defined as

$$\hat{g}(t) = \frac{1}{\pi}\int_{-\infty}^{\infty} \frac{g(\tau)}{t-\tau}d\tau \tag{1.48}$$

which can also be given as (assuming $\otimes$ is the convolution operator)

$$\hat{g}(t) = g(t) \otimes \frac{1}{\pi t} \tag{1.49}$$

Using the Fourier transform, this is rewritten as

$$\hat{G}(f) = G(f) \cdot [-j\text{sgn}(f)] \tag{1.50}$$

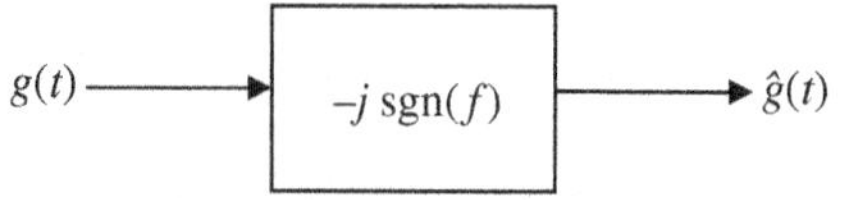

FIGURE 1.51 Hilbert transform transfer function.

Hence the Hilbert transform can be obtained by passing $g(t)$ through the filter shown in Fig. 1.51, with the sign function defined as

$$\text{sgn}(f) = \begin{cases} 1 & f > 0 \\ 0 & f = 0 \\ -1 & f < 0 \end{cases} \tag{1.51}$$

Using this definition in the pre-envelope, the Fourier transform of Eq. (1.47) is shown as

$$G_p(f) = G(f) + j[-j\text{sgn}(f)] \cdot G(f) \tag{1.52}$$

$$G_p(f) = G(f)\{1+\text{sgn}(f)\} \tag{1.53}$$

This is an operation that simply doubles the positive side of $G(f)$ and suppresses the negative side.

$$G_p(f) = \begin{cases} 2G(f) & f > 0 \\ G(0) & f = 0 \\ 0 & f < 0 \end{cases} \tag{1.54}$$

As an example, let's consider the frequency response of the real-valued signal $g(t)$ in Fig. 1.52.

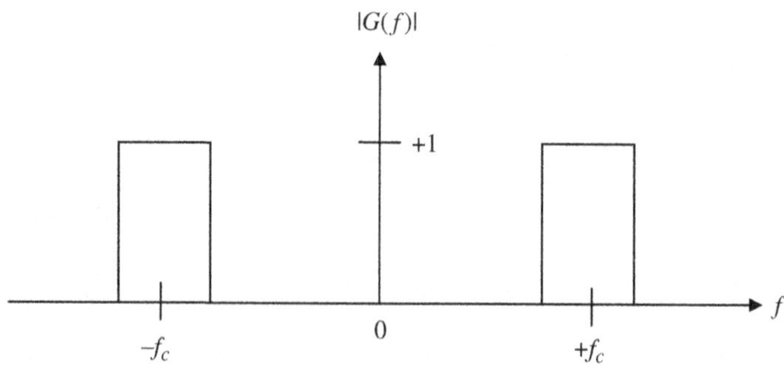

FIGURE 1.52 Frequency response of the real-valued signal $g(t)$.

The pre-envelope spectrum is drawn in Fig. 1.53.

The complex envelope is obtained by spectrally downshifting the pre-envelope to DC (direct current) or zero IF (intermediate frequency) (see Fig. 1.54), which is mathematically represented as

$$\tilde{g}(t) = g_p(t) \cdot e^{-j\omega_c t} \tag{1.55}$$

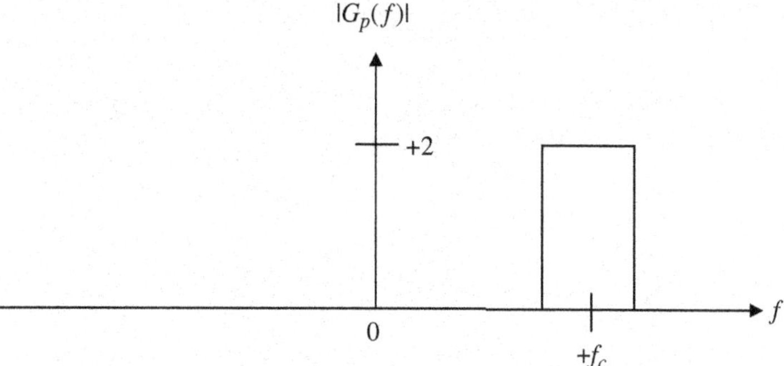

FIGURE 1.53 Pre-envelope frequency response of the real-valued signal $g(t)$.

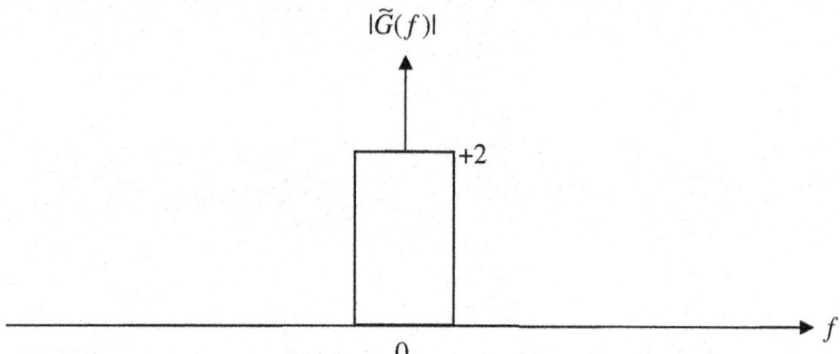

FIGURE 1.54 Complex envelope frequency response of the real-valued signal $g(t)$.

Relationship to Hardware. Next, the complex envelope is related to HW by the use of quadrature modulation and demodulation functions (see Fig. 1.55). We have shown the input to the quadrature modulator is the complex envelope. Also, the output of the quadrature demodulator is the received complex envelope signal (see Fig. 1.56).

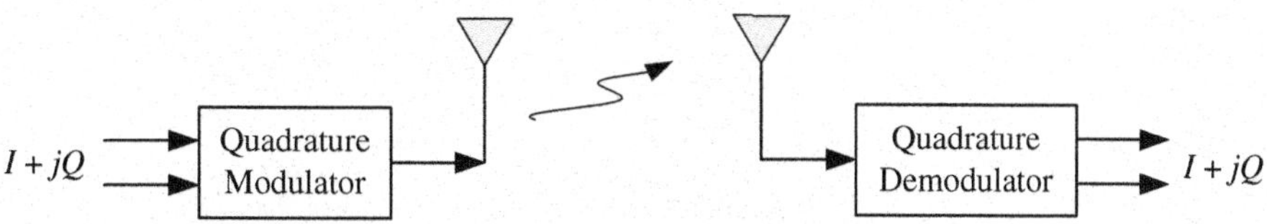

FIGURE 1.55 Spectral up and down conversion using quadrature modulators.

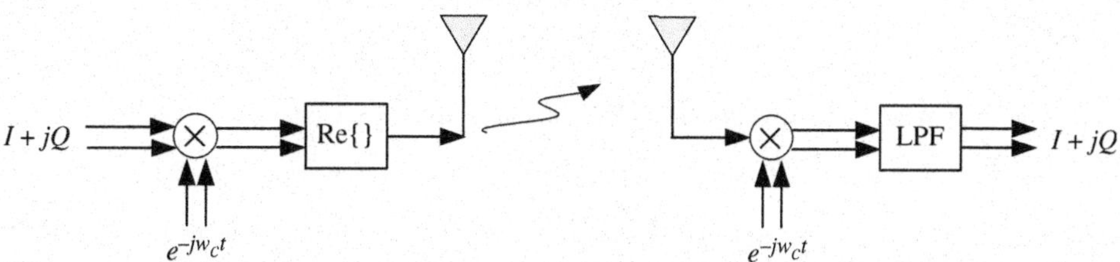

FIGURE 1.56 Spectral up and down conversion using complex envelope theory.

1.5.2 Band Pass Filter (BPF) Example

In this section, we wish to obtain the baseband representation of a band pass system specifically a BPF. Let us follow the band pass signal approach discussed earlier, so we have

$$h(t) = 2h_c(t) \cdot \cos[\omega_c t] - 2h_s(t)\sin[\omega_c t] \tag{1.56}$$

The complex envelope is

$$\tilde{h}(t) = h_c(t) + jh_s(t) \tag{1.57}$$

which can be rewritten as

$$h(t) = \mathrm{Re}\{2\tilde{h}(t) \cdot e^{j\omega_c t}\} \tag{1.58}$$

$$h(t) = \mathrm{Re}\{\tilde{h}(t) \cdot e^{j\omega_c t} + \tilde{h}^* e^{-j\omega_c t}\} \tag{1.59}$$

where * is used to denote the complex conjugate operation.

Using the Fourier transform, we get

$$H(f) = \tilde{H}(f - f_c) + \tilde{H}^*(-f - f_c) \tag{1.60}$$

The complex envelope of the output signal of a band-limited system is the convolution of the impulse response $\tilde{h}(t)$ of the system with the complex envelope $\tilde{g}(t)$ of the input band pass signal.

For sake of completeness, we show the real-valued BPF and then its complex envelope representation (see Fig. 1.57).

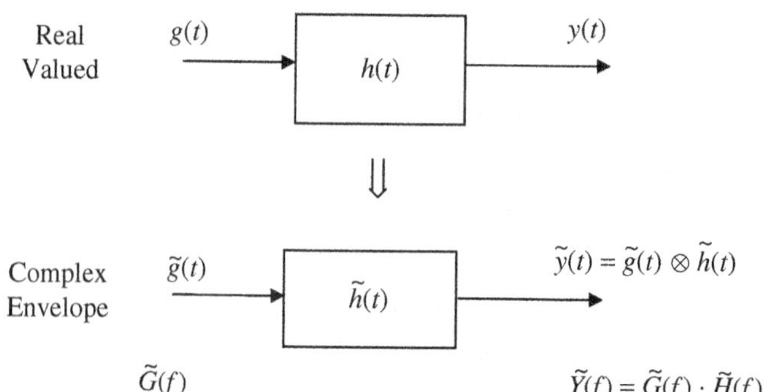

FIGURE 1.57 Complex envelope representation of a BPF.

The baseband output signal is mathematically represented as

$$\tilde{y}(t) = \{h_c(t) + jh_s(t)\} \otimes \{g_c(t) + jg_s(t)\} \tag{1.61}$$

After carrying out the complex arithmetic, we arrive with the following

$$\tilde{y}(t) = \{h_c(t) \otimes g_c(t) - h_s(t) \otimes g_s(t)\} + j\{h_s(t) \otimes g_c(t) + h_c(t) \otimes g_s(t)\} \tag{1.62}$$

The above equation shows that the real-valued BPF, $h(t)$, can be modeled by its analytic equivalent using the block diagram in Fig. 1.58.

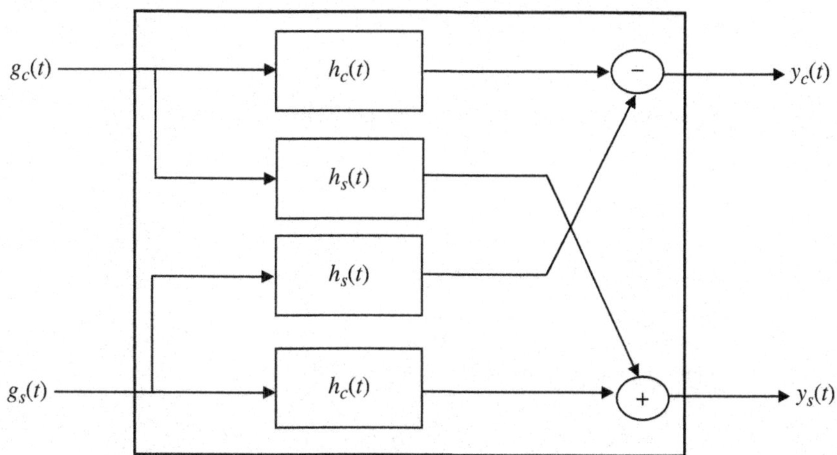

FIGURE 1.58 Complex envelope representation of a BPF, $\tilde{h}(t)$.

1.5.3 Modulation Example

In this section, we will apply the complex envelope theory to a modulation scheme. Let us first consider the real-valued transmitted signal to be

$$g(t) = \mathrm{Re}\{g_p(t)\} \tag{1.63}$$

$$g(t) = \mathrm{Re}\{\tilde{g}(t) \cdot e^{jw_c t}\} \tag{1.64}$$

which can easily be shown to equal

$$g(t) = g_c(t) \cdot \cos[\omega_c t] - g_s(t) \cdot \sin[\omega_c t] \tag{1.65}$$

which is in the canonical form we presented earlier in this topic. Now let us see how we can apply this theory to an angle-modulated signal, say frequency modulation (FM).

$$s(t) = A \cdot \cos\left[2\pi f_c t + 2\pi k_f \int_{-\infty}^{t} m(t)\, dt\right] \tag{1.66}$$

$$s(t) = A \cdot \mathrm{Re}\left\{ s^{j2\pi f_c t} \cdot e^{j2\pi k_f \int_{-\infty}^{t} m(t)\, dt} \right\} \tag{1.67}$$

where k_f = sensitivity factor and $m(t)$ = information modulation signal.

We can quickly write down the complex envelope representation of an FM signal as

$$\tilde{s}(t) = A \cdot e^{j2\pi k_f \int_{-\infty}^{t} m(t)\, dt} \tag{1.68}$$

$$\tilde{s}(t) = A \cdot \cos\left[2\pi k_f \int_{-\infty}^{t} m(t)\, dt\right] + jA \cdot \sin\left[2\pi k_f \int_{-\infty}^{t} m(t)\, dt\right] \tag{1.69}$$

Next let us discuss the complex envelope of a QAM signal.

$$s(t) = A(t) \cdot \cos[2\pi f_c t + \theta(t)] \tag{1.70}$$

or

$$s(t) = A(t) \cdot \mathrm{Re}\left\{e^{j2\pi f_c t} \cdot e^{j\theta(t)}\right\} \tag{1.71}$$

In a similar fashion, we can write down the complex envelope as

$$\tilde{s}(t) = A(t) \cdot e^{j\theta(t)} \tag{1.72}$$

or

$$\tilde{s}(t) = A(t) \cdot \cos[\theta(t)] + jA(t) \cdot \sin[\theta(t)] \tag{1.73}$$

Taking this one step further, for Cartesian coordinate representation,

$$\tilde{s}(t) = a(t) + jb(t) \tag{1.74}$$

and for polar coordinate representation, we have

$$\tilde{s}(t) = \sqrt{a^2(t) + b^2(t)} \cdot e^{j\tan^{-1}\left[\frac{b(t)}{a(t)}\right]} \tag{1.75}$$

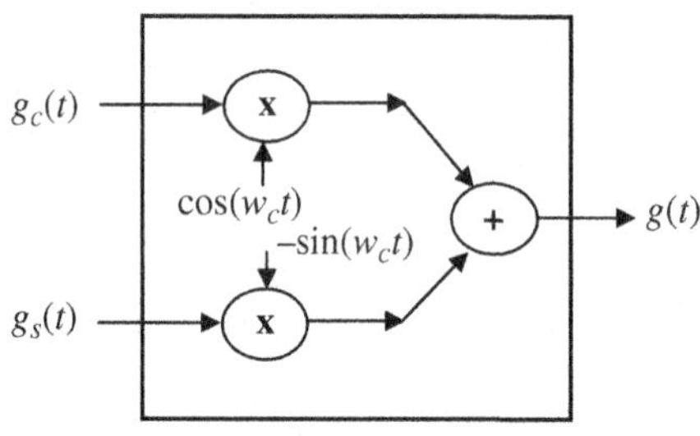

FIGURE 1.59 Quadrature modulator block diagram.

Now let us understand the fact that the transmitted signal is simply the complex envelope multiplied by a complex carrier signal. Pictorially, we have the quadrature modulator shown in Fig. 1.59.

As you can see, we simply form the complex envelope of the signal and then perform a quadrature modulation operation to shift the spectrum to a carrier frequency f_c. Here, we have also managed to accomplish a connection to the actual HW implementation.

Let us revisit the FM signal derived earlier in order to write down the complex envelope as follows

$$g_c(t) = A \cdot \cos\left[2\pi k_f \int_{-\infty}^{t} m(t)\,dt\right] \tag{1.76}$$

$$g_s(t) = A \cdot \sin\left[2\pi k_f \int_{-\infty}^{t} m(t)\,dt\right] \tag{1.77}$$

Then, a commonly used method of generating an FM signal is understood and is given in Fig. 1. 60.

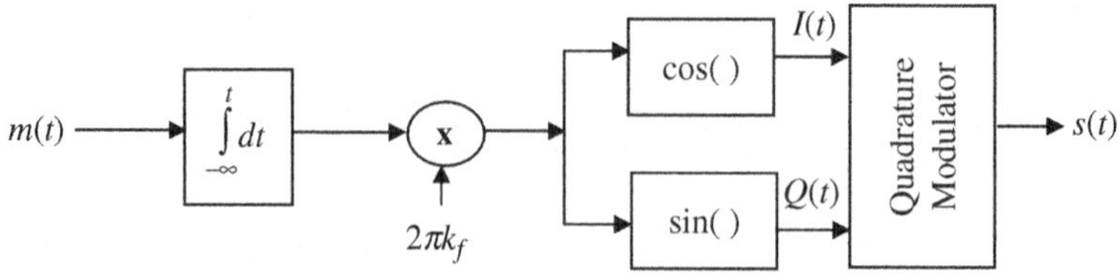

FIGURE 1.60 Baseband implementation of an FM modulator.

The baseband signals, I and Q, will be discussed in more detail in the next chapter when we address FSK. It is important to note that good understanding of the complex envelope of the FSK signal will be essential and useful not only in designing efficient transmission subsystems, but also for signal detection being performed in the receiver.

1.5.4 Zero IF (ZIF) Discussion

In the previous section, we have managed to connect the quadrature modulator device to the complex envelope theory. In this section, we will discuss what can be said for signals that appear at the output of a receiver quadrature demodulator device. An operation such as this is also commonly called "Zero IF (ZIF)" or "Direction Conversion Receiver (DCR)."

Let us present the quadrature demodulator operational theory. Assume the channel has not introduced any other distortion besides additive white Gaussian noise (AWGN), $n(t)$. So we can write the received signal as

$$r(t) = s(t) + n(t) \tag{1.78}$$

We will represent $n(t)$ as a band pass signal since it already encountered a BPF in the receiver. So we have

$$r(t) = A(t) \cdot \cos[\omega_c t + \theta(t)] + N(t) \cdot \cos[\omega_c t + \phi(t)] \tag{1.79}$$

The quadrature demodulator works as shown in Fig. 1.61.

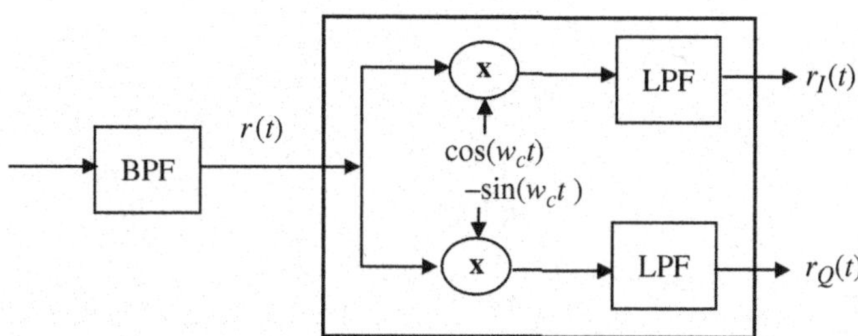

FIGURE 1.61 Quadrature demodulator block diagram.

Using some basic trigonometric identities, we have the received signal mathematically represented as

$$r_I(t) = r(t) \cdot \cos[\omega_c t] \tag{1.80}$$

After substituting for the received signal, we get

$$r_I(t) = \frac{A(t)}{2} \cdot \{\cos[\theta(t)] + \cos[2\omega_c t + \theta(t)]\} + \frac{N(t)}{2} \cdot \{\cos[\phi(t)] + \cos[2\omega_c t + \phi(t)]\} \tag{1.81}$$

The LPF removes the summed frequency components, so we have

$$r_I(t) = \frac{A(t)}{2} \cdot \cos[\theta(t)] + \frac{N(t)}{2} \cdot \cos[\phi(t)] \tag{1.82}$$

$$r_I(t) = \frac{1}{2} \cdot g_c(t) + \frac{1}{2} \cdot n_c(t) \tag{1.83}$$

It is important to observe that in the absence of other impairments, the in-phase components of noise add to the in-phase components of the signal. Similarly, the quadraphase components are given as (after the LPF operation)

$$r_Q(t) = \frac{A(t)}{2} \cdot \sin[\theta(t)] + \frac{N(t)}{2} \cdot \sin[\phi(t)] \tag{1.84}$$

If a QAM signal was transmitted, then we would have

$$r_I(t) = \frac{a(t)}{2} + \frac{n_c(t)}{2} \tag{1.85}$$

and

$$r_Q(t) = \frac{b(t)}{2} + \frac{n_s(t)}{2} \tag{1.86}$$

For the FM example given earlier, we have

$$r_I(t) = \frac{A(t)}{2} \cdot \cos\left[2\pi k_f \int_{-\infty}^{t} m(t)\,dt\right] + \frac{n_c(t)}{2} \tag{1.87}$$

and

$$r_Q(t) = \frac{A(t)}{2} \cdot \sin\left[2\pi k_f \int_{-\infty}^{t} m(t)\,dt\right] + \frac{n_s(t)}{2} \tag{1.88}$$

In the absence of noise, we can obtain the phase of the received signal as

$$\theta(t) = \tan^{-1}\left[\frac{\frac{A(t)}{2} \cdot \sin\left[2\pi k_f \int_{-\infty}^{t} m(t)\,dt\right]}{\frac{A(t)}{2} \cdot \cos\left[2\pi k_f \int_{-\infty}^{t} m(t)\,dt\right]}\right] \tag{1.89}$$

$$\theta(t) = 2\pi k_f \int_{-\infty}^{t} m(t)\,dt \tag{1.90}$$

The instantaneous frequency can be obtained by differentiating the phase to obtain

$$\omega_i(t) = \frac{d}{dt}\theta(t) = 2\pi k_f m(t) \tag{1.91}$$

A block diagram is given in Fig. 1.62.

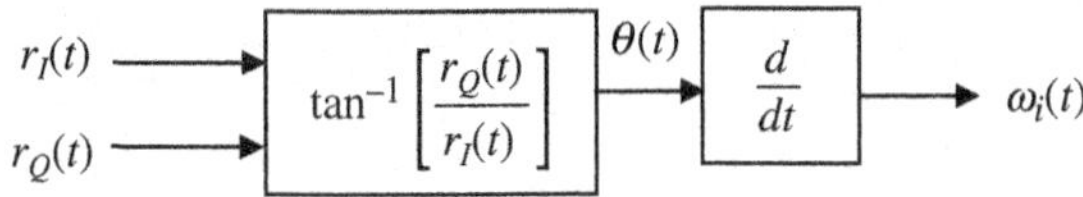

FIGURE 1.62 Polar coordinate system–based discriminator receiver.

1.5.5 Generalization of Tx and Rx Functions

In this section, we hope to build upon the previous sections where we discussed the complex envelope theory to create a universal transmitter and receiver architectural block diagram. A general transmitter is given in Fig. 1.63.

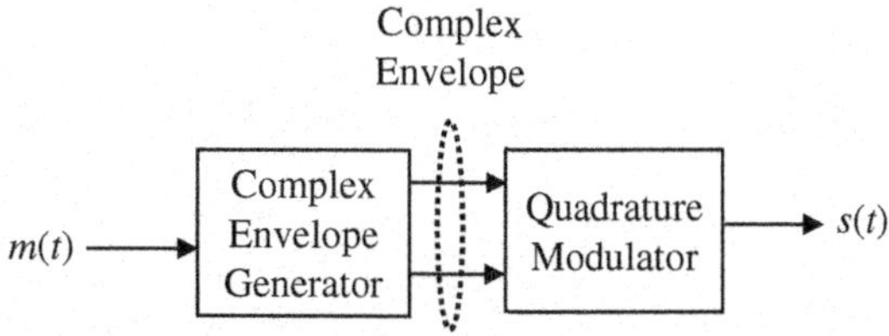

FIGURE 1.63 Universal modulation scheme transmitter.

Eventually we need to reverse the operations performed at the transmitter, this occurs in the receiver as shown in Fig. 1.64. At the transmitter, we have taken the information input and then mapped it to a modulation scheme, which would later be transmitted over the wireless medium. What we have essentially performed is a series of domain transformations. First we have taken the information signal from the information domain and then transformed the signal into the complex envelope domain. This is then spectrally upshifted to the real-valued domain. Hence any modulation scheme can be implemented in the generic transmitter block diagram.

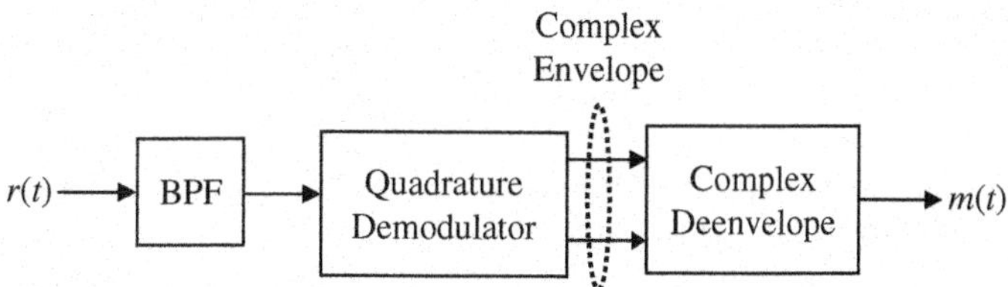

FIGURE 1.64 Universal modulation scheme receiver.

As stated earlier, the receiver must reverse the operations performed at the transmitter. Hence, we first perform the spectrally downshifting operation, and then perform the complex deenvelope operation. Typically, this is called demodulation and/or signal detection in the literature. We have included the generic receiver to show additional insight into what the actual operations mean.

1.6 RECEIVER SENSITIVITY DEFINITIONS

A figure of merit commonly used to evaluate a digital communication system is the probability of error, otherwise known as BER. The BER is defined as the number of bit errors encountered divided by the total number of bits transmitted. The BER is typically plotted on a log-log scale graph, which is a function of a scaled version of the signal to noise ratio (SNR). We would like to show the interrelationship between a few SNR representations [47].

Let us define the received SNR as follows

$$\frac{S}{N} = \frac{E_b}{T_b} \cdot \frac{1}{N} \tag{1.92}$$

where S = signal power and N = noise power.
The amount of noise that enters our decision device in the receiver depends on the BW of the channel selectivity filter; hence we now have

$$\frac{S}{N} = \frac{E_b}{T_b} \cdot \frac{1}{\text{BW} \cdot N_o} \tag{1.93}$$

$$\frac{S}{N} = \frac{E_b}{N_o} \cdot \frac{R_b}{\text{BW}} \tag{1.94}$$

where $\frac{E_b}{N_o}$ = bit energy per noise PSD. This can also be interpreted as saying the SNR normalized by the bit rate.

Let us consider an example of binary signaling, using an SRC filter for the equivalent receive filter response. We now have

$$\frac{S}{N} = \frac{E_b}{N_o} \cdot \frac{R_b}{\frac{R_b}{2} \cdot 2} \tag{1.95}$$

which gives

$$\frac{S}{N} = \frac{E_b}{N_o} \tag{1.96}$$

So far we have provided the mathematical relationships between the SNR and SNR per bit. Let us now present an alternative representation

$$\frac{S}{N} = \frac{S}{kT \cdot \text{BW} \cdot \text{NF}} \tag{1.97}$$

where $kT = -174$ dBm/Hz and NF = the receiver noise figure. This value typically corresponds to the noise generated in the RF front end plus the noise rise due to multiple access interference (MAI).

Lastly, equating Eqs. (1.96) and (1.97), we have

$$\frac{E_b}{N_o} = \frac{S}{kT \cdot R_b \cdot \text{NF}} \tag{1.98}$$

which can be expressed as follows in dB form:

$$\left(\frac{E_b}{N_o}\right)_{\text{dB}} = (S)_{\text{dBm}} - 10 \cdot \log(kT) - 10 \cdot \log(R_b) - 10 \cdot \log(\text{NF}) \tag{1.99}$$

REFERENCES

[1] V. J. MacDonald, "The Cellular Concept," *The Bell System Technical Journal*, Vol. 58, No. 1, Jan. 1979, pp. 15–41.

[2] A. D. Kucar, "Mobile Radio: An Overview," *IEEE Communications Magazine*, Vol. 29, No. 11, Nov. 1991, pp. 72–85.

[3] T. S. Rappaport, "The Wireless Revolutions" *IEEE Communications Magazine*, Vol. 29, Nov. 1991, pp. 52–71.

[4] IS-136 North American Digital Cellular Standard.

[5] M. Eriksson et al., "System Overview and Performance Evaluation of GERAN— The GSM/EDGE Radio Access Network," IEEE 2000, pp. 902–906.

[6] N. R. Sollenberger, N. Seshadri, and R. Cox, "The Evolution of IS-136 TDMA for Third-Generation Wireless Services," *IEEE Personal Communications*, Vol. 6, No. 3, June 1999, pp. 8–18.

[7] D. D. Falconer, F. Adachi, and B. Gudmundson, "Time Division Multiple Access Methods for Wireless Personal Communications," *IEEE Communications Magazine*, Vol. 33, No. 1, Jan. 1995, pp. 50–57.

[8] GSM 05.01, "Digital Cellular Telecommunication System (Phase 2+); Physical Layer on the Radio Path General Description."

[9] GSM 05.02, "Digital Cellular Telecommunication System (Phase 2+); Multiplexing and Multiple Access on the Radio Path."

[10] GSM 05.05, "Digital Cellular Telecommunication System (Phase 2+); Radio Transmission and Reception."

[11] GSM 05.03, "Digital Cellular Telecommunication System (Phase 2+); Channel Coding."

[12] A. Mehrotra, *GSM System Engineering*, Artech House, 1997, New York.

[13] C. Ferrer and M. Oliver, "Overview and Capacity of the GPRS (General Packet Radio Service," IEEE 1998, pp. 106–110.

[14] R. Yallapragada, V. Kripalani, and A. Kripalani, "EDGE: A Technology Assessment," *ICPWC*, 2002, pp. 35–40.

[15] R. Kalden, I. Meirick, and M. Meyer, "Wireless Internet Access Based on GPRS," *IEEE Personal Communications*, Vol. 7, No. 2, April 2000, pp. 8–18.

[16] A. Furuskar, S. Mazur, F. Muller, and H. Olofsson, "EDGE: Enhanced Data Rates for GSM and TDMA/136 Evolution," *IEEE Personal Communications*, Vol. 6, No. 3, June 1999, pp. 56–66.

[17] IS-95 Spread Spectrum Digital Cellular Standard.

[18] 3GPP2, "Physical Layer Standard for cdma2000 Spread Spectrum Systems," Rev. D.

[19] www.3gpp.org

[20] 3GPP Technical Specification TS25.211, "Physical Channel and Mapping of Transport Channels onto Physical Channels (FDD)," Release 7.0.

[21] 3GPP Technical Specification TS25.212, "Multiplexing and Channel Coding (FDD)," Release 7.0.

[22] 3GPP Technical Specification TS25.213, "Spreading and Modulation (FDD)," Release 7.0.

[23] 3GPP Technical Specification TS25.214, "Physical Layer Procedures (FDD)," Release 7.0.

[24] E. Berruto, G. Colombo, P. Monogioudis, A. Napolitano, and K. Sabatakakis, "Architectural Aspects for the Evolution of Mobile Communications Toward UMTS," *IEEE Journal on Selected Areas in Communications*, Vol. 15, No. 8, Oct. 1997, pp. 1477–1487.

[25] J. S. Dasilva, B. Arroyo, B. Barani, and D. Ikonomou, "European Third-Generation Mobile Systems," *IEEE Communications Magazine*, Vol. 34, No. 10, Oct. 1996, pp. 68–83.

[26] IEEE 802.11a, Wireless LAN Specification.

[27] IEEE 802.11b, Wireless LAN Specification.

[28] IEEE 802.11g, Wireless LAN Specification.

[29] B. P. Crow, I. Widjaja, J. G. Kim, and P. T. Sakai, "IEEE 802.11 Wireless Local Area Networks," *IEEE Communications Magazine*, Vol. 35, No. 9, Sept. 1997, pp. 116–126.

[30] Motorola, iDen Specification, August 2000.

[31] M. A. Birchler and S. C. Jasper, "A 64KBps Digital Land Mobile Radio System Employing M-16QAM," ICWC, 1992, pp. 158–162.

[32] Bluetooth Specification Version 2.0 + Enhanced Data Rate, "Architecture & Terminology Overview."

[33] M. Phillips, "Reducing the cost of Bluetooth systems," *Electronics & Communication Engineering Journal*, Vol. 13, No. 5, Oct. 2001, pp. 204–208.

[34] Bluetooth Specification Version 2.0 + Enhanced Data Rate, "Core System Package."

[35] Motorola, Flex Paging Standard.

[36] D. J. Goodman and R. A. Myers, "3G Cellular Standards and Patents," in: *International Conference on Wireless Networks, Communications and Mobile Computing*, IEEE 2005, pp. 415–420.

[37] www.sdrforum.org

[38] A. Gatherer, T. Stetzler, M. McMahan, and Edgar Auslander, "DSP-Based Architectures for Mobile Communications: Past, Present, and Future," *IEEE Communications Magazine*, Vol. 38, No. 1, Jan. 2000, pp. 84–90.

[39] J. A. Wepman, "Analog-to-Digital Converters and Their Applications in Radio Receivers," *IEEE Communications Magazine*, Vol. 33, No. 5, May 1995, pp. 39–45.

[40] W. H. W. Tuttlebee, "Software-Defined Radio: Facets of a Developing Technology," *IEEE Personal Communications*, Vol. 6, No. 2, April 1999, pp. 38–44.

[41] J. Palicot and C. Roland, "A New Concept for Wireless Reconfigurable Receivers," *IEEE Communications Magazine*, Vol. 41, No. 7, July 2003, pp. 124–132.

[42] S. Haykin, "Cognitive Radio: Brain-Empowered Wireless Communications," *IEEE Journal on Selected Areas in Communications*, Vol. 23, No. 2, Feb. 2005, pp. 201–220.

[43] A. Papoulis and S. U. Pillai, *Probability, Random Variables and Stochastic Processes*, McGraw-Hill, 2002, New York.

[44] A. Papoulis, *Signal Analysis*, McGraw-Hill, 1977, New York.

[45] J. G. Proakis, *Digital Communications*, McGraw-Hill, 1989, New York.

[46] S. Haykin, *Communication Systems*, John Wiley & Sons, 1983, New York.

[47] B. Sklar, *Digital Communications Fundamentals and Applications*, Prentice Hall, 1988, New Jersey.

CHAPTER 2
MODULATION THEORY

2.1 MODULATION IMPAIRMENTS

Modulation is defined as a technique of mapping the information signal to a transmission signal (modulated signal) better suited for an operating medium, such as the wireless channel. This mapping can take on any number of forms such as phase variations (PSK), amplitude variations (QAM), and frequency variations (FSK). These variations will be in some proportion to the information signal itself. Regardless of the rules involved in the mapping, the modulated signal will encounter a variety of blocks that will add impairments into the signal. These impairments can take on the form of ISI due to low pass filtering at the transmitter, which is used to meet emissions masks, or due to differential group delay found in the receiver filters, or due to any nonlinearities encountered in both the transmit and receive sections of the communication link.

At the transmit side, we must mention the transmit power amplifier (PA) operating in the efficient, yet nonlinear, region and the frequency conversion blocks. The modulated signal will enter a low pass filter (LPF) in order to suppress the transmission emissions in the adjacent frequency bands. The filter type choice is extremely important since this will shape the transmission bandwidth by slowing down the phase trajectories and, in doing so, may add interference from neighboring modulation symbols. This interference is due to additional memory inserted into the signal and is called ISI—not to mention the ISI caused by the coherence bandwidth of the wireless channel.

The receiver section will also contain filters to perform channel selectivity. The received modulated signal will enter a decision or detection device that will be used in the extraction of the information signal from the modulation. Hence, we must concentrate on overall filter impulse response—by this, we mean the transmitter plus receiver filters.

In Fig. 2.1, we show such an example, where there is a filter used at the transmitter and one used in the receiver. The received signal quality is monitored and measured by viewing the receiver signal constellation diagram or eye diagram. Let us discuss the latter of the two. Here we see, when considering the overall impulse response, the received signal should have as little ISI as possible, in order to use a low-complexity symbol-by-symbol-based detector. A rule of thumb typically used is that the more filtering performed on the transmission signal to reduce emissions, the more ISI will be inserted into the transmitted signal.

In the figure, the transmit filter frequency response $H_T(f)$ can be seen as in cascade with the receive filter frequency response $H_R(f)$. The overall frequency response is given as $H(f)$ and defined as

$$H(f) = H_T(f) \cdot H_R(f) \tag{2.1}$$

In Fig. 2.2, we show an example of a received eye diagram assuming a cascade of transmit and receive filters. This eye diagram shows the ISI present as well as timing jitter present. The ISI is present due to the nonlinear phase filters used in the transmitter and receiver.

The received signal's eye diagram is oversampled at a rate of 8 times the symbol rate. The x-axis shows two symbol times in duration [1].

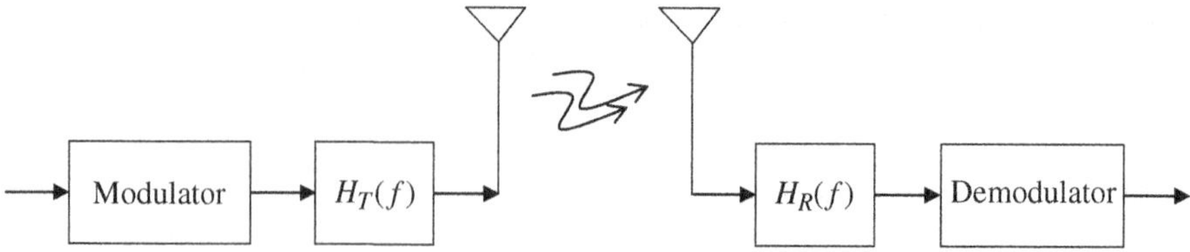

FIGURE 2.1 Overview block diagram emphasizing the filtering operations.

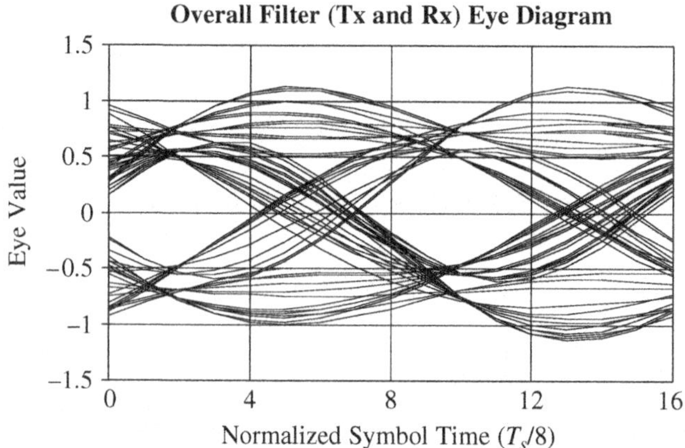

FIGURE 2.2 Example of a received signal's eye diagram.

2.1.1 ISI-Free System Requirements

In order to have an ISI-free system, we require the overall filter response to exhibit zero-ISI properties. Before we present a typically used solution, we will first review the Nyquist theorem which we will leverage in our discussion of ISI.

Nyquist's Theorem. Nyquist's theorem tells us the maximum possible transmission bit rate, given a filter BW, while still achieving ISI-free conditions. Specifically, the maximum data rate in order to achieve ISI-free transmission through an ideal brick wall filter with cutoff frequency (BW = $1/2T_s$) is given as $1/T_s$ bps (see [2] and [3]). Here we pass these impulses through an ideal brick wall filter, in order to provide the transmit pulses (see Fig. 2.3).

The Nyquist filter and theorem have essentially two issues associated with them. The first issue is the impractical implementation of the ideal brick wall (Nyquist) filter. It is well known that band-limited

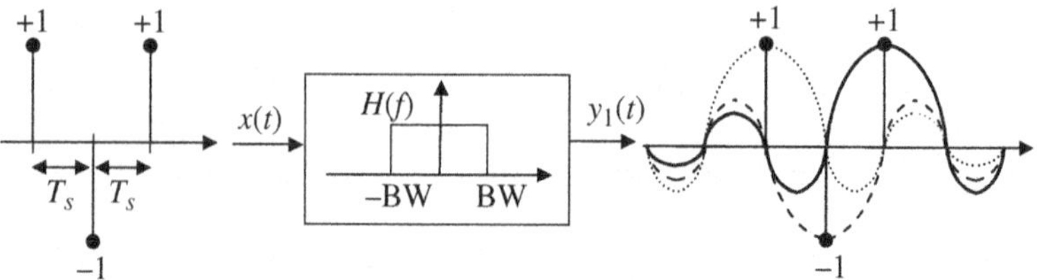

FIGURE 2.3 Nyquist's ISI-free transmission.

signals cannot be time limited [4]. The second issue is the impulse train carrying the information to the input of the Nyquist filter. The first issue can be solved if we use practical filters that have a frequency response that is symmetrical around the 3-dB transition point. A commonly used filter that exhibits the symmetrical property is the raised cosine (RC) filter. The symmetry, in response, is used for providing linear phase and thus constant time delays for all in-band frequency components [5–8].

The RC filter frequency response is given as

$$H(f) = \begin{cases} 1, & 0 < f < \mathrm{BW} - f_S \\ \dfrac{1}{2}\left\{1 - \sin\left[\dfrac{\pi}{2\alpha}\left(\dfrac{f}{\mathrm{BW}} - 1\right)\right]\right\}, & \mathrm{BW} - f_S < f < \mathrm{BW} + f_S \\ 0, & f > \mathrm{BW} + f_S \end{cases} \tag{2.2}$$

which can also be written as

$$H(w) = \begin{cases} 1, & 0 \le w \le \dfrac{\pi}{T_s}(1 - \alpha) \\ \cos^2\left(\dfrac{T_s}{4\alpha}\left[w - \dfrac{\pi(1-\alpha)}{T_s}\right]\right), & \dfrac{\pi}{T_s}(1 - \alpha) \le w \le \dfrac{\pi}{T_s}(1 + \alpha) \\ 0, & w > \dfrac{\pi}{T_s}(1 + \alpha) \end{cases} \tag{2.3}$$

where α = the roll-off factor, BW = the Nyquist bandwidth, and f_S = symbol rate. The roll-off factor α is also used to represent the excess bandwidth with respect to the theoretical minimum BW. For example, $\alpha = 0.5$ can be interpreted to mean that the filter bandwidth is 50% larger than the Nyquist bandwidth BW; in other words, the filter bandwidth = Nyquist BW$(1 + \alpha)$.

The RC frequency responses shown in Fig. 2.4 correspond to roll-off factors of $\alpha = 0$ (ideal brick wall), $\alpha = 0.1$, $\alpha = 0.5$, and $\alpha = 1.0$. As mentioned, the roll-off factor not only defines the response in the transition region, but also defines the BW expansion beyond what Nyquist defined. For example, with $\alpha = 0.5$, the occupied BW is 1.5 times the Nyquist bandwidth.

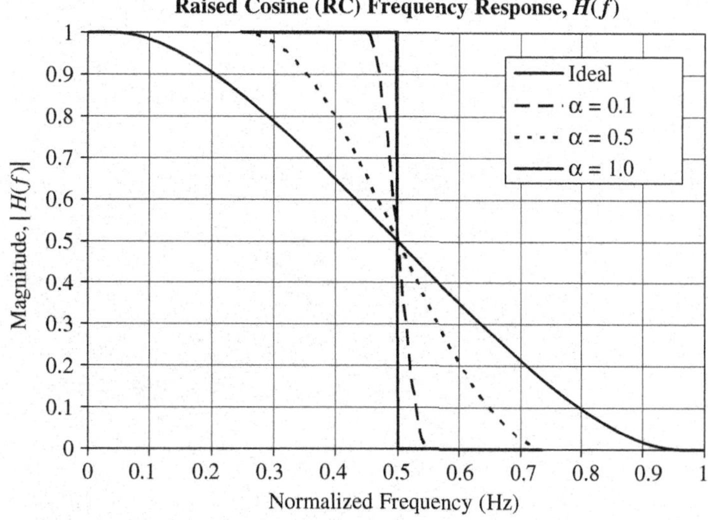

FIGURE 2.4 Raised cosine frequency response.

We define zero ISI or ISI free as having overlapping pulses that do not interfere with one another at the symbol sampling points: $h(nT_S) = 0 \ \forall \ n = \pm 1, \pm 2, \ldots$. An eye diagram of a signal using an RC filter is shown in Fig. 2.5. Here the zero-ISI requirements are clearly visible at the sampling instances.

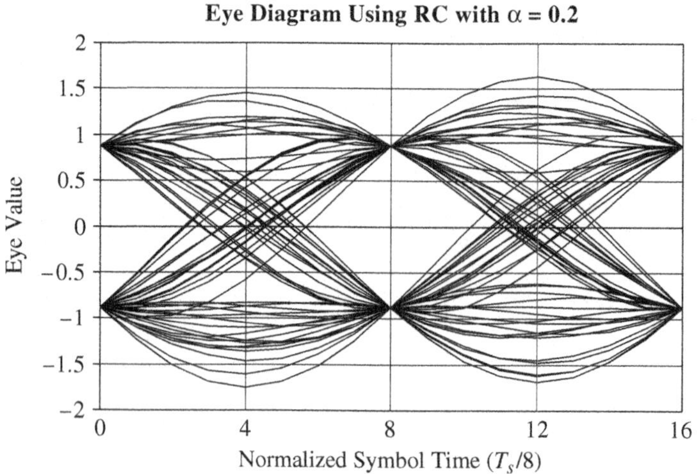

FIGURE 2.5 Raised cosine ($\alpha = 0.2$) eye diagram.

We show two eye diagrams corresponding to two roll-off factors, $\alpha = 0.2$ and $\alpha = 0.9$ (see Figs. 2.5 and 2.6). These eye diagrams show that as we use more bandwidth beyond the Nyquist band, there is less jitter in the transition regions of the symbols. Also the transitional overshoots are larger for the smaller values of the roll-off factor due to the increase in the amplitude of the temporal tails of the impulse response.

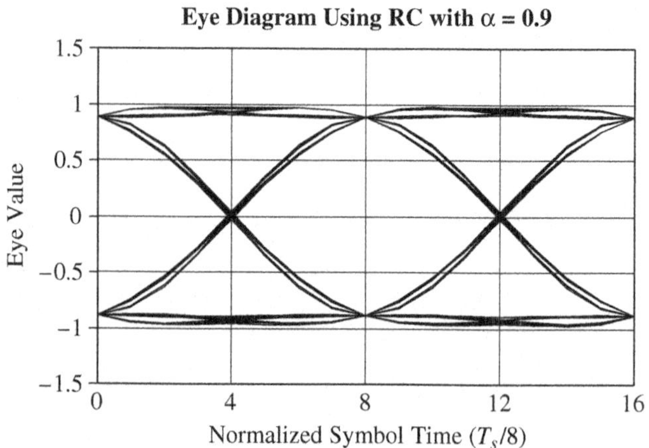

FIGURE 2.6 Raised cosine ($\alpha = 0.9$) eye diagram.

The impulse response for the RC frequency shape is given as

$$h(t) = \frac{\cos\left(\frac{\alpha\pi t}{T}\right)}{1 - \frac{4\alpha^2 t^2}{T^2}} \cdot \frac{\sin\left(\frac{\pi t}{T}\right)}{\frac{\pi t}{T}} \tag{2.4}$$

It is clear when we set $\alpha = 0$, we obtain the impulse response of the ideal brick wall filter. This response is a slowly decaying impulse response. Also, if we set $\alpha = 1$, then this will have the fastest decaying impulse response, but we have sacrificed occupied bandwidth, specifically twice the Nyquist bandwidth, to obtain this feature.

Decomposition Tool. There is still, however, another potential issue with the Nyquist theorem in that transmit impulse responses are used. Let us, for example, assume that the source is a non-return-to-zero (NRZ) pulse stream. If we simply ignore the impulse stream and send the NRZ through the Nyquist filter, we will see ISI present on the transmit signal. Let us explain this through the use of Fig. 2.7. Here, NRZ pulses are sent to the Nyquist brick wall filter. The NRZ pulses can be represented as an impulse stream convolved with the NRZ pulse. Next we perform the Fourier transform and arrive with a sampled version of the (sin x)/x function.

Here we see that we can basically decompose the NRZ data stream into an impulse response stream convolved with a (sin x)/x filter. In comparing this signal to that required by the Nyquist criteria, we see that the (sin x)/x operation is the additional block. One method to get around this is to use an x/sin x compensation or equalizer block to remove the NRZ signal. Now, when we cascade the equalizer with the RC filter and the response of the NRZ signal, we satisfy Nyquist's ISI-free theorem.

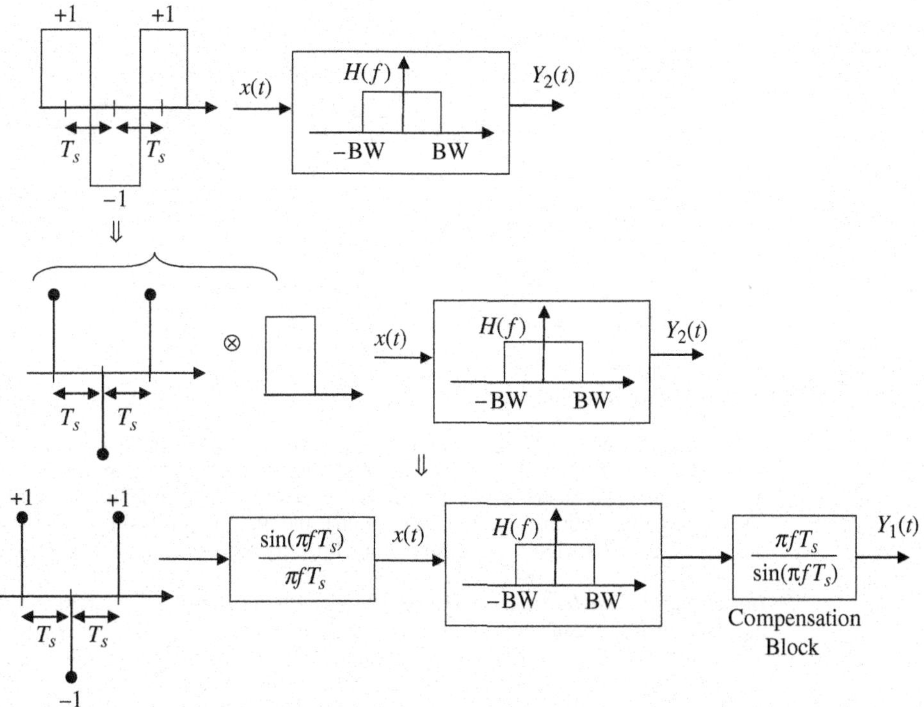

FIGURE 2.7 Decomposition of the NRZ pulse stream to illustrate the Nyquist zero-ISI case.

Square Root Raised Cosine (SRC) Filter. As discussed above, the receiver will also use some sort of filtering and we would really like the zero-ISI criteria to hold in the receiver. Here we discuss the possibility of using the transmit pulse-shaping filter in the receiver. If we used the RC filter in the transmitter, then when another receiver filter is used, ISI will be present in the signal entering the detection device and possibly significantly degrade performance. Another method is to evenly split up the RC filter into two parts: the transmitter will use the SRC filter and the receiver will also use the SRC filter. In this case, the receiver signal acts as a matched filter that is matched to the pulse shaping used in the transmitter. In fact, it is easy to show that maximum received energy can be extracted when an SRC is used at the receiver and the transmitter [9].

The impulse response for the SRC filter is given as

$$h(t) = \frac{\sin\left(\frac{(1-\alpha)\pi t}{T}\right) + \frac{4\alpha t}{T} \cdot \cos\left(\frac{(1+\alpha)\pi t}{T}\right)}{\frac{\pi t}{T}\left[1 - \left(\frac{4\alpha t}{T}\right)^2\right]} \tag{2.5}$$

IS-95 Pulse-Shaping Filter. In Fig. 2.8, we show the IS-95 pulse-shaping filter and relate it to the SRC filter used in other standards [10]. (Solid = SRC and Dashed = IS-95)

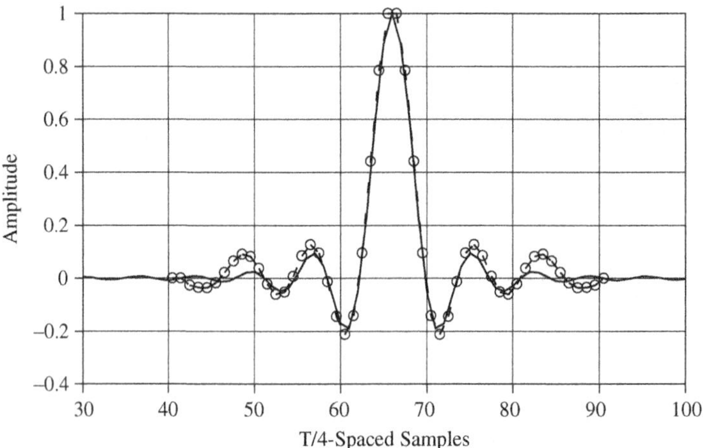

FIGURE 2.8 Pulse-shaping filter comparison.

2.1.2 Transmit Power Amplifier Discussion

In this section, we will address the issues of the transmit signal (usually in the form of Cartesian coordinates) by considering the nonlinearity of the transmit PA. Thus far we have discussed the low pass–filtering requirements of the modulation schemes as they pertain to the transmit emissions mask and receiver implications. Another major issue is the susceptibility of the modulation scheme to nonlinearities. The most significant source of nonlinearity comes from the transmit PA [11].

Input/Output Transfer Characteristics. In Fig. 2.9, we show a model used to include the effects of the transmit PA nonlinearity into the system simulations. First, the complex-valued signal is converted to the polar coordinate system (without any loss of information); the amplitude component enters the two functions. The first function takes the input amplitude and transforms it to an output amplitude based on the PA transfer function. This phenomenon is typically called AM-AM distortion. Similarly, the input amplitude signal enters the phase component and outputs the amount of phase that

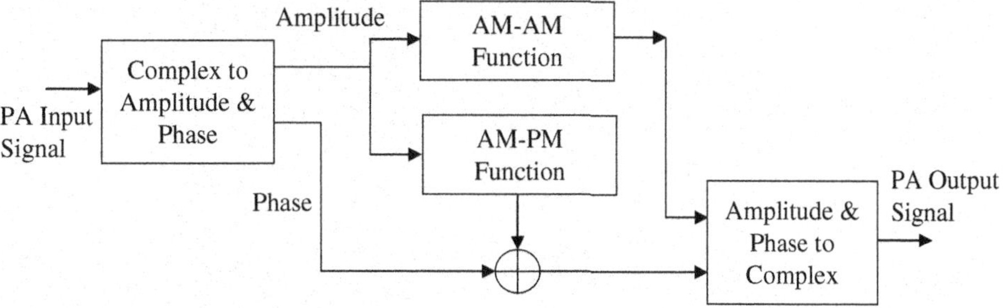

FIGURE 2.9 Transmit power amplifier nonlinear model.

needs to be added to or subtracted from the input phase. This phenomenon is typically called AM-PM distortion. Lastly, as the distortion is inserted into the symbol stream, the waveform is converted back into the Cartesian coordinate system [12].

Next we discuss some typical characteristics of the transmit PA (see Fig. 2.10). We will perform this operation not only in terms of the input and output power, but also in terms of the in-band and out-of-band emissions.

Some noteworthy comments should be made concerning the PA operating point. The larger the desired output power is, the closer we are operating to the nonlinear region of compression. Once the operating point is chosen, the input signal causes variations around this point which is then reflected to the output. The closer the operating point is to the compression region and the larger the amplitude variations that exist on the modulation scheme, the more saturation (or clipping) can occur. However, we desire to operate in this region simply because we require these output power levels and the high efficiency values. Moving the operating point away from the compression region pushes us further into the linear region. This is accomplished by reducing the transmit power as well as the PA efficiency [13].

Spectral Regrowth Issues. We have spent some considerable amount of time presenting the ISI-free conditions and their relationship to frequency emissions. When a modulation signal encounters a nonlinearity, in this case the transmit PA, the signal becomes distorted and its occupied frequency BW increases. The out-of-band spectral components that were once suppressed are now regenerated.

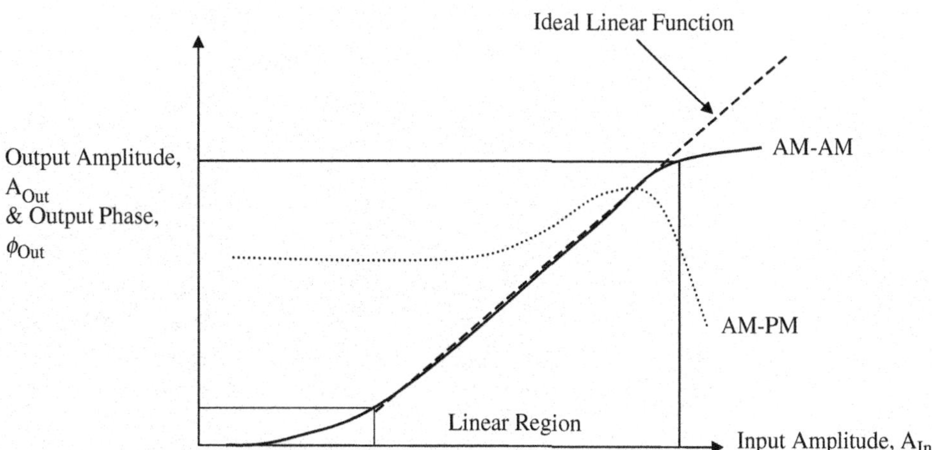

FIGURE 2.10 Typical characteristics of transmit power amplifiers.

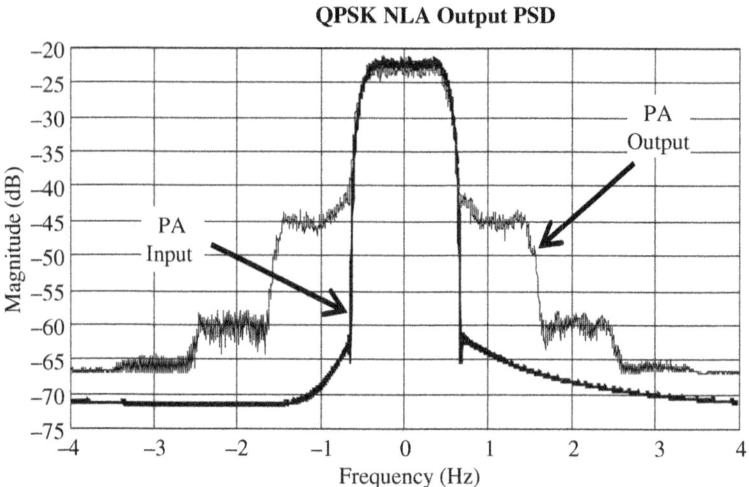

FIGURE 2.11 Spectral regrowth for the QPSK modulation scheme.

This phenomenon is called spectral regrowth. Moreover, the in-band signal is distorted and can lead to Bit Error Rate (BER) degradation, depending on the degree of nonlinearity [14].

Next we give an example of Quaternary Phase Shift Keying (QPSK), Offset QPSK (OQPSK), and $\pi/4$-Shifted Differential QPSK ($\pi/4$-DPSK) modulation schemes.

The first transmit Power Spectral Density (PSD) plot in Fig. 2.11 compares the PA input and output emissions for QPSK. The PA operating point used in these simulation results was intentionally placed in the nonlinear region. The spectral regrowth phenomenon, as well as the significant signal information present outside the main spectral lobe, are clearly visible.

The second transmit PSD plot in Fig. 2.12 compares the PA input and output emissions for OQPSK. For the same PA nonlinear operating point, the spectral regrowth is slightly less than the previous plot.

The last transmit PSD plot compares the PA input and output emission for $\pi/4$-QPSK. Similar results to QPSK can be seen when comparing the PSD plot in Fig. 2.13 to the one presented earlier.

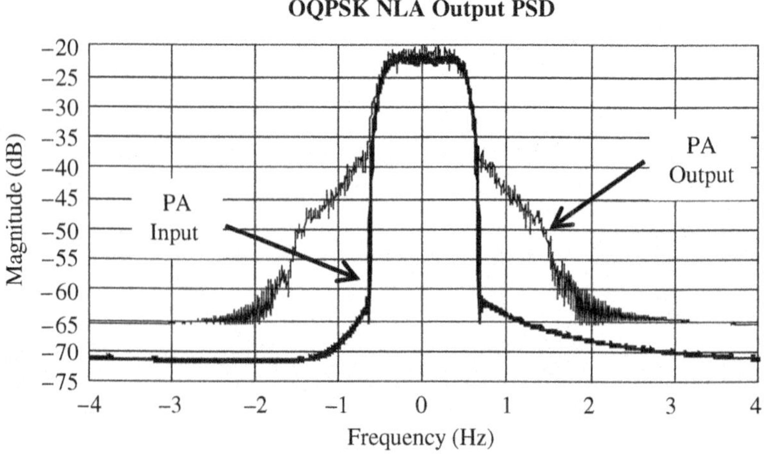

FIGURE 2.12 Spectral regrowth for the OQPSK modulation scheme.

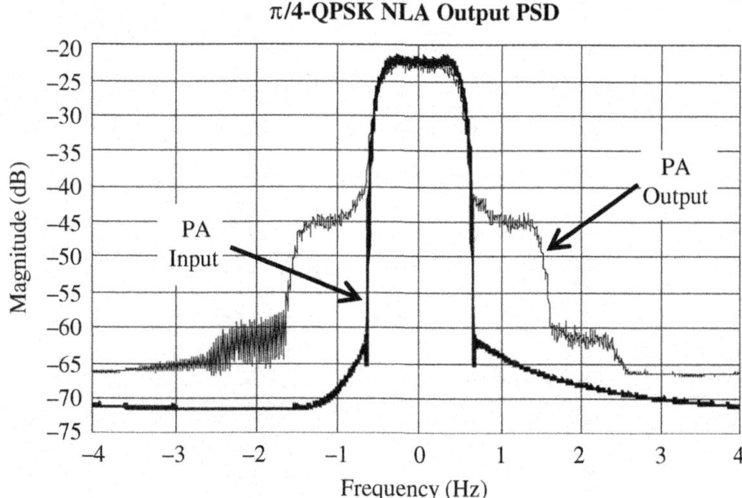

FIGURE 2.13 Spectral regrowth for the π/4-QPSK modulation scheme.

Compensation Techniques. In this subsection, we will present some commonly used PA linearization techniques [13]. As discussed above, spectral regrowth occurs when a modulation scheme with envelope variations enters a PA with a nonlinear transfer function. In the latter sections, we will discuss options communications system designers have when choosing a modulation scheme. So, in this section, we will leave the modulation schemes and concentrate on the PA transfer characteristics [15–17].

The transmit PA nonlinearities are typically modeled as AM-AM and AM-PM transfer characteristics. Above, we used AM-AM to represent the input AM power versus the output PA signal, and AM-PM to represent the input AM power versus the output phase signal. As discussed above, operating in the highly efficient region is desirable, hence either another type of PA is chosen or a linearization technique is performed to reduce the out-of-band emissions. There are essentially two types of linearization techniques available: feed forward and feed backward.

An example of a feed forward technique is called predistortion. Here, the baseband signal is predistorted using a priori information about the characteristics of the PA. Hence, additional distortion is intentionally inserted into the signal such that the PA output signal will have less spectral regrowth. The benefits of this technique are that it is not complex and can be implemented in a cost-effective manner. The disadvantage is that if the PA nonlinear characteristics are assumed time variant, then the linearization technique may not be able to track the changes that occur with aging, temperature, manufacturing sampling, and so forth. Hence, some feedback information path should be used to track the slow time-varying nature of the component. In Fig. 2.14, we provide a block diagram depicting the predistortion technique.

In Fig. 2.15, we provide an example using 16-QAM modulation. These symbols are intentionally distorted in both amplitude and phase, in order to compensate for the PA nonlinear characteristic. The desired output 16-QAM constellation is shown in the figure.

Note that the PA characteristics are typically stored in a form of a lookup table (LUT). However, due to the temperature change, component aging, manufacturing process, carrier frequency, and so forth, the characteristics change. Hence, an adaptive predistortion technique is a good solution. There, the PA output signal is sampled and fed back so as to track time-varying characteristics [18].

The second technique of Cartesian feedback continuously examines the transmitter PA output and then corrects for any nonlinearity in the transmit chain [19]. The performance of this technique is highly dependent on the overall delay in the loop, as well as the gain and phase imbalance in the feedback—not to mention that the PA characteristics also vary with time. Here the PA output signal

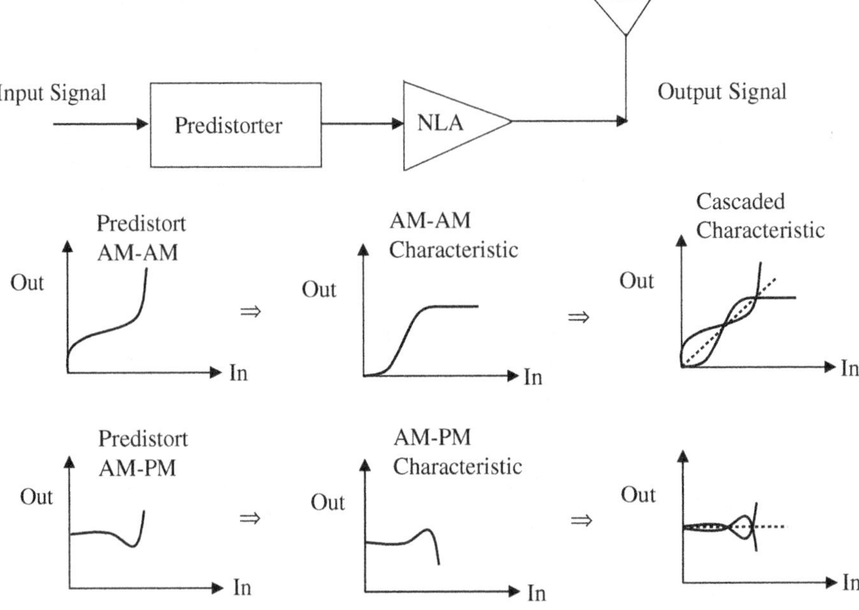

FIGURE 2.14 Predistortion linearization block diagram and example.

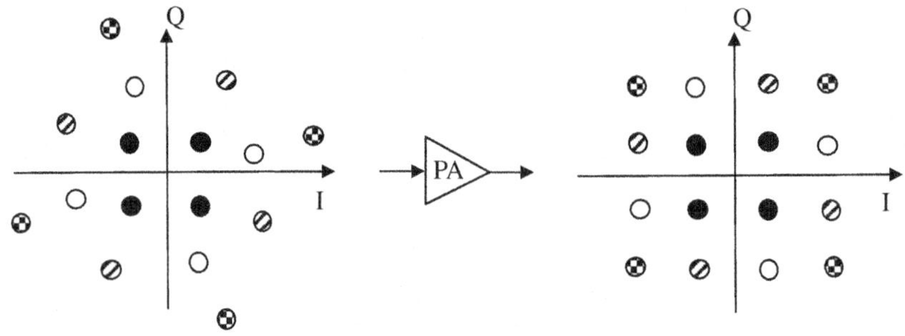

FIGURE 2.15 16-QAM predistortion linearization example.

is sampled and spectrally down converted in order to observe the distortion present on the signal and then a correction algorithm is applied to reduce the spectral regrowth. In Fig. 2.16, we plot the simplified Cartesian feedback linearization technique.

The advantage of this technique is that the performance can be very good. The disadvantage is that additional components are required to sample the PA output signal and then spectrally down convert them to baseband and the digitize them. However, this adaptive method can be used to compensate for time-varying mechanisms. In order to minimize the many variables that can lead to performance degradation, it is highly recommended that the entire technique be performed on a single IC [20, 21].

2.1.3 Gain and Phase Imbalance

In addition to the AM and PM distortion, there exists the possibility of the quadrature modulator being less than perfect. These imperfections will take on the form of gain and phase distortion. Recall the discussion on the complex envelope of a band pass signal in Chap. 1, where we connected the complex

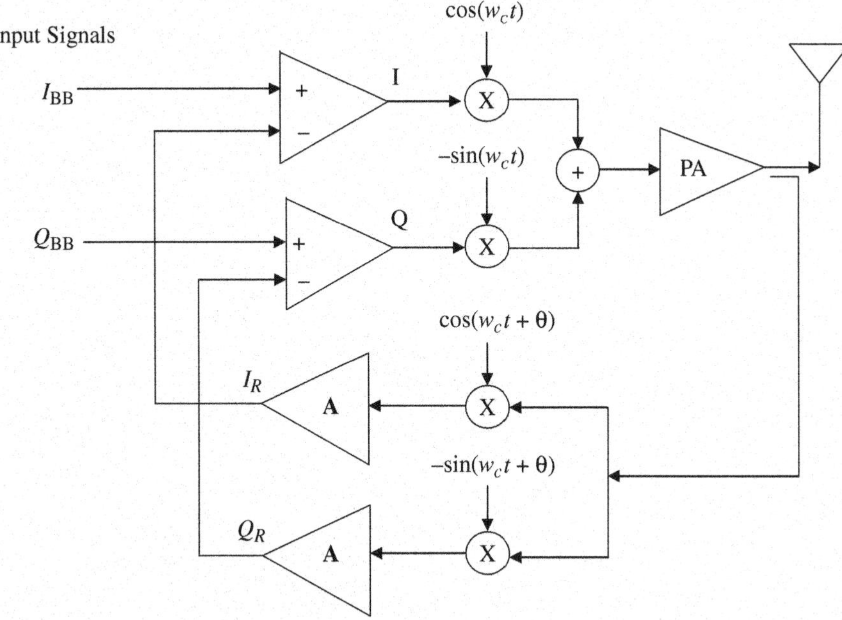

FIGURE 2.16 Cartesian feedback block diagram and example.

envelop to the input of a quadrature modulator which performed the spectral shifting operations. Similarly, the received signal can encounter a quadrature demodulator which acts as a spectral downshifting operator. In either case, the shifting operations can inject gain and phase imbalance, thus corrupting/distorting the signal constellation diagram. A simple gain imbalance impairment model is given in Fig. 2.17.

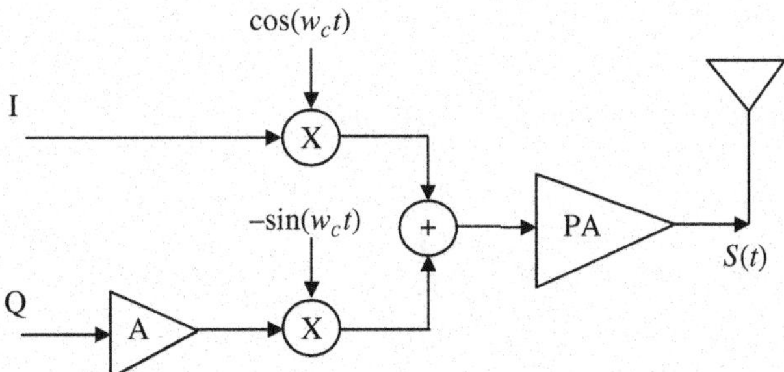

FIGURE 2.17 Gain imbalance model for the quadrature modulator.

Ignoring any nonlinearities in the PA model, the transmit envelope is written as

$$|S(t)| = \sqrt{\cos^2(\theta) + A^2 \sin^2(\theta)} \tag{2.6}$$

Moving our attention to the phase imbalance component, the following model assumes the phase offset between the quadrature sinusoids is not exactly 90 degrees out of phase, and thus, we draw the diagram in Fig. 2.18 to capture this imbalance.

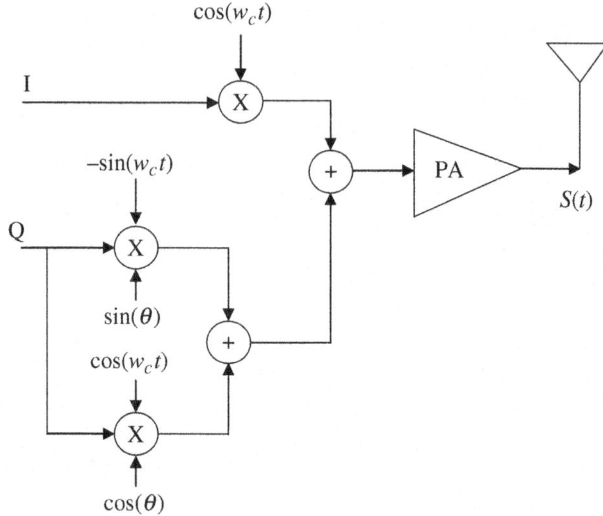

FIGURE 2.18 Phase imbalance model for the quadrature modulator.

The typical quadrature modulator output signal is written as

$$S(t) = I(t)\cos(\omega_c t) - Q(t)\sin(\omega_c t) \tag{2.7}$$

which can be rewritten as

$$S(t) = I(t)\cos(\omega_c t) + Q(t)\cos\left(\omega_c t + \frac{\pi}{2}\right) \tag{2.8}$$

In the absence of having an accurate phase shifter, the transmitted signal is represented as follows, where we have used a trigonometric identity given in the appendix of this book:

$$S(t) = I(t)\cos(\omega_c t) + Q(t)\cos(\omega_c t)\cos(\theta) - Q(t)\sin(\omega_c t)\sin(\theta) \tag{2.9}$$

So, when $\theta = \pi/2$, the modulator converts to the conventional model.

2.2 MODULATION SCHEME MIGRATION

In this section, we will present some commonly used modulation schemes. We will provide a migration path for each modulation scheme plus an overall direction the wireless community is presently taking. Whenever possible, we will compare the modulation schemes on the following merits:

Spectral efficiency (bps/Hz)

Receiver complexity (coherent or noncoherent detection)

BER performance

Transmit PA linearity requirements

We have chosen to begin with BPSK due to its simplicity and superior BER performance (see Fig. 2.19). Here, spectral efficiency can be doubled when going to QPSK, with no sacrifice in BER performance. If we turn our attention to spectral regrowth and receiver complexity, then two options exist. If spectral regrowth is not an immediate concern, then the path on the right can be chosen to select Differential QPSK (DQPSK), in order to have a simple receiver. Now, phase trajectories can be manipulated to attempt to give an additional benefit of lower spectral regrowth through the use of $\pi/4$-QPSK and $\pi/4$-FQPSK.

Returning to the available options, one can choose the OQPSK path where the immediate attention is focused on lowering spectral regrowth. This path can continue with FQPSK, which further modifies the phase trajectories to have a behavior closer to our desired case.

FSK is clearly a modulation choice, and we have decided to insert this next to OQPSK since MSK can be viewed as an extension of OQPSK. In order to focus on transmit emissions, GMSK is also included. When spectral efficiency is not an immediate concern, *M*-ary FSK modulation can surely be used.

All of these modulation schemes eventually encounter the advanced modulation scheme block containing 16-QAM, 64-QAM, and Variable Rate–PSK (see Fig. 2.19). In classical wireline communications, these modulation schemes are in no way viewed as advanced. However, in the wireless medium, higher-order modulation schemes impose many interesting and technically difficult hurdles [22].

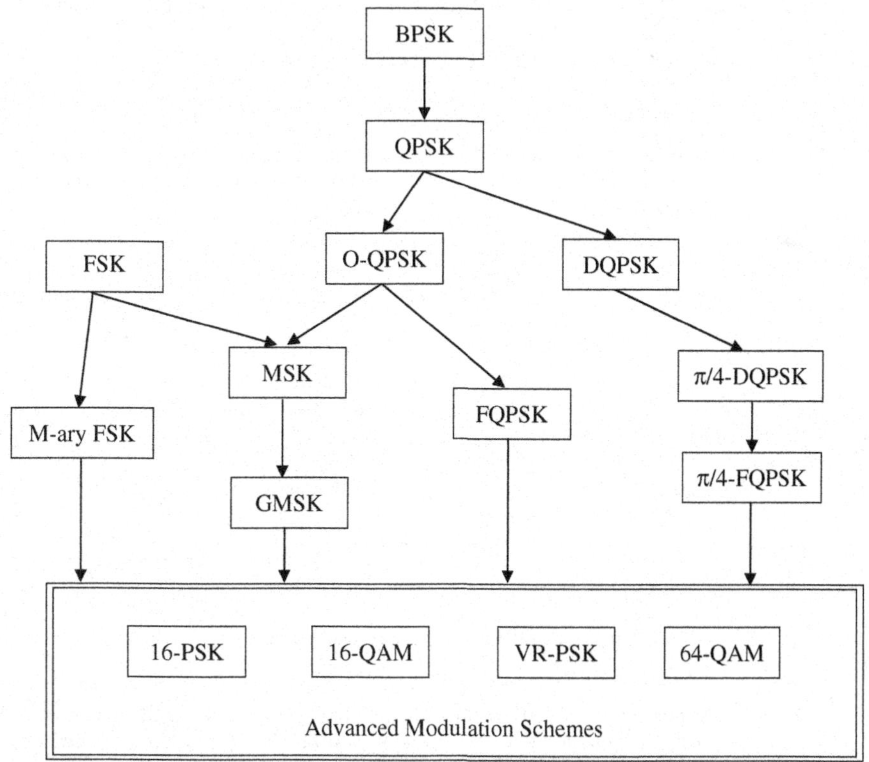

FIGURE 2.19 Modulation scheme migration path.

2.2.1 Binary Phase Shift Keying (BPSK)

In this section, we will present the BPSK modulation scheme. A binary signal is mapped to one of two possible phase changes. This mapping is fully described by a phase state table. For example, let's consider the phase state table in Table 2.1.

TABLE 2.1 BPSK Phase State Table

Data bit	Phase change
0	0
1	π

This table is translated to a signal constellation diagram (see Fig. 2.20). This modulation scheme has the two phase states located 180 degrees out of phase. This is the furthest possible distance that phase states can be separated on this two-dimensional plane. BPSK is also commonly called "antipodal signaling" and

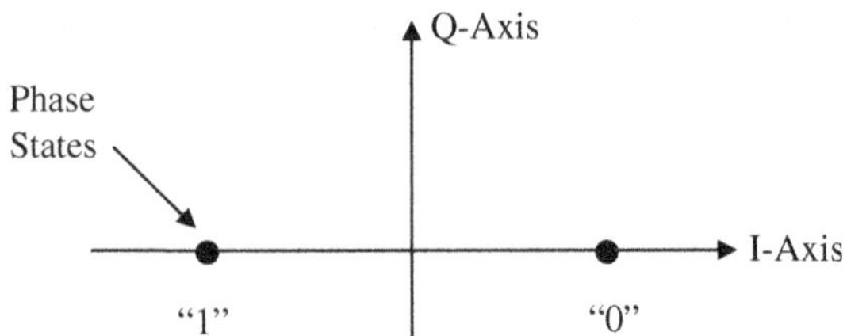

FIGURE 2.20 BPSK signal constellation diagram.

achieves the best probability of error in an AWGN channel [23]. This scheme has a spectral efficiency of 1 bps/Hz.

The transmitted signal is mathematically represented as

$$S(t) = m(t) \cdot \cos(\omega_c t) \tag{2.10}$$

As the constellation shows BPSK is a real-valued signal. A block diagram of the BPSK modulator is shown in Fig. 2.21. The time interval of the information signal $m(t)$ is denoted as T_b, bit time. An LPF was inserted to reduce the out-of-band emissions.

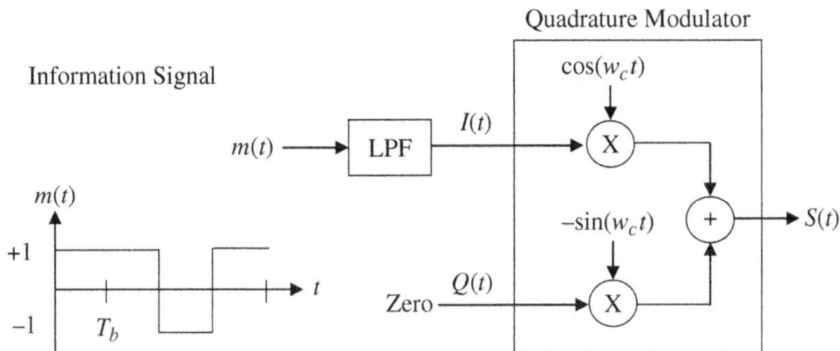

FIGURE 2.21 BPSK modulation transmission block diagram.

As presented in the BPSK modulation block diagram in Fig. 2.21, there is a dimension of the complex envelope domain that was not being utilized. Sending additional information on the Q-channel will double the spectral efficiency of this modulation scheme.

2.2.2 Quaternary Phase Shift Keying (QPSK)

QPSK has four allowable phase changes; prior to considering the associated phase state table where we need to group adjacent bits of information to make a di-bit or a symbol, we would like to present the QPSK modulator block diagram. In Fig. 2.22, 4 bits are drawn that are converted to 2 symbols. The symbols are grouped in terms of the filled and unfilled circles and boxes used.

The information signal $m(t)$ enters a serial-to-parallel (S/P) converter which generates a complex-valued symbol defined as $a(t) + jb(t)$. The S/P converter will take the bits corresponding to even time intervals and place them into $a(t)$, while the bits corresponding to odd time intervals will be placed into $b(t)$. The time duration of $a(t)$ and $b(t)$ is equal to two bit times, $2T_b$, or a symbol time T_s.

If we assume that the information-bearing signal $m(t)$ is a random sequence, then $a(t)$ and $b(t)$ will also appear random. Moreover, there is no interrelationship between the two symbol streams.

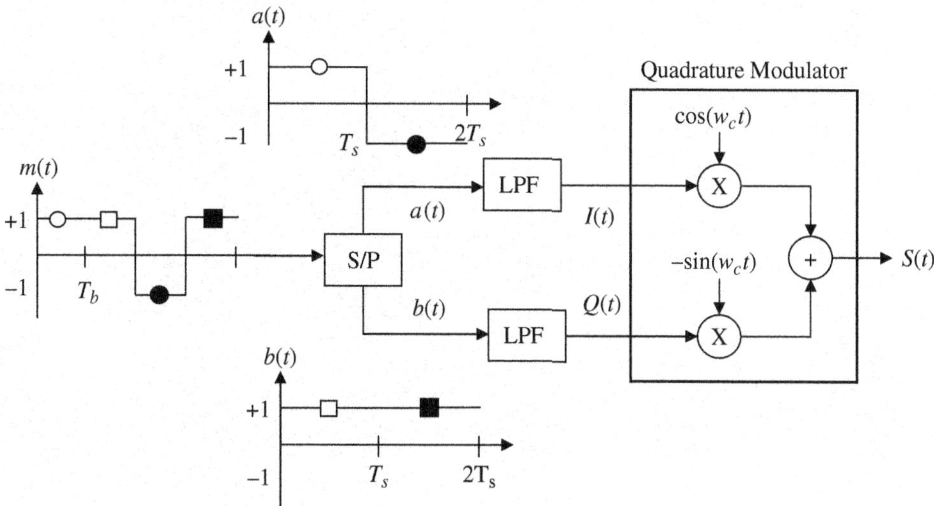

FIGURE 2.22 QPSK modulation transmission block diagram.

The S/P converter is actually creating the complex envelope of the QPSK modulation. The four allowable phase changes are $\phi(t) \in \{0, \pm\pi/2, \pm\pi\}$, and they correspond to the phase states outlined in Table 2.2. The signal constellation diagram is given in Fig. 2.23.

The signal constellation has phase trajectories that correspond to the following absolute phase state table.

The transmitted signal is given as

TABLE 2.2 QPSK Phase State Table

$a(t)$-$b(t)$	$\phi(t)$
0–0	$-3\pi/4$
0–1	$3\pi/4$
1–0	$-\pi/4$
1–1	$\pi/4$

$$S(t) = a(t) \cdot \cos(\omega_c t) - b(t) \cdot \sin(\omega_c t) \tag{2.11}$$

This is also represented as follows, using the complex envelope notation:

$$S(t) = \mathrm{Re}\{[a(t) + jb(t)] \cdot e^{j\omega_c t}\} \tag{2.12}$$

or

$$S(t) = A(t) \cdot \cos[\omega_c t + \phi(t)] \tag{2.13}$$

where the allowable phase changes are calculated as

$$\phi(t) = \tan^{-1}\left[\frac{Q(t)}{I(t)}\right] \tag{2.14}$$

Recall that we have inserted an LPF in order to reduce the out-of-band emissions. This is accomplished by slowing down the phase trajectories as the signal travels from one phase state to another. Using an SRC LPF, we are able to replot the signal constellation diagram. Here, we see the information is mapped to the phase and amplitude change of the carrier signal.

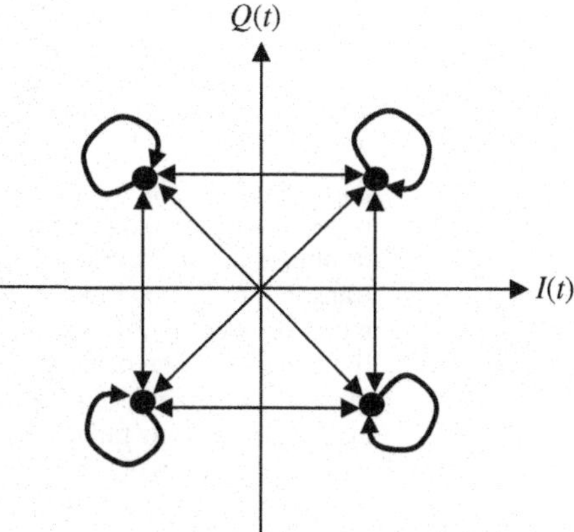

FIGURE 2.23 QPSK signal constellation diagram.

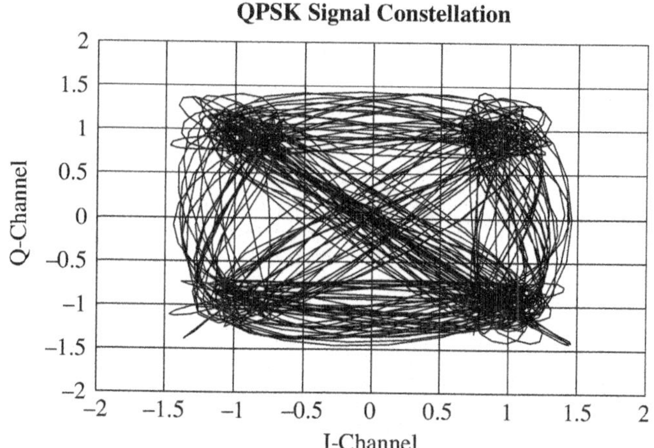

FIGURE 2.24 SRC Filtered ($\alpha = 0.3$) QPSK signal constellation diagram.

The SRC-filtered signal constellation clearly shows the four phase states with ISI present, as well as the phase trajectories that traverse through the origin (see Fig. 2.24).

Let's discuss some properties of the QPSK modulation scheme. The complex envelope can be rewritten in the polar coordinate system as follows:

$$a(t) + jb(t) = \sqrt{a^2(t) + b^2(t)} \cdot e^{j\tan^{-1}\left[\frac{b(t)}{a(t)}\right]} \tag{2.15}$$

where the commonly used terms to describe the amplitude and phase are given as

$$A(t) = \sqrt{a^2(t) + b^2(t)} \tag{2.16}$$

$$\phi(t) = \tan^{-1}\left[\frac{b(t)}{a(t)}\right] \tag{2.17}$$

If we were to plot the amplitude of the QPSK modulation scheme as a function of time, using an RC filter with roll-off factor of $\alpha = 0.2$ and 0.9, the plot in Fig. 2.25 would be observed.

Here we notice that the amplitude will vary significantly and come close to zero at certain times. When the amplitude is close to zero, this corresponds to phase trajectories very close to the origin. A commonly used measurement of amplitude variations is the peak-to-average ratio (PAR). For this particular example, the PAR = 2.86 dB for the roll-off factor $\alpha = 0.2$, and 1.11 dB for roll-off factor $\alpha = 0.9$. The reason behind presenting this statistic is to address the effects of encountering a nonlinearity. The larger the amplitude variation, the more linear operating range is required in the transmit PA. For example, consider the typical PA characteristic in Fig. 2.26, where we have outlined a pair of operating points A and B. When operating the PA at point A, there is a linear gain across the dynamic range of the input signal. The signal was not distorted in the time domain, so the frequency domain exhibits no distortion. However, when operating at point B, the gain is not linear across the dynamic range of the input signal. In fact, there is distortion for the high-amplitude cases, shown in the form of clipping. The signal is distorted in the time domain, and hence, we would expect distortion in the frequency domain.

The PA designers would like to push the operating point as close as possible to the compression region since this is the highly powered-efficient area. However, the communication system designers

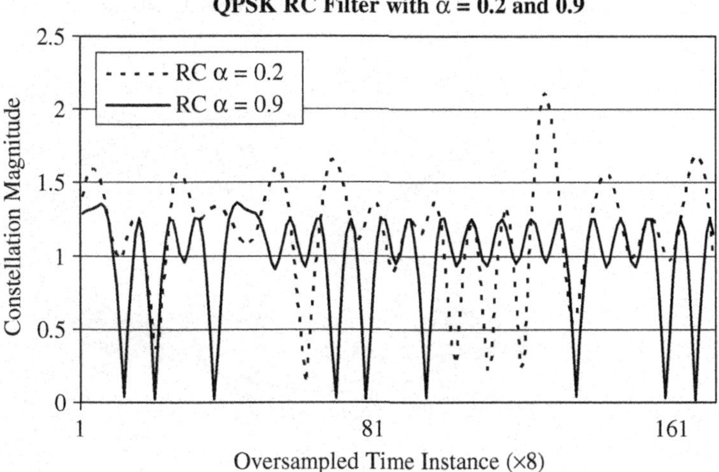

FIGURE 2.25 Amplitude variation of RC-filtered QPSK with roll-off factor $\alpha = 0.2$ and 0.9.

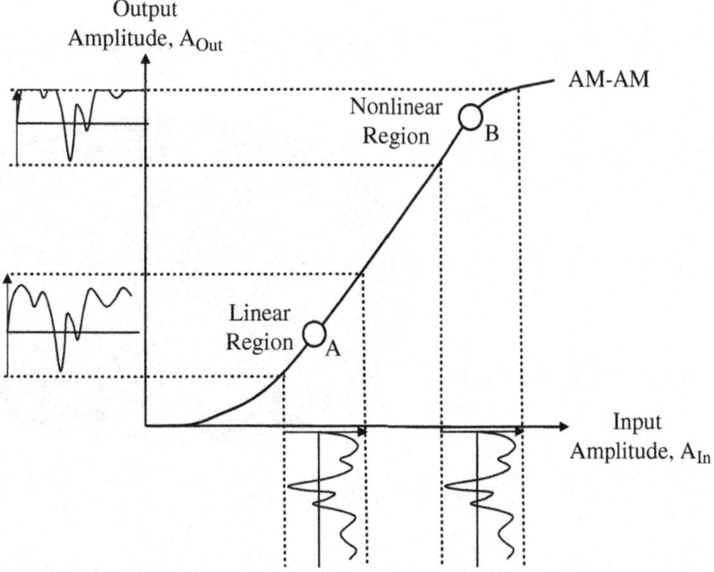

FIGURE 2.26 Transmitter nonlinearity degradation example.

must trade off out-of-band emissions for PA efficiency. Next we show the PA input and output signal constellation diagrams for a particular nonlinearity to emphasize the distortion (see Figs. 2.27 and 2.28).

Based on the constellation plots shown in Fig. 2.28, the distortion is clearly visible in two areas. The first area is for the larger-amplitude phase trajectories; these are mapped to a circle of constant radius. The second area is for the smaller-amplitude phase trajectories (i.e., those traversing close to the origin). Although they have a linear amplitude response, they exhibit nonlinear phase distortion, which also has a contribution to distortion.

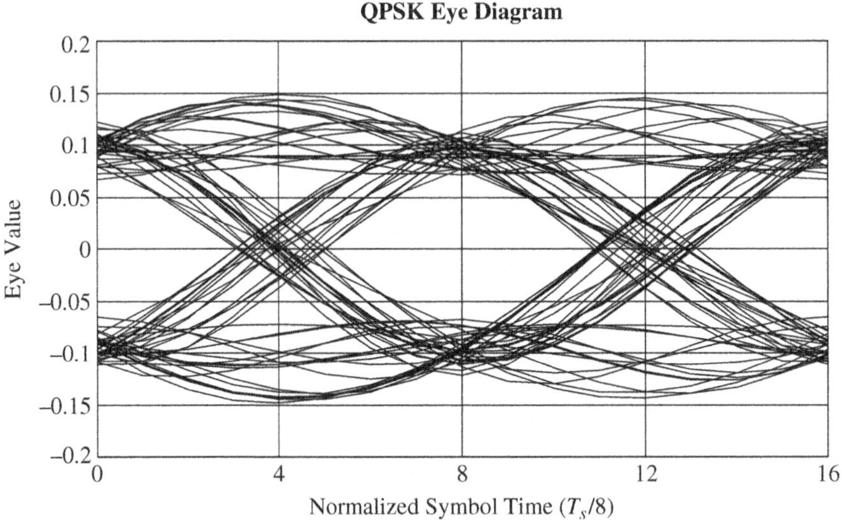

FIGURE 2.27 PA input signal constellation diagram for SRC-filtered QPSK with roll-off factor $\alpha = 0.3$.

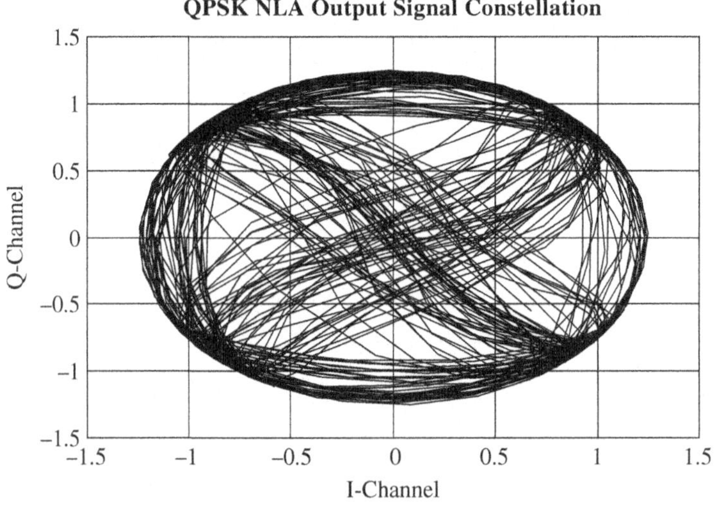

FIGURE 2.28 PA output signal constellation diagram for SRC-filtered QPSK with roll-off factor $\alpha = 0.3$.

As you would expect, this PA output constellation exhibits significant distortion, which is visible in the time domain. The frequency domain must contain spectral spreading and regrowth to dual the temporal variations.

Let us briefly turn our attention to the receiver where we will introduce the demodulation operations. A simplified block diagram of the receiver is shown in Fig. 2.29, where we place emphasis on the channel derotation or compensation aspect.

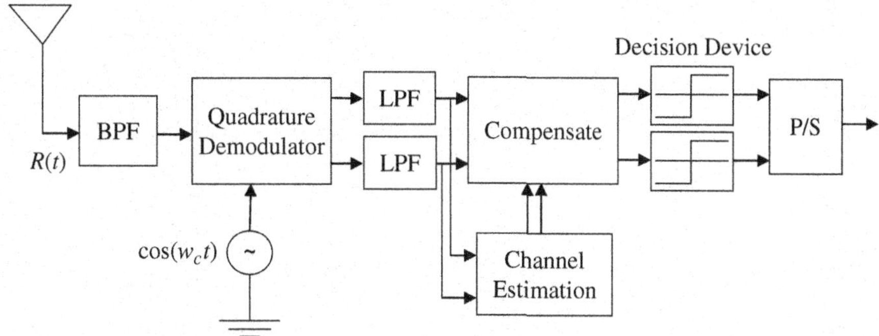

FIGURE 2.29 A simplified block diagram of QPSK demodulation with special emphasis on coherent detection operation.

Let us walk through this receiver; first, the received signal enters a BPF to reduce the noise and adjacent channel interference. This filtered signal is then spectrally down converted to produce the received complex envelope signal. Note that we have previously drawn the quadrature demodulator with LPFs internal to the block. The LPFs shown in Fig. 2.29 are in addition to the quadrature demodulation filtering. The received signal constellation will have a time-varying phase offset (the amplitude will vary as well; but for sake of discussion, let's keep this simple) which will be estimated by the channel estimation (CE) block. Once the phase offset is estimated, it is removed or compensated, so the received signal has ideally zero phase offset. It is important to remove this phase offset since the decision device will make decisions on the received symbol, using a fixed reference plane. Once the symbol is detected, the symbol is then converted to a serial bit stream with the aid of the parallel-to-serial (P/S) converter.

The above procedure of estimating the received phase offset and removing it from the signal prior to detection is called coherent detection. The classical coherent detection receiver creates a local sinusoid at some carrier frequency which is exactly in phase and in frequency locked to the received signal. This operation jointly, spectrally down converts the signal and phase compensates it. This operation could also have been done separately as shown in Fig. 2.29. Here the receiver establishes a local version of the channel phase offset to aid channel phase compensation. This procedure can be complicated and, in some cases, may not be an option to deploy in a commercial product. An alternative solution would be to use noncoherent detection, which has a simpler form of implementation complexity. We must note, however, with coherent detection that the following probability of bit error (P_b) relationship holds true:

$$P_b(\text{BPSK}) = P_b(\text{QPSK}) \tag{2.18}$$

Another issue besides increased receiver complexity is that of phase trajectories traveling through or near the undesirable origin. This property increases the PAR, causes spectral regrowth, and forces the PA to operate in the less-efficient operating region. Let us address the second issue by forcing the phase trajectories to avoid the origin altogether. This is accomplished with the OQPSK modulation scheme.

2.2.3 Offset QPSK (OQPSK)

For this modulation scheme, we assume the receiver implementation complexity is not an issue. OQPSK provides a solution to reducing the spectral regrowth created when the signal enters a non-linearity. This is accomplished by forcing the phase trajectories to avoid the origin. A block diagram of the OQPSK transmitter is shown in Fig. 2.30.

In comparing this block diagram to that presented earlier for QPSK, we see a bit time (T_b) delay was inserted into the complex symbol generation.

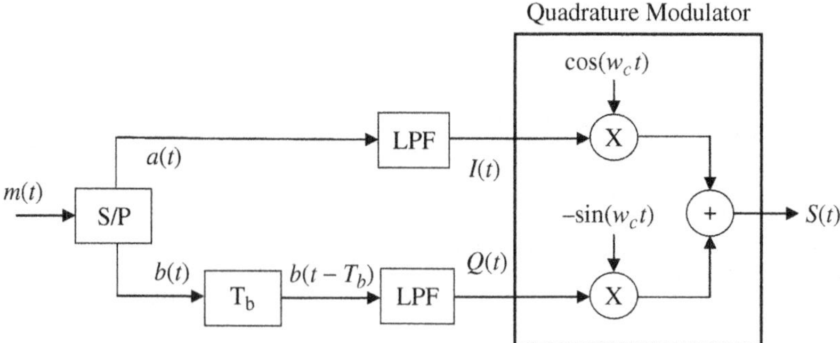

FIGURE 2.30 OQPSK modulation transmission block diagram.

In Fig. 2.31, we show how inserting a bit time delay will force the phase trajectories to avoid the origin. We will start with the complex symbol at the output of the S/P converter.

As we see from the above example, the QPSK problematic phase trajectory corresponds to a phase change of π or 180 degrees. By inserting a bit time delay, OQPSK was able to split up the 180-degree phase change into two $\pi/2$ phase changes and thus avoid the origin.

The OQPSK signal constellation diagram using an SRC with roll-off factor $\alpha = 0.3$ is shown in Fig. 2.32.

The above signal constellation diagram clearly shows the phase trajectories close to the origin have been avoided. This reduces the PAR of the complex envelope signal.

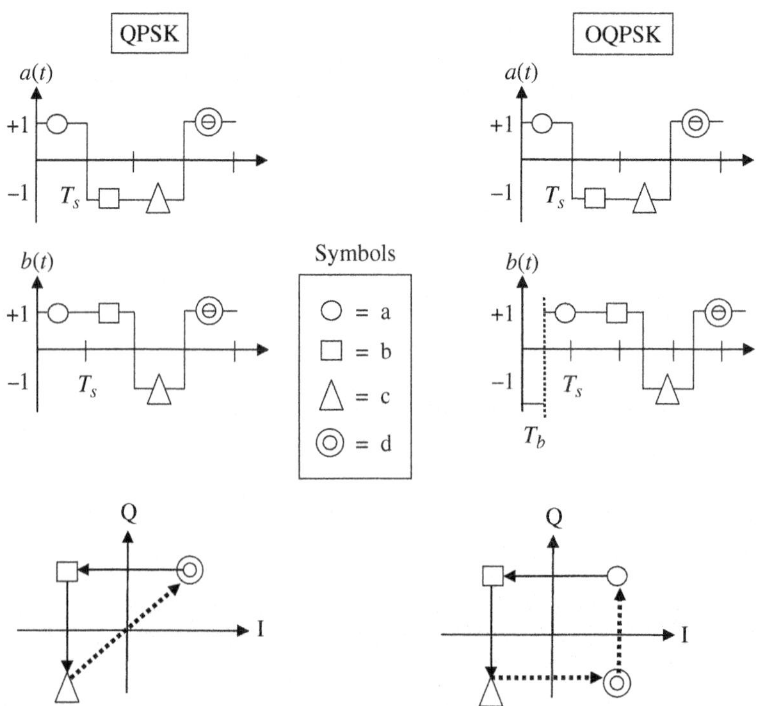

FIGURE 2.31 Comparison of QPSK and OQPSK Phase trajectories.

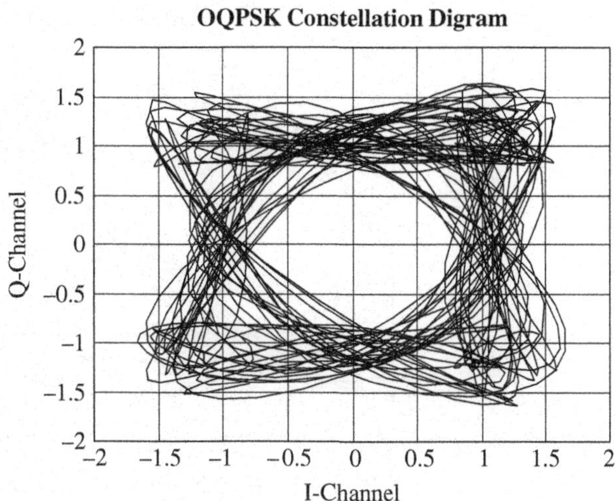

FIGURE 2.32 OQPSK signal constellation diagram.

The signal constellation diagram for the OQPSK modulation signal at the output of a nonlinear PA is given in Fig. 2.33, using an SRC with a roll-off factor $\alpha = 0.3$.

The OQPSK eye diagram at the output of the SRC filter with a roll-off factor $\alpha = 0.3$ is provided in Fig. 2.34. It is a great opportunity to briefly discuss the OQPSK eye diagram using this SRC filter. The RC filter has been slit up into two SRC filters, one at the transmitter and the other at the receiver. The signal at the output of the receiver SRC is ISI free; however, the transmitter SRC filter output signal contains ISI. It is this ISI that is present in the OQPSK eye diagram.

The receiver block diagram for OQPSK is very similar to that presented earlier for QPSK, with the exception that the signals entering the decision devices must be delayed according to the time offsets

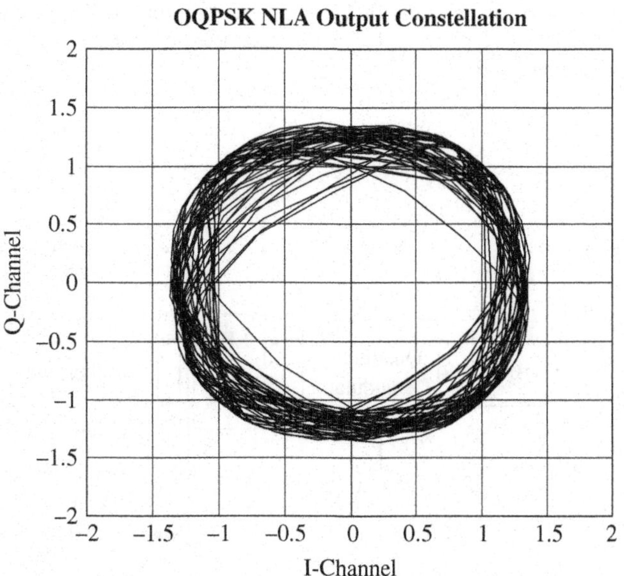

FIGURE 2.33 OQPSK signal constellation at the output of a nonlinear PA.

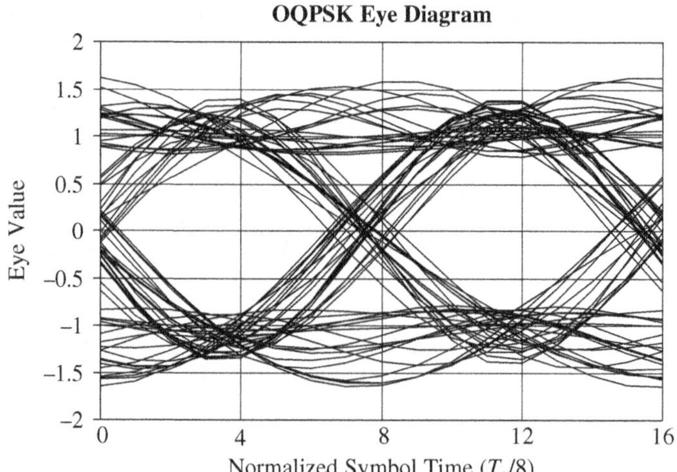

FIGURE 2.34 OQPSK eye diagram at the output of the SRC filter.

inserted in the transmitter. In the OQPSK modulator shown in Fig. 2.30, we inserted a bit time delay into the Q-channel. Now in the receiver, the I-channel must be delayed (or appropriately sampled) in order for the symbol to be time aligned prior to entering the P/S converter. The timing recovery block inserted into Fig. 2.35 will manage the sampling time instances.

Using coherent detection as a method of detecting OQPSK signals in the receiver produces a very interesting result; specifically, the following probability of bit error relationship holds true:

$$P_b(\text{BPSK}) = P_b(\text{QPSK}) = P_b(\text{OQPSK}) \tag{2.19}$$

Hence, this modulation technique will produce the same BER performance as that of QPSK modulation; however, the phase trajectories traveling through or near the origin have been eliminated [24].

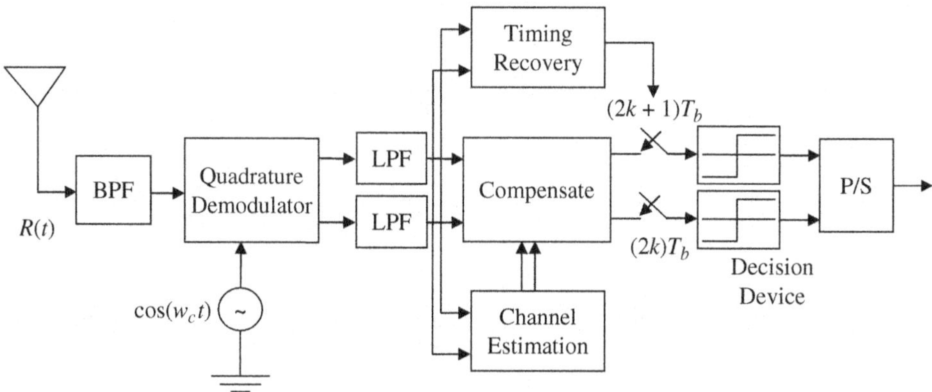

FIGURE 2.35 OQPSK receiver block diagram.

2.2.4 Differential QPSK (DQPSK)

In referencing the modulation scheme migration shown in Fig. 2.19, we now take a path where the transmitter PA nonlinearity is not a concern, but the receiver complexity is a more critical point.

DQPSK inserts the information into phase differences of the signal to be transmitted. In the Cartesian coordinate system, this is represented as (where k denotes the symbol time index)

$$S_k = S_{k-1} \cdot e^{j\Delta\phi_k} \tag{2.20}$$

And in the Polar coordinate system this is represented as (given $S_k = e^{j\theta_k}$) and called phase differential encoding.

$$\theta(k) = \theta(k - T_s) + \Delta\phi(k) \tag{2.21}$$

where $\theta(k)$ is the present phase to be transmitted, $\theta(k - T_s)$ is the previously transmitted phase, and $\Delta\phi(k)$ is the present phase change to be used to transmit the information signal. Let us present an alternative approach and discuss the polar coordinate implementation. Figure 2.36 depicts the DQPSK modulator.

The LUT is essentially performing the operation of a phase state table. An example of such a phase state table or mapping is provided in Table 2.3.

TABLE 2.3 DQPSK Example Phase State Table

$a(t)$-$b(t)$	$\Delta\phi(k)$
0–0	0
0–1	$\pi/2$
1–0	$-\pi/2$
1–1	π

This table has the corresponding signal constellation diagram in Fig. 2.37.

Sometimes, this is referred to as $\pi/2$-DQPSK because the four allowable phase changes are multiples of $\pi/2$. These phase changes are controlled by the phase state table and implemented with the phase differential encoding equation given earlier. In the receiver, we wish to determine the phase change amount on the signal. Since we have a single equation with a single unknown, the answer is trivial and is mathematically represented as

$$\Delta\phi(k) = \theta(k) - \theta(k - T_s) \tag{2.22}$$

This equation is called the phase differential decoding equation. Here, the received phase is subtracted by a delayed version of itself to produce one of the four allowable phase changes. This operation is called differential decoding or differential detection, depending where the equation is implemented in the receiver.

We can best describe the difference between these two descriptions with the receiver block diagrams in Figs. 2.38 and 2.39; both will make use of the polar coordinate system. In the first diagram, we implement differential decoding. This means that coherent detection was applied to the received signal to estimate the symbols that created $\theta(k)$. Then the differential decoding equation is applied on the detected symbols to produce $\Delta\phi(k)$.

Next we discuss the differential detection operation where coherent detection is not performed. Instead, phase differences are applied directly to the received signal. This is a form on noncoherent detection.

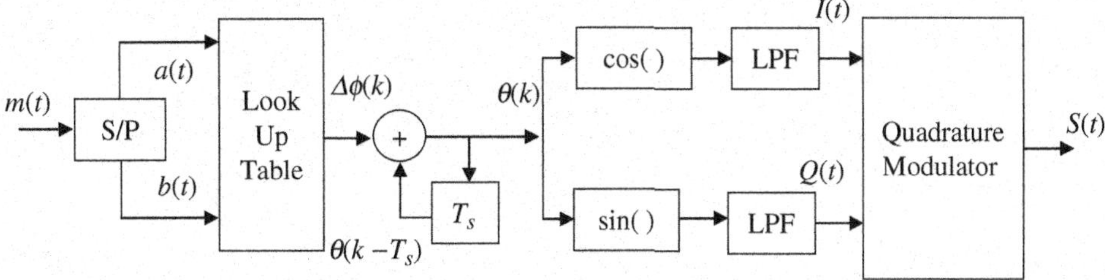

FIGURE 2.36 DQPSK modulator block diagram.

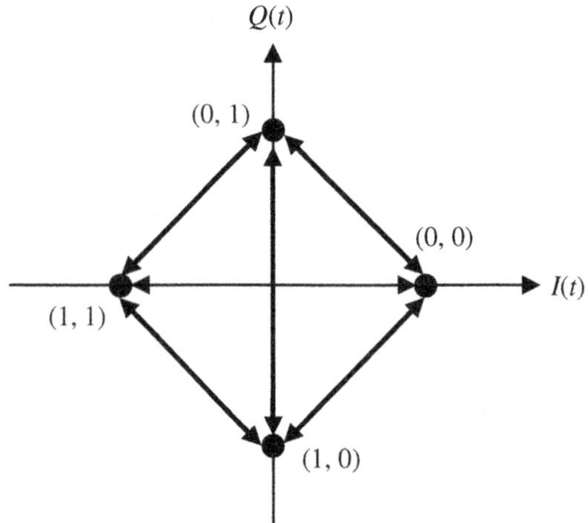

FIGURE 2.37 DQPSK signal constellation diagram.

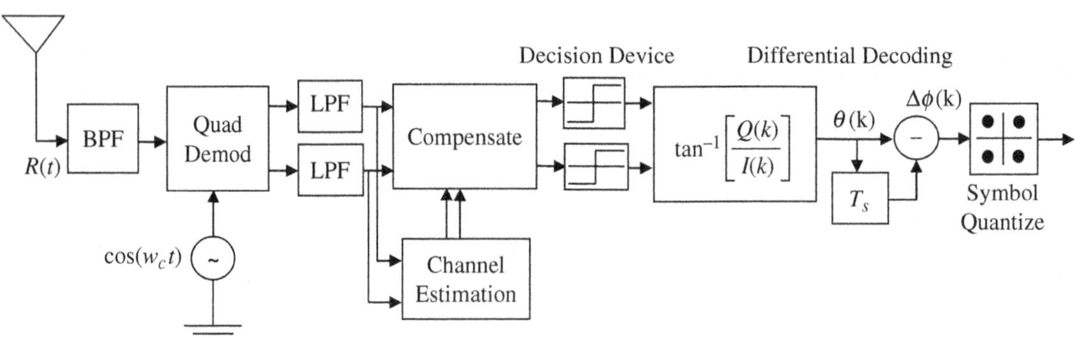

FIGURE 2.38 Receiver block diagram emphasizing differential decoding operations.

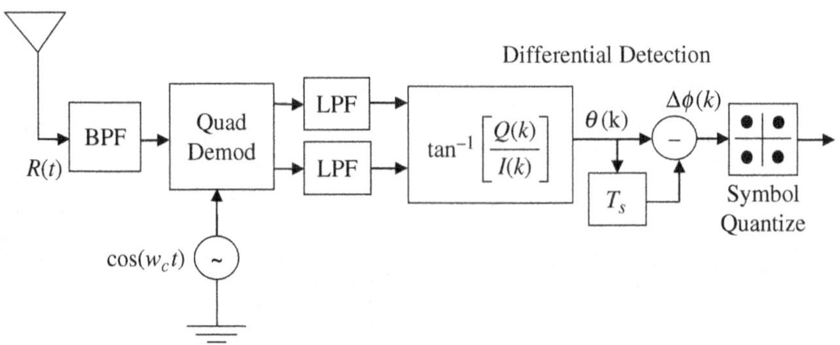

FIGURE 2.39 Receiver block diagram emphasizing differential detection operations.

Based on the differential encoding equation, we were able to intuitively provide/solve the equation to be used in the receiver. A more detailed derivation will be supplied in Chap. 4, where we discuss detection techniques.

It must be mentioned that when differential detection is used, a single error will propagate into two errors since the error will be used as a phase reference for the next symbol. With this in mind, we notice that a receiver would require approximately 2.5 dB more E_b/N_o for differential detection when compared to coherent detection operating in an AWGN channel and maintaining an average BER = 1E-3.

Lastly, we have managed to give the receiver an option to use either coherent detection or differential detection, thus having the possibility of deploying a simple implementation, if so desired. However, we still have a spectral regrowth issue to address, particularly if the transmit PA is operating close to the 1 dB compression point, in other words, in the nonlinear region.

2.2.5 π/4-Shifted Differential QPSK (π/4-DQPSK)

A solution to offer options for the receiver detection and implementation architectures, while avoiding the origin, is π/4-DQPSK. This is very similar to the DQPSK modulation scheme presented earlier, except now the four allowable phase changes are multiples of $\pi/4$. For example, let us consider the phase state table in Table 2.4.

Using this modulation scheme, we will be able to show the unfiltered signal constellation diagram in Fig. 2.40. Please note this constellation was generated using the same DQPSK modulation block diagram presented in the previous section, with the only difference being the LUT contents. The constellation has eight phase states where, at any symbol time instance, only four are allowable. The phase trajectories alternate between "circles" and "boxes."

TABLE 2.4 π/4-DQPSK Phase State Table Definitions

$a(t)$-$b(t)$	$\Delta\phi(k)$
0–0	$-3\pi/4$
0–1	$3\pi/4$
1–0	$-\pi/4$
1–1	$\pi/4$

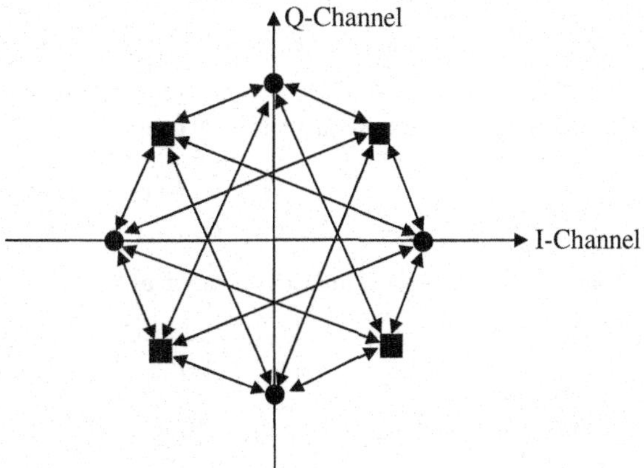

FIGURE 2.40 π/4-DQPSK signal constellation diagram.

Earlier we presented the DQPSK modulator block diagram, using the polar coordinate system; we would now like to this opportunity to present the Cartesian coordinate equivalent approach (see Fig. 2.41).

The above approach, when considering the phase variations, is equivalent to the polar coordinate approach. This is a very simple, robust, and easily implementable architecture. Carrying out the above mathematical operations leads us to the following relationship:

$$I_k = a_k \cdot I_{k-1} - b_k \cdot Q_{k-1} \tag{2.23}$$

$$Q_k = a_k \cdot Q_{k-1} + b_k \cdot I_{k-1} \tag{2.24}$$

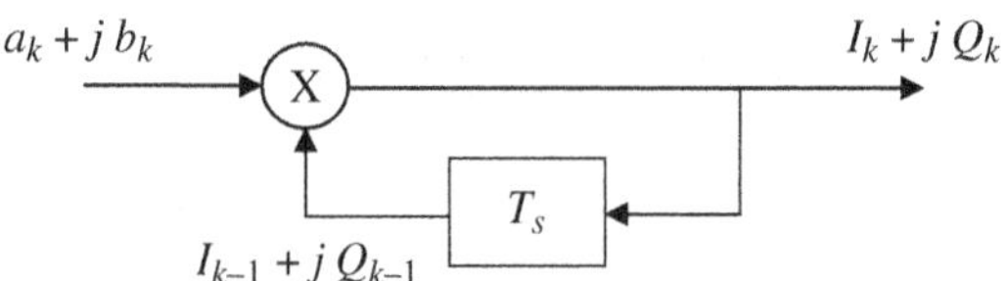

FIGURE 2.41 Conventional baseband DQPSK modulator.

which can be represented in matrix notation as

$$\begin{bmatrix} I_k \\ Q_k \end{bmatrix} = \begin{bmatrix} I_{k-1} & -Q_{k-1} \\ Q_{k-1} & I_{k-1} \end{bmatrix} \cdot \begin{bmatrix} a_k \\ b_k \end{bmatrix} \tag{2.25}$$

Stated another way, the symbols to be presently transmitted are represented in vector notation as $\underline{i}(k)$ and are generated by projecting the present information signal vector $\underline{m}(k)$ onto the phase matrix.

$$\underline{i} = I \cdot \underline{m} \tag{2.26}$$

Observe that the phase matrix determinant is expressed as $\det(I) = I_{k-1}^2 + Q_{k-1}^2 > 0$. Similarly, we can show the demodulator as in Fig. 2.42, using a symbol time (T_s) delay block and a conjugate operation.

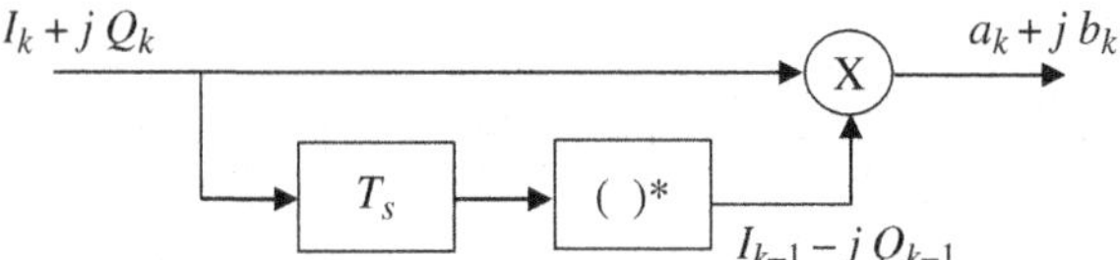

FIGURE 2.42 Conventional DQPSK demodulator.

After carrying out the above mathematical operations, we arrive at the following relationship:

$$a_k = I_k \cdot I_{k-1} + Q_k \cdot Q_{k-1} \tag{2.27}$$

$$b_k = Q_k \cdot I_{k-1} - I_k \cdot Q_{k-1} \tag{2.28}$$

which can also be represented in matrix notation as

$$\begin{bmatrix} a_k \\ b_k \end{bmatrix} = \begin{bmatrix} I_k & Q_k \\ Q_k & -I_k \end{bmatrix} \cdot \begin{bmatrix} I_{k-1} \\ Q_{k-1} \end{bmatrix} \tag{2.29}$$

which can also be written using the previously defined phase matrix:

$$\underline{m} = I^{-1} \cdot \underline{i} \tag{2.30}$$

as

$$\begin{bmatrix} a_k \\ b_k \end{bmatrix} = \begin{bmatrix} I_{k-1} & Q_{k-1} \\ -Q_{k-1} & I_{k-1} \end{bmatrix} \cdot \begin{bmatrix} I_k \\ Q_k \end{bmatrix} \tag{2.31}$$

Note that we have assumed, without loss of generality, that the transmitted signal constellation has a constant envelope, when sampled at the symbol rate.

A brief inspection of the above equations reveals that the real part of the conventional demodulator output consists of the sum of the autocorrelations of the input signal. Also, the imaginary part of the conventional demodulator output consists of the difference of the cross-correlations of the input signal. We will make use of this observation in the later chapters.

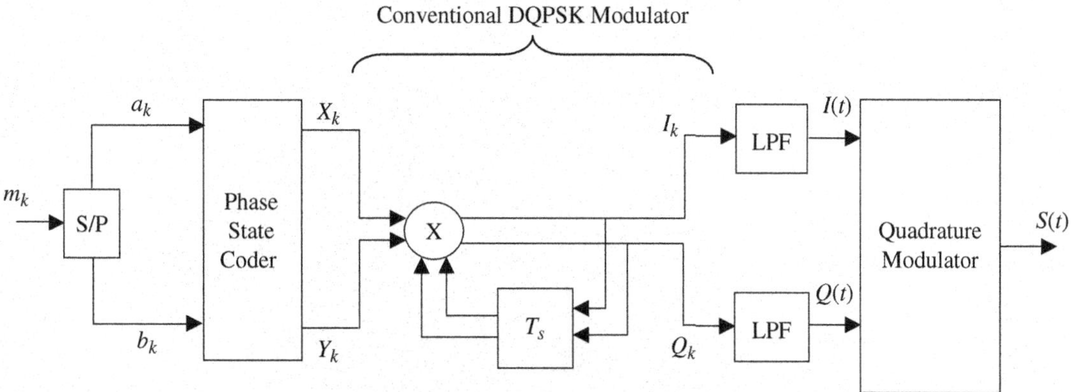

FIGURE 2.43 Block diagram showing use of a phase state coder, as the mechanism to switch between different phase state tables.

Phase State Coder. We can use the easily implementable, conventional modulator for any system that has exactly the same phase state table in Table 2.4. Now if this is not the case, instead of redesigning the modulator, we would like to reuse it and basically change the input signals. This is accomplished through what we will call the phase state coder. A block diagram supporting this idea is shown in Fig. 2.43, using a phase state coder, as the mechanism to switch between different phase state tables as the need arises.

For example, the North American Digital Cellular (NADC) system uses the phase state table in Table 2.5.

TABLE 2.5 $\pi/4$-DQPSK Phase State Table for NADC

$a(t)$-$b(t)$	$\Delta\phi(k)$
0–0	$\pi/4$
0–1	$3\pi/4$
1–0	$-\pi/4$
1–1	$-3\pi/4$

For this example, we notice that quadrants I and III have swapped their phase changes when comparing Table 2.5 to Table 2.4. Hence with the following implementation of the phase state coder, we can extend the use of the conventional modulator through the block diagram in Fig. 2.44 [25]. In fact, the phase state coder for this example consists of invertors. One can easily calculate the phase state coder needed for the Japan Digital Cellular (JDC) system, given its phase state table.

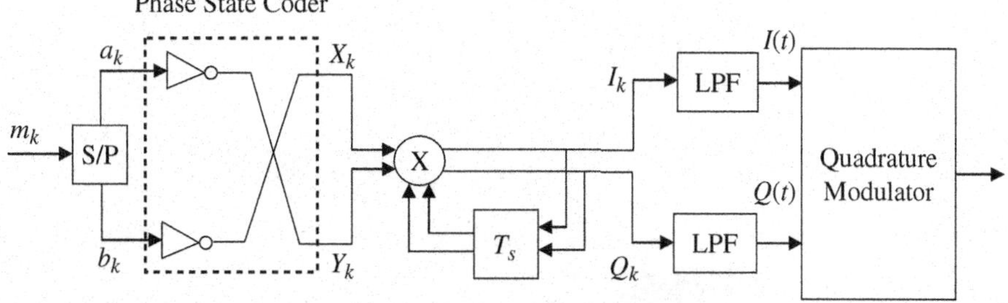

FIGURE 2.44 $\pi/4$-DQPSK modulator using the NADC phase state coder design.

Note that as far as the receiver is concerned, the same demodulator can be used with the change in the symbol quantizer that performs a different symbol demapping operation. Since the phase state table is time invariant and known a priori, this should pose no problem.

Now let us show the filtered version of the signal constellation diagram using an SRC filter (see Fig. 2.45). The constellation diagram clearly shows the presence of ISI on all 8 phase states. It also shows the origin is avoided by the phase trajectories; however, they do come very close.

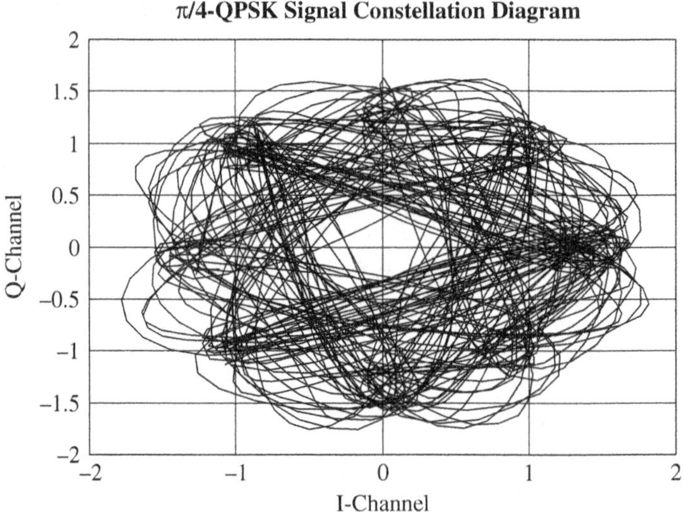

FIGURE 2.45 $\pi/4$-DQPSK signal constellation using SRC with roll-off factor $\alpha = 0.3$.

Also the one-dimensional eye diagram is shown in Fig. 2.46.

Note that the 5-level eye diagram comes from the fact that as we project the two-dimensional constellation onto one dimension (either the I- or Q-axis), there are five states that become pronounced.

In fact, if we were to rotate the constellation by $\pi/8$ in either direction, then the resulting eye diagrams of the I- and Q-channels will consist of 4 levels instead of the previous 5 levels.

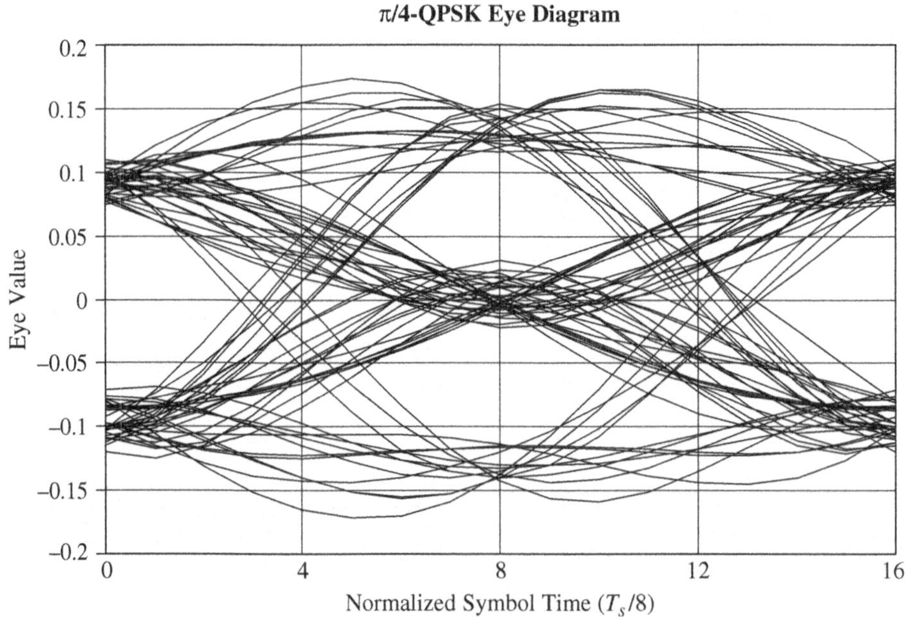

FIGURE 2.46 $\pi/4$-DQPSK eye diagram using SRC with roll-off factor $\alpha = 0.3$.

We would like to take this moment to briefly pause and review what has been discussed so far. We began with BPSK modulation scheme, which required operating the PA in the linear region and coherent detection for bit extraction. We then doubled our spectral efficiency to 2 bps/Hz, by using QPSK. Here the constellation diagram has phase trajectories that crossed the origin and were problematic in that they contribute to the spectral regrowth phenomenon when the modulated signal passes through a nonlinearity. In an effort to avoid the origin, OQPSK was selected. Here the troublesome π trajectories were split up into two $\pi/2$ trajectories, thus avoiding crossing the origin. However, we didn't give many options to the receiver.

Next we mentioned that if spectral regrowth could have been tolerated or less-efficient PAs could have been used, then DQPSK may be another option. Here the phase changes were inserted into the transmitted signal by phase differences. The origin was crossed, but we gave the communication system designer options for the receiver.

In an attempt to recollect the PA high-efficiency benefit while maintaining options at the receiver, $\pi/4$-DQPSK was presented. Here the origin was completely avoided; however, depending on the choice of the LPF used, the phase trajectories came dangerously close.

This last $\pi/4$-DQPSK signal constellation diagram in Fig. 2.47 corresponds to the output of a nonlinear PA operating in the nonlinear region. Notice the constellation amplitude has almost been mapped to a circle of constant radius. Also note that the eight phase states are not readily visible now; hence there is some potential for in-band BER degradation.

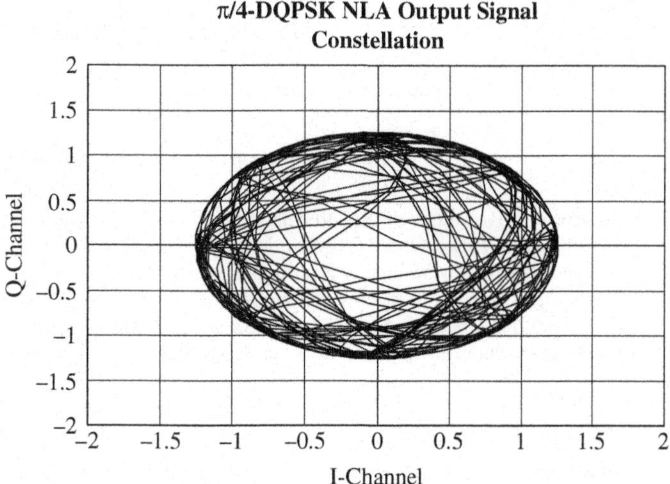

FIGURE 2.47 Nonlinear PA output PSD of $\pi/4$-DQPSK.

2.2.6 Frequency Shift Keying (FSK)

Next we would like to temporarily leave the topic of PSK and discuss the topic of FSK. We do this for a number of reasons. First, FSK is a commonly used modulation scheme and deserves the attention. Second, we can discuss a particular modulation scheme that can essentially be a member of both the PSK and FSK family. Lastly, this will be used as a foundation when we discuss controlling the constellation phase trajectories. Let us start with BFSK (see Fig. 2.48); here we vary the carrier frequency in proportion to the information signal $m(t)$. The transmitted signal is represented as

$$s(t) = A \cdot \cos\left[2\pi f_c t + 2\pi \int_{-\infty}^{t} f_d \cdot m(t)\, dt\right] \tag{2.32}$$

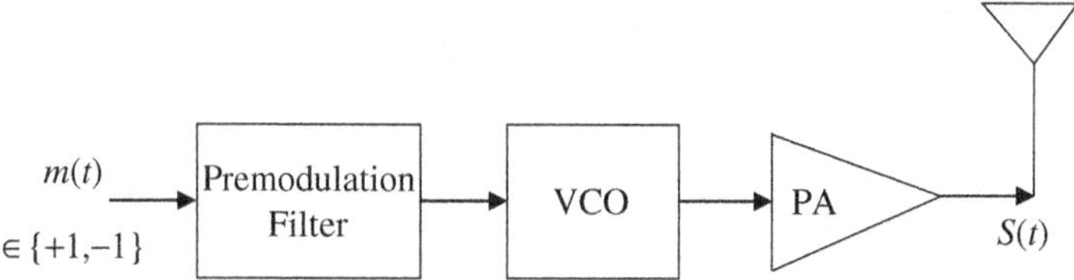

FIGURE 2.48 Conventional FM generation method for BFSK.

which can be generated using the typical FSK transmission method of applying a time-varying signal to a Voltage Controlled Oscillator (VCO). The premodulation filter is inserted in order to reduce the instantaneous phase changes, thus reducing the out-of-band emissions. Sometimes, this filter is called a splatter control filter.

A simplified way of viewing the transmit spectrum is to consider BFSK as consisting of two spectral tones (see Fig. 2.49).

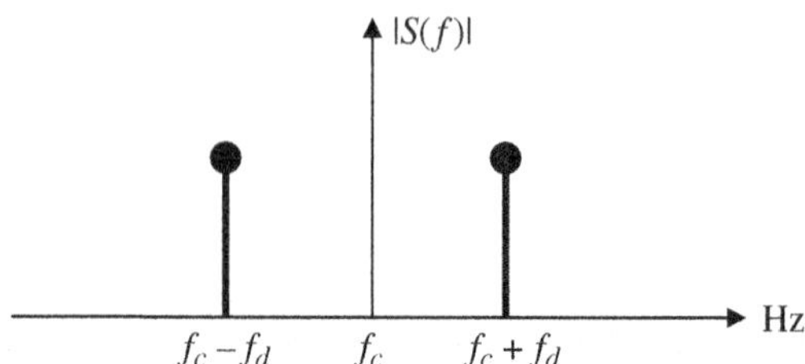

FIGURE 2.49 Simplified view of BFSK modulation scheme.

So, for positive values of the information signal $m(t)$, the transmitted frequency is shifted upward by an amount of f_d Hz, where f_d = frequency deviation. Also for negative values of the information signal, the transmitted frequency is shifted downward by f_d.

A question remains as to how we choose the appropriate frequency deviation for an application. What is commonly accepted is to introduce a variable that represents the normalized frequency deviation of the modulation scheme; the variable is called the modulation index and is defined as

$$\beta = \frac{2f_d}{R_b} \quad \text{[unitless]} \tag{2.33}$$

We will shortly discuss the fact that there are certain values of β that have more benefits over others. We have already presented the classical approach to generate the FSK waveform. However, as the wireless community faces multiple standards using multiple modulation schemes, we would more than likely use a Cartesian coordinate representation, that is, complex envelope approach.

Recall when we introduced the concept of complex envelope in Chap. 1, we already provided the reader with a baseband FM modulator (see Fig. 2.50). Here we redraw it for sake of convenience.

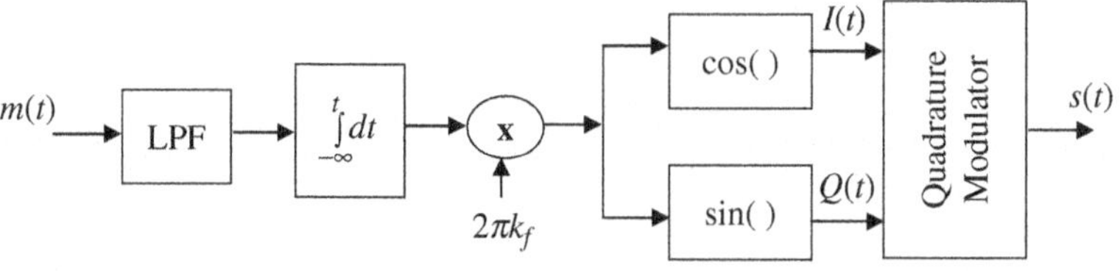

FIGURE 2.50 Baseband FM modulator.

Above we see the input signal is integrated and multiplied by the sensitivity factor in order to get the modulation index β desired. The trigonometric functions sine and cosine are commonly implemented in HW as LUTs; similarly, they are implemented in SW by some series expansion as well as a LUT.

Next we aim to discuss the complex envelope signal in order to help us understand how they are transmitted. Also, we want to be able to better understand the signal output of the receiver quadrature demodulator for circumstances such as when a Zero IF (ZIF) Receiver or Direction Conversion Receiver (DCR) is used.

For sake of discussion, let us consider a case without the premodulation filter present and assume $R_b = 10$ kbps and $f_d = 5$ kHz; then the modulation index is defined as

$$\beta = \frac{2 \cdot 5\,\text{kHz}}{10\,\text{kbps}} = 1 \tag{2.34}$$

With this above information, let's use the random input data in Fig. 2.51as an example.

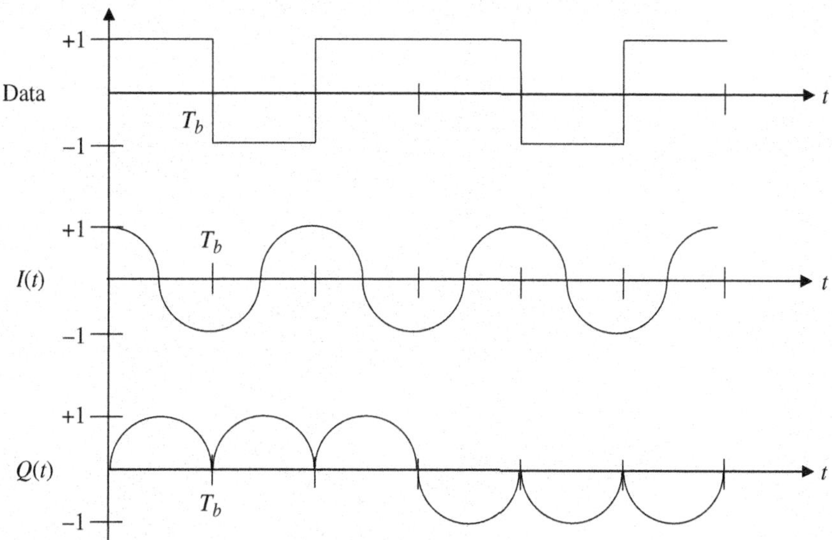

FIGURE 2.51 Unfiltered complex envelope of BFSK with $\beta = 1$.

The above waveforms can also be better understood using the following signal constellation diagram. The amount of phase change encountered in a bit time is equal to

$$\begin{aligned} \Delta\Psi &= 2\pi f_d \cdot T_b \\ \Delta\Psi &= \frac{2\pi f_d}{R_b} \\ \Delta\Psi &= \pi \cdot \beta \end{aligned} \tag{2.35}$$

So, using the parameters form the previous example, $\beta = 1$, we see there are $\pm\pi$ phase changes every bit time, which can also be shown as in Fig. 2.52, where we consider the phase changes of the first two data bits drawn in the previous figure.

In general, the modulation index is an indicator of the amount of phase allowed to change during a bit time T_b.

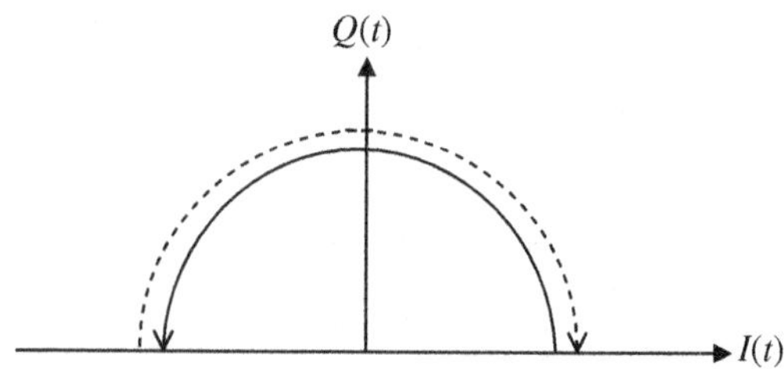

FIGURE 2.52 Phase changes for BFSK with $\beta = 1$.

As previously shown, when not considering the effects of the premodulation filter, the occupied BW becomes theoretically infinite because of the sharp phase reversals observed on the I- and Q-channels. The addition of the premodulation filter slows down the phase transitions and thus makes the transmit PSD more spectrally compact or efficient. In Fig. 2.53, we consider the effects of including an example LPF. This is given by the dashed lines used to emphasize the waveforms in the transition regions.

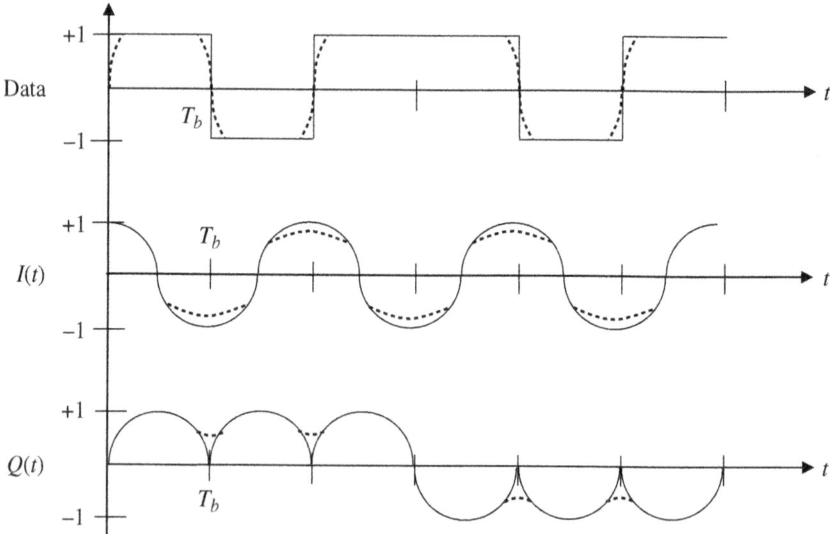

FIGURE 2.53 Filtered complex envelope of BFSK with $\beta = 1$

As we can see, the instantaneous phase reversals that occurred at the bit boundaries have slowed down. For this example, the Q-channel doesn't have the opportunity to go to zero at the bit boundaries. Also, we see the I-channel doesn't reach its maximum value of 1 at those corresponding time instances. We see there is an interchannel relationship or cross-correlation property between the complex signal components.

Continuing along the lines of using the complex envelope theory in presenting the FSK waveform, we now discuss the receiver block diagram (see Fig. 2.54).

An alternative and simplified representation is obtained by carrying out the mathematical operations shown below. The instantaneous frequency variation is the time derivative of the phase signal.

$$w_i(t) = \frac{d}{dt}\left[\tan^{-1}\left(\frac{Q}{I}\right)\right] \tag{2.36}$$

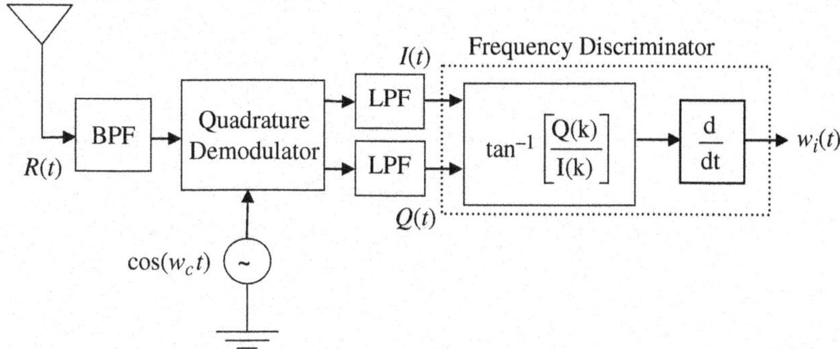

FIGURE 2.54 Baseband frequency discriminator block diagram.

$$w_i(t) = \frac{I(t)\frac{d}{dt}Q(t) - Q(t)\frac{d}{dt}I(t)}{I^2(t) + Q^2(t)} \tag{2.37}$$

Using the above equations, it is easy to draw the alternative block diagram of a frequency discriminator (see Fig. 2.55).

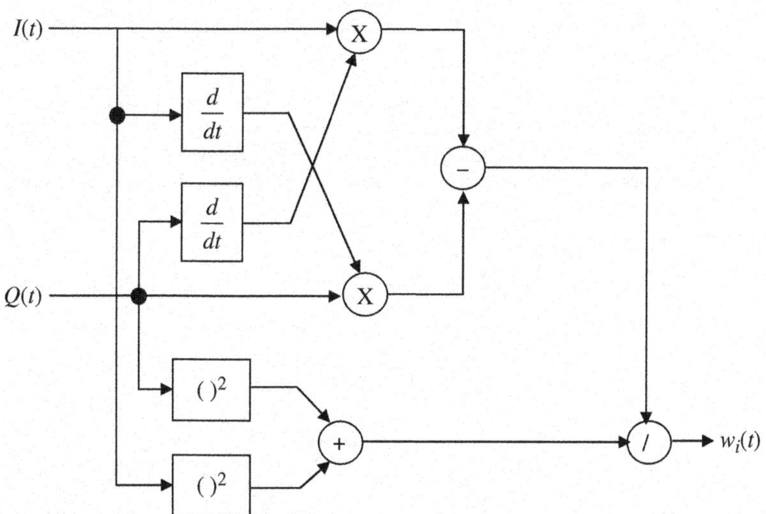

FIGURE 2.55 Alternative representation of the baseband frequency discriminator.

When BFSK does not satisfy the system requirements of your standard, M-ary FSK can be deployed. Let's, for example, consider a 4-level FSK modulation scheme. The typical method described above, which used filtered data to excite the VCO, is shown in Fig. 2.56. Note that the LPF input signal now has an alphabet of size 4.

If we once again allow ourselves to have a simplified viewpoint of the modulation scheme, then the transmit spectrum can be viewed as having four spectral lines, centered around the center carrier frequency (see Fig. 2.57).

We can now present the complex envelope waveforms for a 4-level FSK system with a data rate $R_b = 3.2$ kbps and a frequency deviation $f_d = 1.6$ kHz. In this case, the modulation index is $\beta_1 = 1$

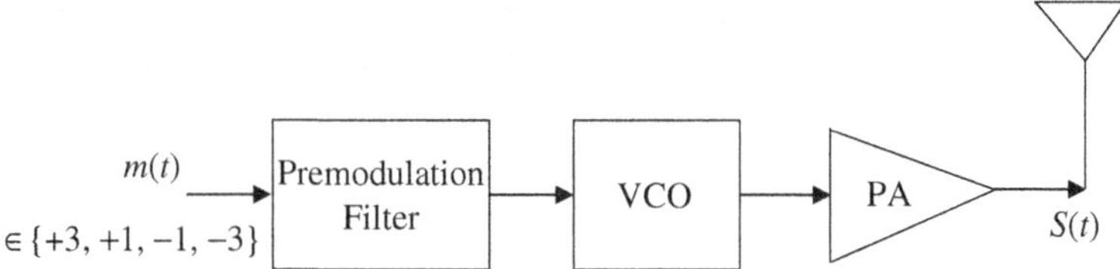

FIGURE 2.56 Conventional FM generation method for 4-level FSK.

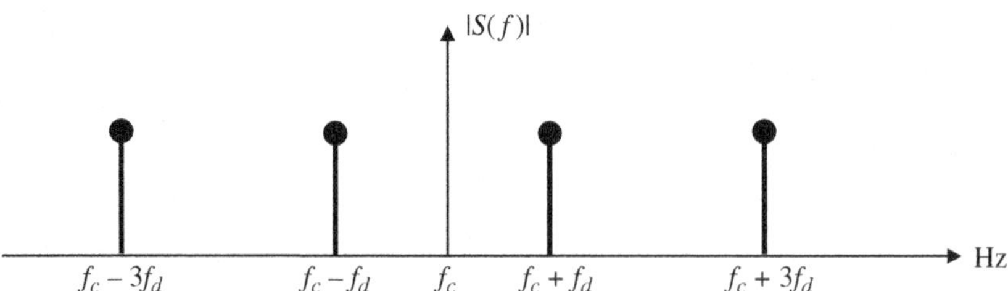

FIGURE 2.57 Simplified view of 4-level FSK modulation scheme.

for the two inner spectral components and $\beta_2 = 3$ for the outer two spectral components. The corresponding baseband waveforms are displayed in Fig. 2.58.

The three waveforms shown in Fig. 2.58 constitute the baseband frequency modulator input data signal, and the analytic output signals. The waveforms are provided for a few values of the 4-level random data.

We were able to present an equation to quantify the separation between the frequency deviations provided by the unitless modulation index parameter β. Next we want to understand if there are certain values of the modulation index that exhibit properties that are in some way better than other modulation indices [26]. The frequency deviations should be separated enough in order to be separable

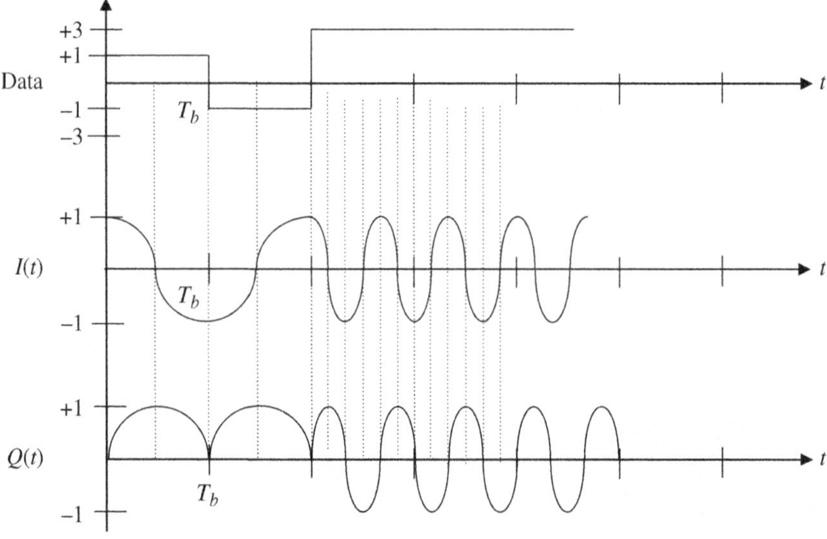

FIGURE 2.58 Unfiltered complex envelope of 4-level FSK with $\beta = 1$ and 3.

in the receiver. At first thought, we would consider waveforms that are orthogonal to one another. Let the two frequency deviations (assuming BFSK) be represented as

$$S_1(t) = \sqrt{2P} \cdot \cos(2\pi f_c t + 2\pi f_d t + \theta)$$
$$S_2(t) = \sqrt{2P} \cdot \cos(2\pi f_c t - 2\pi f_d t) \tag{2.38}$$

where θ has been inserted to represent an arbitrary phase offset between the two FSK signals. If we let these signals be represented as vectors in Cartesian coordinate system, then the orthogonality constraint is viewed as the two vectors being perpendicular to one another. Alternatively, it can be stated as

$$\int_0^{T_b} S_1(t) \cdot S_2(t)\, dt = 0 \tag{2.39}$$

After substituting the previous expressions and carrying out the integration, we arrive with an expression for the correlation of two sinusoids separated in frequency by $\Delta f = f_c + f_d - (f_c - f_d) = 2f_d$. In Fig. 2.59, we plot the cross-correlation function for BFSK modulation as a function of the modulation index [23]. Similarly, we have shown the cross-correlation for BPSK modulation or antipodal signaling, which is a constant value of -1.

Below we see the orthogonality condition is met for modulation indices that are multiples of 0.5. We would like to take this opportunity to relate this finding to what was presented earlier for BPSK. The BPSK modulation scheme has waveforms that are antipodal to one another, that is, the correlation between the 2 possible sinusoids is -1. This is drawn in the Fig. 2.59 as a dashed line. So, it was earlier stated the antipodal signaling provided the best performance in an AWGN channel. Hence modulation indices that produce a correlation factor close to -1 should produce better BER results. In fact, as can be seen from Fig 2.59, choosing $\beta = 0.715$ produces a correlation coefficient of $\rho = -0.217$. This point was made so the reader has a complete picture of the effects of the chosen β on system performance.

The modulation indices of interest, and ones commonly used, are related to satisfying the orthogonality constraint. In particular, we wish to discuss $\beta = 0.5$ case, sometimes called Fast Frequency Shift

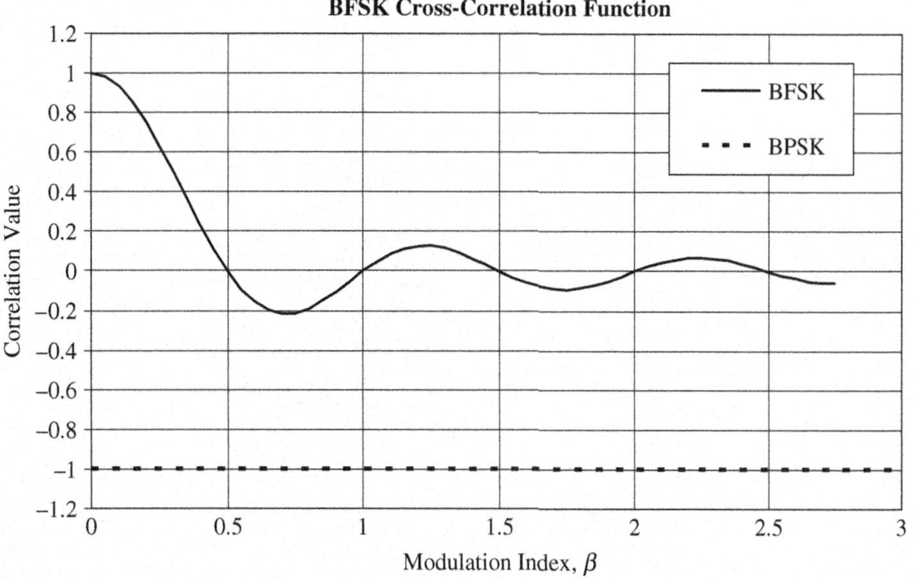

FIGURE 2.59 FSK correlation function.

Keying (FFSK), because this is the closest the two frequency deviations can be placed while still maintaining the principle of orthogonality. This is more often called Minimum Shift Keying (MSK) [27].

We will shortly see that MSK has a number of wonderful properties. It can be generated and detected in a number of ways. Let us first start with the complex envelope baseband approach.

2.2.7 Minimum Shift Keying (MSK)

We have specifically chosen to call attention to the orthogonal BFSK modulation scheme described above. In this section, we will discuss some of the special properties this modulation scheme has. Given a BFSK modulation with index $\beta = 0.5$, we can quickly draw the constellation diagram in Fig. 2.60, knowing the allowable phase change in a bit time is equal to $\beta\pi$.

The corresponding one-dimensional complex envelope waveforms can be drawn based on the above modulation scheme parameters (see Fig. 2.61). These baseband MSK waveforms were generated using the baseband FM modulator presented earlier.

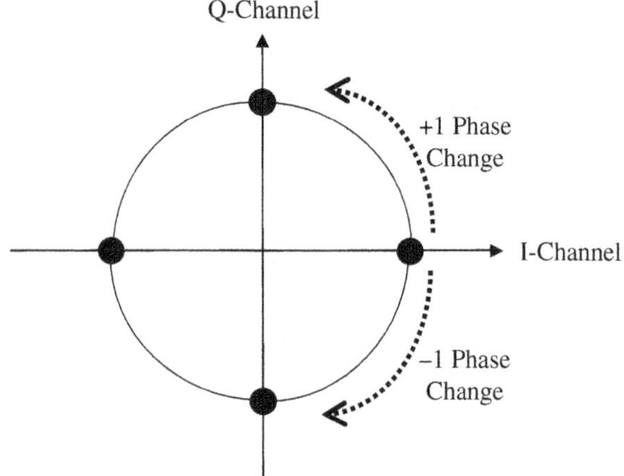

FIGURE 2.60 MSK signal constellation diagram.

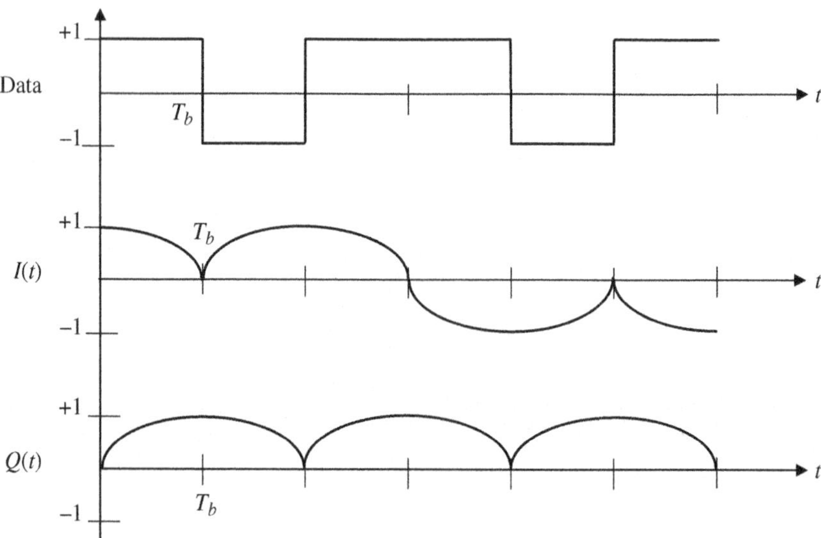

FIGURE 2.61 Unfiltered complex envelope of MSK.

Next we will briefly present another method to generate MSK, called offset bandpass filtering of BPSK [27]. This method consists of first generating a BPSK signal and then sending this signal through a BPF whose center frequency is offset by a specific amount. The transmission block diagram is shown in Fig. 2.62.

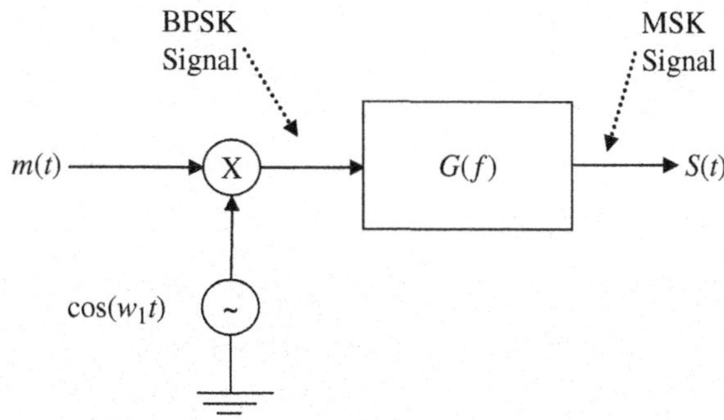

FIGURE 2.62 Offset bandpass filtering method of generating an MSK waveform.

This method will produce an MSK signal $S(t)$ at the output, with an apparent carrier frequency of f_o and deviations $f_1 = f_o - R_b/4$ and $f_2 = f_o - R_b/4$. The impulse response of the conversion filter, $g(t)$, is given as

$$g(t) = 2 \cdot \sin(2\pi f_2 t) \qquad (0 \leq t \leq T_b) \tag{2.40}$$

This filter is offset, in frequency, from the center of the BPSK signal generated at its input. This method was introduced for a few reasons: first, for the sake of completeness to show the special properties this modulation scheme has and, secondly, to introduce the relationship MSK has to the PSK family.

Another method used to generate MSK waveforms is treating this modulation scheme as if it were a member of the PSK family, specifically OQPSK. We will present a method used to generate MSK as treating it as OQPSK with sinusoidal pulse shaping [28]. This can be shown by the modulator block diagram in Fig. 2.63.

In Fig. 2.64, we present the waveforms at various points in the modulator block diagram given above.

In comparing these baseband waveforms to those shown earlier, one can easily see that they are different. Hence if the waveforms are detected using the same receiver, then the resulting bits would be different as well. However, pre- and/or post-processing can be performed to extract the correct information [29].

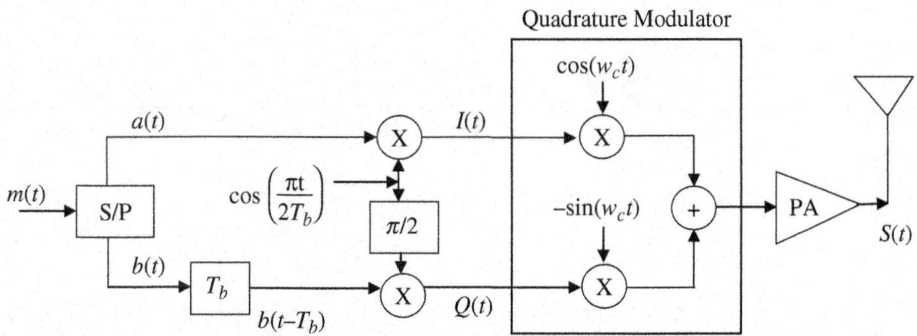

FIGURE 2.63 Offset QPSK method of generating an MSK waveform.

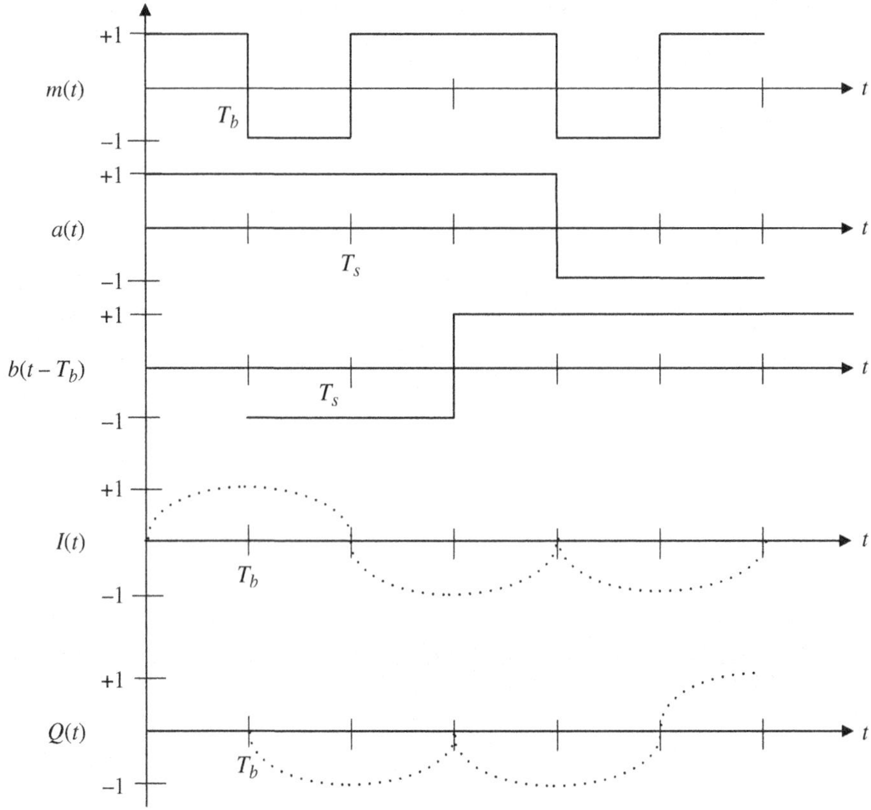

FIGURE 2.64 Baseband waveforms for MSK generated from the Offset QPSK approach.

The receiver for this approach is discussed next (see Fig. 2.65). This receiver's operating principles are easily described by the received complex envelope waveforms that are matched filtered with the pulse shaping used in the transmitter. The outputs are then sampled at the symbol rate, with one symbol stream offset by T_b from the other. This was done to align the two symbol streams so that the decision device can produce the information signal $m(t)$.

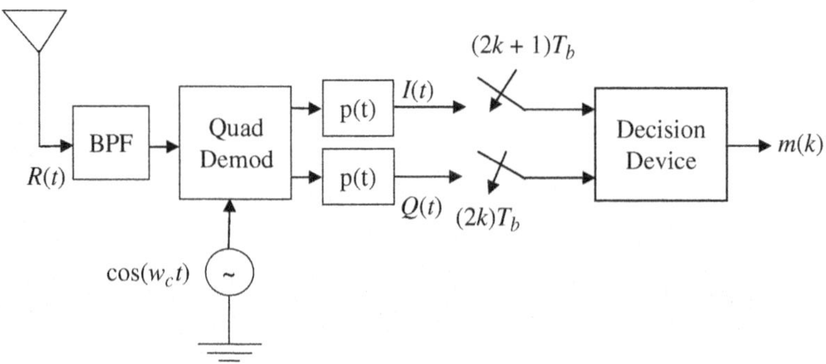

FIGURE 2.65 Receiver block diagram for the MSK waveform generated by the Offset QPSK approach.

In the above receiver block diagram, the matched filter has the following impulse response:

$$p(t) = \cos\left(\frac{\pi t}{2T_b}\right) \qquad (-T_b \le t \le T_b) \tag{2.41}$$

We started this section with a block diagram that contained a premodulation filter. A filter commonly used with MSK is the Gaussian filter GMSK. The Gaussian filter has excellent group delay characteristics. The 3 dB cutoff frequency of this filter is typically normalized by the bit rate to create an equivalent description called BT_b product, where B = 3 dB BW of the premodulation filter and $T_b = 1/R_b$ is the bit time duration. We will close this section by making a few points and observations as to how the BT_b product affects the system performance.

The autocorrelation functions for both MSK and OQPSK with NRZ pulse shaping are shown in Fig. 2.66 [26]. The figure assumes a bit time duration normalized to 1 sec (T = 1 sec).

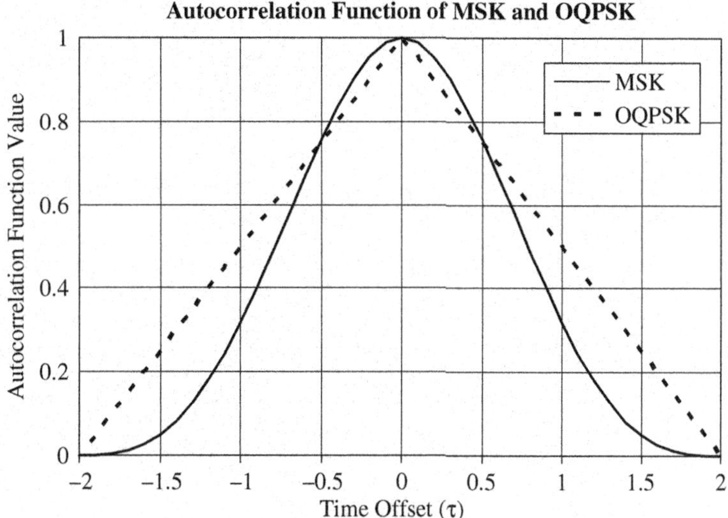

FIGURE 2.66 MSK and OQPSK autocorrelation functions.

The autocorrelation function for MSK is given as [30–33]

$$R_{\mathrm{MSK}}(\tau) = \frac{1}{\pi}\left\{\pi\left(1 - \frac{|\tau|}{2T}\right)\cdot\cos\left(\frac{\pi|\tau|}{2T}\right) + \sin\left(\frac{\pi|\tau|}{2T}\right)\right\} \tag{2.42}$$

The autocorrelation function for OQPSK is given as

$$R_{\mathrm{OQPSK}}(\tau) = 1 - \frac{|\tau|}{2T} \tag{2.43}$$

The PSD is obtained by taking the Fourier transform of the above respective autocorrelation functions (see Fig. 2.67).

The mathematical equations used to plot the above curves are given below:

$$\mathrm{PSD}_{\mathrm{MSK}}(f) = \frac{8\cdot P\cdot T\cdot(1 + \cos[4\pi fT])}{\pi^2\cdot(1 - 16\cdot T^2\cdot f^2)^2} \tag{2.44}$$

$$\mathrm{PSD}_{\mathrm{OQPSK}}(f) = 2\cdot P\cdot T\cdot\left(\frac{\sin[2\pi fT]}{2\cdot\pi\cdot f\cdot T}\right)^2 \tag{2.45}$$

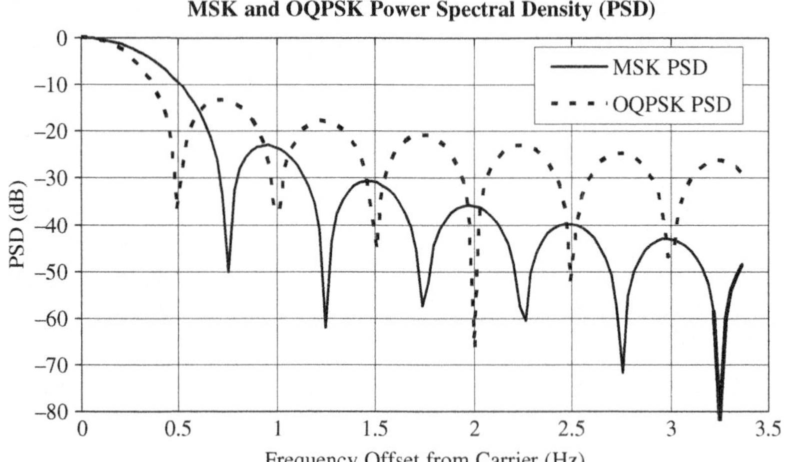

FIGURE 2.67 PSD comparison of MSK with OQPSK modulation schemes.

From the above equations, it is easy to see that MSK spectrum falls off at a rate of f^{-4}, while OQPSK falls off at a rate of f^{-2}.

Gaussian-Filtered MSK (GMSK). In this subsection, we will address the GMSK modulation scheme, specifically the interaction of the Gaussian LPF with the MSK waveform [34]. As mentioned earlier, lowering the BT_b product will reduce the out-of-band emissions; this is clearly shown in Fig. 2.68 when we view their respective PSD plots [35–40].

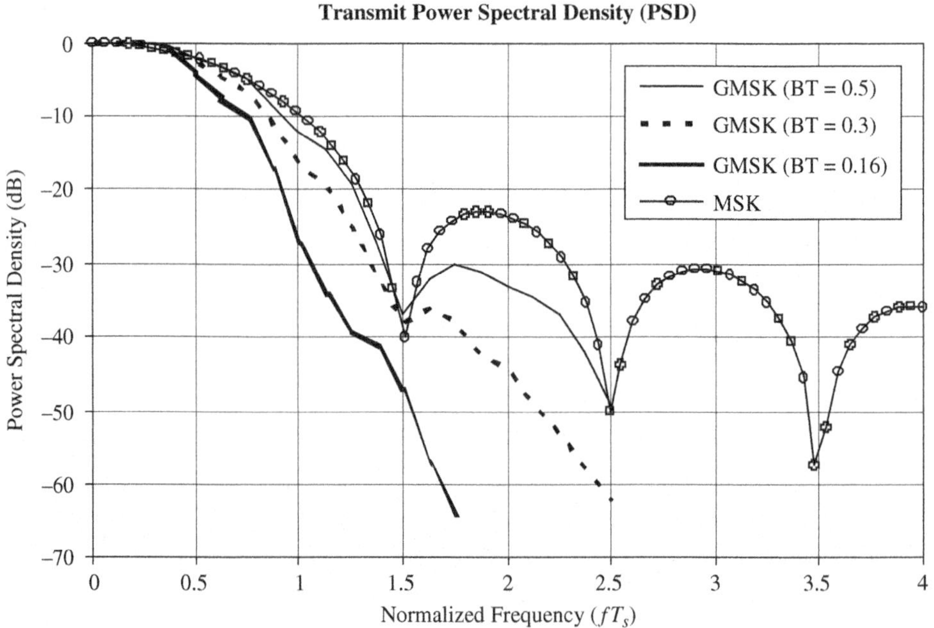

FIGURE 2.68 GMSK PSD plot comparison for various BT_b product values.

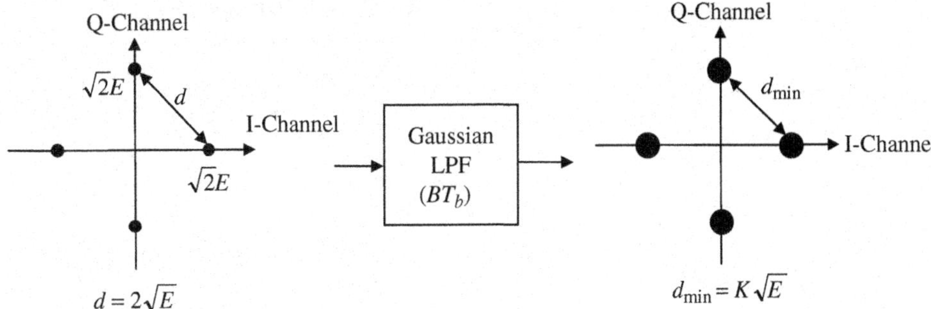

FIGURE 2.69 GMSK filtering impairment.

This spectral compactness is achieved by slowing down the phase transitions. Without the filtering, the phase transitions produced a ±90-degree phase shift in a single bit time duration. After the insertion of the Gaussian LPF, the achievable phase shifts now become $\pm(90 - \alpha)$ degrees, where α increases as BT_b decreases. This occurs because the higher frequency components become reduced in amplitude due to more filtering performed on the information signal and secondly because memory is inserted into the modulation symbol. The term memory used to describe the present phase change is dependent not only on the present information bits but also on the past and filtered bits as well.

An alternative way to describe the same behavior is to consider the effects of the band limiting on the signal constellation diagram (see Fig. 2.69). The constellation on the left shows the unfiltered MSK phase states. When a Gaussian LPF is used, the phase transitions have been slowed down to obtain a spectrally efficient transmit emission spectrum. This inserted memory causes ISI which we have shown as large dark circles to emphasize a scatter of points around each state. As we decrease the BT_b product, the radius of the four large circles increases, resulting in BER performance degradation since the minimum distance d_{min} decreases. The values of K have been investigated for various values of BT_b. (Consult [34] for further details.)

In Fig. 2.70, we plot the I-channel eye diagram of MSK, assuming a 16-time-oversampled signal. This diagram clearly shows the absence of ISI.

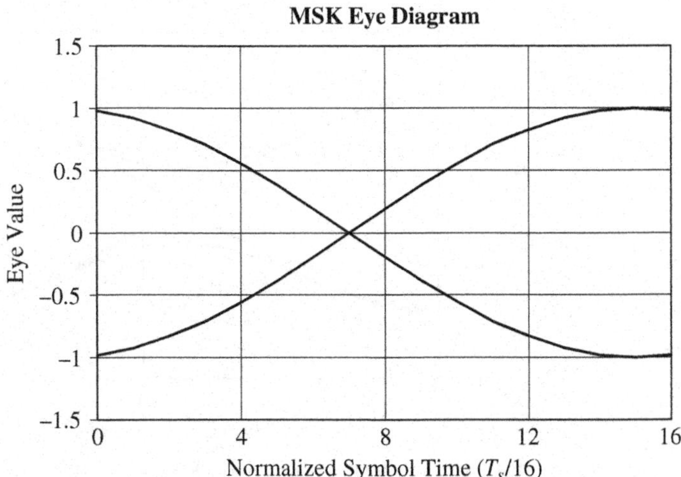

FIGURE 2.70 MSK eye diagram (BT_b = infinity).

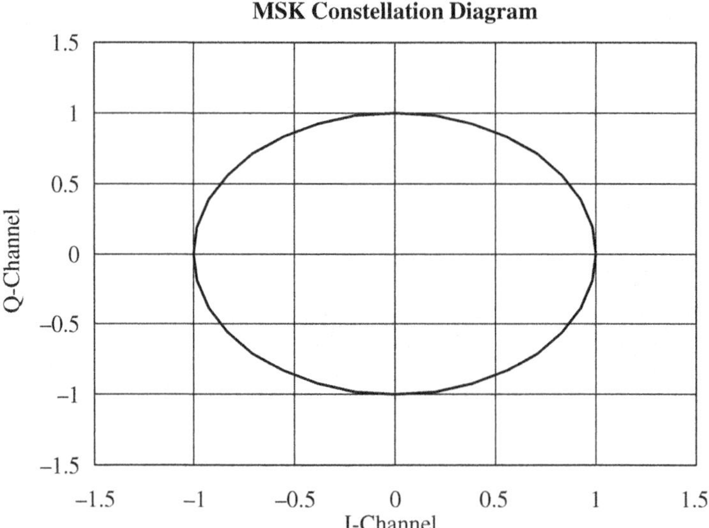

FIGURE 2.71 MSK signal constellation.

The corresponding MSK signal constellation is drawn in Fig. 2.71. Notice this modulation scheme is a constant envelope waveform.

When the BT_b product is reduced to 0.5, notice some ISI is present in the eye diagram when a Gaussian filter GMSK is used (see Fig. 2.72).

A point worth mentioning is as the BT_b product decreases, more filtering is applied to the information signal. This additional filtering essentially slows down the phase trajectories. This in turn causes an increase in ISI and symbol timing jitter. The GMSK eye diagram in Fig. 2.73 is for a BT_b product of 0.3.

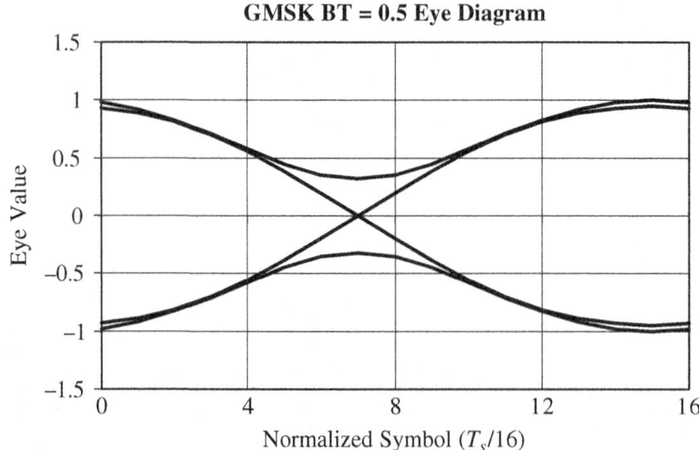

FIGURE 2.72 GMSK eye diagram with $BT_b = 0.5$.

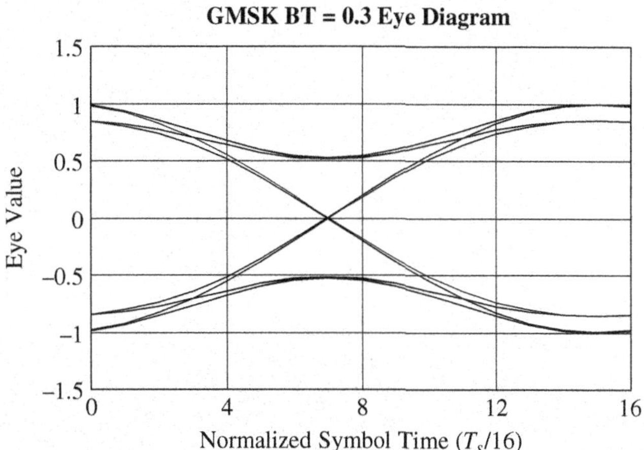

FIGURE 2.73 GMSK eye diagram with $BT_b = 0.3$.

2.2.8 Controlling Phase Trajectories

In this section, we will discuss manipulating the phase trajectories, rather than having them determined by the impulse response of the pulse-shaping filter and the modulation scheme chosen. Specifically, we will briefly introduce what is commonly called Feher's PSK modulation schemes (see [41–43]).

What Dr. Feher realized is that linear modulation schemes require portable devices to operate their transmit PA in its linear region. In this range, the PA is inefficient in terms of power consumption, but is required in order to benefit from the spectral efficiency of linear modulation schemes. Driving the PA into saturation will allow the PA to be more efficient; however, spectral regrowth issues need to be addressed. Using constant envelope modulation schemes allows us to saturate the PA; however, they are inherently spectrally inefficient. So a median ground must be found in order to use the more efficient PAs. Dr. Feher's family of modulation schemes do this by controlling the phase trajectories on the transmit signal constellation diagrams in a number of ways. We will walk the reader through some of the approaches taken and then summarize by providing further reading in the references [44–49].

Below we emphasize the inherent correlation between the I- and Q-channels of a FSK modulated waveform. There are two points to make, first being when one channel is at a minimum, the other is at a maximum. The channels work in concert to produce the modulated waveform. The second point is as one channel decreases (or increases), the other channel increases (or decreases). The combination of the two channels produces a constant envelope. In this section, we will insert this cross-correlation property of the I- and Q-channels into the modulator in order to not only create a spectrally efficient modulation scheme, but also have less spectral regrowth when a nonlinearity is introduced to the modulated waveform (see Figs. 2.74 and 2.75). By controlling the amount of cross-correlation inserted, the envelope variation can be significantly reduced [50].

Feher QPSK (FQPSK). The first approach is called FQPSK-1 [41]. Here the intent is to build on OQPSK and try to reduce the envelope variations, in order to use the more power efficient, nonlinear PA without the tremendous concern around the typical issues surrounding spectral regrowth. Here intersymbol and jitter free (IJF) filters were used in the OQPSK modulation scheme (see Fig. 2.76) [51].

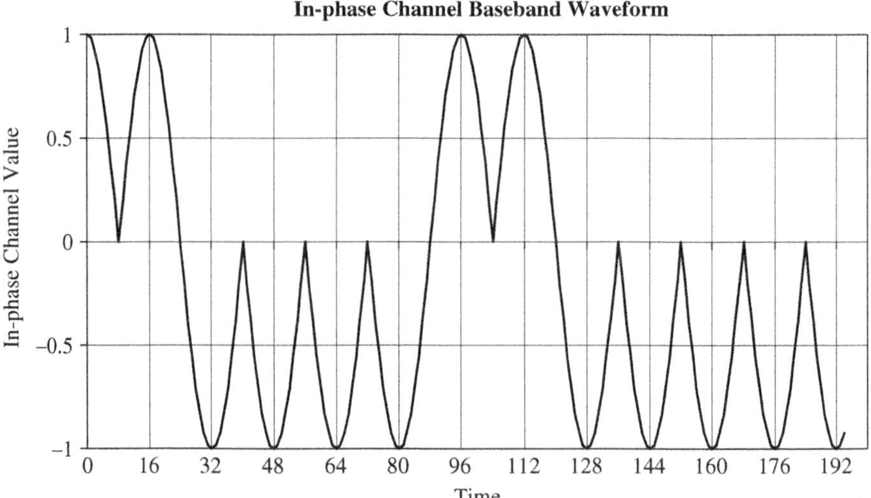

FIGURE 2.74 I-channel waveform.

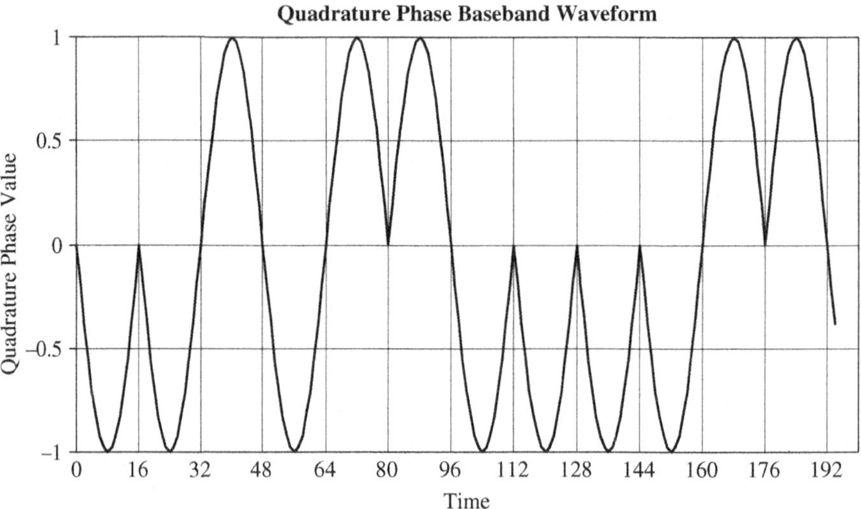

FIGURE 2.75 Q-channel waveform.

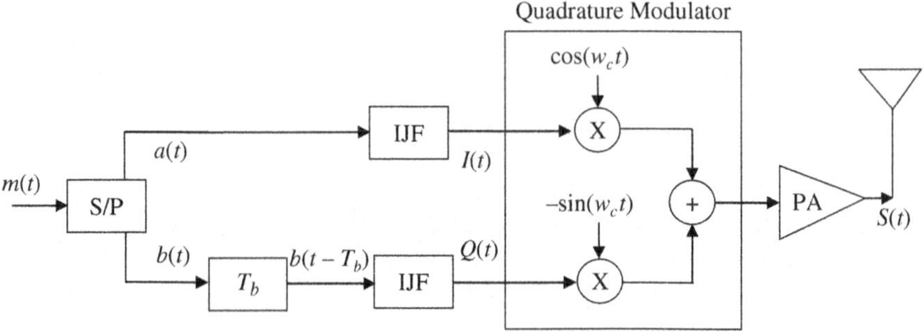

FIGURE 2.76 FQPSK-1 modulation waveform generation block diagram.

The IJF pulse-shaping filters had the following responses for the I- and Q-channels:

$$I(t) = a(n)p(t - nT_s) + a(n - 1)p(t - (n - 1)T_s) \tag{2.46}$$

$$Q(t) = b(n)p(t - (n - 0.5)T_s + b(n - 1)p(t - (n + 0.5)T_s + b(n - 2)p(t - (n + 1.5)T_s) \tag{2.47}$$

where the pulse shaping was given as

$$p(t) = \frac{1}{2}\left[1 + \cos\left(\frac{\pi t}{T_s}\right)\right] \qquad (-T_s \leq t \leq T_s) \tag{2.48}$$

We can also write the equations at the output of the nonlinear PA as follows:

$$S_{out}(t) = \frac{I(t) \cdot \cos(\omega_c t)}{\sqrt{I^2(t) + Q^2(t)}} - \frac{Q(t) \cdot \sin(\omega_c t)}{\sqrt{I^2(t) + Q^2(t)}} \tag{2.49}$$

The IJF-QPSK signal constellation diagram is compared to the IJF-OQPSK constellation (see Fig. 2.77).

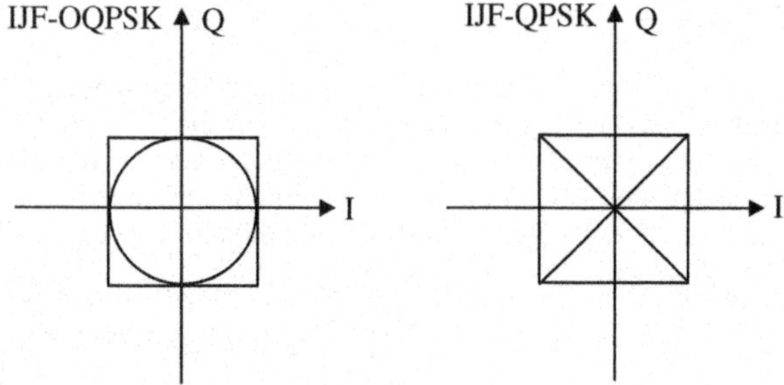

FIGURE 2.77 IJF constellation diagram comparisons.

The IJF criteria discussed above can best be explained with the pictorial description in Fig. 2.78.

Next, an improved concept called FQPSK-KF was applied [43]. The previous approach was used as a stepping stone to the following FQPSK-KF modulation scheme. Here the artificial cross-correlation property is inserted into the FQPSK-1 modulation scheme. A simplified block diagram for this waveform generation is shown in Fig. 2.79.

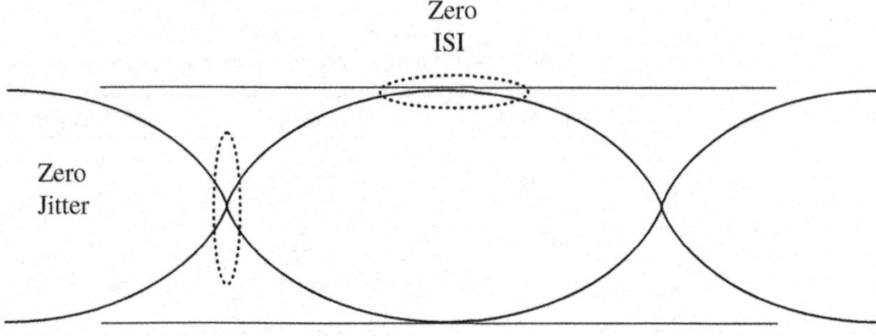

FIGURE 2.78 Description of the IJF filter.

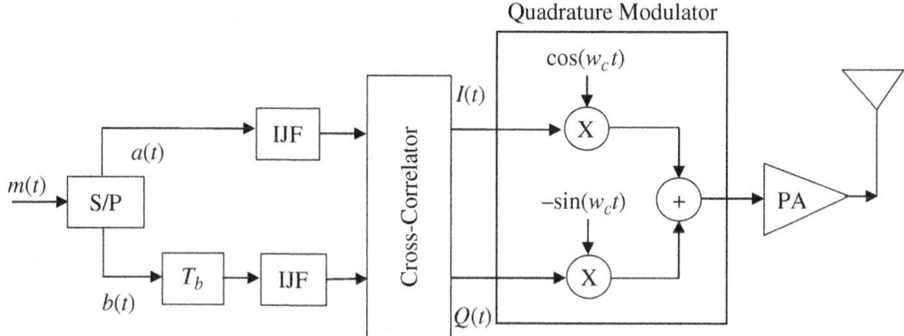

FIGURE 2.79 FQPSK-KF modulation waveform block diagram.

The basic idea behind the cross-correlator block is to produce some desirable behavior of reducing envelope variations. For example, let's consider

- +When the I- (or Q-) channel is zero, then the Q- (or I-) channel is at 1.
- +When the I (or Q) channel is nonzero, the Q- (or I-) channel is reduced to A, where the constraint $(1/\sqrt{2} \leq A \leq 1)$ applies.

Let's now describe the cross-correlation used in [43] next. The output of the IJF filter will be controlled by four transitions plus the two sinusoidal waveforms. Choice of the combinations depends on the present symbol and the preceding symbols. Here we observe two input symbols and then generate an output consisting of half–symbol time waveform.

The four possible transition functions are defined with A $(1/\sqrt{2} \leq A \leq 1)$ as the system design parameter as

$$f_1 = 1 - (1 - A)\cos^2\left(\frac{\pi t}{T_s}\right) \tag{2.50}$$

$$f_2 = 1 - (1 - A)\sin^2\left(\frac{\pi t}{T_s}\right) \tag{2.51}$$

$$f_3 = -1 + (1 - A)\cos^2\left(\frac{\pi t}{T_s}\right) \tag{2.52}$$

$$f_4 = -1 + (1 - A)\sin^2\left(\frac{\pi t}{T_s}\right) \tag{2.53}$$

Based on the 2-input symbols or 4 bits, it is easy to show there are 16 half-symbol combinations that can occur. They are summarized in Table 2.6.

TABLE 2.6 FQPSK Combinations of Waveform Generation

I/Q channel	Q/I channel	No. of combinations
$\pm\cos\left(\frac{\pi t}{T_s}\right)$	$\pm\sin\left(\frac{\pi t}{T_s}\right)$	4
$\pm A\cos\left(\frac{\pi t}{T_s}\right)$	f_1 or f_3	4
$\pm A\sin\left(\frac{\pi t}{T_s}\right)$	f_2 or f_4	4
$\pm A$	$\pm A$	4

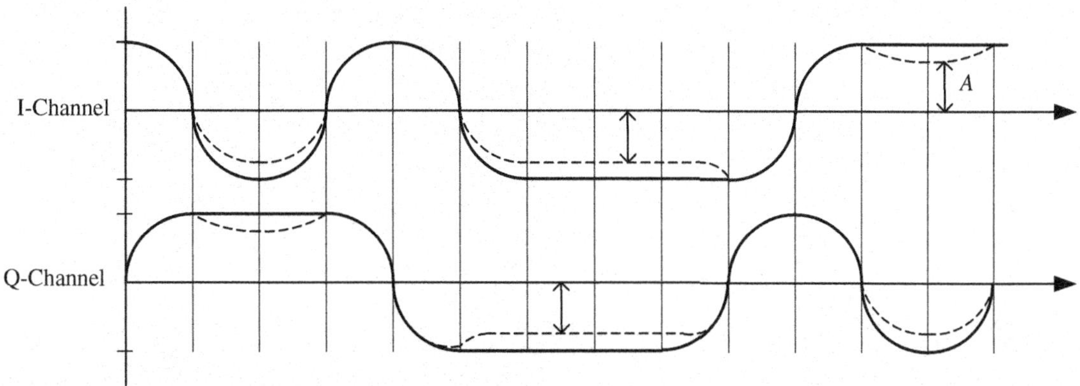

FIGURE 2.80 Cross-correlated baseband waveforms.

The cross-correlated output waveforms are shown in Fig. 2.80 [44]. The solid line corresponds to the IJF-OQPSK modulation and the dashed line corresponds to the Cross-Correlated Phase Shift Keying (XPSK) modulation with the design parameter $A = 0.707107$.

It is important to note that this original technique was presented where the output waveforms were generated at half symbol rate. Other variations were subsequently proposed that generated output waveforms at the symbol rate, reducing possible phase discontinuities that may otherwise arise.

As the system design parameter A decreased, the envelope variations decreased. Hence by varying the parameter A, the cross-correlator can manipulate the constellation diagram to go from IJF-OQPSK to FQPSK-KF and thus change the spectral regrowth effects. Since the introduction of the Feher-patented PSK modulation schemes, much research has been conducted to continue to strive for more power and spectrally efficient modulation schemes. Also alternatives to those presented by Dr. Feher have also emerged, such as those described in [52] and [53].

Numerous publications have demonstrated the FQPSK family of modulation schemes to have significant advantages over QPSK, OQPSK, and MSK and $\pi/4$-DQPSK, especially in the highly nonlinear amplified channel application [54–56].

In an effort to further improve spectral efficiency, a double-jump (DJ)-filtered cross-correlated FQPSK modulation scheme, called DJ-FQPSK, was also proposed [45]. For this case, the cross-correlator output signals are filtered with the frequency response of the square root DJ filter. However, prior to presenting the square root filters, let's present two forms of a DJ filter.

$$H_1(f) = \begin{cases} 1 & , \quad 0 \le f < (1-\alpha)f_n \\ (1-JR) - \dfrac{(1-2JR)(f+\alpha f_n - f_n)}{2\alpha f_n} & , \quad (1-\alpha)f_n \le f < (1+\alpha)f_n \\ 0 & , \quad (1+\alpha)f_n \le f \end{cases} \tag{2.54}$$

where α = the roll-off factor, JR = the jump rate, and f_n = Nyquist frequency.

$$H_2(f) = \begin{cases} 1 & , \quad 0 \le f < (1-\alpha)f_n \\ \dfrac{1}{2} - \dfrac{T_s}{2\alpha}(1-C)\left(f - \dfrac{1}{2T_s}\right) & , \quad (1-\alpha)f_n \le f < (1+\alpha)f_n \\ 0 & , \quad (1+\alpha)f_n \le f \end{cases} \tag{2.55}$$

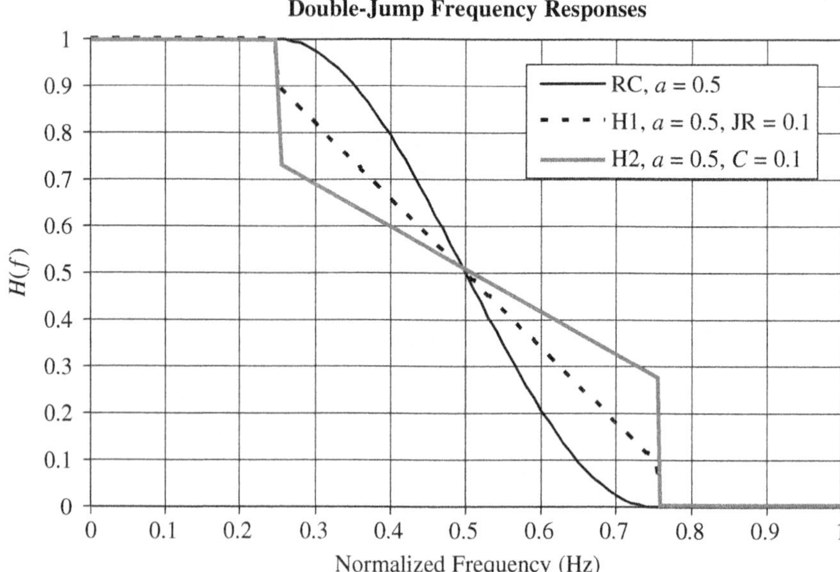

FIGURE 2.81 Double-jump filter frequency response comparison.

Figure 2.81 compares the frequency responses of the RC to that of the DJ filters defined above [57, 58].

Since these filters satisfy the Nyquist criteria of zero ISI, one can imagine applying the same reasoning used for the SRC to the square root DJ filter. A comparison of the square root DJ filters (SDJ) is shown in Fig. 2.82.

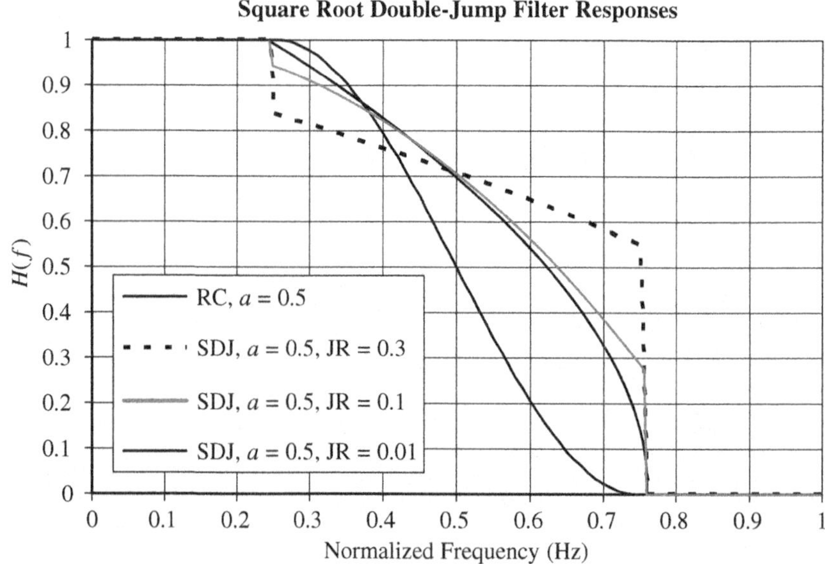

FIGURE 2.82 Square root double-jump frequency response comparisons.

The impulse response of a typical DJ filter is plotted with that of a Nyquist pulse-shaping filter, and both show satisfying the zero-ISI criteria (see Fig 2.83). Notice that the amplitudes of the tails in the DJ filter impulse response are smaller than those of the RC filter [59–61].

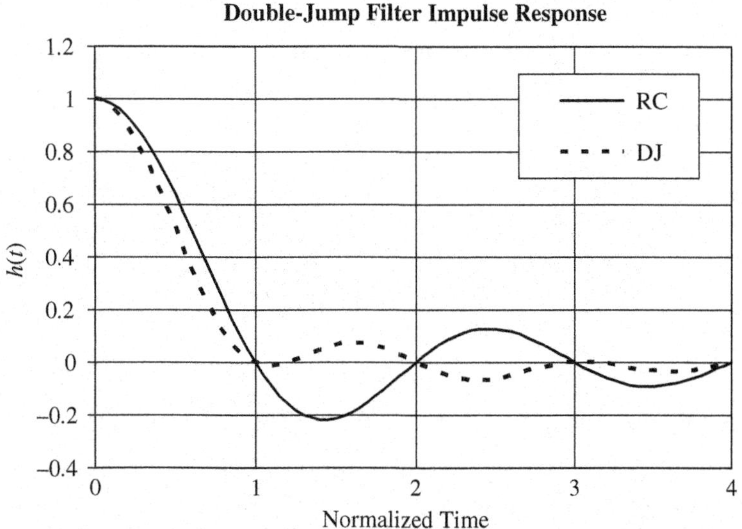

FIGURE 2.83 Double-jump impulse response.

Feher π/4-QPSK. In the previous sections, we have presented the successful technique of cross-correlating the complex envelope in order to achieve high spectral efficiency and high PA efficiency. Next we turn our attention to applying this technique to π/4-QPSK [62]. As previously shown, π/4-QPSK technique has phase trajectories that avoid the origin; however, they come dangerously close, using practical values of roll-off factors in defining the SRC filter. The cross-correlator is applied to π/4-QPSK as shown in Fig. 2.84 [63–66].

Thus, the previous FQPSK family of modulation schemes worked with a coherent demodulator. With this new scheme, differential detection is an option for the receiver now. The cross-correlation operations are scaled and shifted versions of sinusoids and are given below where we have split up the description into three cases:

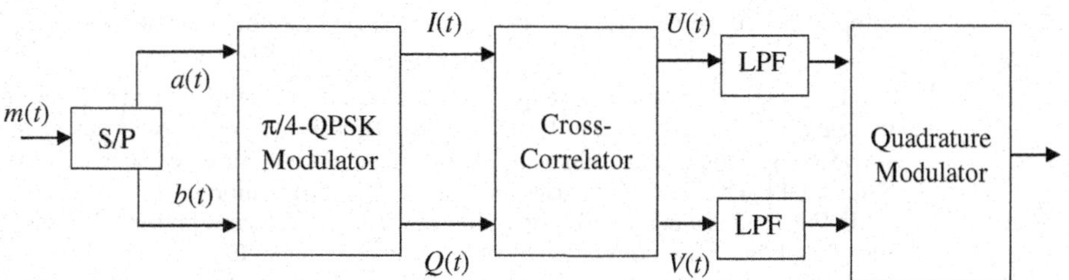

FIGURE 2.84 π/4-FQPSK waveform block diagram.

CASE 1: I(k) I(k − 1) < 0

If $Q(k)\,Q(k-1)$, then

$$U(t) = C_{11} + 0.5C_1\cos(2\pi(t/T_s - k + 1)) \qquad \text{for } (k-1)T_s < t < (k-0.5)T_s$$
$$V(t) = Q(k-1)\cos(\pi(t/T_s - k + 1))$$
$$U(t) = C_{21} - 0.5C_2\cos(2\pi(t/T_s - k + 1)) \qquad \text{for } (k-0.5)T_s < t < kT_s$$
$$V(t) = -Q(k)\cos(\pi(t/T_s - k + 1))$$

where we use the following definitions:

$$C_0 = 2A(I(k-1) + I(k))$$
$$C_1 = I(k-1) - C_0$$
$$C_2 = C_0 - I(k)$$
$$C_{11} = 0.5(I(k-1) + C_0)$$
$$C_{21} = 0.5(C_0 + I(k))$$
$$A = \text{correlation factor (0 to 1)}$$

CASE 2: Q(K) Q(k − 1) > 0, THEN

$$U(t) = A_1 + A_2\cos(\pi(t/T_s - k + 1)) \qquad \text{for } (k-1)T_s < t < kT_s$$
$$V(t) = B_1 + B_2\cos(\pi(t/T_s - k + 1))$$

where the following definitions

$$A_1 = 0.5((I(k-1) + I(k))$$
$$A_2 = 0.5(I(k-1) - I(k))$$
$$B_1 = 0.5(Q(k-1) + Q(k))$$
$$B_2 = 0.5(Q(k-1) - Q(k))$$

CASE 3: Q(k − 1) Q(k) = 0

Simply replace the $I(\sim)$ and $Q(\sim)$ by $Q(\sim)$ and $I(\sim)$ in the above equations, respectively.

As far as demodulation techniques are concerned, both coherent and noncoherent detection can be used and will be discussed in latter chapters. However, as one would expect, since the modulated waveform generation involved a state machine, an optimal receiver, such as a Maximum Likelihood Sequential Estimator, should also make use of this information [67].

In Fig. 2.85 is a comparison of various modulation schemes presented so far. The potential benefits of applying cross-correlation techniques can be dramatic.

Similarly, the comparison can be made with the $\pi/4$-FQPSK modulation scheme shown in Fig 2.86.

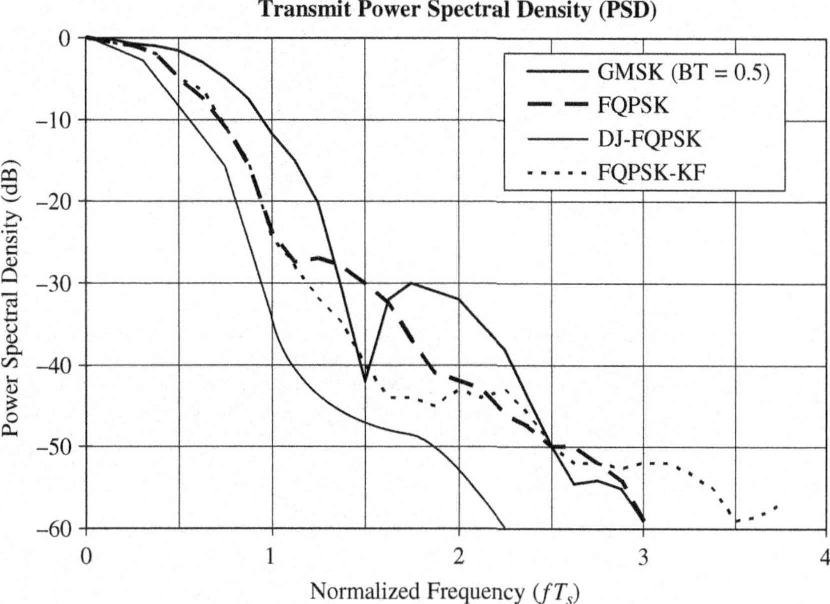

FIGURE 2.85 PSD comparison of FQPSK.

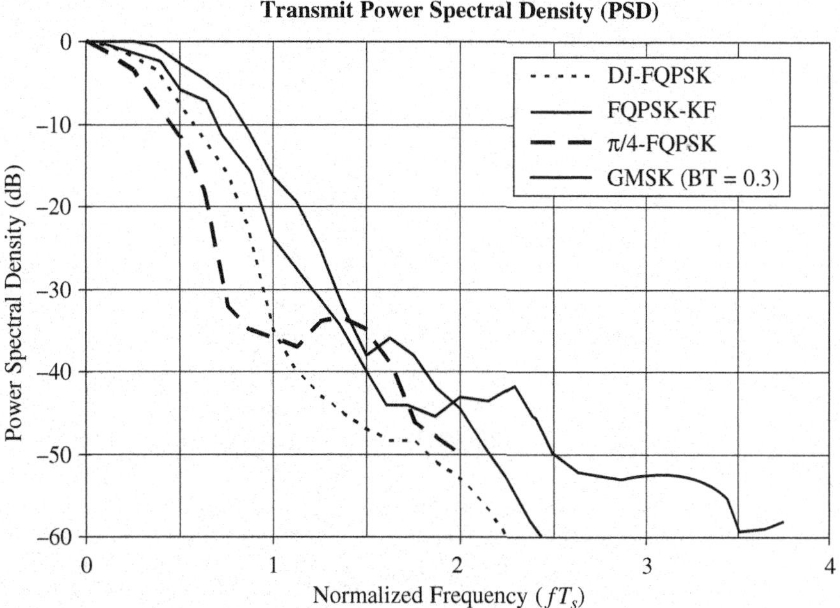

FIGURE 2.86 PSD comparison of $\pi/4$-FQPSK.

There exists an enhanced FQPSK version, where the I- and Q-channel waveforms are calculated at the symbol-by-symbol rate rather than the conventional bit-by-bit rate. This now means the subtle phase discontinuity that existed was removed or smoothed. Also by exploiting the memory inserted into the transmit signal when applying the cross-correlator, the optimum demodulator now becomes a trellis decoder (Viterbi sequence estimator). In continuing with the original FQPSK modulation scheme, the number of states to be used in the trellis decoder is 16 [68–71].

2.2.9 *M*-ary Frequency Shift Keying (FSK)

Earlier we have presented binary FSK and discussed the optimal choices of the modulation index and where the orthogonality principle plays a role. In this subsection, we will discuss increasing the number of modulation levels in FSK to M, which is commonly called M-ary FSK or M-level FSK [72, 73].

First, let us consider the following simple example for illustrative purposes. If a sinusoid is represented as a vector, then the orthogonal binary and ternary FSK modulation schemes can be viewed as in Fig. 2.87, using the Cartesian coordinate system. It is easy to convey this point in two and three dimensions; higher dimensions have decision regions that are more difficult to draw.

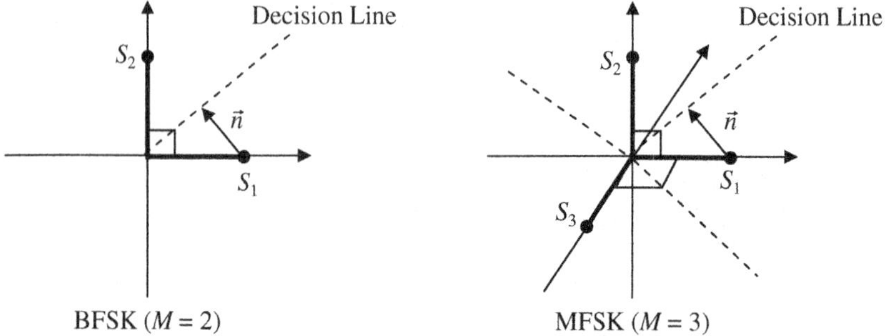

FIGURE 2.87 Comparison of binary and ternary FSK signal constellation diagrams.

Above we see that the distance between any two signal vectors is constant. This in turn means the noise immunity vector $\vec{n}$ remains constant as M increases. In other words, adding new signal vectors to the FSK alphabet does not make the signals vulnerable to small noise vectors [74].

Let us discuss the rate conversions as we introduce the M-ary modulation scheme in Fig. 2.88.

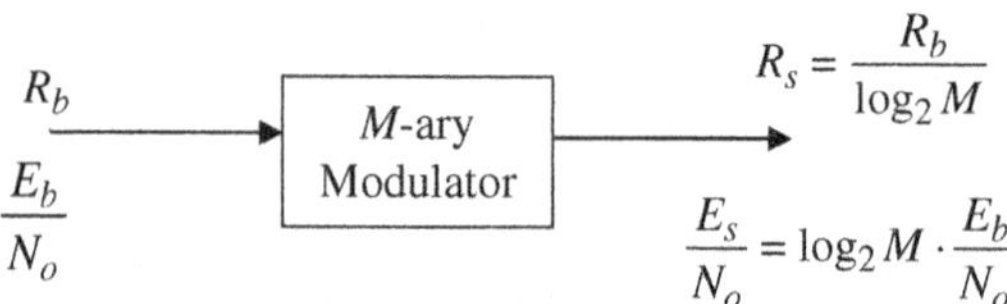

FIGURE 2.88 Data rate conversions for M-ary modulation.

If we were to plot the symbol error rate (SER) versus the SNR, we would observe as more modulation schemes levels are used, the performance degrades (i.e., the error rate performance curve shifts to the right), as shown in Fig. 2.89. Also note the following relationship: $M = 2^k$.

The reason this occurs is that as M increases, there are more neighboring decision regions and thus more ways to make a symbol error. In specific, there are $M - 1$ ways to make a symbol error.

Now to observe the error rate performance improvement, we must use the normalized SNR or E_b/N_o. Recall the following relationship:

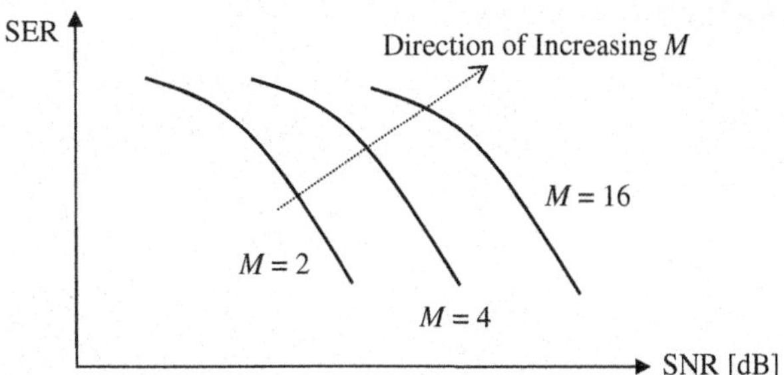

FIGURE 2.89 M-ary FSK symbol error rate performance.

$$\frac{E_b}{N_o} = \frac{S}{N} \cdot \frac{W}{R_b} \tag{2.56}$$

$$\frac{E_b}{N_o} = \frac{S}{N} \cdot \frac{WT_s}{\log_2(M)} \tag{2.57}$$

$$\frac{E_b}{N_o} = \frac{S}{N} \cdot \frac{1}{k} \cdot WT_s = \frac{S}{N} \cdot \frac{1}{k} \cdot \frac{W}{R_s} \tag{2.58}$$

We notice as the bandwidth W increases, the symbol time decreases and their product remains approximately equal to 1. With this observation, we can now show the improvement as the number of modulation levels increases [23].

$$\frac{E_b}{N_o} \cong \frac{S}{N} \cdot \frac{1}{k} \tag{2.59}$$

This improvement can also be seen in the BER curves in Fig. 2.90, where the receiver is assumed to have used ideal coherent detection.

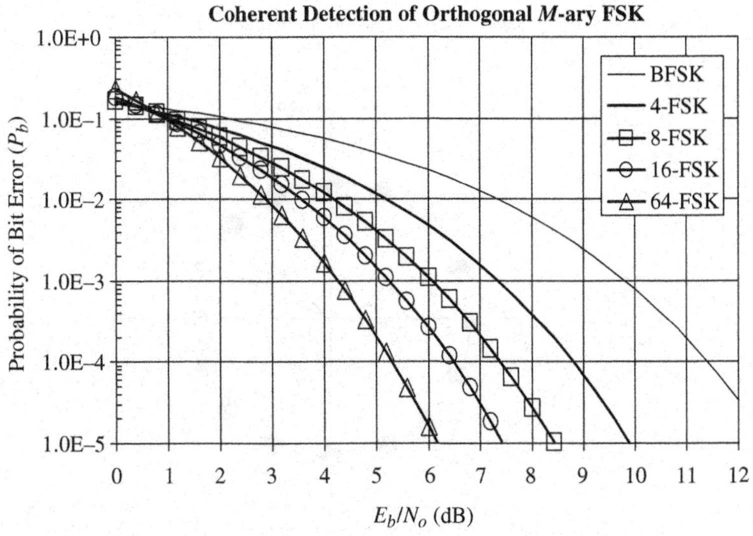

FIGURE 2.90 Alternative representation of M-ary FSK Bit Error Rate performance.

2.2.10 *M*-ary Phase Shift Keying (PSK)

Let us consider the example of 16-PSK, the transmit signal is generated as shown in Fig. 2.91. The input bit stream is grouped into sections of k bits, thus making 2^k possible modulation levels. The input bit rate is R_b bps and is converted to a symbol rate of R_s sps. In this case, $R_s = 1/4\ R_b$, hence occupying less transmission BW.

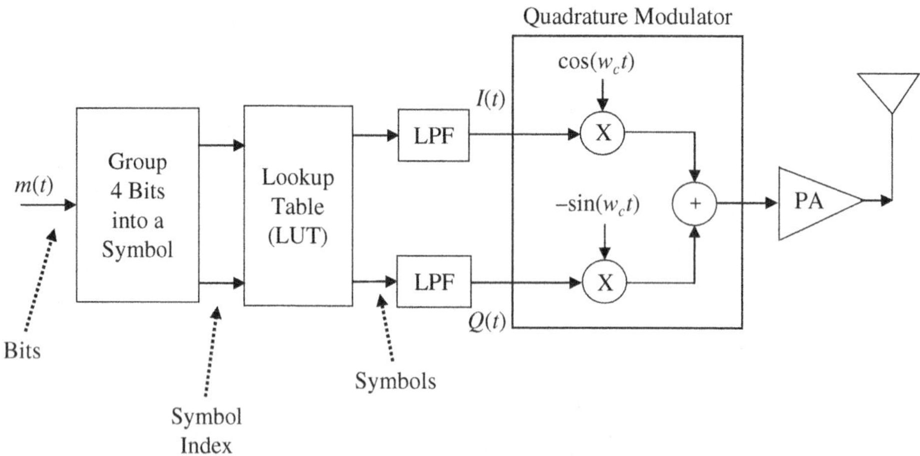

FIGURE 2.91 General 16-PSK waveform generator.

This example modulation scheme is spectrally efficient at 4 bps/Hz. It does, however, have poor BER performance when compared to QPSK, there is approximately 7-dB E_b/N_o degradation. Since there are quite a few modulation levels, a more complicated receiver is required. Similarly, since the modulation scheme is not a member of the constant envelope family, it suffers from spectral regrowth [75–77].

Consider the 16-PSK signal constellation diagram in Fig. 2.92.

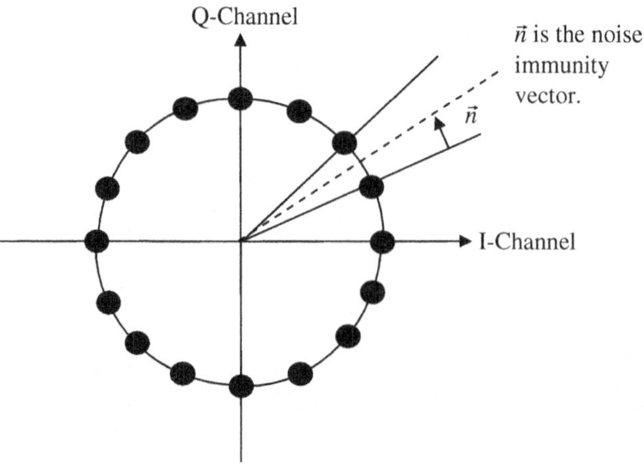

FIGURE 2.92 Unfiltered 16-PSK constellation block diagram.

The more levels (or modulation order) you consider, the smaller the noise vector required to be used before it can introduce an error. Hence, as more modulation levels are added, the BER performance degrades. This is the opposite behavior described for the M-ary FSK modulation scheme.

Figure 2.93 provides the BER performance versus the SNR per bit for M-ary PSK, where $M = 2$, 4, 16, 32, and 64.

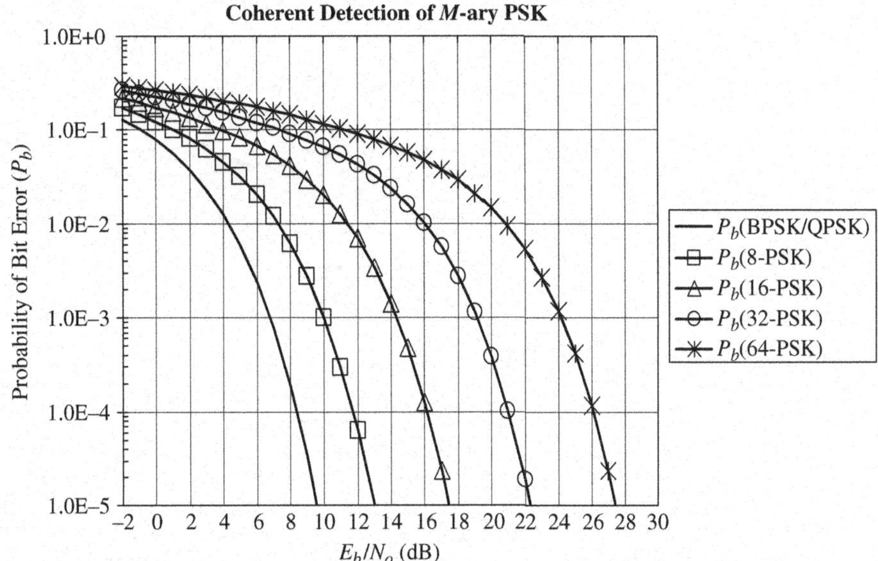

FIGURE 2.93 M-ary PSK Bit Error Rate performance.

We can plot the signal constellation diagram for 16-PSK using an RC LPF with $\alpha = 0.3$. The diagram in Fig. 2.94 shows us the transmit waveform has envelope variations and is therefore not a constant envelope.

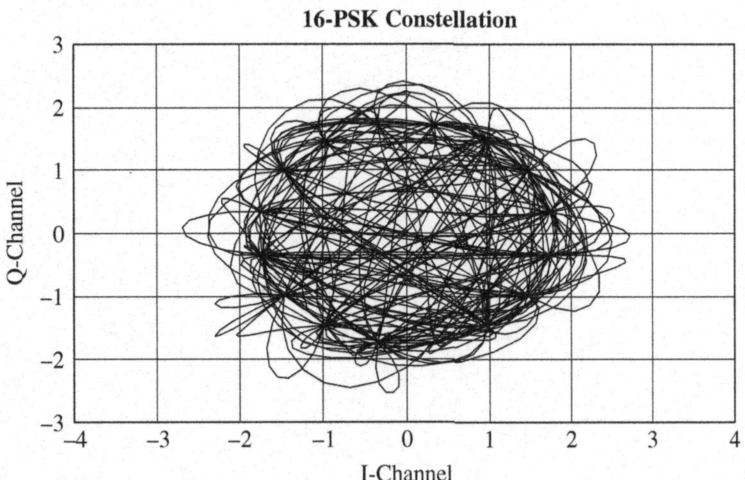

FIGURE 2.94 16-PSK signal constellation using RC with $\alpha = 0.3$.

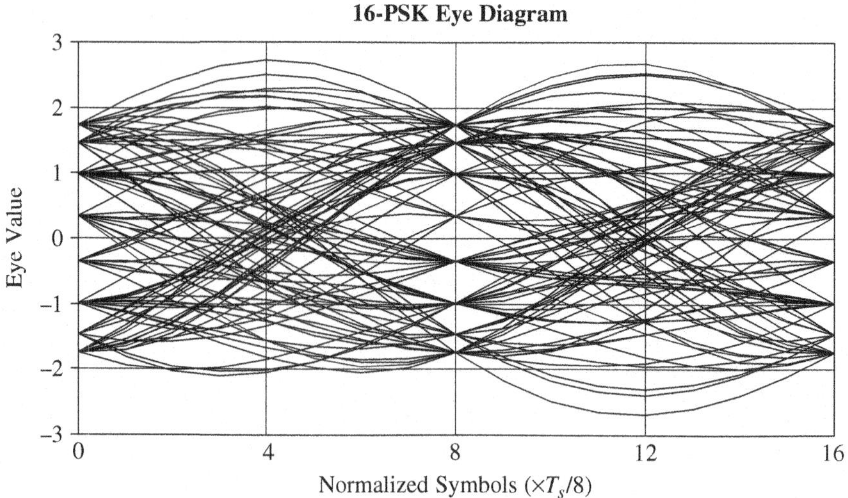

FIGURE 2.95 16-PSK signal constellation using RC with $\alpha = 0.3$.

The eye diagram of the corresponding signal constellation is given in Fig. 2.95. We can clearly see the additional levels required to support the 16-PSK modulation waveform. Also note they are not all equidistant from each other.

We made a point to mention noncoherent detection as a means to simply demodulate M-DPSK modulation waveforms. The BER of M-DPSK ($M = 2, 4$, and 8) is presented in Fig. 2.96. When comparing the BER performance of 8-DPSK to 8-PSK, we notice that noncoherent detection requires approximately 3-dB-higher E_b/N_o for BER = 1E-5 in an AWGN channel.

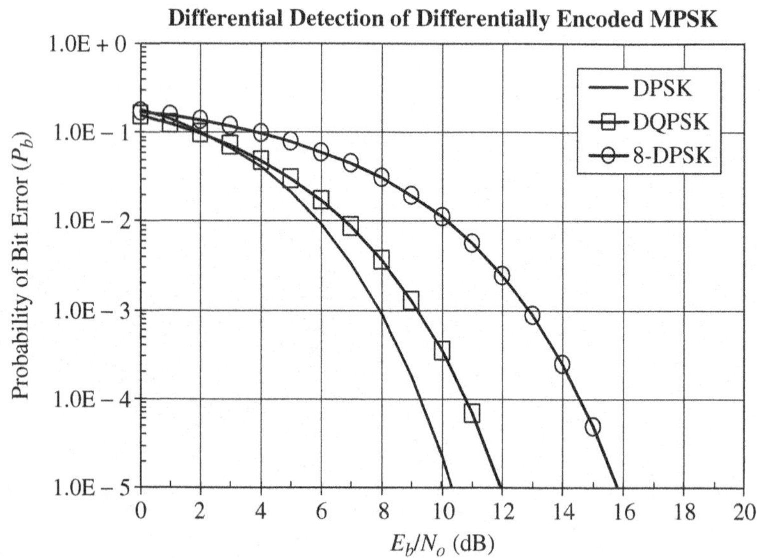

FIGURE 2.96 Differential detection of PSK modulation schemes.

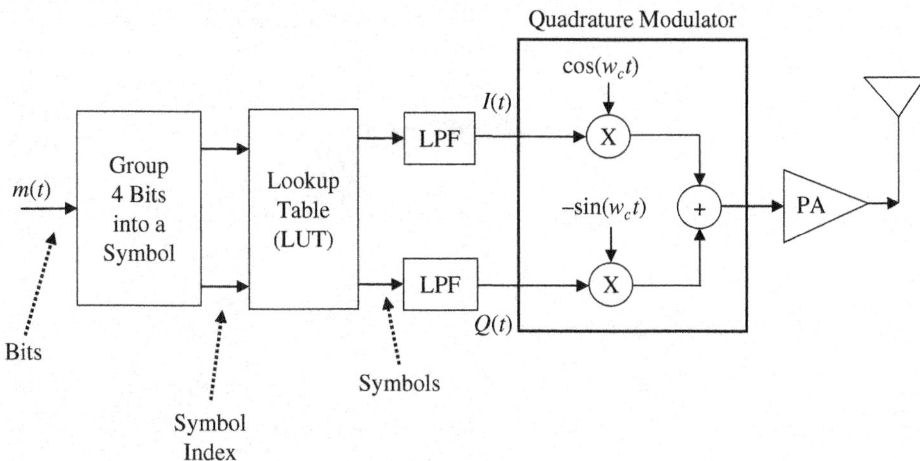

FIGURE 2.97 General 16-QAM waveform generator.

2.2.11 *M*-ary Quadrature Amplitude Modulation (QAM)

Let us consider the example of 16-QAM modulation scheme. The transmit signal is generated where not only is the phase varied, but also the amplitude. The waveform generator shown in Fig. 2.97 is similar to what was presented earlier for the *M*-ary PSK waveforms. The difference is in the contents of the LUT.

This modulation scheme is spectrally efficient with 4 bps/Hz. The BER performance is better than the PSK counterpart, because of the larger noise immunity vector with 16-QAM. In an AWGN channel, 16-QAM is approximately 4 dB better than 16-PSK. Since there are more modulation levels, when compared to QPSK, a more linear and complex receiver is needed to perform demodulation [78, 79].

The signal constellation diagram for 16-QAM is shown in Fig. 2.98. This signal is sometimes called the square QAM constellation. The dashed lines were inserted to represent the decision boundaries.

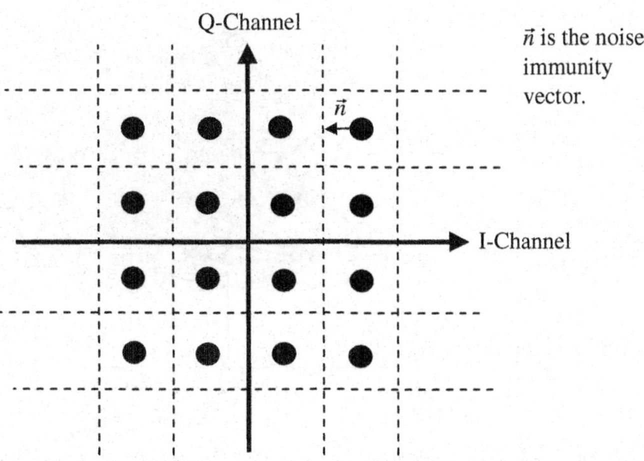

FIGURE 2.98 Unfiltered 16-QAM constellation block diagram.

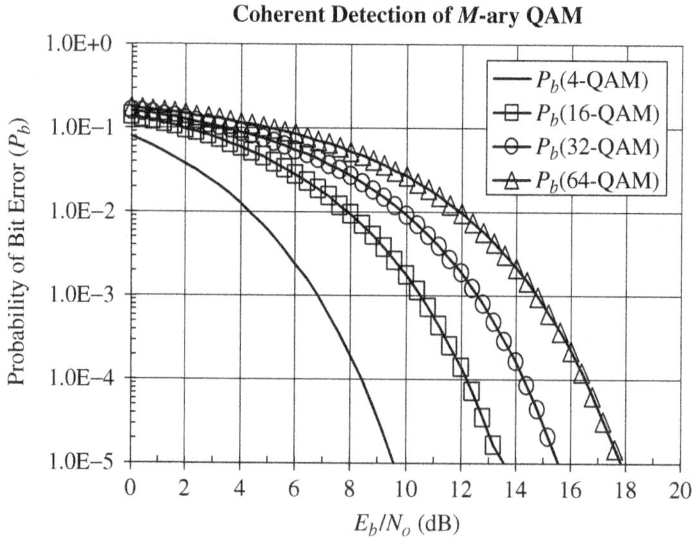

FIGURE 2.99 BER curves for QAM modulation signals.

Next the BER performance results for M-ary QAM is presented in Fig. 2.99, with modulation levels of $M = 4, 16, 32$, and 64.

Next we assume the transmit filters used an RC filter with $\alpha = 0.3$ and plot the signal constellation in Fig. 2.100. The diagram clearly shows the transmit waveform is a nonconstant envelope signal.

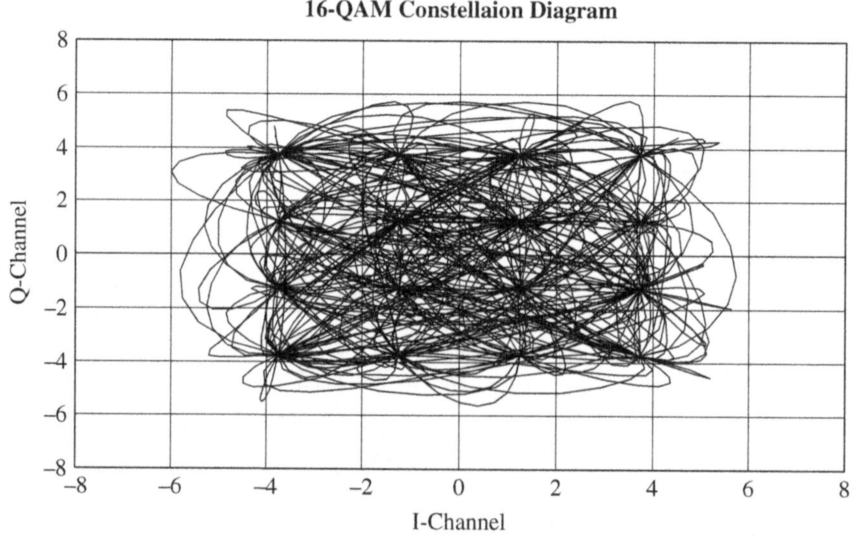

FIGURE 2.100 16-QAM signal constellation diagram using RC with $\alpha = 0.3$.

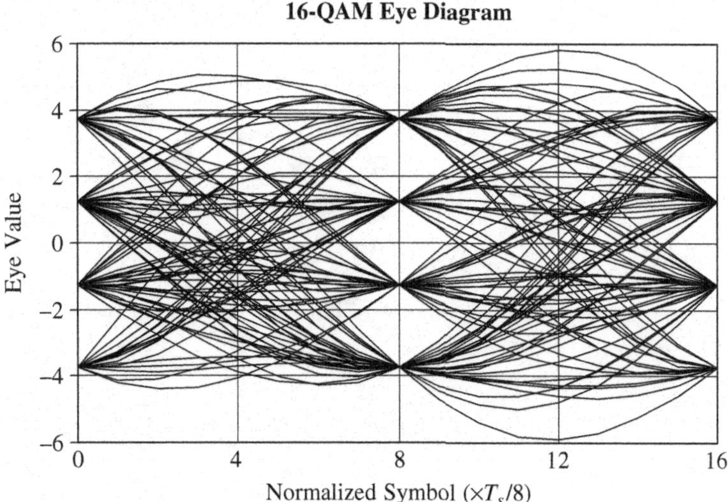

FIGURE 2.101 16-QAM eye diagram using RC with $\alpha = 0.3$.

The eye diagram of the corresponding signal constellation is given in Fig. 2.101. Since an RC filter was used, there is no ISI present in the signal. The 16-QAM modulation has 4 levels on the I-Channel and 4 levels on the Q-Channel, as visible.

Earlier it was mentioned there was a 4-dB advantage of 16-QAM over 16-PSK, and this is expected when we compare the eye diagram in Fig. 2.101, with 4 levels, to the eye diagram for 16-PSK with 8 levels in Fig. 2.95.

Let us provide an MPSK and MQAM performance comparison, specifically the peak-to-average power ratios (PAPRs) of $\pi/4$-DQPSK, 16-PSK, and 16-QAM modulation schemes for two different types of transmit pulse-shaping filters (see Table 2.7).

TABLE 2.7 Modulation Scheme Comparison

Modulation scheme	Ideal PAPR	Raised cosine (with $\alpha = 0.35$)	Root raised cosine (with $\alpha = 0.35$)
$\pi/4$-DQPSK	1	~1.7	~1.48
16-PSK	1	~1.76	~1.59
16-QAM	~1.4	~2.29	~2.11

Note that 16-PSK has less envelope variations when compared to 16-QAM; however, the tradeoff exists between this and BER performance, which has approximately 4-dB degradation in SNR. The ideal column corresponds to not using any filter and thus having an infinite BW.

There has been much work to reduce the PAPR of QAM systems in areas such as bit-to-constellation mapping and NLA linearization techniques. We would like to present an alternative constellation mapping to the conventional square 16-QAM modulation scheme presented in Fig. 2.102. This has also been called the Star-QAM modulation scheme where the 16 states have been evenly split across 2 circles of different radii [9].

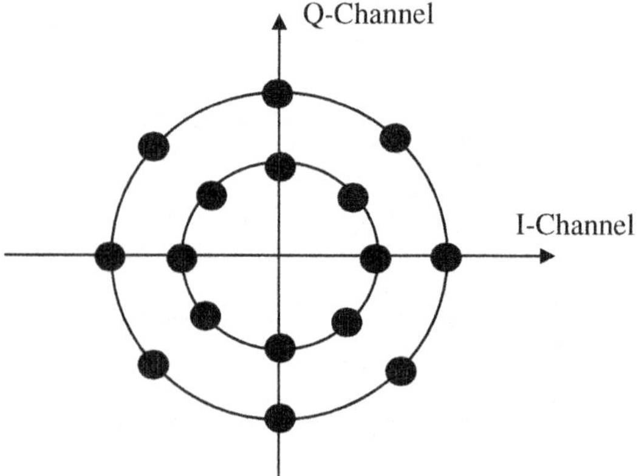

FIGURE 2.102 Star QAM signal constellation diagram.

This solution of constellation mapping provides two immediate benefits. First, it allows for efficient differential encoding and decoding methods, which now give the receiver designer a choice to use coherent detection or noncoherent detection. Second, it allows for less-complicated receive structures, practically speaking, when we consider signal processing functions such as AGC, carrier recovery for coherent detection, constellation stabilization, demapping from states to bits, and so forth.

2.3 MODULATION SCHEME COMPARISONS

In this section, we will provide an interim summary of the modulation schemes provided in this chapter. We say interim since the details of the various choices available to the communication system designer have yet to be discussed and are available in Chap. 4. In the meantime we will present a preview of the comparisons as well as on what grounds they should be compared.

First we will discuss some goals of digital communication system designs. These will in turn create more detailed requirements given in the actual application.

1. Minimize BER
2. Minimize required power, E_b/N_o
3. Maximize transmission bit rate, R_b
4. Minimize occupied BW
5. Minimize system complexity
6. Maximize system efficiency

On many occasions, not all of the above goals can be optimally solved jointly; however, in a practical system, not being optimal in all planes of the above goals will not prove to be catastrophic or even problematic for that matter. As we progress through this book, we will touch each goal when the relevant signal processing topics are discussed. Let us begin with the first goal [80–82].

2.3.1 BER Performance

In this subsection, we will present the BER performance of a few commonly used modulation schemes. We do this since they are most often used as a point of reference and they have simple, mathematically tractable, closed form expressions. The way this figure of merit is presented is BER versus the received SNR. As discussed in Chap. 1, this can also be plotted against any normalized SNR value as well. Since their BER behavior has an exponentially decaying response, it is meaningful to plot their performance on a log scale.

A commonly used reference is the BER of QPSK in an AWGN channel with no ISI present. The equation used is

$$P_e = Q\left(\sqrt{\frac{2E_b}{N_o}}\right) \tag{2.60}$$

where

$$Q(x) = \frac{1}{\sqrt{2\pi}}\int_x^{\infty} e^{-\frac{z^2}{2}}\,dz \tag{2.61}$$

and is plotted in Fig. 2.103, for sake of discussion, where we have highlighted two operating points A and B.

First, in order to satisfy our first goal, all communication system designers strive to have this performance curve move to the left as much as theoretically and practically possible.

Operating point A shows we can have an average BER = 1E-2 for a given received E_b/N_o = 4.2 dB. The average BER can be lowered to 1E-3 for operating point B; however, we require a receiver E_b/N_o = 6.8 dB. Since there is no ISI present in the system (transmit + receive), increasing the transmission power will directly have a positive impact on the performance against noise. Recall earlier, we had called this antipodal signaling performance. Moreover, if we plot the BER versus E_b/N_o, we will have the performance observations (assuming coherent detection).

Now let us add the BER performance of DPSK and FSK for coherent detection and noncoherent detection for further comparisons. Below we present their theoretical equations for sake of discussion, for binary signaling (see [2] and [23]).

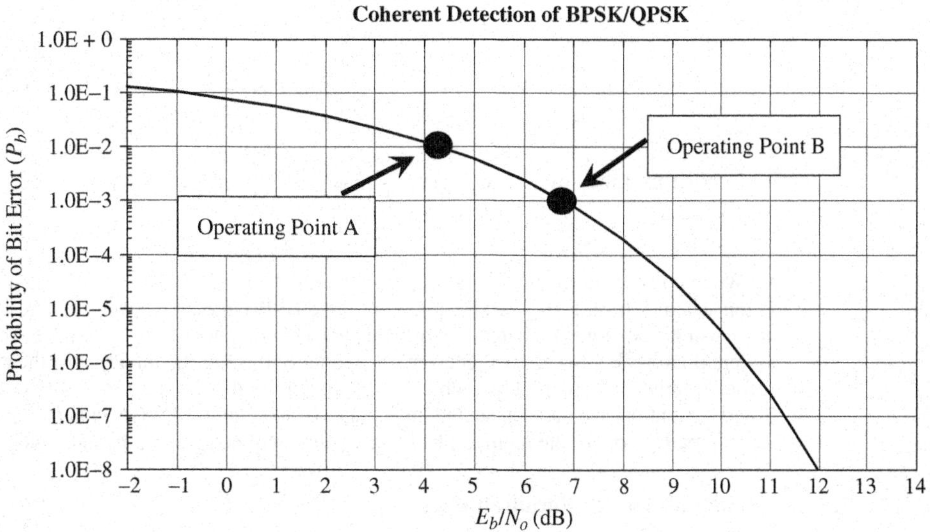

FIGURE 2.103 Bit Error Rate (BER) performance of BPSK/QPSK.

PSK:
$$P_e = Q\left(\sqrt{\frac{2E_b}{N_o}}\right) \tag{2.62}$$

DPSK:
$$P_e = \frac{1}{2} \cdot e^{-\frac{E_b}{N_o}} \tag{2.63}$$

Coherent FSK:
$$P_e = Q\left(\sqrt{\frac{E_b}{N_o}}\right) \tag{2.64}$$

Noncoherent FSK:
$$P_e = \frac{1}{2} \cdot e^{-\frac{E_b}{2N_o}} \tag{2.65}$$

All of the above equations are plotted in Fig. 2.104, for sake of comparisons.

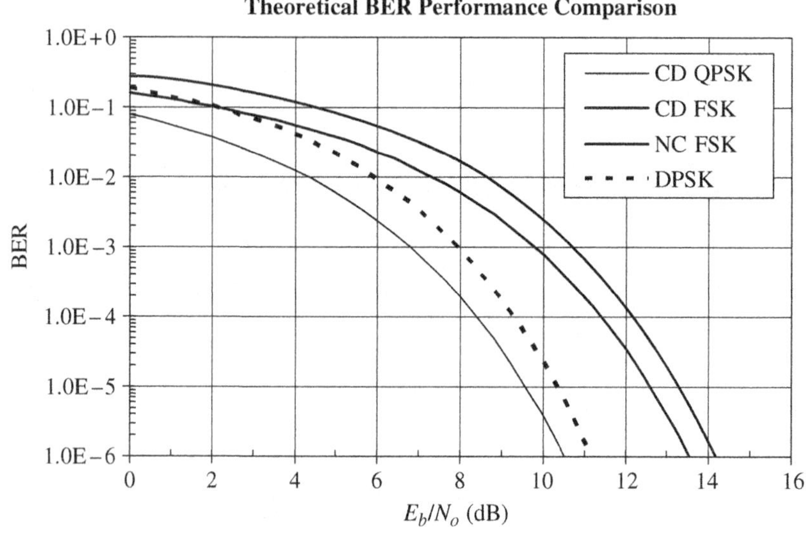

FIGURE 2.104 BER performance comparisons of binary modulation schemes.

When considering binary signaling, we notice that we pay 3 dB of E_b/N_o when switching modulation schemes from PSK to FSK, given the same BER target. Also we see there is a loss of ~1 dB when using a noncoherent detector compared to a coherent detector. This choice would be made in a system if a possibly lower-cost receiver was required. It is not the intention of the author to limit the reader's application to highly mobile scenarios, since there are other wireless applications that deserve attention. We have presented this comparison under the AWGN channel assumptions; a similar comparison under the multipaths fading channel will be given in the latter chapters.

2.3.2 Power versus Bandwidth Requirements

It is not enough to compare modulation schemes only on the merit of BER performance. The communication system designer must also consider the occupied BW required to transmit and effectively communicate and thus obtain the BER results presented in the above subsection. Since we have already presented various transmit PSDs of modulation schemes, in this comparison, we will use BW to denote data rate rather than the spectral domain. A great way to demonstrate this comparison is through the Shannon-Hartely capacity theorem [2].

This theorem is summarized as follows: system capacity perturbed by AWGN is a function of signal power, noise power, and BW and reported as

$$C = W \cdot \log_2\left[1 + \frac{S}{N}\right] \quad \text{(bps)} \tag{2.66}$$

$$\frac{C}{W} = \log_2\left[1 + \frac{E_b}{N_o} \cdot \frac{R_b}{W}\right] \quad \text{(bps/Hz)} \tag{2.67}$$

where we can also add the following definition:

$$\frac{R_b}{W} = \log_2(M) \tag{2.68}$$

We can plot this relationship in what is commonly called a BW–power efficiency plane, as shown in Fig. 2.105 [23]. Here three modulation schemes are plotted: PSK, QAM, and FSK for a BER = 1.0E–3.

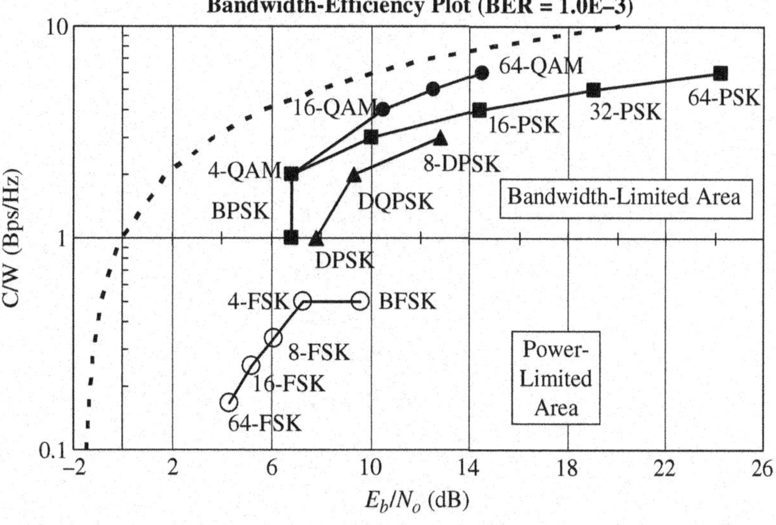

FIGURE 2.105 BW–power efficiency plane characteristics for PSK, QAM, DPSK, and FSK modulation schemes.

So, what can we determine by viewing this BW–power efficiency plot? Well, given a modulation scheme and a targeted BER, the communication system designer can then determine the BW efficiency and the E_b/N_o required to maintain the average BER target.

First, there are two regions highlighted in the plot: BW-limited region and power-limited region. The intention of these two regions is to classify the modulation schemes. For those modulation schemes in the BW-limited regions, we notice they are highly efficient. Hence the communication system designer should choose modulation schemes in the BW-limited region if there is a limitation of the available BW for the wireless system. On the other hand, the modulation schemes in the power limited region should be used when there is a limitation on the received power. More specifically the BW–power efficiency plot shows us if we have a BW constraint then PSK and QAM modulation schemes look promising. Also if we have a Power constraint then FSK modulations schemes look promising since they require lower E_b/N_o to obtain the same average BER.

Let us take a deeper look into the BW-limited region. Recall we have previously discussed the fact that 16-QAM had better SNR (or E_b/N_o) performance requirements than 16-PSK. This is clearly shown in the plot, where notice their normalized capacity of (bps/Hz) are the same for both modulation schemes. Now let's discuss the move from BPSK to QPSK; here we notice they both require ~6.8 dB of E_b/N_o to maintain the target BER. However, we notice our normalized capacity doubled from 1 to 2 bps/Hz. Also we discussed the BER of QPSK was identical to that of antipodal signaling. Besides the receiver complexity, the advantages far outweigh the disadvantages for this particular choice of modulation schemes.

Let's turn our attention into the Power limited region. This shows us that the FSK family of modulation schemes is spectrally inefficient. Recall we present the orthogonality principle in the FSK discussion; we see as we move from 2 level to 4 level FSK, our normalized capacity remains at 0.5 bps/Hz. The reason is we doubled our data rate when going from BFSK to 4-level FSK; however, we also doubled our occupied BW, thus keeping the ratio of the two the same. However, notice that we actually have an improvement in E_b/N_o when going to the higher number of modulation levels.

A more general observation is that as the number of modulation levels increases for PSK/QAM, the required E_b/N_o increases as well. This is because the distance between neighboring constellation points becomes closer as the number of modulation levels increases. The opposite effect is visible for the FSK modulation scheme family, where increasing the number of modulation levels decreases the required E_b/N_o. As discussed earlier, each addition to the FSK modulation alphabet does not allow a smaller noise immunity vector to cause a decision error. The distance between the neighboring FSK symbols is actually constant for the orthogonal signaling case we discussed above.

2.3.3 Some Wireless Standard Choices

In the above sections a comparison of some of the above modulation schemes was presented. First the goals were outlined and then BER, E_b/N_o, and BW were discussed, in relation to the desired effects with respect to the goals. It was shown that antipodal signaling with coherent detection produced the lowest/best BER in the AWGN channel. We noticed that spectral efficiency can be increased by increasing the number of modulation levels of PSK and QAM, at the expense of power (i.e., E_b/N_o). Earlier we discussed modulation schemes whose phase trajectories cross the origin had tremendous spectral regrowth issues. That's why schemes that avoided them had less spectral regrowth.

In this subsection, we will provide a short overview of some of the digital cellular standard's choices in the modulation format.

GSM system uses GMSK, which we saw earlier as a spectrally efficient modulation scheme with the benefit of having an approximate constant envelope waveform. It can be viewed as OQPSK with sinusoidal pulse shaping and can be detected either coherently or noncoherently with a differential detector.

JDC and NADC use $\pi/4$-DQPSK, which is a spectrally efficient modulation scheme, provided the PA is operating in quasi-linear region. It avoids the origin and can be detected noncoherently through the use of a differential detector or limiter discriminator.

In the above-mentioned TDMA schemes, the UL transmission is a major consumer of battery current and thus will play a significant role in determining the MS or UE talk time. It behooves the communication system designer to choose a modulation scheme that not only avoids the origin of the signal constellation diagram, but also resembles a circle as much as possible.

Turning our attention to spread spectrum systems, specifically DS-CDMA systems, we notice that IS-95 uses OQPSK on the UL for the above-mentioned reasons. Many researchers have shown the benefits of OQPSK over $\pi/4$-DQPSK, which is the main reason for its use. Since we are transmitting chips in the CDMA systems and we typically operate with negative chip SNR values, it doesn't make much sense to be able to have an option to differentially detect the chips themselves. This will be discussed in more detail in Chap. 7.

Now CDMA2000 and WCDMA have chosen to use what is called Hybrid-PSK (HPSK). This modulation scheme reduces the number of times the phase trajectories travel close to the origin. Hence, instead of completely avoiding the origin, they have chosen to dramatically limit the number of such occurrences. It accomplishes this task by limiting the phase transitions every other chip time interval. Note this can be extended further.

REFERENCES

[1] R. E. Crochiere and L. R. Rabiner, "Interpolation and Decimation of Digital Signals—A Tutorial Review," *Proceedings of the IEEE*, Vol. 69, No. 3, Mar. 1981, pp. 300–331.

[2] J. G. Proakis, *Digital Communications*, McGraw-Hill, 1989, New York.

[3] Dr. K. Feher, *Wireless Digital Communications Modulation & Spread Spectrum Applications*, Prentice Hall, 1995, New Jersey.

[4] A. Papoulis and S. U. Pillai, *Probability, Random Variables and Stochastic Processes*, 4th Ed., McGraw-Hill, 2002, New York.

[5] N. S. Alagha and P. Kabal, "Generalized Raised-Cosine Filters," *IEEE Transactions on Communications*, Vol. 47, No. 7, July 1999, pp. 989–997.

[6] F. Harris, "On the Design of Pre-Equalized Square-Root Raised Cosine Matched Filters for DSP Based Digital Receivers," *IEEE*, 1993, pp. 1291–1295.

[7] T. Tsukahara, K. Shiojima and M. Ishikawa, "A GHz-Band IF Si Chip-Set and Dielectric Root-Nyquist Filter for Broadband Wireless Systems," *IEEE Wireless Communications* Conference, 1997, pp. 32–37.

[8] G. S. Deshpande and P. H. Wittke, "Optimum Pulse Shaping in Digital Angle Modulation," *IEEE Transactions Communications*, Vol. 29, No. 2, Feb. 1981, pp. 162–168.

[9] W. T. Webb and L. Hanzo, *Modern Quadrature Amplitude Modulation: Principles and Applications for Fixed and Wireless Channels*, IEEE Press, 1994, New York.

[10] 3rd Generation Partnership Project 2 (3GPP2), "Medium Access Control (MAC) Standard for cdma2000 Spread Spectrum Systems," Sept. 2005, Release D.

[11] J. Boccuzzi, "Performance Evaluation of Non-Linear Transmit Power Amplifiers for North American Digital Cellular Portables," *IEEE Transactions on Vehicular Technology*, Vol. 44, No. 2, May 1995, pp. 220–228.

[12] M. C. Jeruchim, P. Balaban, and K. S. Shanmugan, *Simulation of Communication Systems*, Plenum Press, 1992, New York.

[13] S. C. Cripps, *RF Power Amplifiers for Wireless Communications*, Artech House, 1999, Massachusetts.

[14] S. Hischke and J. Habermann, "New Results on the Effects of Nonlinear Amplifiers on DOQPSK and π/4-DQPSK Signals," *IEEE*, 1998, pp. 386–390.

[15] S. Ono, N. Kondoh, and Y. Shimazaki, "Digital Cellular System with Linear Modulation," *IEEE*, 1989, pp. 44–49.

[16] M. R. Heath and L. B. Lopes, "Assessment of Linear Modulation Techniques and Their Application to Personal Radio Communications Systems," in: *Fifth International Conference on Mobile Radio and Personal Communications*, Coventry, UK, 1989, pp. 112–116.

[17] J. Hammuda, Spectral Efficiency of Digital Cellular Mobile Radio Systems," *IEEE*, 1995, pp. 92–96.

[18] J. K. Cavers, "The Effect of Quadrature Modulator and Demodulator Errors on Adaptive Digital Predistorters for Amplifier Linearization," *IEEE Transactions on Vehicular Technology*, Vol. 46, No. 2, May 1997, pp. 456–466.

[19] Y. Akaiwa and Y. Nagata, "Highly Efficient Digital Mobile Communications with a Linear Modulation Method," *IEEE Journal on Selected Areas in Communications*, Vol. 5, No. 5, June 1987, pp. 890–895.

[20] P. Garcia, J. de Mingo, A. Valdovinos, and A. Ortega, "An Adaptive Digital Method of Imbalances Cancellation in LINC Transmitters," *IEEE Transactions on Vehicular Technology*, Vol. 54, No. 3, May 2005, pp. 879–888.

[21] M. Matsui, T. Nakagawa, K. Kobayashi, and K. Araki, "Compensation Technique for Impairments on Wideband Quadrature Demodulators for Direct Conversion Receivers," *IEEE*, 2004, pp. 1677–1681.

[22] J. D. Oetting, "A Comparison of Modulation Techniques for Digital Radio," *IEEE Transactions on Communications*, Vol. 27, No. 12, Dec. 1979, pp. 1752–1762.

[23] B. Sklar, *Digital Communications Fundamentals and Applications*, Prentice Hall, 1988, New Jersey.

[24] J. He, C. G. Englefield, and P. A. Goud, "Performance of SRC-Filtered ODQPSK in Mobile Radio Communications," *IEEE*, 1993, pp. 668–671.

[25] J. Boccuzzi, US Patent #5,438,592, "π/4-DQPSK Phase State Encoder/Decoder."

[26] M. K. Simon, S. M. Hinedi, and W. C. Lindsey, *Digital Communication Techniques Signal Design and Detection*, Prentice Hall, 1995, New Jersey.

[27] R. E. Zeimer and C. R. Ryan, "Minimum Shift Keyed Modem implementations for High Data Rates," *IEEE Communications Magazine*, Vol. 21, No. 7, 1983, pp. 28–37.

[28] S. Pasupathy, "Minimum Shift Keying: A Spectrally Efficient Modulation," *IEEE Communications Magazine*, Vol. 17, No. 4, 1979, pp. 14–22.

[29] C. C Powell and J. Boccuzzi, United States Patent #5,309,480, "Non-Coherent Quadrature Demodulation of MSK Signals."

[30] D. H. Morais and K. Feher, "The Effects of Filtering and Limiting on the Performance of QPSK, Offset QPSK and MSK Systems," *IEEE Transactions on Communications*, Vol. 28, No. 12, Dec. 1980, pp. 1999–2009.

[31] D. H. Morais and K. Feher, "Bandwidth Efficiency and Probability of Error Performance of MSK and Offset QPSK Systems," *IEEE Transactions on Communications*, Vol. 27, No. 12, Dec. 1979, pp. 1794–1801.

[32] S. A. Gronemeyer and A. L. McBride, "MSK and Offset QPSK Modulation," *IEEE Transactions on Communications*, Vol. 24, No. 8, Aug. 1976, pp. 809–820.

[33] M. C. Austin, M. U. Chang, D. F. Horwood, and R. A. Maslov, "QPSK, Staggered QPSK and MSK—A Comparative Evaluation," *IEEE Transactions on Communications*, Vol. 31, No. 2, Feb. 1983, pp. 171–182.

[34] K. Murota and K. Hirade, "GMSK Modulation for Digital Radio Telephony," *IEEE Transactions on Communications*, Vol. 29, No. 7, July 1981, pp. 1044–1050.

[35] R. De Buda, "Coherent Demodulation of Frequency-Shift Keying with Low Deviation Ratio," *IEEE Transactions on Communications*, Vol. 20, June 1972, pp. 429–435.

[36] M. R. Heath and L. B. Lopes, "Variable Envelope Modulation techniques for Personal Communications," *in: Second IEEE National Conference on Telecommunications*, York, UK, *April* 1989, pp. 249–254.

[37] C.-E. Sundberg, "Continuous Phase Modulation," *IEEE Communications Magazine*, Vol. 24, No. 4, April 1986, pp. 25–38.

[38] J. B. Anderson and C.-E. W. Sundberg, "Advances in Constant Envelope Coded Modulation," *IEEE Communications Magazine*, Vol. 29, No. 12, Dec. 1991, pp. 36–45.

[39] M. Ishizuka and Y. Yasuda, "Improved Coherent Detection of GMSK," *IEEE Transactions on Communications*, Vol. 32, No. 3, March 1984, pp. 308–311.

[40] I. Korn, "Generalized Minimum Shift Keying," *IEEE Transactions on Information Theory*, Vol. 26, No. 2, March 1980, pp. 234–238.

[41] C. Brown and K. Feher, "Cross-Correlated Correlative FQPSK Modulation Doubles the Capacity of PCS Networks," *IEEE*, 1996, pp. 800–804.

[42] K. Feher et al., US Patent #4,567,602; US Patent #4,664,565; US Patent #5,491,457; US Patent #5,784, 402; US Patent #6,198,777.

[43] Y. Guo and K. Feher, "A New FQPSK Modem/Radio Architecture for PCS and Mobile Communications," IEEE 1994, pp. 1004–1008.

[44] S. Kato and K. Feher, "XPSK: A New Cross-Correlated Phase-shift Keying Modulation Technique," *IEEE Transactions on Communications*, Vol. 31, No. 5, May 1983, pp. 701–707.

[45] Y. Guo and K. Feher, "A New FQPSK Modem/Radio Architecture for PCS and Mobile Satellite Communications," *IEEE Journal on Selected Areas in Communications*, Vol. 13, No. 2, Feb. 1995, pp. 245–353.

[46] J. Borowski and K. Feher, "Nonobvious Correlation Properties of Quadrature GMSK, A Modulation Technique Adopted Worldwide," *IEEE Transactions on Broadcasting*, Vol. 41, No. 2, June 1995, pp. 69–75.

[47] J. Wang and K. Feher, "Spectral Efficiency Improvement Techniques for Nonlinearly Amplified Mobile Radio Systems," IEEE 1988, pp. 629–635.

[48] C. Brown and K. Feher, "Reconfigurable Digital Baseband Modulation for Wireless Computer Communication," *IEEE*, 1995, pp. 610–616.

[49] C. Brown and K. Feher, "Cross-Correlated Correlative Encoding: An Efficient Modulation Method," *IEEE Transactions on Broadcasting*, Vol. 43, No. 1, March 1997, pp. 47–55.

[50] S. Pasupathy, "Correlative Coding: A Bandwidth-Efficient Signaling Scheme," *IEEE Communications Society Magazine*, Vol. 15, No. 4, July 1977, pp. 4–11.

[51] J. C. Y. Huang, K. Feher, and M. Gendron, "Techniques to Generate ISI and Jitter-Free Bandlimited Nyquist Signals and a Method to Analyze Jitter Effects," *IEEE Transactions on Communications*, Vol. 27, No. 11, Nov. 1979, pp. 1700–1711.

[52] T. J. Hill, "A nonproprietary, constant envelope variant of shaped offset QPSK (SOQPSK) for improved spectral containment and detection efficiency," *Proceedings of the IEEE Military Communications Conference*, Vol. 1, Oct. 2000, pp. 347–352.

[53] L. Li and M. K. Simon, "Performance of Coded OQPSK and MIL-STD SOQPSK with Iterative Decoding," *IEEE Transactions on Communications*, Vol. 52, No. 11, Nov. 2004, pp. 1890–1900.

[54] P. S. K. Leung and K. Feher, "F-QPSK—Superior Modulation Technique for Mobile and Personal Communications," *IEEE Transactions on Broadcasting*, Vol. 39, No. 2, June 1993, pp. 288–294.

[55] T. Le-Ngoc and K. Feher, "Performance of IJF-OQPSK Modulation Schemes in a Complex Interference Environment," *IEEE Transactions on Communications*, Vol. 31, No. 1, Jan. 1983, pp. 137–144.

[56] H. Mehdi and K. Feher, "FBPSK, Power and Spectrally Efficient Modulation for PCS and Satellite Broadcasting Applications," *IEEE Transactions on Broadcasting*, Vol. 42, No. 1, March 1996, pp. 27–31.

[57] L. E. Franks, "Further Results on Nyquist's Problem in Pulse Transmission," *IEEE Transactions on Communications Technology*, Vol. 16, No. 2, April 1968, pp. 337–340.

[58] L. E. Franks, "Pulses Satisfying the Nyquist Criterion," *Electronics Letters*, Vol. 28, No. 10, May 1992, pp. 951–952.

[59] E. Panayirci and N. Tugbay, "Class of Optimum Signal Shapes in Data Transmission," *IEEE Proceedings*, Pt. F, Vol. 135, No. 3, June 1988, pp. 272–276.

[60] N. Tugbay and E. Panayirci, "Energy Optimization on Band-Limited Nyquist Signals in the Time Domain," *IEEE Transactions on Communications*, Vol. 35, No. 4, April 1987, pp. 427–434.

[61] F. Amoroso, "The Bandwidth of Digital Data Signals," *IEEE Communications Magazine*, Vol. 18, No. 6, Nov. 1980, pp. 13–24.

[62] M. Yu and K. Feher, "π/4-FQPSK: An Efficiency Improved, Standardized π/4-DQPSK Compatible Modulation/Nonlinearly Amplified RF Wireless Solution," *IEEE Transactions on Broadcasting*, Vol. 42, No. 2, June 1996, pp. 95–101.

[63] D. S. Dias and K. Feher, "Baseband Pulse Shaping for π/4 FQPSK in Nonlinearly Amplified Mobile Channels," *IEEE Transactions on Communications*, Vol. 42, No. 10, Oct. 1994, pp. 2843–2852.

[64] J.-S. Lin and K. Feher, "Noncoherent Limiter-Discriminator Detection of Standardized FQPSK and OQPSK," IEEE 2003, pp. 795–800.

[65] M. Yu and K. Feher, "An Improved π/4-DQPSK Compatible Feher's π/4-FQPSK Nonlinearly Amplified Modulation," *IEEE*, 1995, pp. 226–230.

[66] H. C. Park, K. Lee, and K. Feher, "Continuous Phase Modulation for F-QPSK-B Signals," *IEEE Transactions on Vehicular Technology*, Vol. 56, No. 1, Jan. 2007, pp. 157–172.

[67] T. Le-Ngoc and S. B. Slimane, "IJF-OQPSK Modulation Schemes with MLSE Receivers for Portable/Mobile Communications," *IEEE*, 1992, pp. 676–680.

[68] M. Cho and S. C. Kim, "Non-Coherent Detection of FQPSK Signals Using MLDD," in: *Proceedings of the 7th Korea-Russia Symposium, KORUS 2003*, Ulsan University, Ulsan, Korea, pp. 314–318.

[69] J. S. Lin and K. Feher, "Noncoherent Limiter-Discriminator Detection of Standardized FQPSK and OQPSK," *IEEE*, 2003, pp. 795–800.

[70] M. K. Simon and D. Divsalar, "A Reduced Complexity Highly Powered/Bandwidth Efficient Coded FQPSK System with Iterative Decoding," *IEEE*, 2001, pp. 2204–2210.

[71] J. A. McCorduck and K. Feher, "Study of Alternative Approaches to Non-Coherent Detection of FQPSK and Other Quadrature Modulated Systems," *IEEE*, 2002, pp. 680–684.

[72] A. Linz and A. Hendrickson, "Efficient Implementation of an I-Q GMSK Modulator," *IEEE Transactions on Circuits and Systems-II, Analog and Digital Signal Processing*, Vol. 43, No. 1, Jan. 1996, pp. 14–23.

[73] M. C. Austin and M. U. Chang, "Quadrature Overlapped Raised Cosine Modulation," *IEEE Transactions on Communications*, Vol. 29, No. 3, March 1981, pp. 237–249.

[74] B. Sklar, "Defining, Designing, and Evaluating Digital Communication Systems," *IEEE Communications Magazine*, Vol. 31, No. 11, Nov. 1993, pp. 92–101.

[75] P. J. Lee, "Computation of the Bit Error Rate of Coherent M-ary PSK with Gray Bit Mapping," *IEEE Transactions on Communications*, Vol. 34, No. 5, May 1986, pp. 488–491.

[76] D. Subasinghe and K. Feher, "Error Floors in Digital Mobile Radio Systems Due to Delay Distortion," *IEEE*, 1988, pp. 194–198.

[77] J. C.-I. Chuang, "Comparison of Coherent and Differential Detection of BPSK and QPSK in a Quasi-Static Fading Channel," *IEEE*, 1988, pp. 749–755.

[78] S. Oshita, "Optimum Spectrum for QAM Perturbed by Timing and Phase Errors," *IEEE Transactions on Communications*, Vol. 35, No. 9, Sept. 1987, pp. 978–980.

[79] H. Yamamoto, "Advanced 16-QAM Techniques for Digital Microwave Radio," *IEEE Communications Magazine*, Vol. 19, No. 3, May 1981, pp. 36–45.

[80] G. D. Forney, Jr., R. G. Gallager, G. R. Lang, F. M. Longstaff, and S. U. Qureshi, "Efficient Modulation for Band-Limited Channels," *IEEE Journal on Selected Areas in Communications*, Vol. 2, No. 5, Sept. 1984, pp. 632–647.

[81] W. T. Webb, "Modulation Methods for PCNs," *IEEE Communications Magazine*, Vol. 30, No. 12, Dec. 1992, pp. 90–95.

[82] S. Murakami, Y. Furuya, Y. Matsuo, and M. Sugiyama, "Optimum Modulation and Channel Filters for Nonlinear Satellite Channels," *IEEE Transactions on Communications*, Vol. 27, No. 12, Dec. 1979, pp. 1810–1818.

CHAPTER 3
WIRELESS MULTIPATH CHANNEL

In this chapter the wireless propagation phenomenon between the transmitting antenna(s) and the receiving antenna(s) will be presented. The model derived is commonly called the wireless multipath channel model.

First additive white Gaussian noise (AWGN) is presented due to its simple elegance and ability to support closed form expressions and mathematical tractability. Rayleigh fading is discussed along with its companion, Rician multipath fading. Next the wireless channel bandwidth is introduced thus starting a discussion on frequency selective fading (FSF). These natural forms of interference are complimented with the system-made forms of interference such as adjacent channel interference (ACI) and co-channel interference (CCI).

Taking a more macroscopic view of the wireless channel model will lead to describing how the separation between the transmitter and receiver, called path loss or signal attenuation, affects the received signal power.

As discussed in Chap. 1, complex envelope theory was introduced with a few purposes in mind, one being efficient computer simulation. With this comes a presentation of a few simulation models to emulate the multipath fading phenomenon.

Lastly, the time varying nature of the wireless channel bandwidth (BW) is discussed by means of the birth (or creation) and death (or deletion) of multipaths being received. As a means of a demonstration vehicle, an application to DS-CDMA is given in terms of the implications of this time variation to the performance of the RAKE receiver.

3.1 ADDITIVE WHITE GAUSSIAN NOISE

This first section will introduce additive interference in the form of white Gaussian noise (WGN). The interference will be present for all frequency channels and that's the reason behind using the adjective, white. This interference is indeed a random process whose realization can best be described by a density function called Gaussian. This interference is not part of the transmitted signal and is commonly called noise.

3.1.1 Mathematical Representation

The AWGN wireless channel model can best be described with the aid of Fig. 3.1, with $s(t)$ being the transmitted modulation scheme, possibly nonlinearly amplified.

Here $n(t)$ is indeed a random variable and is described by the following probability density function (PDF).

$$p_N(n) = \frac{1}{\sigma_N\sqrt{2\pi}} \cdot e^{\frac{-(n-m_N)^2}{2\sigma_N^2}} \tag{3.1}$$

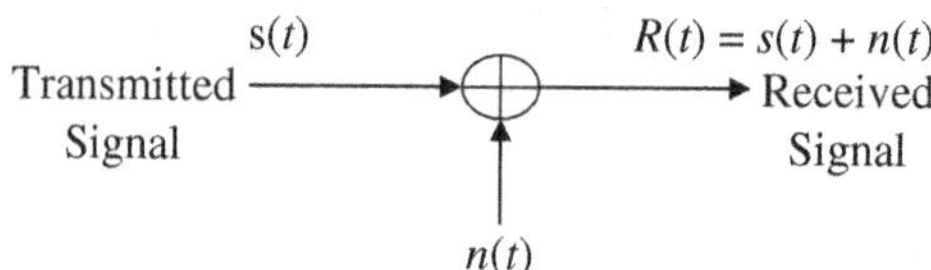

FIGURE 3.1 Graphical representation of the AWGN channel.

where m_N = mean value and σ_N = standard deviation of the noise signal.

As discussed above, this noise has components for all the frequency channels and is shown in Fig. 3.2, where N_o is the noise power spectral density.

The above plot is centered around the carrier frequency and is sometimes referred to as the double sided density since the region of interest is governed by the channel BW.

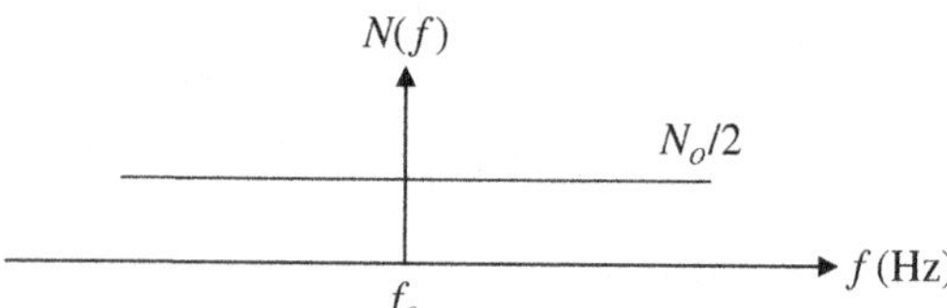

FIGURE 3.2 White Gaussian noise (WGN) power spectral density.

3.1.2 Statistical Properties

Some of the statistical properties of the AWGN random samples are the mean and variance given as

$$E\{n\} = m_N \tag{3.2}$$

$$\sigma_N^2 = E\{(n - m_N)^2\} \tag{3.3}$$

Given the AWGN flat spectral density, we can simply write down the autocorrelation as the inverse FFT of $N(f)$, shown in Fig. 3.3.

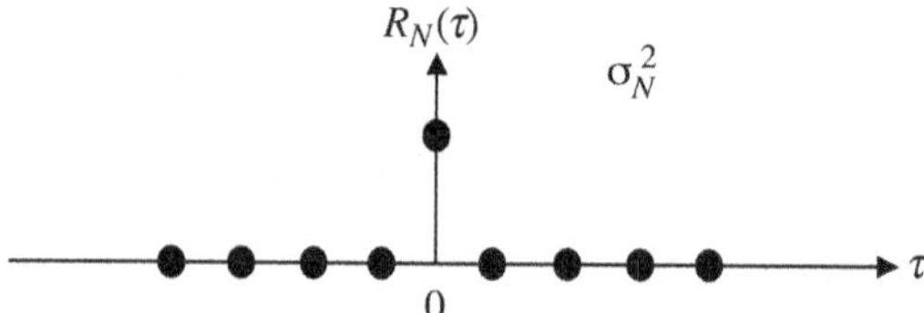

FIGURE 3.3 WGN autocorrelation function.

An alternative way to explain the AWGN noise variance is to define the correlation as

$$R_N(\tau) = E\{n(t) \cdot \overline{n}(t + \tau)\} \tag{3.4}$$

The overbar is used to denote the complex conjugate operation. Hence, we see the AWGN samples are only correlated with themselves and not correlated for time offsets other than zero.

3.2 RAYLEIGH MULTIPATH FADING PHENOMENON

In this section we will describe the receiver signal variations typically encountered in a wireless environment. These signal variations are random and are accurately modeled using a Rayleigh distribution [1–5].

3.2.1 Mathematical Representation

Signals transmitted between transmit and receive antennas will undergo various signal power fluctuations. These signal variations become visible through the use of an antenna where several paths will be summed together. These paths enter the antenna after having traveled different routes and encountered reflections and diffractions due to the surrounding obstructions. The received signal will experience random fluctuations in both amplitude and phase. The random amplitude fluctuations can be observed by recording the signal envelope as the antenna is physically moved as a function of distance. These recordings have been shown to exhibit a Rayleigh distribution. Given r to denote the received signal's envelope or amplitude then the random phenomenon is described as

$$p(r) = \frac{r}{\sigma^2} \cdot e^{\frac{-r^2}{2\sigma^2}} \qquad (0 \leq r) \tag{3.5}$$

where σ^2 is the average signal power.

It is well known that the envelope of a complex Gaussian signal has a Rayleigh distribution [6]. The probability that the received signal's envelope is below a specified value, say R, is given by the CDF below

$$P(r \leq R) = \int_0^R p(r)\,dr \tag{3.6}$$

$$P(r \leq R) = 1 - e^{\frac{-R^2}{2\sigma^2}} \tag{3.7}$$

In an effort to provide more quantitative analysis the mean value of the Rayleigh random variable is given by

$$E\{r\} = \int_0^\infty r \cdot p(r)\,dr \tag{3.8}$$

$$E\{r\} = \sigma\sqrt{\frac{\pi}{2}} \tag{3.9}$$

The variance of the Rayleigh distribution is given by

$$\sigma_r^2 = E\{r^2\} - [E\{r\}]^2 \tag{3.10}$$

$$\sigma_r^2 = \int_0^\infty r^2 p(r)\,dr - \sigma^2\frac{\pi}{2} \tag{3.11}$$

$$\sigma_r^2 = \sigma^2\left(2 - \frac{\pi}{2}\right) \tag{3.12}$$

The multipath fading model can be represented as multiplicative distortion and is shown in Fig. 3.4.

A plot emphasizing the wide signal variations is made possible with Fig.3.5.

This distribution has often been used to describe small scale signal variations in a multipath fading environment. The term small scale is used to emphasize the widely varying phenomenon occurring at

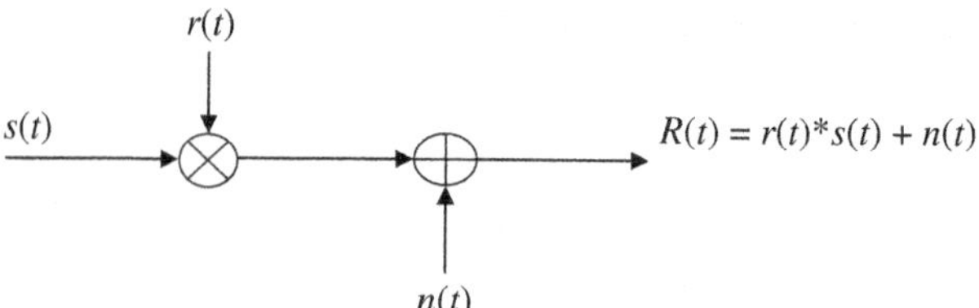

FIGURE 3.4 Graphical representation of multipath fading with noise.

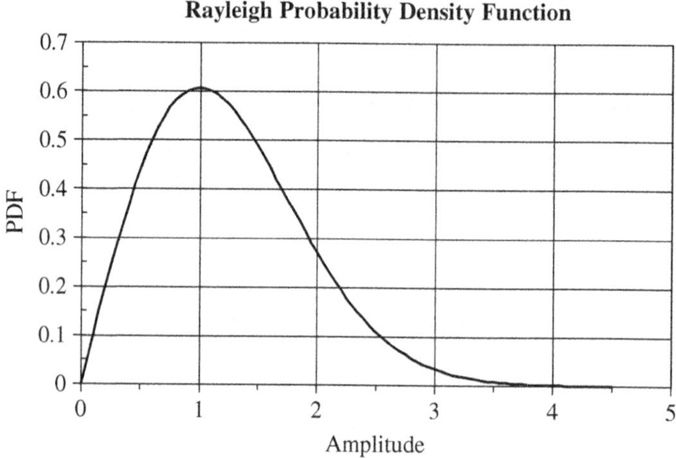

FIGURE 3.5 Rayleigh probability density function plot (normalized power = 1).

a fraction of a carrier wavelength apart; specifically fades can occur about every half wavelength apart in distance. These variations are caused by the local surrounding environment of the receiver and not from signal variations that occurred a significant distance apart, say 2 km away [7].

When signals from various routes sum together interesting observations can be made. Depending on the angle of arrival between the signals or their relative phase offsets/differences the signal can sum to create constructive interference, sometimes called a maxima or a peak. Similarly, these summing signals can add to create destructive interference, sometimes called a minima or a fade or a null. As described above these peaks and fades can vary considerably.

Let us take a top view of the multipath fading description just given as we draw an example. The sum of the multipath rays produces the Rayleigh phenomenon. Instead of gathering reflection coefficient data for various obstructions such as concrete, metal, and so on. It is practically accepted to use a statistically equivalent ray whose envelope is Rayleigh distributed and whose phase is uniformly distributed (assuming an omni directional receiving antenna). This is shown in Fig. 3.6 as a bold dashed line.

Let us introduce the notion of vehicular movement into the channel. The faster the vehicle travels, the more often the signal will encounter both peaks and nulls. On the contrary, the slower the vehicle travels, the less often the signal will encounter both peaks and nulls. Moreover, when a null is encountered the signal will actually reside there longer with a slower vehicle speed, as apposed to the faster vehicle speed.

The channel model discussion was presented from the point of view of the mobile station [8]. We can apply the theorem of reciprocity and reverse the directions of the propagation arrows to arrive at an uplink model. However, care must be taken when modeling multiple receive antennas on the uplink due to the fact that the surrounding environment of the base station is different than that of the mobile station [9].

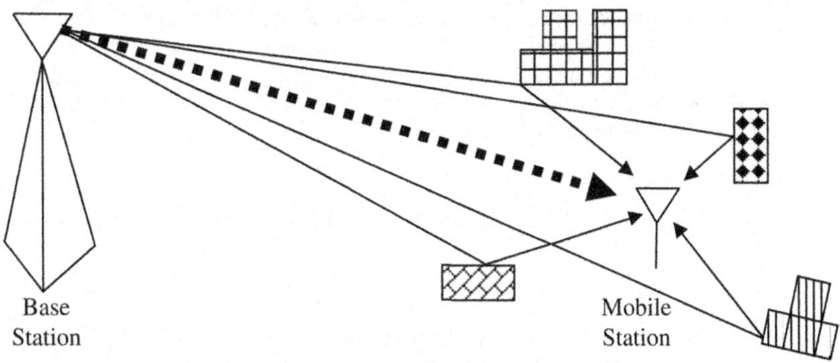

FIGURE 3.6 High-level view of the multipath propagation phenomenon.

3.2.2 Statistical Properties

Two quantitative metrics of interest to us in the design of wireless digital communications, deal with the peaks and fades described above. In particular, there is interest to know the number of times per second a fade with a certain depth (in dB) is actually encountered. Once this rate is known, the next quantity of interest is the amount of time the signal actually spends below this null. Both of the parameters will become more apparent and useful in the chapters that follow.

The first quantity discussed above is called the Level Crossing Rate (LCR) which is defined as the expected rate at which the envelope crosses a specified signal level, R, in the positive direction. It is mathematically defined as

$$N_R = \int_0^\infty \dot{r} p(R, \dot{r})\, d\dot{r} \tag{3.13}$$

where N_R is the level crossing rate, R is the specified signal level, and $\dot{r}$ is the time derivative of the envelope. The results of the integration are presented below where the actual derivation can be seen in [1], it is omitted here for sake of convenience.

$$N_R = \sqrt{2\pi} \cdot f_m \cdot \rho \cdot e^{-\rho^2} \tag{3.14}$$

where $\rho = \dfrac{R}{R_{\text{RMS}}}$ and f_m = maximum Doppler shift or frequency which is defined as

$$f_D = f_m \cdot \cos(\theta) = \frac{v}{\lambda} \cdot \cos(\theta) \tag{3.15}$$

where v = vehicular speed, λ is the wavelength of the carrier frequency, θ is the angle of arrival with respect to the velocity vector traveling in the same direction as the vehicle. Assuming an omni directional antenna in the receiver, the arriving multipaths have the opportunity to experience a wide range of angles. This leads to a Doppler spread since each arriving ray will have its own frequency shift associated with it. However, the aggregate sum of all the multipaths will exhibit a behavior that is typically called Doppler spread. Assuming a uniformly distributed angle of arrival, the Doppler frequency response or Doppler spectrum can be written as

$$S(f) = \frac{3\sigma^2}{2\pi \cdot f_m} \cdot \frac{1}{\sqrt{1 - \left(\dfrac{f - f_c}{f_m}\right)^2}} \tag{3.16}$$

where f_c = the carrier frequency. Now returning our attention to the LCR, we notice the quantity is proportional to the max Doppler shift. In other words, as the vehicle speed increases, the average number of times we encounter a normalized signal fade, R (in dB), increases. That is we encounter the faded signal more often because the temporal distance between these nulls has decreased due to mobility.

The second quantity discussed above is called the average fade time duration (AFTD) which is defined as the average time the received signals' envelope is below the normalized specific signal level, R. The mathematical definition is given as

$$\bar{\tau} = \frac{1}{TN_R} \cdot \sum_i \tau_i \tag{3.17}$$

$$\bar{\tau} = \frac{1}{N_R} \cdot P(r \leq R) \tag{3.18}$$

where T = the total time under consideration. The resulting expression is given as

$$\bar{\tau} = \frac{e^{\rho^2} - 1}{\rho \cdot f_m \cdot \sqrt{2\pi}} \tag{3.19}$$

A quick observation of this result shows the AFTD is inversely proportional to the vehicle speed or Doppler frequency. For example, for fast fading vehicle speeds, the average time the signal is below the specified value is small when compared to the case when a slow moving vehicle is considered.

Given the above relationships we will choose two values of Doppler frequencies to calculate some typical LCR and AFTD quantities. Let us assume two Doppler frequencies, f_m = 200 Hz and f_m = 50 Hz. Also assume our interest lies in the statistics centered around a normalized 20 dB fade. Below we plot both of the quantities of interest where we extract the 2 points to compare against.

Using the plot and the previous LCR definitions (see Fig. 3.7), it is easy to extract the LCR for a 20-dB fade. For maximum Doppler spreads of 200 and 5 Hz, we encounter this fade level for 49.6 and 1.24 times per second, respectively.

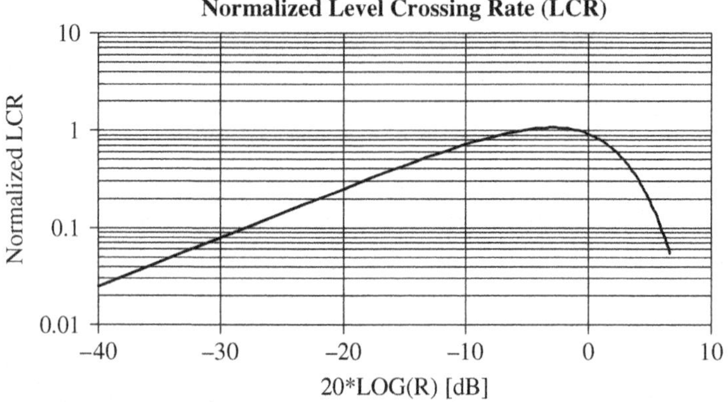

FIGURE 3.7 Level Crossing Rate (LCR) plot.

Using Fig. 3.8 along with the previous definition of the AFTD, we can extract the average time we are below the 20-dB fading level. For maximum Doppler spreads of 200 and 5 Hz, the average time durations are 200 μsec and 8 msec, respectively.

Moreover, we plot the temporal variations of the received signal assuming a Rayleigh fading channel (see Fig. 3.9). This plot aims to show that the time spent below the specified fading level varies in duration, with the average value given above.

An equally important component of the multipath fading phenomenon is the phase variation. The Rayleigh fading component will vary the signal envelope depending on fades and peaks present. The uniform phase component will rotate the signal depending on the aggregate sum of the arriving signals. Note

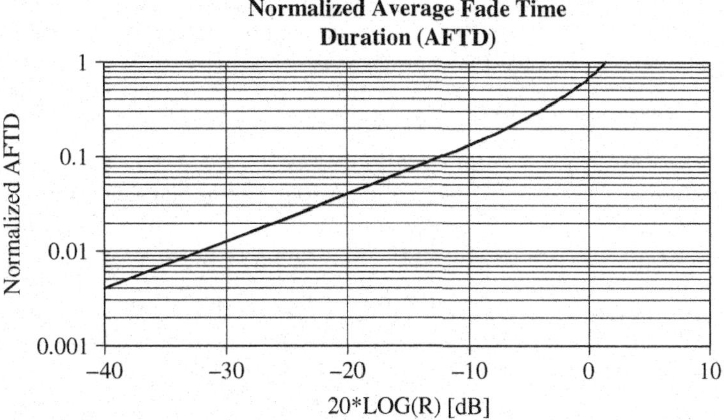

FIGURE 3.8 Average fade time duration plot.

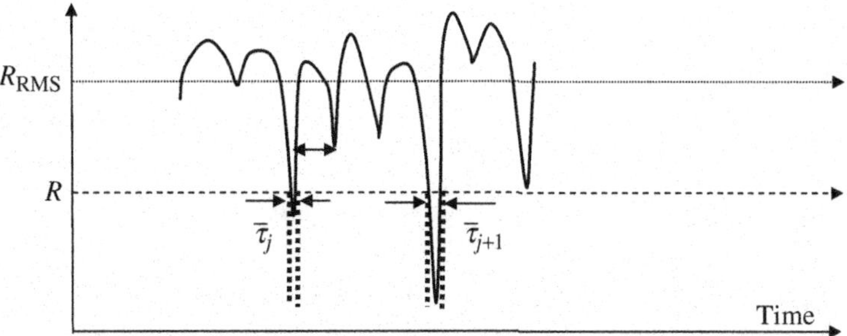

FIGURE 3.9 Rayleigh fading temporal variations to emphasize the AFTD.

although the Rayleigh and phase varying components have been called out or discussed separately, it is quite common and acceptable to simply describe both effects when the phrase multipath fading is used. The best way to describe this phenomenon is to transmit a Quaternary Phase Shift Keying (QPSK) signal and emphasize the received signal's constellation diagram at different points in time. Here the emphasis is on both the amplitude variations and their associated phase shifts. This is described in Fig. 3.10.

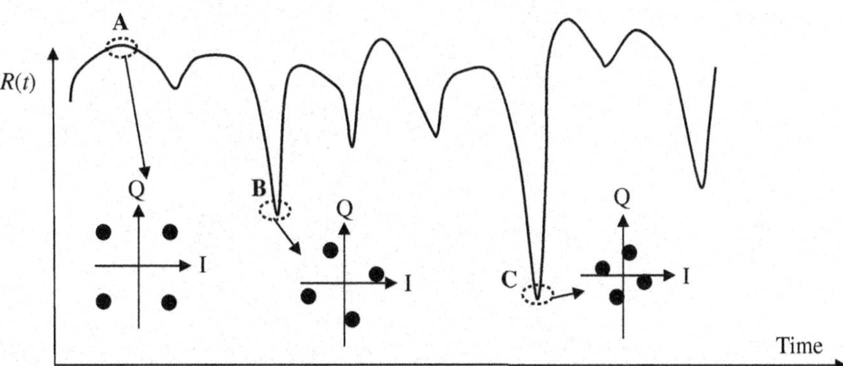

FIGURE 3.10 QPSK example to demonstrate the Rayleigh fading phenomenon.

Figure 3.10 shows the received signal constellation diagram for QPSK after encountering the multipath faded channel. The point worthy of mention is that as the signal encounters a fade not only does the signal amplitude decrease significantly, but also the phase changes abruptly. The reduced signal level or equivalently reduction in signal to noise ratio (SNR), makes estimating this abrupt phase change challenging (see Fig. 3.11).

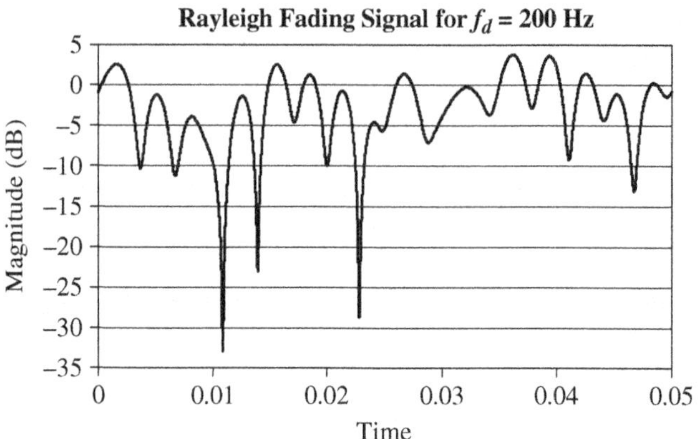

FIGURE 3.11 Temporal amplitude variations of Rayleigh fading.

3.3 RICIAN MULTIPATH FADING PHENOMENON

Here there is a dominant stationary signal or path making a significant contribution to the received signal power such as a line of sight propagation path. Such is the case when the base station (BS) is visible to the terminal, the small scale variations are now described with the help of the Rician distribution.

3.3.1 Mathematical Representation

The Rician distribution is given by the following:

$$p(r) = \frac{r}{\sigma^2} \cdot e^{\frac{-(r^2 + A^2)}{2\sigma^2}} \cdot I_o\left(\frac{rA}{\sigma^2}\right) \qquad (0 \leq A), (0 \leq r) \tag{3.20}$$

where $I_o(x)$ is the modified Bessel function of the first kind and zeroth order, $A^2/2$ is proportional to the power of the dominant signal.

A question arises as to how dominant or significant does the signal path need to be? The answer lies in what is commonly called the K factor and is defined below

$$K = 10^*\log\left[\frac{A^2}{2\sigma^2}\right] \qquad \text{(dB)} \tag{3.21}$$

This factor describes the ratio between the power of the dominant signal and the power (or variance) of the multipath signal. For example, let us assume there is no dominant signal, in this case $A = 0$ and $K = -\infty$ and this distribution now takes on the form of Rayleigh. On the other hand, assume there is no multipath signal, in this case $\sigma^2 = 0$ and $K = \infty$, here the distribution approaches the form of an impulse (or Dirac delta function). A plot of this distribution is given in Fig. 3.12 for various values of K factor.

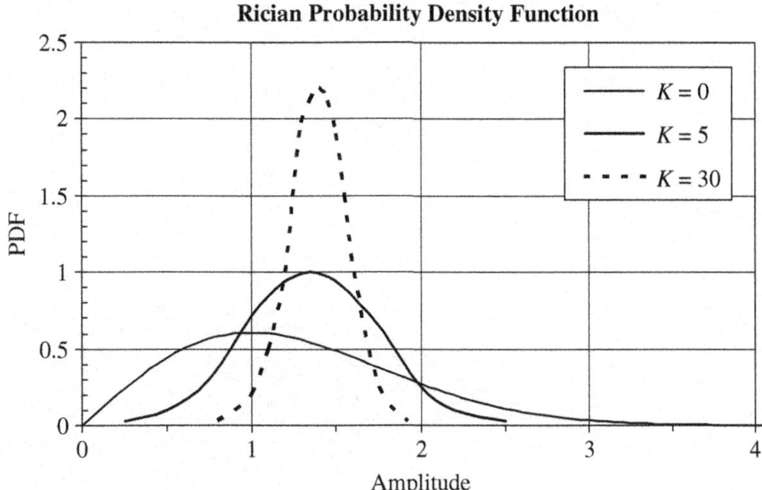

FIGURE 3.12 Rician distribution plot.

From Fig. 3.12 we see as the power of the line of sight increases the envelope variations become less and less. The number of deep fades encountered reduces thus reducing the instantaneous errors and improving the overall average Bit Error Rate (BER) performance.

3.3.2 Statistical Properties

The mean of the Rician distributed random variable is given as

$$E\{r\} = \sigma\sqrt{\frac{\pi}{2}} \cdot \left[(1 + K)I_o\left(\frac{K}{2}\right) + KI_1\left(\frac{K}{2}\right)\right] \cdot e^{-k/2} \tag{3.22}$$

The CDF of the Rician signal is the probability the received signal envelope is below a specific value, say R.

$$P(r \leq R) = \int_0^R p(r)\,dr \tag{3.23}$$

If we define a normalized signal level as

$$\rho = \frac{R}{\sqrt{A^2 + 2\sigma^2}} \tag{3.24}$$

Then the CDF is written as

$$P(r \leq R) = 1 - Q(\sqrt{2K}, \sqrt{2(K + 1)\rho^2}) \tag{3.25}$$

where $Q(a, b)$ is the Marcum Q function defined below

$$Q(a, b) = \int_b^\infty x \cdot e^{\frac{-(x^2 + a^2)}{2}} \cdot I_o(ax)\,dx \tag{3.26}$$

Earlier the LCR for Rayleigh fading was presented, next we present the LCR for Rician fading. The LCR is given by the following expression:

$$L_R = \sqrt{2\pi(K+1)} \cdot f_m \cdot \rho \cdot e^{-K-(K+1)\rho^2} \cdot I_o\left(2\rho\sqrt{K(K+1)}\right) \tag{3.27}$$

where we have defined the normalized signal level as

$$\rho = \frac{R}{R_{\text{RMS}}} \tag{3.28}$$

As a sanity check by forcing $K = 0$ will produce exactly the same result given earlier for the Rayleigh fading case. In Fig. 3.13 we plot the LCR for various values of K.

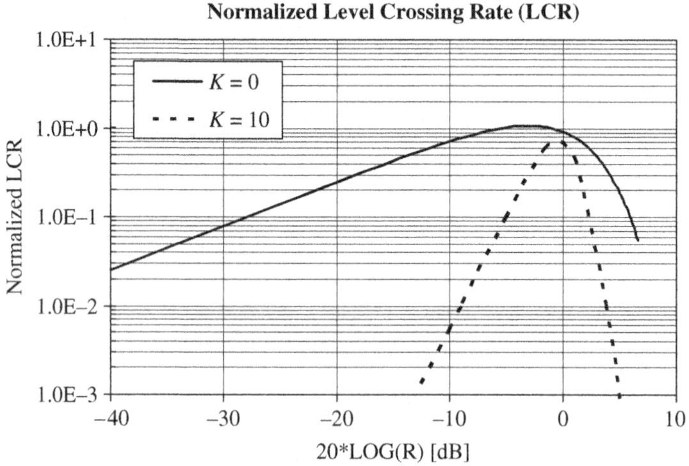

FIGURE 3.13 Rician LCR comparison for various K factors.

Similarly we can provide the AFTD for the Rician-faded signal, it is given by the following expression [10, 11].

$$\text{AFTD} = \frac{1 - Q(\sqrt{2K}, \sqrt{2(K+1)\rho^2})}{\sqrt{2\pi(K+1)} \cdot f_m \cdot \rho \cdot e^{-K-(K+1)\rho^2} \cdot I_o(2\rho\sqrt{K(K+1)})} \tag{3.29}$$

Lastly, for the sake of completeness the PDF of Nakagami faded signals is presented.

$$p(r) = \frac{2m^m}{\Gamma(m)\cdot\Phi^m} \cdot r^{2m-1} \cdot e^{\frac{-mr^2}{\Phi}} \tag{3.30}$$

where $\Phi = E\{r^2\}$ and m = Nakagami parameter, $m \geq 1/2$ and $\Gamma(m)$ is a gamma function.

The corresponding CDF is given as follows, using the following definition: $\rho = r/\Phi$.

$$P(r) = \frac{\Gamma(m, m\rho^2)}{\Gamma(m)} \tag{3.31}$$

where the incomplete gamma function is defined below.

$$\Gamma(a, b) = \int_0^b x^{a-1} \cdot e^{-x}\,dx \tag{3.32}$$

The level crossing rates and average fade time durations can also be derived, we simply show the results below for the Nakagami channel.[12]

$$\text{LCR} = f_d \cdot \sqrt{2\pi} \cdot \frac{m^{m-1/2}}{\Gamma(m)} \cdot \rho^{2m-1} \cdot e^{-m\rho^2} \tag{3.33}$$

$$\text{AFTD} = \frac{\Gamma(m, m\rho^2)}{f_d \cdot \rho^{2m-1} \cdot \sqrt{2\pi^{2m-1}}} \cdot e^{m\rho^2} \tag{3.34}$$

3.4 FREQUENCY SELECTIVE FADING

In the previous sections we have lumped the rays that arrive at or about the same time into a single statistically equivalent multipath. This single multipath doesn't discriminate against carrier frequencies and it is often called frequency flat fading. In this section we will discuss a frequency selective fading channel or more specifically the bandwidth of the wireless channel.

3.4.1 Power Delay Profile Methodology

Figure 3.14 depicts a scenario whereby the signals received by the terminal's antenna have arrived by propagating or traveling through distinct propagation paths. Here distinct is used with reference to the multipaths time of arrival. This scenario shows local scatterers surrounding the Mobile Station

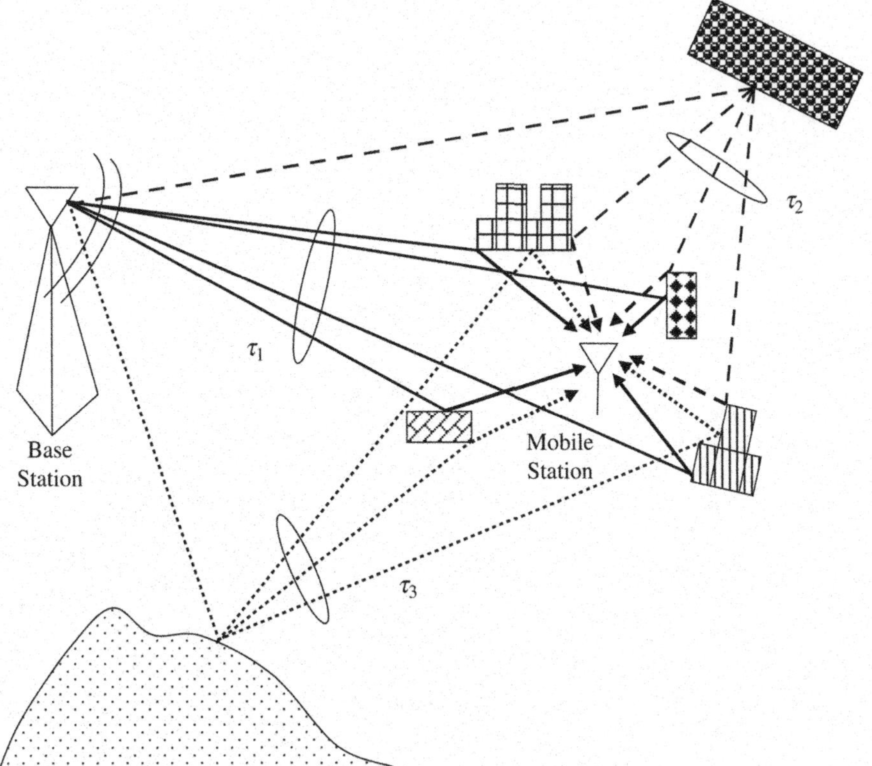

FIGURE 3.14 Frequency selective fading scenario.

(MS) being illuminated by the signal transmitted by the BS. Also shown is the presence of large, distinct reflectors that can essentially act as point sources with respect to the MS position.

The plot of signal power versus time of arrival is called the power delay profile (PDP). The above multipath propagation scenario is modeled by a three-ray model that has the following PDP with $P(\tau_k) = \alpha_k^2$, denoting the average power of the kth ray (see Fig. 3.15).

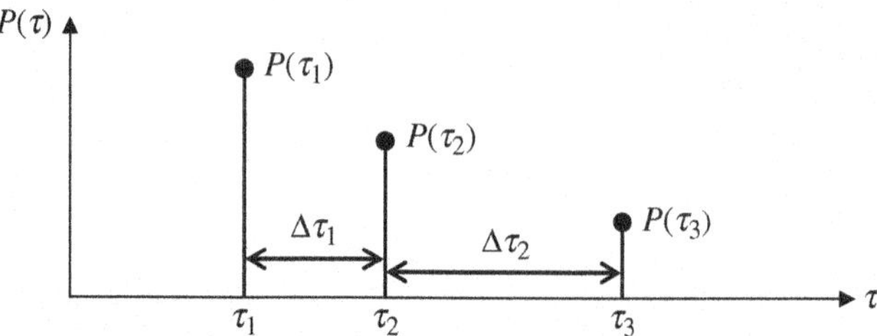

FIGURE 3.15 A power delay profile example.

The times of arrival and their associated powers are modeled as random variables. The significant multipaths arriving at times, τ_1, τ_2, and, τ_3 are sometimes called echoes. The corresponding FSF channel model can be drawn as shown in Fig. 3.16.

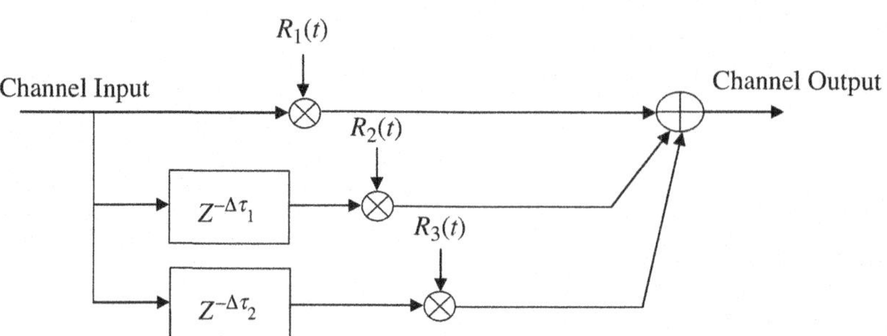

FIGURE 3.16 Three-ray frequency selective fading channel model.

Where three independent and identically distributed random variables, R_1, R_2, and, R_3 are used to model the Rayleigh fading process of each respective echo. It is common practice to assume the rays arriving at distinct times are independent of each other.

In order to assist in the derivation of mathematically closed form expressions, some mathematical representations of delay spread profiles are given as follows:[13]

Exponential
$$P(\tau) = \frac{A}{\tau_{\text{RMS}}} \cdot e^{-\frac{B\tau}{\tau_{\text{RMS}}}} \tag{3.35}$$

Double spike
$$P(\tau) = \partial(\tau) + \partial(\tau - K \cdot \tau_{\text{RMS}}) \tag{3.36}$$

Gaussian
$$P(\tau) = \frac{1}{\tau_{\text{RMS}}\sqrt{2\pi}} \cdot e^{-\frac{\tau^2}{2\tau_{\text{RMS}}^2}} \tag{3.37}$$

The procedure used to capture this information is to transmit a pulse at some location in the area of interest and then travel through this area of interest capturing the transmitted pulse plus the echoes that arrive. These echoes are scaled and time-shifted versions of the transmitted pulse. This phenomenon is also called time dispersion besides delay spread and less frequently called timing jitter [14], since this describes the position of the start of the received pulse. The time dispersion effect is visible in that the channel widens the time duration of the pulse so the received pulse is wider than the transmitted one. Now the comparison of this time dispersion to the transmitted symbol time is the same as the comparison of the channel BW (or coherence BW) relative to the occupied BW of the modulated signal, this will be discussed in the next section.

In an effort to quantitatively compare various PDPs and provide a means to support system design analysis, various parameters are defined, which in some extent vaguely quantify the multipath channel of interest. The mean excess delay is defined as the first moment of the PDP and is given below [15–17].

$$\bar{\tau} = \frac{\sum_k P(\tau_k) \cdot \tau_k}{\sum_k P(\tau_k)} \tag{3.38}$$

$$\bar{\tau} = \frac{\sum_k \alpha_k^2 \cdot \tau_k}{\sum_k \alpha_k^2} \tag{3.39}$$

The root mean square (RMS) delay spread is defined as the square root of the second central moment of the PDP and is given by

$$\tau_{\text{RMS}} = \sqrt{\overline{\tau^2} - (\bar{\tau})^2} \tag{3.40}$$

With the following definition

$$\overline{\tau^2} = \frac{\sum_k P(\tau_k) \cdot \tau_k^2}{\sum_k P(\tau_k)} = \frac{\sum_k \alpha_k^2 \cdot \tau_k^2}{\sum_k \alpha_k^2} \tag{3.41}$$

These parameters assume the first ray is arriving at $\tau_0 = 0$. As we will soon see the units of these parameters are expressed in nanoseconds (nsec) for the indoor environment and microseconds (μsec) for the outdoor environment.

3.4.2 Coherence Bandwidth (CBW)

The frequency selective nature of the propagation channel arises when there exists dominant multipath at significant time of arrival offsets. This next parameter is an attempt to statistically measure the BW of the wireless channel. The CBW is defined as the separation of frequencies over which the propagation channel has the same fading statistics (in other words range of frequencies where the channel is considered to be frequency flat). Here the frequency separation between two carriers is increased until the fading observed by each carrier begins to vary differently. This measured similarity between the two carriers is defined where the correlation function is above 0.5 and is given below [1].

$$\text{CBW} = \frac{1}{2\pi \cdot \tau_{\text{RMS}}} \tag{3.42}$$

Note other definitions exist depending on the choice of the correlation function and relative importance to the system designer. This shows the CBW is inversely proportional to the RMS delay spread of the channel.

Let's consider the following two-ray multipath PDP example (see Fig. 1.17).

Given the PDP below, the mean excess delay is given as

$$\bar{\tau} = \frac{(1) \cdot (0) + (0.1) \cdot (2)}{1 + 0.1} = \frac{0.2}{1.1} = 0.18\ \mu\text{sec}$$

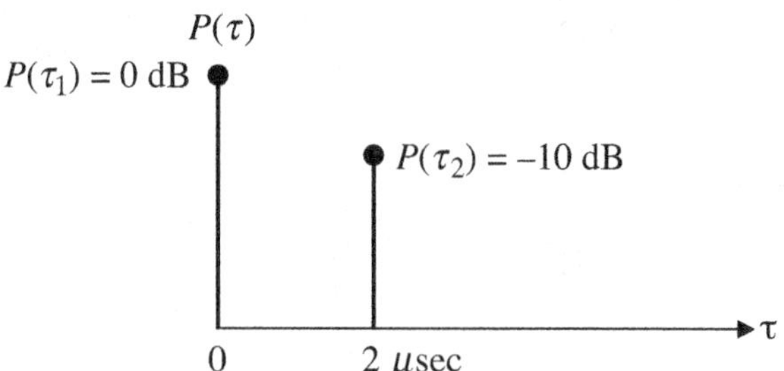

FIGURE 3.17 Two-ray multipath PDP example.

And the RMS delay spread is calculated as

$$\overline{\tau^2} = \frac{(1)\cdot(0)^2 + (0.1)\cdot(2)^2}{1 + 0.1} = \frac{0.4}{1.1} = 0.36\,\mu\text{sec}$$

$$\tau_{\text{RMS}} = \sqrt{\frac{0.4}{1.1} - \left[\frac{0.2}{1.1}\right]^2} = 0.575\,\mu\text{sec}$$

We can now write down the CBW for this type of channel as

$$\text{CBW} = \frac{1}{(2\pi)\cdot(0.575\,\mu\text{sec})} \cong 277\,\text{kHz}$$

If the transmit occupied BW is larger than or similar to 277 kHz, then the transmitted signal will encounter an FSF channel. In other words, the received signal will have obvious spectral variations (peaks and nulls) within its occupied BW, thus causing some frequency components to experience attenuation while others will not. Alternatively, the frequency selective fading nature of the channel can be easily detected by comparing the transmitted waveform (either chip, symbol, or bit time) to the total time dispersion of the channel. Similarly, the RMS delay spread is compared against the waveform time interval and is of significant value; then the channel is FSF. We will present some typical values of the CBW in the sections that follow once the measured PDPs are presented and discussed. Below we present an example illustrating this point by considering a two-ray channel model. The first case has the time difference between the two echoes to be much smaller than the symbol time interval. The second case has a transmitted time interval that is smaller than the delay spread of the channel (see Fig. 3.18).

Above we have taken an extreme for the FSF case, where the waveform time interval is shorter than the delay spread. As long as the delay spread is a significant portion of the waveform symbol interval, the transmitted signal will encounter a frequency selective channel. Figure 3.18 essentially describes the FSF criteria in the time domain. Next we present the similar description in the frequency domain.

Now that we have a quantitative expression for the BW of the wireless channel, we wish to compare it to the occupied BW of the transmitted signal. If the transmitted BW is much larger than the CBW then the received signal will experience FSF. On the other hand, if the transmitted BW is much smaller than the CBW, then the received signal will experience frequency flat fading. It is actually the same channel; however, how the varying baud rates relate to this gives rise to the above used notation.

Let us assume a PDP with the following exponential profile.

$$P(\tau) = \frac{1}{2\pi\cdot\tau_{\text{RMS}}}\cdot e^{-\frac{\tau}{\tau_{\text{RMS}}}} \tag{3.43}$$

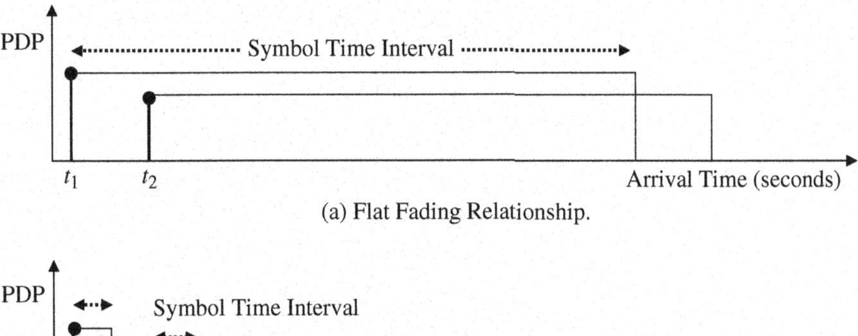

FIGURE 3.18 Frequency selective fading example.

Then the envelope correlation coefficient of two signals separated in time by Δt seconds and in frequency by Δf Hz is equal to the following [1].

$$\rho(\Delta f, \Delta t) = \frac{J_o^2(2\pi \cdot f_m \cdot \Delta t)}{1 + [2\pi \cdot \Delta f]^2 \cdot \tau_{\text{RMS}}} \tag{3.44}$$

This is a very interesting result since it allows us to gain insight into a few properties. For example, let us consider two signals whose frequency separation is varied, and comparisons are made at the same time interval, for example, $\Delta t = 0$.

$$\rho(\Delta f, 0) = \frac{1}{1 + [2\pi \cdot \Delta f]^2 \cdot \tau_{\text{RMS}}} \tag{3.45}$$

A plot of this function is given in Fig. 3.19 for two values of the RMS delay spread (2 μsec and 100 nsec).

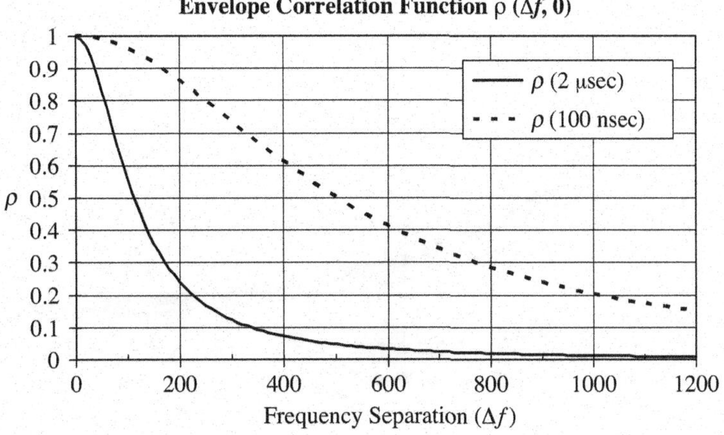

FIGURE 3.19 Envelope correlation plot for two signals separated by Δf Hz.

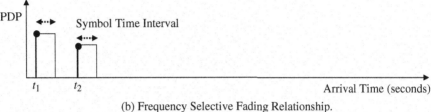

This plot shows the envelope correlation decreases as the frequency separation between the two signals increases. Also the rate of decay for the correlation function increases as the CBW decreases.

Next consider the example of 2 time samples of a single signal, whose time separation is varied, and comparisons are made at the same frequency, for example, $\Delta f = 0$.

$$\rho(0, \Delta t) = J_o^2(2\pi \cdot f_m \cdot \Delta t) \tag{3.46}$$

A plot of this temporal correlation for various Doppler frequencies is given in Fig. 3.20.

As previously discussed, there can be a few different definitions used for the CBW of the channel. In Fig. 3.21 we show two of them for sake of comparison.

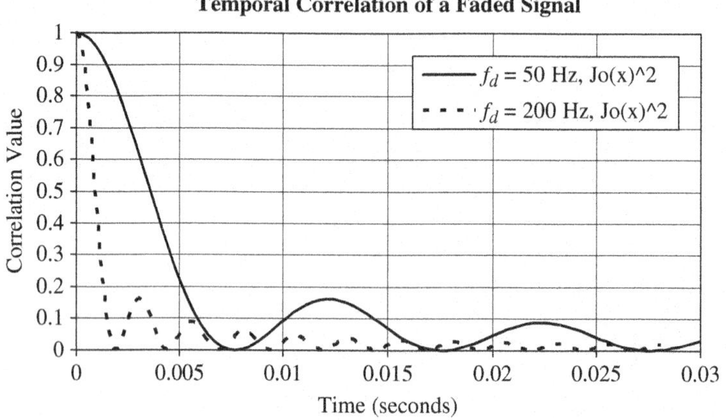

FIGURE 3.20 Temporal correlation of two time samples.

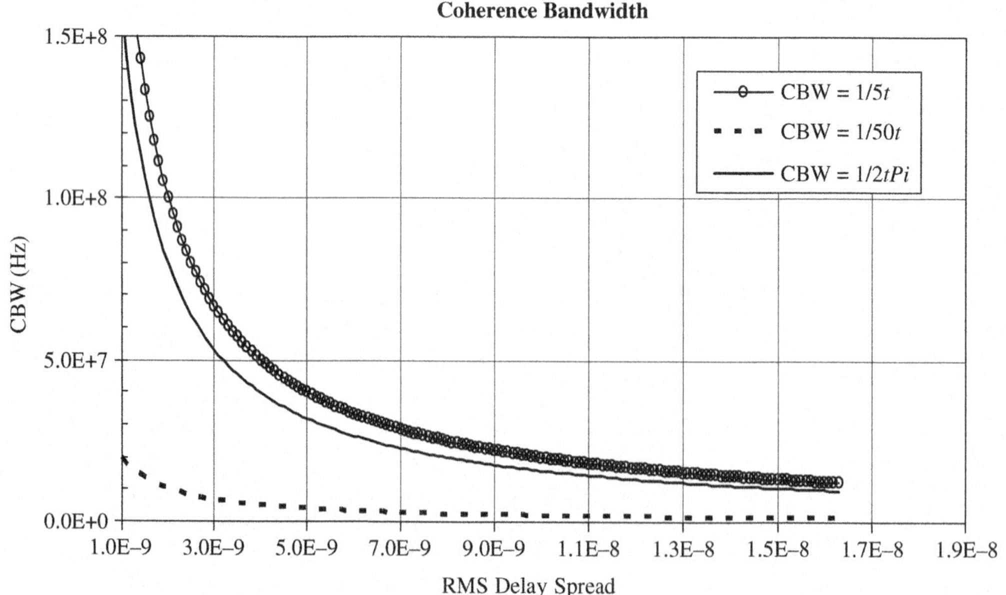

FIGURE 3.21 Coherence bandwidth comparisons.

The plot shows significant deviation between the definitions when the delay spread is very small.

3.4.3 Coherence Time (CT)

The PDP presented earlier was used to describe the FSF nature of the wireless channel by defining the CBW in the vicinity of the receiver. The signal envelope statistical variations were also presented earlier. However, they do not describe the time varying phenomenon of the wireless channel. This will be accomplished by relating the CT to the Doppler spread.

The Doppler spread is the spectral widening of the carrier signal caused by the time variations of the channel. Recall the max Doppler frequency is dependent on the carrier frequency, vehicle speed, and angle of arrival with respect to the velocity vector (direction of movement).

The CT is inversely proportional to the Doppler Spread [1, 7].

$$\mathrm{CT} \propto \frac{1}{f_m} \tag{3.47}$$

It is defined as the time duration in which two samples have a strong correlation. If the symbol time duration is much less than the CT then the channel will not change over its transmission. If the CT is defined as the time where the autocorrelation function is greater than 0.5, then we have the following relationship (with the maximum Doppler spread defined as $f_m = v/\lambda$).

$$\mathrm{CT} \cong \frac{9}{16\pi \cdot f_m} \tag{3.48}$$

Given the normalized time correlation function of

$$\rho(\tau) = J_o(\tau) \tag{3.49}$$

The corresponding spectrum is equal to

$$S(f) = \int_{-\infty}^{\infty} \rho(\tau) \cdot e^{-j\omega\tau} d\tau \tag{3.50}$$

$$S(f) = \frac{1}{\pi f_m} \cdot \frac{1}{\sqrt{1 - \left(\frac{f}{f_m}\right)^2}} \qquad |f| \le f_m \tag{3.51}$$

This is equivalent to the Doppler spectrum given earlier. The Fourier transform of the temporal correlation produces the Doppler spectrum [18].

3.4.4 Indoor Delay Spread Measurements

In this section a brief overview of some of the delay spread measurements for an indoor environment will be provided. [19, 20] First we present a measured PDP taken at 2.4 GHz band in Australia (see Fig. 3.22) [21].

The first point to make here is to observe the units of the excess delayed multipaths are expressed in nanoseconds. As you would expect there are plenty of propagation measurements at various carrier frequencies made for differing bandwidths. We will first give a summary of comparisons made between 900 MHz and 1.9 GHz since this is of immediate interest. Comparisons to other frequencies will be provided latter in this section.

In residential homes, office building and factories τ_{RMS} statistical information was collected and separately discussed in terms of line of sight (LOS) and non-line of sight (NLOS) cases. For presentation

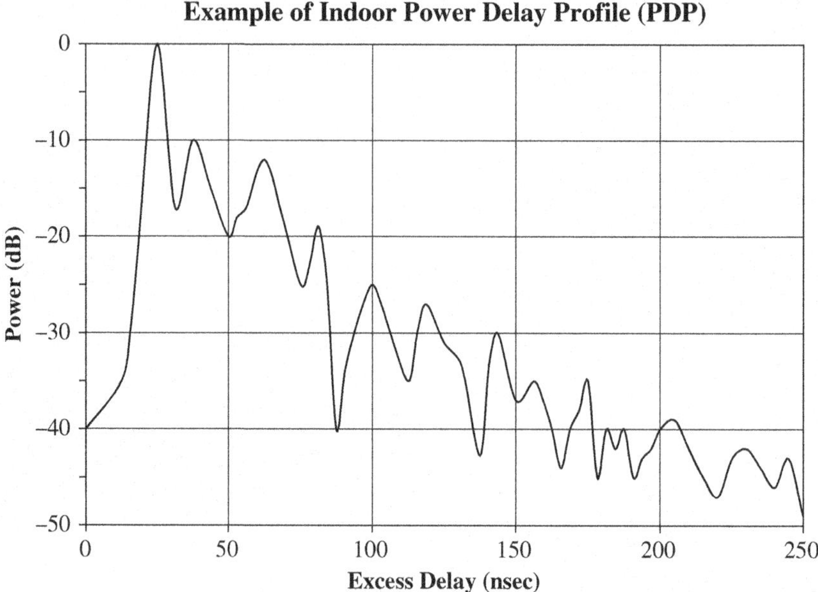

FIGURE 3.22 Example of a measured indoor PDP.

TABLE 3.1 Comparison of Delay Spread Values for 900 MHz and 1.9 GHz Frequency Bands

Delay spread	LOS	NLOS
τ_{RMS}	<100 nsec	<300 nsec

purposes, a summary can be provided and shown in Table 3.1 when considering the indoor propagation measurements.

In considering the contributions listed above we noticed the measurement parameters were not all exactly the same, such as carrier frequency (f_c), BS antenna height (h_{BS}), mobile antenna height (h_{MS}), and cell size. No attempt has been made to normalize the published data to a given set of parameters because the results overwhelmingly were in agreement in terms of their conclusions. An interesting parameter deserving special attention is the carrier frequency. The frequencies considered in these results were: 450 MHz, 850 MHz, 1.3 GHz, 1.7 GHz, 4 GHz, and 98 GHz, to name a few. For some measurements conducted, no significant statistical difference in delay spread was found at these frequencies. What this indicates is that τ_{RMS} may not, by itself, limit the choice of a higher carrier frequency operating within a large building. We have chosen to show an example where the measured delay spread statistics for a particular environment showed relatively no difference for a few carrier frequencies chosen (specifically 850 MHz and 1.7 GHz). This is shown in Fig. 3.23 [22].

The above results correspond to an office building in New York City (NYC) that measured approximately 61 m by 61 m in size.

However, there were also other published measurements that concluded τ_{RMS} to decrease significantly as the carrier frequency increased. Below we reproduce results supporting such statements in Figs. 3.24 and 3.25 [23].

The above-measured propagation results were taken in a university in Canada. The legend used above shows I = 37.2 GHz and J = 893 MHz. The higher carrier frequency had the smaller delay spread statistics in the channel.

The following measurements were taken in a university in the United Kingdom [24].

It was shown from these two plots how the delay spread decreases as the carrier frequency increases. As we will show later in this chapter, the higher frequencies have larger propagation path loss, and thus for the same cell size more path loss occurs.

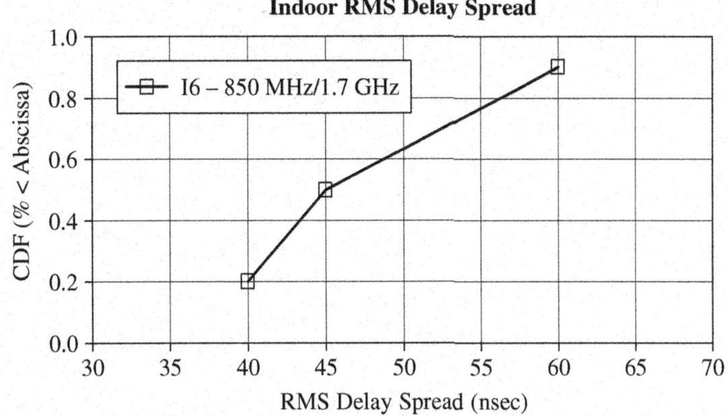

FIGURE 3.23 Indoor delay spread example showing frequency independence.

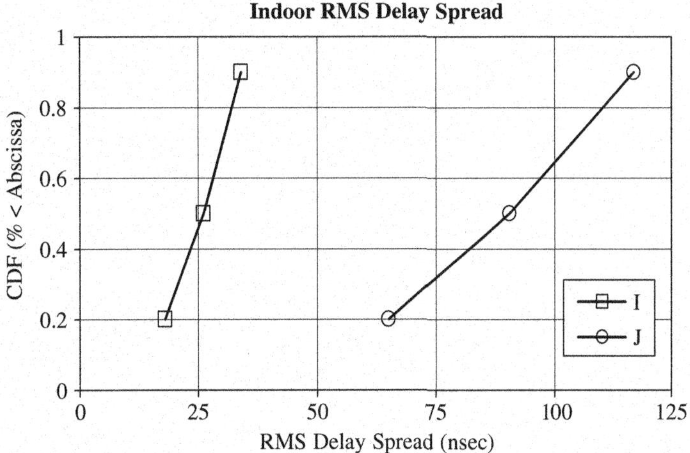

FIGURE 3.24 Indoor delay spread statistics for some measurements made in Canada.

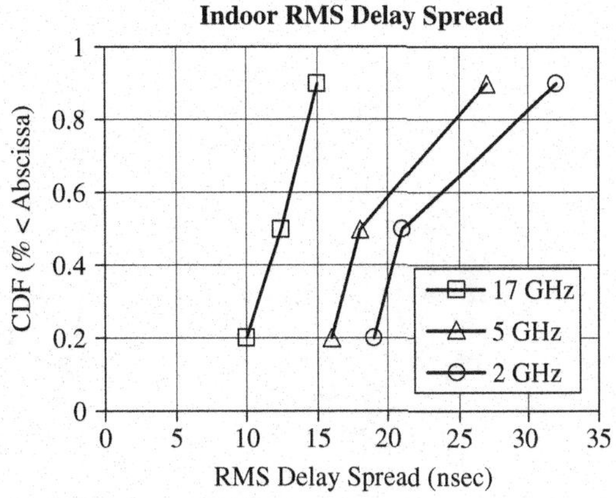

FIGURE 3.25 Indoor delay spread statistics for measurements made in the United Kingdom.

As mentioned above there have been many propagation measurements from various sites and at a variety of carrier frequencies. We have collected information from the published data and have plotted the RMS delay spread as a function of carrier frequency, this is shown in Fig. 3.26.

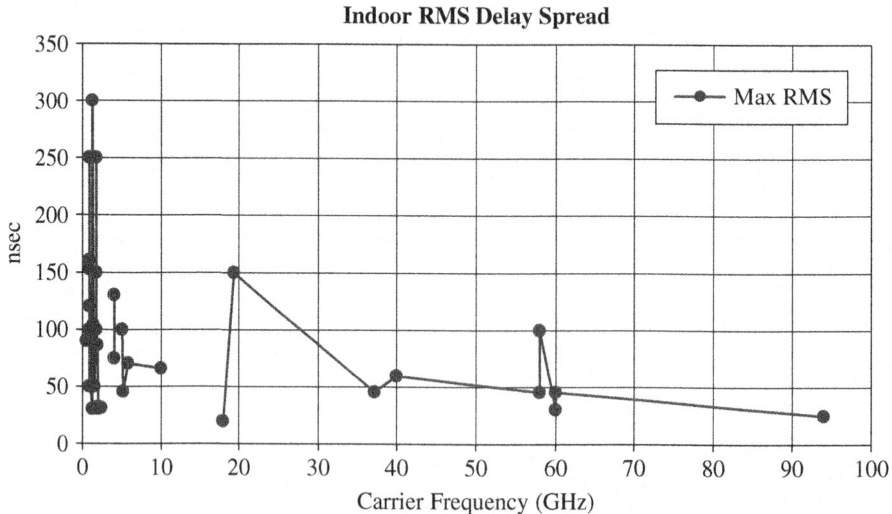

FIGURE 3.26 Summary of various propagation measurements.

From a bird's eye view there seems to be a slight slope indicating the RMS delay spread decreases as the carrier frequency increases. We must mention a sample support issue exists precluding us from drawing conclusions with high statistical confidence. We nevertheless present the findings for informative purposes (see Fig. 3.27).

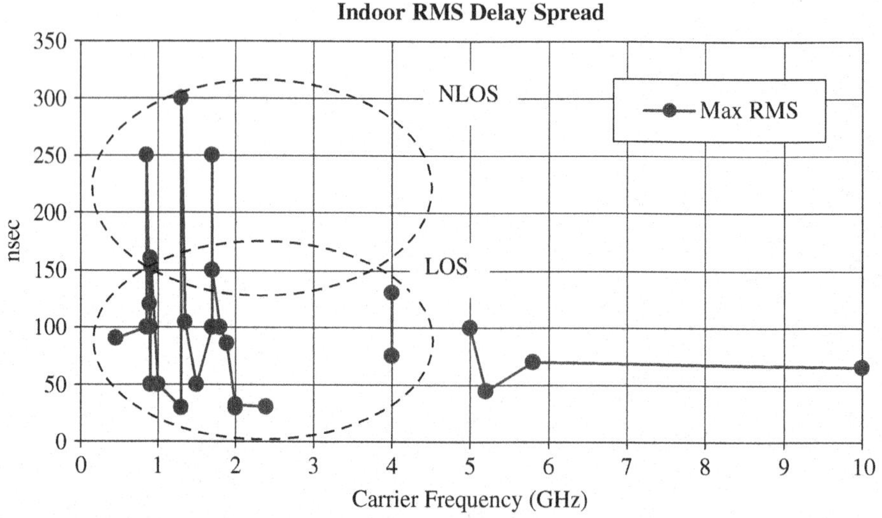

FIGURE 3.27 Summary of various propagation measurements (zoom in).

If we zoom in around the lower frequency range, we notice the data has significant variability. For sake of simplicity and continuity we will group them into LOS and NLOS areas. Where the NLOS cases have the larger values shown in the group and the LOS cases have the smaller values. Note an overlap in the measurements. Now that we have zoomed in, we can see there are more data points available in this region than at the higher frequency regions. Lastly, recall the above data points have been made in different buildings, made up of different construction material and in different countries (i.e., USA, Canada, Netherlands, Sweden, China, UK, etc.).

It should be mentioned that there exists a range of τ_{RMS} delay spread measurements that exist in the literature. One example is some measurements published $\tau_{\text{RMS}} = 30$ nsec while others publish values greater than 100 nsec. The first explanation is that the actual environment will vary (residential, factory, office building) not to mention factors such as indoor furniture, wall and floor construction, and so on. For those measurements with very large values of delay spread then special attention should be given to the outside surrounding buildings (possibly mountains). In other words, outside buildings and mountains can act as reflectors and thus increasing the multipath arrival time and in turn increasing τ_{RMS}.

Another rather interesting question is how does the delay spread vary as a function of distance or antenna separation. Hashemi [25] had showed there is a linear dependence (over a region considered) between τ_{RMS} and path loss and is supported in Figs. 3.28 and 3.29.

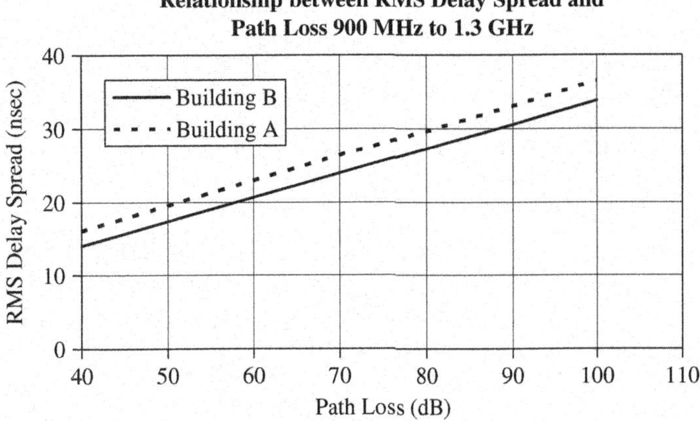

FIGURE 3.28 Delay spread and path loss relationship.

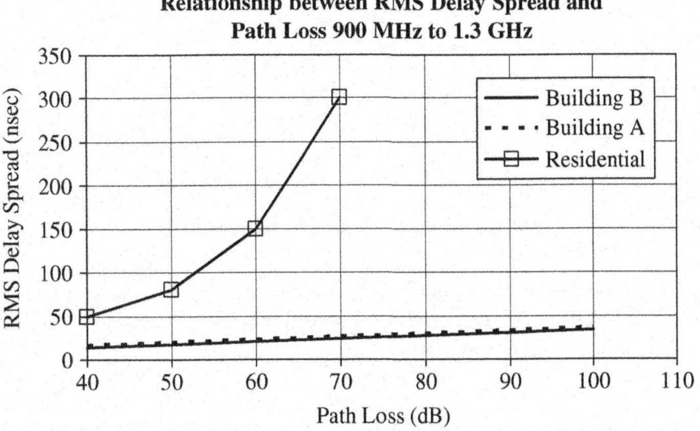

FIGURE 3.29 Delay spread and path loss relationship.

The above results were taken in Canada. Here we see as τ_{RMS} decreases, the path loss also decreased. Now since path loss is a function of antenna separation, one can also conclude that for these particular set of measurement that τ_{RMS} is indeed a function of distance. It is worth mentioning that there have also been measurements made where this argument is not supported. Here reports have been made stating delay spread was not correlated with distance.

It was also shown that a relationship other than linear exists between the path loss and RMS delay spread, see Fig. 3.29. The measured distances were larger in the residential results.

Generally speaking, we can prepare a simplified model that relates RMS delay spread to distance, with the use of Eqs. (3.52) and (3.53).

$$\tau_{\text{RMS}} = X \cdot e^{-Y \cdot PL} \tag{3.52}$$

$$PL = A + B \cdot \log(d) \tag{3.53}$$

Combining these equations gives us the following relationship

$$\tau_{\text{RMS}} = X \cdot e^{-Y\{A + B \cdot \log(d)\}} \tag{3.54}$$

where one can curve fit the above equations to come up with close approximations to the variables used within.

The last point we want the reader to be aware of is the actual procedure or methodology used in obtaining the τ_{RMS} results. Also we will neglect the differences in multipath arrival time bin widths and focus more on the threshold value used to consider if multipaths are valid to be used in the τ_{RMS} calculation. The typical range has been 10 dB to 20–30 dB below the main path power. In the next figure we present more results showing the effects of the choice/value of threshold on the τ_{RMS} calculated. These measured results were taken in Canada. It is expected as the multipath threshold increases, more multipaths will be used in the calculation of τ_{RMS} and thus increasing the value (see Fig. 3.30).

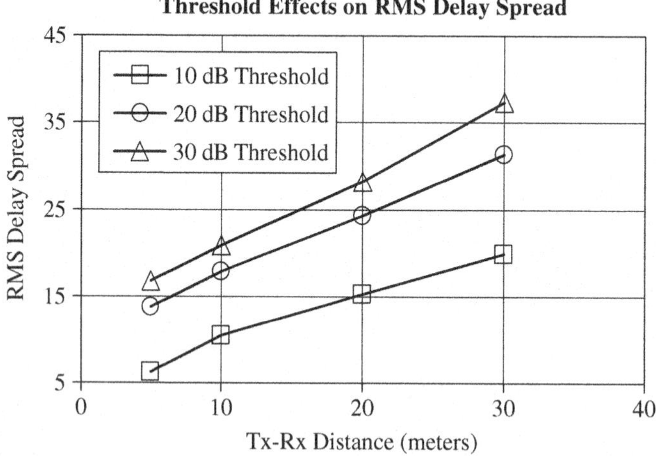

FIGURE 3.30 Delay spread relationship with the data collection threshold.

This plot shows two key points: first as antenna separation increases, so does τ_{RMS} and secondly as the threshold increases the delay spread increases. Notice the seemingly linear relationship that can exist between both variables.

As was previously mentioned the relationship between τ_{RMS} and path loss or distance was noticeable and significant for the propagations measurements made in Hashemi paper. There is also evidence to show that this relationship is insignificant or weak, thus supporting results published in other studies.

In Figs. 3.31, 3.32, and 3.33, we provide a variety of curves from various measurement campaigns taken throughout the world, such as USA, Canada, UK, Singapore, China, Japan, Korea, and the Netherlands. Moreover, a wide range of carrier frequencies were investigated. What we do see is there seems to be two groups of curves possibly due to the LOS and NLOS classification, as well as the actual building construction material that varies across the world [26–56].

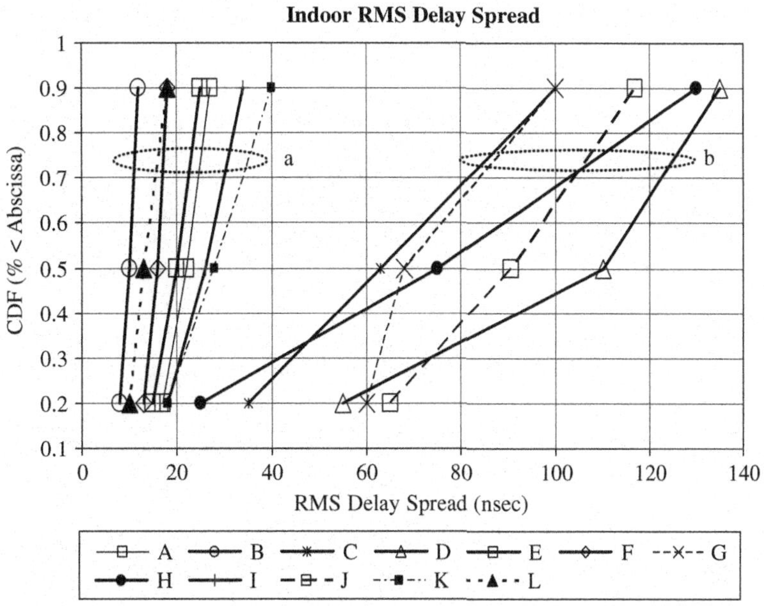

FIGURE 3.31 Summary of some measured indoor delay spread values.

We have named the two groups shown in Fig. 3.31 as "a" and "b." The maximum RMS delay spread for groups "a" and "b" are 40 nsec and 135 nsec, respectively.

In Fig. 3.31 we have labeled the curves with the following legend indicating the country where the measurements were performed and the carrier frequency: (a) = UK, 5 GHz; (b) = UK, 17 GHz [24, 57, 58, 59]; (c) = USA, 1.3 GHz/4 GHz; (d) = USA, 1.3 GHz [60]; (e) = Netherlands, 11.5 GHz; (f) = Japan, 60 GHz [61]; (g) = Singapore, 56 GHz [62, 63]; (h) = UK, 19.37 GHz; (i) = Canada, 37.2 GHz; (j) = Canada, 893 MHz [23, 64–66]; (k) = China, 900 MHz [67]; (l) = Canada, 40 GHz.

In order to make the figures more legible, we have decided to separate the curves into two plots to emphasize the point (see Figs. 3.32 and 3.33). The first plot is shown below, here it is clear that two significant families exist. The worse case mean delay spread is approximately 40 nsec (CBW = 398 MHz). These measured results correspond to measurement campaigns conducted in the UK, USA, Japan, Singapore, Canada, and China regions.

This second plot also shows two families present. Here the maximum delay spread value increased to approximately 135 nsec (CBW = 1.18 MHz). These measured results correspond to campaigns conducted in the USA, Netherlands, UK, and Canada regions.

In this section a summary of some of the propagation measurement campaigns conducted across the world was provided. As discussed earlier no attempt has been made to normalize the measured data. Delay spread results as a function of frequency and distance was summarized. When an extensive comparison was made about the available data, strong behavioral similarity existed for the RMS delay spread.

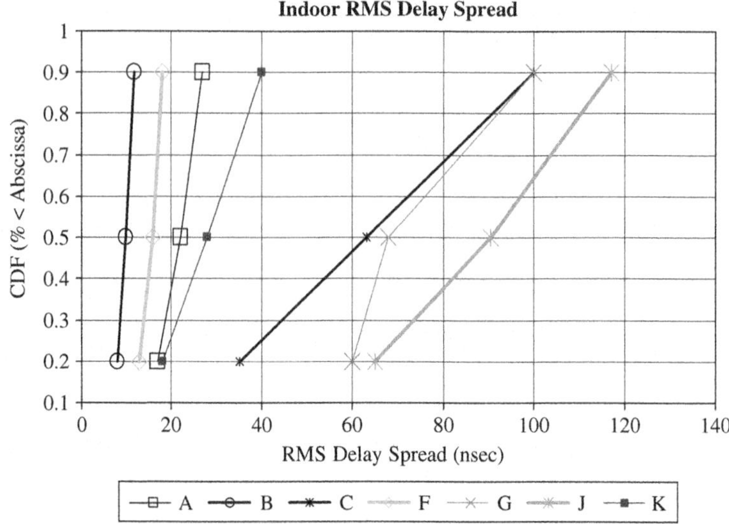

FIGURE 3.32 Partial summary of some measured indoor delay spread values.

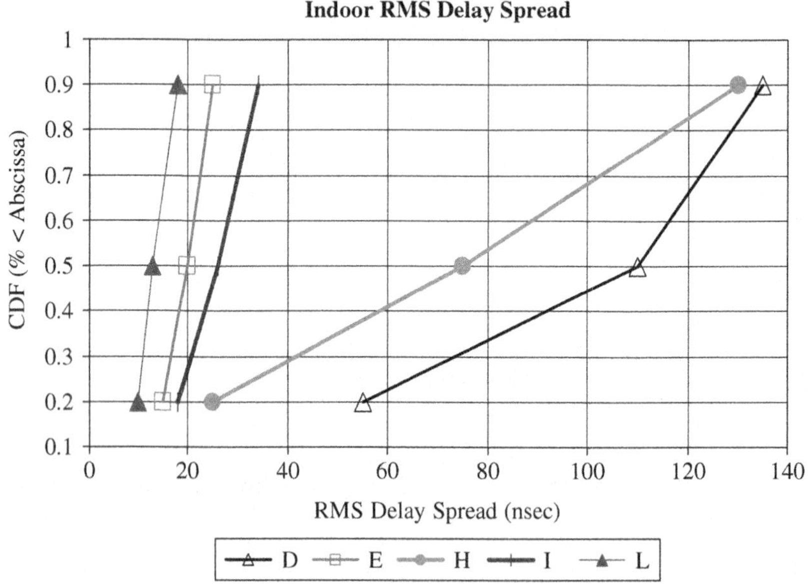

FIGURE 3.33 Partial summary of some measured indoor delay spread values.

3.4.5 Outdoor Delay Spread Measurements

In this section a brief overview of some of the delay spread measurements for an outdoor environment will be presented. First we provide a measured PDP shown in Fig. 3.34 [68, 69].

The first point to make here is to observe the units of the excess delay spread in microseconds. Propagation measurements were made for both macrocellular and microcellular environments for a variety of carrier frequencies. As noted above, measured environments varied significantly, specifically

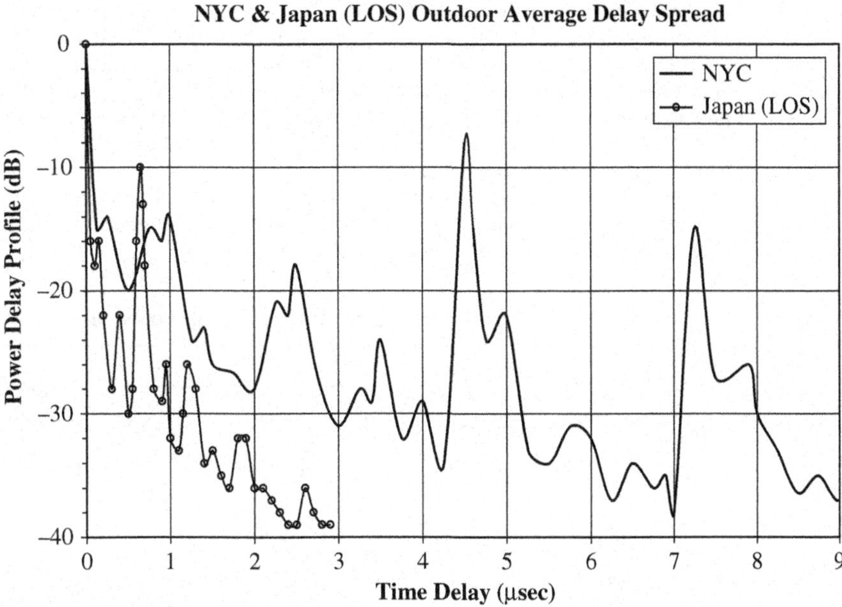

FIGURE 3.34 Example PDP of an outdoor channel.

speaking carrier frequency, BW, cell radius, country, environment, h_{BS}, h_{MS}, and so on. Similarly, no attempt has been made to normalize the propagation measurements; they are presented and some observations are discussed. The first question of concern is given the popular frequency bands used; is there a difference in delay spread? In considering the 850 MHz and 1.9 GHz frequency bands, the measurements revealed the RMS delay spread was statistically equivalent. In Fig. 3.35, we plot some results originally published by Devasirvantham in the Red Bank, NJ area [70, 71].

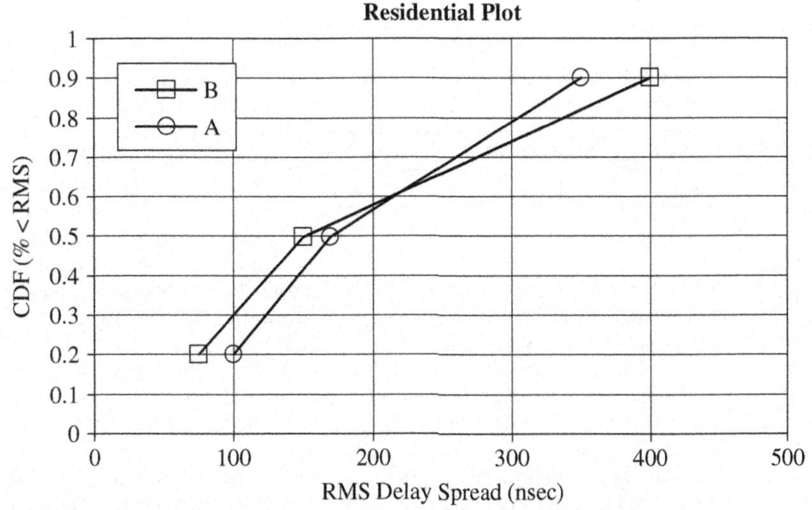

FIGURE 3.35 CDF of the measured τ_{RMS} values in outdoor residential areas.

As this plot shows there is no significant difference of τ_{RMS} between the two carrier frequencies investigated. Figure 3.35 has the corresponding legend: A = 850 MHz and B = 1.9 GHz.

The next question concentrates on the relationship between delay spread and distance. There has been an increasing body of literature showing there is a strong relationship. We will present some propagation results revealing as the cell radius (or *Tx-Rx* distance) decreases, the τ_{RMS} also decreases. The following results are from St. Louis, Missouri and the Red Bank, NJ areas shown by the dashed and solid lines, respectively [72].

Figure 3.36 shows, as the transmit and receive separation distance increases, the largest τ_{RMS} value increases. Lastly, these measured results were in the 1.9 GHz frequency band for the 20, 50, and 90% CDF values.

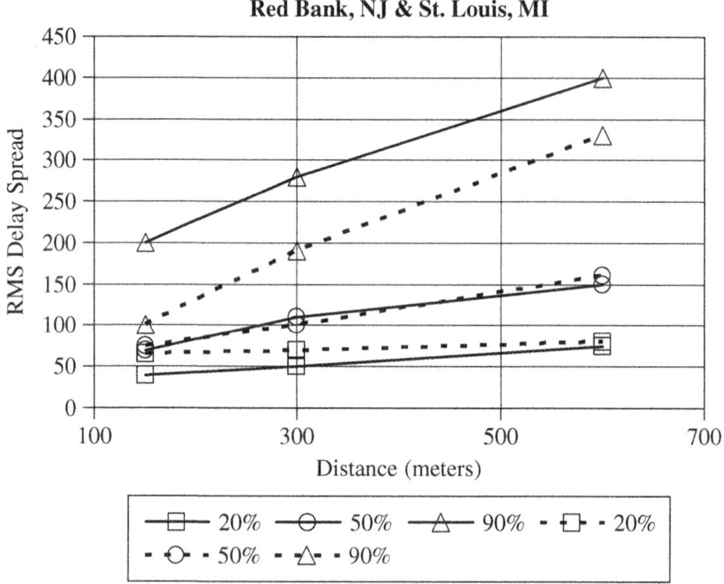

FIGURE 3.36 Delay spread statistics depending on distance.

TABLE 3.2 RMS Delay Spread Values for Different Environments

Urban	Suburban/Rural	Mountainous
<8 μsec	<2.2 μsec	<12 μsec

The next question we wanted to answer had to do with how the RMS delay spread varied from not only country to country but also from one environment to another. An excellent summary is provided in [73] where they have chosen three types of classifications: mountainous, urban, and suburban/rural. The results from this contribution are summarized in Table 3.2 for 95% CDF.

We have taken the liberty to collect a reasonable sample size of propagation measurements and plotted them together, shown in Fig. 3.37. Here we see the majority of the measurements have a τ_{RMS} less than 500 nsec, with a few results that are less than 3 μsec.

The legend used in Fig. 3.37 plot is given as: (a) = Red Bank, N.J. (850 MHz/1.9 GHz); (b) = St. Louis, MI (1.9 GHz); (c) = NYC, N.Y. (910 MHz); (d) = Computer Simulation, 900 MHz; (e) = Australia (1.9 GHz); (f) = Switzerland (900 MHz); (g) = Colorado (1.9 GHz); (h) = Toronto (910 MHz); and (i) = four US cities (900 MHz).

From the references used in Fig. 3.37, we can essentially see three sections present, indicating small, medium, and large delay spread. We selected a few measurements and show them in Fig. 3.38.

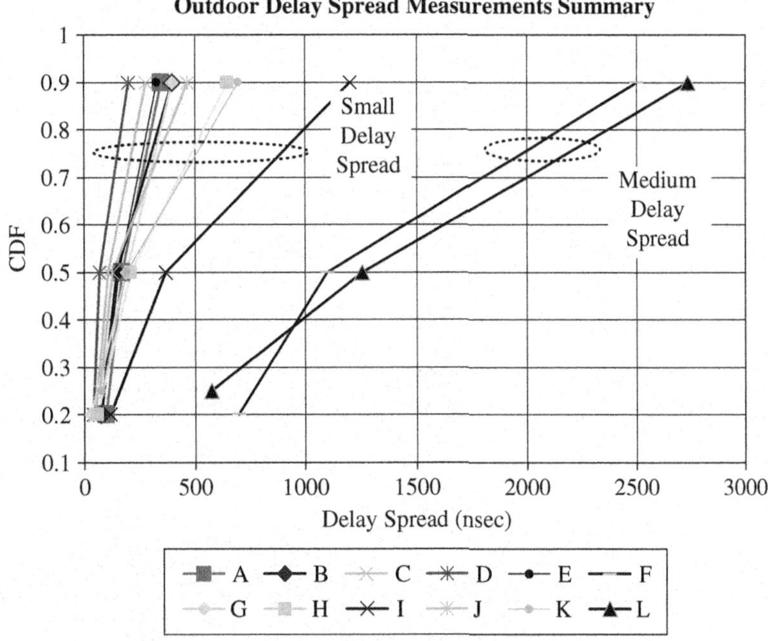

FIGURE 3.37 Summary of outdoor propagation measurements.

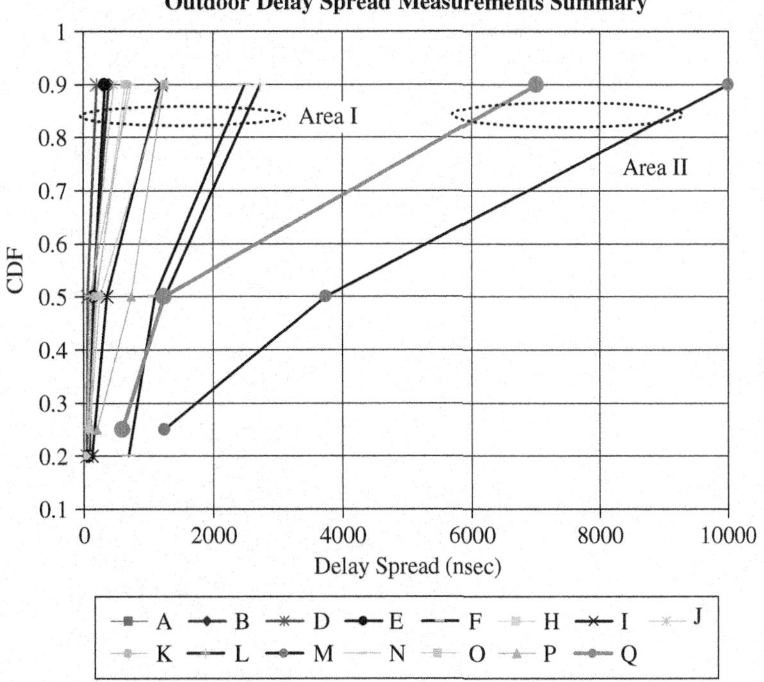

FIGURE 3.38 Additional summary of outdoor propagation measurements.

These measurements include some mountainous environments. We can easily see the mountainous regions approached the 10 μsec point. The results presented were taken from the following references: [74–88]

As discussed above no attempt was made to normalize the data since it was measured in widely varying environments and countries, not to mention measurement test setups as well. Having said this, we can see there are a least two areas/sections that deserve special attention. Specifically areas I and II can correspond to urban/suburban and mountainous. The mountainous section shows τ_{RMS} as much as 10 μsec. Where as the other section clearly shows an overlap between the urban and suburban environments. In fact, this can also be seen in the [73] reference. A closer look can reveal section I can be split up into two parts: A and B, where B can be associated with possibly urban and macrocellular cell sizes. Having said this, there has also been some results in Germany and presented in Table 3.3 [89].

TABLE 3.3 Comparison of Micro- and Macrocell Results

Parameter	Microcellular	Macrocellular
τ_{RMS}	<2 μsec	<8 μsec
τ_{excess}	<6 μsec	<16 μsec

TABLE 3.4 Comparison between East Coast and West Coast τ_{RMS} Measurements in USA

	East coast	West coast
τ_{RMS} (max)	7.5 μsec	25.5 μsec

Staying along these lines of environment classification, we have two items remaining: the first item deals with results taken in US cities and compare two cities on the West Coast to two cities on the East Coast [90]. This is summarized in Table 3.4.

This difference is mainly due to the more dispersive nature of the northern California region. We have found that the West Coast German cities are less dispersive than the West Coast US cities. Also note that the above published measurements are not exhaustive, but are meant to convey patterns.

The second item remaining is the choice of the PDP threshold to be used to remove unwanted noise and interference. Below we present some results that show as the threshold is lowered, to effectively include more echoes in the RMS calculator, the excess delay spread increases (see Fig. 3.39).

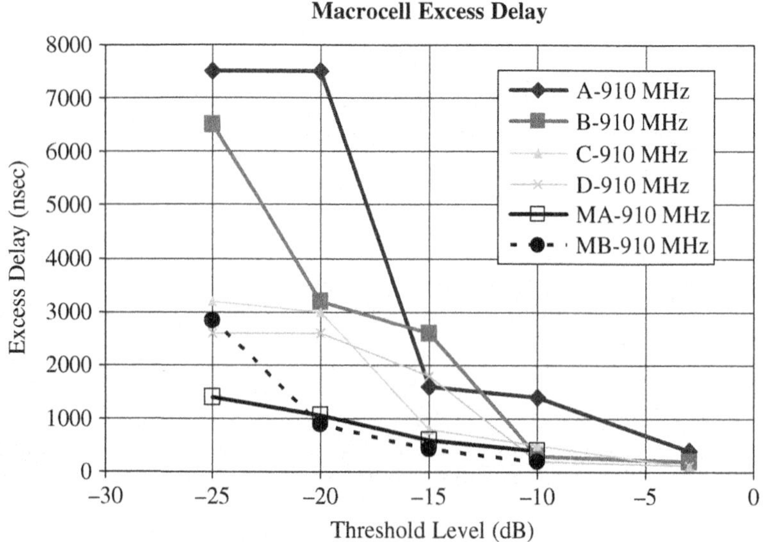

FIGURE 3.39 Delay spread dependency with data collection threshold.

In fact, the results were taken for both the micro- and macrocellular environments operating in the 910 MHz frequency band.

The above characteristic is useful knowledge especially when comparing the measured data presented herein. This will allow comparisons to be made on the same as well as fair playing ground.

Let's continue with a discussion on the PDP using Fig. 3.40 to support our point. We have shown earlier to τ_{RMS} delay spread was not only dependent on the surrounding environment but also on distance (ΔL) from the BS. Let's consider Fig. 3.40 which consists of essentially three regions that source the scattering signals: local scattering, large building scattering, and mountainous scattering [91].

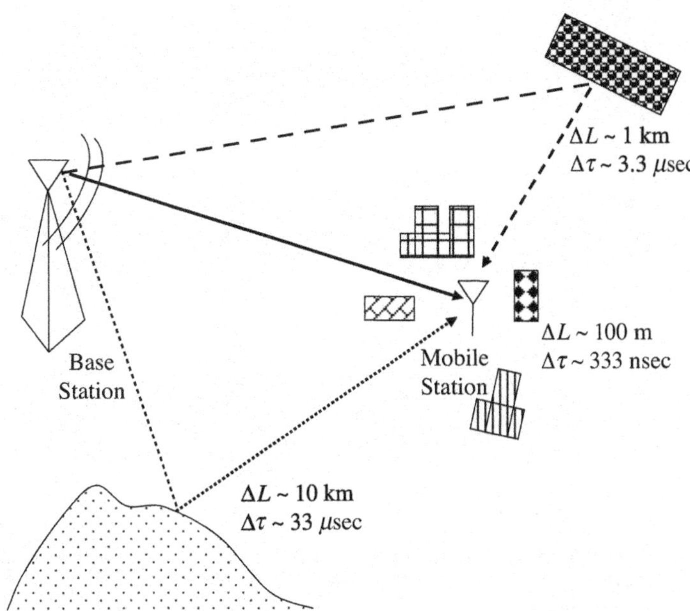

FIGURE 3.40 General delay spread characterization for three scattering regions.

What is important to note here is that signals scattered from large building and mountains will be rescattered by objects near the mobile resulting in a cascade of multipath arrivals at the MS antenna. This clustering concept was earlier proposed by Saleh and Valenzuela in [92].

3.5 MAN-MADE SYSTEM INTERFERENCE

Up until now the channel model has been developed based on natural phenomenon. By this we mean the wave pattern generated due to buildings, hills, mountains, and so on. Now a component of the channel model is presented whereby the source is from the communication system designer. In particular this section will address co-channel interference (CCI) and adjacent channel interference (ACI).

3.5.1 Co-Channel Interference

CCI arises from another signal operating in the same frequency band, at the same time but with a different transmitting waveform. This comes from frequency reuse in cellular systems. For example, let us consider a cellular system (drawn with hexagonal shapes) with a frequency reuse pattern of 7 as shown in Fig. 3.41.

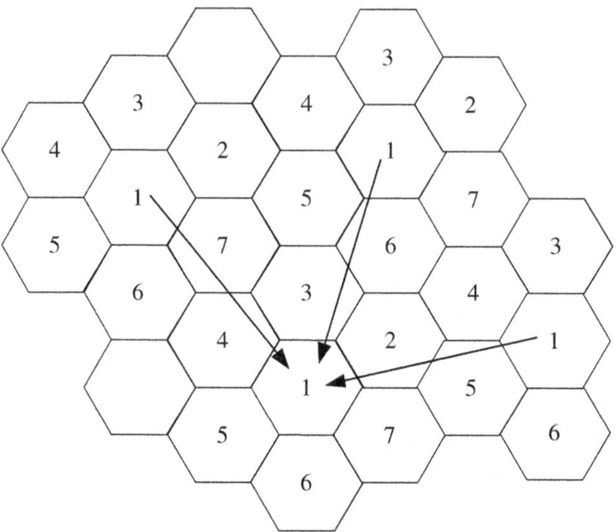

FIGURE 3.41 Cell layout assuming a reuse pattern of 7.

Here we see all cells operating on frequency f_j will interfere with one another (given they transmit at the same time). Smaller reuse factors exist, for example 3 or 1, where one would expect this interference will increase. This is a typical scenario faced by time division multiple access (TDMA) systems. In code division multiple access (CDMA) systems, however, a reuse factor of 1 is typically used as well as usage of different PN codes for other cell separation.

A simple model for CCI is given as follows, assuming the above cellular example with a frequency flat fading channel.

$$r(t) = s(t) \cdot h(t) + \sum_{k=1}^{6} s_k(t) \cdot h_k(t) + n(t) \tag{3.55}$$

where $s(t)$ = desired signal, $h(t)$ = channel response of desired signal, $s_k(t)$ = kth interfering signal, $h_k(t)$ = channel response of the kth interfering signal, $n(t)$ = AWGN.

This simple model assumes a single modulated interferer or user present in the frequency reuse cells. This is a nonrealistic scenario, hence Eq. (3.55) should be modified to include an average number of users in each cell with proper power distribution. It is clear as the number of users increases in each cell that the aggregate sum can approach a Gaussian distribution. In fact, some capacity analyses have modeled this other user noise as Gaussian or Colored noise. A few assumptions exist and are given in [93].

A simplified model for this type of interference is given in Fig. 3.42 for the downlink communication. Here we have chosen to show the top (or strongest in power) 6 interfering signals.

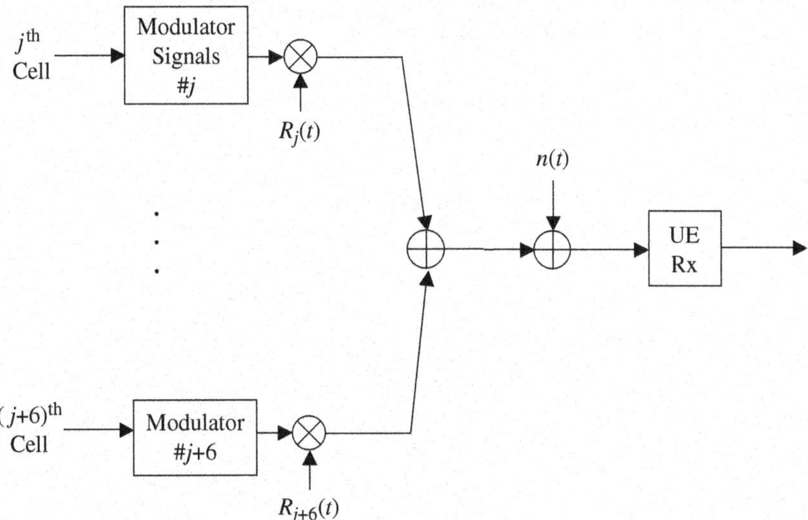

FIGURE 3.42 Simplified CCI simulation model.

These interfering signals are typically modeled as the aggregate sum of the transmitted signals from that particular cell. Note this number can be either decreased or increased based on the system under consideration and measured/simulated results. Also shown is that each interferer is independently faded.

3.5.2 Adjacent Channel Interference

ACI arises from other signals that transmit on an adjacent frequency channel at the same time the desired signal is transmitting. In both TDMA and CDMA systems the spectrum available to the service provider will be segmented. This means that adjacent channels can be of either similar radio access technology or a different one. Moreover, depending on the choice of the radio access technology, modulation scheme, and emissions mask, channels beyond adjacent can contribute to this form of unwanted interference, let us provide a picture given in Fig. 3.43.

Below we see an alternative ACI can potentially create enough in-band interference in the desired signal, thus causing performance degradation. A simplified model of ACI is considering a single side of the transmitted signal and is given below (see Fig. 3.44).

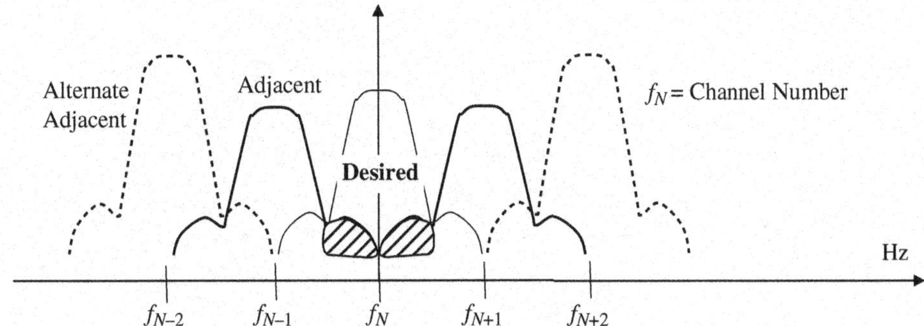

FIGURE 3.43 Transmit spectra for adjacent and alternate adjacent channel interference.

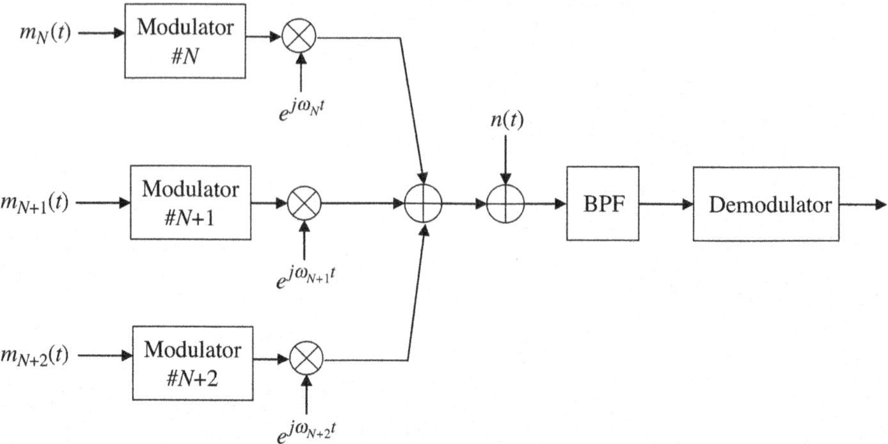

FIGURE 3.44 Simplified ACI simulation model.

As discussed above this is a form of interference that is created by the system designer. Frequency bands can be placed closer together to increase the overall system capacity; however, greater in-band distortion and performance degradation can occur if spectral regrowth is not properly addressed or managed. Recall our earlier attention that spectral regrowth occurs when the transmitting signal encounters a nonlinear, transmit power amplifier.

3.6 PROPAGATION PATH LOSS

In this section a comparison of methods used for predicting path loss is given. Here path loss is defined as the difference between the received power and the transmitted power. The methods presented herein are based on both empirical data and mathematical formulation [94, 95].

3.6.1 Free Space Path Loss

When the medium/environment between the transmitter and receiver has no obstructions, but simply a direct line of sight, then the free space propagation model can be used to predict the received signal power. Typically path loss models predict the received power decreases as the difference between the transmit and receive distance is raised to a power. The free space path loss model is defined as

$$P_{Rx} = P_{Tx} \cdot G_{Tx} \cdot G_{Rx} \cdot \left[\frac{\lambda}{4\pi d}\right]^2 \tag{3.56}$$

where P_{Rx} = predicted receive power, P_{Tx} = transmit power, G_{Tx} = gain of transmitting antenna, G_{Rx} = gain of receiving antenna, λ = carrier wavelength, d = distance separation between the transmitter and receiver. Above we see there is a decrease in power by 6 dB for each doubling of distance.

We can actually rewrite the free space path loss equation to equal the following (making use of $\lambda f = c$):

$$P_{Rx} = P_{Tx} \cdot G_{Tx} \cdot G_{Rx} \cdot \left[\frac{c}{4\pi d f}\right]^2 \tag{3.57}$$

where c = speed of light (meters per second).

Now we can see there is a decrease in power by 6 dB when either the distance or the carrier frequency is doubled. Since we are typically interested in the effects of the power exponent, we can write down the normalized free space path loss as follows:

$$L_{FS} = \frac{P_{Rx}}{P_{Tx} \cdot G_{Tx} \cdot G_{Rx}} = \left[\frac{c}{4\pi df}\right]^2 \tag{3.58}$$

Assuming we wish to represent this free space path loss as a positive value then the above equation can be inversed and expressed below in dB form, where distance is measured in km and the carrier frequency is expressed in MHz.

$$L_{FS}(\text{dB}) = 32.4 + 20 \cdot \log(d_{km}) + 20 \cdot \log(f_{MHz}) \tag{3.59}$$

As discussed above this path loss model assumes the channel has no obstructions between the transmitter and receiver; in reality there are obstructions and the resulting receiver power will be less. In other words, path loss predicted by this model is overly optimistic.

3.6.2 Hata Path Loss Model

The Hata model is a mathematical representation of extensive field measurements conducted by Okumura in Japan [96]. He provides attenuation curves relative to the propagation path loss in free space. The measurements were made for frequencies in the range of 150 MHz to 1.9 GHz and for transmit-receive distances between 1–100 km. Hence to use this model the free space path loss is first created and then an adjustment parameter is added $a(f, d)$ to consider the relative offsets and lastly a terrain parameter is added, G_{Terr}, to take into account the type of terrain. The Okumura model is given as

$$L_{Okumura}(\text{dB}) = L_{FS} + a(f, d) - G_{Terr} \tag{3.60}$$

Plots of the above correction factors are given in [96].

Okumura also found relationships of the BS antenna height and of the MS antenna height. Specifically the BS varies at a rate of 20 dB/decade and the MS varies at a rate of 10 db/decade. Now the complete model is given as

$$L_{Okumura}(\text{dB}) = L_{FS} + a(f, d) - G(h_{BS}) - G(h_{MS}) - G_{Terr} \tag{3.61}$$

where the above antenna height gains are represented as

$$G(h_{BS}) = 20 \cdot \log\left(\frac{h_{BS}}{200}\right) \quad h_{BS} > 30\,\text{m} \tag{3.62}$$

$$G(h_{MS}) = \begin{cases} 10 \cdot \log\left(\dfrac{h_{MS}}{3}\right) & h_{MS} \leq 3\text{ m} \\ 20 \cdot \log\left(\dfrac{h_{MS}}{3}\right) & 10\text{ m} > h_{MS} > 3\text{ m} \end{cases} \tag{3.63}$$

Now that we have presented the baseline data, we can now begin our discussion on the Hata model [97]. The Hata model is based on the Okumura measurements. The model starts with average path loss in the urban environment and then adjustment factors are added for other environments. The Hata model for the urban environment is given as (in the 150 MHz to 1.5G Hz frequency range).

$$\begin{aligned} L_{Hata}^{Urban}(\text{dB}) = {} & 69.55 + 26.16\log(f_c) - 13.82\log(h_{BS}) - a(h_{MS}) \\ & + [44.9 - 6.55\log(h_{BS})] \cdot \log(d) \end{aligned} \tag{3.64}$$

where f_c = carrier frequency in MHz, h_{BS} is the BS antenna height in units of meters (30–200 m), h_{MS} is the MS antenna height in units of meters (1–10 m), d is the distance separation in units of km (1–20 km), a(h_{MS}) is the correction factor for the MS antenna height.

For small- to medium-sized cities the MS antenna height correction factor becomes.

$$a(h_{\mathrm{MS}}) = (1.1 \cdot \log(f_c) - 0.7) \cdot h_{\mathrm{MS}} - (1.56 \cdot \log(f_c) - 0.8) \tag{3.65}$$

Also for large cities the MS antenna height correction is given by

$$a(h_{\mathrm{MS}}) = \begin{cases} 8.29 \cdot (\log(1.54 \cdot h_{\mathrm{MS}})^2 - 1.1\ \mathrm{dB} & f_c < 300\ \mathrm{MHz} \\ 3.2 \cdot (\log(11.75 \cdot h_{\mathrm{MS}})^2 - 4.97\ \mathrm{dB} & f_c \geq 300\ \mathrm{MHz} \end{cases} \tag{3.66}$$

Next, to obtain the path loss in a suburban environment, we first start with the path loss for the urban environment and then add an adjustment parameter

$$L_{\mathrm{Hata}}^{\mathrm{suburban}}(\mathrm{dB}) = L_{\mathrm{Hata}}^{\mathrm{urban}}(\mathrm{dB}) - 2\left[\log\left(\frac{f_c}{28}\right)\right]^2 - 5.4 \tag{3.67}$$

This approach can also be applied to the open area environment as is given as

$$L_{\mathrm{Hata}}^{\mathrm{rural}}(\mathrm{dB}) = L_{\mathrm{Hata}}^{\mathrm{urban}}(\mathrm{dB}) - 4.78 \cdot [\log(f_c)]^2 - 18.33 \cdot \log(f_c) - 40.98 \tag{3.68}$$

The Hata model predicts the path loss relatively close to not only the original Okumura data but also to some other models used [7]. This will be discussed later in this section.

3.6.3 Modified Hata Path Loss Model

There has been much work performed to extend the model for carrier frequencies up to 2 GHz. The modified Hata model is given as follows [17].

$$\begin{aligned} L_{\mathrm{Mod-Hata}}^{\mathrm{urban}}(\mathrm{dB}) &= 46.3 + 33.9 \cdot \log(f_c) - 13.82 \cdot \log(h_{\mathrm{BS}}) \\ &\quad - a(h_{\mathrm{MS}}) + [44.9 - 6.55 \cdot \log(h_{\mathrm{BS}})] \cdot \log(d) + C_M \end{aligned} \tag{3.69}$$

where $C_M = 0$ dB for medium cities and suburban areas, $= 3$ dB for metropolitan areas.

The valid range of these modified parameters is now

$$\begin{aligned} 1.5\ \mathrm{GHz} &\leq f_c \leq 2\ \mathrm{GHz} \\ 30\ \mathrm{m} &\leq h_{\mathrm{BS}} \leq 200\ \mathrm{m} \\ 1\ \mathrm{m} &\leq h_{\mathrm{MS}} \leq 10\ \mathrm{m} \\ 1\ \mathrm{km} &\leq d \leq 20\ \mathrm{km} \end{aligned} \tag{3.70}$$

3.6.4 Bertoni-Walfrisch Path Loss Model

In this section we will approach the path loss discussion from a different angle. Previous models were based on measured data where mathematical representations were formulated that would best describe the behavior observed in the measured data. This present model will actually start with a mathematical description of the environment and then provide the prediction of the path loss. As we will soon see, there is reasonable agreement between these two approaches [91].

A block diagram showing the underlying mechanisms of the mathematical model is presented in Fig. 3.45.

In Fig. 3.45 various variables have been shown and next we attempt to define them. R = distance between the transmitter and receiver, b = separation of building rows, λ = wavelength of carrier frequency, α = angle to last roof tip, r = distance to the MS from the building rooftop, θ diffracting angle to the MS, and x = distance from building to mobile location.

The total path loss is the sum of the following individual components.

$$L_{\mathrm{TOT}} = L_{\mathrm{FS}} + \text{multiple screen diffraction} + \text{diffraction to the ground level} \tag{3.71}$$

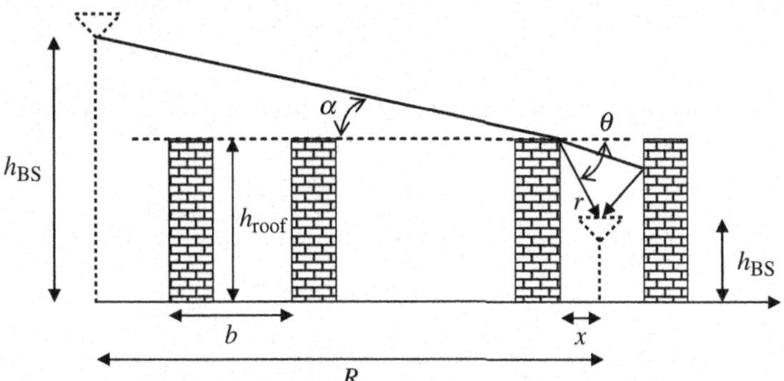

FIGURE 3.45 Mathematical propagation model.

Next we discuss each component separately. The free space loss is given as

$$L_{FS} = \left[\frac{\lambda}{4\pi R}\right]^2 \tag{3.72}$$

The multiple screen diffraction is the reduction in the rooftop signal at the building row just before the subscriber as a result of the propagation past the previous rows of buildings.

$$L_{MSD} = (2.3474)^2 \cdot \left[\alpha\sqrt{\frac{b}{\lambda}}\right]^{1.8} \tag{3.73}$$

$$\alpha = a\tan\left[\frac{h_{BS} - h_{roof}}{R}\right] \cong \frac{h_{BS} - h_{roof}}{R} \tag{3.74}$$

And the diffraction to the ground level is the reduction due to the diffraction of the rooftop fields down to the street level.

$$L_{DS} = \frac{\lambda}{2\pi^2 r} \cdot \left(\frac{1}{\theta} - \frac{1}{2\pi + \theta}\right)^2 \tag{3.75}$$

$$\theta = \mathrm{a}\tan\left(\frac{h_{roof} - h_{MS}}{\lambda}\right) \text{ and } r = \sqrt{(h_{roof} - h_{MS})^2 + x^2} \tag{3.76}$$

The combined path loss is given as

$$L_{TOT} = \left(\frac{\lambda}{4\pi R}\right)^2 \cdot (2.3474)^2 \cdot \left[\alpha\sqrt{\frac{b}{\lambda}}\right]^{1.8} \cdot \frac{\lambda}{2\pi^2 r} \cdot \left(\frac{1}{\theta} - \frac{1}{2\pi + \theta}\right)^2 \tag{3.77}$$

Note the overall distance exponent is 3.8, which is remarkably close to the range of values typically chosen between 3 and 4.

With certain building environment assumptions such as $h_{MS} = 1.5$ m, $h_{BS} = 30$ m, $b = 60$ m, $h_{roof} = 10$ m we can write an approximation as

$$L_{TOT} = 53.7 + 21 \cdot \log(f_c) + 3.8 \cdot \log(R) \tag{3.78}$$

$$L_{Hata} = 49.2 + 26.2 \cdot \log(f_c) + 35.2 \cdot \log(R) \tag{3.79}$$

This close agreement supports the assumption that propagation takes place over the buildings, with diffraction from the rooftop to the ground level. We will make various comparisons about measurements published across the world. Their details can be found in references: [98–123].

First we present a summary of the published results for a microcell environment. We will define microcells as those cells that have a radius of less than 1 km. In viewing the following microcell path loss summary we can make a few observations (see Fig. 3.46). First we can see a large variation in the published results mainly due to differences in countries, BS antenna height, MS antenna height, surrounding

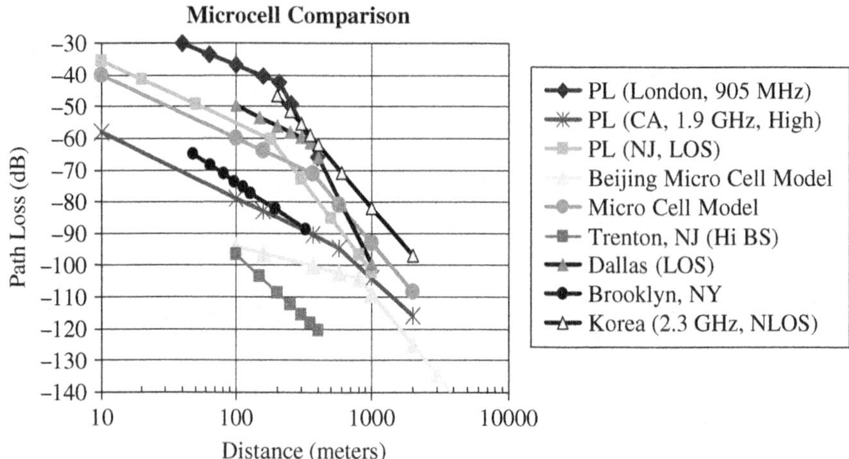

FIGURE 3.46 Microcell path loss summary.

environments, and so forth. Another observation that can be made is that there is a two-slope response. The point where the two-sloped lines meet is called the "knee" or "break point" of the curve. This break point can be calculated as shown in [91] and can be approximated for high carrier frequencies as

$$d_{\text{break}} = \frac{4 \cdot h_{\text{BS}} \cdot h_{\text{MS}}}{\lambda} \tag{3.80}$$

The typical values of this break point occur in the hundred of meters range.

We have chosen to plot a variety of measurements [123, 125]. These measurements include countries such as UK, USA, China, and Korea. We have included additional results from various USA cities: California, Trenton, N.J., and Brooklyn, N.Y. In this microcell case we can see for a distance of 500 m the difference in path loss is between −65 to −105, in other words approximately 40 dB difference. This observation has been made neglecting measurement results from Trenton, N.J., which provides the worse case losses.

In order to further emphasize the above observations we have collected a subset and have shown them below. As an example, the breakpoint for the New Jersey results was found to be approximately 200 m. In Fig. 3.47 we show a curve labeled microcell model. This model has a breakpoint set to 100 m (arbitrarily set) to convey the two-slope behavior. The model is given as generally as

$$PL = \begin{cases} A + B \cdot \log(d) & d \le d_{\text{break}} \\ A + B \cdot \log(d_{\text{break}}) + C \cdot \log\left(\frac{d}{d_{\text{break}}}\right) & d > d_{\text{break}} \end{cases} \tag{3.81}$$

where we have used the following values to plot the curve above

$$A = -20$$

$$B = -20$$

$$C = -50$$

$$d_{\text{break}} = 100 \text{ m}$$

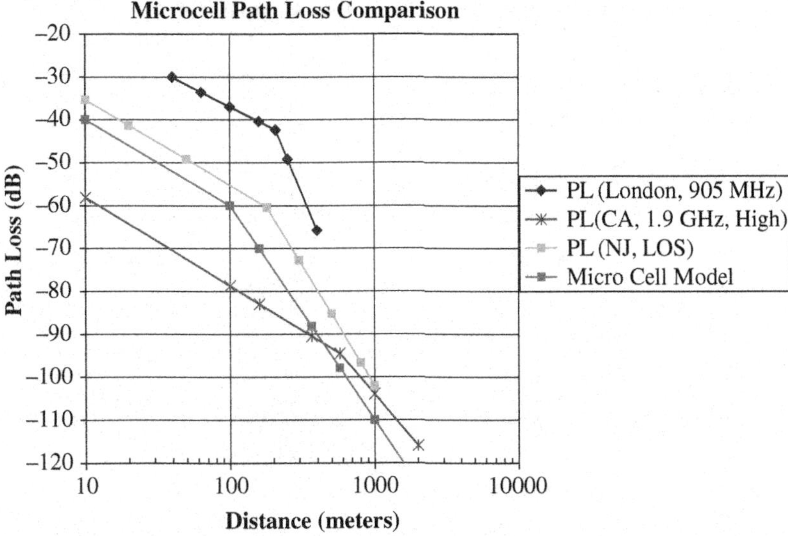

FIGURE 3.47 Microcell path loss summary.

Next we move our attention to the macrocell environment where we have defined macrocells having a radius greater than 1 km. At this point the breakpoint would have occurred already so the curves are a straight line (when plotted on log scale). The following plots differ from presentations made in other texts in that we have included additional countries in the comparison [7]. From a high level we see the path loss curves have somewhat similar slopes but differing in the initial path loss offset. The major exception being the path loss results pertaining to the NYC measurements. The macrocell path loss summary is given below (see Fig. 3.48).

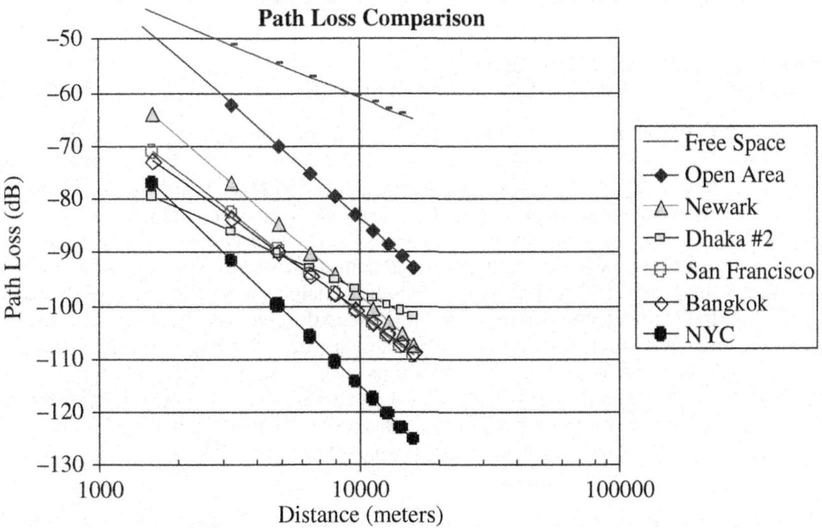

FIGURE 3.48 Macrocell path loss summary.

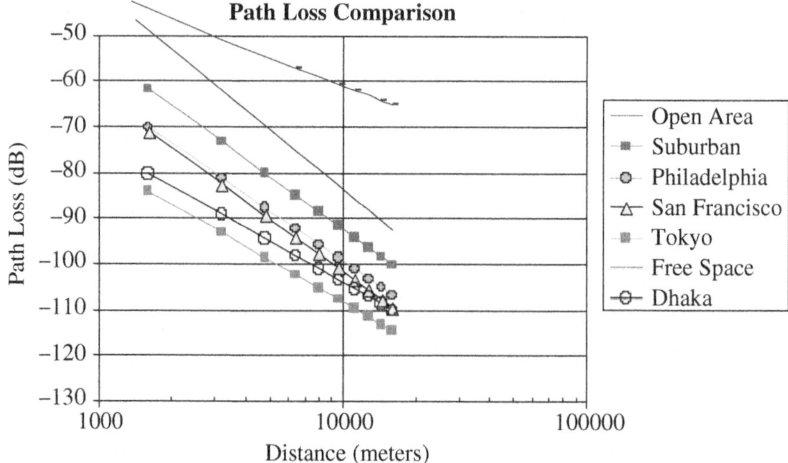

FIGURE 3.49 Macrocell path loss summary.

TABLE 3.5 Path Loss Exponent Comparison

Location	Path loss exponent
Open	4.35
Suburban	3.84
Newark, NJ	4.31
Philadelphia, PA	3.68
NYC	4.8
Tokyo, Japan	3.05
Free space	2
Dhaka, India	2.98
Bangkok	3.6

These results show the macrocell path loss of Bangkok was similar to San Francisco, CA with Newark, N.J. being not far behind.

Next additional results were provided, these results show San Francisco, CA has a macrocell path loss that is close to Philadelphia, PA. Tokyo, Japan had the worse path loss from this group (see Fig. 3.49).

Next we list the distance exponent for various cities used in this comparison (see Table 3.5).

3.7 SHADOWING DISCUSSION

In this section we will discuss the third mechanism used to characterize the wireless channel, namely log-normal shadowing. Shadowing occurs when the LOS is obstructed between the BS and MS due to either large buildings, terrain, and so on. The presence of these obstructions vary the mean of the received signal that is highly dependent on the environment [91, 126, 127].

The change in the mean received signal can result due to variations in the building height. As we have seen earlier in the path loss sections, diffraction from the rooftop to ground level contributes to the path loss. So variations in the building heights will cause random variations in this diffraction loss to the MS. These variations are also due to different rooftop constructions (triangular versus flat top) and not to mention the occasional absence of a building (typically found in lots and intersections). This mean variation can be found by averaging the small scale fading over a few to many wavelengths. Recall the fast multpath (or small scale) fading occurs in fractions of a wavelength while the large scale (or slow) fading occurs over a few to many wavelengths.

The log-normal shadowing is expressed as

$$p(x) = \frac{1}{\sigma\sqrt{2\pi}} \cdot e^{-\frac{(x - x_m)^2}{2\sigma^2}} \tag{3.82}$$

where x = 10*log(X) measured in units of dB, x_m = area average measured in dB, X = the received signal level, σ = standard deviation measured in dB.

The typical values of the standard deviation (σ) of the log-normal shadowing is in the range of 5–12 dB.

This log-normal fading has been suggested to be a result of a cascade of events along the propagation path. Each event multiplies the signal by a randomly varying amount. Now when expressed in dBs, the product of random events becomes the sum of random variables, which approaches a Gaussian distribution [6].

Suburban environments would have a larger σ dues to the large variations in the environment, where as urban would have a lower σ, but recall the path loss would be higher for this urban environment.

We can address the question of how does the log-normal fading vary with carrier frequency. A relationship was fitted to Okumura's data and is given as follows for the macrocell case [128].

$$\sigma = 0.65 \cdot [\log(f_c)]^2 - 1.3 \cdot \log(f_c) + A \tag{3.83}$$

with

$$A = \begin{cases} 5.2 & \text{urban} \\ 6.6 & \text{suburban} \end{cases}$$

It is worth mentioning that the log-normal standard deviation difference between 900 MHz and 1.9 GHz has been shown to be approximately 1 dB. Figure 3.50 plots the above equation.

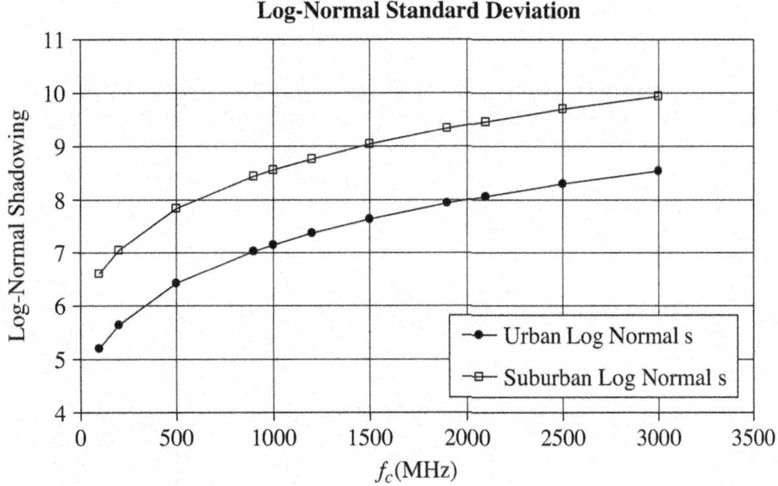

FIGURE 3.50 Log-normal shadowing dependency on carrier frequency.

3.7.1 Mathematical Representation

In this section we will present a method used to model the log-normal shadowing [129]. Here WGN source is low pass filtered by a first order low pass filter (LPF) (IIR structure).

$$x(k + 1) = \alpha \cdot x(k) + (1 - \alpha) \cdot n(k) \tag{3.84}$$

where $n(k)$ = the zero mean Gaussian random variable, α = low pass filter parameter.

The LPF parameter controls the spatial decorrelation length of the log-normal shadowing. Specifically we can write the autocorrelation of the shadowing signal as

$$R_{xx}(k) = \sigma^2 \cdot \alpha_d^{\frac{vT}{D}|k|} \tag{3.85}$$

where α_d = the correlation between two samples separated by a spatial distance of D meters assuming a vehicle speed of v m/sec, and T = simulation sampling interval.

For example, in a suburban environment σ = 7.5 dB with a correlation of approximately 0.82 at a distance of 100 m. This was for the 900 MHz frequency band. Now for 1.7 GHz, another value reported was σ = 4.3 dB with a correlation of approximately 0.3 at a distance of 10 m. These values are chosen to adjust the correlation of the generated shadowing signal. Typical decorrelation lengths are 5 m for urban and 300 m for suburban [91]. The application of the log-normal fading is shown in Fig. 3.51.

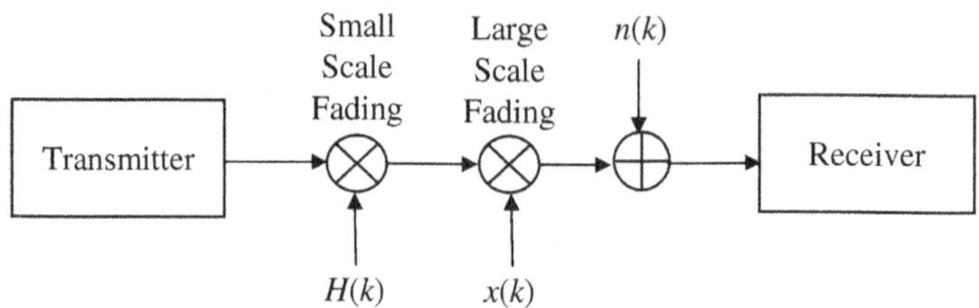

FIGURE 3.51 Channel model including log-normal shadowing.

The slow fading generation model is shown in Fig. 3.52.

A question remains as to how long does one average the received signal power to determine the local mean variations. Some work has been published showing keeping the averaging interval greater than 20 wavelengths is fine [130].

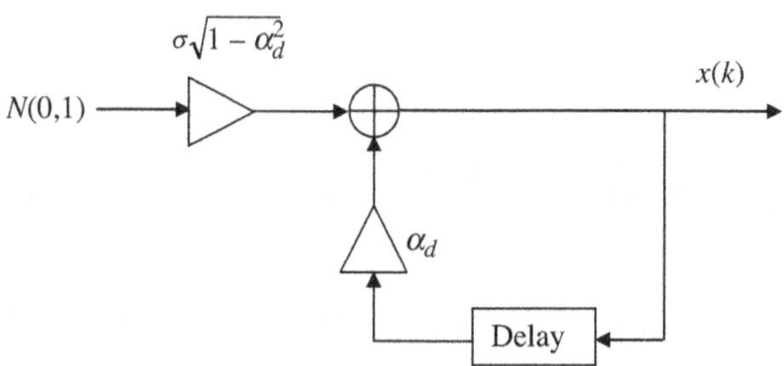

FIGURE 3.52 Log-normal shadowing generation block diagram.

3.8 MULTIPATH FADING SIMULATION MODELS

In this section we will provide a few methods that are typically used to simulate (and in some cases emulate in test equipment) the multipath, small scale fading. We will begin by discussing the classical Jakes model [1], next we present a modified Jakes model [130–134] to overcome potential short comings of the model in certain applications, and lastly we present a probabilistic model to generate the fading waveform.

Regardless of the model chosen the simulation methodology is shown in Fig. 3.53. The assumption in the figure is that the channel is frequency flat. The frequency selective fading models can be built using the flat fading model.

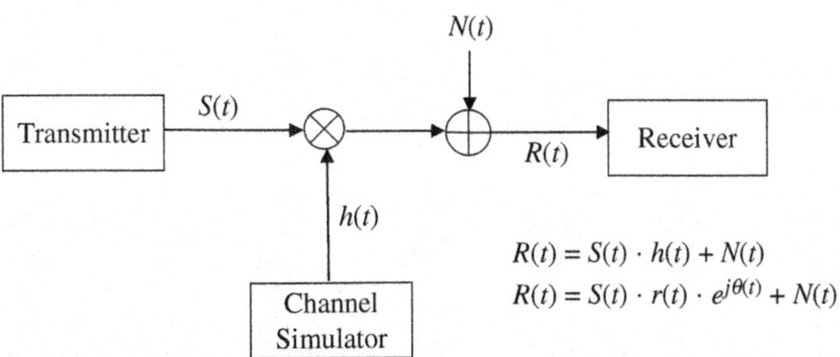

FIGURE 3.53 Simplified communication system.

3.8.1 Jakes Fading Model

The classical model used to generate multipath fading is the Jakes model [1]. This model assumes the local scatterers are uniformly distributed on a circle of constant radius around the mobile (see Fig. 3.54).

The Jakes model includes the effects of each arriving ray as a sinusoid with a certain frequency shift. This frequency shift is due to the Doppler principle. These shifts are due to the incoming rays

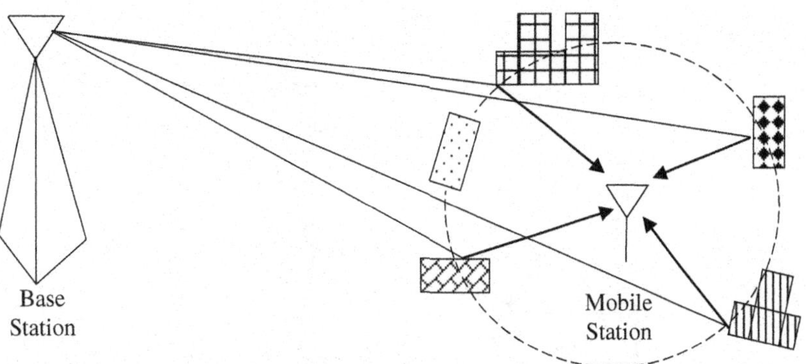

FIGURE 3.54 Classical Jakes multipath fading model.

having an angle of arrival with respect to the MS velocity vector. The circle of constant radius is an approximation of the surrounding scatterers/reflectors the MS would indeed see in reality, but its simplicity allows for robust and insightful channel models.

Mathematical Representation. The mathematical development begins with the superposition of plane waves expressed in the low pass equivalent form ($H(t) = h_I(t) + jh_Q(t)$).

$$H(t) = \frac{1}{\sqrt{N}} \sum_{n=1}^{N} e^{j(2\pi \cdot f_m \cdot t \cdot \cos(\alpha_n) + \phi_n)} \tag{3.86}$$

With N = number of paths, f_m = maximum Doppler phase offset, $\alpha_n = 2\pi n/N$.

Assuming $N/2$ is an odd number, the use of an omnidirectional antenna and the angle of arrivals are uniformly distributed, the complex envelop of a Rayleigh multipath fading channel is given as (with M denoting the number of frequency components)

$$h_I(t) = 2\sum_{n=1}^{M}\cos[\beta_n]\cdot\cos\left[2\pi f_n t\right] + \sqrt{2}\cos[\alpha]\cdot\cos\left[2\pi f_m t\right] \tag{3.87}$$

$$h_Q(t) = 2\sum_{n=1}^{M}\sin[\beta_n]\cdot\cos\left[2\pi f_n t\right] + \sqrt{2}\sin[\alpha]\cdot\cos\left[2\pi f_m t\right] \tag{3.88}$$

where the following variables are defined.

$$M = \frac{1}{2}\left(\frac{N}{2} - 1\right); \qquad \beta_n = \frac{\pi n}{M}; \qquad f_n = \frac{v}{\lambda}\cos\left[\frac{2\pi n}{M}\right]; \qquad f_m = \frac{v}{\lambda} \tag{3.89}$$

The Jakes Rayleigh fading channel model sums discrete sinusoids corresponding to discrete rays with differing angle of arrivals or Doppler frequencies. The block diagram is shown in Fig. 3.55.

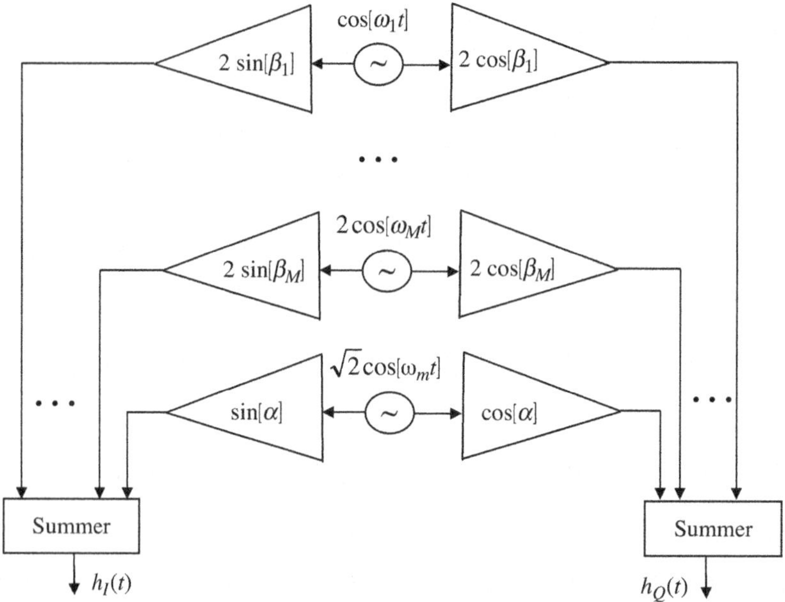

FIGURE 3.55 The classical Jakes multipath fading simulator generating the complex envelope of the multipath fading channel.

The phases β_n are chosen to reasonably approximate an aggregate phase with uniform distribution, whose value is $1/2\pi$. The Jakes model assumes N equal strength rays with uniformly distributed arrival angles.

The Jakes model can also be written as (in nonnormalized form)

$$H_k(t) = s\sum_{n=1}^{M}[\cos(\beta_n) + j\sin(\beta_n)]\cdot\cos(2\pi f_n\cdot\cos(\alpha_n t) + \theta_{nk}) + \sqrt{2}\cos(2\pi f_m t + \theta_{0k}) \tag{3.90}$$

With $\alpha_n = \frac{2\pi n}{M}$ and $\beta_n = \frac{\pi n}{M+1}$. Now to generate multiple waveforms we make use of the following phase offset for each of the kth multipath waveform.

$$\theta_{nk} = \frac{\pi n}{M+1} + \frac{2\pi}{M+1} \cdot (k-1) \tag{3.91}$$

The simulated channel should produce a reasonable approximation to the Rayleigh distribution. If M is very large then we can make use of the central limit theorem (CLT) to state that $H(t)$ is a complex Gaussian random process and so $|H(t)|$ is Rayleigh.

As discussed earlier in this chapter the autocorrelation of the generated multipath fading signal should closely approximate, $J_o(w_m t)$. It has been shown by many authors that this accuracy improves with increasing the number of sinusoids. A very good approximation can be had for M greater than 32 frequency sinusoids.

As the need for wideband channel modeling increases, it is important to be able to generate multipath faded signals that are uncorrelated. We can extend the Jakes model to create up to M multipath faded signals using the same sinusoids. Here the nth sinusoids is given an additional phase shift of $\gamma_{nj} + \beta_{nj}$ for ($j = 1, \ldots, M$). These values can be determined by imposing the additional requirement that the multipath faded signals are uncorrelated, (or nearly uncorrelated as possible). Additionally by using two quadrature sinusoids per offset, the use of the phase shifters to perform the $\gamma_{nj} + \beta_{nj}$ phase shift can be eliminated. The extended Jakes model is shown in Fig. 3.56.

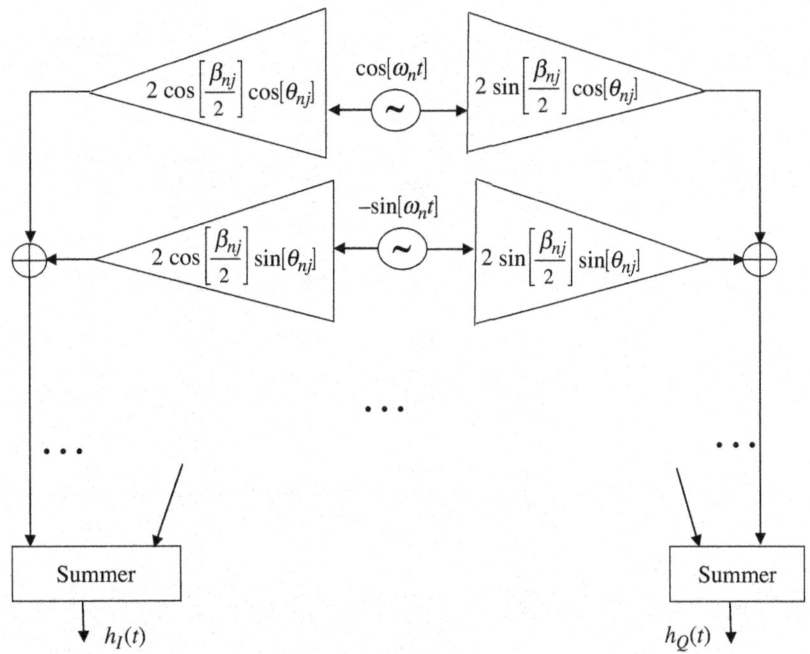

FIGURE 3.56 Extending the Jakes multipath fading model generating the complex envelope multipath channel signal.

With the above figure, use these definitions ($j = 1, \ldots, M$) and ($n = 1, \ldots, M$)

$$\theta_{nj} = \beta_{nj} + \gamma_{nj}; \qquad \beta_{nj} = \frac{\pi n}{M+1} \qquad \gamma_{nj} = \frac{2\pi(j-1)}{M+1} \tag{3.92}$$

In the presentation of the Jakes model we discussed how it can reasonably model the Rayleigh multipath fading time variation and the autocorrelation function. However, we did not discuss the cross-correlation of the different multipath faded signals. It is well known that the issue with the Jakes model is that there can be significant cross-correlation between the different generated signals [131, 132].

And when these correlated multipath signals are used to characterize a RAKE receiver demodulating many multipaths or a spatial diversity receiver, then the system BER results will be slightly worse than the truly uncorrelated model and thus bias results in a possibly conservative manner.

3.8.2 Modified Jakes Fading

In this section we will discuss a suggested method to modify the Jakes model to have the desired cross-correlation properties. These models will prove to be useful in FSF channels and antenna diversity receivers. We will briefly present a few modified Jakes models. First we will start with the one proposed by [135].

By adjusting the multipath arrival angles and the use of some special sequences we have

$$H_k(t) = \sqrt{\frac{2}{N_o}} \cdot \sum_{n=1}^{N_o} A_k(n) \cdot \{(\cos[\beta_n] + j\sin[\beta_n]) \cdot \cos[\omega_m \cdot \cos(\alpha_n t) + \theta_{nk}]\} \tag{3.93}$$

With θ_n = independent random phases that are uniformly distributed and the following:

$$N_o = 4; \qquad \beta_n = \frac{\pi n}{N_o} \qquad \alpha_n = \frac{2\pi n}{N_o} - \frac{\pi}{N_o} \tag{3.94}$$

In this model orthogonal functions (Walsh-Hadamard sequences, $A_k(n)$) weight the sinusoids before summing to generate the uncorrelated waveforms.

The next model we will review is the Li & Huang [136]. Here the complex envelope is written as

$$h_{Ik}(t) = 2\sum_{n=0}^{N_o-1} \cos[\omega_m \cdot \cos(\alpha_{nk}) + \phi^I_{nk}] \tag{3.95}$$

$$h_{Qk}(t) = 2\sum_{n=0}^{N_o-1} \sin[\omega_m \cdot \sin(\alpha_{nk}) + \phi^Q_{nk}] \tag{3.96}$$

where ϕ^I_{nk} and ϕ^Q_{nk} are uniformly distributed $[0, 2\pi]$ acting as noise seeds with $k = 0, \ldots M\text{-}1$, $\alpha_{nk}=$ the nth arrival angle of the kth fader.

$$N_o = \frac{N}{2}; \qquad \omega_{nk} = \omega_m \cdot \cos[\alpha_{nk}]; \qquad \alpha_{nk} = \frac{2\pi n}{N} + \frac{2\pi k}{MN} + \alpha_{oo} \tag{3.97}$$

The last model is the [137] model. Here he builds on the [135] model using two Walsh-Hadamard sequences in the following manner.

$$H_k(t) = \sqrt{\frac{1}{N_o}} \sum_{n=0}^{N_o-1} (A_{k1}(n) + jA_{k2}(n)) \cdot \cos[\omega_m \cdot \cos(\alpha_{nk}) + \theta_n] \tag{3.98}$$

With A_1 and A_2 being the different orthogonal weighting functions and the following phases

$$\alpha_{nk} = \frac{2\pi n}{N} + \frac{2\pi k}{MN} \tag{3.99}$$

All of the above modifications are successful in reducing the cross-correlation among different multipaths. Placing implementation complexity aside for the moment, any of the above-mentioned methods proves to be useful.

3.8.3 Low Pass Filtering of WGN

In this subsection we will discuss another method used to generate Rayleigh multipath fading signals. Here we begin with a complex Gaussian noise source that has a zero mean. It is well known that the magnitude of such a complex signal produces a Rayleigh distribution while its phase is uniformly distributed. In order to introduce the presence of a maximum Doppler spread, we must pass this noise source through a low pass filter whose cutoff frequency is equal to the maximum expected Doppler shift. Note the insertion of an LPF introduces correlation between adjacent samples. A block diagram is shown in Fig. 3.57.

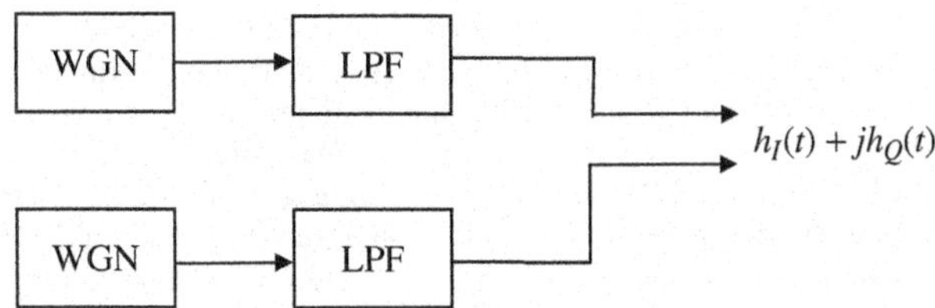

FIGURE 3.57 Low pass filtering approach to generate Rayleigh multipath fading.

As shown earlier in this chapter, ideally the LPF will have a pole or asymptote at the maximum Doppler frequency when considering the Doppler spectrum, $S(f)$.

The choice of the LPF is critical in order to obtain the desired autocorrelation properties discussed above. An example of an LPF using an finite impulse response (FIR) structure is given in Fig. 3.58.

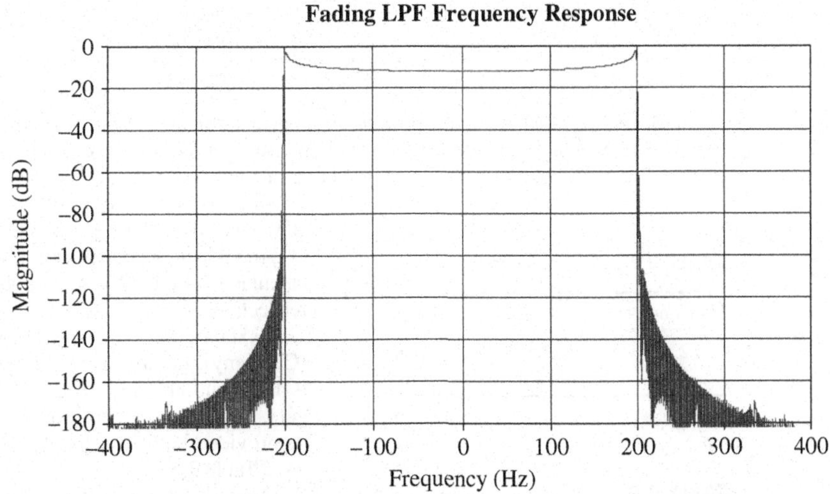

FIGURE 3.58 LPF response of the WGN filters.

Notice there is a peaking behavior around the maximum Doppler frequency in order to approximate the ideal power spectral density (PSD) response given earlier (see Eq. 3.51).

Note, for sake of simplicity, we have chosen to present the FIR structure of the LPF. Certainly the infinite impulse response (IIR) structures can be used and in all cases reduce the simulation run time because of the reduced number of taps required in representing the desired multipath frequency response.

3.9 MULTIPATH BIRTH/DEATH DISCUSSION

In the above sections we have discussed the phenomenon that occurs when the transmission BW is larger than the channel CBW, namely FSF. For this wideband channel we have presented measured results and quantified the time dispersion or delay spread. Since the channel is a random variable, various statistics were presented to describe the PDP. In reality this PDP changes and an important parameter to discuss is the average lifetime of a multipath. In a related issue, we wish to better understand the actual birth and death (B/D) of the multipaths that we have used to calculate the delay spread. In other words the CBW varies as a function of time [138–141].

The approach we have taken was to discuss the multipath B/D from the perspective of a DS-CDMA system, where each finger in the RAKE receiver is actively demodulating a distinguishable or noticeable multipath. Specifically we will discuss the finger lifetime, where a finger is defined as having its mean power within 10 or 15 dB of the mean power of the peak path.

We will now discuss some of the publicly available data for finger (or multipath) lifetime [138]. First some results were measured in Helsinki, Finland at a carrier frequency of 2.154 GHz. These results showed that the second RAKE finger was always used along the specific measurement route taken. The CDF of the number of finger distances was presented and we will summarize their results for the 90% (assuming fingers used are within 10 dB of the peak value), urban environment (see Table 3.6).

TABLE 3.6 Spatial Lifetime of Finger Tracking Multipaths (Finland)

Finger number	BW = 5 MHz
2	35 m
3	6 m

From these values we can suggest the finger lifetimes assuming a $v = 30$ km/hr vehicle speed and thus arriving to the following observations. Here we translate the spatial lifetime of a finger into a temporal lifetime to gain some insight.

$$\text{Lifetime of finger \# 2} = \frac{(35\,\text{m})(60)(60)}{30{,}000} = 4.2\ \text{sec}$$

$$\text{Lifetime of finger \# 3} = \frac{(6\,\text{m})(60)(60)}{30{,}000} = 0.72\ \text{sec}$$

In using this translation we can state the finger/multipath lifetime is dependent on vehicle speed. The faster the vehicle speed, the shorter the multipath lifetime. Another observation is that the lifetime of finger number three is shorter than that of finger two.

Other results indicated the following for a 5 MHz BW for CDF = 90% (see Table 3.7).

TABLE 3.7 Lifetime Distance

Finger number	Route A	Route B
2	3.5 m	12 m
3	1 m	3 m

And for the second RAKE finger, the lifetimes were calculate to be 0.42 and 1.44 sec for routes A and B, respectively, for $v = 30$ km/hr.

Next based on data from Seoul, Korea, and Munich (Germany), the average lifetime of the second finger is 15 m and about 6.5 m for the third and fourth fingers, assuming a 10 dB threshold. The 90% percentile is provided below for the 900 MHz carrier frequency (see Table 3.8).

TABLE 3.8 Lifetime Distance as a Function of the Threshold (Korea and Germany)

	Threshold	
Finger number	10 dB	15 dB
2	65 m	90 m
3	27 m	40 m
4	13 m	21 m

Now that we have the lifetime that there are N or more fingers above the threshold, let us convert the distance values into time.

Consider the lifetime distance of 15 m and a vehicle speed of $v = 30$ km/h, this gives us an average lifetime of

$$\text{lifetime} = \frac{\text{distance}}{\text{velocity}} \cong 1.8\ \text{sec}$$

As you would expect from the above, equations if we increased the vehicle speed to 100 and 250 km/hr, then the lifetime would become 540 and 216 msec, respectively.

In comparing these results to those represented earlier, it can be seen that latter multipath lifetime values measured in Korea/Germany are larger than those measured in Helsinki. As we will discuss this important point in the later chapters, it is worthy of a short discussion in this chapter.

The point in this topic is that there is a lifetime of a multipath and it varies from one geographic location (country) to another. This imposes a great design constraint for CDMA terminals. They must be able to operate in any (or close to) environment it encounters in real life. We will see how important a multipath searcher becomes to combat the birth and death of multipaths.

REFERENCES

[1] W. C. Jakes, *Microwave Mobile Communications*, IEEE Press, 1993, New York.

[2] P. A. Bello, "Measurement of Random Time-Variant Linear Channels," *IEEE Transactions on Information Theory*, Vol. IT-15, No. 4, July 1969, pp. 469–475.

[3] R. H. Clarke, "A Statistical Theory on Mobile-Radio Reception," *The Bell System Technical Journal*, July-Aug. 1968, pp. 957–1000.

[4] S. O. Rice, "Statistical Properties of a Sine Wave Plus Noise," *Bell System Technical Journal*, Vol. 27, Jan. 1948, pp. 109–157.

[5] S. O. Rice, "Mathematical Analysis of Random Noise," *Bell System Technical Journal*, Vol. 23, July 1944, pp. 282–332.

[6] A. Papoulis and S. U. Pillai, *Probability, Random Variables and Stochastic Processes*, McGraw Hill, 2002, New York.

[7] W. C. Y. Lee, *Mobile Communication Engineering: Theory and Applications*, McGraw-Hill, 1998, New York.

[8] J. Toftgard, S. N. Hornsleth, and J. B. Andersen, "Effects on Portable Antennas on the Presence of a Person," *IEEE Transactions on Antennas and Propagation*, Vol. 41, No. 6, June 1993, pp. 739–746.

[9] W. C. Y. Lee, "Effect on Correlation between Two Mobile Radio Base-Station Antennas," *IEEE Transactions on Communications*, Vol. 21, No. 11, Nov. 1973, pp. 1214–1224.

[10] G. L. Stuber, *Principles of Mobile Communication*, Kluwer Academic Publishers, 1996, Massachusetts.

[11] N. C. Beaulieu and X. Dong, "Level Crossing Rate and Average Fade Duration of MRC and EGC Diversity in Rician Fading," *IEEE Transactions on Communications*, Vol. 51, No. 5, May 2003, pp. 722–726.

[12] A. Abdi, K. Wills, H. A. Barger, M. S. Alouini and M. Kaveh, "Comparison of the Level Crossing Rate and Average Fade Duration of Rayleigh, Rice and Nakagami Fading Models with Mobile Channel Data," *IEEE VTC* 2000, pp. 1850–1857.

[13] K. Feher, *Wireless Digital Communications: Modulation & Spread Spectrum Applications*, Prentice Hall, 1995, New Jersey.

[14] J. C-I. Chuang, "The Effects of Time Delay Spread on Portable Radio Communications Channels with Digital Modulation," *IEEE Journal on Selected Areas in Communications*, Vol. SAC-5, No. 5, June 1987, pp. 879–889.

[15] T. S. Rappaport, "Indoor Radio Communications for Factories of the Future," *IEEE Communications Magazine*, May 1989, pp. 15–24.

[16] D. M. J. Devasirvatham, "Multipath Time Delay Spread in the Digital Portable Radio Environment," *IEEE Communications Magazine*, June 1987, Vol. 25, No. 6, pp. 13–21.

[17] T. S. Rappaport, *Wireless Communications: Principles and Practice*, Prentice Hall, 1996, New Jersey.

[18] J. G. Proakis, *Digital Communications*, McGraw-Hill, 1989, New York.

[19] J. M. Keenan and A. J. Motley, "Radio Coverage in Buildings," *Br. Telecom Technical Journal*, Vol. 8, No. 1, Jan. 1990, pp. 19–24.

[20] M. R. Heath, "Propagation Measurements at 1.76 GHz for Digital European Cordless Telecommunications," *IEEE* Global Telecommunications Conference, 1990, pp. 1007–1012.

[21] H-J. Zepernick and T. A. Wysocki, "Multipath Channel Parameters for the Indoor Radio at 2.4 Ghz ISM Band," *IEEE*, 1999, pp. 190–193.

[22] D. M. J. Devasirvatham, R. R. Murray, and C. Banerjee, "Time Delay Spread Measurements at 850 MHz and 1.7 GHz inside a Metropolitan Office Building," *Electronic Letters*, Vol. 25, No. 3, Feb. 1989, pp. 194–196.

[23] L. Talbi and G. Y. Delisle, "Comparison of Indoor Propagation Channel Characteristics at 893 MHz and 37.2 GHz," *IEEE Vehicular Technical Conference 2000*, pp. 689–694.

[24] P. Nobles and F. Halsall, "Delay Spread and Received Power Measurements within a Building at 2 GHz, 5 GHz and 17 GHz," *IEEE*, 1997, *10th International Conference on Antenna and Propagation*, No. 436, pp. 2.319–2.324.

[25] H. Hashemi, "The Indoor Radio Propagation Channel," *IEEE Proceedings*, Vol. 81, No. 7, July 1993, pp. 943–968.

[26] H. Hashemi and D. Tholl, "Statistical Modeling and Simulation of the RMS Delay Spread of Indoor Radio Propagation Channels," *IEEE Transactions on Vehicular Technology*, Vol. 43, No. 1, Feb. 1994, pp. 110–120.

[27] C. Bergljung and P. Karlsson, "Propagation Characteristics for Indoor Broadband Radio Access Networks in the 5 GHz Band," *IEEE*, 1998, pp. 612–616.

[28] R. J. C. Bultitude, R. F. Hahn, and R. J. Davies, "Propagation Considerations for the Design of an Indoor Broad-Band Communications System at EHF," *IEEE Transactions on Vehicular Technology*, Vol. 47, No. 1, Feb. 1998, pp. 235–245.

[29] A. Chandra and A. Kumar and P. Chandra, "Comparative Study of Path Losses from Propagation Measurements at 450 MHz, 900 MHz, 1.36 GHz and 1.89 GHz in the corridors of a Multifloor Laboratory-cum-Office Building," *IEEE VTC* 1999, pp. 2272–2276.

[30] A. Affandi, G. El Zein, and J. Citerne, "Investigation of Frequency Dependence of Indoor Radio Propagation Parameters," *IEEE VTC* 1999, pp. 1988–1992.

[31] D. M. J. Devasirvatham, "A Comparison of Time Delay Spread and Signal Level Measurements Within Two Dissimilar Office Buildings," *IEEE Transactions on Antennas and Propagation*, Vol. AP-35, No. 3, March 1987, pp. 319–324.

[32] D. M. J. Devasirvatham, "Multipath Time Delay Jitter Measured at 850 MHz in the Portable Radio Environment," *IEEE Journal on Selected Areas in Communications*, Vol. SAC-5, No. 5, June 1987, pp. 855–861.

[33] D. M. J. Devasirvatham, "Time Delay Spread and Signal Level Measurements of 850 Mz Radio Waves in Building Environments," *IEEE Transactions on Antennas and Propagation*, Vol. AP-34, No. 11, Nov. 1986, pp. 1300–1305.

[34] D. M. J. Devasirvatham, M. J. Krain, D. A. Rappaport, and C. Banerjee, "Radio Propagation Measurements at 850 MHz, 1.7 GHz, and 4 GHz inside Two Dissimilar Office Buildings," *Electronic Letters*, Vol. 26, No. 7, March 1990, pp. 445–447.

[35] J. Medbo, H. Hallenberg, and J-E. Berg, "Propagation Characteristics at 5 GHz in Typical Radio-LAN Scenarios," *IEEE*, 1999, pp. 185–189.

[36] S. Y. Seidel, T. S. Rappaport, M. J. Feuerstein, K. L. Blackard, and L. Grindstaff, "The Impact of Surrounding Buildings on Propagation for Wireless In-Building Personal Communications System Design," *IEEE* 1992, pp. 814–818.

[37] F. Lotse, J-E. Berg, and R. Bownds, "Indoor Propagation Measurements at 900 MHz," *IEEE*, 1993, pp. 629–632.

[38] T. S. Rappaport, "Characterization of UHF Multipath Radio Channels in Factory Buildings," *IEEE Transactions on Antennas and Propagation*, Vol. 37, No. 8, Aug. 1989, pp. 1058–1069.

[39] T. S. Rappaport and C. D. McGillem, "UHF Multipath and Propagation Measurements in Manufacturing Environments," *IEEE*, 1988, pp. 825–831.

[40] T. S. Rappaport, S. Y. Seidel, and K. Takamizawa, "Statistical Channel Impulse Response Models for Factory and Open Plan Building Radio Communication System Design," *IEEE Transactions on Communications*, Vol. 39, No. 5, May 1991, pp. 794–807.

[41] T. S. Rappaport and C. D. McGillem, "UHF Fading in Factories," *IEEE Journal on Selected Areas in Communications*, Vol. 7, No. 1, Jan. 1989, pp. 40–48.

[42] R. Ganesh and K. Pahlavan, "Statistical Modeling and Computer Simulation of Indoor Radio Channel," *IEEE Proceedings*, Vol. 138, No. 3, June 1991, pp. 153–161.

[43] R. Ganesh and K. Pahlavan, "On the Modeling of Fading Multipath Indoor Radio Channels," *IEEE Global Telecommunications Conference*, 1989, pp. 1346–1350.

[44] A. A. M. Saleh and R. A. Valenzuela, "A Statistical Model for Indoor Multipath Propagation," *IEEE Journal on Selected Areas in Communications*, Vol. SAC-5, No. 2, Feb. 1987, pp. 128–137.

[45] R. J. C. Bultitude, P. Melancon, H. Zaghloul, G. Morrison, and M. Prokki, "The Dependence of Indoor Radio Channel Multipath Characteristics on Transmit/Receive Ranges," *IEEE Journal on Selected Areas in Communications*, Vol. 11, No. 7, Sept. 1993, pp. 979–990.

[46] H. Hashemi, "Impulse Response Modeling of Indoor Radio Propagation Channels," *IEEE Journal on Selected Areas in Communications*, Vol. 11, No.7, Sept. 1993, pp. 967–978.

[47] P. F. M. Smulders and A. G. Wagemans, "Wideband Indoor Radio Propagation Measurements at 58 GHz," *Electronic Letters*, Vol. 28, No. 13, June 1992, pp. 1270–1272.

[48] G. J. M. Janssen, P. A. Stigter and R. Prasad, "Wideband Indoor Channel Measurements and BER Analysis of Frequency Selective Multipath Channel at 2.4, 4.75, and 11.5 GHz," *IEEE Transactions on Communications*, Vol. 44, No. 10, Oct. 1996, pp. 1272–1288.

[49] M. R. Williamson, G. E. Athanasiadou, and A. R. Nix, "Investigating the Effects on Antenna Directivity on Wireless Indoor Communication at 60 GHz," *IEEE*, 1997, pp. 635–639.

[50] S-C. Kim, H. L. Bertoni, and M. Stern, "Pulse Propagation Characteristics at 2.4 GHz Inside Buildings," *IEEE Transactions on Vehicular Technology*, Vol. 45, No. 3, Aug. 1996, pp. 579–592.

[51] A. Bohdanowicz, G. J. M. Janssen, and S. Pietrzyk, "Wideband Indoor and Outdoor Multipath Channel Measurements at 17 GHz," *IEEE Vehicular Technology Conference 1999*, pp. 1998–2003.

[52] P. Hafezi, D. Wedge, M. A. Beach, and M. Lawton, "Propagation Measurements at 5.2 GHz in Commercial and Domestic Environments," *IEEE*, 1997, pp. 509–513.

[53] A. F. AbouRaddy, S. M. Elnoubi, and A. El-Shafei, "Wideband Measurements and Modeling of the Indoor Radio Channel at 10 GHz Part II: Time Domain Analysis," *National Radio Science Conference*, Feb. 1998, pp. B14.1–B14.8.

[54] J-H. Park, Y. Kim, Y-S. Hur, K. Lim, and K-H. Kim, "Analysis of 60 GHz Band Indoor Wireless Channels with Channel Configurations," *IEEE*, 1998, pp. 617–620.

[55] A. M. D. Turkmani and A. F. de Toledo, "Radio Transmission at 1800 MHz into and within Multistory Buildings," *IEEE Proceedings*, Vol. 138, No. 6, Dec. 1991, pp. 577–584.

[56] A. A. Arowojolu, A. M. D. Turkmani, and J. D. Parson, "Time Dispersion Measurements in Urban Microcellular Environments," *IEEE* Vehicular Technology Conference, 1994, pp. 150–154.

[57] P. Nobles, D. Ashworth, and F. Halsall, "Propagation Measurements in an Indoor Radio Environment at 2, 5, and 17 GHz," *IEEE* Colloquium on High Bit Rate UHF/SHF Channel Sounders—Technology and Measurement, 1993, pp. 4.1–4.6.

[58] P. Nobles, D. Ashworth, and F. Halsall, "Indoor Radiowave Propagation Measurements at Frequencies up to 20 GHz," *IEEE* Vehicular Technology Conference, 1994, pp. 873–877.

[59] P. Nobles, and F. Halsall, "Delay Spread Measurements within a Building at 2 GHz, 5 GHz, and 17 GHz," *IEEE* Propagation Aspects of Future Mobile Systems, 1996, pp. 8.1–8.6.

[60] D. A. Hawbaker and T. S. Rappaport, "Indoor Wideband Radiowave Propagation Measurements at 1.3 GHz and 4 GHz," *Electronic Letters*, Vol. 26, No. 21, Oct. 1990, pp. 1800–1802.

[61] T. Manabe, Y. Miura, and T. Ihara, "Effects of Antenna Directivity, and Polarization on Indoor Multipath Propagation Characteristics at 60 GHz," *IEEE Journal on Selected Areas in Communications*, Vol. 14, No. 3, April 1996, pp. 441–448.

[62] S. Sumei, "Broadband Measurements of Indoor Wireless LAN Channels at 18 GHz Using Directive Antennas," International Conference on Communication Technology, 1996, pp. 474–477.

[63] S. P. T. Kumar, B. Farhang-Boronjeny, S. Uysal, and C. S. Ng, "Microwave Indoor Radio Propagation Measurements and Modeling at 5 GHz for Future Wireless LAN Systems," Asia Pacific Microwave Conference (APMC 1999), pp. 606–609.

[64] L. Talbi, "Experimental Comparison of Indoor UHF and EHF Radio Channel Characteristics," Canadian Conference on Electrical and Computer Engineering, 2000, pp. 265–269.

[65] L. Talbi, "Effect of Frequency Carrier on Indoor Propagation Channel," *Electronic Letters*, Vol. 36, No. 15, July 2000, pp. 1309–1310.

[66] L. Talbi and G. Y. Delisle, "Experimental Characterization of EHF Multipath Indoor Radio Channels," *IEEE Journal on Selected Areas in Communications*, Vol. 14, No. 3, April 1996, pp. 431–440.

[67] G. Ke, Q. Jianzhong, and Z. Yuping, "Measurement of UHF Delay Spread in Building," International conference on Microwave and Millimeter Wave Technology, 1998, pp. 178–180.

[68] D. C. Cox and R. P. Leck, "Distributions of Multipath Delay Spread and Average Excess Delay for 910 MHz Urban Mobile Radio Paths," *IEEE Transactions on Antenna and Propagation*, Vol. AP-23, No. 2, March 1975, pp. 206–213.

[69] A. Yamaguchi, K. Suwa, and R. Kawasaki, "Received Signal Level Characteristics for Wideband Radio Channel in Microcells," IEEE Personal, Indoor and Mobile Radio Communications Conference, 1995, pp. 1367–1371.

[70] D. M. J. Devasirvatham and R. R. Murray, "Time Delay Spread Measurements at Two Frequencies in a Small City," IEEE Military Communications Conference, 1995, pp. 942–946.

[71] D. M. J. Devasirvatham, "Radio Propagation Studies in a Small City for Universal Portable Communications," IEEE Vehicular Technology Conference, 1988, pp. 100–104.

[72] D. M. J. Devasirvatham, R. R. Murray, and D. R. Wolter, "Time Delay Spread Measurements in a Wireless Local Loop Test Bed," IEEE Vehicular Technology Conference, 1995, pp. 241–245.

[73] L. J. Greenstein, V. Erceg, Y. S. Yeh, and M. V. Clark, "A New Path-Gain/Delay-Spread Propagation Model for Digital Cellular Channels," *IEEE Transactions on Vehicular Technology*, Vol. 46, No. 2, May 1997, pp. 477–485.

[74] T. S. Rappaport and S. Y. Seidel, "900 MHz Multipath Propagation Measurements in Four United States Cities," *Electronic Letters*, Vol. 25, No. 15, July 1989, pp. 956–958.

[75] D. C. Cox, "Delay Doppler Characteristics of Multipath Propagation at 910 MHz in a Suburban Mobile Radio Environment," *IEEE Transactions on Antenna and Propagation*, Vol. AP-20, No. 5, Sept. 1972, pp. 625–635.

[76] C. Cheon, H. L. Bertoni, and G. Liang, "Monte Carlo Simulation of Delay and Angle Spread in Different Building Environments," IEEE Vehicular Technology Conference, 2000, pp. 49–56.

[77] H. L. Bertoni, P. Pongsilamanee, C. Cheon, and G. Liang, "Sources and Statistics of Multipath Arrival at Elevated Base Station Antenna," IEEE Vehicular Technology Conference, 1999, pp. 581–585.

[78] G. T. Martin and M. Faulkner, "Delay Spread Measurements at 1890 MHz in Pedestrian Areas of the Central Business District in the City of Melbourne," IEEE Vehicular Technology Conference, 1994, pp. 145–149.

[79] J. P. de Weck, P. Merki, and R. W. Lorenz, "Power Delay Profiles Measured in Mountainous Terrain," IEEE Vehicular Technology Conference, 1988, pp. 105–112.

[80] J. A. Wepman, J. R. Hoffman, and L. H. Lowe, "Characterization of Macrocellular PCS Propagation Channels in the 1850–1990 MHz Band," International Conference on Universal Personal Communications, 1994, pp. 165–170.

[81] E. S. Sousa, V. M. Jovanovic, and C. Daigneault, "Delay Spread Measurements for the Digital Cellular Channel in Toronto," IEEE Trans. On Vehicular Tech., 1992, pp. 3.4.1–3.4.6.

[82] D. C. Cox and R. P. Leck, "Correlation Bandwidth and Delay Spread Multipath Propagation Statistics for 910 MHz Urban Mobile Radio Channels," *IEEE Transactions on Communications*, Vol. COM-23, No. 11, Nov. 1975, pp. 1271–1280.

[83] G. T. Martin and M. Faulkner, "Wide Band PCS Propagation Measurements in Four Australian Cities," *IEEE 10th International Conf. on Antennas and Propagations*, April 1997, pp. 2.199–2.202.

[84] D. C. Cox, "910 MHz Urban Mobile Radio Propagation: Multipath Characteristics in New York City," *IEEE Transactions on Communications*, Vol. COM-21, No. 11, Nov. 1973, pp. 1188–1194.

[85] R. J. C. Bultitude and G. K. Bedal, "Propagation Characteristics on Microcellular Urban Mobile Radio Channels at 910 MHz," *IEEE Journal on Selected Areas in Communications*, Vol. 7, No. 1, Jan. 1989, pp. 31–39.

[86] R. J. C. Bultitude and G. K. Bedal, "Propagation Characteristics on Microcellular Urban Mobile Radio Channels at 910 MHz," *IEEE Journal on Selected Ares in Communications*, 1988, pp. 152–160.

[87] S. Kozono and A. Taguchi, "Mobile Propagation Loss and Delay Spread Characteristics with a Low Base Station Antenna on an Urban Road," *IEEE Transactions on Vehicular Technology*, Vol. 42, No. 1, Feb. 1993, pp. 103–109.

[88] K. Siwiak and H. L. Bertoni, and S, M. Yano, "Relation between Multipath and Wave Propagation Attenuation," *Electronics Letters*, Vol. 39, No. 1, Jan. 2003, pp 142–143.

[89] S. Y. Seidel, R. Singh, and T. S. Rappaport, "Path Loss and Multipath Delay Statistics in Four European Cities for 900 MHz Cellular and Microcellular Communications," *Electronic Letters*, Aug. 1990, Vol. 26, Issue 20, pp. 1713-1715.

[90] T. S. Rappaport, S. Y. Seidel, and R. Singh, "900-MHz Multipath Propagation Measurements for U.S. Digital Cellular Radiotelephone," *IEEE Transaction on Vehicular Technology*, Vol. 39, No. 2, May 1990, pp. 132–139.

[91] H. L. Bertoni, *Radio Propagation for Modern Wireless Systems*, Prentice Hall, 2000, New Jersey.

[92] A. A. M. Saleh and R. A. Valenzuela, "A Statistical Model for Indoor Multipath Propagation," *IEEE Journal on Selected Areas in Communications*, Vol. SAC-5, No. 2, Feb. 1987, pp. 128–137.

[93] J. Liberti, Jr. and T. S. Rappaport, *Smart Antennas for Wireless Communications*, Prentice Hall, 1999, New Jersey.

[94] T-S. Chu and L. J. Greenstein, "A Quantification of Link Budget Differences between the Cellular and PCS Bands," *IEEE Transactions on Vehicular Technology*, Vol. 48, No. 1, Jan. 1999, pp. 60–65.

[95] A. Bahai, M. V. Clark, V. Erceg, L. J. Greenstein, and A. Kasturia, "Link Reliability for IS-54/136 Handsets with Difference Receiver Structures," *IEEE Transactions on Vehicular Technology*, Vol. 48, No. 1, Jan. 1999, pp. 213–223.

[96] Y. Okumura et al., "Field Strength and Its Variability in UHF and VHF Land Mobile Radio Service," Review of the Electrical Communication Laboratory, Vol. 16, 1968, pp. 825–873.

[97] M. Hata, "Empirical Formula for Propagation Loss in Land Mobile Radio Services," *IEEE Transactions on Vehicular Technology*, Vol. VT-29, No. 3, Aug. 1980, pp. 317–325.

[98] C. Chrysanthou and H. L. Bertoni, "Variability of Sector Averaged Signals for UHF Propagation in Cities," *IEEE Transactions on Vehicular Technology*, Vol. 39, No. 4, Nov. 1990, pp. 352–358.

[99] J. Walfisch and H. L. Bertoni, "A Theoretical Model of UHF Propagation in Urban Environments," *IEEE Transactions on Antennas and Propagation*, Vol. 36, No. 12, Dec. 1988, pp. 1788–1796.

[100] T. K. Sarkar, Z. Ji, K. Kim, A. Medouri, and M. Salazar-Palma, "A Survey of Various Propagation Models for Mobile Communication," *IEEE Antennas and Propagation Magazine*, Vol. 45, No. 3, June 2003, pp. 51–81.

[101] P. E. Mogensen, P. Eggers, C. Jensen, and J. B. Anderson, "Urban Area Radio Propagation Measurements at 955 and 1845 MHz for Small and Micro Cells," *GLOBECOM* 1991, pp. 1297–1302.

[102] M. J. Feuerstein, K. L. Blackard, T. S. Rappaport, S. Y Seidel and H. H. Xia, "Path Loss, Delay Spread and Outage Models as Functions of Antenna Height for Microcellular System Design," *IEEE Transactions on Vehicular Technology*, Vol. 43, No. 3, Aug. 1994, pp. 487–498.

[103] R. Grosskopf, "Prediction of Urban Propagation Loss," IEEE Transactions on Antennas and Propagation, Vol. 42, No. 5, May 1994, pp. 658–665.

[104] A. J. Dagen, A. Iskandar, C. Banerjee, G. Nease, J. Vancraeynest, N. Chan, and P. Campanella, "Results of Propagation Loss and Power Delay Profile Measurements in an Urban Microcellular Environment," IEEE Personal, Indoor, and Mobile Radio Communications Conference, 1992, pp. 417–425.

[105] D. M. J. Devasirvatham, V. Banerjee, R. R. Murray, and D. A. Rappaport, "Two-Frequency Radiowave Propagation Measurements in Brooklyn," International Conference on Universal Personal Communications, 1992, pp. 23–27.

[106] J. P. P. do Carmo, "Behavior of a Mobile Communication Link in an Urban Environment: A Test Case," *International Crimean Conference Microwave and Telecommunication Technology, 2002*, pp. 225–228.

[107] K. Taira, S. Sekizawa, G. Wu, H. Harada, and Y. Hase, "Propagation Loss Characteristics for Microcellular Mobile Communications in Microwave Band," International Conference on Universal Personal Communications, 1996, pp. 842–846.

[108] J. H. Whitteker, "Measurements of Path Loss at 910 MHz for Proposed Microcell Urban Mobile Systems," *IEEE Transactions on Vehicular Technology*, Vol. 37, No. 3, Aug. 1988. pp. 125–129.

[109] A. J. Rustako, Jr., V. Erceg, R. S. Roman, T. M. Willis, and J. Ling, "Measurements of Microcellular Propagation Loss at 6 GHz and 2 GHz Over Non-Line-of-Sight Paths in the City of Boston," *IEEE GLOBECOM*, 1995, pp. 758–763.

[110] S. Aguirre, K. C. Allen, and M. G. Laflin, "Signal Strength Measurements at 915 MHz and 1920 MHz in an Outdoor Microcell Environment," International Conference on Universal Personal Communications, 1992, pp. 16–22.

[111] R. C. Bernhardt, "The Effect of Path Loss Models on the Simulated Performance of Portable Radio Systems," *IEEE GLOBECOM*, 1989, pp. 1356–1360.

[112] A. M. D. Turkmani and A. F. de Toledo, "Characterization of Radio Transmissions Into and Within Buildings," International Conference on Antennas and Propagation, 1993, pp. 138–141.

[113] S. Todd, M. E.Tanany, G. Kalivas, and S. Mahmoud, "Indoor Radio Path Loss Comparison Between the 1.7 GHz and 37 GHz Bands," *IEEE ICUPC* 1993, pp. 621–625.

[114] B. Y. Hanci and I. H. Cavdar, "Mobile Radio Propagation Measurements and Tuning the Path Loss Model in Urban Areas at GSM-900 Band in Istanbul-Turkey," IEEE Vehicular Technology Conference, 2004, pp. 139–143.

[115] A. B. M. S. Hossain and R. Ali, "Propagation-Path Losses Characterization for 900 MHz Cellular Communication in Dhaka City," International Conference on Communications, 2003, pp. 6–8.

[116] Q. Cao and M. Zhang, "Mobile Radio Propagation Studies at 900 MHz in Suburban Beijing," International Symposium on Antennas, Propagation, and EM Theory, 2003, pp. 549–552.

[117] M. E. Hughes, W. J. Tanis II, A. Jalan, and M. Kibria, "Narrowband Propagation Characteristics of 880 MHz and 1922 MHz Radio Waves in Macrocellular Environments," IEEE ICUPC Conference 2003, pp. 610–615.

[118] L. Melin, M. Ronnlund, and R. Angbratt, "Radio Wave Propagation A Comparison Between 900 and 1800 MHz," *IEEE*, 1993, pp. 250–252.

[119] E. Green, "Path Loss and Signal Variability Analysis for Microcells,"International Conference on Mobile Radio and Personal Communications, 1989, pp. 38–42.

[120] G-S. Bae and H-K. Son, "A Study of 2.3 GHz Bands Propagation Characteristic Measured in Korea," IEEE Vehicular Technology Conference, 2003, pp. 995–998.

[121] L. Piazzi and H. L. Bertoni, "Achievable Accuracy of Site-Specific Path-Loss Predictions in Residential Environments," *IEEE Transactions on Vehicular Technology*, Vol. 48, No. 3, May 1999, pp. 922–930.

[122] H. L. Bertoni, W. Honcharenko, L. R. Maciel, and H. H. Xia, "UHF Propagation Prediction for Wireless Personal Communications," *Proceedings of the IEEE*, Vol. 82, No. 9, Sept. 1994, pp. 1333–1359.

[123] V. Erceg, S. Ghassemzadeh, M. Taylor, D. Li, and D. L. Schilling, "Urban/Suburban Out-of-Sight Propagation Modeling," *IEEE Communications Magazine*, Jun 1992, pp. 56–61.

[124] M. V. S. N. Prasad and R. Singh, "Terrestrial Mobile Communication Train Measurements in Western India," *IEEE Transactions on Vehicular Technology*, Vol. 52, No. 3, May 2003, pp. 671–682.

[125] J. E. J. Dalley, M. S. Smith, and D. N. Adams, "Propagation Losses Due to Foliage at Various Frequencies," *National Conference on Antennas and Propagations*, March 1999, pp. 267–270.

[126] M. Gudmundson, "Correlation Model for Shadow Fading in Mobile Radio Systems," *Electronics Letters*, Vol. 27, No. 23, Feb. 1991, pp. 2145–2146.

[127] M. M. Zonoozi, P. Dassanayake, and M. Faulkner, "Mobile Radio Channel Characterization," IEEE International Conference on Information Engineering, 1995, pp. 403–406.

[128] S. R. Saunders, *Antennas and Propagation for Wireless Communication Systems*, John Wiley & Sons, 1999, New York.

[129] A. J. Coulson, A. G. Williamson, and R. G. Vaughan, "A Statistical Basis for Log-normal Shadowing Effect in Multipath Fading Channels," *IEEE Transactions on Communications*, Vol. 46, No. 4, April 1998, pp. 494–502.

[130] A. Urie, "Errors in Estimating Local Average Power of Multipath Signals," *Electronics Letters*, Vol. 27, No. 4, Feb. 1991, pp. 315–317.

[131] M. Patzold and F. Laue, "Statistical Properties of Jakes' Fading Channel Simulator," *Vehicular Technology Conference 1998*, pp. 712–718.

[132] T. Fulghum and K. Molnar, "The Jakes Fading Model Incorporating Angular Spread for a Disk of Scatterers," *IEEE Vehicular Technology Conference 1998*, pp. 489–493.

[133] A. Stephenne and B. Champagne, "Effective Multipath Vector Channel Simulator for Antenna Array Systems," *IEEE Transactions on Vehicular Technology*, Vol. 49, No. 6, Nov. 2000, pp. 2370–2381.

[134] Y. Li and Y. L. Guan, "Modified Jakes' Model for Simulating Multiple Uncorrelated Fading Waveforms," IEEE International Conference on Communications, 2000, pp. 46–49.

[135] P. Dent, G. E. Bottomley, and T. Croft, "Jakes Fading Model Revisited," *Electronics Letters*, Vol. 29, No. 13, 1993, pp. 1162–1163.

[136] Y. Li and X. Huang, "The Simulation of Independent Rayleigh Faders," *IEEE Transactions on Communications*, Vol. 50, No. 9, Sept. 2002, pp. 1503–1514.

[137] Z. Wu, "Model of Independent Rayleigh Faders," *Electronics Letters*, Vol. 40, No. 15, July 2004, pp. 949–951.

[138] H. M. El-Sallabi, H. L. Bertoni, and P. Vainikainen, "Experimental Evaluation of Rake Finger Life Distance for CDMA Systems," *IEEE Antennas and Wireless Propagation Letters,* Vol. 1, 2002, pp. 50–52.

[139] C. Cheon and H. L Bertoni, "Fading of Wide Band Signals Associated with Displacement of the Mobile in Urban Environments," *Vehicular Technology Conference,* 2002, pp. 1–5.

[140] A. S. Khayrallah and G. E. Bottomley, "Rake Finger Allocation in the DS-CDMA Uplink," IEEE Vehicular Technlogy Conference, 2005, pp. 122–126.

[141] H. M. El-Sallabi, H. L. Bertoni and P. Vainikainen, "Channel Characterization for CDMA Rake Receiver Design for Urban Environment," IEEE International Conference on Communications, 2002, pp. 911–915.

CHAPTER 4
MODULATION DETECTION TECHNIQUES

In this chapter we will present a wide variety of detection techniques to be used in the wireless digital communications area. Generally speaking, they are applicable to just about any digital communication system. These techniques will comprise both coherent detection and noncoherent detection principles. For illustrative purposes, we have chosen to provide detectors for Differential Quaternary Phase Shift Keying (DQPSK) and Minimum Shift Keying (MSK) modulation schemes. These techniques can also be extended to work for the higher order modulation schemes presented earlier in Chap. 2. Each technique will be described by mathematical terms and accompanied with block diagrams. Whenever possible, theoretical and/or simulation performance results will be provided. Lastly, a comparison of some of the present techniques will be provided and several conclusions will be drawn based on not only performance but also implementation complexity.

4.1 DIFFERENCE BETWEEN PRACTICE AND THEORY

When a signal is transmitted it will encounter the wireless multipath channel discussed in Chap. 3. The received signal will have the following sources of degradation: sample timing offset, carrier frequency offset, and carrier phase offset. The timing offset will be taken care of by the timing recovery mechanism to be discussed in the latter chapters. The frequency offset will be taken care of by the automatic frequency control (AFC) mechanism to be discussed in later chapters as well. In this chapter we will address the phase offset degradation.

If one were to blindly attempt to recover the information from the received symbols without compensating for the phase offset present on the signal, then performance degradation will surely result. If the receiver estimates the phase offset and compensates for it by derotating the received signal prior to entering the detector then this is called coherent detection. On the other hand if no attempt is made to estimate the received signal's phase offset and we extract the information from the received signal, this is called noncoherent detection.

In digital communications there exist a few system design goals that we must mention prior to presenting the choices available in the receiver, specifically modulation detection [1–8]. These goals are listed below (nonexhaustive):

- Maximize transmission bit rate, R_b
- Minimize Bit Error Rate, BER
- Minimize required power, E_b/N_o
- Minimize required system bandwidth, BW
- Maximize system utilization/capacity, C

- Minimize system complexity
- Minimize power consumption

Being able to optimally choose the system values for each of the above parameters is a challenging task. The suggested approach is to consider the system as a whole: transmitter, plus channel plus receiver. In this case certain system design trade-offs can be made in order to benefit the entire communication link.

4.2 COHERENT DETECTION

In a typical consumer electronic product, the received signal is spectrally down converted to base band. Due to inaccuracies in the implementation and signal processing algorithms, residual frequency and phase offsets will exist. A general block diagram of a coherent receiver is given below assuming a super heterodyne architecture (see Fig. 4.1). First, note the following block diagram is simplified

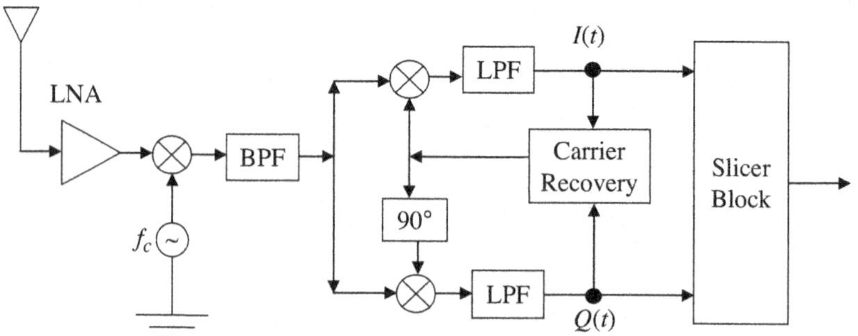

FIGURE 4.1 General receiver block diagram.

with the intention of conveying the present message. The received signal is first amplified by a low noise amplifier (LNA). It is then spectrally shifted to an intermediate frequency (IF) where the band pass filter (BPF) will extract the desired information signal. Then the signal is brought down to DC or Zero IF (ZIF) so the signal processing algorithms can operate on the baseband signal.

The general principle is that the received signal is down converted to baseband producing the complex envelope signal, $I(t) + j\,Q(t)$. This complex envelope will have a residual frequency and phase offset. The carrier recovery algorithm will estimate this offset and adjust the locally generated carrier in such a way as to reduce this offset, ideally this frequency/phase offset would be zero. In essence we are derotating the baseband signal to compensate for this residual offset introduced by inaccuracies in both the transmitter and receiver, as well as the distortion introduced by the channel. Due to the time varying elements in the communication link, this phase offset is also time varying and should be tracked. Figure 4.1 shows the frequency/phase offset compensation occurring in the analog domain, we should mention that more accuracy can be obtained if this operation is performed in the digital domain.

This can best be explained by a simple diagram shown in Fig. 4.2. Assume the received signal is $r(t)$ that contains not only the modulation, $\theta(t)$, but also a time invariant phase offset due to the wireless channel, ϕ.

Here the carrier recovery (CR) algorithm estimates the received signal's carrier frequency, f_c, and phase offset, $\hat{\phi}$. Notice any frequency offset produces a "phase roll" in the signal. For sake of simplicity assume a zero frequency offset ($\Delta f = 0$) so we have the following mathematical representation

$$d(t) = \text{Re}\left\{e^{j[\theta(t)+\phi_e]}\right\} \qquad \phi_e = \phi - \hat{\phi} \tag{4.1}$$

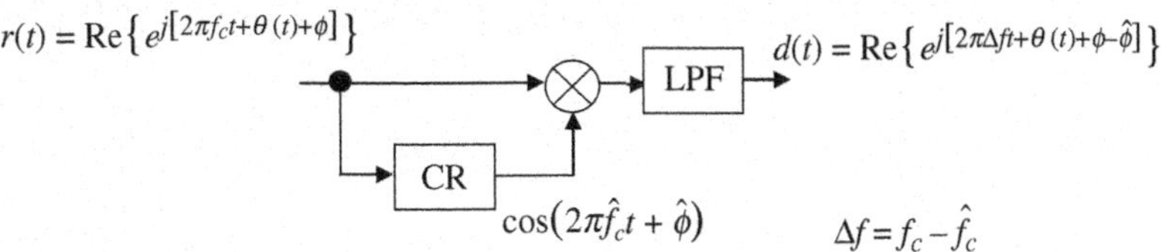

FIGURE 4.2 Coherent detection principle.

If we assume that QPSK was transmitted then we would receive the following constellation diagram (see Fig. 4.3). The received constellation is rotated by the amount equal to the phase error (assumed constant for this example).

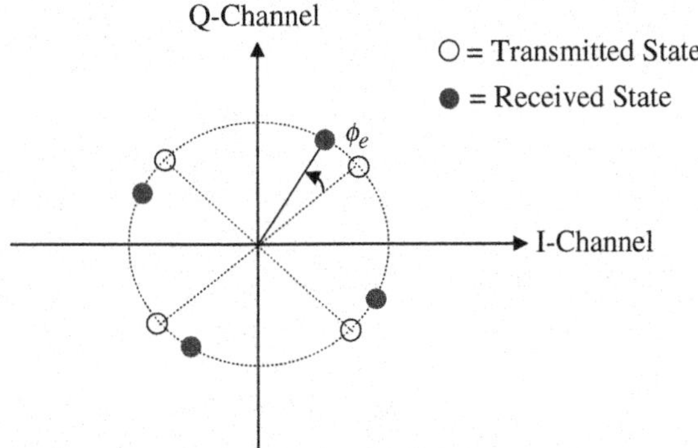

FIGURE 4.3 QPSK constellation offset due to phase error.

As you can clearly see from Fig. 4.3, the residual/uncompensated phase offset will push the phase states closer to the decision boundary. This will make states more susceptible to noise and thus degrade BER performance.

In the next subsections we will begin by introducing some IF techniques used to perform coherent detection and then present some baseband techniques which perform with a higher level of accuracy and are more commonly deployed in commercial products. These are applicable to both time division multiple access (TDMA) and code division multiple access (CDMA) systems.

The IF techniques are discussed first for historical reasons in that the initial techniques used for carrier recovery in the digital cellular area were more focused on the IF signal (analog section). As time and technology progressed more accurate signal processing algorithms were invented and applied in the baseband, complex envelope domain (digital section) [9–11].

4.2.1 Costas Loop

This first coherent detection technique builds upon the Costas loop receiver shown below [12]. The received signal is directly converted to ZIF by the quadrature demodulator operation. The locally generated carrier signal is adjusted by low pass filtering the product of the baseband I- and Q-signals (see Fig. 4.4).

The received signal is represented as follows, where $\theta(t)$ is the modulation signal and $\phi(t)$ is the phase offset.

$$r(t) = A(t)\cos[2\pi f_c t + \theta(t) + \phi(t)] \tag{4.2}$$

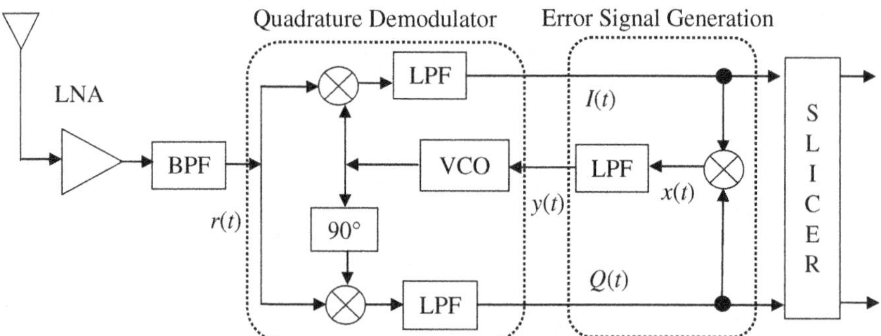

FIGURE 4.4 Costas loop receiver block diagram.

The corresponding baseband signals are, where $\hat{\phi}(t)$ is the locally generated phase offset.

$$I(t) = \frac{1}{2}A(t)\cos[\theta(t) + \phi(t) - \hat{\phi}(t)] \tag{4.3}$$

$$Q(t) = \frac{1}{2}A(t)\sin[\theta(t) + \phi(t) - \hat{\phi}(t)] \tag{4.4}$$

The output of the mixer/multiplier is given as

$$x(t) = I(t) \cdot Q(t)$$

$$x(t) = \frac{1}{4}A^2(t)\sin[2\theta(t) + 2\phi(t) - 2\hat{\phi}(t)] \tag{4.5}$$

Next we apply the low pass filter (LPF) to remove the modulation signal to produce

$$y(t) = K \cdot A^2(t)\sin[2\phi(t) - 2\hat{\phi}(t)] \tag{4.6}$$

where K = gain constant (overall). Hence the signal that controls the Voltage Control Oscillator (VCO) is dependent on the phase difference between the received signal and the locally generated signal. For this example the error signal driving the VCO is proportional to the phase error doubled.

A modified version of the Costas loop is given in Fig. 4.5, in fact some performance results for Gaussian Filter MSK (GMSK) modulation are given in [13] and referred to as a closed loop coherent detector.

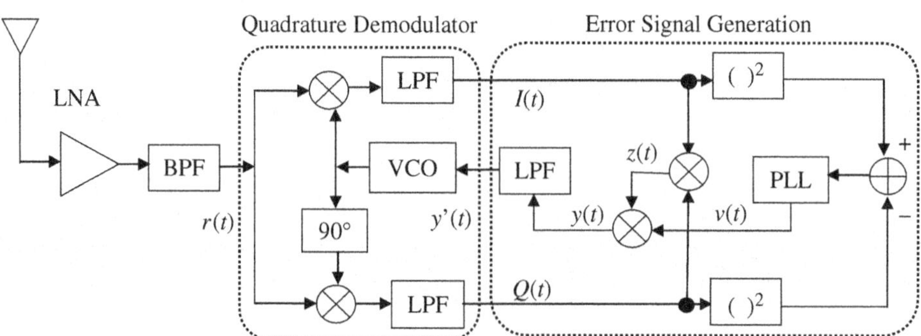

FIGURE 4.5 Modified Costas loop (closed loop) receiver block diagram.

The baseband signals are given as

$$I(t) = A(t)\cos[\theta(t) + \phi(t) - \hat{\phi}(t)] \tag{4.7}$$

$$Q(t) = A(t)\sin[\theta(t) + \phi(t) - \hat{\phi}(t)] \tag{4.8}$$

The first mixer/multiplier output is

$$z(t) = \frac{1}{2} \cdot A^2(t) \cdot \sin[2\theta(t) + 2\phi(t) - 2\hat{\phi}(t)] \tag{4.9}$$

The subtractor output error signal component is

$$I^2(t) - Q^2(t) = A^2(t) \cdot \cos[2\theta(t) + 2\phi(t) - 2\hat{\phi}(t)] \tag{4.10}$$

The second mixer/multiplier output is

$$y(t) = z(t) \cdot v(t)$$

$$y(t) = A^4(t) \cdot K \cdot \sin[4\theta(t) + 4\phi(t) - 4\hat{\phi}(t)] \tag{4.11}$$

The LPF removes the modulation and we now have the following

$$y'(t) = K \cdot A^4(t) \cdot \sin[4\phi(t) - 4\hat{\phi}(t)] \tag{4.12}$$

Here we have the error signal that is proportional to the phase error, but this time quadrupled. This error signal can then enter a phase divider operation, making the error signal proportional to the phase error and thus changing the carrier recovery loop characteristics.

4.2.2 Frequency Doubling (or Quadrupling)

In this section we will make use of trigonometric identities in order to remove the modulation from the received signal in order to observe the residual frequency/phase error [10]. The basic principle is as follows; first let's consider a Binary Phase Shift Keying (BPSK) symbol.

$$r(t) = A(t) \cdot \cos[\omega_c t + \pi \cdot a(t) + \phi(t)] \tag{4.13}$$

where the bit of information belongs to the following alphabet: $a(t) \in \{0,1\}$ and $\phi(t)$ = time varying phase offset. By squaring and applying DC removal (by using a high pass filter) we obtain the resulting signal

$$r^2(t) = \frac{1}{2} \cdot A^2(t) \cdot \cos[2w_c t + 2\pi \cdot a(t) + 2\phi(t)] \tag{4.14}$$

$$r^2(t) = \frac{1}{2} \cdot A^2(t) \cdot \cos[2w_c t + 2\phi(t)] \tag{4.15}$$

By squaring this noise-free received signal, we obtain a signal centered at twice the carrier frequency. Additionally, the modulation has been removed simply leaving the phase offset term. The similar idea can be applied to QPSK symbols, except this time we need to have a $4x$'s multiplier to remove the 4 possible phase shifts. To make the discussion more interesting we will apply this technique to $\pi/4$-DQPSK modulation. Please recall the transmitted signal is represented as following (when considering the 8 phase states).

$$s(t) = A(t) \cdot \cos[\omega_c t + \theta(t)] \tag{4.16}$$

where we can approximate the $\pi/4$-DPQSK modulation when the modulation information is $\theta(t) \in \left\{0, \frac{\pi}{4}, \frac{\pi}{2}, \frac{3\pi}{4}, \pi, -\frac{\pi}{4}, -\frac{\pi}{2}, -\frac{3\pi}{4}\right\}$. Let us calculate the maximum frequency deviation, f_d, in a symbol time, T_s, for the above phase changes. $\pi/4$-DQPSK modulation only allows 4 possible phase changes per symbol time which we will make use of to calculate the four corresponding frequency changes. The first 2 phase changes are given as

$$\theta_1 = \omega_1 T_s = \pm\frac{\pi}{4} \tag{4.17}$$

$$2\pi f_{d1} T_s = \pm\frac{\pi}{4} \tag{4.18}$$

$$f_{d1} = \pm\frac{1}{8T_s} \tag{4.19}$$

And the second 2 phase changes are given a follows:

$$\theta_2 = \omega_2 T_s = \pm\frac{3\pi}{4} \tag{4.20}$$

$$2\pi f_{d2} T_s = \pm\frac{3\pi}{4} \tag{4.21}$$

$$f_{d2} = \pm\frac{3}{8T_s} = 3f_{d1} \tag{4.22}$$

The above analysis shows the four allowable phase changes can be interpreted as four allowable frequency changes: $\pm R_s/8,\ \pm 3R_s/8$ (Hz).

We can therefore approximate the $\pi/4$-DQPSK as a 4-level frequency shift keyed (FSK) modulation scheme [14]. We will make use of this representation in the following mathematical coherent analysis. Modeling $\pi/4$-DQPSK as 4-Level FSK we can write the above equations as

$$s(t) = A(t) \cdot \cos[\omega_c t + a(t) \cdot 2\pi \cdot f_{d1}] \tag{4.23}$$

where $a(t)$ are the symbols to be transmitted and given as $a(t) \in \{\pm 1, \pm 3\}$.

The idealized transmit spectrum can be shown as in Fig. 4.6, assuming each phase change generates a single spectral tone. In reality there will be many additional frequency components, but for illustrative purposes let us assume they create a single spectral tone.

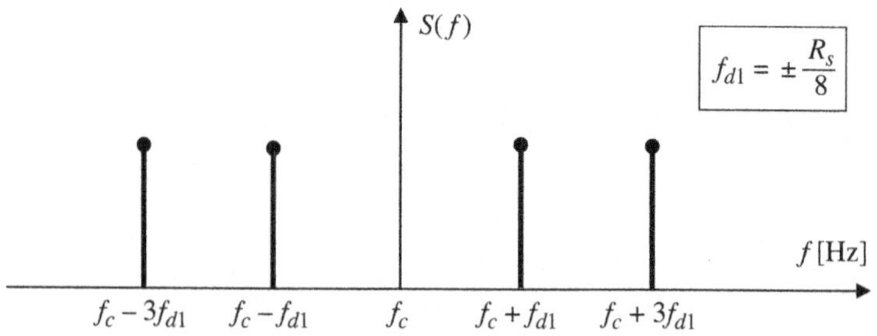

FIGURE 4.6 Ideal 4-level FSK transmit spectrum.

The approach is to generate a local frequency such that when mixed with the received signal will generate a spectral line (or tone) that can be used to estimate the received signal phase offset. A block diagram of one such technique is given in Fig. 4.7, the performance results for GMSK are given in [15] where this technique was called the open loop coherent detector.

A similar receiver block diagram is shown in Fig. 4.8 for DQPSK modulation, where it also provides an example of generating the $^1/_2$ symbol clock.

Above the received signal is quadrupled and multiplied by a half symbol clock to generate a signal centered at 4x's the carrier frequency. The carrier x's 4 signal is extracted by a BPF and then

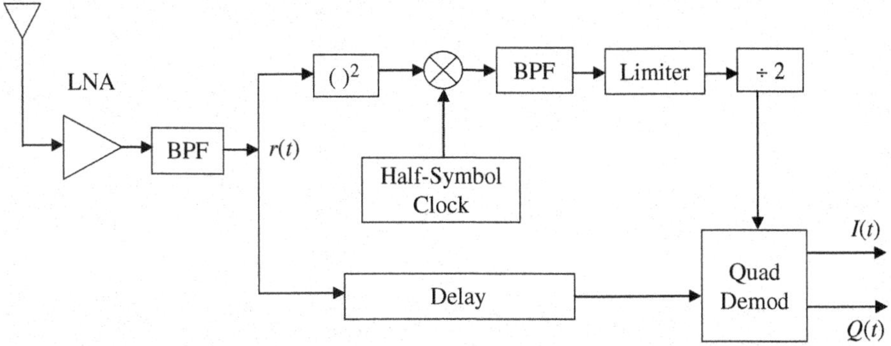

FIGURE 4.7 Coherent open loop receiver block diagram for GMSK.

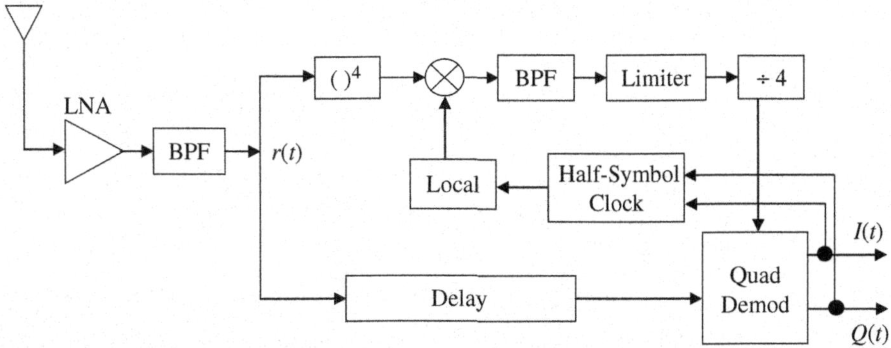

FIGURE 4.8 Coherent receiver block diagram based on quadrupling principles.

divided by 4 to obtain the carrier frequency prior to entering the quadrature demodulator. Note a limiter can be inserted into this signal path to help remove the effects of amplitude modulation/variation. Also the correct estimate of the time delay through the BPF and dividing circuits is critical for best system performance. The delay block is used to time align the received signal with the estimated coherent reference signal.

Now let's actually follow the signal path as it passes through this receiver. After the quadrupler we have the following mathematical representation

$$s^4(t) = \frac{1}{4} \cdot A^4(t) \cdot \left\{1 + 2 \cdot \cos[2\omega_c t + 2\theta(t)] + \frac{1}{2} + \frac{1}{2} \cdot \cos[4\omega_c t + 4\theta(t)]\right\} \tag{4.24}$$

Since we have modeled the $\pi/4$-DQPSK modulation as 4-level FM, let us concentrate on the signal centered at $4x$'s the carrier frequency.

$$s_o(t) = K(t) \cdot \cos[4\omega_c t + 4\theta(t)] \tag{4.25}$$

$$s_o(t) = K(t) \cdot \cos\left[4\omega_c t + 4 \cdot a(t) \cdot 2\pi \cdot \frac{1}{8T_s}\right] \tag{4.26}$$

$$s_o(t) = K(t) \cdot \cos\left[4\omega_c t + \frac{a(t) \cdot \pi}{T_s}\right] \tag{4.27}$$

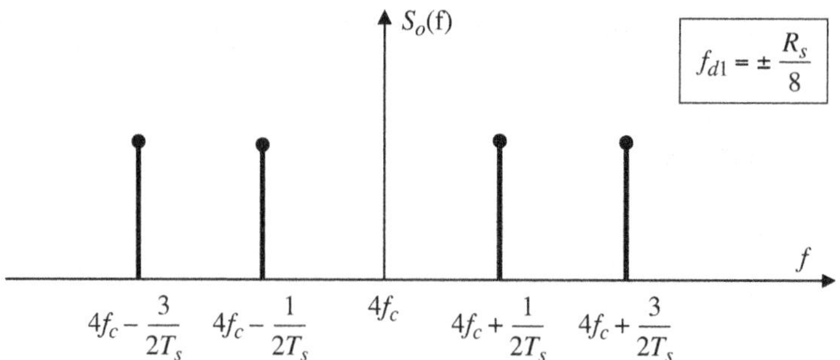

FIGURE 4.9 Ideal 4-level FSK spectrum at the quadrupler output.

The spectrum of this signal can be approximated as follows, when continuing with the above assumption of the spectral tones (see Fig. 4.9).

Next we generate a local half symbol clock signal to be used with the quadrupler output signal in the following fashion

$$y(t) = s^4(t) \cdot \cos[4\theta(t)] \tag{4.28}$$

$$y(t) = s^4(t) \cdot \cos\left[a(t) \cdot 2\pi \cdot \frac{1}{2T_s}\right] \tag{4.29}$$

Using the above equation for the quadrupler output we arrive with the following:

$$y(t) = c(t) \cdot \cos[4\omega_c t] + c(t) \cdot \cos[4\omega_c t + 8\theta(t)] + \text{other terms} \tag{4.30}$$

We BPF this signal to only allow the spectral line at $4x$'s the carrier frequency to pass through and this signal can be further hard limited to remove the amplitude variations (possibly due to fading). Now in order for us to use this signal for demodulation, we require a divide by 4 operation to generate a spectral line at the carrier frequency. This resulting signal is then used in the quadrature demodulation operation.

We have presented three techniques that use the received signal to essentially generate a spectral component that can be used as a locally generated reference. First, a few major issues exist with this family of coherent detection techniques, namely the received signal can be very noisy and thus degrade performance. Second, the received signal can experience a large Doppler spread due to fast multipath fading. This frequency spread translates to phase spread making it difficult to obtain a single value for the phase offset. Third, we can be operating in a FSF channel in which case we will be receiving multiple copies (echoes) of the transmitted signal. In this case the question naturally arises as to which multipath do we try to demodulate? The obvious answer would be all of them, hence we should leave this to the topic of equalization to be discussed latter in this book.

In Fig. 4.10 we plot performance of coherent detection and noncoherent detection in a flat fading channel for $\pi/4$-DQPSK modulation using the quadrupling technique discussed above. Here we clearly see the particular disadvantages of this technique when compared to others [16]. The coherent detection receiver used the open loop technique described above and the noncoherent detection receiver used the differential detector (which will be explained in the later sections). The theoretical BER curves are also provided for sake of reference.

What Fig. 4.10 conveys is that for fast fading channels, the irreducible error floor of coherent detection can be higher than that of noncoherent detection. A point worth mentioning is that the open loop technique is sensitive to various parameters such as BPF, bandwidth (BW), and the accuracy of the delay block. Being somewhat sloppy in choosing these parameters can lead to results that are worse than noncoherent detection.

The main shortcoming of this type of coherent detection (phase estimation) technique is that it is attempting to generate a spectral tone (frequency component) with the same frequency and phase offset

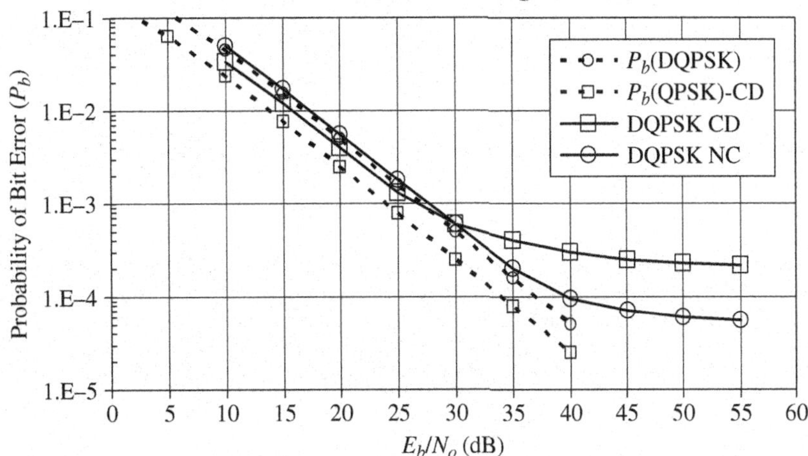

FIGURE 4.10 Comparison of some simulated coherent and noncoherent detection techniques.

as the received signal. For relatively clean channel conditions, the performance results seem reasonable. However, for high signal to noise ratio (SNR) and high Doppler frequency channels there is a spectral spreading phenomenon occurring and thus have more uncertainty (or estimation error) in the coherent detection operations [17].

The last point to make on this family of detectors is that more robust architectures can be used such as decision directed or adaptive estimation. But the argument against taking such a path is this increase in complexity is rather unnecessary as there are other techniques that operate directly on the baseband *I*- and *Q*-signals, thus virtually eliminating the influence of the analog tolerances in the receiver design.

4.2.3 Pilot Symbol Aided Detection (PSAD)

As shown and discussed above, many authors accurately present coherent detection performance results to be better than noncoherent detection performance results. Closed form expressions used to support this statement assume the locally generated demodulation signal is not only in exact frequency and phase alignment with the received signal, but also has an SNR of infinity. Practically speaking, the SNR is much lower and can dramatically affect the performance results if a very noisy reference is used in the receiver. The three above-mentioned methods presented in this chapter thus far derive the phase reference from the received signal itself blindly. For these cases if the receiver is of poor quality then the estimated phase reference will be noisy and degrade performance.

A way to combat the blind shortcoming, pilot symbols are periodically inserted into the data stream in a time-multiplexed fashion [18, 19]. This allows the receiver to accurately estimate the phase reference during these known symbol time intervals. These estimates are combined with other pilot symbol fields and/or use decision directed updates to estimate the phase reference over the unknown random data portion of the time slot. Suffice it to say that this technique is modulation scheme agnostic [20–24]. A very simple diagram showing this concept is shown in Fig. 4.11 from the perspective of the transmitter.

Above we can see that the pilot symbols are periodically inserted into the transmit data stream so the receiver can make use of them. A system trade-off exists here: placing the pilot symbols too close together will reduce the overall system efficiency and data throughput but will allow the receiver to accurately estimate the phase reference over the majority of the time slot. On the other hand, placing

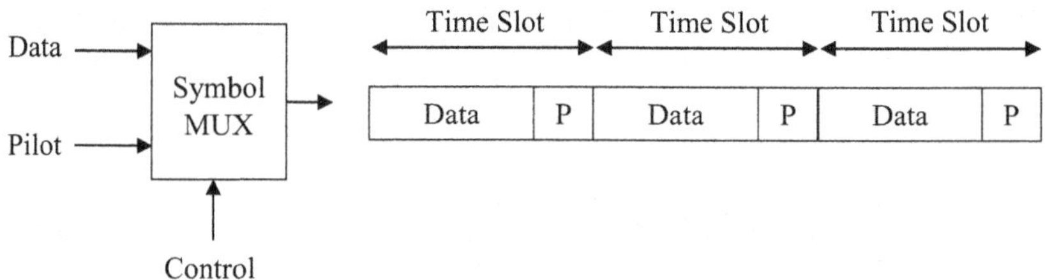

FIGURE 4.11 Pilot symbol inserted transmission example (time multiplexed).

them too far apart will reduce the accuracy of the phase reference especially for the condition when the channel is time varying. Hence they must be placed close enough to sample the desired maximum rate of change such that time variations can be accurately estimated (and tracked). Below we overlay the time varying channel on the transmitted signal to help express this idea of using time division multiplexed (TDM) pilot symbols to sample the time varying channel. In Fig. 4.12 we have defined the

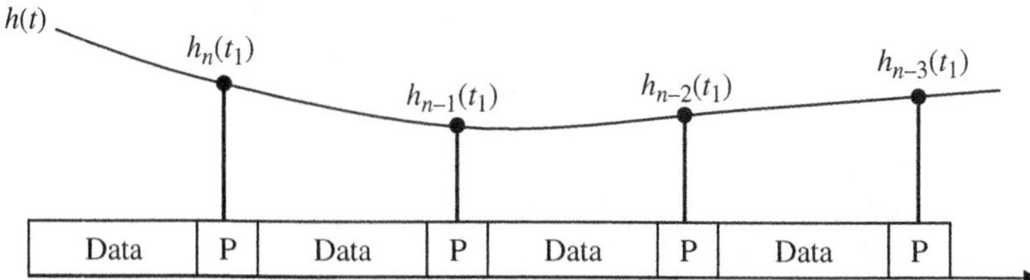

FIGURE 4.12 Pilot symbol aided channel estimation.

channel samples as $h_{n-k}(t_p)$ which is equal to the channel response during the pilot symbol that corresponds to the kth time slot and the pth symbol position within the kth time slot.

Once the channel is estimated it can be applied to the remainder of the time slot assuming a very slowly time varying channel or for faster varying channels, intermediate values can be interpolated for the data portion of the time slot.

The signal processing functions to be performed at the receiver side are shown in Fig. 4.13. Here the position of the pilot symbols in the time slot is assumed to be known, which is a reasonable

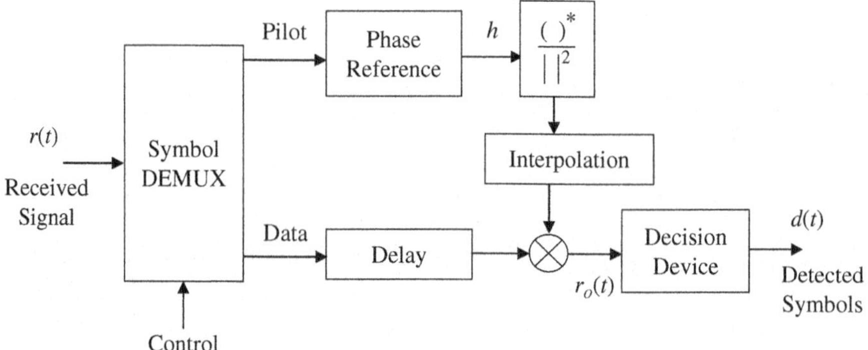

FIGURE 4.13 Pilot symbol aided demodulation block diagram.

assumption due to time synchronization. The phase reference is estimated during this time interval and applied during the data portion of the time slot. The delay block is inserted into the data signal path in order to time-align the interpolated estimate with the data stream. Since the pilots symbols are inserted in a TDM fashion a few time slots are required to collect reliable channel estimates before they can be applied during the data portion of the time slot.

Let us say for sake of example, that only one pilot symbol is transmitted every time slot. And the received signal during this time interval is given as

$$r(t) = p \cdot h(t) + n(t) \tag{4.31}$$

It is easy to estimate the channel, $h(t)$, by using the following step: multiply the received signal by the conjugate of the pilot symbol, p^*.

$$r_p(t) = r(t) \cdot p^* = p \cdot p^* \cdot h(t) + n(t) \cdot p^* \tag{4.32}$$

$$r_p(t) = |p|^2 \cdot h + n_p(t) \tag{4.33}$$

If we assume the pilot symbol energy is normalized to unity, then we can rewrite this as follows, where $n_p(t)$ is the noise signal modified by the pilot signal.

$$r_p(t) = h(t) + n_p(t) \tag{4.34}$$

And if we further assume a reasonable SNR value such that the noise component contributes only marginally then we have the following approximation

$$r_p(t) \cong \hat{h}(t) \tag{4.35}$$

Now if we assume phase reference is time invariant during the time slot then we can apply this estimate to the data portion of the time slot.

$$r_d(t) = d(t) \cdot h(t) + n(t) \tag{4.36}$$

$$r_d(t) \cdot r_p^*(t) = d(t) \cdot h(t) \cdot \hat{h}^*(t) + n(t) \cdot \hat{h}^*(t) \tag{4.37}$$

$$r_d(t) \cdot r_p^*(t) = d(t) \cdot |\hat{h}(t)|^2 + n_h(t) \tag{4.38}$$

Hence phase derotation was accomplished by the use of the complex conjugation operation on the channel estimate. Assume we normalize by the power of the channel and the SNR is at a reasonable value then we have the following:

$$r_o(t) \cong d(t) \tag{4.39}$$

This compensated symbol clearly shows the amplitude and phase distortion (introduced by the channel) has been removed. We have chosen to normalize the channel estimate for the purposes of supporting our explanation; however is it not a requirement. The decision to perform such an operation should be based on the signal processing functions that follow the detector, such as forward error correction, and so on.

If the channel is time varying then using the channel estimate throughout the entire time slot is incorrect and can prove to cause detrimental effects on system performance. In order to obviate the risk of degraded performance a solution is to update the channel estimate as it is applied in the time slot. This is accomplished with the assistance of two blocks: Delay first-in first-out (FIFO) and Interpolation. The Delay FIFO buffer allows us to use noncausal signal processing algorithms, by this we mean using the channel estimates of not only the past and present, but also the future to make a more accurate estimate of the present channel. Hence this delay buffer allows us to have more accurate channel estimates to be used in the time slot. The length of the FIFO is also a measure of the degree of noncausality used in the receiver. The second block uses the past, present, and future estimates to create a time varying channel estimate. Here the estimate is updated every symbol position in order to track the time variations.

In Fig. 4.14 we show some more details of the interpolation, where we use four channel estimates to aid the coherent detector. These four channel estimates, h_1, h_2, h_3, and h_4, are updated every time slot.

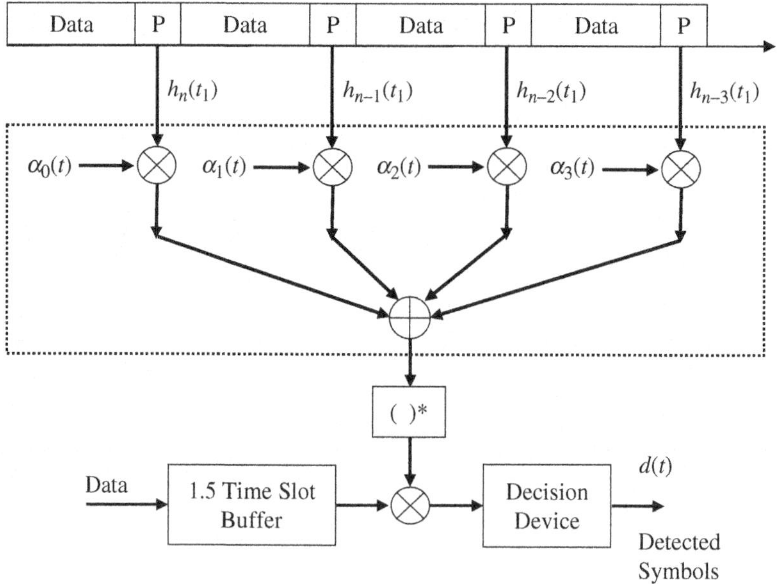

FIGURE 4.14 Noncausal channel estimation example.

They all enter the interpolation block that has time varying weights: $\alpha_0(t)$, $\alpha_1(t)$, $\alpha_2(t)$, and $\alpha_3(t)$, where they are updated every symbol time, $t = kT_s$. These weights depend on where the data symbol is within the time slot.

As far as interpolation goes, various techniques exist such as linear, cubic-spline, polynomial based, and so on. For the fast time varying examples discussed we have found the linear interpolator to perform the worst and a polynomial based interpolator to perform the best.

A last point necessary to mention is that PSAM is a member of a more general family called pilot aided modulation (PAM). We have presented a solution using time division multiplexed (TDM) pilot symbols, solutions also exist for Frequency Division Multiplexing (FDM) pilot tones and lastly CDM pilot channels. We would like to briefly discuss the CDM solution and provide more details in later chapter that discusses WCDMA. The general idea is to transmit a continuous pilot signal that would be used to coherently demodulate the data signal [25]. This is also applicable to Orthogonal Frequency Division Multiplexing (OFDM)-based waveforms, where pilot symbols are generally time and frequency multiplexed into the transmitted OFDM symbol.

4.2.4 Multiple Bit Observation (MBO) of MSK—Coherent Detection

In this section we will build upon the assumption that carrier recovery has been performed. Once the carrier frequency and phase offsets have been removed, the conventional next step would be to use a symbol- (or bit-) based detector. What we will present in this section is the performance gain in widening the number of symbols (or bits) used to perform a decision on a single information signal. We will maintain the naming convention of MBO [26]. This technique differs from the multiple symbol differential detection (MSDD), to be discussed latter, in that as the observation window is made larger, we still make a decision on a single bit of information rather than a sequence of bits.

In particular, we will observe n bits of a continuous phase FSK (CPFSK) waveform and perform a decision on 1 bit, specifically the first bit in this observation window. For sake of simplicity let us use binary CPFSK waveform with modulation index, $\beta = 0.5$ or MSK modulation.

$$s(t) = A(t) \cdot \cos\left[\omega_c t + \frac{\pi t}{2T} \cdot a_1 + \theta_1\right] \qquad (0 \leq t \leq T) \tag{4.40}$$

Here a_1 = data $\{+1, -1\}$ and θ_1 = phase at the beginning of the observation interval. We can also write down the FSK waveform during the ith bit time as

$$s(t) = A(t) \cdot \cos\left[\omega_c t + \frac{\pi}{2T} \cdot a_i\left(t - (i-1)T\right) + \sum_{j=1}^{i-1} \frac{\pi}{2} \cdot a_j + \theta_1\right] \tag{4.41}$$

In the following interval $(i-1)T \leq t \leq iT$

Since we have used the coherent detection receiver we will assume θ_1 is known to us and therefore we will set it to zero for simplification. We now define the n-tuple signal as $s(t, a_1, A_k)$ where A_k represents a particular data sequence of length $n - 1$ bits $\{a_2, a_3, \ldots, a_n\}$, a_1 is the first bit of the n-tuple signal that we are trying to estimate.

We can write the a posteriori probability of the received signal, r given a_1 and A_k as follows (assuming n symbols or bits) [27]:

$$p(r|a_1, A_k) = C \cdot e^{\frac{2}{N_o}\int_0^{nT} r(t)s(t,a_1,A_k)dt} \tag{4.42}$$

The n-tuple data sequence is unknown and can be removed from the above equation in the following manner.

$$p(r|a_1) = E_{A_k}\{p(r|a_1, A_k)\} \tag{4.43}$$

$$p(r|a_1) = \int p(r|a_1,A_k) \cdot p(A_k)dA_k \tag{4.44}$$

where the density function of the data bits, $f(\alpha_i)$, is denoted below and $\partial(x)$ is a Dirac delta function that equals 1 only when $x = 0$.

$$p(A_k) = \prod_{i=2}^{n} f(\alpha_i) = \prod_{i=2}^{n}\left[\frac{1}{2} \cdot \partial(\alpha_i - 1) + \frac{1}{2} \cdot \partial(\alpha_i + 1)\right] \tag{4.45}$$

Looking over all the possible sequences of A_k, which amounts to 2^{n-1}

$$p(r|a_1) = \int_{a_2}\int_{a_2} \cdots \int_{a_n} p(r|a_1, A_k)p(A_k)dA_k \tag{4.46}$$

$$p(r|a_1) = \frac{1}{2^{n-1}} \cdot \sum_{i=1}^{2^{n-1}} p(r|a_1, A_i) \tag{4.47}$$

In order to decide whether $a_1 = +1$ or $a_1 = -1$ we will present the optimal coherent detector in the form a likelihood ratio test, L. This represents the ratio of the probability of received sequence is correct given the first bit is a 1 and the probability of received sequence is correct given the first bit is a -1.

$$L = \frac{p(r|1)}{p(r|-1)} = \frac{p(r|a_1 = 1)}{p(r|a_1 = -1)} \tag{4.48}$$

Substituting the above equations we arrive with the following:

$$L = \frac{\int_A e^{\frac{2}{N_o}\int_0^{nT} r(t)\cdot s(t,1,A)dt} f(A)dA}{\int_A e^{\frac{2}{N_o}\int_0^{nT} r(t)\cdot s(t,-1,A)dt} f(A)dA} \tag{4.49}$$

Assuming the data bits are independent with a density function given above, we can now rewrite the likelihood ratio test as follows:

$$L = \frac{e^{\frac{2}{N_o}\int_0^{nT} r(t)\cdot s(t,1,A_1)dt} + \cdots + e^{\frac{2}{N_o}\int_0^{nT} r(t)\cdot s(t,1,A_m)dt}}{e^{\frac{2}{N_o}\int_0^{nT} r(t)\cdot s(t,-1,A_1)dt} + \cdots + e^{\frac{2}{N_o}\int_0^{nT} r(t)\cdot s(t,-1,A_m)dt}} \tag{4.50}$$

where $m = 2^{n-1}$ representing the number of combinations to test against in the observation window of length n. In other words, the receiver correlates the received waveform with each of the "m" possible transmitted signals beginning with $a_1 = 1$. A similar series of operations are performed for each of the "m" possible transmitted signal beginning with $a_1 = -1$.

$$L = \frac{\sum_{k=1}^{2^{n-1}} e^{\frac{2}{N_o}\int_0^{nT} r(t)\cdot s(t,1,A_k)dt}}{\sum_{k=1}^{2^{n-1}} e^{\frac{2}{N_o}\int_0^{nT} r(t)\cdot s(t,-1,A_k)dt}} \tag{4.51}$$

A block diagram is shown in Fig. 4.15. Note the integrations are performed over nT bits time intervals. As mentioned earlier n bits are observed and a decision is made on the first bit. Then the observation window shifts by 1 bit and a new likelihood ratio is calculated, and so on and so forth.

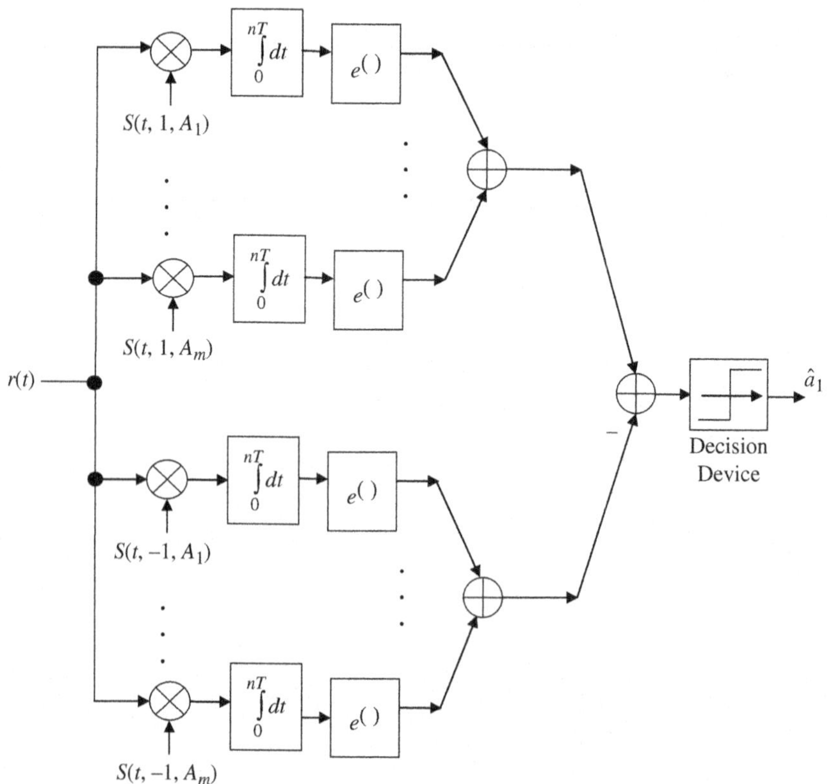

FIGURE 4.15 Multiple bit observation-based coherent detection block diagram.

The upper part of the detector represents the likelihood the first bit in the n-tuple sequence is a "1," while the lower part represents the likelihood the first bit in the sequence is a "−1." The bit most likely transmitted, a_1, is determined by the subtraction operation and a hard decision device.

4.3 NONCOHERENT DETECTION OF DQPSK

In this section we will present the various noncoherent detection choices available to the system designer. We have chosen to use the DQPSK modulation scheme as an example; note these techniques can easily be extended for other modulation schemes.

Recall the classical noncoherent detector makes no attempt to estimate the received signal's phase offset. As a result of this assumption the received constellation diagram will have a time varying phase offset that precludes directly using symbol threshold(ing) on the received signal.

4.3.1 Differential Detection

In this section we will present the first noncoherent technique called differential detection (DD). Since no attempt is made in estimating the received phase offset, phase differences will be made in an effort to reconstruct the transmitted signal. Three variations of this technique will be presented, all of which have been shown to produce similar BER performance under the same operating conditions [28–35].

Baseband Approach of DD. This first variation will be the baseband implementation of DD. The following block diagram shows the noncoherent receiver using the DD (see Fig. 4.16). Let's start with

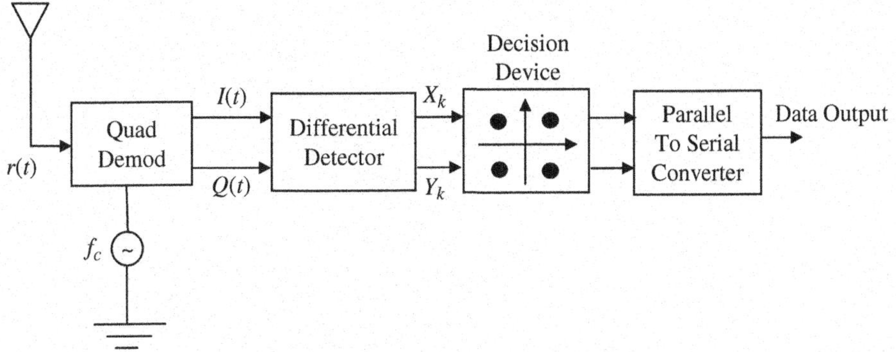

FIGURE 4.16 Noncoherent receiver block diagram.

a brief description of the following block diagram. The received signal is immediately spectrally down converted to baseband. The assumption is the locally generated carrier frequency is exactly equal to the received carrier frequency; however, the phase difference is not resolved. These complex symbols ($I + jQ$) enter the DD which creates detected symbols ($X + jY$). A hard binary decision is made on each of the real and imaginary parts of the DD output. These binary decisions then enter a P/S converter block which essentially reverses the operations performed by the transmitter by time division demultiplexing the real and imaginary decisions into a single bit stream.

The actual operations performed within the DD module assuming a complex-valued waveform are given in Fig. 4.17. The received signal is first delayed by a symbol time, T_s, and then conjugated, this result is then multiplied by the received signal. Essentially we are using the previous received symbols as a reference signal to demodulate the present received symbols.

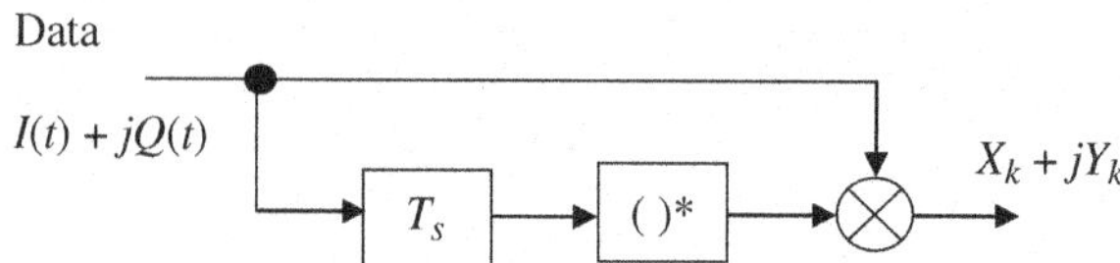

FIGURE 4.17 Differential detection mathematical operations.

The input to the differential detector is assumed to be given as follows, where we have conveniently written the input in the Cartesian as well as Polar coordinate system.

$$I(t) + jQ(t) = A(t) \cdot e^{j\theta(t)} \tag{4.52}$$

The DD output signal is given as follows assuming the Cartesian coordinate system and the time index k is in units of symbol times (T_s).

$$X_k + jY_k = \{I_k + jQ_k\} \cdot \{I_{k-1} - jQ_{k-1}\} \tag{4.53}$$

$$X_k + jY_k = \{I_k \cdot I_{k-1} + Q_k \cdot Q_{k-1}\} + j\{Q_k \cdot I_{k-1} - I_k \cdot Q_{k-1}\} \tag{4.54}$$

In order to provide some insight into the DD operations, let's discuss the following. If we were to perform the ensemble average of X_k then it would consist of the sum of the in-phase and quadrature phase autocorrelations, evaluated with a lag equal to a symbol time. Hence this provides a measure of how correlated the present symbol is to the previous one. Performing the same operations on Y_k produces the sum of the in-phase and quadrature phase cross correlations, evaluated with a lag equal to the symbol time. The above mathematical equations are simply drawn in a block diagram form (see Fig. 4.18) [36–38].

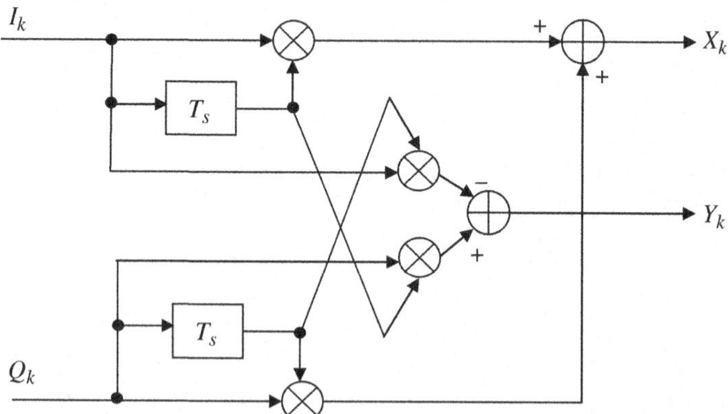

FIGURE 4.18 Baseband differential detector block diagram in Cartesian coordinate system.

Alternatively we can show the DD output signal in the Polar coordinate system.

$$X(t) + jY(t) = \{A(t) \cdot e^{j\theta(t)}\} \cdot \{A(t - T_s) \cdot e^{j\theta(t-T_s)}\}^* \tag{4.55}$$

$$X(t) + jY(t) = A(t) \cdot A(t - T_s) \cdot e^{j[\theta(t)-\theta(t-T_s)]} \tag{4.56}$$

$$X(t) + jY(t) = A(t) \cdot A(t - T_S) \cdot e^{j\Delta\phi(t)} \tag{4.57}$$

$$X(t) + jY(t) = A(t) \cdot A(t - T_s) \cdot \{\cos[\Delta\phi(t)] + j\sin[\Delta\phi(t)]\} \tag{4.58}$$

Here we see the output signal amplitude is proportional to the power of the input signal. Moreover, the output phase is the difference between contiguous phases. This can be shown in Fig. 4.19.

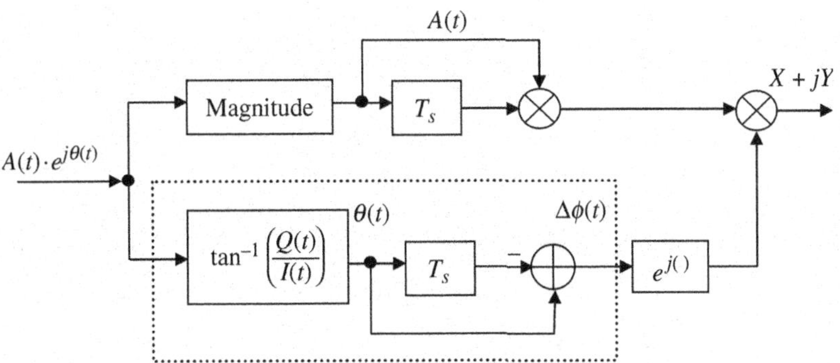

FIGURE 4.19 Baseband differential detector block diagram in Polar coordinate system.

In fact, if the amplitude variations were not important or could be ignored, then the dashed box in Fig. 4.19 is the polar coordinate representation of the DD. It sufficiently carries the modulated waveform information and so is suitable for detection.

Assuming the input signal is a $\pi/4$-DQPSK modulated signal then the DD input complex envelope signal has a constellation diagram (see Fig. 4.20). Recall, due to the zero Inter-Symbol Interference (ISI) the states shown have assumed a pulse shaping filter that satisfied the Nyquist condition. This constellation diagram has eight phase states where the four allowable phase changes are denoted by the solid and dashed arrows.

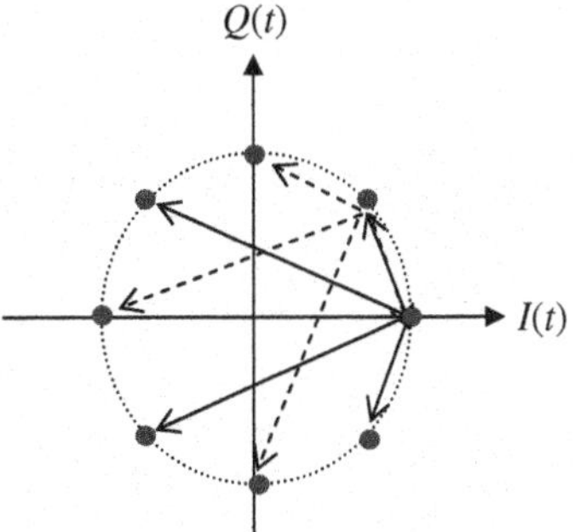

FIGURE 4.20 $\pi/4$-DQPSK modulation constellation state diagram.

The DD output signal has a constellation diagram shown in Fig. 4.21 with 4 states. Each state (or symbol) corresponds to 2 bits where this bit-to-symbol mapping was discussed earlier in Chap. 2, with the help of the phase state table.

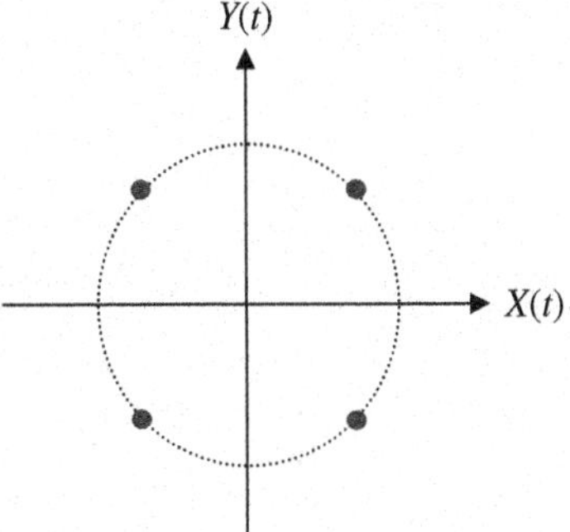

FIGURE 4.21 $\pi/4$-DQPSK differential detector output constellation state diagram.

IF Approach of DD. The second variation will be the IF implementation of the DD. The following block diagram captures its functionality (see Fig. 4.22) [39]. Similar to the baseband DD where the delayed version of the received signal is used as a reference for the present received signal, the IF approach follows suit. The phase difference operations will be shown below with the assistance of the equations that follow.

Let us assume the signal at the output of the BPF shown in the block diagram is given as (in the absence of noise)

$$r(t) = A(t) \cdot \cos[\omega_c t + \theta(t)] \tag{4.59}$$

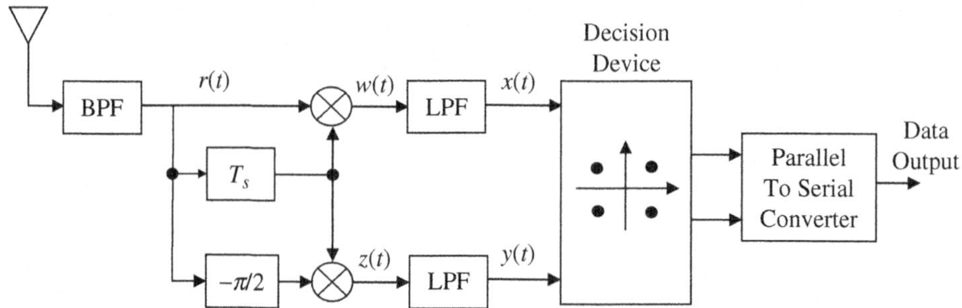

FIGURE 4.22 IF differential detection block diagram.

The output of the first multiplier is

$$w(t) = A(t) \cdot A(t - T_s) \cdot \cos[\omega_c t + \theta(t)] \cdot \cos[\omega_c(t - T_s) + \theta(t - T_s)] \tag{4.60}$$

And after the top LPF we have

$$x(t) = A(t) \cdot A(t - T_s) \cdot \frac{1}{2} \cdot \cos[\omega_c T_s + \theta(t) - \theta(t - T_s)] \tag{4.61}$$

Similarly, the output of the bottom multiplier is

$$z(t) = A(t) \cdot A(t - T_s) \cdot \cos[\omega_c(t - T_s) + \theta(t - T_s)] \cdot \sin[\omega_c t + \theta(t)] \tag{4.62}$$

After the bottom LPF we have

$$y(t) = A(t) \cdot A(t - T_s) \cdot \frac{1}{2} \cdot \sin[\omega_c T_s + \theta(t) - \theta(t - T_s)] \tag{4.63}$$

Now if we assume $\omega_c T_s = k \cdot 2\pi$ where $k = 0, 1, 2, \ldots$, then the output of the IF DD is the same as that of the baseband DD. What we have basically done was to jointly perform frequency down conversion and noncoherent DD in a single operation. The baseband signals are given as

$$x(t) = A(t) \cdot A(t - T_s) \cdot \frac{1}{2} \cdot \cos[\Delta\phi(t)] \tag{4.64}$$

$$y(t) = A(t) \cdot A(t - T_s) \cdot \frac{1}{2} \cdot \sin[\Delta\phi(t)] \tag{4.65}$$

These are the same equations presented earlier in the baseband DD section. At this point a general comment should be made on this type of receiver architecture. With this approach the receiver performance is fixed to be equal to that of noncoherent detection or differential detection. In the previous section the detector could have been replaced with a coherent detector, if so desired. Another point to mention is that this joint frequency down conversion and detection is performed in the analog domain, hence inaccuracies/tolerances should be considered on the overall system performance.

Discriminator with Integrate and Dump (I&D) Detection. The third variation will be the discriminator with integrate and dump implementation of a noncoherent detector. The following block diagram shows the operations (see Fig. 4.23). These operations are very similar to what a paging receiver and an AMPS demodulator would use [40–42].

The received signal at the output of the BPF is given as follows (assuming no noise is present):

$$r(t) = A(t) \cdot \cos[\omega_c t + \theta(t)] \tag{4.66}$$

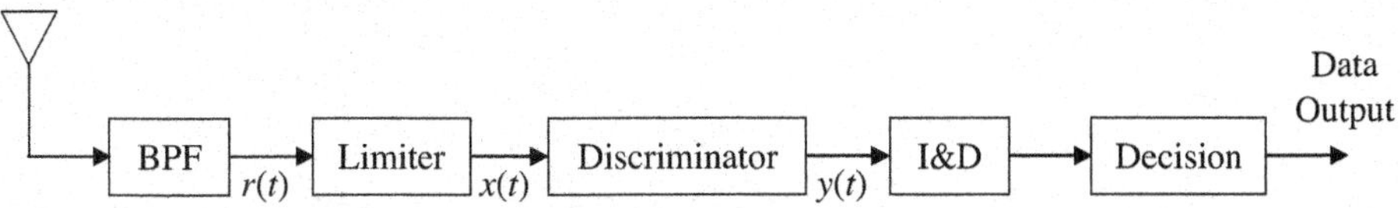

FIGURE 4.23 Limiter-discriminator w/I&D block diagram.

The limiter output is given below (assuming the limiter has not distorted the signal's phase, but only amplitude).

$$x(t) = K \cdot \cos[\omega_c t + \theta(t)] \tag{4.67}$$

The model for the discriminator is given as

$$\omega_i(t) = \frac{d}{dt}\{\theta_i(t)\} \tag{4.68}$$

Then we have the output written as follows, where $m(t)$ is the transmitted information signal.

$$y(t) = \frac{d}{dt}\{x(t)\} = -\left(\omega_c + \frac{d}{dt}\theta(t)\right) \cdot \sin[\omega_c t + \theta(t)] \tag{4.69}$$

Looking only at the envelope of this signal gives the following:

$$y(t) = \frac{d}{dt}\left\{2\pi k_f \int_0^t m(t)\,dt\right\} \tag{4.70}$$

$$y(t) = 2\pi k_f \cdot m(t) \tag{4.71}$$

Assuming our $\pi/4$-DQPSK reference modulation scheme, then the discriminator output resembles the following eye diagram where we have identified the phase offset changes to obtain a certain eye opening (see Fig. 4.24). A symbol time duration is shown.

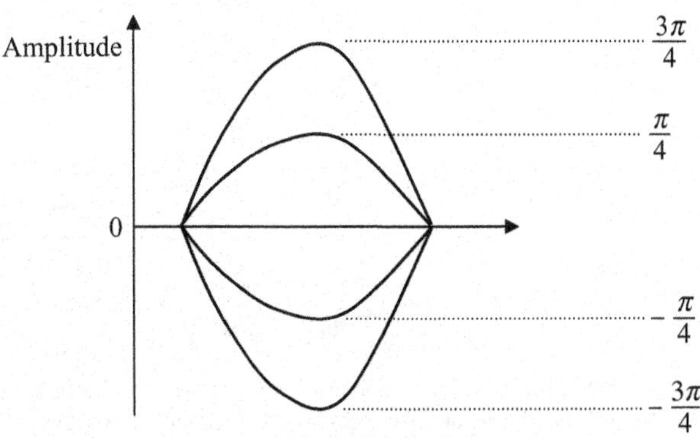

FIGURE 4.24 $\pi/4$-DQPSK discriminator output eye diagram.

Now after the I&D operations we have the modified eye diagram shown in Fig. 4.25 where we have also shown the phase offset change associated with an eye opening.

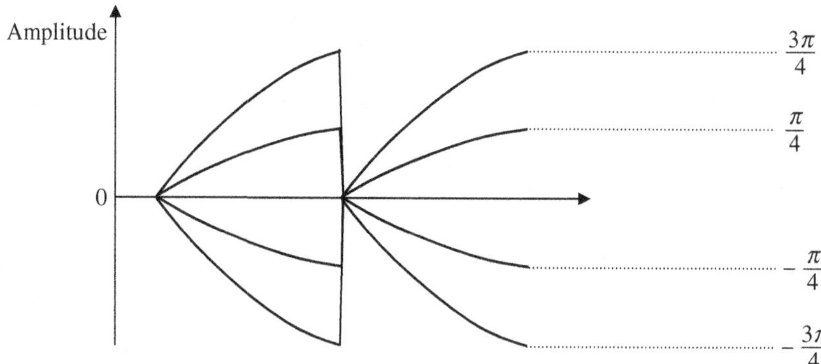

FIGURE 4.25 $\pi/4$-DQPSK I&D output eye diagrams.

With these eye diagrams it is easy to see a multilevel thresholding operation is required in the decision process. The I&D operation after the discriminator is there to behave as a post detection filter to improve performance.

As a side note, the discriminator can also be implemented in the Polar coordinate system as follows where the quadrature demodulator performed spectral down conversion with ideal knowledge of the carrier frequency and not its phase (see Fig. 4.26).

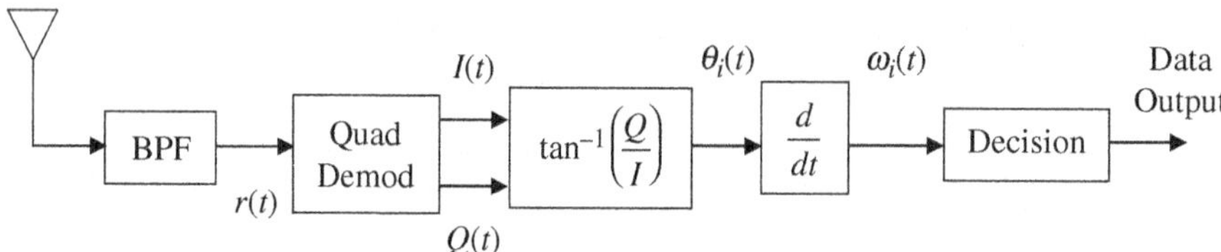

FIGURE 4.26 Baseband implementation of the discriminator detector.

The instantaneous phase of the received signal is given as

$$\theta_i(t) = \tan^{-1}\left(\frac{Q(t)}{I(t)}\right) \tag{4.72}$$

And the instantaneous frequency is given as

$$\frac{d}{dt}\theta_i(t) = \frac{I(t)\cdot\frac{d}{dt}Q(t) - Q(t)\cdot\frac{d}{dt}I(t)}{I^2(t) + Q^2(t)} \tag{4.73}$$

$$\frac{d}{dt}\theta_i(t) = \omega_i(t) \tag{4.74}$$

Various methods can be used to implement the differentiator. One such method is a capacitor in the analog domain, also the reader is referred to [43] for various methods to be performed in the digital domain.

Now let us step back to briefly comment on the BER performance of coherent and noncoherent detection. In comparing the signal processing operations required to perform the PSAM and DD, the former is more complex. Hence in using the differential detector we have simplified the receiver architecture but sacrificed BER performance (or E_b/N_o)[1]. Figure 4.27 shows the BER performance in

[1] Some authors use the phrase "SNR per bit" as an alternative to E_b/N_o.

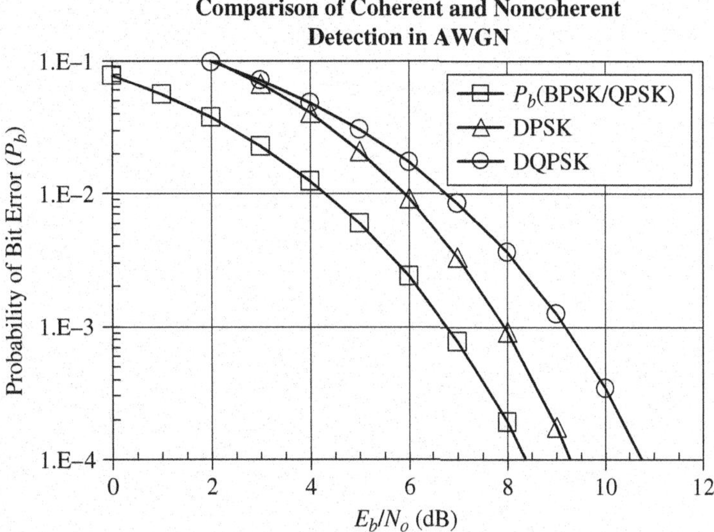

FIGURE 4.27 Comparison of coherent and noncoherent detection in AWGN.

an AWGN (static) channel for BPSK, QPSK, DPSK, and DQPSK modulation schemes. Coherent detection was used for BPSK/QPSK and noncoherent detection was used for DPSK/DQPSK.

Here we see that we have sacrificed 2.5 dB of E_b/N_o when the receiver phase offset was neglected, not estimated nor compensated. The mathematical equations used to plot Fig. 4.27 are given in Appendix 4A.

Similarly, we can plot the BER performance in a flat fading (dynamic) channel below for BPSK, QPSK, DPSK, and DQPSK modulation schemes (see Fig. 4.28).

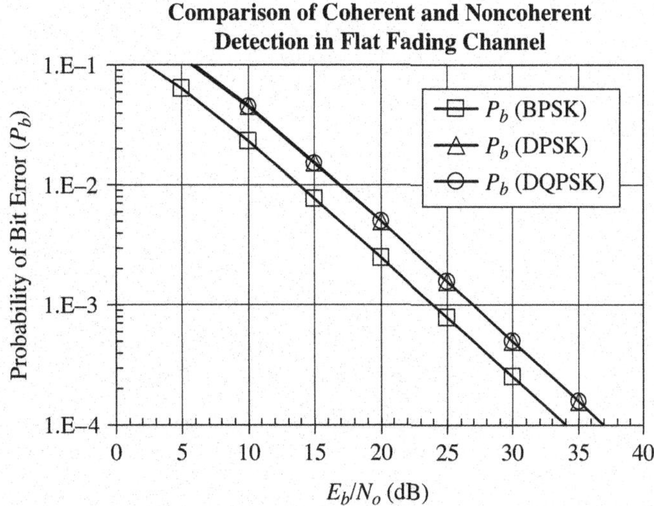

FIGURE 4.28 Comparison of coherent and noncoherent detection in Rayleigh fading.

Here we see approximately 3 dB penalty in E_b/N_o when we use noncoherent detection instead of coherent detection. In many communication systems this loss is significant and must be made up (or compensated for) elsewhere in the link budget. On the other hand, this loss must be discussed in the context of implementation complexity. In some cases the more critical issue may be receiver complexity and in this case the price paid is performance. However as discussed above, this loss in E_b/N_o must be considered in the overall communication system link budget. Various parameters can be optimized to help reduce or eliminate this loss such as: coding gain, output power, noise figure, antenna gain, cell size, and so on.

In the next noncoherent detection technique we show how this 3 dB performance gap can be made smaller by increasing the complexity of this simple noncoherent receiver. In other words, by widening the observation window from 2 symbols for the conventional differential detector to a larger value, performance gain can be obtained. Hence we begin to close the performance gap shown in Fig. 4.28.

4.3.2 Multiple Symbol DD

In this noncoherent technique the observation window used to estimate the transmitted signal is increased resulting in desirable performance improvement [44–51]. The conventional DD discussed in the previous section has an observation window of 2 symbols. In other words, we observe 2 symbols to make a decision on 1 symbol. The extension to this is to use N symbols to make a decision on N-1 symbols. Let us now formulate the problem and rederive the decision metric to be used in the receiver (given earlier in [52]).

Let us assume the transmitted signal is given as (notice the constant envelop notation)

$$s(k) = e^{j\phi(k)} \tag{4.75}$$

where $\phi(k)$ is the modulation phase at the time instant k, for our $\pi/4$-DQPSK modulation scheme the phases are

$$\phi(k) = \frac{\pi}{4} \cdot m \qquad (m = 0, 1, 2, \ldots, 7) \tag{4.76}$$

The received signal is given as follows (assuming an AWGN channel)

$$r(k) = s(k) \cdot e^{j\theta(k)} + n(k) \tag{4.77}$$

where the channel has inserted a phase offset of $\theta(k)$ and we have further assumed the phase is constant over the observation window and $n(k)$ is the AWGN component of the received signal. This obviously depends on the actual system design, but in general is a descent assumption.

$$r(k) = e^{j[\phi(k)+\theta(k)]} + n(k) \tag{4.78}$$

We can write the a posteriori probability of r given s and θ as follows (assuming a sequence of length N samples)

$$P(r|s, \theta) = \frac{1}{(2\pi\sigma_n^2)^N} \cdot e^{-\frac{\|r - s \cdot e^{j\theta}\|^2}{2\sigma_n^2}} \tag{4.79}$$

where σ_n^2 = the noise variance. Please note we have dropped the dependency on the time variable, k, in the above equations as well as the equations that follow, for the simple reason of not cluttering the equations.

The above expression can be expanded and rewritten to equal.

$$P(r|s,\theta) = \frac{1}{\left(2\pi\sigma_n^2\right)^N} \cdot e^{-\frac{1}{2\sigma_n^2}\left\{\sum_{i=0}^{N-1}\left(|r_{k-i}|^2 + |s_{k-i}|^2\right) - 2\left|\sum_{i=0}^{N-1} r_{k-i} \cdot s_{k-i}^*\right| \cdot \cos[\theta - \alpha]\right\}} \tag{4.80}$$

where the following phase variable has been defined.

$$\alpha = \tan^{-1}\left\{\frac{\operatorname{Im} g\left[\sum_{i=1}^{N-1} r_{k-i}\cdot s_{k-i}^{*}\right]}{\operatorname{Re}\left[\sum_{i=1}^{N-1} r_{k-i}\cdot s_{k-i}^{*}\right]}\right\} \tag{4.81}$$

Since this is a noncoherent technique, the channel phase offset is unknown to us, so we wish to remove it from the above equation in the following manner:

$$P(r|s) = E_\theta\{P(r|s,\theta)\} \tag{4.82}$$

$$P(r|s) = \int_{-\pi}^{\pi} P(r|s,\theta)p(\theta)d\theta \tag{4.83}$$

Here we have assumed the channel phase offset is a uniformly distributed random variable with a probability density function (PDF) equal to $p(\theta)$. The details of the integration are left to the reader and so we simply state the result. ($I_o(x)$ = modified, zeroth order Bessel function)

$$P(r|s) = \frac{1}{\left(2\pi\sigma_n^2\right)^N}\cdot e^{-\frac{1}{2\sigma_n^2}\sum_{i=0}^{N-1}\left(|r_{k-i}|^2+|s_{k-i}|^2\right)}\cdot I_o\left(\frac{1}{\sigma_n^2}\left|\sum_{i=0}^{N-1} r_{k-i}\cdot s_{k-i}\right|\right) \tag{4.84}$$

In order to obtain the maximum likelihood estimate (MLE) we make use of the following approximation:

$$\ln[I_o(x)] \cong \frac{x^2}{4} \tag{4.85}$$

Hence maximizing the a posteriori probability of r given s is equivalent to maximizing the following expression:

$$\max_{\phi}\left|\sum_{i=0}^{N-1} r_{k-i}\cdot s_{k-i}^{*}\right|^2 \tag{4.86}$$

$$\max_{\phi}\left|\sum_{i=0}^{N-1} r_{k-i}\cdot e^{-j\phi_{k-i}}\right|^2 \tag{4.87}$$

We now have an MLE metric that is dependent on the possibilities of the transmitted phases. We wish to have a term resembling the differentially encoded data. This is accomplished as follows. First we notice that adding a phase ambiguity, say θ_a, to all the estimated phases has an identical decision rule.

$$\max_{\phi}\left|\sum_{i=0}^{N-1} r_{k-i}\cdot e^{-j[\phi_{k-i}+\theta_a]}\right|^2 \tag{4.88}$$

where θ_a is assumed to be uniform. If we let $\theta_a = -\phi_{k-N+1}$ then the metric becomes

$$\max_{\phi}\left|\sum_{i=0}^{N-1} r_{k-i}\cdot e^{-j[\phi_{k-i}-\phi_{k-N+1}]}\right|^2 \tag{4.89}$$

And assuming differential encoding of phases, $\Delta\phi_{k-i} = \phi_{k-i} - \phi_{k-i-1}$, the decision rule becomes

$$\max_{\Delta\phi}\left|\sum_{i=0}^{N-1} r_{k-i}\cdot e^{-j\sum_{m=0}^{N-i-2}\Delta\phi_{k-i-m}}\right|^2 \tag{4.90}$$

Here we use an observation of N symbols to make a block (or joint) decision of N–1 symbols. An important point to make here is, as the observation window (N) approaches infinity, the performance of the maximum likelihood differential detection (MLDD) noncoherent detection technique approaches coherent detection with differential decoding [52].

Let us say, for example, we have an interest to set $N = 2$, then the decision rule is

$$\max_{\Delta\phi}\left|\sum_{i=0}^{1} r_{k-i}\cdot e^{-j\sum_{m=0}^{-i}\Delta\hat{\phi}_{k-i-m}}\right|^2 \tag{4.91}$$

$$\max_{\Delta\phi}\left|r_k\cdot e^{-j\Delta\hat{\phi}_k}+r_{k-i}\right|^2 \tag{4.92}$$

$$\max_{\Delta\phi}\left\{\left|r_{k-i}\right|^2+\left|r_k\right|^2+2\mathrm{Re}\left(r_k\cdot r_{k-1}^*\cdot e^{-j\Delta\hat{\phi}_k}\right)\right\} \tag{4.93}$$

The decision rule can be further written as follows, where we have removed the first two terms since they are independent of $\Delta\hat{\phi}k$.

$$\max_{\Delta\hat{\phi}_k}\mathrm{Re}\left\{r_k\cdot r_{k-1}^*\cdot e^{-j\Delta\hat{\phi}_k}\right\} \tag{4.94}$$

Now what does this say: We will try all the possible transmitted phases in the form of $\Delta\phi_k$ and then choose the phase that produces the largest metric value. We perform an exhaustive search over all possible phase changes and then choose the one that was most likely transmitted. This can be shown in the following block diagram (see Fig. 4.29). The differential detector output symbol is compared

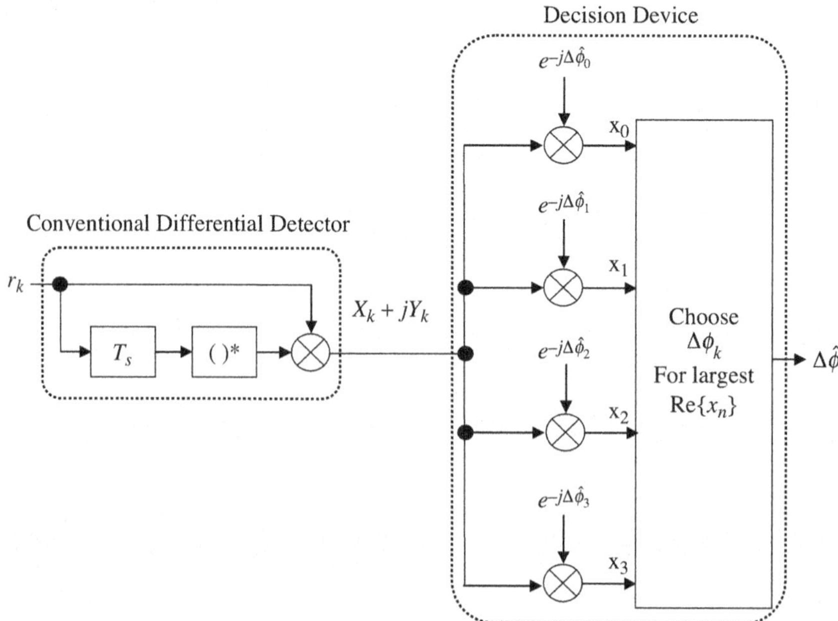

FIGURE 4.29 Multiple symbol DD ($N = 2$ symbols) receiver block diagram.

against all possible phase changes by rotating the symbol to the positive real axis. This is essentially performing a nearest neighbor operation.

Here we first perform a phase difference operation and then we exhaustively compare this received phase difference to all the possible transmitted phase differences. Let us say we have transmitted a $\pi/4$-DQPSK modulation then the possible received candidates are

$$\Delta\hat{\phi}_0 = -\frac{\pi}{4}, \Delta\hat{\phi}_1 = \frac{3\pi}{4}, \Delta\hat{\phi}_2 = \frac{\pi}{4}, \Delta\hat{\phi}_3 = -\frac{3\pi}{4} \tag{4.95}$$

If $\pi/4$ was the actual transmitted symbol then the $\text{Re}\{x_2\}$ would be the largest and thus we would choose$\Delta\hat{\phi}_2$. The DD output symbol is compared against the complex conjugate of all the possible phase changes. This was accomplished in order to have a projection onto the real axis.

An alternative way to view this is that the block diagram has two components: a DD and a decision part. The first dashed block is, in fact, a conventional DD while the second dashed block is performing the operations of a decision. Earlier we presented a block diagram for the DD where the decision device was drawn to actually perform the nearest neighbor function. Here the same function is essentially performed.

What this shows us is that the conventional DD is indeed a max a posteriori probability detector. Up until this point the conventional DD was derived from basic intuition given the differential encoding rule used at the transmitter. There we had one equation with one unknown and so the detector rules were trivial to write down.

We must now present the decision rule where we observe 3 symbols and make a joint decision on 2 symbols in the observation window. We can simply write down the decision rule to be

$$\max_{\Delta\hat{\phi}_k, \Delta\hat{\phi}_{k-1}} \text{Re}\{r_k \cdot r_{k-1}^* \cdot e^{-j\Delta\hat{\phi}_k} + r_{k-1} \cdot r_{k-2}^* \cdot e^{-j\Delta\hat{\phi}_{k-1}} + r_k \cdot r_{k-2}^* \cdot e^{-j[\Delta\hat{\phi}_k + \Delta\hat{\phi}_{k-1}]}\} \tag{4.96}$$

Here we see the metric consists of three parts. The first part corresponds to an ordinary DD using an observation window of 2 symbols. The second part also corresponds to an ordinary DD except the inputs have been delayed by 1 symbol. The third part corresponds to a 2-symbol DD using an observation window of 3 symbols. In other words, we basically have two DDs that are operating in a serial fashion with the third DD computing the sum of the individual DD outputs. Alternatively stated, there are two conventional DDs ($N = 2$) operating on symbols of consecutive time intervals. The third detector is basically calculating a metric that is the sum of the individually calculated metrics. All the possible combinations are exhaustively compared and the sequence with the largest projection on the real axis is chosen.

A note to make here is that this block decoder performs an exhaustive search over all possible DQPSK sequences of length N, to provide the MLE of the transmitted symbol. Since this is QPSK the number of searches is 4^{N-1}. For 3-symbol observation window this equals 16 combinations. As we will soon see at the end of this chapter, increasing N will improve performance at the expense of increase complexity. As complexity grows the system designer must evaluate system performance to determine if rather a coherent detector is required. A block diagram for this MLDD using 3-symbol observation window is given in Fig. 4.30.

4.3.3 Decision Feedback Differential Detection (DF-DD)

In this noncoherent technique we will make use of past detected symbols in order to improve the detection of the present symbol [53–55]. This was shown in [56] where the resulting decision metric equations are rewritten here to establish a baseline in understanding and to be used in the performance comparison study that follows. Lastly, it forms an alternative representation of DD/decoding. Let us assume the transmitted phases are given as.

$$\phi_n = \phi_{n-l} + \sum_{i=0}^{l-1} \Delta\phi_{n-i} \tag{4.97}$$

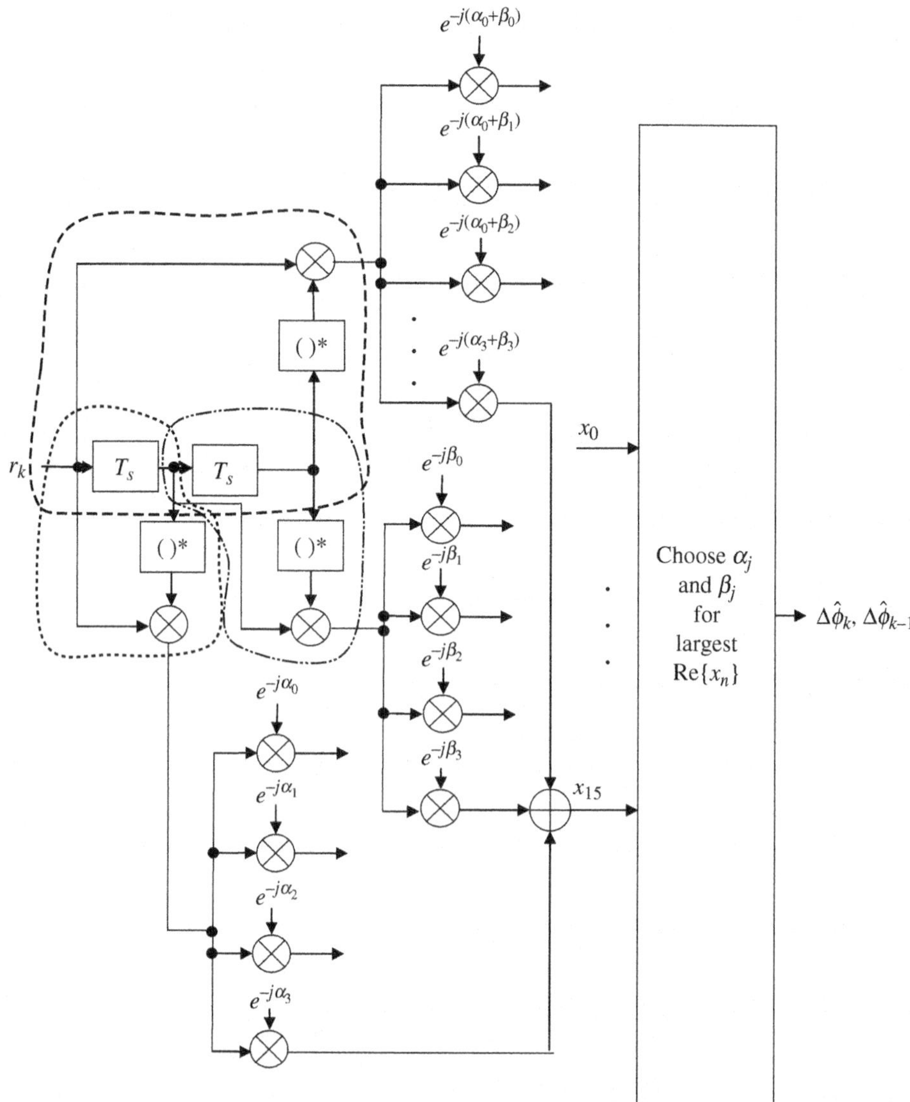

FIGURE 4.30 Multiple symbol DD (N = 3 symbols) receiver block diagram.

The output of an l-symbol DD is (ψ_n = received phase at the nth symbol time) given as the sum of l–1 phase differences.

$$\Delta\psi_n(l) = \psi_n - \psi_{n-l} \tag{4.98}$$

$$\Delta\psi_n(l) = \sum_{i=0}^{l-1} \Delta\phi_{n-i} \tag{4.99}$$

We wish to maximize the a posteriori probability given as (where R = noise covariance matrix and U = noise vector).

$$p(\Delta\psi|\Delta\phi) = \frac{1}{(2\pi)^{l/2} \cdot |R|^{1/2}} \cdot e^{-\frac{1}{2} \cdot U^T \cdot R^{-1} \cdot U} \tag{4.100}$$

The steps of the derivation are shown in [56] and left out for the sake of convenience. The decision rule becomes the following:

$$\Delta\hat{\phi}_n = \min_{\Delta\phi_n} \sum_{l=1}^{L} [\mu_l]^2 \tag{4.101}$$

$$\Delta\hat{\phi}_n = \min_{\Delta\phi_n} \sum_{l=1}^{L} \left[\Delta\psi_n(l) - \Delta\phi_n - \sum_{i=1}^{l-1} \Delta\hat{\phi}_{n-i} \right]^2 \tag{4.102}$$

For $L = 2$ symbols we have the following:

$$\Delta\hat{\phi}_n = \min\{\mu_1^2 + \mu_2^2 - \mu_1 \cdot \mu_2\} \tag{4.103}$$

For $L = 3$ symbols we have the following:

$$\Delta\hat{\phi}_n = \min\left\{ \mu_1^2 + \mu_2^2 + \mu_3^2 - \frac{2}{3} \cdot (\mu_1 \cdot \mu_2 + \mu_1 \cdot \mu_3 + \mu_2 \cdot \mu_3) \right\} \tag{4.104}$$

Ignoring the cross terms in the above decisions will allow us to approximate the $L = 2$ solution with the following:

$$\Delta\hat{\phi}_n \cong \min\{\mu_1^2 + \mu_2^2\} \tag{4.105}$$

$$\Delta\hat{\phi}_n \cong \min\{[\Delta\psi_n(1) - \Delta\phi_n]^2 + [\Delta\psi_n(2) - \Delta\phi_n - \Delta\phi_{n-1}]^2\} \tag{4.106}$$

We can alternatively show an iterative solution based on a 1-symbol DD using the following $\{\Delta\psi_n(2) = \Delta\psi_n(1) + \Delta\psi_{n-1}(1)\}$

$$\mu_2 = \Delta\phi_n + \Delta\hat{\phi}_{n-1+} - \Delta\psi_n(1) - \Delta\psi_{n-1}(1) \tag{4.107}$$

So the DF-DD block diagram for $L = 2$ symbols is shown in Fig. 4.31. Each of the four $\Delta\phi_n$ candidates is searched.

4.3.4 Nonredundant Error Correction (NEC)

In this section we will present the NEC demodulation technique for $\pi/4$-DQPSK modulation [57, 58]. Note with very simple modifications this technique can be used for MDPSK and MSK waveforms [59]. This principle is similar to other error correction techniques, a syndrome is calculated to determine the presence of an error. If it is present then it is corrected. The only difference is that the transmitter did not insert redundancy or parity bits into the data stream; here we work directly on the waveform and make use of differential encoding properties to aid error correction. A significant amount of technical information is available in the literature that discusses this special property [60, 57, 61–63].

Assuming the phase symbols are differentially encoded, then the transmitted signal is (at the ith symbol interval) equal to.

$$\theta_i = \theta_{i-1} + \Delta\phi_i \tag{4.108}$$

where the four allowable phase changes are given as

$$\Delta\phi_i = \frac{\pi}{4} \cdot a_i \qquad a_i \in \{1,3,5,7\} \tag{4.109}$$

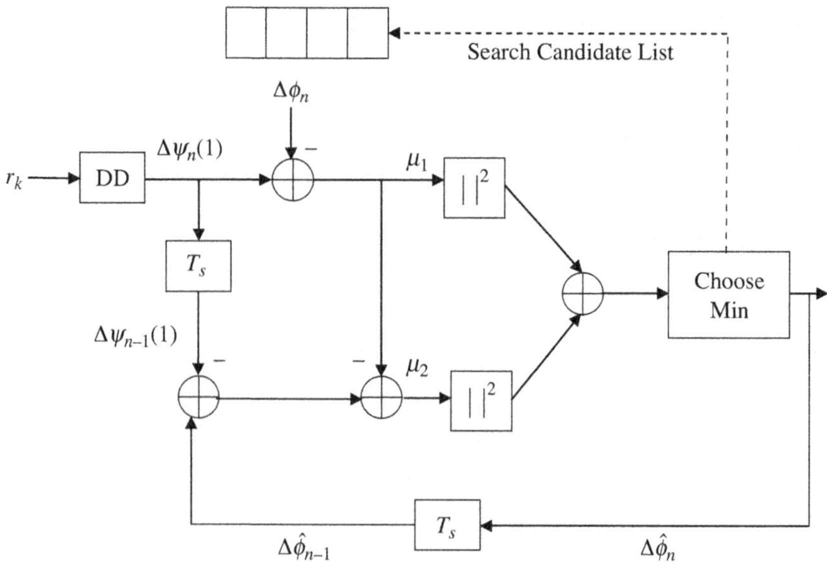

FIGURE 4.31 DF-DD block diagram for $L = 2$ symbols.

Now the output of a general kth order DD at the ith symbol interval is written as

$$r_{k,i} = \frac{\pi}{4} \cdot \left[\sum_{j=0}^{k-1} a_{i-j}\right] \bmod 8 = \theta_i - \theta_{i-k} \tag{4.110}$$

Let us consider the 1st and 2nd order DD output as

$$r_{1,i} = \frac{\pi}{4} \cdot a_i \tag{4.111}$$

$$r_{2,i} = \frac{\pi}{4} \cdot [a_i + a_{i-1}] \bmod 8 \tag{4.112}$$

Neglecting the $\pi/4$ constant multiplier then the output of the kth order DD is given as (in the absence of any errors)

$$d_{k,i} = \left[\sum\nolimits_{j=0}^{k-1} a_{i-j}\right] \bmod 8 \tag{4.113}$$

where

$$d_{k,i} = \begin{cases} 0,2,4,6 & k = \text{even} \\ 1,3,5,7 & k = \text{odd} \end{cases} \tag{4.114}$$

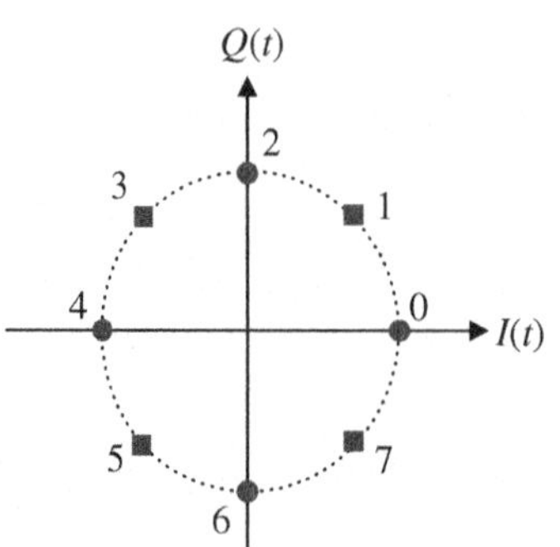

FIGURE 4.32 $\pi/4$-DQPSK signal constellation diagram mapping.

The constellation is quantized to the following rule where we have assigned a value to each of the possible phase states (see Fig. 4.32).

The differential detector output signal is composed of the transmitted symbol plus the error signal described as follows:

$$r_{k,i} = d_{k,i} + e_{k,i} \tag{4.115}$$

If the differential detector output is error-free than the error, $e_{k,i}$ would equal 0.

In order to calculate the syndromes we will sum the $(k+1)^{\text{th}}$ order DD output and the k successive 1st order DD outputs as

$$s_{k,i} = \left[\sum_{j=0}^{k} r_{1,i-j} - r_{k+1,i}\right] \bmod 8 \tag{4.116}$$

$$s_{k,i} = \left[\sum_{j=0}^{k}(d_{i-j} + e_{1,i-j}) - \sum_{j=0}^{k} d_{i-j} + e_{k+1,i}\right] \bmod 8 \tag{4.117}$$

$$s_{k,i} = \left[\sum_{j=0}^{k} e_{1,i-j} - e_{k+1,i}\right] \bmod 8 \tag{4.118}$$

For a single-error correction detector the syndromes become

$$s_{1,i} = [e_{1,i} + e_{1,i-1} - e_{2,i}] \bmod 8 \tag{4.119}$$

$$s_{1,i-1} = [e_{1,i-1} - e_{2,i-1}] \bmod 8 \tag{4.120}$$

The error would be able to be corrected with the following rule inside look up Table 4.1 (LUT) where $N \in \{1,3\}$.

A single error correcting NEC block diagram is shown in Fig. 4.33, where we have chosen to show the three components of the receiver. These consist of differential detectors, syndrome generators, and error detector and correction circuit.

Let us attempt to not only provide insight into the above receiver, but also to compare the present governing mechanism to the receivers given earlier. The 1NEC receiver is essentially comparing the same differentially detected symbols in the decision process. The signal path across the top of the figure computes

TABLE 4.1 Single-Error Correction NEC Table Look-up

$S_{1,i}$	$S_{1,i-1}$	$e_{1,i-1}$
0	0	0
0	N	0
N	0	0
N	N	N

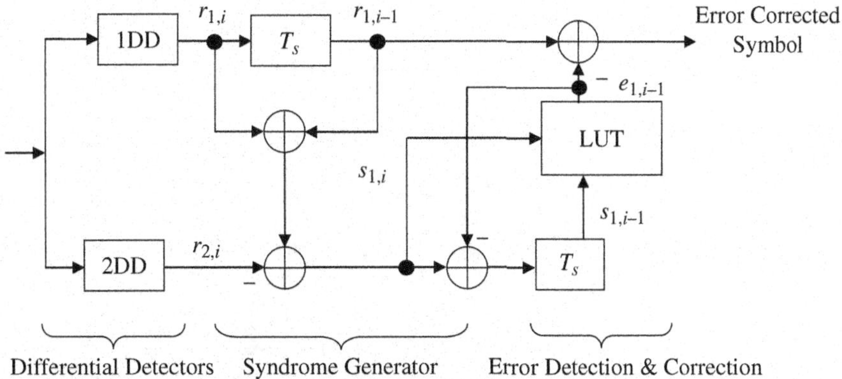

FIGURE 4.33 Single error correcting NEC block diagram.

the sum of adjacent phases, the signal path across the bottom of the figure computes the phase difference of nonadjacent symbols. Very simply put: The sum of adjacent phase differences should add to the difference of the summed phases. When this condition is met, there is no need to believe an error is present in the group of symbols. Although we must mention it is possible to have errors inserted into the group of observed symbols that would be undetectable. When this condition is not met, then the syndromes will be used to detect the errors so they can be corrected.

It has been shown that for an MDPSK signal using DD of order 1 to L, the output is a code sequence of a rate $1/L$ convolutional code which has L–1 error correction capability. Above we have presented the single-error correction NEC, we now wish to show the double-error correction NEC. Two errors can be detected and corrected using the following six syndromes.

$$s_{1,i} = (e_{1,i} + e_{1,i-1} - e_{2,i}) \bmod 8 \tag{4.121}$$

$$s_{1,i-1} = (e_{1,i-1} + e_{1,i-2} - e_{2,i-1}) \bmod 8 \tag{4.122}$$

$$s_{1,i-2} = (e_{1,i-2} - e_{2,i-2}) \bmod 8 \tag{4.123}$$

$$s_{2,i} = (e_{1,i} + e_{1,i-1} + e_{1,i-2} - e_{3,i}) \bmod 8 \tag{4.124}$$

$$s_{2,i-1} = (e_{1,i-1} + e_{1,i-2} - e_{3,i-1}) \bmod 8 \tag{4.125}$$

$$s_{2,i-2} = (e_{1,i-2} - e_{e,i-2}) \bmod 8 \tag{4.126}$$

Once again we would like to point out modulo 8 operations, which was due to the fact that $\pi/4$-DQPSK has 8 phase states. As we will encounter in the next section for GMSK the operations performed consist of modulo 2 operations.

The double error correcting NEC block diagram is shown in Fig. 4.34. Here we show the six syndromes are being calculated and used to determine the error signal $e_{i,j-2}$.

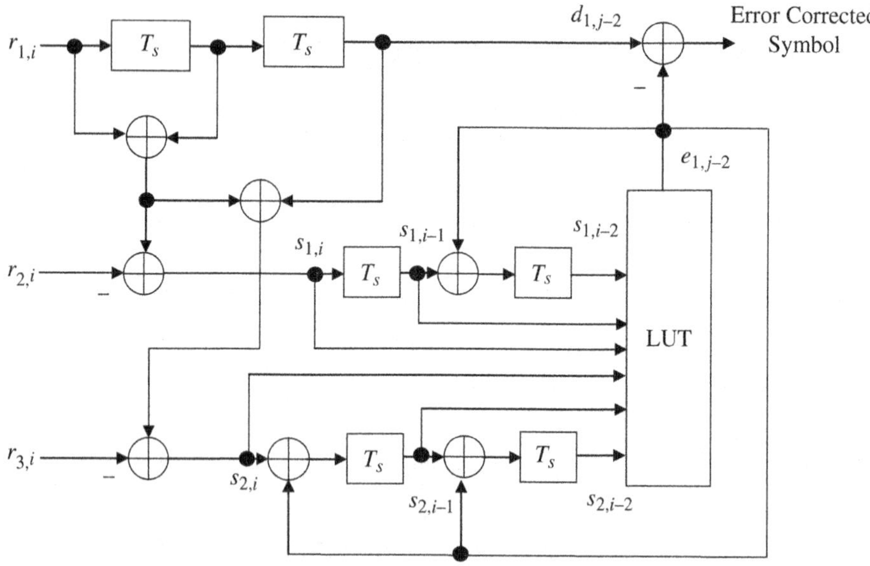

FIGURE 4.34 Double-error correcting NEC block diagram.

The look-up table used in the error pattern calculation for the double-error correction NEC is shown in Table 4.2 $N \in \{1,3\}$ and $M \in \{0,1,3\}$.

In this subsection we have presented a single-error correction (1NEC) and double-error correction (2NEC) receiver. In fact, higher error correcting receivers exist and have been omitted from this chapter due to presentation complexity as well as they can be obtained by extending the techniques presented herein.

TABLE 4.2 2NEC Table Look-Up

$S_{1,i}$	$S_{1,i-1}$	$S_{1,i-2}$	$S_{2,i}$	$S_{2,i-1}$	$S_{2,i-2}$	$e_{1,j-2}$
M	N	N	$M+N$	N	N	N
M	$M+N$	N	$M+N$	$M+N$	N	N
M	N	N	N	N	N	N
0	$N-M$	N	N	N	N	N
0	N	$N-M$	N	N	N	N
0	N	N	$N-M$	N	N	N
0	N	N	N	$N-M$	N	N
0	N	N	N	N	$N-M$	N

Prior to moving to the next section, where MSK will be addressed, we would like to provide a brief summary of the noncoherent detection techniques presented. We began with the conventional DD, as well as its alternative representations (i.e., IF DD, limiter discriminator w/I&D), then increased the observation window used in the detector. This introduced the MLDD, DF-DD, and the NEC family of detectors.

4.4 NONCOHERENT DETECTION OF MSK

In the following subsections we will present various noncoherent detection techniques used to demodulate MSK waveforms. We will first start with the conventional DD to exploit some inherent properties of MSK. Then widen the observation window used to make a decision on the transmitted symbol. We will see that the noncoherent detection techniques previously presented for DQPSK waveforms can be easily modified to be used for the MSK waveforms [64].

4.4.1 Differential Detection

In this section we will show how to use the symbol-based DD to detect MSK waveforms [65, 66]. Recall Chap. 2 discussed MSK modulation to be a member of the BFSK family of modulation schemes with a modulation index, β, equal to $^1/_2$. And for sake of review the allowable phase change in a 1 symbol time interval was

$$\Phi = \omega T_s = 2\pi \cdot \Delta f \cdot T_s \tag{4.127}$$

$$\Phi = \beta\pi = \frac{\pi}{2} \tag{4.128}$$

The MSK signal is expressed as [with data bit defined as $a_k \in \{-1, +1\}$ and the peak frequency deviation given as ($f_d = R_b/4 = 1/4T$) where R_b = bit rate].

$$s(t) = \cos\left[\omega_c t + \frac{\pi}{2T} \cdot a_k \cdot t\right] \tag{4.129}$$

This provides us with a phase change described by

$$\theta(t) = \frac{\pi}{2T} \cdot a_k \cdot t \tag{4.130}$$

Then the phase difference between consecutive MSK symbols is written as

$$\theta(t) - \theta(t - T) = \frac{\pi}{2} \cdot a_k \tag{4.131}$$

This shows us that the phase change of the MSK modulation scheme is either +90° or −90° depending on the information bit to be transmitted.

Hence we can model the transmitter with the following differential encoding rule.

$$\theta_k = \theta_{k-1} + \Delta\phi_k \tag{4.132}$$

where using the previous definitions provide us with the following phase difference notation.

$$\Delta\phi_k = \frac{\pi}{2} \cdot a_k \tag{4.133}$$

The conventional differential detector will be described as following in Polar coordinates

$$\theta_k - \theta_{k-1} = \Delta\phi_k \tag{4.134}$$

And as follows in the Cartesian coordinate system

$$I_k + jQ_k = \cos[\Delta\phi_k] + j\sin[\Delta\phi_k] \tag{4.135}$$

Under the ideal conditions the imaginary component will carry the information while the real component would be constant. However, due to the presence of receiver frequency offsets, the waveform will rotate and thus the modulation information will spill over from the imaginary component to the real component [67].

A block diagram of a receiver using the DD to extract the information bits from MSK waveforms is given in Fig. 4.35. The assumption used is that there is no frequency uncertainty present.

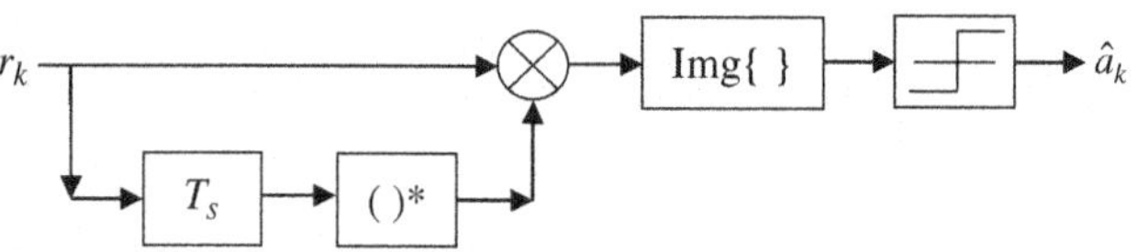

FIGURE 4.35 Differential detection-based receiver for MSK.

Discriminator Detection w/I&D of MSK. In this section we will show the discriminator detection of MSK modulation. The discriminator detector is mathematically the same as presented early in this chapter. A block diagram of a GMSK-based communication system is shown in Fig. 4.36 in the complex envelope domain.

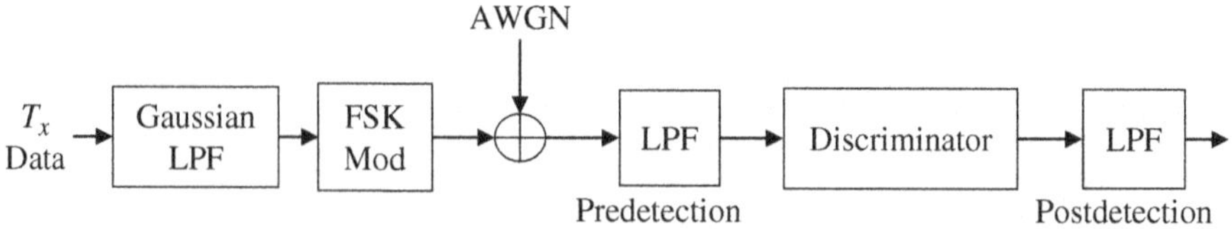

FIGURE 4.36 GMSK-based communication system block diagram.

Please refer to the discriminator equations presented in Section 4.3.1. The major difference is that for MSK, the limiter discriminator output signal would ideally consist of 2 levels, since only 2-phase changes are allowed. However, we immediately follow the above statement with this caution. Recall in Chap. 2, the BT product controlled the spectral efficiency of GMSK, but this came at the expense of inserting intersymbol interference into the transmitted signal. Hence as the BT product reduces the number of distinct levels at the limiter discriminator output begins to be less obvious [68].

Moreover, techniques such as equalizers can be used to not only remove the ISI intentionally introduced at the transmitter, but also the ISI inserted by the FSF wireless channel [69–72].

4.4.2 Nonredundant Error Correction of MSK

In this section we will present a noncoherent technique that is very similar to that used in the $\pi/4$-DQPSK section given earlier in this chapter [73, 74]. Generalizing the DD for an mth order detector results in the following:

$$r_{m,i} = \left[\sum_{j=0}^{m-1} d_{i-j}\right] \mod 2 \tag{4.136}$$

Recall the conventional DD corresponds to $r_{1,i}$. The mth order DD output represents the parity check sum of m successive transmitted data. When noise is present the mth order detector output is given as

$$r_{m,i} = \left[\sum_{j=0}^{m-1} d_{i-j} + e_{m,i}\right] \mod 2 \tag{4.137}$$

$e_{m,i}$ = error symbol which has a value of 1 when an error is present otherwise it is zero. The syndrome is calculated in the same way as in the DQPSK example and is the mod 2 sum of the $(m + 1)$ order detector and m successive first-order detector outputs, the syndrome $s_{m,i}$ is given as

$$s_{m,i} = \left[\sum_{j=0}^{m} r_{1,i-j} + r_{m+1,i}\right] \mod 2 \tag{4.138}$$

$$s_{m,i} = \left[\sum_{j=0}^{m} (d_{i-j} + e_{m,i-j}) + \sum_{j=0}^{m} d_{i-j} + e_{m+1,i}\right] \mod 2 \tag{4.139}$$

$$s_{m,i} = \left[\sum_{j=0}^{m} e_{1,i-j} + e_{m+1,i}\right] \mod 2 \tag{4.140}$$

For the single-error correction NEC the two syndromes are given as

$$s_{1,i} = e_{1,i} + e_{1,i-1} + e_{2,i} \tag{4.141}$$

$$s_{1,i-1} = e_{1,i-1} + e_{2,i-1} \tag{4.142}$$

A block diagram for the single-error correction NEC (1NEC) receiver is given in Fig. 4.37. Note that the LUT shown in Fig. 4.37 can be also modeled by an AND logical gate.

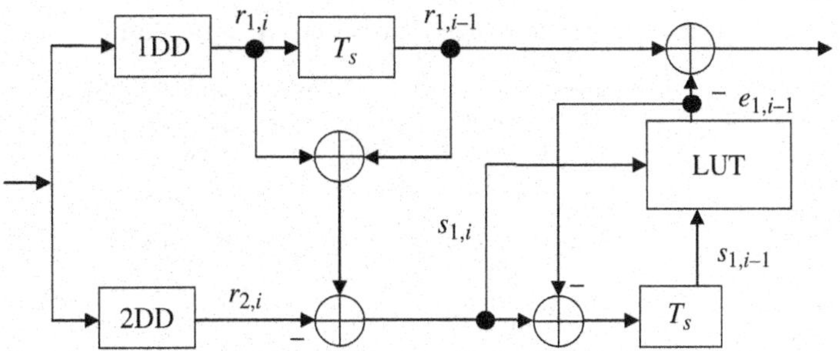

FIGURE 4.37 Single-error correction NEC for MSK modulation.

Note this similarity of the above NEC receiver block diagram for MSK to that of $\pi/4$-DQPSK previously shown. The exception being the modulo mathematical operations.

Next we present the double-error correction NEC receiver. Here the corresponding syndromes are given as

$$s_{1,i} = (e_{1,i} + e_{1,i-1} - e_{2,i}) \bmod 2 \tag{4.143}$$

$$s_{1,i-1} = (e_{1,i-1} + e_{1,i-2} - e_{2,i-1}) \bmod 2 \tag{4.144}$$

$$s_{1,i-2} = (e_{1,i-2} - e_{2,i-2}) \bmod 2 \tag{4.145}$$

$$s_{2,i} = (e_{1,i} + e_{1,i-1} + e_{1,i-2} - e_{3,i}) \bmod 2 \tag{4.146}$$

$$s_{2,i-1} = (e_{1,i-1} + e_{1,i-2} - e_{3,i-1}) \bmod 2 \tag{4.147}$$

$$s_{2,i-2} = (e_{1,i-2} - e_{e,i-2}) \bmod 2 \tag{4.148}$$

where $e_{1,i\text{-}3}$ and $e_{1,i\text{-}4}$ have been eliminated in the previous decoding interval. Arranging the syndromes in the following order

$$S_i = \begin{Bmatrix} S_{1,i} & S_{1,i-1} & S_{1,i-2} \\ S_{2,i} & S_{2,i-1} & S_{2,i-2} \end{Bmatrix} \tag{4.149}$$

The pattern detector is looking for the occurrence of the following nine patterns. This was suggested in [59] and is duplicated here for sake of convenience.

$$\begin{gathered}
\begin{Bmatrix} 0 & 1 & 1 \\ 1 & 1 & 1 \end{Bmatrix};\begin{Bmatrix} 1 & 0 & 1 \\ 0 & 0 & 1 \end{Bmatrix};\begin{Bmatrix} 1 & 1 & 1 \\ 0 & 1 & 1 \end{Bmatrix}; \\
\begin{Bmatrix} 0 & 1 & 0 \\ 1 & 1 & 1 \end{Bmatrix};\begin{Bmatrix} 0 & 0 & 1 \\ 1 & 1 & 1 \end{Bmatrix};\begin{Bmatrix} 1 & 1 & 1 \\ 1 & 1 & 1 \end{Bmatrix}; \\
\begin{Bmatrix} 0 & 1 & 1 \\ 1 & 1 & 0 \end{Bmatrix};\begin{Bmatrix} 0 & 1 & 1 \\ 1 & 0 & 1 \end{Bmatrix};\begin{Bmatrix} 0 & 1 & 1 \\ 0 & 1 & 1 \end{Bmatrix}
\end{gathered} \tag{4.150}$$

When they are detected then $e_{1,i\text{-}2} = 1$ otherwise it is equal to zero. This error correction can be performed by modulo 2 adding $e_{1,i\text{-}2}$ to $r_{i,i\text{-}2}$.

The double-error correction NEC (2NEC) block diagram for MSK modulation scheme is shown in Fig. 4.38.

We can make the same comment for MSK as we did earlier for the DQPSK modulation scheme. It was that higher-order error correction capabilities exist that essentially build upon what we presented in this subsection. Moreover, this increase in error correction capability comes at the expense of an increase in complexity.

4.4.3 Multiple Bit Observation of MSK—Noncoherent Detection

In this section we will continue along the lines of widening the observation window used to make a decision of what was transmitted. Here we create a $2n$-tuple waveform A_k which is equal to $\{a_1, a_2, \ldots, a_n, \ldots, a_{2n+1}\}$. We will rewrite the expression of CPFSK for sake of discussion as

$$s(t) = \cos\left[\omega_c t + \frac{a_i \pi}{2T} \cdot (t - (i-1)T) + \frac{\pi}{2} \cdot \sum_{j=1}^{i-1} a_j + \theta\right] \tag{4.151}$$

Since we are in the noncoherent section we will assume that the phase offset, θ, is unknown to us and that is uniformly distributed between $+\pi$ and $-\pi$. We will create a local $2n$-tuple waveform that will be correlated against what was received and make a decision on the center bit of this observation

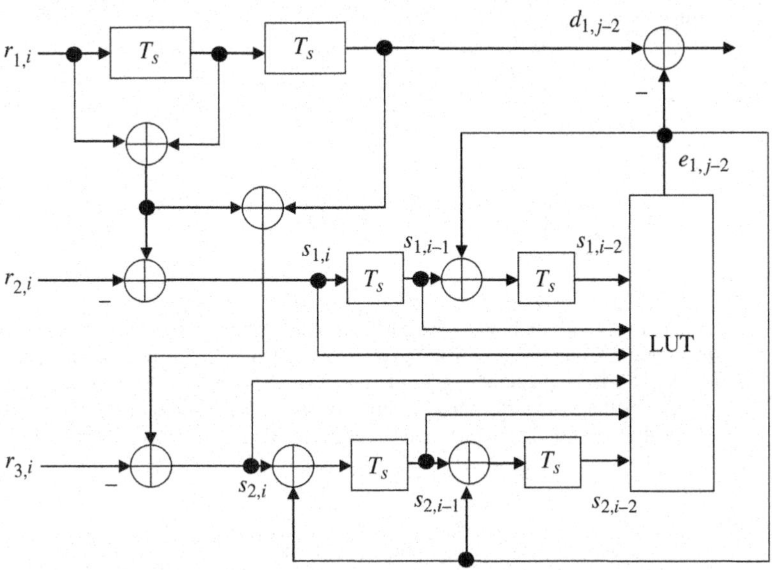

FIGURE 4.38 Double-error correction NEC for MSK modulation.

window. This position is different from what was presented in the coherent detection section, which was the first bit in the observation interval. Here we choose the center bit position because the cross correlation between the received and locally generated waveform is the highest at this position [26]. As we have done earlier, we will present the likelihood ratio test as the ratio of probability of the received data symbol = 1 to that with the received symbol = −1.

$$L = \frac{\int_{\theta}\int_{A} f(A)\cdot f(\theta)\cdot e^{\frac{2}{N_o}\int r(t)\cdot s(t,1,A,\theta)dt}\, d\theta\, dA}{\int_{\theta}\int_{A} f(A)\cdot f(\theta)\cdot e^{\frac{2}{N_o}\int r(t)\cdot s(t,-1,A,\theta)dt}\, d\theta\, dA} \tag{4.152}$$

where $f(A)$ is the density function of the data bits, this was given earlier in the discussion of $\pi/4$-DQPSK. Which becomes (assuming $m = 2^{2n}$) equal to the following equation once we evaluate the integral over the data sequences.

$$L = \frac{\int_{\theta}\sum_{k=1}^{m} e^{\frac{2}{N_o}\int r(t)\cdot s(t,1,A_k,\theta)dt} f(\theta)\cdot d\theta}{\int_{\theta}\sum_{k=1}^{m} e^{\frac{2}{N}\int r(t)\cdot s(t,-1,A_k,\theta)dt} f(\theta)\cdot d\theta} \tag{4.153}$$

This ratio is rewritten, after mathematical manipulations, to remove the dependency on the received signal's phase (which again is assumed to be uniformly distributed).

$$L = \frac{\sum_{i=1}^{m} I_o\left(\frac{2z_{1,i}}{N_o}\right)}{\sum_{i=1}^{m} I_o\left(\frac{2z_{-1,i}}{N_o}\right)} \tag{4.154}$$

With

$$z_{1,i} = \sqrt{\left(\int^{(2n+1)T} r(t) \cdot s(t,1,A_i,0)\,dt\right)^2 + \left(\int^{(2n+1)T} r(t) \cdot s(t,1,A_i,\pi/2)\,dt\right)^2} \qquad (4.155)$$

$$z_{-1,i} = \sqrt{\left(\int^{(2n+1)T} r(t) \cdot s(t,-1,A_i,0)\,dt\right)^2 + \left(\int^{(2n+1)T} r(t) \cdot s(t,-1,A_i,\pi/2)\,dt\right)^2} \qquad (4.156)$$

Note the above results reveal that in place of performing coherent detection, we have performed correlation operations on both axes of the complex coordinate system (i.e., real and imaginary). These correlations are squared and summed to remove the dependency of the received phase offset on the decision metric. This is a typical method used to perform the noncoherent detection of FSK waveforms [75, 76].

A block diagram performing the above mathematical operations are shown in Fig. 4.39, where instead of actually performing the ratio, we have decided to analyze the difference of the two calculated metrics.

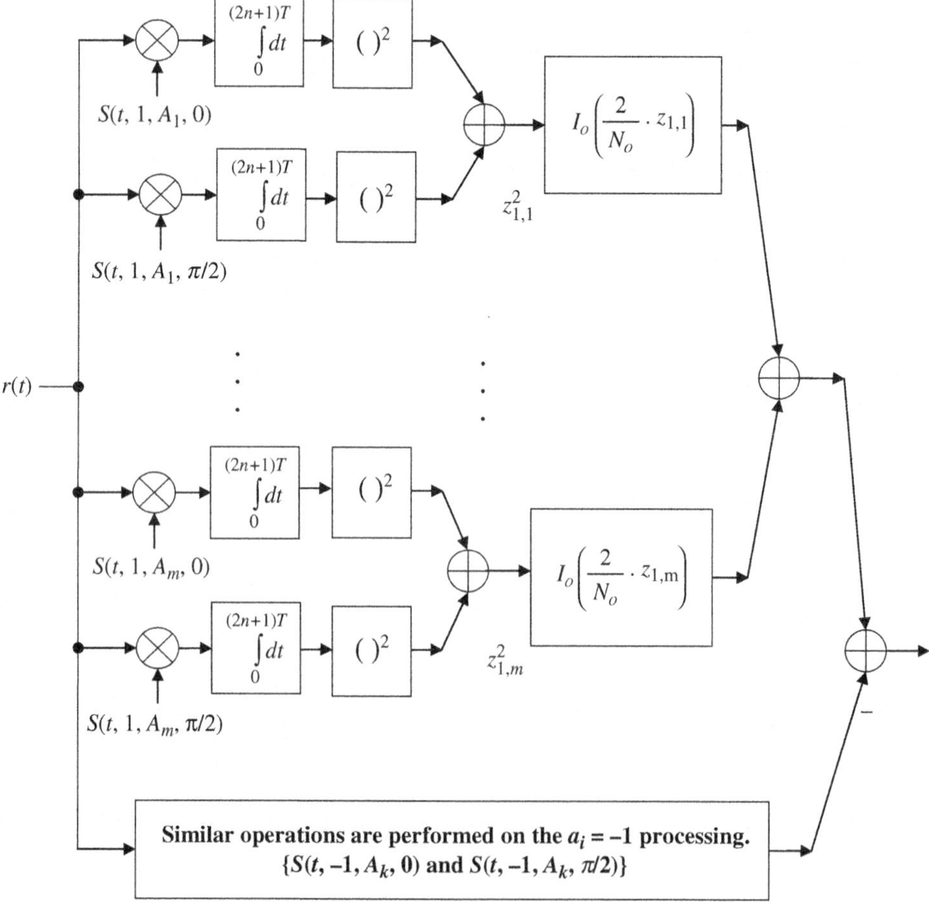

FIGURE 4.39 Multiple-bit observation-based noncoherent detection block diagram.

4.5 BER PERFORMANCE COMPARISONS

In this section we will present the BER performance of a few of the previously mentioned modulation schemes using some of the above-mentioned demodulation techniques. We will begin with the BER performance of the coherent and noncoherent detection receivers in AWGN channel. Next their performance in a flat fading channel is presented and lastly their performance in a two-ray, frequency selective Rayleigh fading channel is provided [77–97, 51].

4.5.1 AWGN Performance

The results in this section were obtained either from computer simulations and/or from their theoretical description. The mathematical equations used to plot the corresponding figures are shown in Appendix 4A. Here we will present the BER performance of DQPSK and FSK (MSK) modulation schemes. First, we compare the performance of the multiple symbol detectors such as MLDD, NEC, and DF-DD as shown in Fig. 4.40.

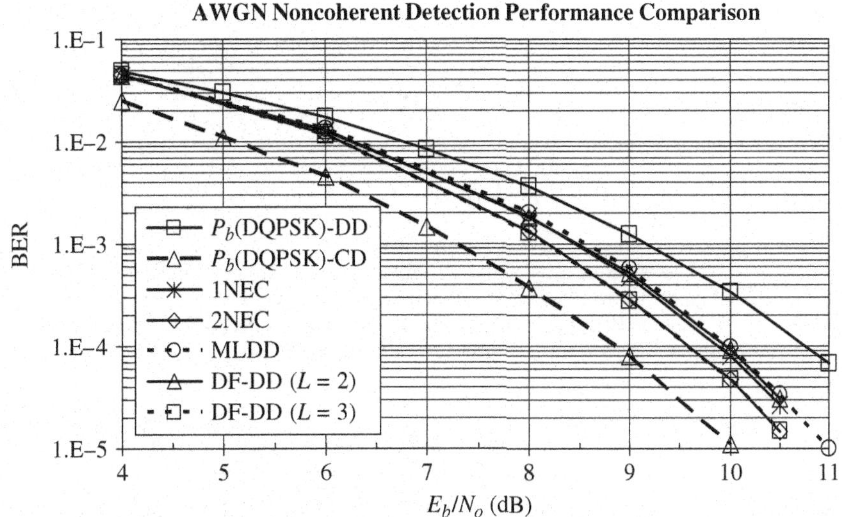

FIGURE 4.40 Noncoherent detection performance comparison for AWGN channel.

So when comparing the noncoherent technique to the coherent technique (with differential decoding), there is approximately 2 dB difference in performance. We can make up approximately 1.3 dB of this difference when using noncoherent detectors that either correct 2 errors or have an observation window of 4 symbols. We can see in going to a wider observation window improved performance can be obtained. There is a gain of approximately 0.9 dB (considering a BER = 1E-4) when using the MLDD, 1NEC, or DF-DD (L = 2 symbols) detectors. Moreover, there is an additional 0.4 dB improvement when using the 2NEC and DF-DD (L = 3 symbols) detectors. A rather interesting result of our investigation reveals that we can classify the performance of these detectors by their observation window only. By this we mean the performance of the 3-symbol window detectors, that is, MLDD, 1NEC, and DF-DD (L = 2 symbols) all perform approximately the same (within 0.1 dB of each other). This now means we choose the observation window size based on the desired performance and then choose the receiver detection technique based on implementation complexity. Note this work was performed in [98].

Next the NEC performance plots are shown below. Here we have used the Rician channel model with Rician factor, $K = 10$ dB and a Doppler spread of 60 Hz. Here we see using the single-error correction receiver produces approximately 1 dB of gain at BER = 1E-4 (see Fig. 4.41).

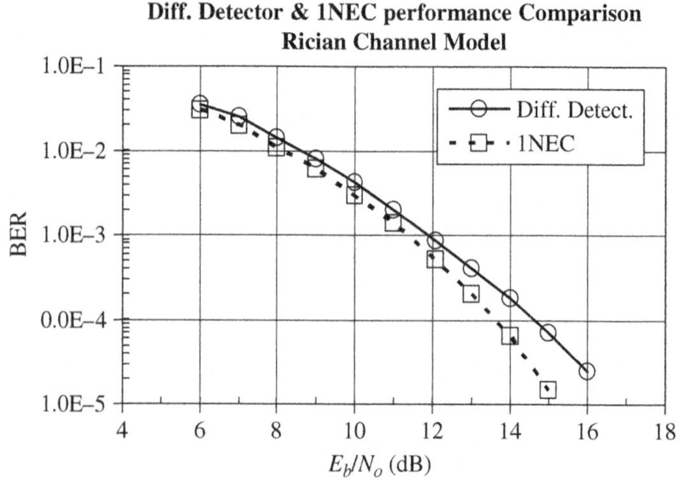

FIGURE 4.41 Single-error correcting NEC receiver performance in Rician fading channel model.

If we were to focus on the MLDD performance improvement for DQPSK and 8DPSK modulation schemes, we would want to show the following two figures (Figs. 4.42 and 4.43).

First the BER performance for DQPSK is presented. The results show when increasing the observation window from 2 symbols (which is already the case for the differential detector) to 3 symbols

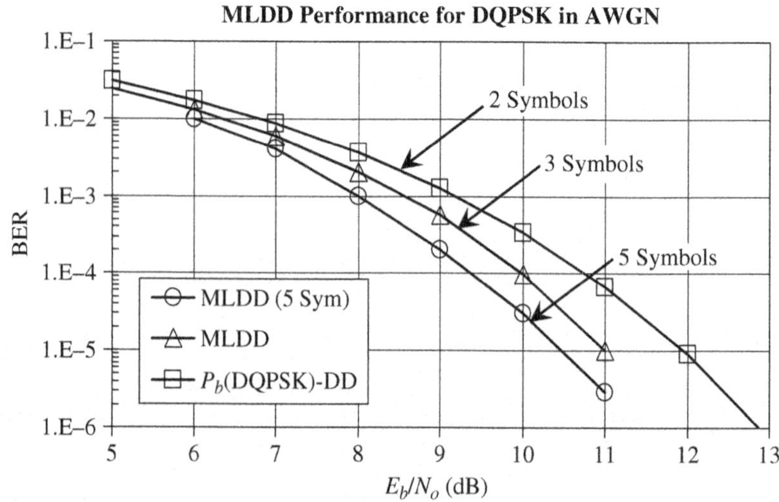

FIGURE 4.42 MLDD detector performance for DQPSK in an AWGN channel.

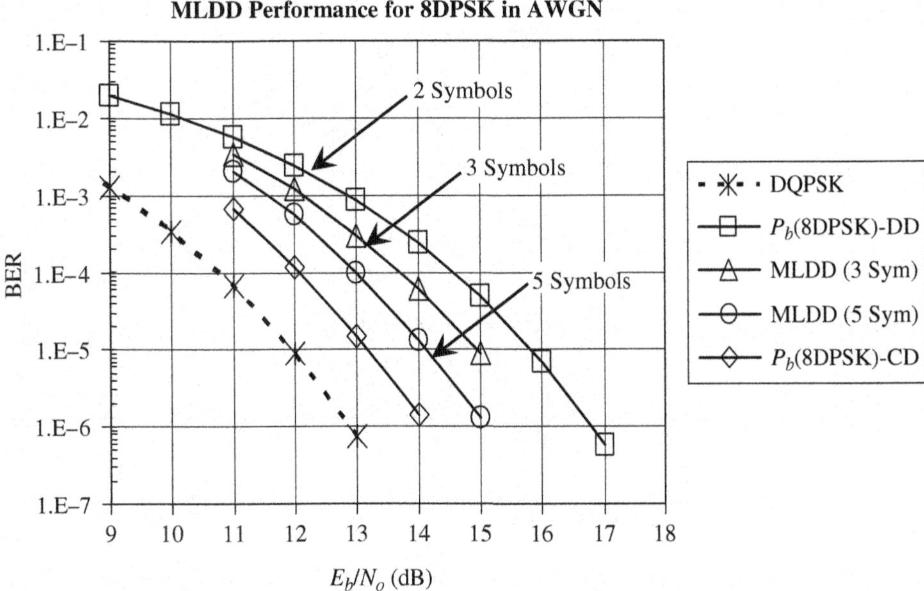

FIGURE 4.43 MLDD detector performance for 8DPSK in an AWGN channel.

(for MLDD) there is approximately a gain of 1 dB. Moreover, if this window was increased to 5 the overall performance gain would be increased to 1.5 dB. Thus we can further close the performance gap between noncoherent detection and coherent detection with differential decoding (see Fig. 4.42).

Next the BER performance for 8DPSK is presented. The results show when increasing the observation window from 2 symbols to 3 symbols there is a gain of 0.8 dB. An additional gain of 0.7 dB can be seen when we further increased the observation window to 5 symbols, totaling 1.5 dB. This has closed the 2.5 dB performance gap between coherent and noncoherent detection to approximately 1 dB. The differential detector performance of DQPSK is also shown in Fig. 4.43 for sake of reference.

Lastly, the performance of the MBO detectors are shown here for CD using the upper bounds defined in [26] for the theoretical coherent detection curve. A point worth noting is that these results assumed a BFSK modulation index, $\beta = 0.715$ (for reasons already discussed in Chap. 2) (see Fig. 4.44).

Here we see increasing the observation window from 1 bit to 2 bits gives approximately 2.4 dB of gain. Increasing this further to 3 bits gives an additional gain of approximately 1 dB. Moreover, a 5-bit observation window gives another 0.3 dB. Hence we can achieve approximately 3.7 dB of performance gain in using the coherent detection-based MBO.

The noncoherent (NC) performance is shown in Fig. 4.45, where we see using either a 3-bit or 5-bit observation window dramatically improved performance over the single-bit demodulator. Also similar comments can be made regarding the BFSK modulation index value, $\beta = 0.715$.

Increasing the observation window to 3 bits provides approximately 3.25 dB of gain. Further increasing this window to 5 bits provides another 0.9 dB of gain. Hence the total gain shown is approximately 4.2 dB. A very interesting observation, made earlier in [26], is that the MBO can improve BER performance to be better than that of coherent detection of QPSK (or antipodal signaling).

Some public results are worthy of mention here where they show the BER performance dependency on the cutoff frequency of the LPF [99]. Here we see that as the BT product becomes smaller, the performance degrades more rapidly as shown in [99]. In fact, there is approximately 1 dB of loss for BER = 1E-4 for BT = 0.25 (see Fig. 4.46).

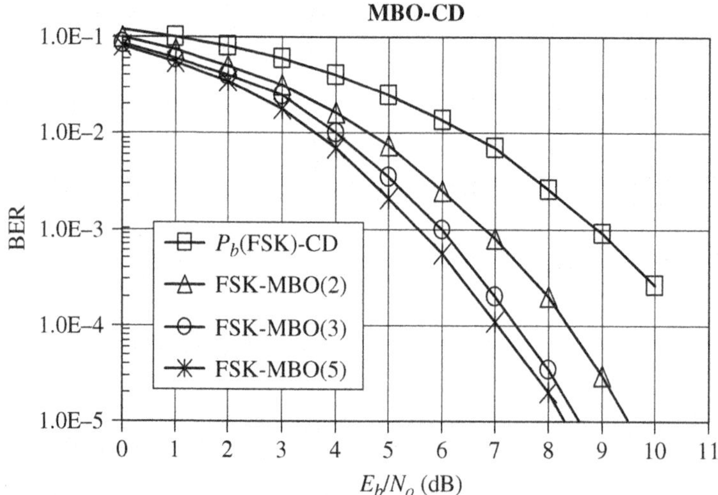

FIGURE 4.44 Coherent detection using MBO for FSK in an AWGN channel.

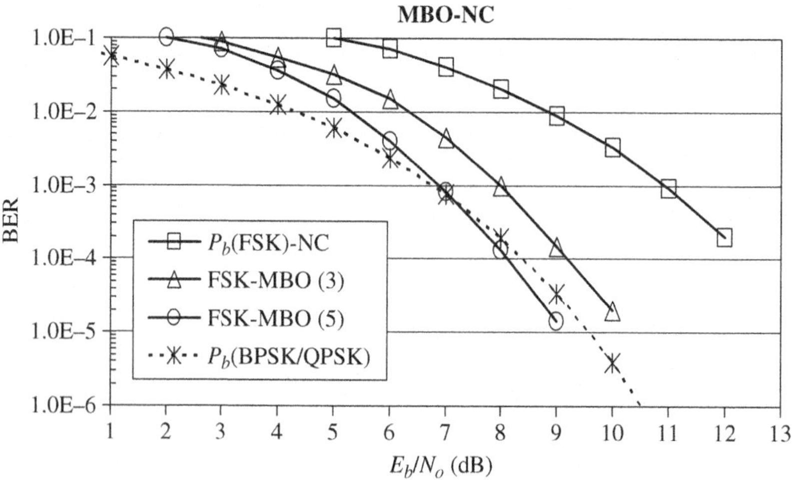

FIGURE 4.45 Noncoherent MBO detector performance for FSK in an AWGN channel.

TABLE 4.3 GMSK BER Performance Approximation Parameters

Modulation	Alpha value
Coherent QPSK	2
MSK	1.7
GMSK (BT = 0.25)	1.36
GMSK (BT = 0.2)	1

We also have plots for BER approximations of the various BT product curves using the following familiar formula [39].

$$P_e = Q\left(\sqrt{\alpha \cdot \frac{E_b}{N_o}}\right) \tag{4.157}$$

With the use of Table 4.3, there is good agreement to the simulations performed over the BER range of interest.

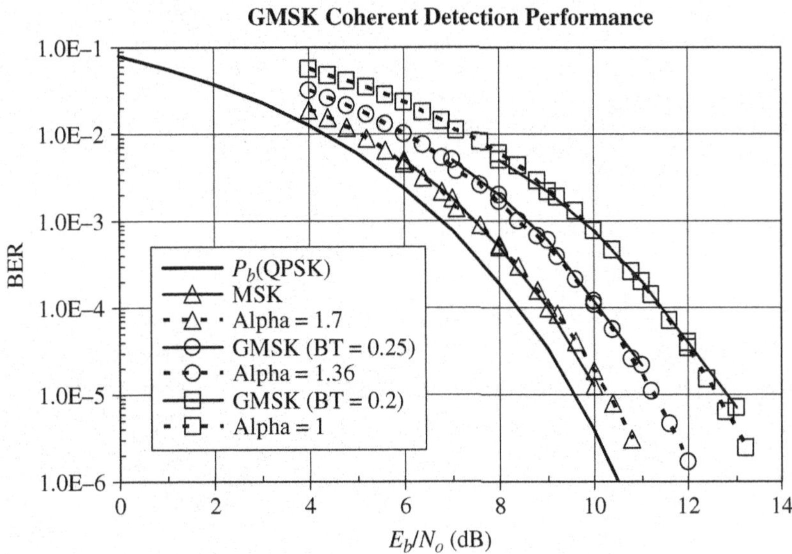

FIGURE 4.46 GMSK coherent detection for various values of BT products.

4.5.2 Multipath Fading Performance

In this section we will present the BER performance of DQPSK using the DD technique in a flat Rayleigh fading channel. Below we plot the theoretical BER performance results of coherent and noncoherent detection for QPSK, DQPSK, and FSK modulation schemes (see Fig. 4.47). Here we see

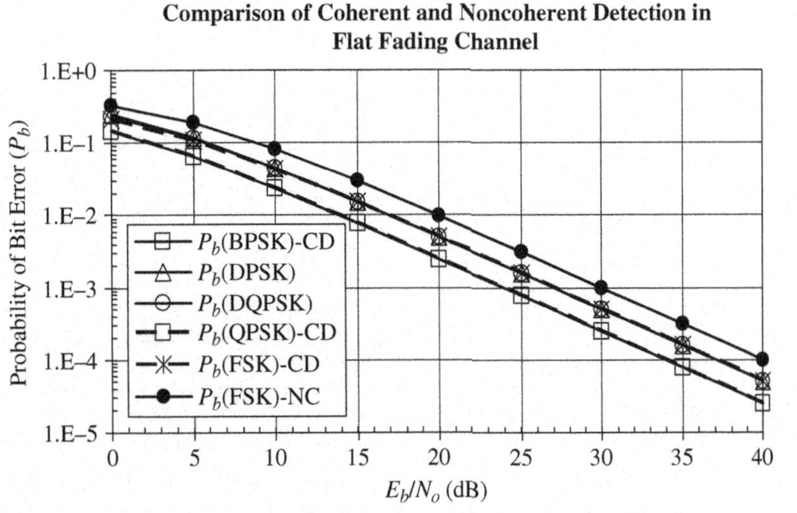

FIGURE 4.47 Theoretical coherent and noncoherent detection performance comparison in a flat fading channel.

there is approximately 3 dB of loss when going from QPSK coherent detection to DQPSK noncoherent detection. Similarly, there is a loss of 3 dB when going from FSK coherent detection to FSK noncoherent detection. As discussed earlier these equations are provided in Appendix 4A for reference.

Next simulation results are presented for DQPSK modulation using a noncoherent differential detector in a flat Rayleigh fading channel (see Fig. 4.48). The theoretical BER performance curve is

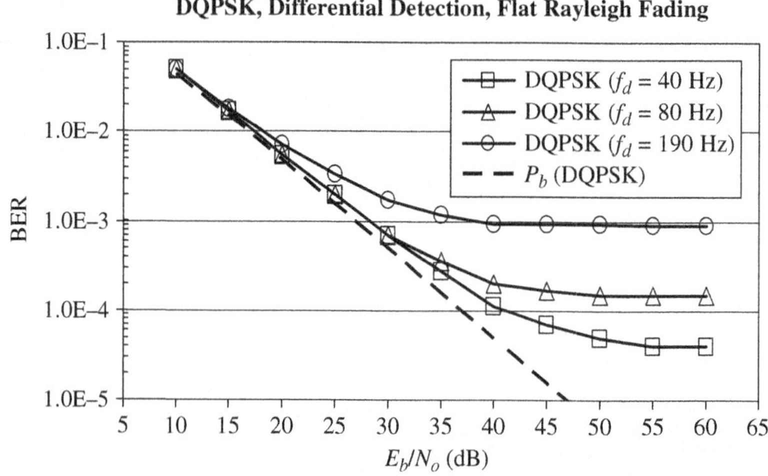

FIGURE 4.48 Noncoherent detection performance of DQPSK in a flat Rayleigh fading channel.

drawn by a dashed line and used for sake of reference. The channel model was previously described in Chap. 3. Three Doppler frequencies were used: f_d = 40 Hz, f_d = 80 Hz, and f_d = 190 Hz. Here we see as f_d increases the performance degrades to a point where any further increase in E_b/N_o will result in no BER improvement. This behavior is commonly described as an irreducible error floor (IBER).

4.5.3 Frequency Selective Fading Performance

In this section we will provide simulation results for the DD in a two-ray, Rayleigh fading channel. The two-ray channel model will be characterized by the following three parameters [100, 101].

1. Doppler spread (Hz)
2. Delay spread (s)
3. Power profile (dB)

This can be shown in Fig. 4.49, where D represents the average power of the desired ray and U represents the average power of the undesired ray.

It is common practice to normalize the time separation of the two rays by the modulation symbol time (τ/T_s). Also the ratio of the average power of the first arriving ray to that of the second arriving ray will be expressed in dB, and used to identify the degree of channel frequency selectivity present. This was discussed earlier in Chap. 3.

$$\left(\frac{D}{U}\right)_{\text{dB}} = (D)_{\text{dB}} - (U)_{\text{dB}} \tag{4.158}$$

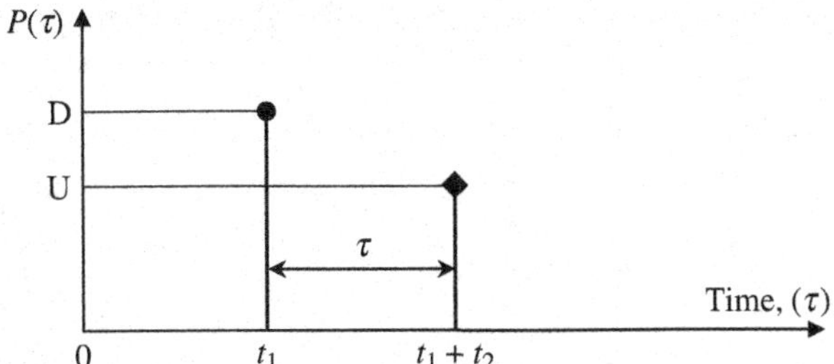

FIGURE 4.49 Two-ray power delay profile example.

Note for D/U equal to either +infinity or −infinity, the channel is also described as a flat fading channel. As we will soon see, the example of a two-ray model was used due to its simplicity and usefulness in conveying a message.

The first plot shows the BER performance of DQPSK using the differential detector in a channel model with Doppler spread of 80 Hz and a time delay spread of $\tau = 0.25T_s$. The curves are drawn for various values of D/U ratios. When the two rays are of equal average power, an IBER = 2.5E-2 results. As the second ray decreases in power the performance improves. This is because the DD is a symbol-based detector that performs very well in a frequency nonselective channel. Other techniques exist that should be used when the user is experiencing a frequency selective fading channel (see Fig. 4.50).

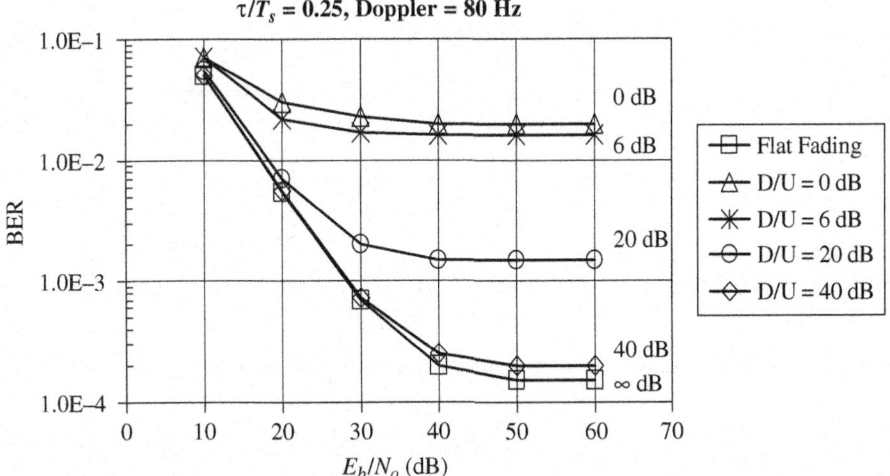

FIGURE 4.50 Differential detection performance of DQPSK in a two-ray ($\tau/T_s = 0.25$) channel.

This next plot shows the BER performance of DQPSK using the differential detector for a Doppler spread of 80 Hz and a time delay spread of $\tau = 0.5T_s$. Similar observations can be made here, as the D/U ratio decreases the BER performance degrades (see Fig. 4.51).

The time separation of the two rays was further increases to $\tau = T_s$ and we plot the BER performance using the differential detector. As the D/U ratio decreases we see the BER performance degrades and the BER floor increases (see Fig. 4.52).

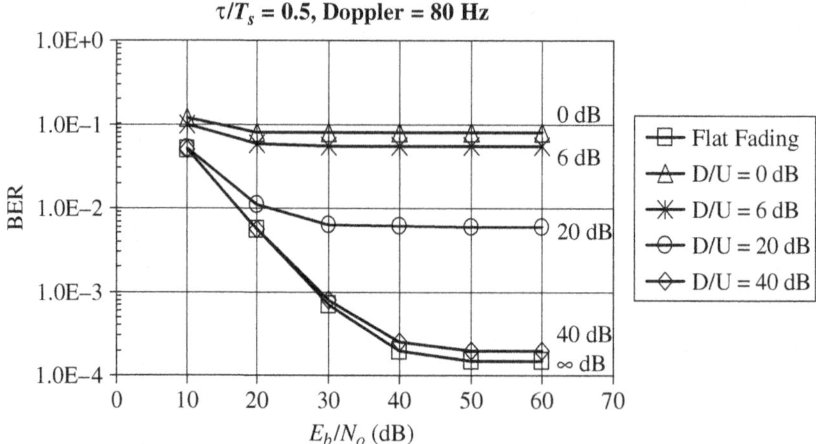

FIGURE 4.51 Differential detection performance of DQPSK in a two-ray ($\tau/T_s = 0.5$) channel.

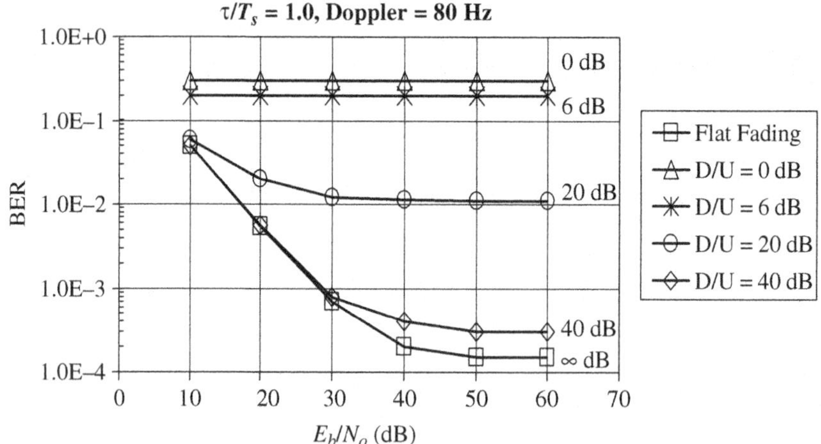

FIGURE 4.52 Differential detection performance of DQPSK in a two-ray ($\tau/T_s = 1.0$) channel.

A short observation of the above three FSF performance figures is presented. These FSF performance results have showed us in the presence of a second ray (even 20 dB below the average power of the first arriving ray) can cause significant degradation in BER performance. This confirms the earlier statements that the DD is a symbol-based detector and does not perform any explicit operations to compensate for ISI.

Figure 4.53 plots the results of Figs. 4.50 to 4.52 in this subsection in an effort to provide some insight into the behavior as D/U ratio is varied. Here we observe as D/U ratio increases the IBER gradually improves at first until the D/U becomes greater than approximately 6 dB. Once this D/U ratio threshold is passed, the IBER rapidly improves to become much closer to the value corresponding to the frequency flat channel.

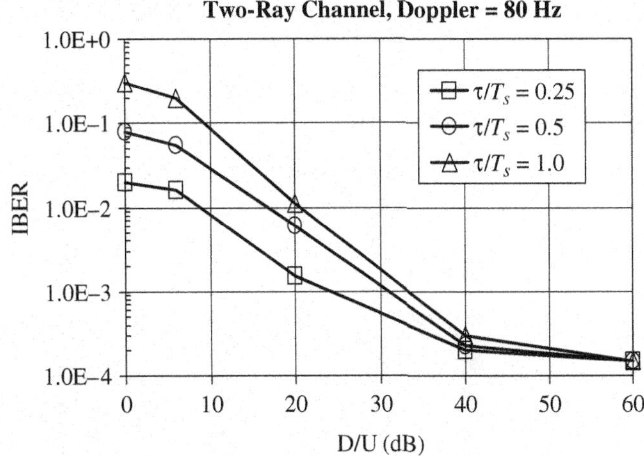

FIGURE 4.53 Irreducible BER performance of differential detection for DQPSK.

Figure 4.54 shows the differential detector BER performance for time-delay spread values of τ/T_s = 0.0, 0.25, 0.5, and 1.0. This figure clearly shows as soon as the second ray is introduced into the channel, the BER dramatically increases by approximately two orders of magnitude.

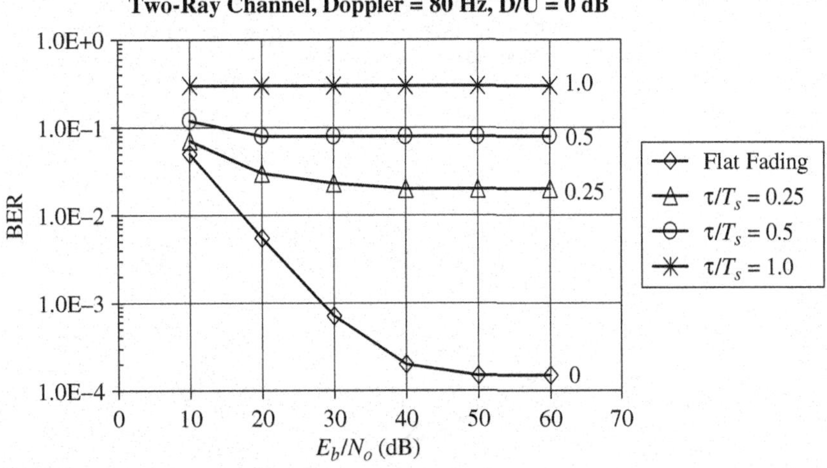

FIGURE 4.54 Differential detection BER performance comparison for various time-delay spread values (for D/U = 0 dB).

Alternatively we can plot the results of Figs. 4.50 to 4.53 in this subsection as a function of time-delay spread. This was done in order to provide the reader with a different perspective on the performance. This plot shows as the second ray is introduced into the channel model, BER performance dramatically degrades for D/U ratio values of 0 dB, 6 dB, and 20 dB. Notice the performance

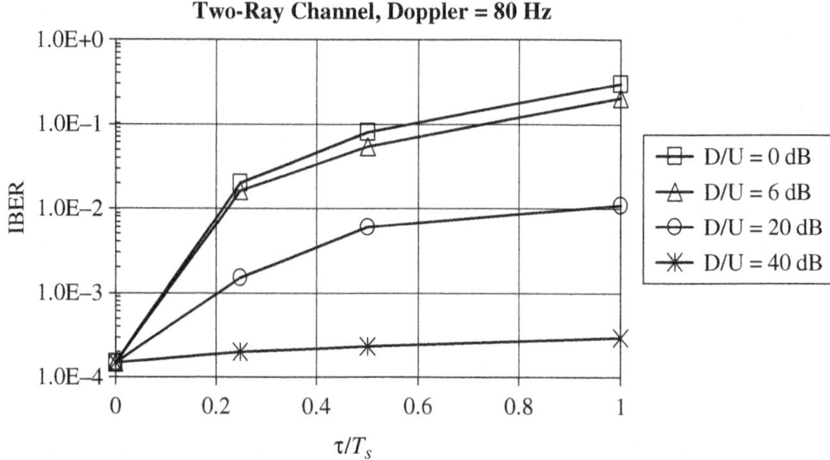

FIGURE 4.55 Differential detection time-delay spread performance for various values of D/U ratios.

degradation due to D/U = 40 dB is minimal, in other words, very close to that of a frequency flat channel (see Fig. 4.55).

This last plot shows the BER versus the D/U ratio for 2 values of time-delay spread (τ/T_s = 0.25 and 1.0) and 2 values of Doppler spreads (f_d = 80 Hz and 190 Hz). The f_d = 190 Hz Doppler spread results were drawn with dashed lines. Here we see for the two-ray, equal average power channel model, the BER is independent of Doppler, but dependent on the time-delay spread. For large values of the D/U ratio, the BER is dependent on the Doppler spread only (as shown earlier). Now for values of D/U ratio in between there exists a transition region that goes from time-delay spread dependency to Doppler spread dependency (see Fig. 4.56).

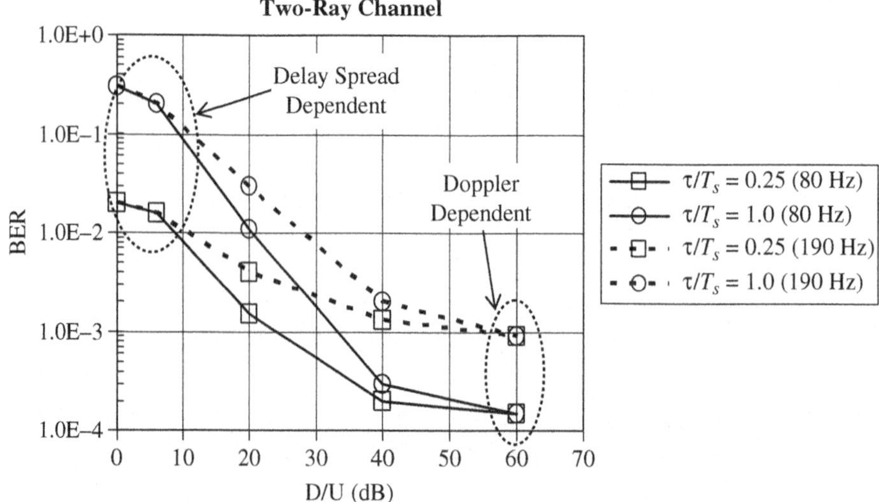

FIGURE 4.56 Differential detection performance summary.

APPENDIX 4A: MATHEMATICAL BER EXPRESSIONS

In this section we will provide a concise summary of the mathematical equations used to plot the above figures as well as to provide insight into the modulation scheme and detection techniques. The equations are a function of the following:

$$\gamma_b = \frac{E_b}{N_o} = \frac{\text{Signal Power}}{\text{(Noise Density)(Bit Rate)}} \tag{4A.1}$$

The mathematical equations are presented below.

AWGN, BPSK

$$P(\gamma_b) = \frac{1}{2}\cdot \mathrm{erfc}\left(\sqrt{\gamma_b}\right) \tag{4A.2}$$

AWGN, BFSK

$$P(\gamma_b) = \frac{1}{2}\cdot \mathrm{erfc}\left(\sqrt{\frac{\gamma_b}{2}}\right) \tag{4A.3}$$

Multipath, BPSK

$$P(\gamma_b) = \frac{1}{2}\cdot \left[1 - \sqrt{\frac{\gamma_b}{1+\gamma_b}}\right] \tag{4A.4}$$

Multipath, BFSK, Coherent

$$P(\gamma_b) = \frac{1}{2}\cdot \left[1 - \sqrt{\frac{\gamma_b}{2+\gamma_b}}\right] \tag{4A.5}$$

Multipath, DPSK

$$P(\gamma_b) = \frac{1}{2\cdot(1+\gamma_b)} \tag{4A.6}$$

Multipath, BFSK noncoherent

$$P(\gamma_b) = \frac{1}{2+\gamma_b} \tag{4A.7}$$

Multipath, QPSK

$$P(\gamma_b) = \frac{1}{2}\cdot \left\{1 - \frac{C}{\sqrt{2-C^2}}\right\} \tag{4A.8}$$

where

$$C = \sqrt{\frac{2\gamma_b}{1+2\gamma_b}} \tag{4A.9}$$

Multipath, DQPSK

$$P(\gamma_b) = \frac{1}{2}\cdot \left\{1 - \frac{C}{\sqrt{2-C^2}}\right\} \tag{4A.10}$$

where

$$C = \frac{2\gamma_b}{1+2\gamma_b} \tag{4A.11}$$

AWGN, DPSK, Noncoherent

$$P(\gamma_b) = \frac{1}{2}\cdot e^{-\gamma_b} \tag{4A.12}$$

AWGN, BFSK, Noncoherent

$$P(\gamma_b) = \frac{1}{2}\cdot e^{-\frac{\gamma_b}{2}} \tag{4A.13}$$

REFERENCES

[1] B. Sklar, "Defining, Designing and Evaluating Digital Communication Systems," *IEEE Communications Magazine*, Nov. 1993, pp. 92–101.

[2] O. Belce, "Comparison of Advanced Modulation Schemes for LEO Satellite Downlink Communications," *IEEE*, 2003, pp. 432–437.

[3] J. C-I Chuang, "Comparison of Coherent and Differential Detection of BPSK and QPSK in a Quasi-static Fading Channel," *IEEE Conference on Communications*, 1988, pp. 749–755.

[4] F. Adachi and M. Sawahashi, "Performance Analysis of Various 16 Level Modulation Schemes Under Rayleigh Fading," *Electronics Letters*, Vol. 28, No. 17, Aug. 1992, pp. 1579–1581.

[5] M. P. RistenBatt, "Alternatives in Digital Communications," *Proceedings of the IEEE*, Vol. 61, No. 6, June 1973, pp. 703–721.

[6] K. Pahlavan and A. H. Levesque, "Wireless Data Communications," *Proceedings of the IEEE*, Vol. 82, No. 9, Sept. 1994, pp. 1398–1430.

[7] T. S. Chu and L. J. Greenstein, "A Quantification of Link Budget Differences between the Cellular and PCS Bands," *IEEE Transactions on Vehicular Technology*, Vol. 48, No. 1, Jan. 1999, pp. 60–65.

[8] P. A. Moore, "Demodulating an Angle-Modulated Signal," United States Patent # 4,766,392, Aug. 1988.

[9] K. K. Clarke and D. T. Hess, "Frequency Locked Loop FM Demodulator," *IEEE Transactions on Communications Technology*, Vol. COM-15, No. 4, Aug. 1967, pp. 518–524.

[10] S. Komaki, O. Kurita, and T. Memita, "QPSK Direct Regenerator with a Frequency Tripler and a Quadrupler," *IEEE Transactions on Communications*, Vol. COM-27, No. 12, Dec. 1979, pp. 1819–1828.

[11] P. Y. Kam and T. M. Cheong, "Analysis of *M*th Power Carrier Recovery Structure for MPSK," *Conference on Information, Communications and Signal Processing*, Sept. 1997, pp. 1496–1500.

[12] C. R. Cahn, "Improving Frequency Acquisition of a Costas Loop," *IEEE Transactions on Communications*, Vol. COM-25, No. 12, Dec. 1977, pp. 1453–1459.

[13] S. H. Goode, "A Comparison of Gaussian Minimum Shift Keying to Frequency Shift Keying for Land Mobile Radio," *IEEE Vehicular Technology Conference*, 1984, pp. 136–141.

[14] S. H. Goode, H. L. Kazecki, and D. W. Dennis, "A Comparison of Limiter-Discriminator, Delay and Coherent Detection for $\pi/4$-DQPSK," *IEEE Vehicular Technology Conference*, 1990, pp. 687–694.

[15] S. H. Goode, "An Open Loop Technique for the Detection of Minimum Shift Keyed Signals," *IEEE Vehicular Technology Conference*, 1986, pp. 116–121.

[16] D. Makrakis, A. Yongacoglu, and K. Feher, "Novel Receiver Structures for Systems Using Differential Detection," *IEEE Transactions on Vehicular Technology*, Vol. VT-36, No. 2, May 1987, pp. 71–77.

[17] K. Otani, K. Daikoku, and H. Omori, "Burst Error Performance Encountered in Digital Land Mobile Radio Channel," *IEEE Transactions on Vehicular Technology*, Vol. VT-30, No. 4, Nov. 1981, pp. 156–161.

[18] J. K. Cavers, "An Analysis of Pilot Symbol Assisted QPSK for Digital Mobile Communications," *IEEE GLOBECOM*, 1990, pp. 928–933.

[19] S. Sampei and T. Sunaga, "Rayleigh Fading Compensation Method for 16QAM In Digital Land Mobile Radio Channels," *IEEE Vehicular Technology Conference*, 1989, pp. 640–646.

[20] J. K. Cavers, "An Analysis of Pilot Symbol Assisted Modulation for Rayleigh Fading Channels," *IEEE Transactions on Vehicular Technology*, Vol. 40, No. 4, Nov. 1991, pp. 686–693.

[21] J. K. Cavers, "An Analysis of Pilot Symbol Assisted 16QAM for Digital Mobile Communications," *IEEE Vehicular Technology Conference*, 1991, pp. 380–385.

[22] L. Tong, B. M. Sadler, and M. Dong, "Pilot Assisted Wireless Transmissions," *IEEE Signal Processing Magazine*, Nov. 2004, pp. 12–25.

[23] B. R. Tomiuk, N. C. Beaulieu and A. A. Abu-Dayya, "Maximal Ratio Combining with Channel Estimation Errors," *IEEE Conference on Communications, Computers and Signal Processing*, 1995, pp. 363–366.

[24] T. M. Schmidl, A. G. Dabak, and S. Hosur, "The Use of Iterative Channel Estimation (ICE) to Improve Link Margin in Wideband CDMA Systems," *IEEE Vehicular Technology Conference*, 1999, pp. 1307–1311.

[25] S. Min and K. B. Lee, "Channel Estimation Based on Pilot and Data Traffic Channels for DS/CDMA Systems," *IEEE GLOBECOM*, 1998, pp. 1384–1389.

[26] W. P. Osborne and M. B. Luntz, "Coherent and Noncoherent Detection of CPFSK," *IEEE Transactions on Communications*, Vol. COM-22, No. 8, Aug. 1974, pp. 1023–1036.

[27] J. G. Proakis, *Digital Communications*, McGraw-Hill, 1989, New York.

[28] C. L. Liu and K. Feher, "Noncoherent Detection of π/4-QPSK Systems in CCI-AWGN Combined Interference Environment," *IEEE*, 1989, pp. 83–94.

[29] W. C. Lindsey and M. K. Simon, "On The Detection of Differentially Encoded Polyphase Signals," *IEEE Transactions on Communications*, Vol. COM-20, No. 6, Dec. 1972, pp. 1121–1128.

[30] E. Arthurs and H. Dym, "On The Optimum Detection of Digital Signals in the Presence of White Gaussian Noise—A Geometric Interpretation and a Study of Three Basic Data Transmission Systems," *IRE Transactions on Communications Systems*, Dec. 1962, pp. 336–372.

[31] P. Fines and A. H. Aghvami, "Fully Digital M-ary PSK and M-ary QAM Demodulators for Land Mobile Satellite Communications," *Electronics & Communication Engineering Journal*, Dec. 1991, pp. 291–298.

[32] K. Defly, M. Lecours, and N. Boutin, "Differential Detection of the OQPSK Signal: Coding and Decoding," *IEEE International Conference on Communications*, 1989, pp. 1660–1664.

[33] F. Adachi, "Error Rate Analysis of Differentially Encoded and Detected 16 APSK Under Rician Fading," *IEEE Transactions on Vehicular Technology*, Vol. 45, No. 1, Feb. 1996, pp. 1–11.

[34] K. Kiasaleh and T. He, "Performance of DQPSK Communication Systems Impaired by Mixed Imbalance, Timing Error and Rayleigh Fading," *IEEE International Conference on Communications*, 1996, pp. 364–368.

[35] G. Chrisikos, "Analysis of 16-QAM over a Nonlinear Channel," *IEEE Personal, Indoor and Mobile Radio Communications*, 1998, pp. 1324–1329.

[36] C. C. Powell & J. Boccuzzi, "Performance of π/4-QPSK Baseband Differential Detection under Frequency Offset Conditions," *IEEE GLOBECOM*, 1991, pp. 526–530.

[37] S. Chennakeshu and G. J. Saulnier, "Differential Detection of /4-Shifted-DQPSK For Digital Cellular Radio," *IEEE Vehicular Technology Conference*, 1991, pp. 186–191.

[38] H. Furukawa, K. Matsuyama, T. Sato, T. Takenaka, and Y. Takeda, "A /4-Shifted DQPSK Demodulator for Personal Mobile Communications System," *IEEE Personal, Indoor and Mobile Radio Communications*, 1992, pp. 618–622.

[39] T. S. Rappaport, *Wireless Communications: Principles and Practice*, Prentice Hall.

[40] J. S. Lin and K. Feher, "Noncoherent Limiter-Discriminator Detection of Standardized FQPSK and OQPSK," *IEEE Wireless Communications and Networking Conference*, 2003, pp. 795–800.

[41] J. P. Fonseka, "Limiter Discriminator Detection of M-ary FSK Signals," *IEE Proceedings*, Vol. 137, No. 5, Oct. 1990, pp. 265–272.

[42] Y. Akaiwa and Y. Nagata, "Highly Efficient Digital Mobile Communications with a Linear Modulation Method," *IEEE Journal on Selected Areas in Communications*, Vol. SAC-5, No. 5, June 1987, pp. 890–895.

[43] A. V. Oppenheim and R. W. Schaher, *Discrete-Time Signal Processing*, Prentice Hall, 1989, New Jersey.

[44] M. Cho and S. C. Kim, "Non-Coherent Detection of FQPSK Signals using MLDD," *Proceedings of the 7th Korea-Russia Symposium*, KORUS 2003, pp. 314–318.

[45] L. Li and M. K. Simon, "Performance of Coded OQPSK and MIL-STD SOQPSK with Iterative Decoding," *IEEE Transactions on Communications*, Vol. 52, No. 11, Nov. 2004, pp. 1890–1900.

[46] D. Divsalar and M. K. Simon, "Maximum-Likelihood Differential Detection of Uncoded and Trellis Coded Amplitude Phase Modulation over WGN and Fading Channels-Metrics and Performance," *IEEE Transactions on Communications*, Vol. 42, No. 1, Jan. 1994, pp. 76–89.

[47] F. Adachi and M. Sawahashi, "Viterbi-Decoding Differential Detection of DPSK," *Electronics Letters*, Vol. 28, No. 23, Nov. 1992, pp. 2196–2198.

[48] W. G. Phoel, "Improved Performance of Multiple-Symbol Differential Detection of Offset QPSK," *IEEE Wireless Communications and Networking Conference*, 2004, pp. 548–553.

[49] M. K. Simon and D. Divsalar, "Multiple Symbol Partially Coherent Detection of MPSK," *IEEE Transactions on Communications*, Vol. 42, No. 2/3/4, Feb./March/April 1994, pp. 430–439.

[50] R. Schober, I. Ho, and L. Lampe, "Enhanced Multiple-Bit Differential Detection of DOQPSK," *IEEE Transactions on Communications*, Vol. 53, No. 9, Sept. 2005, pp. 1490–1497.

[51] M. K. Simon, "Multiple-Bit Differential Detection of Offset QPSK," *IEEE Transactions on Communications*, Vol. 51, No. 6, June 2003, pp. 1004–1011.

[52] D. Divsalar and M. K. Simon, "Multiple Symbol Differential Detection of MPSK," *IEEE Transactions on Communications*, Vol. 38, No. 3, March 1990, pp. 300–308.

[53] F. Adachi and M. Sawahashi, "Decision Feedback Differential Detection of 16-DAPSK Signals," *Electronics Letters*, Vol. 29, No. 16, Aug, 1993, pp. 1455–1457.

[54] F. Adachi and M. Sawahashi, "Decision Feedback Differential Detection of Differentially Encoded 16APSK Signals," *IEEE Transactions on Communications*, Vol. 44, No. 4, April 1996, pp. 416–418.

[55] F. Adachi, "BER Analysis of 2PSK, 4PSK, and 16QAM with Decision Feedback Channel Estimation in Frequency-Selective Slow Rayleigh Fading," *IEEE Transactions on Vehicular Technology*, Vol. 48, No. 5, Sept. 1999, pp. 1563–1572.

[56] F. Adachi and M. Sawahashi, "Decision Feedback Differential Phase Detection of M-ary DPSK Signals," *IEEE Transactions on Vehicular Technology*, Vol. 44, No. 2, May 1995, pp. 203–210.

[57] J. Yang and K. Feher, "An Improved $\pi/4$-QPSK with Nonredundant Error Correction for Satellite Mobile Broadcasting," *IEEE Transactions on Broadcasting*, Vol. 37, No. 1, March 1991, pp. 9–15.

[58] H. C. Schroeder and J. R. Sheehan, "Nonredundant Error Detection and Correction System," United States Patent # 3,529,290, Sept. 1970.

[59] T. Masamura, "Intersymbol Interference Reduction for Differential MSK by Nonredundant Error Correction," *IEEE Transactions on Vehicular Technology*, Vol. 39, No. 1, Feb. 1990, pp. 27–36.

[60] S. Samejima, K. Enomoto, and Y. Watanabe, "Differential PSK System with Nonredundant Error Correction," *IEEE Journal on Selected Areas in Communications*, Vol. SAC-1, No. 1, Jan. 1983, pp. 74–81.

[61] D. P. C. Wong and P. T. Mathiopoulos, "Nonredundant Error Correction Analysis and Evaluation of Differentially Detected $\pi/4$-Shift DQPSK Systems in a Combined CCI and AWGN Environment," *IEEE Transactions on Vehicular Technology*, Vol. 39, Feb. 1990, pp. 35–48.

[62] D. P. C. Wong and P. T. Mathiopoulos, "Nonredundant Error Correction DQPSK for the Aeronautical-Satellite Channel," *IEEE Transactions on Aerospace and Electronic Systems*, Vol. 31, No. 1, Jan. 1995, pp. 168–181.

[63] J. Yang and K. Feher, "Nonredundant Error Correction for /4-QPSK in Mobile Satellite Channels," IEEE Vehicular Technology Conference, 1991, pp. 782–787.

[64] M. Saitou, M. Kawabata, and Y. Akaiwa, "Direct Conversion Receiver for 2- and 4-Level FSK Signals," *IEEE Conference on Universal Personal Communications*, 1995, pp. 392–396.

[65] H. Leib and S. Pasupathy, "Error-Control Properties of Minimum Shift Keying," *IEEE Communications Magazine*, Jan. 1993, pp. 52–61.

[66] K. Hirade, M. Ishizuka, F. Adachi, and K. Ohtani, "Error-Rate Performance of Digital FM with Differential Detection in Land Mobile Radio Channels," *IEEE Transactions on Vehicular Technology,* Vol. VT-28, No. 3, Aug. 1979, pp. 204–212.

[67] I. Tezcan and K. Feher, "Performance Evaluation of Differential MSK (DMSK) Systems in an ACI and AWGN Environment," *IEEE Transactions on Communications*, Vol. COM-34, No. 7, July 1986, pp. 727–733.

[68] M. Hirono, T. Miki, and K. Murota, "Multilevel Decision Method for Band-Limited Digital FM with Limiter-Discriminator Detection," *IEEE Journal on Selected Areas in Communications*, Vol. SAC-2, No. 4, July 1984, pp. 498–508.

[69] D. L. Schilling, E. A. Nelson, and K. K. Clarke, "Discriminator Response to an FM Signal in a Fading Channel," *IEEE Transactions on Communications Technology*, Vol. COM-15, No. 2, April 1967, pp. 252–263.

[70] D. L. Schilling, E. A. Nelson, and K. K. Clarke, Correction to "Discriminator Response to an FM Signal in a Fading Channel," *IEEE Transactions on Communications Technology*, Oct. 1967, pp. 713–714.

[71] D. K. Schilling, E. Hoffman, and E. A. Nelson, "Error Rates for Digital Signals Demodulated by an FM Discriminator," *IEEE Transactions on Communications Technology*, Vol. COM-15, No. 4, Aug. 1967, pp. 507–517.

[72] M. K. Simon and C. C. Wang, "Differential versus Limiter-Discriminator Detection of Narrow-Band FM," *IEEE Transactions on Communications*, Vol. COM-31, No. 11, Nov. 1983, pp. 1227–1234.

[73] Y. Han and J. Choi, "DMSK System with Nonredundant Error Correction Capability," *IEEE GLOBECOM*, 1991, pp. 770–774.

[74] T. Masamura, S. Samejima, Y. Morihiro, and H. Fuketa, "Differential Detection of MSK with Nonredundant Error Correction," *IEEE Transactions on Communications*, Vol. COM-27, No. 6, June 1979, pp. 912–918.

[75] H. Taub and D. L. Schilling, *Principles of Communication Systems*, McGraw Hill, 1986, New York.

[76] K. S. Shanmugam, *Digital and Analog Communication Systems*, John Wiley & Sons, 1979, New York.

[77] M. G. Shayesteh and A. Aghamohammadi, "On the Error Probability of Linearly Modulated Signals on Frequency-Flat Ricean, Rayleigh, and AWGN Channels," *IEEE Transactions on Communications*, Vol. 43, No. 2/3/4, Feb/Mar/April 1995, pp. 1454–1466.

[78] N. Ekanayake, "Performance of M-ary PSK Signals in Slow Rayleigh Fading Channels," *Electronic Letters*, Vol. 26, No. 10, May 1990, pp. 618–619.

[79] D. A. Shnidman, "Evaluation of the Q-Function," *IEEE Transactions on Communications*, March 1974, Vol. 22, No. 3, pp. 342–346.

[80] S. A. Rhodes, "Effect of Noisy Phase Reference on Coherent Detection of Offset-QPSK Signals," *IEEE Transactions on Communications*, Vol. COM-22, No. 8, Aug. 1974, pp. 1046–1055.

[81] P. C. Jain, "Error Probabilities in Binary Angle Modulation," *IEEE Transactions on Information Theory*, Vol. IT-20, No. 1, Jan. 1974, pp. 36–42.

[82] R. F. Pawula, "Generic Error Probabilities," *IEEE Transactions on Communications*, Vol. 47, No. 5, May 1999, pp. 697–702.

[83] A. Abrardo, G. Benelli, and G. R. Cau, "Multiple-Symbol Differential Detection of GMSK for Mobile Communications," IEEE Transactions on Vehicular Technology, Vol. 44, No. 3, Aug. 1995, pp. 379–389.

[84] R. Matyas, "Effect of Noisy Phase References on Coherent Detection of FFSK Signals," *IEEE Transactions on Communications*, Vol. COM-26, No. 6, June 1978, pp. 807–815.

[85] P. J. Lee, "Computation of the Bit Error Rate of Coherent M-ary PSK with Gray Code Bit Mapping," *IEEE Transactions on Communications*, Vol. COM-34, No. 5, May 1986, pp. 488–491.

[86] T. A. Schonhoff, "Symbol Error Probabilities for M-ary CPFSK Coherent and Noncoherent Detection," *IEEE Transactions on Communications*, June 1976, pp. 644–652.

[87] J. G. Proakis, "Probabilities of Error for Adaptive Reception of M-Phase Signals," *IEEE Transactions on Communications Technology*, Vol. COM-16, No. 1, Feb. 1968, pp. 71–81.

[88] S. Park, D. Yoon, and K. Cho, "Tight Approximations for Coherent MPSK Symbol Error Probability," *Electronics Letters*, Vol. 39, No. 16, Aug. 2003, pp. 1220–1222.

[89] M-S. Alouini, and A. J. Goldsmith, "A Unified Approach for Calculating Error Rates of Linearly Modulated Signals over Generalized Fading Channels," *IEEE Transactions on Communications*, Vol. 47, No. 9, Sept. 1999, pp. 1324–1334.

[90] H. Zhang and T. A. Gulliver, "Error Probability for Maximum Ratio Combining Multichannel Reception of M-ary Coherent Systems over Flat Ricean Fading Channels," WCNC 2004, *IEEE Communications Society*, pp. 306–310.

[91] C. M. Chie, "Bounds and Approximations for Rapid Evaluation of Coherent MPSK Error Probabilities," *IEEE Transactions on Communications*, Vol. COM-33, No. 3, Mar. 1985, pp. 271–273.

[92] J. Sun and I. S. Reed, "Performance of MDPSK, MPSK, and Noncoherent MFSK in Wireless Rician Fading Channels," *IEEE Transactions on Communications*, Vol. 47, No. 6, June 1999, pp. 813–816.

[93] J. J. Koo and K. D. Barnett, "Improved Bounds for Coherent M-ary PSK Symbol Error Probability," *IEEE Transactions on Vehicular Technology*, Vol. 46, No. 2, May 1997, pp. 396–399.

[94] W-P. Yung, "Probability of Bit Error for MPSK Modulation with Diversity Reception in Rayleigh Fading and Log-Normal Shadowing Channel," *IEEE Transactions on Communications*, Vol. 38, No. 7, July 1990, pp. 933–937.

[95] J. W. Craig, "A New, Simple and Exact Result for Calculating the Probability of Error for Two-Dimensional Signal Constellations," *IEEE MILCOM* 1991, pp. 571–575.

[96] S. Chennakeshu and J. B. Anderson, "A New Symbol Error Rate Calculation for Rayleigh Fading Multichannel Reception of MPSK Signals," *IEEE Vehicular Technology Conference*, 1993, pp. 758–761.

[97] P. Y. Kam, "Tight Bounds on Rician-Type Error Probabilities and Some Applications," *IEEE Transactions on Communications*, Vol. 42, No. 12, Dec. 1994, pp. 3119–3128.

[98] P. Petrus and J. Boccuzzi, "Comparison of $\pi/4$-DQPSK Non-Coherent Detection Schemes for PCS Systems," *AT&T Bell Laboratories Technical Memorandum*, 1994.

[99] K. Murota and K. Hirade, "GMSK Modulation for Digital Radio Telephony," *IEEE Transactions on Communications*, Vol. COM-29, No. 7, July 1981, pp. 1044–1050.

[100] P. A. Bello and B. D. Nelin, "The Effect of Frequency Selective Fading on the Binary Error Probabilities of Incoherent and Differentially Coherent Matched Filter Receivers," *IEEE Transactions on Communications Systems*, June 1963, Vol. 11, No. 2, pp. 170–186.

[101] T. A. Schonhoff, "Symbol Error Probabilities for M-ary CPFSK Coherent and Noncoherent Detection," *IEEE Transactions on Communications*, June 1976, Vol. 24, No. 6, pp. 644–652.

CHAPTER 5
PERFORMANCE IMPROVEMENT TECHNIQUES

In this chapter we will build upon the topics previously discussed. Various modulation techniques were introduced followed by a presentation of different channel models. Demodulation techniques were also discussed and their corresponding performance curves were compared. In many applications the resulting system performance based on the above material is insufficient to provide a reliable communication link. Hence, in this chapter, we will present performance improvement techniques that would allow the system designer to be able to create a reliable communication system. We will begin by presenting Forward Error Correction (FEC) codes and their ability to improve performance. Next we will discuss antenna diversity techniques, both at the transmit and receive sides of the communication link. Lastly, we will apply these techniques and discuss the overall system gains in example link budget calculations.

5.1 FORWARD ERROR CORRECTION CODES

In this section we will discuss FEC codes. Here parity bits of information are inserted into the transmit data stream. The receiver makes use of these parity bits to better detect the data bits within a particular block or group of bits. A conventional-coded system block diagram is shown in Fig. 5.1 [1].

The raw information is introduced to the FEC encoder at an uncoded rate of R_u bits/sec, this corresponds to the uncoded data rate. Since the encoder adds parity bits to the data bits, the encoded output data rate is given by R_c bits/sec, with $R_c > R_u$. The coded bits enter the modulator to perform the complex envelop domain transformation and create symbols, R_s symbols/sec.

5.1.1 Interleaving/De-interleaving

As shown in Chap. 3, multipath fading can be characterized by the following parameters: Doppler spread, frequency selectivity, average fade time duration (AFTD), and Level Crossing Rate (LCR). Let us, for the moment, assume that errors only occur when the channel is in a deep fade. In this case the errors would be bunched together in groups, exhibiting a bursty behavior. If the length of the burst of errors is greater than the error correction capability of the FEC decoder, then performance improvements will be greatly diminished [2]. A method to combat this behavior is to use interleaving. The purpose of interleaving is to randomize (or distribute) the bursty errors caused by the multipath channel (or any other channel with memory, as well as Spectral Conversion (RF) impairments). This is best described with the help of Fig. 5.2.

Prior to discussing the benefits of interleaving we would like to present its disadvantage.[1] These operations introduce time delays into the system. This time delay is the actual time required to perform

[1] This is contrary to the effective presentation principles, where the benefits are discussed first.

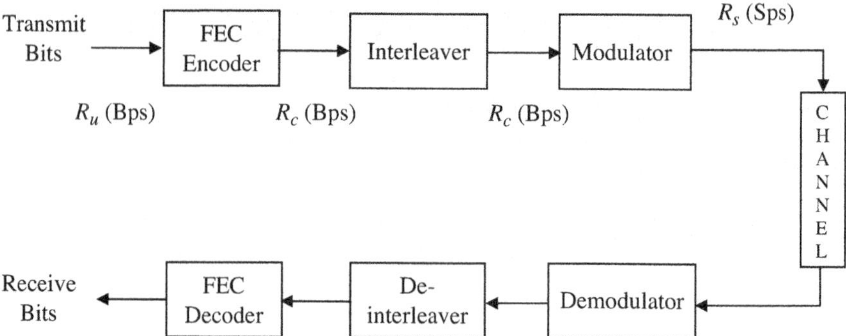

FIGURE 5.1 Coded communication system block diagram.

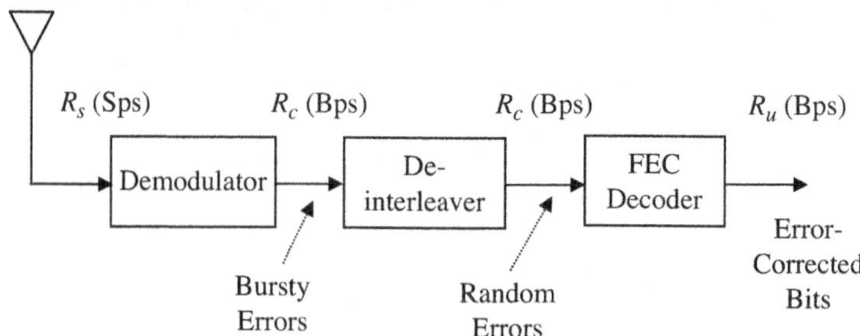

FIGURE 5.2 Coded communication system receive block diagram.

the interleaving plus the time required to perform de-interleaving. This increase in time delay can be detrimental especially in a voice application.

Interleaving improves the performance of a communication system. It is a form of time diversity that is used to mitigate the effects of bursty errors. A bit stream is considered to be ideally interleaved when consecutive received bits are independent, in other words, not affected by the same burst of errors. This is accomplished when adjacent bits of the transmitter are separated by more than the AFTD of the channel. Similarly, a bit stream is considered to be partially interleaved when consecutive received bits are affected by the same burst of errors.

Block Type. The first type of inter/de-interleaving technique to be discussed is called block interleaving. Here a block of data is stored and waiting to be transmitted. The block size is sometimes described the same way the sizes of matrices are defined.

A typical block interleaver is shown below (Fig. 5.3) and contains N rows and M columns, describing an N by M ($N \times M$) matrix. The number of rows, N, is commonly called "interleaver depth." This value is chosen based on the FEC code word size and the error protection needed in the communication system link. Each element (or cell) of the $N \times M$ matrix is either a bit or a symbol, depending on its location in the communication system.

In this example the FEC encoder output bit stream is inserted into each row until all N rows have been filled. At this point the coded data to be transmitted is read out by the columns.

Now turning our attention to the receiver we describe the block de-interleaving operations. Here a matrix of the same $N \times M$ size is constructed by writing the received bits into each column. Once the matrix is full, data is read out of each row (see Fig. 5.4). This de-interleaved bit stream enters the FEC decoder.

One can easily see the additional time delays the information signal incurred as a result of the interleaving and de-interleaving. However, the benefits of using this randomizing approach can be

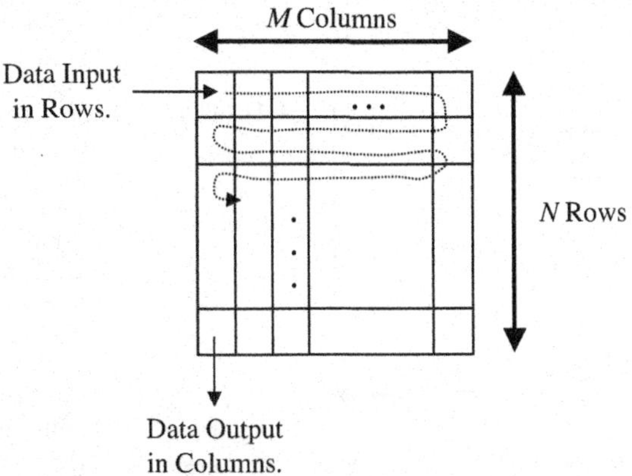

FIGURE 5.3 Block interleaver example.

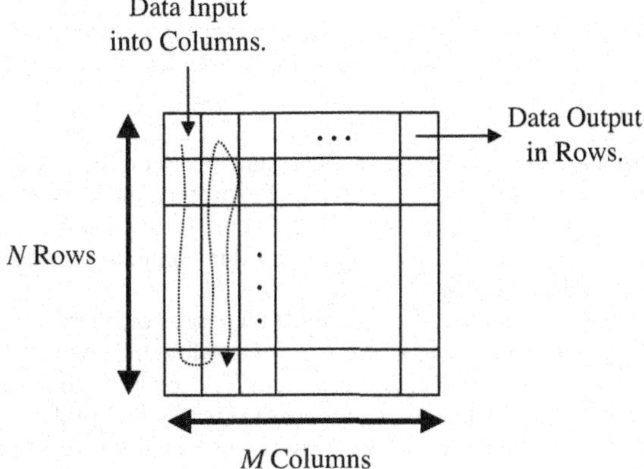

FIGURE 5.4 Block de-interleaver example.

tremendous. The worse case time delay when considering both the interleaving and de-interleaving operations is given by $2NM$ bits.

Let us consider the following bursty error example (see Fig. 5.5). The transmitted bit stream consists of concatenated columns as shown below (C_{ij} = ith column and jth row element).

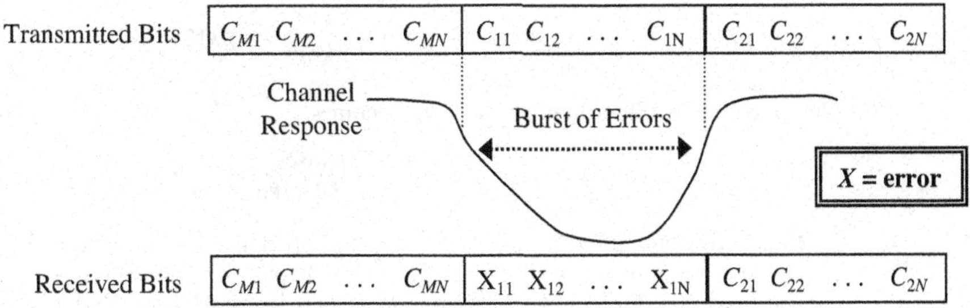

FIGURE 5.5 Bursty error example.

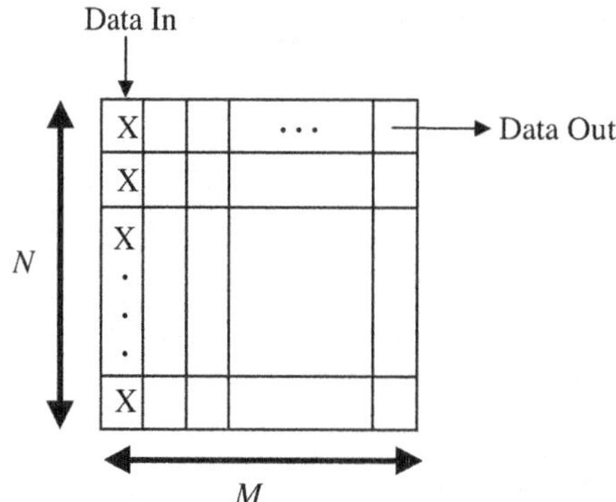

FIGURE 5.6 De-interleaving the bursty error example.

As mentioned earlier, multipath fading can produce strong correlation from one symbol time to the next, especially for low Doppler spreads. It is for this reason, as well as the type of demodulator used in the receiver, that the errors tend to be bursty in nature. The receiver takes the data and writes them into columns. In this particular example shown in Fig. 5.5, there is a burst of N bits. Hence a complete column consists of erroneous (or unreliable) data bits. As the de-interleaver matrix is complete, the data is read out in rows as shown in Fig. 5.6.

The de-interleaver data output bit stream will resemble as shown in Fig. 5.7.

Here we can clearly see that a burst of N errors was redistributed to having a single error, every

$X_{11}\ C_{21} \ldots\ C_{M1}$	$X_{12}\ C_{12} \ldots\ C_{M2}$	$C_{13}\ C_{23} \ldots\ C_{M3}$

FIGURE 5.7 De-interleaver output data stream for the bursty error example.

M bits apart. Hence using a simple single error correcting FEC code would correct and compensate the channel burst of errors. We have redistributed the bursty channel errors into random errors in the code words. This would hopefully be in the error correction capability of the FEC scheme chosen. Block-type interleaving methods are used in communications systems, for example in the 3GPP cellular block interleaving.

In designing interleavers one must take into consideration the largest burst of errors the communication system can reliably handle. It can often happen that the interleaver cannot fully randomize the bursty errors and thus present the FEC decoder with an input signal that contains a number of errors that is larger than the error correction capability of the code word. In this case you either accept the errors and hope the system operating procedures (i.e., upper layers) mitigate this or increase the block size at the expense of longer time delays or finally use a higher error correcting FEC code. Retransmission techniques are also available. In this case if an error is determined the same packet of data can be retransmitted and potentially combined with the previous packet (i.e., Chase combining) or certain data and additional parity bits are retransmitted and combined with the previous packet (i.e., incremental redundancy combining). As you would expect, there are quite a few options available to the communication system designer.

Let us consider a 3.2-kbps binary FSK modem using an FEC code. Our objective is to protect our information against a 10-msec burst of errors. So in designing the interleaver we proceed as follows. The bit time duration is defined as

$$T_b = \frac{1}{R_b} = \frac{1}{3.2\,\text{kbps}} = 0.312\,\text{msec} \tag{5.1}$$

Our protection depth or interleaver depth becomes

$$\text{Depth} = \frac{10\,\text{msec}}{0.132\,\text{msec}} = 33\,\text{bits} \tag{5.2}$$

Hence we would propose using a 32 × M size block interleaving scheme. The value of M will be discussed more in the following sections of this chapter, since this is related to the error correction capability of the FEC code. A last point to make here is that the combined delay of the interleaver and de-interleaver is not considered to be a variable that should be minimized in this example.

Convolutional Type. The convolutional interleaving operation consists of B parallel delay elements (e.g., FIFOs). As a new bit to be transmitted in inserted, the switch moves to the next element to be used. An example of such an interleaver is shown in Fig. 5.8 [3].

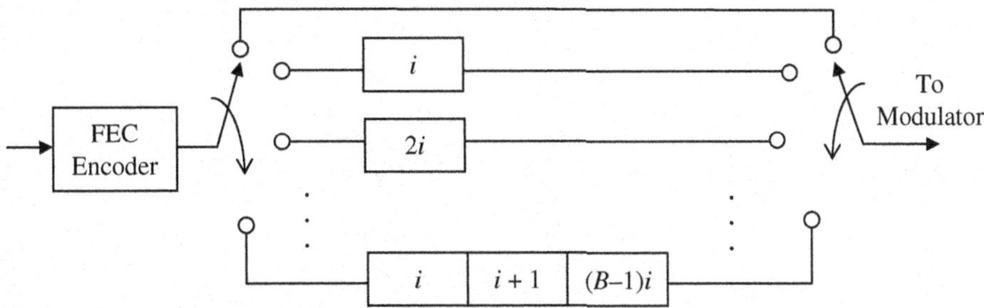

FIGURE 5.8 Convolutional interleaving block diagram.

The interleaver's switches move from one position to the next as new bits are introduced to the block. The de-interleaver will essentially reverse these operations. Next we will present the de-interleaver block diagram shown in Fig. 5.9.

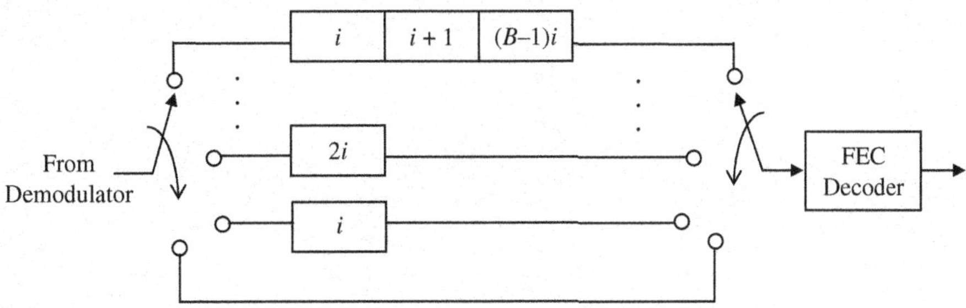

FIGURE 5.9 Convolutional de-interleaving block diagram.

Depending on the actual FEC code used, the convolutional interleaving can have the advantages of using less memory (or delay) elements than the block interleaver. Additionally, this technique can have a lower propagation time delay.

An alternative representation of the convolutional interleaver is shown in Fig. 5.10, where it is created by diagonally splitting the $B \times N$ block interleaver matrix into two halves. Essentially this is creating two triangular matrices.

In comparing this technique to the block interleaver given in the previous section, we see the time delay of the convolutional interleaver is $N(B - 1)$ bit times versus the $2BN$ that would be required for the block interleaver. This time difference is less than half between these two techniques.

Let us consider an example of $B = 3$ symbols, here the memory (or delay) elements are updated every time a new symbol is inserted. Initially 3 symbols are introduced to the interleaver (NA = not available) (see Fig. 5.11).

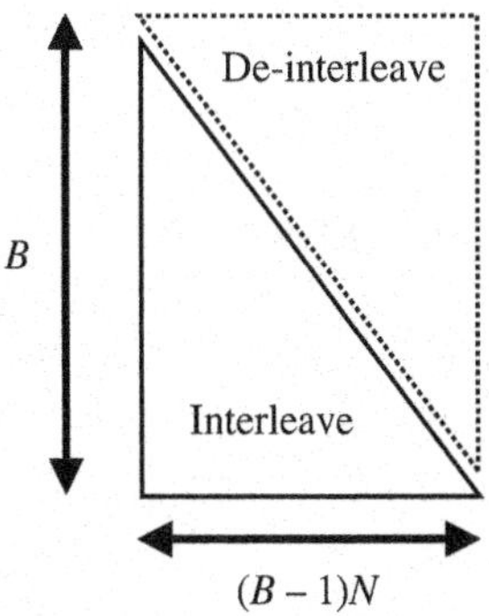

FIGURE 5.10 Convolutional interleaver decomposition.

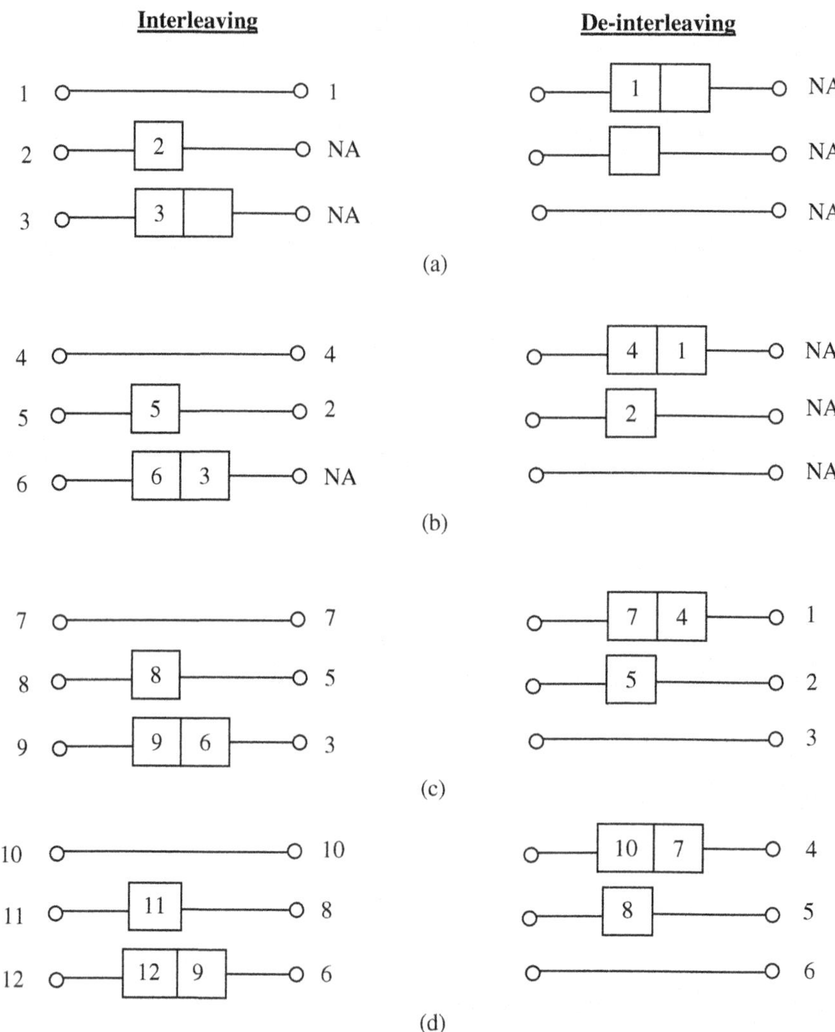

FIGURE 5.11 $N = 3$ symbol example of convolutional interleaving.

5.1.2 Block Codes

We will begin the discussion of block codes starting with linear block codes [4]. Here we group the incoming bits into "words" which then enter what is commonly called a FEC encoder. This coder essentially treats the input word as a vector (of size k bits) and applies an encoding rule across this vector to produce an output "code word." The FEC encoder has the following notation (n,k), where k is the number of input bits that make up a word entering the FEC encoder and n is the number of output bits representing the error protected code word. Here the integer value of n is always greater than that of k (see Fig. 5.12) [5].

Hence the FEC encoder provides a mapping of the 2^k possible k-tuple words to the 2^n possible n-tuple code words. The FEC encoder rule has each one of the 2^k words assigned to one of the 2^n code

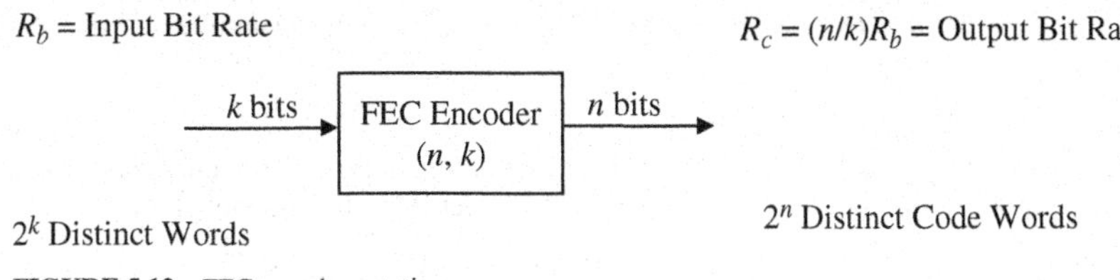

FIGURE 5.12 FEC encoder notation.

words. The mapping is a one-to-one (i.e., unique) and this transformation is linear. This means code words outside the 2^n code words cannot be created by the addition of valid code words.

The input bit rate is the uncoded rate, denoted above as R_b (bits/sec). The encoder output-coded bit rate is n/k times the input bit rate. This ratio is inversely related to what is commonly called the code rate.

Pictorially speaking, there are 2^n code words, of which 2^k are used. The set of all n-tuple code words is typically called a vector space. The entire coded vector space is denoted as circles in Fig. 5.13. In fact, each circle represented an n-tuple codeword. The encoder input-to-output mapping rule selects a subset of the vector space to use as its code-word pool. This subset (or sometimes called subspace providing some criteria are met) is shown in Fig. 5.13 by the black circles [1].

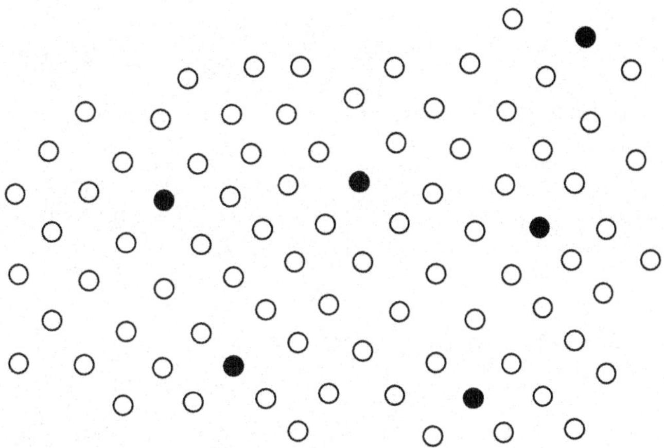

FIGURE 5.13 n-tuple code word vector space.

Prior to moving into the linear block code encoding rules, let us discuss various coding definitions. We list the most commonly used ones below [6].

Hamming weight, $w(c)$. This measurement is the number of nonzero bit locations in the code word vectors, c.

Hamming distance, $d(c_1,c_2)$. This measurement is the number of elements (or bits) in which the two code words, c_1 and c_2, differ.

Minimum distance, d_{min}. This measurement is the smallest value of the Hamming distance, $d(c_i,c_j)$ for i not equal to j.

Error correcting capability, t. This measurement is the number of bit errors that the FEC code word can correct.

$$t = \left\lfloor \frac{d_{\min} - 1}{2} \right\rfloor \tag{5.3}$$

Error detecting capability, e. This measurement is the number of bit errors the FEC code word can detect.

$$e = d_{\min} - 1 \tag{5.4}$$

Let us present the mathematical foundation necessary to understand the FEC encoding operations. Consider the information bits at the encoder input to be denoted as.

$$\vec{x} = \left[x_1, x_2, \ldots, x_k\right] \tag{5.5}$$

And the output of the FEC encoder is a vector defined below.

$$\vec{C} = \left[c_1, c_2, \ldots, c_n\right] \tag{5.6}$$

The generation of the code word is represented as

$$\vec{C} = \vec{x} \cdot G \tag{5.7}$$

where G is commonly called the FEC generation matrix of the code. This provides a one-to-one mapping of the input and output words. The elements of this matrix are given below.

$$G = \begin{bmatrix} g_{11} & g_{12} & \cdots & g_{1n} \\ g_{21} & g_{22} & \cdots & g_{2n} \\ \vdots & & \ddots & \\ g_{k1} & \cdots & & g_{kn} \end{bmatrix} \tag{5.8}$$

For what is typically called systematic codes, the generator matrix can be modified and will have the following form.

$$G = \begin{bmatrix} p_{11} & p_{12} & \cdots & & 1 & 0 & 0 & \cdots \\ p_{21} & p_{22} & & & 0 & 1 & 0 & \cdots \\ \vdots & & \ddots & & \vdots & & \ddots & \\ p_{k1} & \cdots & p_{k(n-k)} & & 0 & & \cdots & 1 \end{bmatrix} \tag{5.9}$$

Which can be rewritten as follows

$$G = [P \quad I_k] \tag{5.10}$$

where P is the parity matrix of size $k \times (n - k)$ and I is the identity matrix of size $k \times k$.

We have a resulting systematic code word because the generator matrix generates a linear block code where the last k bits of each code word are identical to the information bits to be transmitted (e.g., the FEC encoder input) and the remaining $(n - k)$ bits of each code word represent a linear combination of the k information bits at the input.

BCH, Hamming, and Cyclic Codes. Bose-Chadhuri-Hocquenghem (BCH) codes are extensions to the Hamming codes, and because of this reason we wish to first discuss Hamming codes [5]. Hamming codes are characterized by the following FEC encoder constraints

$$(n, k) = (2^m - 1, 2^m - 1 - m) \qquad (m = 2, 3, \ldots) \tag{5.11}$$

These codes have a minimum distance of 3 and are capable of correcting all single errors or detecting 2 or fewer errors within the block of data. Please recall our previous definitions.

$$t = \left\lfloor \frac{d_{\min} - 1}{2} \right\rfloor = 1 \text{ error} \tag{5.12}$$

$$e = d_{\min} - 1 = 2 \text{ errors} \tag{5.13}$$

Since this is a single-error correcting code, syndrome decoding can be used rather nicely to perform the error correction [6]. In general, a t-error correcting (n, k) linear block code is capable of correcting a total of 2^{n-k} error patterns.

Now turn our attention to the BCH codes. As discussed earlier they are extensions to the Hamming codes in that multiple bit error correction capability exists. The most commonly used BCH code words have a block length of $n = 2^m - 1$ where $m = 3, 4, \ldots$

Let us consider an example of (6, 3) BCH codes. This BCH code has an error correction capability of $t = 1$, since $d_{min} = 3$. The codes can be constructed in the following manner. Consider the (6,3) generator matrix as follows, written in systematic form.

$$G = \begin{bmatrix} 1 & 0 & 0 & 1 & 1 & 0 \\ 0 & 1 & 0 & 1 & 0 & 1 \\ 0 & 0 & 1 & 0 & 1 & 1 \end{bmatrix} = [I_3 \ \ P] \tag{5.14}$$

Recall the code word is generated as follows:

$$\vec{C} = \vec{x} \cdot G = [x_1 \ \ x_2 \ \ x_3] \cdot G \tag{5.15}$$

The encoder block diagram can be easily drawn from the previous equation and is given in Fig 5.14.

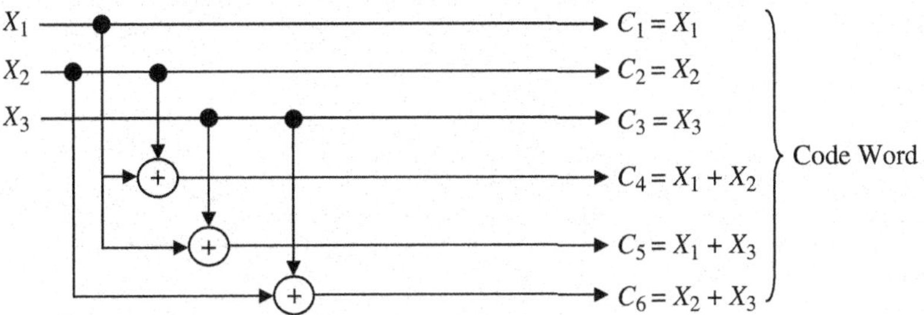

FIGURE 5.14 A (6,3) block code encoder details.

Alternatively we can easily construct the (7,4) code word from the following generator matrix of the encoder (written in systematic form).

$$G = \begin{bmatrix} 1 & 0 & 1 & 1 & 0 & 0 & 0 \\ 1 & 1 & 1 & 0 & 1 & 0 & 0 \\ 1 & 1 & 0 & 0 & 0 & 1 & 0 \\ 0 & 1 & 1 & 0 & 0 & 0 & 1 \end{bmatrix} = [P \ \ I_4] \tag{5.16}$$

Similarly, we can draw the encoder used to generate the output code word as shown in Fig. 5.15.

Let us discuss the FEC encoder in more detail. Recall the generator matrix for a BCH (7,4) code word was given as

$$G = \begin{bmatrix} 1 & 0 & 1 & 1 & 0 & 0 & 0 \\ 1 & 1 & 1 & 0 & 1 & 0 & 0 \\ 1 & 1 & 0 & 0 & 0 & 1 & 0 \\ 0 & 1 & 1 & 0 & 0 & 0 & 1 \end{bmatrix} \tag{5.17}$$

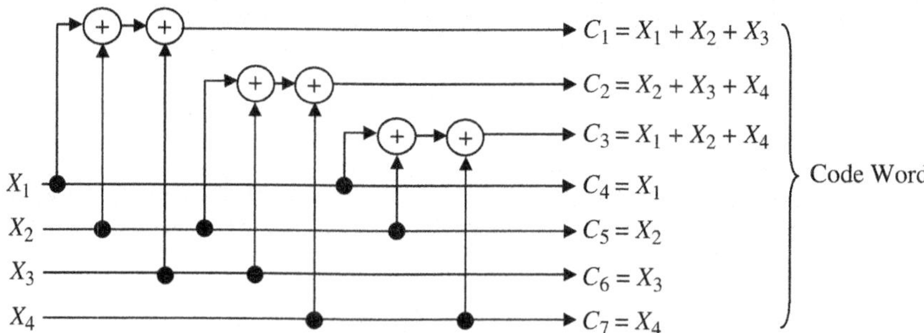

FIGURE 5.15 A (7,4) block code encoder details.

We wish to discuss how the generator matrix was created. Much work has been done to search for generator polynomials that produce good results. We will simply refer the reader to widely published tables that list generator polynomials for various encoders. For example consider Table 5.1 as a starting point [7].

TABLE 5.1 BCH Primitive Polynomials

n	k	$G(x)$	$G(x)$ *octal*	t
7	4	$X^3 + X + 1$	13	1
15	11	$X^4 + X + 1$	23	1
32	26	$X^5 + X^2 + 1$	45	1
63	57	$X^6 + X + 1$	103	1
127	120	$X^7 + X + 1$	211	1
255	247	$X^8 + X^4 + X^3 + X^2 + 1$	435	1

These generator polynomials will be used as encoding rules governing the operations of the FEC encoder. A note to make here is that the generator polynomials listed in Table 5.1 are primitive polynomials.

The generator polynomial for (7,4) code is given as $g(x) = x^3 + x + 1$ and in binary field this is represented as

$$\vec{g} = [1\ 1\ 0\ 1] \tag{5.18}$$

We can create code words by performing cyclic shifting of the generator polynomial as follows:

$$\begin{aligned} \vec{g}_0 &= [1\ 1\ 0\ 1] \\ \vec{g}_1 &= [1\ 1\ 1\ 0] \\ \vec{g}_2 &= [0\ 1\ 1\ 1] \\ \vec{g}_3 &= [1\ 0\ 1\ 1] \end{aligned} \tag{5.19}$$

In fact, performing a third shift will result in the time reversal of $\vec{g}_0$. Now we can take this generator polynomial and create the generator matrix as follows (where T is used to denote the transpose operation).

$$G = [\vec{g}_1^T\, \vec{g}_2^T\, \vec{g}_0^T I_{n-k}] = [PI_{n-k}] \tag{5.20}$$

Note the columns of P can be arranged in any order without affecting the distance property of the code. This was a specific example for (7,4) code, generally speaking elementary row operations on the matrix G can produce the systematic form (see Table 5.2) [8].

The next group of error correcting codes to be discussed are cyclic codes. These codes satisfy the following property: if $C = [C_{n-1}C_{n-2}...C_1C_0]$ is a code word of a cyclic code, then $C = [C_{n-2}C_{n-3}...C_0C_{n-1}]$ is also a code word. In other words, all cyclic shifts of C are code words. These codes have wonderful properties that enable efficient encoding/decoding operations. Thus making it practically possible to implement long codes with many code words.

The generator polynomial of (n,k) cyclic code is a factor of $p^n + 1$ and is given as

$$g(p) = p^{n-k} + g_{n-k-1}p^{n-k-1} + \cdots + g_1 p + 1 \tag{5.21}$$

TABLE 5.2 Primitive Polynomials for BCH Codes

n	k	t	$G(x)$ octal
7	4	1	13
15	11	1	23
15	7	2	721
15	5	3	2467
31	26	1	45
31	21	2	3551
31	16	3	107657
31	11	5	5423325
31	6	7	313365047
63	57	1	103
63	51	2	12471
63	45	3	1701317
63	39	4	166623567
127	120	1	211
127	113	2	41567
127	106	3	11554743
255	247	1	435
255	239	2	267543
255	231	3	156720665

The message vector can be expressed as

$$x(p) = x_{k-1}p^{k-1} + x_{k-2}p^{k-2} + \cdots + x_1 p + x_0 \tag{5.22}$$

The code word to be transmitted is then given as

$$C(p) = x(p) \cdot g(p) \tag{5.23}$$

For the (7,4) code we have been discussing, the generator polynomial is given as $g(p) = p^3 + p + 1$. This polynomial can be translated to a generator matrix given below. Note that this matrix form is for a nonsystematic code, but as mentioned earlier this matrix can be transformed to become systematic.

$$G = \begin{bmatrix} 1 & 0 & 1 & 1 & 0 & 0 & 0 \\ 0 & 1 & 0 & 1 & 1 & 0 & 0 \\ 0 & 0 & 1 & 0 & 1 & 1 & 0 \\ 0 & 0 & 0 & 1 & 0 & 1 & 1 \end{bmatrix} \tag{5.24}$$

A systematic code word can be generated in the following way: multiply the message polynomial, $x(p)$, by p^{n-k}, then divide the result by the generator polynomial, $g(p)$, to obtain the remainder, $r(p)$. Lastly, add the remainder to the shifted data sequence. This is shown mathematically below

$$\frac{p^{n-k}x(p)}{g(p)} = z(p) + \frac{r(p)}{g(p)} \tag{5.25}$$

Equivalently stated as

$$p^{n-k}x(p) = z(p)g(p) + r(p) \tag{5.26}$$

The division is accomplished using a feedback shift register structure shown in Fig. 5.16. For the first k bit times the switch is in the #1 position so the message bits are sent to the output, since we are considering a systematic coder. Then the switch is moved to the number 2 position where the feedback path is disconnected and the parity bits are read out.

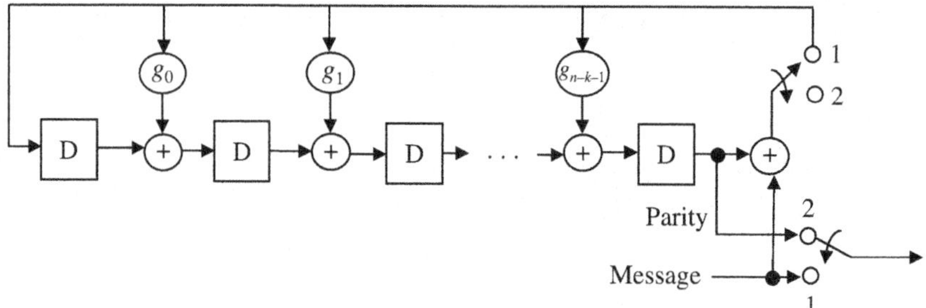

FIGURE 5.16 General cyclic code generator.

Now, for the example of the (7,4) code where the generator polynomial was set to $g(p) = p^3 + p + 1$, the code word generation is given in Fig. 5.17.

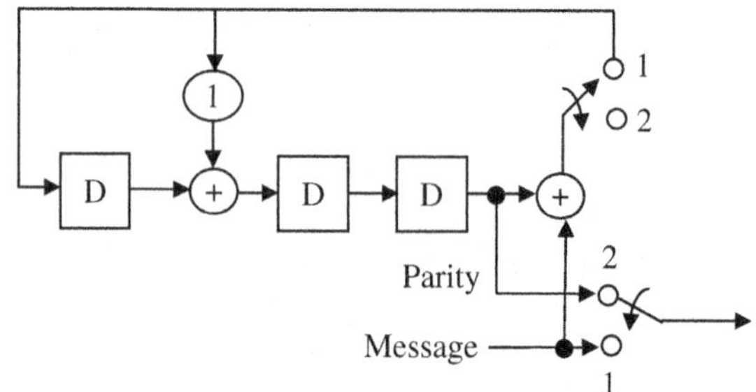

FIGURE 5.17 Cyclic code generator for the (7,4) code.

Syndrome-Based Decoding. As discussed above, for single-error correcting codes, syndrome decoding can be easily used. Let us define the "parity check matrix," H, as satisfying the following desirable property of orthogonality

$$G \cdot H^T = 0 \tag{5.27}$$

This orthogonality constraint results in the following "parity check matrix"

$$H = [I_{n-k}P^T] \tag{5.28}$$

This matrix has the following two properties: First no column can be all zeros and second, all columns must be unique. This matrix can also be rewritten as follows:

$$H^T = \begin{bmatrix} I_{n-k} \\ P \end{bmatrix} \tag{5.29}$$

Let us assume the received vector, $\vec{r}$, is the sum of the transmitted code word, $\vec{C}$, and the error pattern, $\vec{e}$. Here we have

$$\vec{r} = \vec{C} + \vec{e} \tag{5.30}$$

Let us define the syndrome, S, as

$$S = \vec{r} \cdot H^T \tag{5.31}$$

The syndrome is a result of a parity check performed on the received code word to determine whether $\vec{r}$ is a valid member of the code word set. If the syndrome is zero, then $\vec{r}$ is a member and/or may have no errors. If the syndrome is not zero then $\vec{r}$ is not a member of the vector space and it has errors. We can construct the parity check matrix for the (7,4) example we provided above, it is given below

$$H = [I_3\ P^T] = \begin{bmatrix} 1 & 0 & 0 & 1 & 1 & 1 & 0 \\ 0 & 1 & 0 & 0 & 1 & 1 & 1 \\ 0 & 0 & 1 & 1 & 1 & 0 & 1 \end{bmatrix} \tag{5.32}$$

Suppose the transmitted vector is given as

$$\vec{C} = [1\ 0\ 1\ 1\ 0\ 0\ 0] \tag{5.33}$$

And the received vector is given as

$$\vec{r} = [1\ 0\ 1\ 1\ 0\ 1\ 0] \tag{5.34}$$

Based on our previous discussion we can rewrite this as follows:

$$\begin{aligned} \vec{r} &= \vec{C} + \vec{e} \\ \vec{r} &= [1\ 0\ 1\ 1\ 0\ 0\ 0] + [0\ 0\ 0\ 0\ 0\ 1\ 0] \end{aligned} \tag{5.35}$$

The syndrome, S, to be calculated is given as

$$S = \vec{r} \cdot H^T = [\,1\ 0\ 1\ 1\ 0\ 1\ 0\,] \cdot \begin{bmatrix} 1 & 0 & 0 \\ 0 & 1 & 0 \\ 0 & 0 & 1 \\ 1 & 0 & 1 \\ 1 & 1 & 1 \\ 1 & 1 & 0 \\ 0 & 1 & 1 \end{bmatrix} \tag{5.36}$$

$$S = [\,1\ 1\ 0\,]$$

Searching the H matrix to find a column that matches S, we find the sixth column matches. So we have determined the error vector is

$$\hat{e} = [\,0\ 0\ 0\ 0\ 0\ 1\ 0\,] \tag{5.37}$$

Next we add the locally generated error vector to the received vector in order to correct the error present and recover the transmitted code word.

$$\begin{aligned} \vec{r} &= \vec{C} + \vec{e} \\ \vec{r} + \hat{e} &= \vec{C} + \vec{e} + \hat{e} \end{aligned} \tag{5.38}$$

$$\vec{C} = [1\ 0\ 1\ 1\ 0\ 0\ 0\,] + [0\ 0\ 0\ 0\ 0\ 1\ 0] + [0\ 0\ 0\ 0\ 0\ 1\ 0]$$

Below we will show the syndrome is directly related to the error vector and not actually the received vector.

$$\begin{aligned} S &= \vec{r} \cdot H^T = (\vec{C} + \vec{e}) \cdot H^T \\ S &= \vec{C} \cdot H^T + \vec{e} \cdot H^T \end{aligned}$$

$$S = \vec{x} \cdot G \cdot H^T + \vec{e} \cdot H^T$$
$$S = \vec{e} \cdot H^T \tag{5.39}$$

In the above derivation, the last line is obtained using the orthogonality constraint given earlier. Here we see the syndrome is dependent on the error vector.

We can now outline the procedure for correcting the single errors using the syndrome decoding method, it is as follows:

Step 1: Given $\vec{r}$ calculate the syndrome, S.

Step 2: Find the column in H (or the row in H^T) whose value matches the syndrome. This determines the position of the error.

Step 3: Generate this error vector, $\hat{e}$ and add it to the received code word.

$$\vec{r} = \vec{C} + \vec{e}$$
$$\vec{r} + \hat{e} = \vec{C} + \vec{e} + \hat{e} \tag{5.40}$$
$$\vec{r} + \hat{e} = \vec{C}$$

Note the transmitted code word is extracted only if the locally generated error vector, $\hat{e}$, matches the channel generated error vector, $\vec{e}$.

In Fig. 5.18, we provide a generalized block diagram of the error correction method based on syndrome calculation for the (7,4) code. The received code word enters the syndrome calculation block. The syndrome then enters the error pattern detector, which generates the local error vector. Lastly, errors are corrected by the modulo 2 sum of the error vector and the received code word.

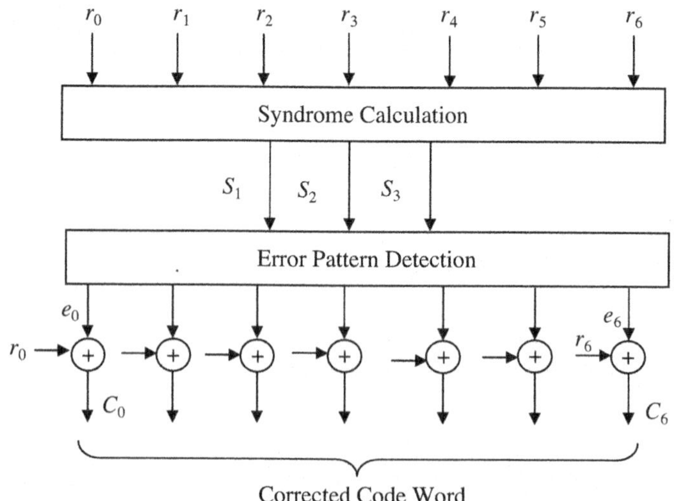

FIGURE 5.18 Syndrome-based error correction block diagram for the (7,4) code.

When the code word is cyclic, the calculation of the syndrome can be performed by a shift register, very similar to what was performed in the encoder discussion above. Recall the received code word, $y(p)$, is written as follows:

$$y(p) = C(p) + e(p) \tag{5.41}$$

$$y(p) = x(p)g(p) + e(p) \tag{5.42}$$

$$\frac{y(p)}{g(p)} = z(p) + \frac{r(p)}{g(p)} \tag{5.43}$$

The division of $y(p)$ by generator polynomial $g(p)$ is carried out by a shift register structure, a general division is shown in Fig. 5.19 [5].

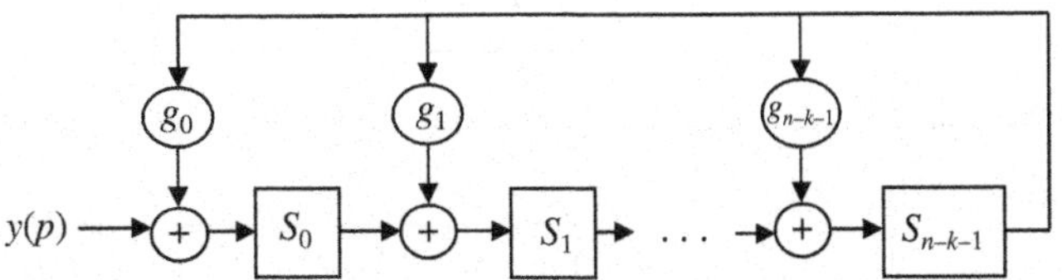

FIGURE 5.19 General syndrome calculation for cyclic codes.

As we have done previously we will use the (7,4) code as an example to illustrate the syndrome calculation. In Fig. 5.20, we show a method of calculating the syndrome.

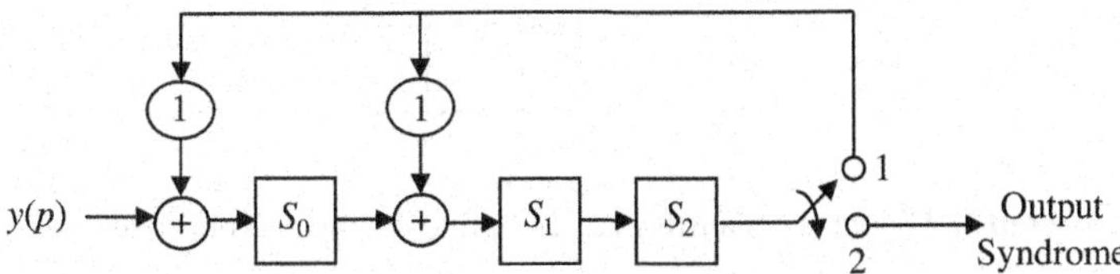

FIGURE 5.20 Syndrome calculation for (7,4) cyclic code.

The syndrome calculation procedure is relatively simple. After n shifts the contents of the registers is equal to the syndrome. For the first n bits the switch is in position 1. Then the switch is placed into position 2 for the next $n - k$ bits in order to extract the syndrome from the registers. It may be premature to conclude the syndrome is a characteristic of the transmitted code word, but recall we have previously shown this to not be the case. In fact, the syndrome is a characteristic of the error pattern.

Another topic related to decoding of the FEC code words is erasure decoding [9]. The FEC decoding operation can be improved if the decoder can recognize a bit of questionable reliability and then declare that bit as an erasure. Here a bit position is identified and an exhaustive search among the possible combinations of bits is conducted.

Recall the number of detectable errors is given by $d_{\text{min}} - 1$, then a pattern of β or fewer erasures can be corrected if the following condition is met

$$\beta \leq d_{\text{min}} - 1 \tag{5.44}$$

Consider this practical example, the received signal strength (i.e., RSSI) is sampled and used as channel measurement information to identify unreliable (or questionable) bits. The purpose here is to weaken their contribution in the decoding phase, because these bits are most likely in error. Once β erasures are selected then the FEC decoding operations are conducted for 2^β iterations.

In fact, this is called *"Chase's Second Algorithm"* and is given below

Step 1: Determine the β bits (actual positions within the code word) with the lowest signal strength.

Step 2: Apply 2^β patterns to the erasure bits and decoding is performed for each candidate.

Step 3: Select the most likely candidate or maximize the a posteriori probability that the received code word is correct.

A block diagram presenting the operation is presented below for the (7,4) code example with $\beta = 2$ erasures. Let's review these steps using an example.

Step 1: Identify the bit positions with the lowest signal strength, here we assume they correspond to the following bit locations within the code word.

		X			X	

Step 2: Apply the patterns to the code words (the number of pattern is $2^\beta = 4$)

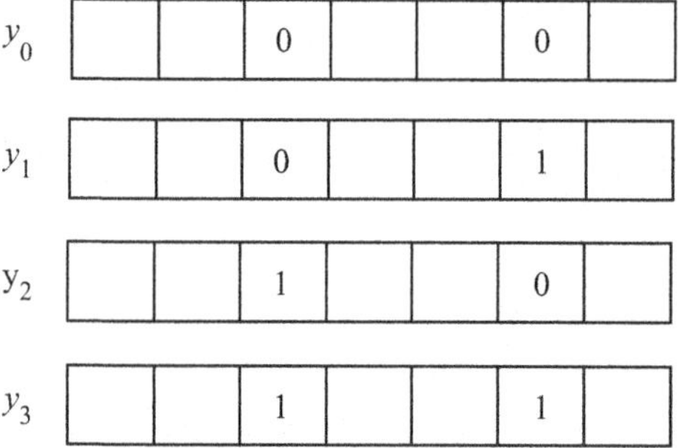

Step 3: Apply the exhaustive search to find the most likely code word transmitted ($j = 0, 1, 2, 3, 4$).

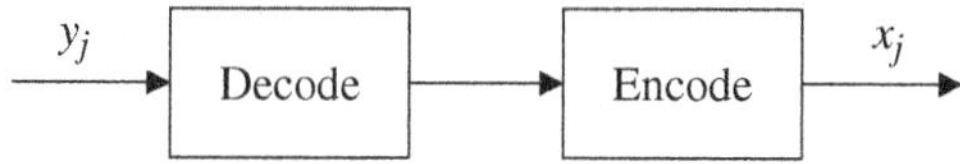

Our goal is to maximize the following equation:

$$\min_j \sum_{i=1}^{7} \gamma_i (y_{i,j} + x_{i,j}) \tag{5.45}$$

Let's provide some insight into the above operations. Let γ_i be equal to the signal strength corresponding to the bit in the ith position in the code word. We select the two smallest values (possibly provided they are below some threshold) and declare them as erasure bits. We create four candidates, which is essentially an exhaustive search of the possible transmitted bits. This in turn provides performance close to that of the maximum likelihood (ML) decoding. Each candidate is decoded and then re-encoded. The resulting two code words, x and y, are XOR-ed together and weighed by their corresponding signal strength. We wish to choose a code word candidate that minimizes the changes in the bit positions where the signal strength is large. We have made use of a relatively safe and reliable assumption that errors don't occur often in the high signal areas.

An alternative way to view error correction is to recall the transmit encoder mapping, where only 2^k n-tuple code words were used. We can place rings of radius equal to the error correction capability of the code word, say t, around valid code words. The significance of this ring is to be able to make the following statement: Any received code word that falls within the circle, can be corrected to the intended code word (see Fig. 5.21).

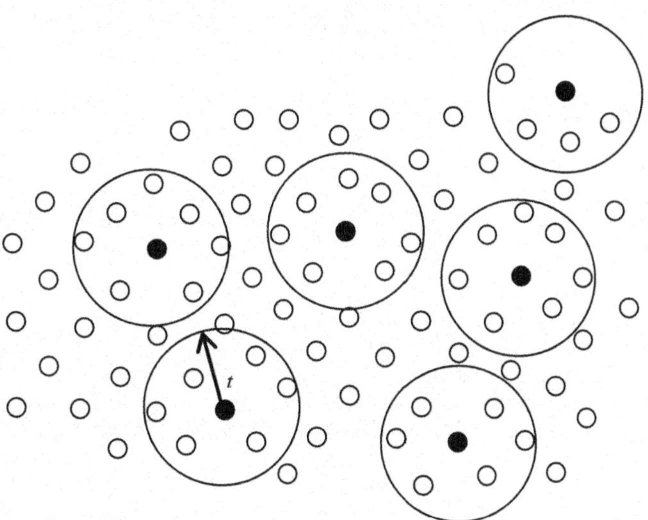

FIGURE 5.21 Error correcting pictorial view.

Let us now discuss the probability of making an erroneous decision. The probability of m errors in a block of n bits is denoted as, $P(m, n)$ and given below where p = the decoder input bit error probability.

$$P(m,n) = \binom{n}{m} \cdot p^m \cdot (1 - p)^{n-m} \tag{5.46}$$

The probability of a code word being in error is upper-bounded as

$$P_m \le \sum_{m=t+1}^{n} P(m,n) \tag{5.47}$$

$$P_m \le \sum_{m=t+1}^{n} \binom{n}{m} \cdot p^m \cdot (1 - p)^{n-m} \tag{5.48}$$

The corresponding bit error probability is given as

$$P_b = \frac{1}{n} \sum_{i=t+1}^{n} i \cdot \binom{n}{i} \cdot p^i \cdot (1 - p)^{n-i} \tag{5.49}$$

The number of code words, N_{cw}, in a sphere of radius, t, is given as

$$N_{cw} = \sum_{i=0}^{t} \binom{n}{i} \tag{5.50}$$

Given the earlier example of the (7,4) code, we know there are 2^7 possible code words of which only 2^4 are used. Hence the number of code words in a sphere of radius $t = 1$ is given as $N_{cw} = 8$.

In an effort to bring this subsection on block codes to some closure, we will briefly present an overview of the block codes discussed. We started by simply providing generator matrices and polynomials for certain block codes. It is out of the scope of this book to discuss how they were obtained. Suffice it to say, these encoding rules or mapping were taken from classical reference books and journal papers. We presented some examples of Hamming codes, BCH codes, and cyclic codes. Syndrome decoding principles were presented for the FEC decoder operations. Lastly, an example of a very powerful technique called erasure decoding was presented.

5.1.3 Convolutional Codes

In this subsection we will discuss an FEC encoding technique using convolutional codes [10]. A convolutional code is described by a set of rules by which the encoding of k data bits into n-coded data bits is defined. Convolutional codes are characterized by three parameters: k = the number of input data bits, n = the number of output data bits, and K = the number of input data bits involved in the calculation of the current output coded bits.

The ratio of k/n is typically called the code rate, this ratio determines the amount of additional redundancy inserted into the code word. Since k is always less than n, the code rate will always be less than unity. The smaller the code rate the more parity bits are inserted into the data stream. Two differences over the block codes are noted as: The value of n does not define the length of the code word and the convolutional encoder has memory. What this says is that the convolutionally encoded bit stream is not only a function of the present input data bit, but also a function of the previous $K - 1$ input bits.

The name "convolutional" is used because the output of this FEC encoder can be viewed as the convolution of the input bit stream and the impulse response of the encoder. Which we will show is a time invariant polynomial.

State Machines and Trellis Diagrams. In this section we will present a few methods that we will use to not only help describe the convolutional codes, but also help analyze them. The first method we will use to describe the code is generator matrix or polynomials. Below we show an encoder with rate = 1/2 and a constraint length of K = 3. The corresponding generator polynomials are given as follows [11]:

$$g_1(x) = 1 + x + x^2 \tag{5.51}$$

$$g_2(x) = 1 + x^2 \tag{5.52}$$

The block diagram of the $R = {}^1/_2$, $K = 3$ encoder is shown in Fig. 5.22.

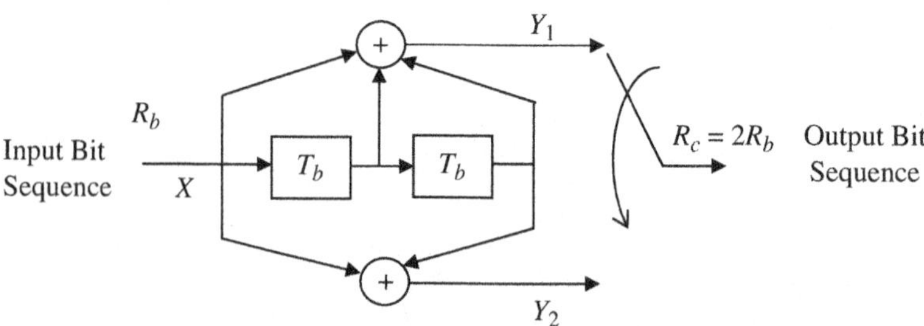

FIGURE 5.22 Rate ${}^1/_2$, $K = 3$, convolutional encoder.

Data is presented to the encoder 1 bit at a time at a rate of R_b Bps. Since the constraint length = 3, we can see that 3 bits are used to create the output bit sequence. The modulo 2 addition operations are controlled by the generator polynomial. These polynomials have special properties (i.e., possess no catastrophic states) and have been computer searched. Some of the details will be discussed later.

This is a rate = ${}^1/_2$ encoder, which means for every input bit presented to the encoder, 2 output bits are generated. The output encoded bit rate, R_c, is twice the input bit rate, R_b. Let us assume the input message sequence is defined as m, and given by m = 1 1 0 1 1. The states of the encoder are shown in Fig. 5.23 for the first 5-bit streams.

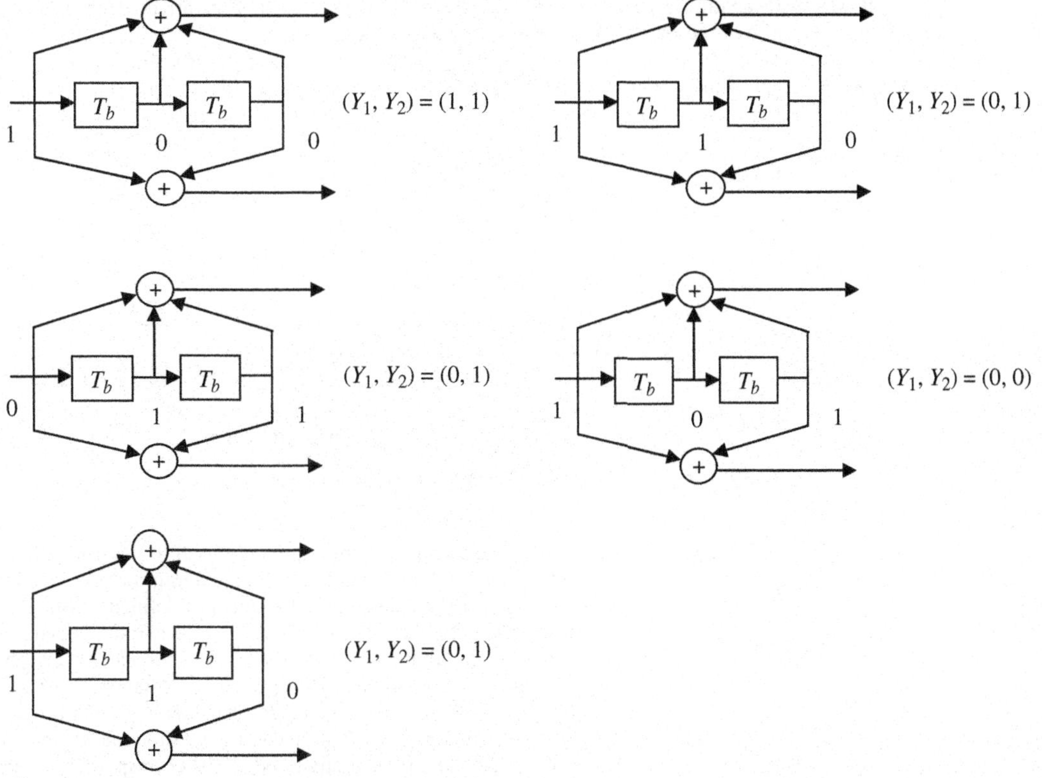

FIGURE 5.23 Rate $^1/_2$, convolutional code state machine.

The input/output bit sequence can be presented in a concise form (see Fig. 5.24).

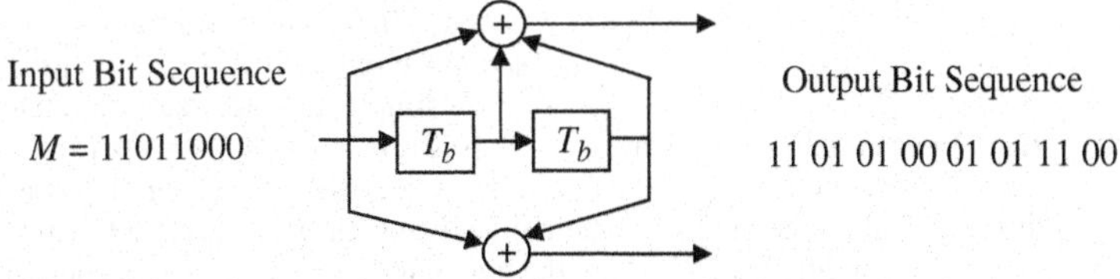

FIGURE 5.24 Convolutional encoder input and output bit stream.

We mentioned earlier in this section that the output can be viewed as the convolution of the input bit stream with the impulse response of the convolutional encoder. In Fig. 5.25 we present the impulse response for the encoder using a particular example.

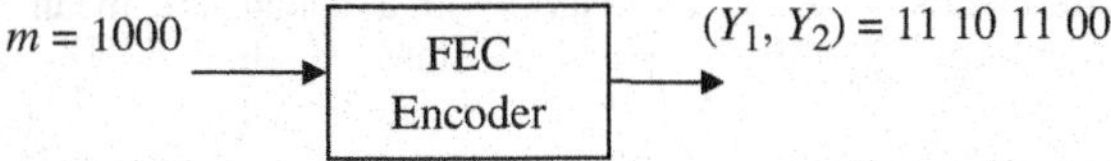

FIGURE 5.25 Rate = $^1/_2$, $K = 3$ convolutional encoder impulse response.

It was also mentioned that the FEC encoder polynomial was time invariant. Hence once the impulse response of the encoder is known, it can be used at any time.

Since convolutional codes are linear we will show the output to be a superposition of the impulse responses, simply shifted in time. Consider the following example (see Fig. 5.26).

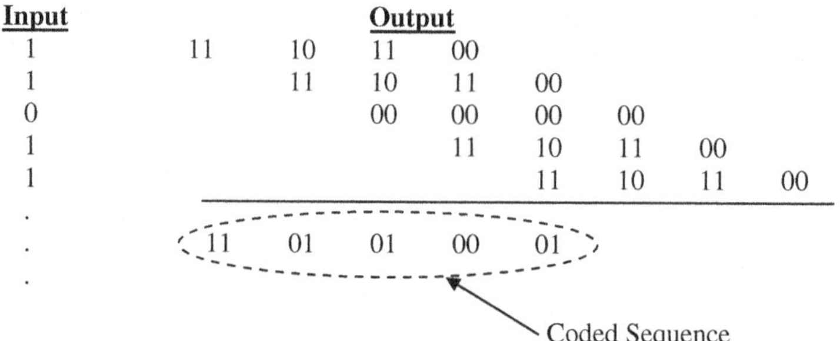

FIGURE 5.26 Convolutional encoder output by overlapping impulse responses.

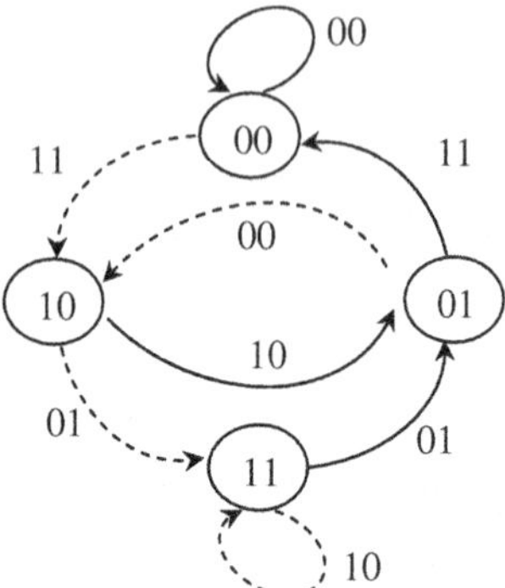

FIGURE 5.27 State diagram representation of the convolutional encoder.

An alternative method commonly used to describe and analyze convolutional codes is the state diagram or state machine. The state of the convolutional encoder is defined as the $K - 1$ right-most bits used to calculate the output coded bits. Continuing with the previous $R = {}^1/_2$, $K = 3$ coder we can show there are $2^{(3-1)} = 4$ states. The encoder state diagram is shown in Fig. 5.27, where the solid line corresponds to a state transition when the input bit $= 0$. The state diagram shows all the possible transitions and various output-coded bits corresponding to their respective input bits [1].

When the input bit is a 1 the corresponding state transition is shown by a dashed line. Similarly an input bit of 0 produces state transitions shown by a solid line. The output-coded bits are written next to the state transition line. For example, consider the case where the present state is 11 and a 0 bit is input into the encoder. This will move the state to 01 and output to a coded bit sequence of 01.

A simple way to track the state transitions is to label each state. We can define each state as $a = \{0,0\}$, $b = \{1,0\}$, $c = \{0,1\}$, $d = \{1,1\}$.

The next method used to analyze the encoder is the tree diagram. This basically adds the notion of a time line to the state diagram, where time increases as we traverse the tree diagram toward the right side of the page. In Fig. 5.28 we show the tree diagram of the FEC encoder with $R = {}^1/_2$ and $K = 3$.

Here we show the first 5 bits of the input and all the possible output combinations. For the input sequence of 11011 we have outlined the corresponding path in the tree diagram. The ability to show a time line is also disadvantageous. By this we mean as the number of input bits increases the size of the tree diagram dramatically and drastically grows out of control.

The last method used to analyze and describe the convolutional encoder is the trellis diagram. This trellis diagram comes about by viewing the repetitive structure of the tree diagram. By viewing the tree diagram we can see the states repeat after K branches (or transitions). Because of the property we can still maintain the time line aspect with a manageable means of description. The trellis diagram is shown in Fig. 5.29 with the number of states equaling $2^{(K-1)} = 2^2 = 4$ states.

Keeping the previous definitions for state transitions (i.e., solid line corresponds to an input bit $=$ 0 and a dashed line corresponds to an input $= 1$) we can outline the path corresponding to the input sequence of 11011 and this is shown in Fig. 5.30.

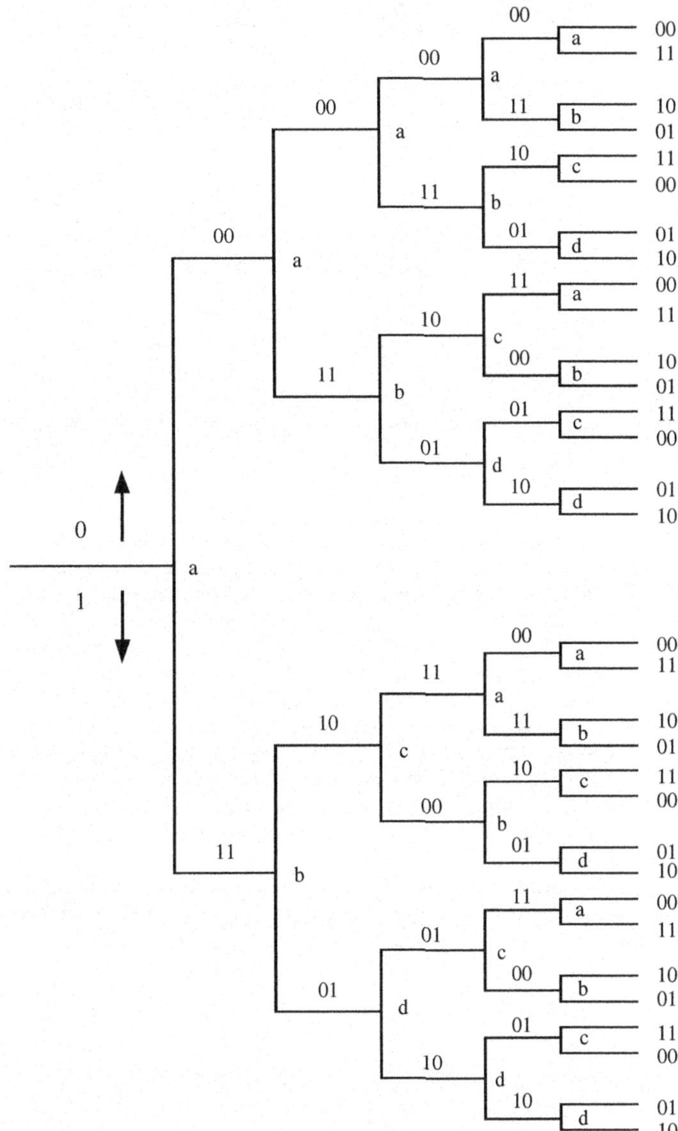

FIGURE 5.28 Convolutional encoder tree diagram representation.

Note each state has a valid transition to certain states, this behavior is controlled by the polynomials used in the encoder. For this particular example, after K input bit durations each state can be entered from only two other states transitions.

Viterbi Algorithm (VA). We now turn our attention to the decoding operations for convolutional codes. Specifically we will discuss an implementation of an ML decoder, commonly referred to as the Viterbi Algorithm [12, 13]. We chose the ML decoder in order to achieve the minimum probability

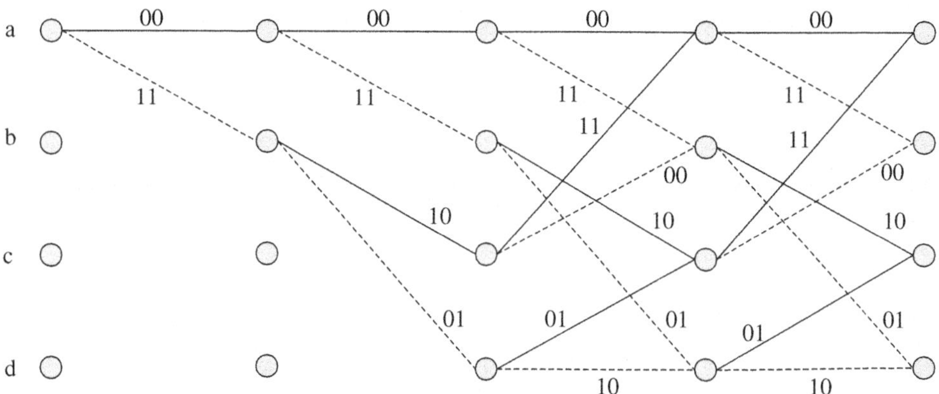

FIGURE 5.29 Trellis diagram representation of the $R = 1/2$ and $K = 3$ convolutional code.

of error. The likelihood functions are given as $P(r|x^k)$ where r = received sequence of information and x^k = one of the possible transmitted sequences of information. The goal of the ML decoder is to choose a particular transmitted sequence that maximizes the likelihood function. This is accomplished by exhaustively comparing or searching all the possible code words that could have been transmitted. Hence for each code word sequence a likelihood value is associated with it. Invoking assumptions of a memoryless channel and additive white Gaussian noise (AWGN) allows the decoder to accumulate likelihood values for each path. The likelihood function can be interpreted as a measure of similarity between all the trellis paths entering each state at time t_k and the received signal at time t_k.

As one can expect the complexity of such a brute force application can quickly grow. One such simplification is to discard paths that are "unlikely," this type of decoder is still optimal. When two paths enter the same state, the most likely path is chosen, and this path is called the surviving path. Each state selects its surviving path. In 1969, Omura showed that the VA is an ML decoding technique and thus optimal [14, 15].

In the next two subsections we will present two versions of the VA, the first operating on hard decisions and the second operating on soft bit values. In the latter version the actual decision on the transmitted bits was made during the VA operations.

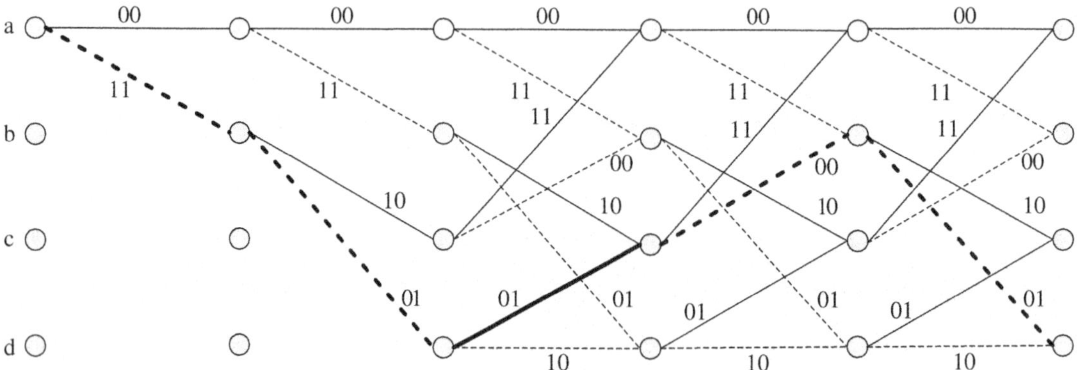

FIGURE 5.30 Trellis diagram path transitions for the 11011 input bit sequence.

Hard Decision Decoding. In this section, we will make binary decisions on the bit sequence entering the Viterbi algorithm, a block diagram showing these operations is provided (see Fig. 5.31). Note that we have purposely excluded the de-interleaving operations in order to not clutter the block diagram. Here binary decisions are made prior to performing the decoding operations, this type of decoding is called hard decision decoding.

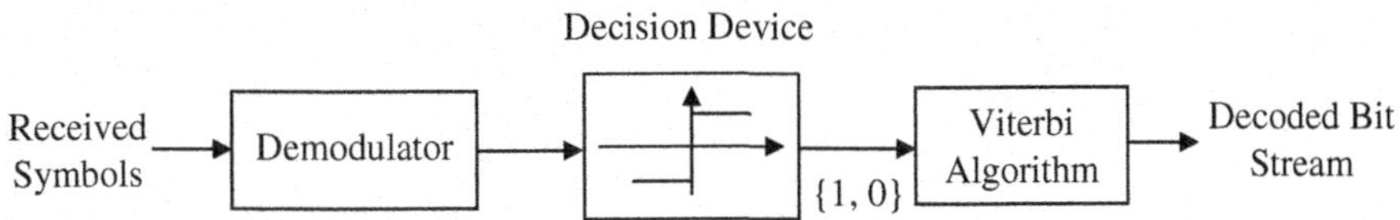

FIGURE 5.31 Hard decision decoding overview.

As discussed above, the VA calculates a measure of similarity between all the states at time equal to t_k and the received code word at time t_k. When binary decisions have been made, the Hamming distance can be used to measure the similarity.

Below we assume the received sequence is given as r = 11 01 01 00 01 and we will build the trellis diagram for each input-coded sequence that enters the Viterbi algorithm. The first 2 received coded bits are compared against all possibly transmitted bits and the Hamming distance is calculated for each possibility. The next 2 received coded bits are also compared against all combinations as well. The exhaustive comparison is shown in Fig. 5.32, where we have placed the Hamming distances next to the state transitions.

The first trellis diagram compares "11" against the possible transmitted bits. The Hamming distance between the locally generated bits and the bits that would have been transmitted, if a 0 was encoded, is a 2. Similarly, the Hamming distance when comparing the received bits to those generated if a 1 was encoded, is a 0. It is very premature to estimate the first encoded bit; however, you can see that, so far, it has a value of 1.

The second trellis diagram compares "01" against the possible transmitted bits. Here four Hamming distances are calculated for each state transition.

The third trellis diagram compares "01" against the possible transmitted bits, this time eight Hamming distances are calculated for each state transition. At this point we need to make a decision on what information to keep going forward in the decoding process.

As a result of comparing the third group of coded bits against all possible code words, each state now has two trellis paths entering them. As stated above we wish to discard the least likely path and select the most likely path, called the survivor path. At this point, the accumulated Hamming distance for each candidate is calculated and the trellis path with the smallest Hamming distance entering that state is selected, this is shown in Fig. 5.33. The trellis diagram presents the accumulated Hamming distance at each state, this is shown as the boxed value.

So the accumulated Hamming distance for states *a*, *b*, *c*, and *d* are equal to 3, 3, 0, and 2, respectively. Now each state has only a single state transition path entering it. The general procedure at this time is to calculate a Hamming distance and update the accumulated Hamming value. Next compare all state transitions and select the path with the smallest accumulated Hamming distance. This procedure is simplified to perform the following operations: Add the metric, compare the state paths, and select the survivor transition. In short, these functions are add, compare, and select (ACS). Hence it is easy to see how the VA can be implemented using these repetitive operations [16].

Now we show the next group of received bits entering the Viterbi decoder and once again select the surviving transition path for each state. This is accomplished by discarding the path with the largest accumulated Hamming distance (see Fig. 5.34).

At this point we have compared four input groups of coded bits against all possible combinations and accumulated the Hamming distances for each state. If we are forced to make a decision, at this moment, on the transmitted bit stream we would choose the trellis path corresponding to the smallest accumulated Hamming distance. For this example we chose the trellis path that terminates at the state

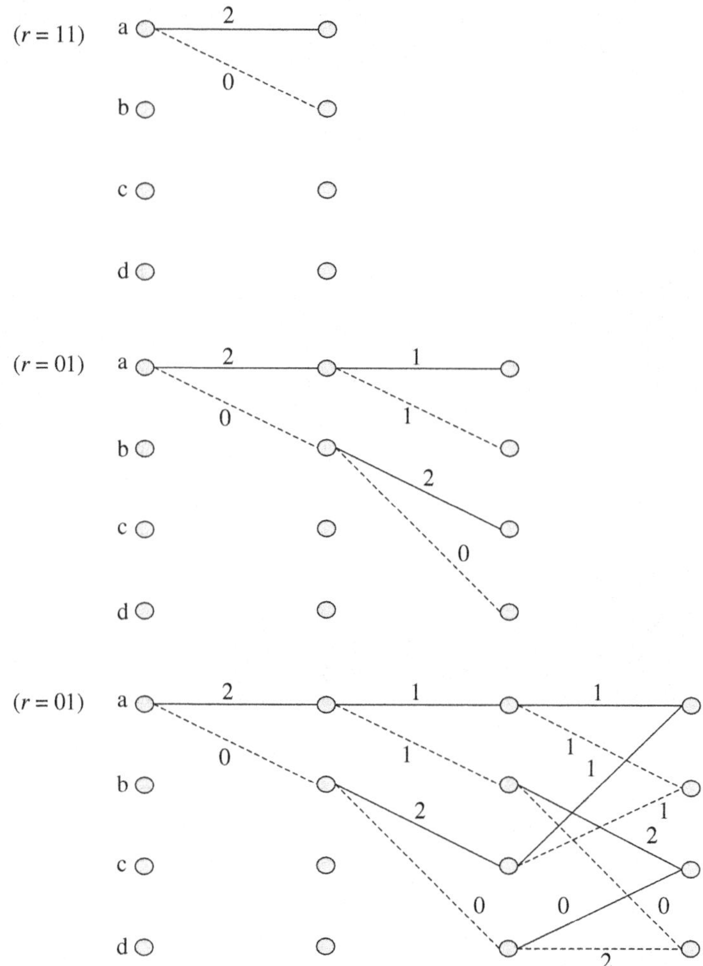

FIGURE 5.32 State transitions of the Viterbi decoding algorithm.

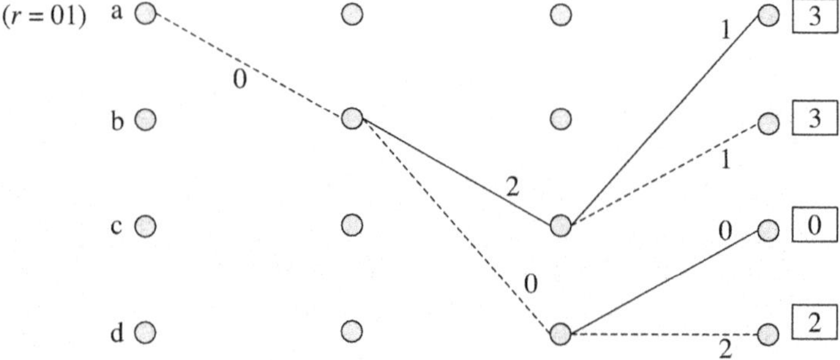

FIGURE 5.33 Survivor state transitions of the Viterbi decoding algorithm for $r = 11\ 01\ 01$.

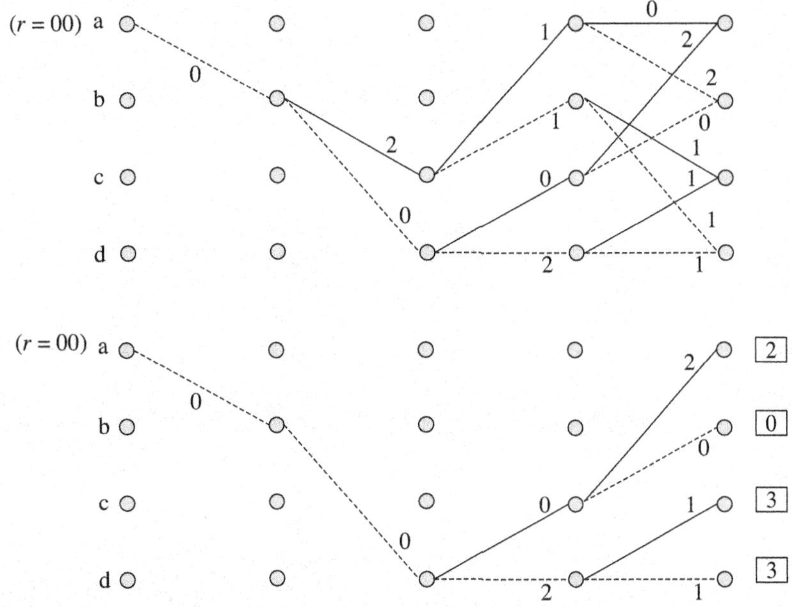

FIGURE 5.34 Survivor state transitions of the Viterbi decoding algorithm for r = 11 01 01 00.

labeled "b" and follow the path back in time to the first transmitted bit, which is 1. In fact, for illustrative purposes, a decision on the second bit can also be made at this time, which also equals 1. Hence the reader can see the operations that are necessary to perform decoding.

In a typical application, the entire frame of data needs to be decoded and so we would continue to accumulate the Hamming distances, compare the joining (or merging) paths, and then select the surviving path. The procedure repeats until the entire data is decoded. Note, it is common to force the very last state to be zero by inserting K zeros at the end of the frame input to the FEC encoder. Now if tentative or final decisions are required prior to decoding the entire frame, then a good rule-of-thumb is to wait for at least five times the constraint length before attempting to make a decision on the bits that were first transmitted. It is generally good practice to give all the accumulated metrics enough time to acquire their statistical distance measurements before one attempts to make any decisions on the transmitted bits. A final point to make about this trellis-decoding operation is that we have presented an approach where a single trellis path is chosen and its associated bit stream is passed to the next signal processing function. We could have maintained a list of such trellis paths where they would be ranked in some order. This type of decoding is called list decoding.

Soft Decision Decoding. In this section we will present a decoder that allows us to distinguish between small amplitude bits and large amplitude bits. In other words, we wish to reduce the influence of questionable bits on the decoding procedure. Here bits that are more reliable impact the decoding decision more than those that are unreliable (or less reliable). This can be accomplished with the help of the following block diagram (see Fig. 5.35).

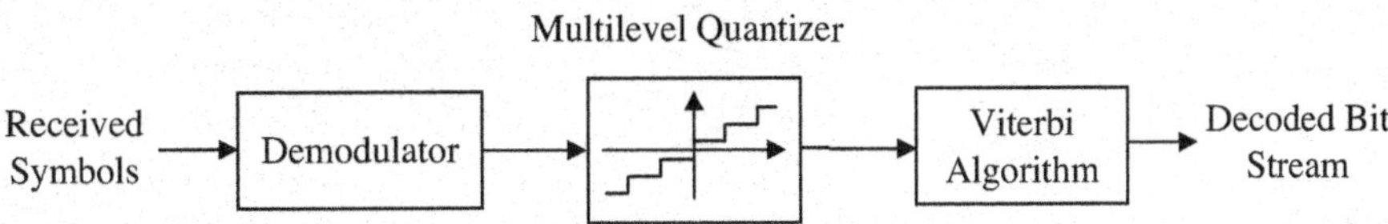

FIGURE 5.35 Soft decision decoding overview.

Here the demodulator output amplitudes are preserved with the use of the multilevel quantizer. When the bit amplitudes are preserved prior to entering the FEC decoder, this is called soft decision decoding. Because of this multiple-quantized bit representation of a code word, the Hamming distance no longer applies as a useful branch metric to us. In fact, the following metric is a suggestion for soft decision decoding

$$\text{metric} = \sum_{k} x_k \cdot y_k \tag{5.53}$$

where y_k is the multilevel-quantized soft symbol output and x_k is the locally generated code word bit stream that takes on the values of $+1$ or -1. By using this metric we have essentially de-emphasized the small amplitude symbols and emphasized the large amplitude symbols in the decision process [17, 18].

Let us take a moment to recall the aim in using FEC codes was to improve system performance [19, 20]. This will be accomplished in the following discussion of coding gain. Very simply, coding gain is defined as follows (for a particular value of BER).

$$\text{Coding gain}_{\text{dB}} = \left(\frac{E_b}{N_o}\right)_{\text{dB}} - \left(\frac{E_b}{N_o}\right)_{\text{coded}} \tag{5.54}$$

Recall the following formula:

$$\frac{E_b}{N_o} = \frac{S}{N} \cdot \frac{\text{BW}}{R_b} \tag{5.55}$$

For uncoded Quaternary Phase Shift Keying (QPSK) modulation with square-root raised cosine filtering at both the transmitter and the receiver we have the following:

$$\left(\frac{E_b}{N_o}\right)_{\text{uncoded}} = \text{SNR} - 3\,\text{dB} \tag{5.56}$$

Consider a coded QPSK system with a rate $R = {}^1/_2$ convolutional code and the coded bit energy (defined as energy of the encoded bits), we now have

$$\left(\frac{E_c}{N_o}\right) = \text{SNR} - 3\,\text{dB} \tag{5.57}$$

Then our uncoded equation becomes

$$\begin{aligned} \frac{E_b}{N_o} &= \frac{1}{R} \cdot \frac{E_c}{N_o} \\ \frac{E_b}{N_o} &= 2 \cdot \frac{E_c}{N_o} \\ \frac{E_b}{N_o} &= \text{SNR} \end{aligned} \tag{5.58}$$

In the above discussion, uncoded refers to the bit energy at the output of the FEC decoder [21, 22].

Below we plot the Bit Error Rate (BER) for $R = {}^1/_2$ convolutional code for various constraint values, K, using a hard decision decoder in an AWGN channel. Figure 5.36 shows the BER performance of Binary Phase Shift Keying (BPSK) using an ideal coherent detector. It also shows the BER performance for $K = 4$, 6, and 8. Focusing on the BER = 1E-3 target, then the coding gains are 1.3 dB, 1.8 dB, and 2.4 dB, respectively. Also a table of some commonly used convolutional codes is given below (see Table 5.3). Here code polynomials are provided for $R = {}^1/_2$ and $R = {}^1/_3$ code rates and for constrait lengths of $K = 3$ to $K = 9$ [23].

We would like to compare the performance of the hard-decision and soft-decision decoders discussed above. This will be accomplished by the following BER performance curves where we compare their respective performance in an AWGN channel.

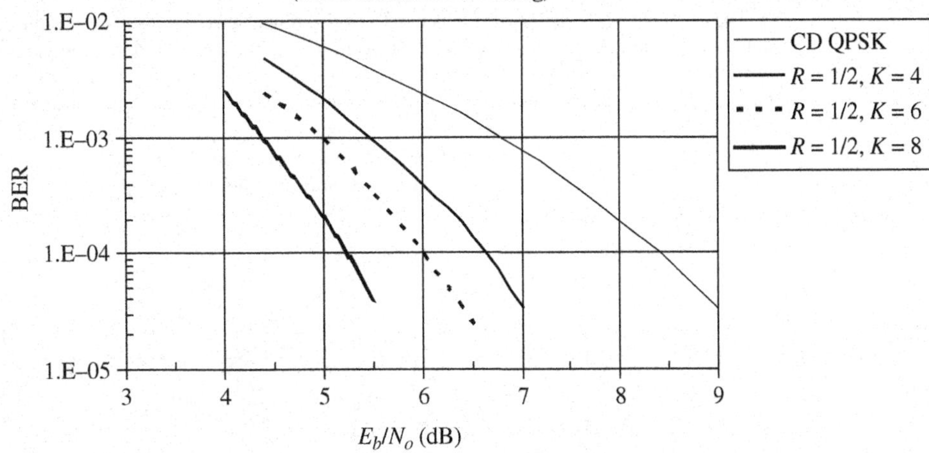

FIGURE 5.36 Performance comparison of various constraint length codes.

TABLE 5.3 Commonly Used Convolutional Codes

Code rate	Constraint length	Code	Free distance
$^1/_2$	3	111 101	5
$^1/_2$	4	1111 1011	6
$^1/_2$	5	10111 11001	7
$^1/_2$	6	101111 110101	8
$^1/_2$	7	1001111 1101101	10
$^1/_2$	8	10011111 11100101	10
$^1/_2$	9	110101111 100011101	12
$^1/_3$	3	111 111 101	8
$^1/_3$	4	1111 1011 1101	10
$^1/_3$	5	11111 11011 10101	12
$^1/_3$	6	101111 110101 111001	13
$^1/_3$	7	1001111 1010111 1101101	15
$^1/_3$	8	11101111 10011011 10101001	16

The ideal antipodal BPSK curve is used as a reference. Two values of constraint lengths were simulated, $K = 5$ and $K = 7$. The additional gain in using soft-decision decoding is approximately 2 dB over hard-decision decoding. For example, consider $K = 7$ and a BER = 1E-3 as the point of interest, the coding gain is approximately 4 dB for soft-decision decoding (see Fig. 5.37).

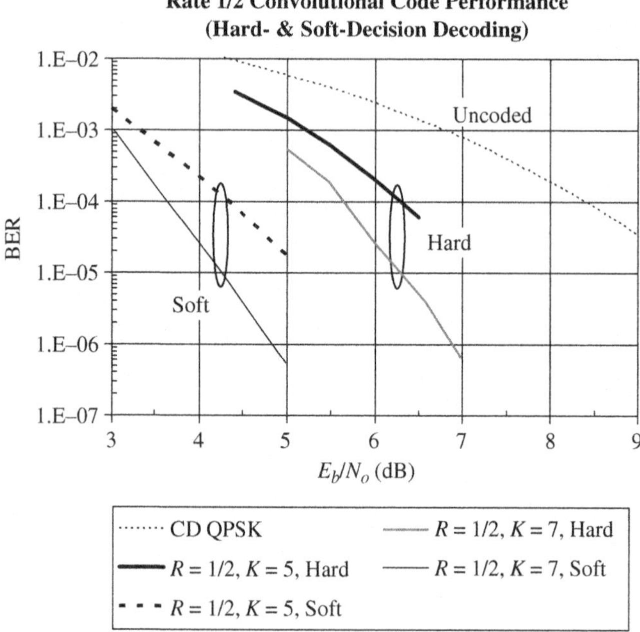

FIGURE 5.37 Performance comparison of hard- and soft-decision decoding.

Next we plot the coding gain of rate $^1/_2$ and $^1/_3$ convolutional codes as a function of BER, with constraint length, $K = 7$ using soft-decision decoding operating in an AWGN channel [23]. Coding gain plots can be seen in (see Fig. 5.38) [7].

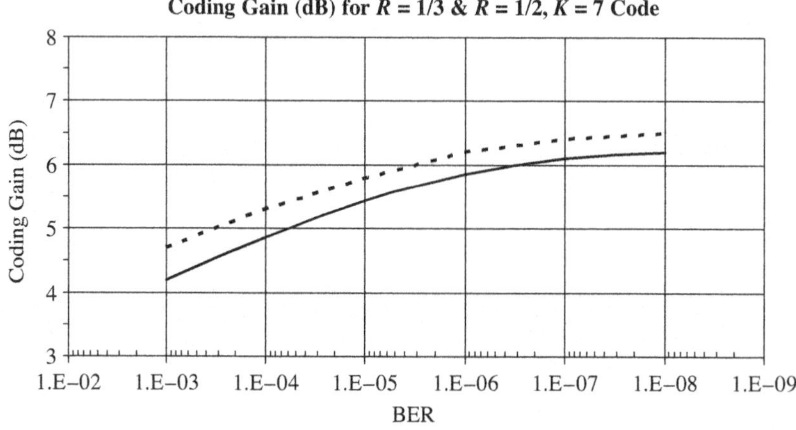

FIGURE 5.38 Coding gains of convolutional codes.

The plot shows us as the BER of interest decreases, the coding gain increases nonlinearly. For the BER values shown, coding gains as much as approximately 6.5 dB are possible. In an effort to provide an interim summary for the reader, we will comment on the two types of decoders presented. First, the hard-decision decoder is presented. Here a binary decision device is placed before the VA in order to make a decision on the received bit stream. In this case the VA can use the Hamming distance as the path metric. The drawback of this technique is that reliable and unreliable bits are equally weighed in the decoding process. Soft decision aims to solve this shortcoming by preserving the amplitude of the received symbol. As we have shown in Fig. 5.37, this provides approximately 2 dB of additional coding gain over hard decision.

Puncture Coding. The concept of puncturing the encoder comes into play when the communication system cannot afford a powerful low-code rate coding scheme. System limitations such as transmission bandwidth may preclude the system designer from using a lower code rate. Alternatively, this application can be approached from another perspective. Let's suppose a data packet is transmitted and it is received with an error, the transmitter can either retransmit the same packet or increase the redundancy information in order to help the receiver decode the packet error-free. In the High Speed Downlink Packet Access (HSDPA) requirements in the 3GPP standard, this is called Hybird-ARQ.

For example, if we assume our system can only support $R = {}^3/_4$ code rate then a "good" punctured code that translates the $R = {}^1/_2$ encoder into a $R = {}^3/_4$ encoder is shown in Fig. 5.39.

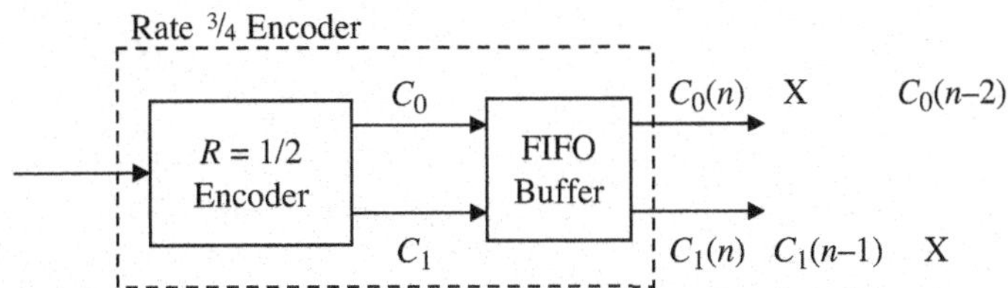

FIGURE 5.39 Rate $^1/_2$ encoder punctured to supply a rate $^3/_4$ encoding.

Here the encoder rate can be shown as 3 bits enter the $R = {}^1/_2$ encoder, thus producing 6 output bits, where 2 of them are not transmitted.

$$\text{Rate} = \frac{\text{input}}{\text{ouptut}} = \frac{3\,\text{bits}}{(6 - 2)\,\text{bits}} = \frac{3}{4} \tag{5.59}$$

From the above block diagram it is apparent that $C_0(n-1)$ and $C_1(n-2)$ are not transmitted. In fact, the place holding operation is also not transmitted. Note one cannot simply pick any combination of bits to puncture and expect the same overall BER performance. Various studies were performed to investigate puncturing patterns that resulted in good performance; some are presented in [24]. Since the transmitter is not transmitting the punctured bits, we are essentially introducing errors into the data stream at the transmitter; this is from the rate $^1/_2$ perspective.

The receiver simply inserts "null bits" into these punctured locations; these null bits are generally forced to have zero value. The receive block diagram showing the depuncturing operations is shown in Fig. 5.40.

The advantages in using this punctured coding technique are given below

- Allows for use of a general rate $^1/_2$ Viterbi decoder to detect a wide range of punctured codes. For example, $R = {}^3/_4, {}^7/_8$, and so on.
- This punctured BER performance is very close to the nonpunctured BER for this $R = {}^3/_4$ system example.
- We can use the less complex Viterbi decoder for rate $^1/_2$ versus the more complex $R = {}^3/_4$ Viterbi decoder.

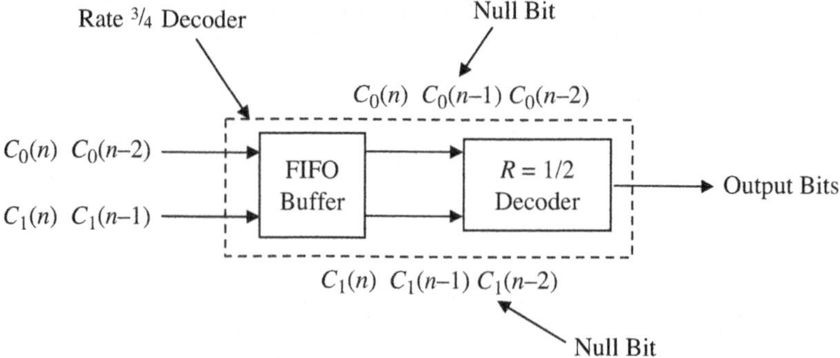

FIGURE 5.40 Rate $^1/_2$ decoder operating as a punctured rate $^3/_4$ code.

One disadvantage in using this punctured coding technique is that the truncation path length increases as the punctured code rate decreases. For example, start with $R = {}^1/_2$ and $K = 7$ code, then the following rules-of-thumb hold true.

Code rate	Truncation path length (T)
$R = {}^1/_2$	>5K bits
$R = {}^3/_4$	>10K bits
$R = {}^7/_8$	>15K bits

Another disadvantage is that soft decision decoding must be used to retrieve the lost information caused by puncturing at the transmitter. The 2-dB gain is needed to compensate for the intentionally induced errors to meet the code rate requirements.

In Fig. 5.41 we present the BER performance results for the punctured $R = {}^3/_4$ and $R = {}^7/_8$ codes. There was no difference in performance for the $R = {}^3/_4$ code when either 75 bits or 100 bits was used

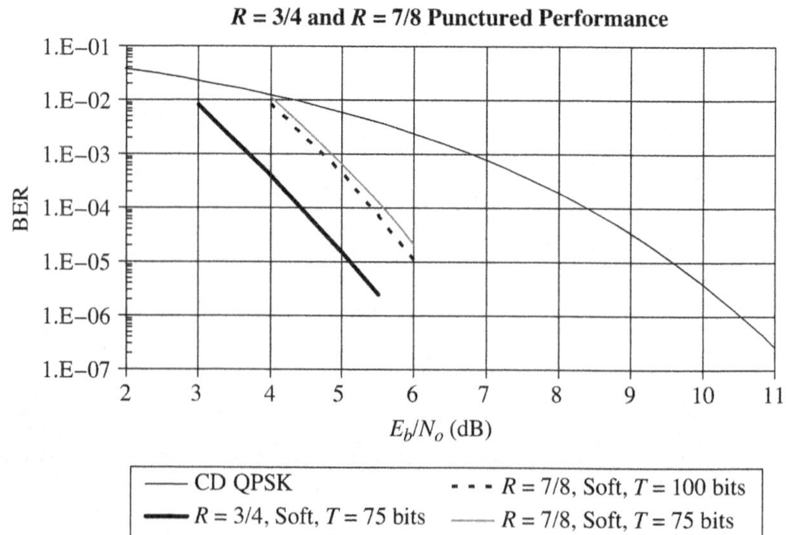

FIGURE 5.41 $R = {}^3/_4$ and $R = {}^7/_8$ punctured code performance in AWGN channel.

for the truncation path length value. However, for the $R = {}^7/_8$ code we can see a slight degradation when we used the 75 bits truncation path length. The measured degradation was approximately 0.2 dB. Also note as the code rate approaches unity, the BER performance improvement diminishes since less parity bits are inserted into the transmit data stream. As this occurs the error correction capability reduces.

It is important to note the $R = {}^3/_4$ and $R = {}^7/_8$ curves were generated by puncturing the same $R = {}^1/_2$ encoder. Similarly, the decoders used the same VA as $R = {}^1/_2$ with the addition of inserting null bits. Hence a significant amount of decoder reuse can be observed with puncturing techniques.

5.1.4 Reed-Solomon (RS) Codes

In 1960, Irving Reed and Gus Solomon introduced RS codes in their paper entitled, *Polynomial Codes over Certain Finite Fields* [25]. RS codes are nonbinary cyclic codes which have burst error correcting capabilities [26]. The nonbinary or code symbols consist of m-bit symbols. The notation used in the RS code is (n,k) where n = number of code symbols in the code word and k = number of data symbols to be encoded. The following relationship holds true

$$(n,k) = (2^m - 1, 2^m - 1 - 2t) \tag{5.60}$$

where m = number of bits used to create a RS symbol and t is the symbol error correcting capability of the code. This leads to mathematical operations that satisfy the rules of the finite fields known as Galois Fields (GF). The number of parity bits is equal to $n - k$ and is equal to $2t$. Alternatively we can write

$$t = \left\lfloor \frac{n - k}{2} \right\rfloor \tag{5.61}$$

An important point to make here is that codes satisfying this relationship are optimal for any code of the same length and dimension. This is sometimes called maximal distance (MD) codes. Also, defining the code minimum distance as $d_{\min} = n - k + 1$, then we can rewrite the error correcting capability as

$$t = \left\lfloor \frac{d_{\min} - 1}{2} \right\rfloor \tag{5.62}$$

What this error-correcting capability property shows us is that in order to correct t symbols we need to have $2t$ parity symbols. Some intuitive reasons given by others is that for each error, one parity symbol is used to locate the error and the other parity symbol is used to find its correct value.

RS Encoder. Let's first discuss the RS encoder, the generating polynomial is represented as

$$g(x) = g_0 + g_1x + g_2x^2 + \cdots + g_{2t-1}x^{2t-1} + g_{2t}x^{2t} \tag{5.63}$$

$$g(x) = (x + \alpha)\cdot(x + \alpha^2)\cdots(x + \alpha^{2t}) \tag{5.64}$$

Recall the degree of the generator polynomial is equal to the number of parity symbols. Alternatively, the generator polynomials for a t error-correcting code must have as roots $2t$ consecutive powers of α. This is shown below.

$$g(x) = \prod_{j=1}^{2t}(x + \alpha^j) \tag{5.65}$$

Keeping along the lines of block codes, we will discuss the systematic form of the encoder. This is accomplished by shifting the message polynomial, $m(x)$, into the rightmost k stages of a code word register and then appending a parity polynomial, $p(x)$ in the leftmost n-k stages. The parity polynomial is obtained by

$$\frac{x^{n-k}\cdot m(x)}{g(x)} = q(x) + \frac{p(x)}{g(x)} \tag{5.66}$$

where $p(x)$ is defined as the remainder polynomial. Now we can write the code word, $c(x)$, as

$$c(x) = p(x) + x^{n-k} \cdot m(x) \tag{5.67}$$

The encoder is similar to that presented for binary cyclic codes, except the mathematical operations are now contained in $GF(2^m)$ dimension, instead of GF(2) for the binary codes.

RS Decoder. Next we move our attention to decoding RS codes. Assume the transmitted code word is corrupted by errors such that the received signal is given as

$$r(x) = c(x) + e(x) = p(x) + x^{n-k} \cdot m(x) + e(x) \tag{5.68}$$

where the error sequence is written as

$$e(x) = e_0 + e_1 x + \cdots + e_{n-1} x^{n-1} \tag{5.69}$$

$$e(x) = \sum_{j=0}^{n-1} e_j \cdot x^j \tag{5.70}$$

Here the received code word has $2t$ unknowns, t of them are for the error locations and the other t are for the error values themselves. Contrast this to the binary case where only the error locations are unknown.

The transmitted code word is given as

$$c(x) = m(x) \cdot g(x) \tag{5.71}$$

It can be shown that the roots of $g(x)$ are also the roots of $c(x)$. Evaluating $r(x)$ at each of the roots of $g(x)$ will produce zero when it is a valid code word. Hence the syndrome symbols, S_i, are computed as (for $i = 1, \ldots, n - k$)

$$S_i = r(x)|_{x=\alpha^i} = r(\alpha^i) \tag{5.72}$$

$$S_i = c(\alpha^i) + e(\alpha^i) \tag{5.73}$$

$$S_i = e(\alpha^i) \tag{5.74}$$

So far we have used the syndromes to determine if the received code word is valid or if an error is present. If an error is present, we next need to determine its location. Let's assume there are p errors in the code word, then the error polynomial is

$$e(x) = e_1 x^1 + e_2 x^2 + \cdots + e_p x^p \tag{5.75}$$

We can define the error-locator polynomials as, $\sigma(x)$

$$\sigma(x) = (1 + \beta_1 x) \cdot (1 + \beta_2 x) \cdots (1 + \beta_p x) \tag{5.76}$$

where the roots of $\sigma(x)$ are given as $\frac{1}{\beta_1}, \frac{1}{\beta_2}, \ldots, \frac{1}{\beta_p}$ where β are the error location numbers. Using the same notation used in [1] and [25], the error-locator polynomial can also be written as

$$\sigma(x) = 1 + \beta_1 x^1 + \beta_2 x^2 + \cdots \beta_p x^p \tag{5.77}$$

The reciprocal of the roots of $\sigma(x)$ are the error-location numbers of the error pattern $e(x)$. Once these roots are located, the error locations will be known. The roots are determined by exhaustively using each of the field elements into $\sigma(x)$.

$$\begin{aligned} \sigma(\alpha^0) &= \sigma(x)\big|_{x=\alpha^0} \\ &\vdots \\ \sigma(\alpha^{n-1}) &= \sigma(x)\big|_{x=\alpha^{n-1}} \end{aligned} \tag{5.78}$$

This determines the error locations and provides us with the pattern of the error polynomial. Next we need to get the error values; the syndrome are used with the roots of the error-location polynomial

$$\begin{aligned} S_1 &= r(\alpha) = e_1\beta_1 + e_2\beta_2 + \cdots + e_t\beta_t \\ S_2 &= r(\alpha^2) = e_1\beta_1^2 + e_2\beta_2^2 + \cdots + e_t\beta_t^2 \\ &\vdots \\ S_p &= r(\alpha^p) = e_1\beta_1^p + e_2\beta_2^p + \cdots + e_t\beta_t^p \end{aligned} \tag{5.79}$$

We solve for the error values of e_j from the above equations. A number of algorithms are available to perform the above tasks, one specifically is the Chien search algorithm. Lastly, once we know the location and value of the error polynomial, error correction still remains to occur. Here the locally generated error sequence is added to the received code word sequence and the decoder operations are complete. Let us summarize the steps used in the decoding procedure.

1. Calculate the syndromes (to determine if an error is present).
2. Create the error-locator polynomials.
3. Use the roots of (**2**) to find the location of the errors.
4. Determine the error values (using syndromes and roots of error-locator polynomial).
5. Create an error polynomial and add it to the received code word.

As an example of a RS code used in some communications systems, the RS code used in Cellular Digital Packet Data (CDPD) is a RS(63, 47) operating in a finite field of GF(64), which means that $m = 6$ bits. This gives an error correction capability of

$$t = \left\lfloor \frac{63 - 47}{2} \right\rfloor = 8 \tag{5.80}$$

At this point it is worth mentioning that practical wireless systems use RS codes in the context of concatenated coding. In this example, the outer code is the RS code while the inner code is the convolutional code (CC). A very simple block diagram is shown in Fig. 5.42.

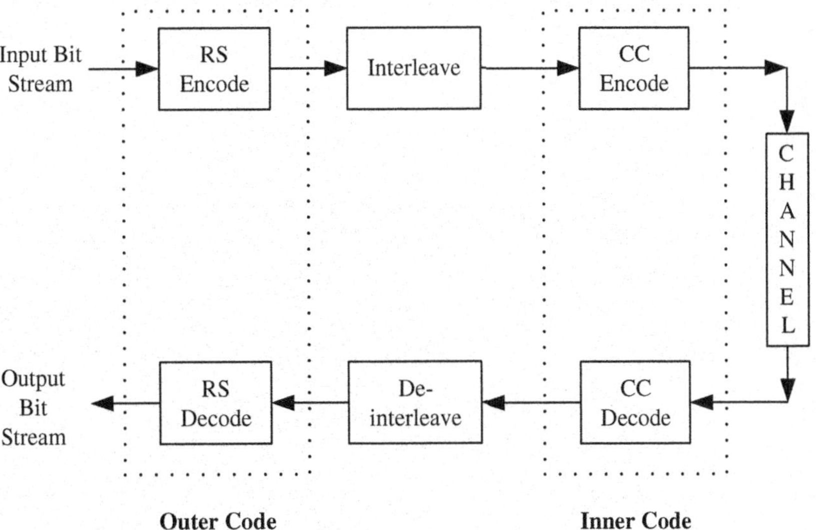

FIGURE 5.42 Concatenated coding block diagram using CC + RS codes.

A reason to introduce a bursty error correcting code such as RS to the CC is as follows. The limit on the coding gain is the ability of the CC to perform well under low signal to noise ratio (SNR) conditions. After an investigation of this shortcoming it was shown that in the low SNR region, when the VA makes errors, the errors tend to occur in bursts [23]. Hence we can redistribute these errors with an interleaver, then a RS can be placed at the outer code and be used to correct bursts of errors produced by the VA in the low SNR regions. The burst of errors occur because the VA basically looses "synchronization" and requires more time to reliably accumulate path metrics [27]. We have plotted the length of the bursts in Appendix 5A, taken from [23], for the sake of continuity. The performance of this concatenated coding technique is presented in the next section. It worth mentioning that the original proposals for the FEC technique of the 3G systems used this concatenated coding approach and then changed to using only turbo codes.

5.1.5 Turbo Codes

This next FEC technique was first presented in 1993 by Berrou, Glavieux, and Thitmajshima and is called turbo codes [28, 29]. They quickly acquired the fame status because these codes achieved performance near the Shannon limit. They released results for a rate $^1/_2$ turbo code achieving a BER = 1E-5 for an E_b/N_o = 0.7 dB, using simple component codes and very large interleavers.

Turbo codes can be viewed as concatenated codes where the decoding is performed in an iterative fashion. Here soft values (or decisions) are passed from the output of one decoder to the input of another decoder. This procedure is repeated (or iterated) where each iteration would improve the performance by providing more confidence into the encoded block of bits.

By the late 1990s, turbo codes were widespread and being used in a variety of digital communication systems. They include Deep Space Communications (CCSDS), 3G Digital Cellular systems (3GPP and CDMA2000), digital video broadcasting (MediaFLO), and WiMax.

Much research has been performed in better understanding and improving the performance of these turbo codes. We have decided to present an encoder, provide details on the decoder, and lastly show some performance results that have gained this FEC technique worldwide attention [30–38].

Turbo Encoder. The particular code structure discussed is called parallel concatenation. Note serial concatenation code structures are available; we refer the reader to [39] for further information on this topic. Here two component codes operate on the same input bit stream. A turbo encoder is shown in Fig. 5.43, where the component codes are of recursive systematic convolutional (RSC) codes.

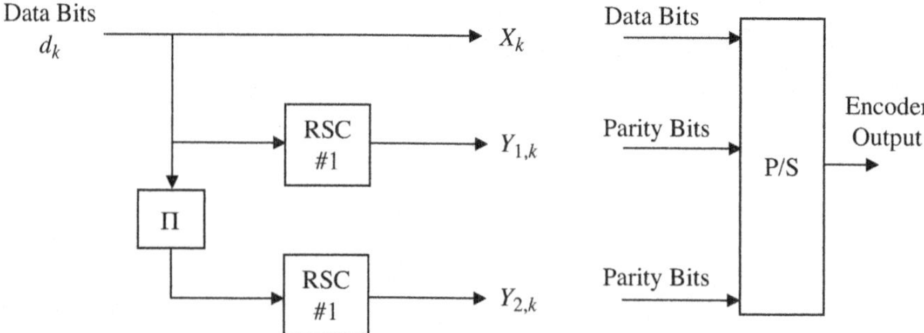

FIGURE 5.43 Parallel concatenated turbo encoder.

As a side note more than two component codes (i.e., RSC) can be used in the encoder and they are called multiple turbo codes. Please refer to [40] for further details. What's interesting in this encoder is the same data bits are encoded twice but with a different arrangement. The first component encoder operates directly on the input data bits, while the second component encoder operates on an interleaved version of the input data bits. We have used the commonly accepted notation of using the Greek letter, Π, for the interleaving operations. The turbo encoder operates on blocks of input data bits, say, N. Hence the turbo encoder is characterized by the block size, N, the interleaver details, the number of component codes, the RSC code, and lastly the parallel to serial (P/S) function.

Recall the rate $^1/_2$ encoder for a nonsystematic convolutional code was given earlier and redrawn for sake of comparing this encoder type to the systematic type (see Fig. 5.44).

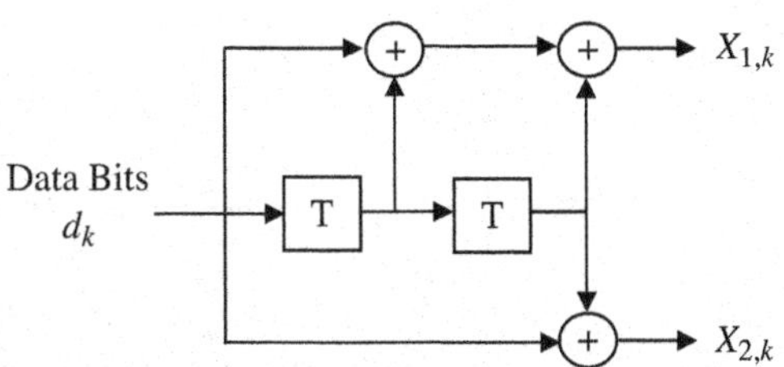

FIGURE 5.44 Nonsystematic, $R = {^1/_2}$, $K = 3$ convolutional encoder.

The nonsystematic rate $^1/_2$ code can be made systematic and one form is the RSC given in Fig. 5.45. Here input data bits are made available to the output and the parity bits are calculated in a recursive fashion.

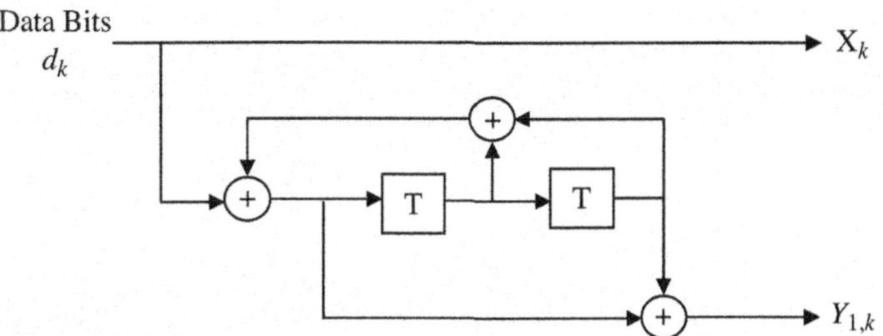

FIGURE 5.45 Systematic, $R = {^1/_2}$, $K = 3$ convolutional encoder.

We can place the RSC codes in the turbo encoder block diagram given earlier to create a more detailed discussion. In the diagram given in Fig. 5.46 we see the RSC do not pass the input data bits again since they were already sent to x_k. Also puncturing can be applied to the output bits to obtain various code rates. In this case, it would occur prior to the parallel to serial operation.

The purpose of the interleaver is to randomize the input data bits in order to decorrelate the input of the two encoders. The reason is if the input to the two encoders are uncorrelated then after the first decoder, a portion of the residual errors can be corrected by the second decoder. Hence you would expect going to a larger interleaver size can drive us closer to our low SNR goal. A disadvantage is that larger interleaver sizes require more turbo iterations and larger memory in the turbo decoder. Lastly, we should also mention the delays caused by the large interleaving and de-interleaving blocks. These are both interesting and challenging design trade-offs.

Turbo Decoder. Let us first start by mentioning there are a few choices available to the system designer concerning the turbo decoder method to use. It is not our intention to compare all of them; we will concentrate on the maximum a posteriori probability (MAP) decoder. As discussed above, the

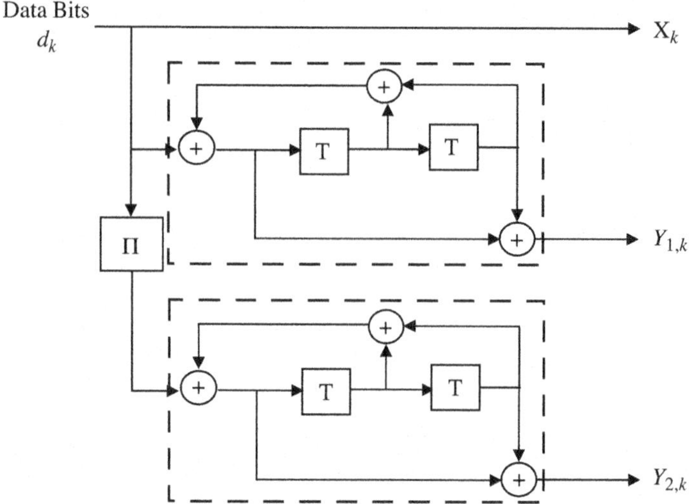

FIGURE 5.46 Block diagram of a rate $^1/_3$ turbo encoder ($K = 3$).

interleaver is placed between the input of the two encoders in order to decorrelate their respective inputs. The decoders make use of this property when iterating. The decoders operate in a concatenated fashion where the output of one decoder is the input to the second decoder. After each iteration errors are corrected.

The soft decisions will be passed from one decoder to another decoder in the form of log-likelihood ratios defined as (with y_j's representing the received signals).

$$L_m = \log\left(\frac{P[d = 1|y_0,y_m]}{P[d = 0|y_0,y_m]}\right) \qquad (m = 1, 2) \tag{5.81}$$

Which can be rewritten as

$$L_m = \log\left(\frac{P[y_0|d = 1]}{P[y_0|d = 0]}\right) + \log\left(\frac{P[y_m|d = 1]}{P[y_m|d = 0]}\right) + \log\left(\frac{P[d = 1]}{P[d = 0]}\right) \tag{5.82}$$

$$L_m = \Lambda_S + \Lambda_{Em} + \Lambda \tag{5.83}$$

The log likelihood consists of three parts: the systematic, extrinsic, and a priori information.

A block diagram of a turbo decoder is given in Fig. 5.47.

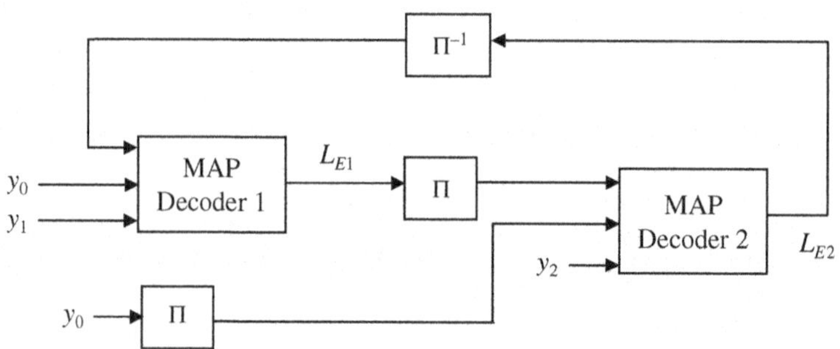

FIGURE 5.47 General block diagram of a turbo decoder.

The first MAP decoder uses the systematic and first parity bit, along with a priori estimates to come up with soft decisions on the transmitted bit sequence in the block of N bits. This extrinsic information is interleaved along with the systematic bits to be used by the second MAP decoder. The second MAP decoder output extrinsic information is de-interleaved and becomes the a priori probability for the first MAP decoder. This process or iteration begins over again.

Many publications exist that compare performance of various decoders such as MAX Log MAP, SOVA, and so on. The reader should refer to the reference section for a list of some of these comparisons.

Using this Log MAP decoder and two component codes, we can create a $R = {}^1/_2$, $K = 5$ turbo code with random interleaver of size = 65,536. The simulated BER performance as a function of E_b/N_o is shown in Fig. 5.48.

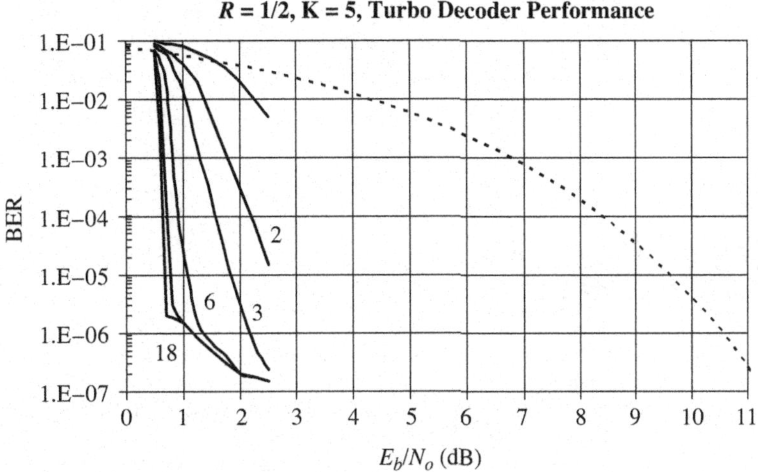

FIGURE 5.48 Turbo decoder BER performance in AWGN.

Here we notice a few interesting observations: specifically as the number of iterations increases, the BER performance improves. Also the incremental improvement decreases with each additional iteration increment. Lastly, we see an error floor or asymptote beginning to take form at least for BER = 1E-5 at an E_b/N_o of approximately 0.7 dB. Lastly, for this BER, there is a coding gain of approximately 9 dB. This asymptote behavior can be expressed using the following approximation [29].

$$P_b \cong D_{\text{free}} \cdot Q\left(\sqrt{2 \cdot d_{\text{free}} \cdot R \cdot \frac{E_b}{N_o}}\right) \tag{5.84}$$

where D_{free} is the average number of ones on the minimum free distance, R is the code rate, d_{free} is the code minimum free distance (this depends on the interleaver and generator polynomials). This bound determines the code performance at high SNR.

As one can see from the turbo decoder's performance (see Fig. 5.48), there is significant performance improvement that can be achieved. It is not readily apparent how these improvements can be obtained by other FEC techniques provided earlier in this chapter. It is for these reasons why next generation digital cellular systems have adopted the turbo coding principle.

In an effort to provide some closure on the performance comparison of the FEC codes presented, in Fig. 5.49 we compare the performance of the convolutional code to that of concatenated coding and lastly to the turbo coding.

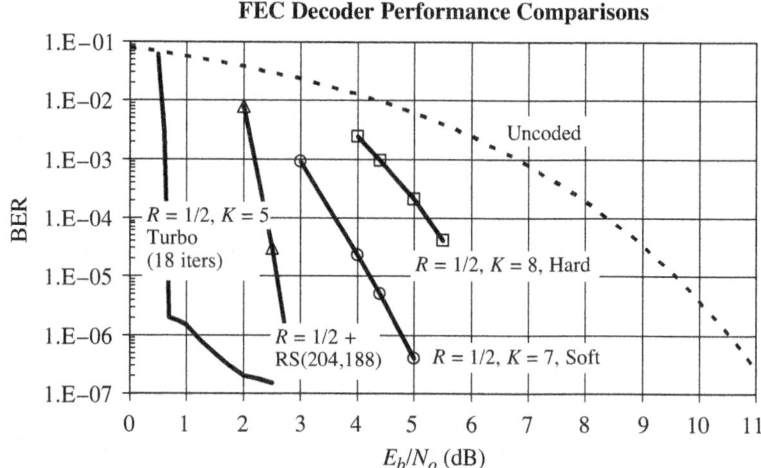

FIGURE 5.49 A comparison of various FEC techniques in an AWGN channel.

5.2 RECEIVE SPATIAL ANTENNA DIVERSITY COMBINING TECHNIQUES

In the "Multipath Fading Channel" Chapter 3 we discussed how severe the wireless channel can become. Thus far we have assumed a single transmit antenna and a single receive antenna in the communication system. Performance degradation was observed when the received signal encountered a fade or destructive interference. In this section, we introduce a technique where multiple-receive antennas are used to aid the detection or demodulation of the transmitted signal. The level of performance improvement highly depends on the signals at each of the receive antennas being independent and uncorrelated from each other. This can be accomplished by spatially separating the antennas from each other. Generally speaking, antenna spacing of half a wavelength is adequate; however, it has been shown that the antenna spacing is dependent on the local scatterers present near the receiver. It is for this reason that antenna distance separation at the BS is larger than that at the MS.

A few multiple-antenna solutions exist, we present them in Table 5.4.

TABLE 5.4 Multiple-Antenna Combining Comparisons

Technique	Complexity	Performance
Switch/selection	Low	Good
Equal gain combining	Middle	Better
Maximal ratio combining	High	Best
Optimal combining	Very high	Very best

Table 5.4 orders the multiple-receive antenna techniques in terms of implementation complexity as well as link level performance. The simplest technique is the switch/selection (SS) which is labeled to have low complexity and good performance. The complexity column is with respect to the techniques listed within the table. It is the lowest since complete multiple-receiver chains are not required to accomplish the operations. As far as performance is concerned, it is given a "good" status even though it doesn't produce the best performance; it still provides considerable gains over the single receive antenna case [41–54].

Now on the other end of the performance spectrum, the discussion becomes more interesting. In an interference-free environment the performance of the maximal ratio combining (MRC) and optimal combining (OC) become theoretically equivalent. And special care must be given to the implementation of the OC since it can perform worse than MRC under certain scenarios and implementations.

5.2.1 Switch/Selection(SS)

This first technique selects one of M diversity receivers which has the largest SNR and this receiver is then used in the detection and demodulation process [55]. In order to aid in the derivations characterizing the signal we assume each diversity receiver is uncorrelated. The multipath channel is Rayleigh distributed having a mean power, b_0, the probability density function (PDF) is given as

$$p(r_i) = \frac{r_i}{b_0} \cdot e^{-\frac{r^2}{2b_0}} \qquad (r_i \geq 0) \tag{5.85}$$

where r_i is the received signal envelope of the ith antenna. Let us further assume that each antenna has the same average noise power, N_o. The local mean SNR per antenna is given as

$$\gamma_i = \frac{r_i^2}{2N_o} \tag{5.86}$$

The mean SNR per antenna is given as

$$\Gamma = \frac{b_0}{N_o} \tag{5.87}$$

A simplified block diagram presenting the SS diversity receiver is shown in Fig. 5.50. There are a total of M receive antennas with a certain spatial separation. Placing the antennas close together will result in correlated fading between antennas and would reduce the overall performance gains. For the uncorrelated case, the antennas are separated in distance far enough to have independent fading on each antenna.

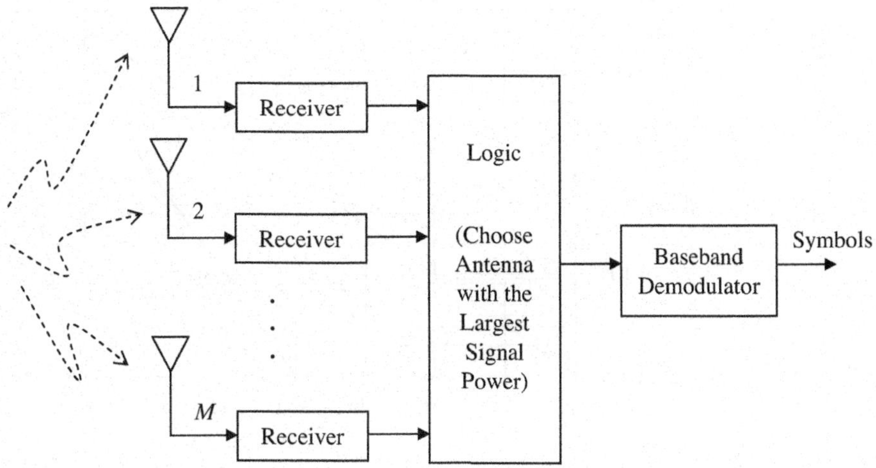

FIGURE 5.50 Switch/selection diversity method.

As discussed above, the main idea is to select the best one of the available M antennas to be used in the baseband for demodulation. This means that $M - 1$ antennas are not used during this time interval. Earlier we stated given M antennas, the probability that all M antennas are in a deep fade is very small. However, for large values of M, there is a good chance that a "few" antennas would be usable for demodulation. Hence not using them is equivalent to wasting potential performance improvement.

Note, the antenna selection can occur prior to the baseband demodulation, as shown above; this is called predetection diversity. Also the antenna selection can occur after baseband demodulation, this means there will be M baseband sections and we will choose one of their outputs to input to the demodulator. If each of the M antennas performed their own demodulation and the selection was made on their respective output symbols, then this is called postdetection diversity. In this last scenario the received symbol sequence can be selected by received SNR and/or maximum eye opening.

The PDF of the local SNR per antenna is given as

$$p(\gamma_i) = \frac{1}{\Gamma} \cdot e^{-\frac{\gamma_i}{\Gamma}} \tag{5.88}$$

Selecting one out of two antennas provides performance improvement because the probability that both antennas are in a deep fade is very low, p^2. In other words, there is a good chance that at least one antenna would have a useful signal to be used for demodulation.

The probability that each of the local SNR values, γ_i, are simultaneously below or equal to γ_s is expressed as

$$P[\gamma_i \cdots \gamma_M \leq \gamma_s] = P[\gamma_i \leq \gamma_s]^M \tag{5.89}$$

The cumulative distribution function (CDF) is written as

$$P_M(\gamma_s) = \left(1 - e^{-\frac{\gamma_s}{\Gamma}}\right)^M \tag{5.90}$$

In Fig. 5.51 we plot the CDF of the SNR for an M branch selection diversity receiver, where the x-axis has been normalized by the mean SNR. Here we see a tremendous improvement when going from $M = 1$ to $M = 2$ antennas. The improvement continues with each additional antenna; however,

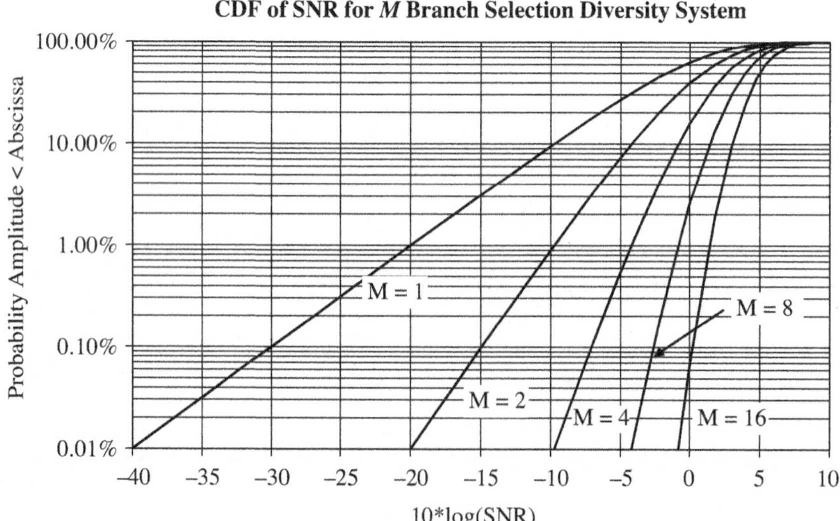

FIGURE 5.51 CDF SNR for a selective diversity system.

it diminishes as M grows. For a single antenna case, we see that approximately 1% of the time the SNR is less than or equal to -20 dB. In going to $M = 2$ antennas, this value decreases to approximately 0.01% of the time. Hence increasing the number of antennas used in the receiver increases the SNR and in doing so reduces the system's BER.

This can also be looked at under a different light, as the more antennas that are used in the receiver the less visible are the deep fades and thus reducing the variability of the received SNR.

The mean SNR of the selected signal is

$$E\{\gamma_s\} = \Gamma \cdot \sum_{k=1}^{M} \frac{1}{k} \tag{5.91}$$

PDF of the mean SNR of the selected signal is given as

$$p(\gamma_s) = \frac{M}{\Gamma} \cdot \left(1 - e^{-\frac{\gamma_s}{\Gamma}}\right)^{M-1} \cdot e^{-\frac{\gamma_s}{\Gamma}} \tag{5.92}$$

We have purposely omitted the steps used to derive the above equations; they can be found in [55]. Our main purpose here is to present the results and provide additional insight.

5.2.2 Equal Gain Combining (EGC)

In this next spatial diversity technique all M antennas will be used to obtain a signal to input into the baseband for demodulation. Here all M received signals are phase aligned (or co-phased) prior to being combined. A block diagram emphasizing this is shown in Fig. 5.52.

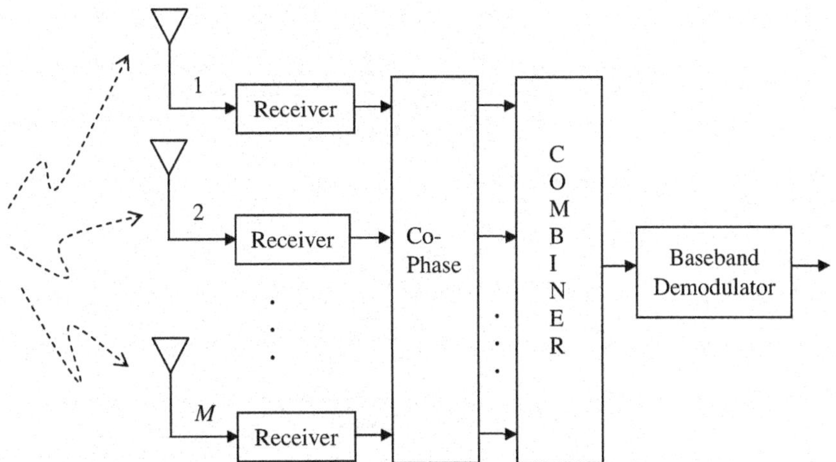

FIGURE 5.52 Equal gain diversity combining method.

Let us consider QPSK modulation that is transmitted and encounters flat Rayleigh fading channel with AWGN. Due to uncorrelated fading the constellation on all the M antennas will have various phase offsets. Co-phasing essentially performs phase derotation and aligns the constellations so that they can be combined coherently.

The mean SNR of EGC output is given as

$$E\{\gamma_s\} = \Gamma \cdot \left[1 + (M-1) \cdot \frac{\pi}{4}\right] \tag{5.93}$$

When comparing this result to that of the SS equation presented earlier, the EGC performance shows each additional antenna provides a larger incremental gain for EGC rather over the SS method.

5.2.3 Maximal Ratio Combining (MRC)

In the previous section, EGC made use of all the M antennas to present a signal to the demodulator. This essentially weighted all antennas equally, regardless of their respective SNR. So what this next operation accomplishes is to emphasize the signals with high SNR and de-emphasizes signals with low SNR. A block diagram of this spatial combing technique is shown in Fig. 5.53.

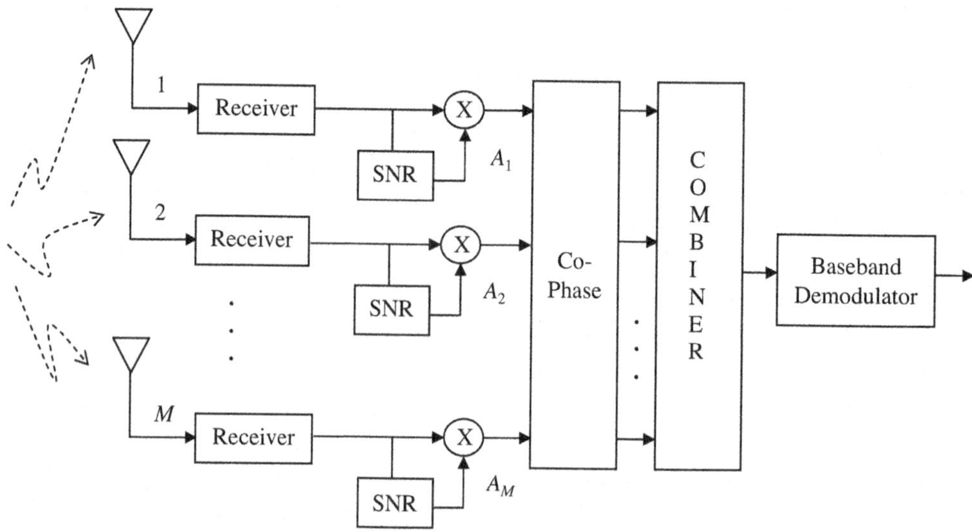

FIGURE 5.53 Maximal ratio combining diversity method.

The PDF of the combiner output SNR is given as

$$p(\gamma) = \frac{1}{(M-1)!} \cdot \frac{\gamma^{M-1}}{\Gamma^{M}} \cdot e^{-\frac{\gamma}{\Gamma}} \tag{5.94}$$

The CDF of the combiner output of SNR is

$$P(\gamma) = \int_{0}^{\gamma_A} p(x)\, dx \tag{5.95}$$

$$P(\gamma) = 1 - e^{-\frac{\gamma}{\Gamma}} \cdot \sum_{k=1}^{M} \frac{\left(\frac{\gamma}{\Gamma}\right)^{k-1}}{(k-1)!} \tag{5.96}$$

The mean SNR at the output of the MRC combiner is given as

$$E\{\gamma\} = \sum_{i=1}^{M} \Gamma = M\Gamma \tag{5.97}$$

When comparing the incremental SNR improvement of MRC to that of SS and EGC, each additional antenna has a greater impact with MRC.

A plot of the CDF of this MRC technique is shown in Fig. 5.54.

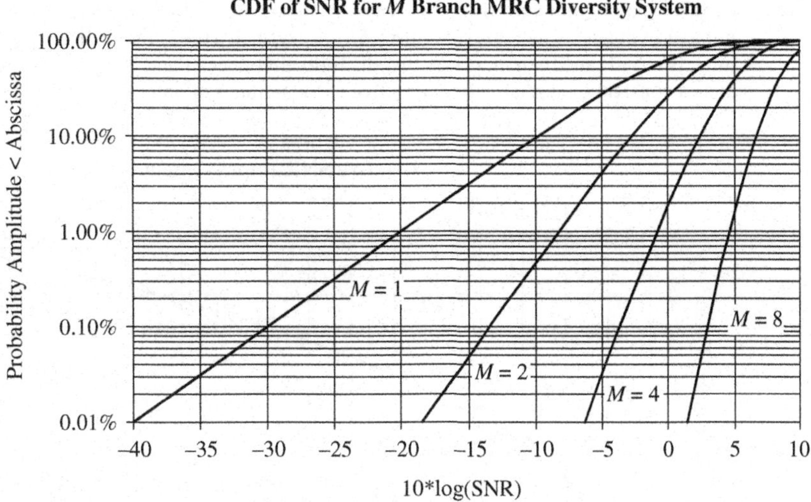

FIGURE 5.54 Output CDF of MRC diversity combining.

Let us compare the results of MRC to those given earlier for SS diversity for the $M = 4$ antennas case. The SS diversity curves show that for approximately 0.5% of the time the SNR is less than or equal to -5 dB. However, for MRC diversity, the probability dramatically decreased to approximately 0.03% of the time.

A plot of all the mean SNR expressions for the SS, EGC, and MRC diversity techniques is given in Fig. 5.55. This shows the potential gains with each additional antenna used in the receiver.

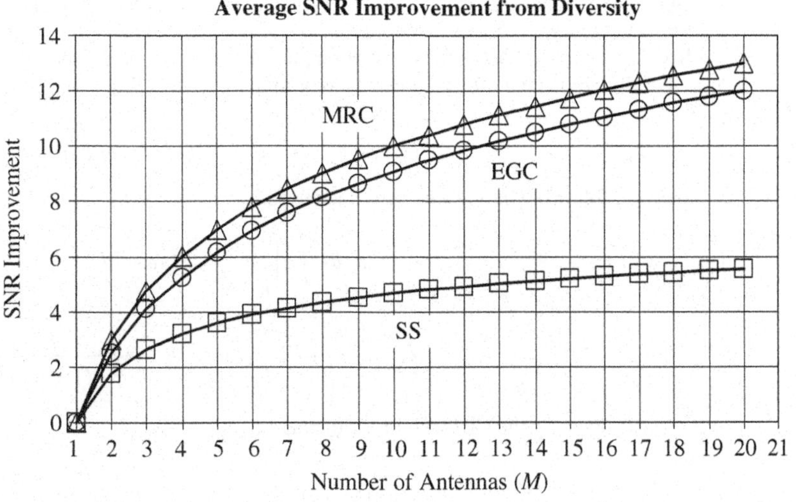

FIGURE 5.55 Mean diversity output SNR comparison.

Here we can clearly see, as M increases the performance gap between the MRC and the SS continues to increase dramatically. While the performance gap between the MRC and EGC stays approximately the same with slight performance improvements.

An alternative viewpoint is given below. Here we compare the performance gain of MRC over that of EGC and plot this improvement factor. We also consider the improvement of MRC over SS as the number of antennas increases. Figure 5.56 clearly shows the gain of MRC over EGC starts at approximately 0.5 dB and then increases to 1 dB for the number of antennas shown. In fact, it can be looked at as a constant gain relatively insensitive to the number of antennas used in the receiver.

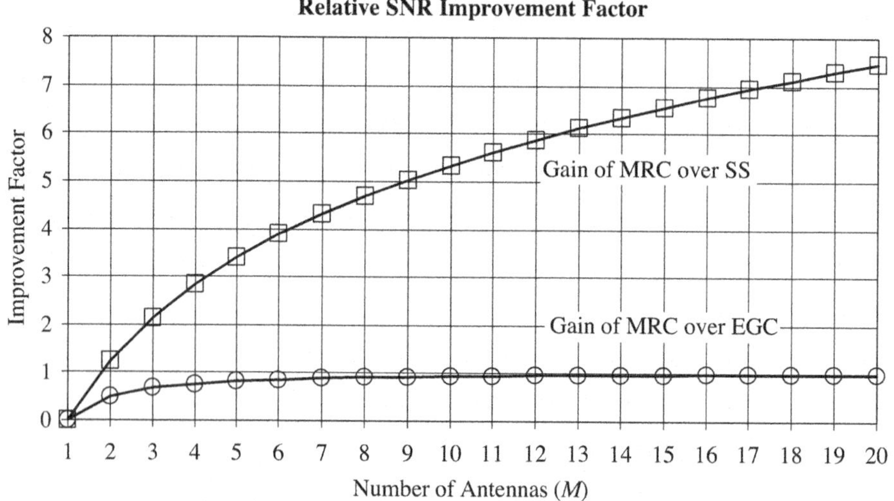

FIGURE 5.56 Relative performance improvement of MRC.

Diversity Performance Summary. Let us take a moment to recall the above results have assumed uncorrelated signals, $\rho = 0$, on each antenna. Here ρ is the magnitude of the complex cross-covariance function of two faded Gaussian signals. Here we fix $M = 2$ and compare the SS to the MRC for various values of the correlation present across the antennas.

For the SS diversity the CDF can be rewritten as

$$P(\gamma_s) = 1 - e^{-\frac{\gamma_s}{\Gamma}}(1 - Q(a,b) + Q(b,a)) \tag{5.98}$$

where the following definitions have been used.

$$Q(a, b) = \int_b^\infty e^{-\frac{1}{2}(a^2+x^2)} I_o(ax)\,x\,dx \tag{5.99}$$

$$a = \sqrt{\frac{2\gamma_s}{\Gamma(1 + \rho^2)}} \quad \text{and} \quad b = \sqrt{\frac{2\gamma_s}{\Gamma(1 - \rho^2)}} \tag{5.100}$$

For MRC diversity the CDF can be written as

$$P(\gamma_R) = 1 - \frac{1}{2\rho}\left[(1 + \rho)\cdot e^{-\frac{\gamma_R}{\Gamma(1 + \rho)}} - (1 - \rho)\cdot e^{-\frac{\gamma_R}{\Gamma(1 - \rho)}}\right] \tag{5.101}$$

One can plot the CDF undercorrelated conditions to notice the effects.

We would like to present BER performance curves comparing MRC and SC (selection combining). It is worth mentioning that various expressions exist for the theoretical BER; below a few commonly used closed form expressions for MRC diversity combining are given. They are provided below for sake of convenience, the first is provided by [56].

$$P_b = \frac{1}{2}\left[1 - \frac{\mu}{\sqrt{2-\mu^2}}\sum_{k=0}^{M-1}\binom{2k}{k}\cdot\left(\frac{1-\mu^2}{4-2\mu^2}\right)^k\right] \tag{5.102}$$

With the help of the following variable definition for coherent PSK modulation

$$\mu = \sqrt{\frac{\gamma}{\gamma+1}} \tag{5.103}$$

and $\gamma = E_b/N_o$. If the E_b/N_o per antenna is required then this can be modified to be the following $\gamma = M \cdot E_b/N_o$. Similarly, the following definition is used for DPSK modulation performance investigations.

$$\mu = \frac{\gamma}{\gamma+1} \tag{5.104}$$

Another commonly used form of the MRC BER is given by [39] and shown below

$$P_b = \left(\frac{1-\mu}{2}\right)^M \sum_{k=0}^{M-1}\binom{M-1+k}{k}\cdot\left(\frac{1+\mu}{2}\right)^k \tag{5.105}$$

Also note that various approximations to the above two expressions exist for large and small SNR; they are not listed here.

The first set of performance curves correspond to the BPSK family of modulation schemes and is shown in Fig. 5.57 for M = 1 to 4 antennas. It is clearly shown by adding another antenna performance

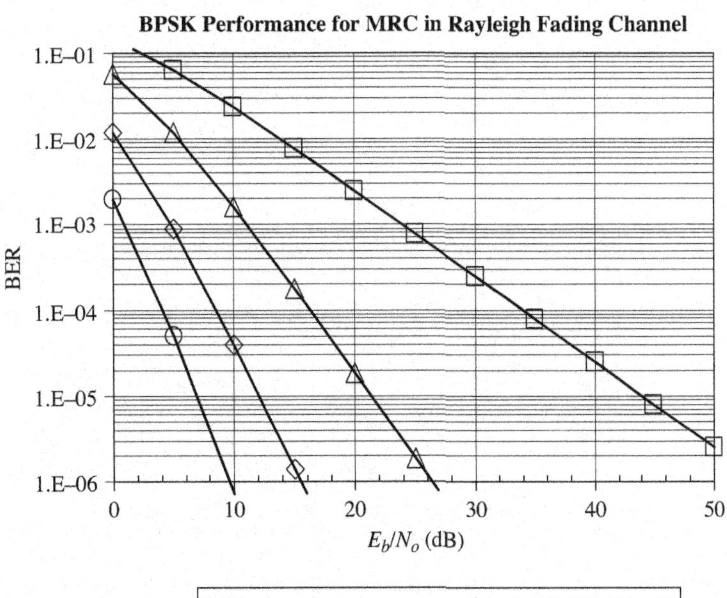

FIGURE 5.57 BPSK BER performance for MRC combining.

improvement can be obtained. Moreover, with each additional antenna, the performance improvement becomes smaller and smaller. In any case, for the $M = 2$ antenna example there is more than 10 dB in improvement in E_b/N_o for a BER = 1E-3.

An interesting observation from Fig. 5.57 shows the BER performance is linearly, dependent on the inverse of the SNR raised to the power of the diversity order.

Next we plot the DPSK modulation scheme BER performance (see Fig. 5.58). The mathematical equations provided earlier have variables that can be changed to plot the performance for either BPSK or DPSK modulation scheme. This family of curves also shows the performance improvement gradually increases as more and more antennas are added to the receiver.

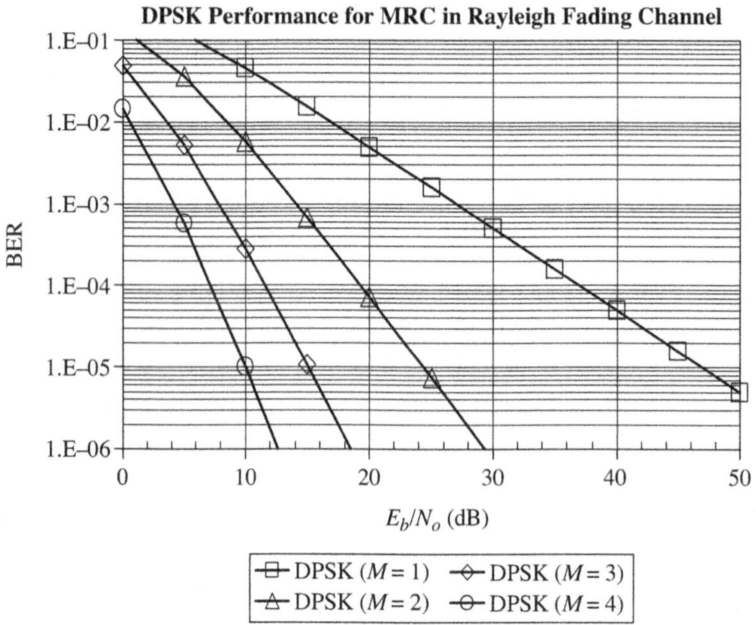

FIGURE 5.58 DPSK BER performance for MRC combining.

Next we wish to show the BER performance gain when comparing MRC and SC diversity combining techniques for BPSK modulation. Earlier in this chapter, we have shown the combiner output mean SNR gain for both of these techniques. It was also shown that as the number of receiver antennas increases the performance difference between EGC and MRC remains relatively the same, while it increases when comparing MRC to SC. First we wish to provide the equations used to plot the curves that follow. The BER performance for MRC is given by [57].

$$P_b = \frac{1}{2}\left[1 - \left(\frac{\gamma}{\gamma + 1}\right)^{1/2} \sum_{k=0}^{M-1} \binom{2k}{k} \cdot \frac{1}{[4(1 + \gamma)]^k}\right] \tag{5.106}$$

The BER performance for selection combining (SC) is given by

$$P_b = \frac{1}{2}\left[1 - \sum_{k=1}^{M} \binom{M}{k} (-1)^{k-1} \left(\frac{\gamma}{\gamma + k}\right)^{1/2}\right] \tag{5.107}$$

Given these two equations we can compare the performance of MRC and SC for $M = 2$ and 4 antennas in Fig. 5.59.

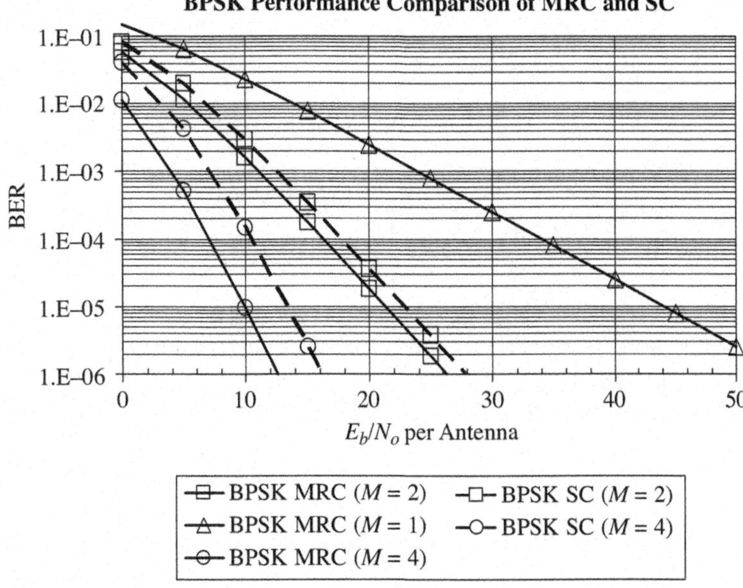

FIGURE 5.59 BPSK BER performance comparison of MRC and SC combining.

Above we can see for the $M = 2$ antenna diversity receiver, the performance difference between the MRC and SC is approximately 1.25 dB. Even with this difference approximately 10 dB of gain is still achievable for the BER = 1E-3 target. This plot also shows us as the number of antennas, M, increases the performance difference between MRC and SC increases. For the particular values chosen above, the performance difference between SC and MRC for the $M = 4$ antennas example is approximately 3 dB. This is an alternative way to show the difference in performance between MRC and SC increases as the number of receive antennas increases.

Lastly, an alternative form of the MRC performance is given below (continuing with the above variable definitions)

$$P_b = \frac{1}{2} \cdot \left[1 - \mu \sum_{k=0}^{M-1} \binom{2k}{k} \cdot \left(\frac{1 - \mu^2}{4} \right)^k \right] \tag{5.108}$$

The alternative form of the SC performance is given below with $\alpha = 1/\gamma$.

$$P_b = \frac{M}{2} \sum_{k=0}^{M-1} \binom{M-1}{k} \cdot (-1)^k \cdot \frac{1}{1+k} \cdot \left[1 - \frac{1}{\sqrt{1 + \alpha(1+k)}} \right] \tag{5.109}$$

Earlier we discussed that the switching selection diversity receiver suffered from complexity since only a single antenna was used at a time for the M available antennas. Next we present the results of some work performed by [58] where more than 1 antenna is used in the demodulation process. We will keep the same notation which was SC2 corresponds to keeping the largest 2 antenna signals while SC3 corresponds to keeping the largest 3 antennas signals. We will first present the mathematical BER equations and then discuss their results. The SC2 performance is given as

$$P_b = \frac{M(M-1)}{2} \left\{ \frac{1}{2} \left[1 - \frac{1}{\sqrt{1+\alpha}} - \frac{\alpha}{2(1+\alpha)\sqrt{1+\alpha}} \right] + \sum_{k=1}^{M-2} \binom{M-2}{k} \cdot (-1)^k \cdot V(k) \right\} \tag{5.110}$$

where

$$V(k) = \frac{1}{2+k} - \frac{1}{k \cdot \sqrt{1+\alpha}} + \frac{2}{k \cdot (k+2) \cdot \sqrt{1 + \frac{\alpha \cdot (2+k)}{2}}} \tag{5.111}$$

The SC3 performance is given as

$$P_b = \frac{M(M-1)(M-2)}{4} \cdot \left\{ \frac{1}{3}\left[1 - \mu \sum_{k=0}^{2} \binom{2k}{k} \cdot \left(\frac{1-\mu^2}{4}\right)^k\right] + \sum_{k=1}^{M-3} \binom{M-3}{k} \cdot (-1)^k \cdot G(k) \right\} \tag{5.112}$$

where $G(k)$ is defined as

$$G(k) = \frac{k-3}{k^2}(1-\mu) - \frac{\mu}{2k}(1-\mu^2) + \frac{9}{k^2(3+k)} \cdot \left(1 - \frac{1}{\sqrt{1 + \alpha\left(1 + \frac{k}{3}\right)}}\right) \tag{5.113}$$

The derivations are provided in [58] and are omitted here. A block diagram showing this technique is shown in Fig. 5.60.

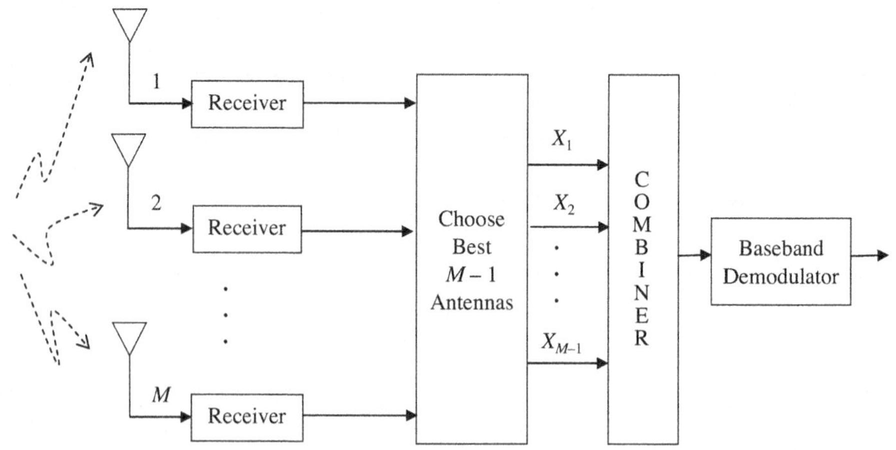

FIGURE 5.60 Modified selection combining diversity $\{SC(M-1)\}$ method.

We begin the performance comparison by presenting the BPSK BER results for the SS/SC diversity technique for $M = 1$ to 4 antennas. The curves show us that greater than 10 dB of improvement can be obtained when going from $M = 1$ to $M = 2$ antennas (see Fig. 5.61).

We have discussed and shown the shortcomings of using the SS diversity technique. Figure 5.62 plots the performance of MRC, SS, and SC2 for the case that there are $M = 3$ antennas available in the receiver. We notice that MRC performs the best while there is a performance gap when viewing

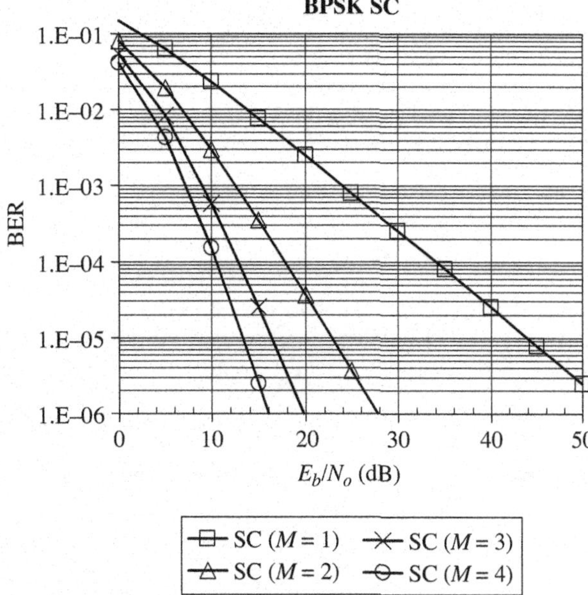

FIGURE 5.61 BPSK BER performance for SC combining.

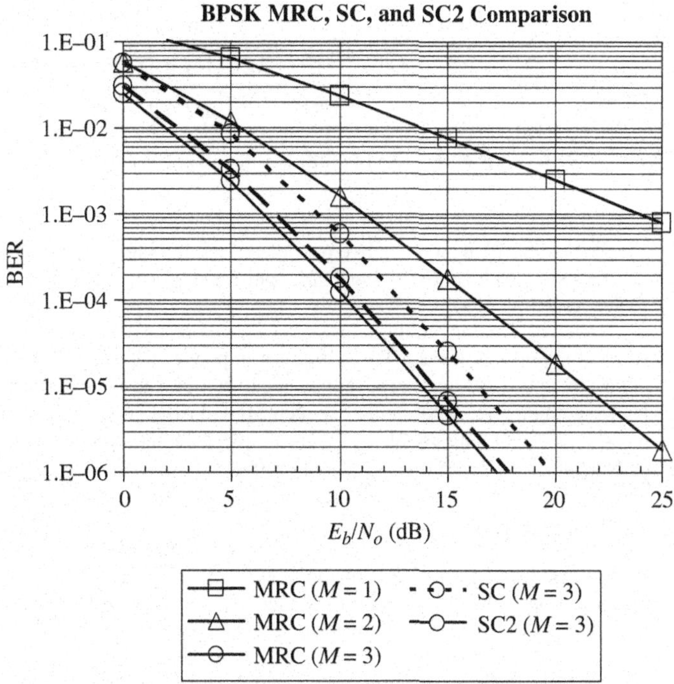

FIGURE 5.62 BPSK BER performance for MRC, SC, and SC2 combining.

SS of approximately 3 dB. Using the SC2 diversity receiver almost the entire gap in performance can be omitted with approximately 1 dB performance difference remaining. What this shows is using 2 out of 3 antennas produces results closer to MRC receiver using $M = 3$ antennas.

Next we increase the number of antennas available to $M = 4$ and compare the performance of MRC, SS, and SC3. This scenario clearly shows the benefits of being able to choose the best 3 out of 4 antennas. One can see the additional gain of using MRC over SC3 is very small, approximately 0.5 dB (see Fig. 5.63).

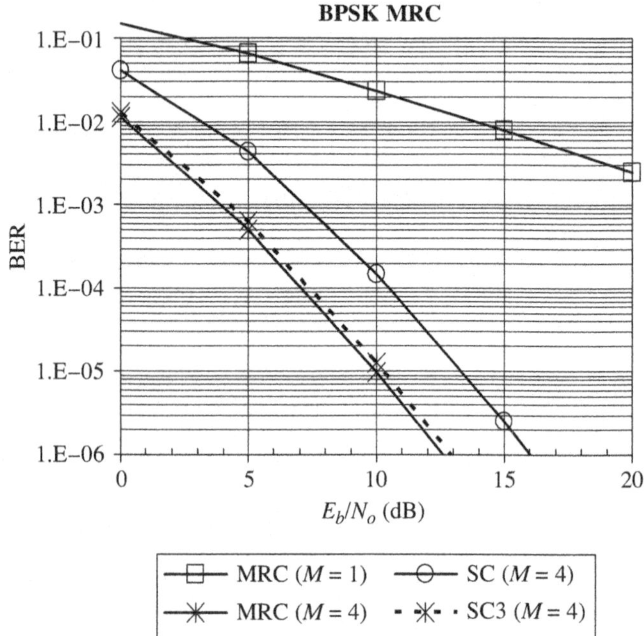

FIGURE 5.63 BPSK BER performance for MRC combining.

Next we have taken the above results and plotted them together to emphasize the points made. It is clear using the best $M - 1$ out of M antennas performs very close to the M antenna MRC diversity system (see Fig. 5.64).

What we have done in this subsection was to present the theoretical performance of the four diversity techniques mentioned. The difference in performance between EGC and MRC was shown to be approximately 1 dB and because their receivers were similar we purposely omitted results for EGC. Hence comparisons were made between SS and MRC for a variety of antennas. This gap in performance was filled with the SC2/SC3 techniques which used the best $M - 1$ of M antennas. Above we showed as M increases this gap becomes narrower.

5.2.4 Optimum Combining (OC)

In the previous three subsections, three antenna combining/selection techniques were presented, all of which improve the system performance. One of the major assumptions used in these sections was that interference was at least, negligible. However, for interference-limited systems or systems that get peak traffic interference, the above techniques are, in fact, not optimal. In this section, we will present the optimal combining receiver and show how this receiver can have a subset, the MRC receiver discussed above. Optimal combining is a form of smart antenna technology. There are basically

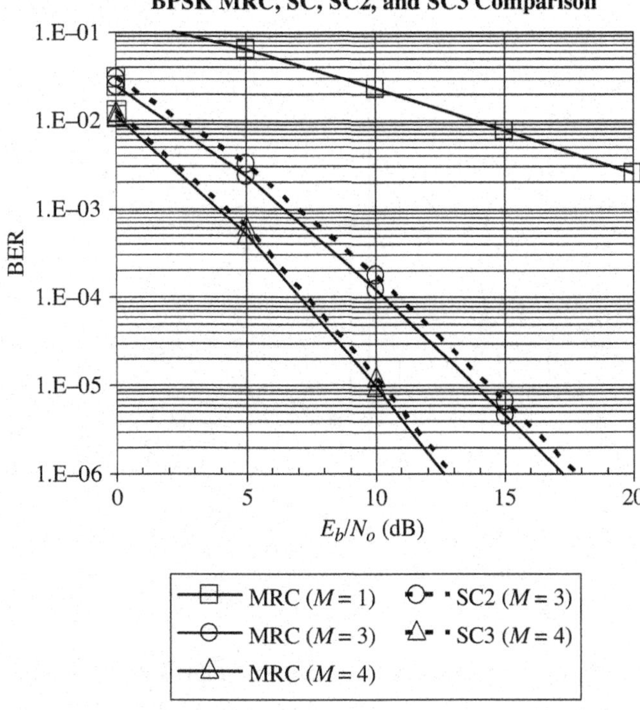

FIGURE 5.64 BPSK BER performance for MRC, SC, SC2, and SC3 combining.

two categories: The first is called beam forming (BF), here the antennas are physically located close to each other. Here the weights are controlled to form antenna patterns (or beams) in the direction of the desired signal and nulls in the direction of the interfering signals. The second category is optimal combining; here the antennas are physically located far apart where antenna patterns are avoided, but array signal processing rules apply [59–62].

Prior to presenting the mathematical equations describing the optimal combining antenna weights, we would like to present the following adaptive antenna array (AAA) block diagram (see Fig. 5.65).

In the above block diagram we see the antenna weights are adaptively controlled by either minimizing or maximizing the chosen cost function. Two commonly used cost functions are Minimum Mean Squared Error (MMSE) and Maximal Signal to Interference plus Noise Ratio (MSINR). It has been shown that in a CCI limited environment, OC performs better than MRC [63]. The received signal vector is given as

$$\underline{x} = [x_1\, x_2 \ldots x_M]^T \tag{5.114}$$

where T is used to denote the transpose operation. The weight vector is defined as, $\underline{w}$, so the AAA output signal as

$$y = \underline{w}^* \cdot \underline{x} \tag{5.115}$$

where * is used to denote the complex conjugate transpose operation.

MMSE and MSINR Cost Functions. In this section we will present the two cost functions used to derive the array weights, namely MSINR and MMSE. The array weights must be adaptively controlled to track the time varying, multipath fading channel.

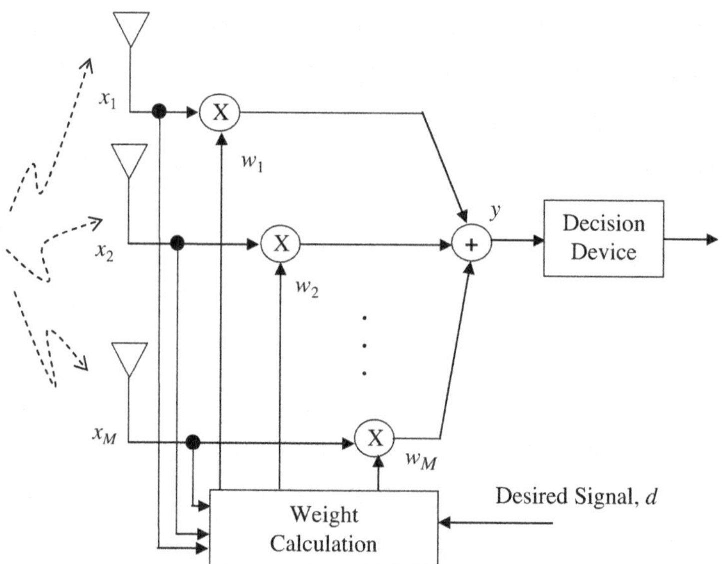

FIGURE 5.65 Adaptive antenna array block diagram.

First, the array weights that minimize the MSE will be denoted as $\underline{w}_{MMSE}$ and are defined as follows:

$$\min_{\underline{w}} E\left\{\left\|\underline{w}^* \cdot \underline{x} - d\right\|^2\right\} \tag{5.116}$$

where d is the desired signal. The solution to the above criterion leads to the following equation for the antenna weights:

$$\underline{w}_{\text{MMSE}} = R_{xx}^{-1} \cdot \underline{r}_{xd} \tag{5.117}$$

where the received signal's covariance matrix and cross-correlation vector are defined and given as below.

$$R_{xx} = E\{\underline{x} \cdot \underline{x}^*\} \tag{5.118}$$

$$\underline{r}_{xd} = E\{\underline{x} \cdot d^*\} \tag{5.119}$$

In practice we have to estimate these parameters using the received signal since the channel is time varying. We can replace the expectation operator by a sample mean estimator, which gives us the following estimate of the covariance matrix.

$$\hat{R}_{xx} = \frac{1}{N}\sum_{i=1}^{N} \underline{x}(t + t_i) \cdot \underline{x}^*(t + t_i) \tag{5.120}$$

And the following estimate of the cross-correlation vector

$$\hat{\underline{r}}_{xd} = \frac{1}{N}\sum_{i=1}^{N} \underline{x}(t + t_i) \cdot d^*(t + t_i) \tag{5.121}$$

We have used N to denote the number of time samples used in the estimate calculation. The use of these estimates is sometimes referred to as Sample Matrix Inversion (SMI) or Direct Matrix Inversion (DMI) [59, 64–66]. Hence the estimated MMSE array weights are given as

$$\hat{\underline{w}}_{\text{MMSE}} = \hat{R}_{xx}^{-1} \cdot \hat{\underline{r}}_{xd} \tag{5.122}$$

The second cost function that we will analyze maximizes the array output SINR given as follows:

$$\max_{\underline{w}} \frac{\underline{w}^* \cdot R_{ss} \cdot \underline{w}}{\underline{w}^* \cdot R_{I+N} \cdot \underline{w}} \tag{5.123}$$

where R_{ss} is the desired signal's covariance matrix and R_{I+N} is the interference plus noise covariance matrix. This leads to the following equation for the adaptive antenna array weights.

$$\underline{w}_{\text{MSINR}} = R_{I+N}^{-1} \cdot \underline{r}_{xd} \tag{5.124}$$

The received signal's interference plus noise covariance matrix is given as follows (with the variable t is omitted for sake of convenience):

$$\hat{R}_{I+N} = \frac{1}{N}\sum_{i=1}^{N}(\underline{x} - \hat{\underline{h}} \cdot d) \cdot (\underline{x} - \hat{\underline{h}} \cdot d)^* \tag{5.125}$$

where $\hat{\underline{h}}$ his an estimate of the desired signal's channel vector. It is commonly accepted to replace this channel estimate with the cross-correlation vector given above. This can now be rewritten as

$$\hat{R}_{I+N} = \frac{1}{N}\sum_{i=1}^{N}(\underline{x} - \hat{\underline{r}}_{xd} \cdot d) \cdot (\underline{x} - \hat{\underline{r}}_{xd} \cdot d)^* \tag{5.126}$$

Hence the estimated MSINR array weights are given as

$$\hat{\underline{w}} = \hat{R}_{I+N}^{-1} \cdot \hat{\underline{r}}_{xd} \tag{5.127}$$

Covariance Matrix Eigen Spectra Decomposition. In this section, we will discuss some eigen spectral properties of the covariance matrix used in the array weight calculation. These properties will not only provide insight into the array weight mechanisms, but also attempt to improve system performance. First the covariance matrices are Hermitian, they are normal matrices and positive definite. Using the Eigen Spectral Decomposition (ESD) theorem we can rewrite the received signal's covariance matrix as follows [67–70]:

$$R = \sum_{i=1}^{M} \lambda_i \cdot \underline{v}_i \cdot \underline{v}_i^* \tag{5.128}$$

where λ_i's are the eigenvalues of the matrix, R, and $\underline{v}_i$ are the associated eigen vectors.

We know the vector space of the covariance matrix, $\underline{v}$, consists of M linearly independent vectors. We can go one step further to classify them into vector subspaces: the signal subspace, V_s, and the noise subspace, V_N. The total vector space can be mathematically written as $V = V_S + V_N$. The covariance matrix can be written as

$$R = \sum_{i=1}^{N_s} \lambda_i \cdot \underline{v}_i \cdot \underline{v}_i^* + \sum_{i=N_s+1}^{M} \lambda_i \cdot \underline{v}_i \cdot \underline{v}_i^* \tag{5.129}$$

where the first summation on the right-hand side corresponds to the signal subspace and the second summation corresponds to the interference plus noise subspace. The MMSE weights can be expressed as follows where we used the ESD on $\hat{R}_{xx}$.

$$\hat{\underline{w}}_{\text{MMSE}} = \left[\sum_{i=1}^{M} \frac{1}{\lambda_i} \cdot \underline{v}_i \cdot \underline{v}_i^*\right] \cdot \hat{\underline{r}}_{xd} \tag{5.130}$$

or equivalently as

$$\hat{\underline{w}}_{\text{MMSE}} = \left[\sum_{i=1}^{N_s} \frac{1}{\lambda_i} \cdot \underline{v}_i \cdot \underline{v}_i^* + \sum_{i=N_s+1}^{M} \frac{1}{\lambda_i} \cdot \underline{v}_i \cdot \underline{v}_i^*\right] \cdot \hat{\underline{r}}_{xd} \tag{5.131}$$

where the variable N_s is used to denote the dimension of the signal subspace of the $\hat{R}_{xx}$ covariance matrix.

The MSINR array weights can be calculated in a similar manner, except for this case, the ESD was performed on $\hat{R}_{I+N}$ matrix, thus giving us

$$\underline{\hat{w}}_{\text{MSINR}} = \left[\sum_{i=1}^{N_I} \frac{1}{\lambda_i} \cdot \underline{v}_i \cdot \underline{v}_i^* + \sum_{i=N_I+1}^{M} \frac{1}{\lambda_i} \cdot \underline{v}_i \cdot \underline{v}_i^* \right] \cdot \hat{\underline{r}}_{xd} \tag{5.132}$$

where N_I is the dimension of the interference vector space of $\hat{R}_{I+N}$.

From this spectral decomposition many insightful array weights can be derived and used under different conditions. Below we will list three of them:

1. *MMSE signal subspace.* Here we use only the eigen vectors that correspond to the signal subspace, which is of size N_s. The reasoning behind this weight is as follows. The cross-correlation vector is a member of the signal subspace and as such is collectively orthogonal to the interference + noise subspace. Hence the second summation would result in contributing ideally a value of zero. In practice, it is not zero, but rather a small value and so by forcing the property performance improvements can be observed.

$$\underline{\hat{w}}_{\text{MMSE-SS}} = \left[\sum_{i=1}^{N_s} \frac{1}{\lambda_i} \cdot \underline{v}_i \cdot \underline{v}_i^* \right] \cdot \underline{r}_{xd} \tag{5.133}$$

2. *MSINR noise subspace.* This technique is also referred to as the Eigen Cancellor ($\underline{\hat{w}}_{MSINR-EC}$) [71]. The reasoning behind this array weight is as follows. Since we are dealing with the R_{I+N} covariance matrix the noise subspace is collectively orthogonal to the interference subspace. Hence in using this noise vector space we can attempt to cancel/suppress the interference that belongs to the interference subspace. That is why the second half of the summation is used.

$$\underline{\hat{w}}_{\text{MSINR-NS}} = \left[\sum_{i=N_I+1}^{M} \frac{1}{\sigma_n^2} \underline{v}_i \cdot \underline{v}_i^* \right] \cdot \underline{r}_{xd} \tag{5.134}$$

3. *MSINR weighted subspace.* Here we use the entire vector space of the covariance matrix, but they are weighted by an eigenvalue modified by the function, $f(x)$ [72]. The reasoning behind this array weight is as follows. Reduced rank techniques tend to degrade faster as the number of interferers increases or the dimension is incorrectly estimated. We wish to retain the entire vector space but weigh then appropriately. By this we mean, each eigenvalue will be inspected and we will determine whether it is associated to a vector in the interference subspace or noise subspace. If they belong to the noise subspace, we will force their respective eigenvalues to equal, σ_n^2.

$$\underline{\hat{w}}_{\text{MSINR-WSS}} = \left[\sum_{i=1}^{M} \frac{1}{f(\lambda_i) \cdot \lambda_i} \cdot \underline{v}_i \cdot \underline{v}_i^* \right] \underline{r}_{xd} \tag{5.135}$$

Interference Study. In this section, we will present simulation results of some of the abovementioned array weights. The received signal is represented as

$$\underline{x} = \underline{h} \cdot d + \sum_{i=1}^{P} \underline{h}_{I_i} \cdot d_i + \underline{n} \tag{5.136}$$

where $\underline{h}$ is an $M \times 1$ desired signal channel vector, $\underline{h}_{I_i}$ is the ith interfering signal channel vectors, and d is the desired signal, d_i is the ith interference signal, and $\underline{n}$ is an $M \times 1$ noise vector. P is used as the number of equal power CCI interferers in the channel. We can define the SINR as follows with INR = interference to noise ratio.

$$\text{SINR} = \frac{\text{SNR}}{1 + \sum_{j=1}^{P} \text{INR}} \tag{5.137}$$

Assuming the desired signal, interfering signals and noise are all independent, we can write the expression for the signal plus interference plus noise covariance matrix.

$$R_{xx} = \sigma_d^2 \cdot \underline{hh}^* + \sum_{i=1}^{P} \sigma_{I_i}^2 \cdot \underline{h}_{I_i} \cdot \underline{h}_{I_i}^* + \sigma_n^2 I \tag{5.138}$$

For sake of completeness the corresponding MMSE and MSINR array weights are given below.

$$\underline{w}_{\text{MMSE}} = \left[\sigma_d^2 \cdot \underline{h}\,\underline{h}^* + \sum_{i=1}^{P} \sigma_{I_i}^2 \cdot \underline{h}_{I_i} \underline{h}_{I_i}^* + \sigma_n^2 I \right]^{-1} \cdot \sigma_d^2 \cdot \underline{h} \tag{5.139}$$

$$\underline{w}_{\text{MSINR}} = \left[\sum_{i=1}^{P} \sigma_{I_i}^2 \cdot \underline{h}_{I_i} \underline{h}_{I_i}^* + \sigma_n^2 I \right]^{-1} \cdot \sigma_d^2 \cdot \underline{h} \tag{5.140}$$

The first results we present are for the $M = 5$ antennas with $P = 1$ CCI. We plot the BER versus the estimator window size (N) for the MMSE, Ideal MSINR, MSINR-WSS, and genie aided knowledge about the MSINR dimension of the noise plus interference subspaces.

Here we have two points to make: For the smaller values of the window size the plot shows how MSINR can dramatically improve performance when compared to MMSE. Here some a priori value of P is assumed. The second point is that the MSINR-WSS technique produced BER results very close to the case when the rank of the R_{I+N} matrix (MSINR-SS) was known (see Fig. 5.66).

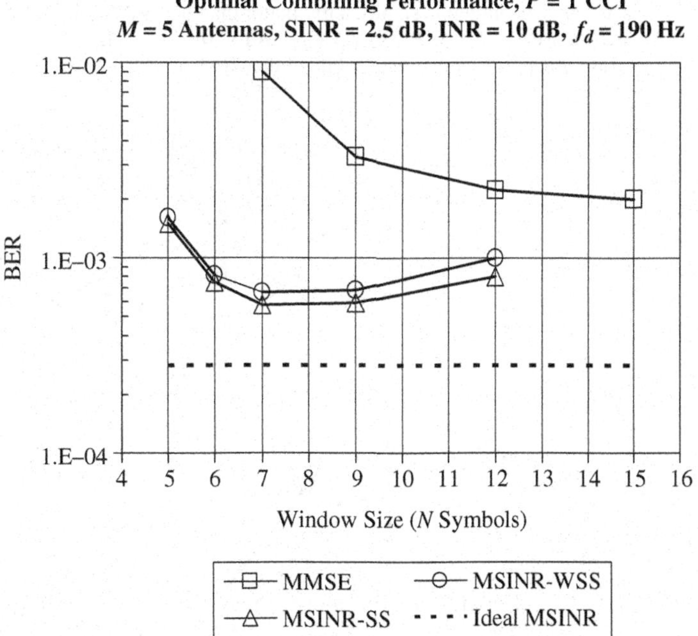

FIGURE 5.66 Optimal combining performance with $P = 1$ CCI.

Next we plot the BER versus the number of CCI, P, using $M = 5$ antenna receiver. MMSE showed small improvement or degradation as P increased. Also we can see how our WSS with known subspace dimensions compares the case where the rank of the covariance matrix is determined to the case when the estimated subspace dimension (MSINR-WSS).

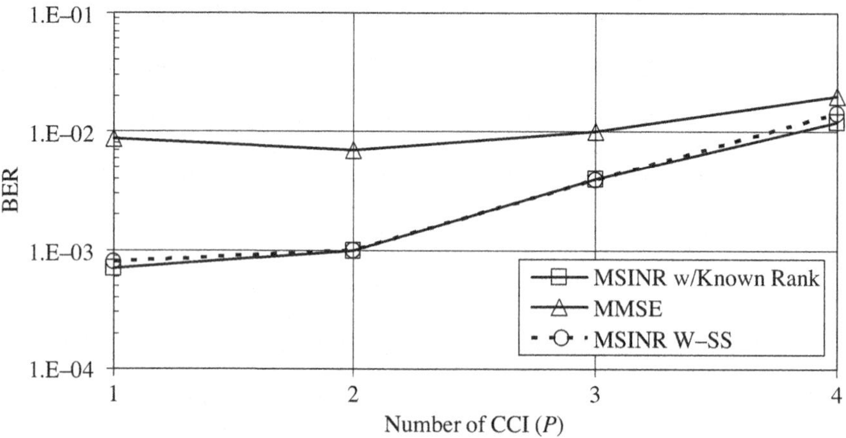

FIGURE 5.67 Optimal combining performance for various number of CCIs.

Figure 5.67 shows us that the MSINR-WSS weights accurately reflect the rank of the covariance matrix. This technique is significantly better than using the conventional MMSE array weights.

Lastly, we plot the BER versus for SINR for $M = 2$ and $M = 5$ antennas. Here we see $M = 5$ antennas case, the performance improvement increases as M increases (see Fig. 5.68).

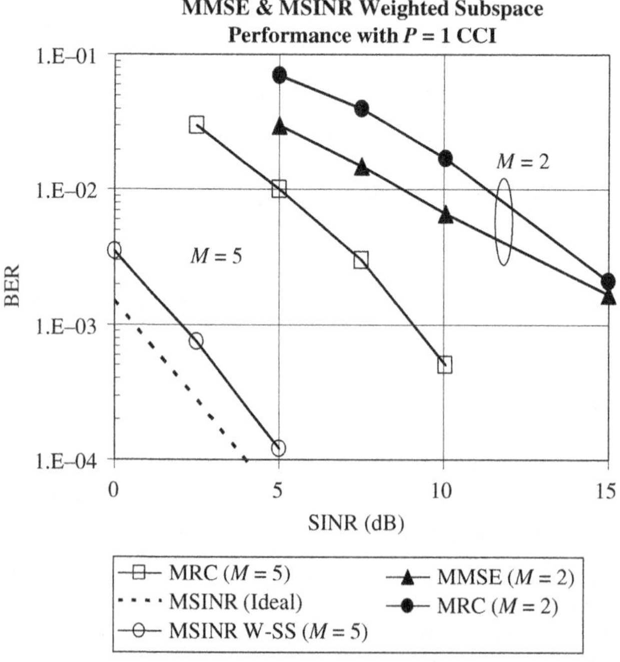

FIGURE 5.68 AAA BER performance comparison for $M = 2$ & $M = 5$ antennas.

5.3 TRANSMIT SPATIAL ANTENNA DIVERSITY TECHNIQUES

In this section, we will discuss a technique called transmit diversity. In the previous sections, we have shown how using receive diversity can dramatically improve system performance. This improvement comes at the expense of receiver complexity (i.e., power consumption, size, cost, and so on). Transmit diversity aims to provide the similar benefits of having receive diversity, but will accomplish this by transmitting with multiple antennas [73–77].

In 1998, a simple space-time block code was presented by Alamouti, showing that 2 transmit antennas and 1 receive antenna provide the same improvement order of MRC with 1 transmit and 2 receive antennas [78]. This technique was later adopted into the WCDMA UMTS and IS-2000 digital cellular standards and is named Space-Time Transmit Diversity (STTD).

5.3.1 Space-Time Block Code (STBC)

As the name STBC implies, this technique involves grouping bits to be transmitted into a block of data. This block will then be transmitted by being spread out over time and space, hence the name Space-Time.

Space-Time Transmit Diversity. In this section, we will present the technique originally proposed by Alamouti and later adopted by WCDMA UMTS. This is sometimes called open loop transmit diversity (for reasons we will see later). Here we have 2 transmit antennas and group the input symbols into a block of 2. The STTD encoding operations performed at the transmitter is shown in Fig. 5.69.

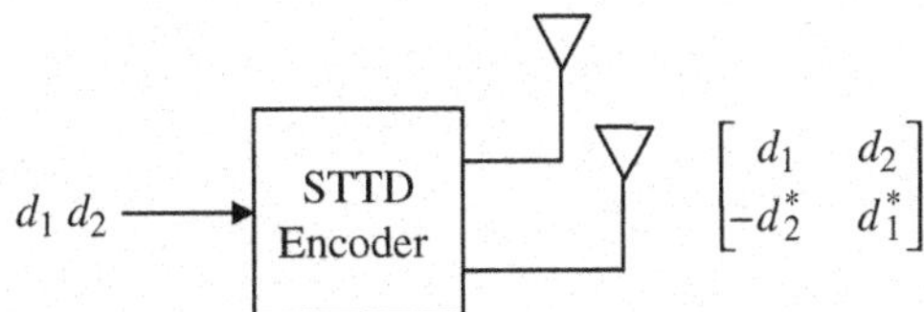

FIGURE 5.69 STTD encoding operations for $M = 2$ antennas.

The transmitted signal can be written in matrix form as

$$S = \begin{bmatrix} s_1 \\ s_2 \end{bmatrix} = \begin{bmatrix} d_1 & d_2 \\ -d_2^* & d_1^* \end{bmatrix} \tag{5.141}$$

It is very easy to see the determinant is defined as $\det(S) = |d_1|^2 + |d_2|^2$, and for nonzero data symbols, allows the signal matrix to be nonsingular.

Assuming a single receive antenna, we can draw the transmit diversity communication system block diagram shown in Fig. 5.70.

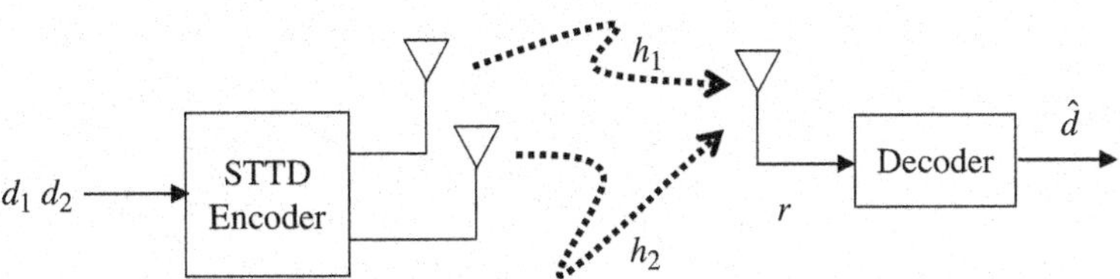

FIGURE 5.70 STTD encoder and decoder communication block diagram.

Let r_1 and r_2 be the 2 received symbols in 2 consecutive symbol intervals and are given below assuming $h_1(t) \cong h_1(t - T_S)$.

$$r_1(t) = d_1(t) \cdot h_1(t) - d_2^*(t) \cdot h_2(t) + n_1(t) \qquad @\ t = nT_s \tag{5.142}$$

$$r_2(t) = d_2(t) \cdot h_1(t) + d_1^*(t) \cdot h_2(t) + n_2(t) \qquad @\ t = (n+1)T_s \tag{5.143}$$

Here we assumed the multipath fading is constant over 2 consecutive symbols and can represent the above equations in matrix notation.

$$\underline{r} = \begin{bmatrix} r_1(t) \\ r_2(t) \end{bmatrix} = \begin{bmatrix} d_1(t) & -d_2^*(t) \\ d_2(t) & d_1^*(t) \end{bmatrix} \cdot \begin{bmatrix} h_1(t) \\ h_2(t) \end{bmatrix} + \begin{bmatrix} n_1(t) \\ n_2(t) \end{bmatrix} \tag{5.144}$$

$$\underline{r} = S^T \cdot \underline{h} + \underline{n} \tag{5.145}$$

Before we present the STTD decoder equations, we would like to state the decoder consists of two functions: channel estimation and linear combiner, as shown in Fig. 5.71.

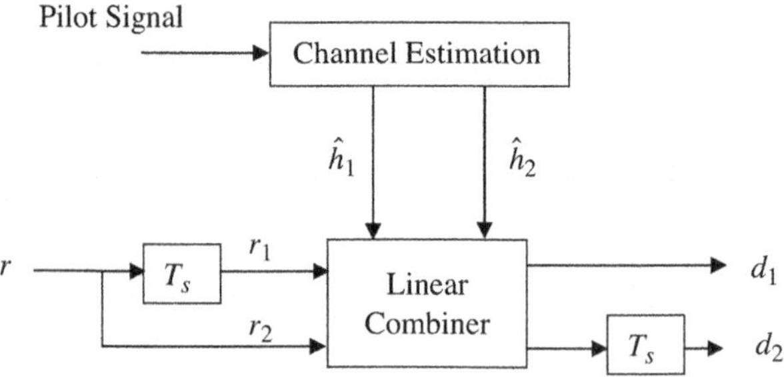

FIGURE 5.71 STTD decoder receiver.

For the sake of this discussion, we will assume the channel has been estimated using any reliable technique. The particular implementation of WCDMA utilizes pilot symbols that are orthogonal across both transmit antennas. We will have the specifics of this in a later chapter where we discuss the WCDMA application.

The linear combiner makes use of the consecutive received symbols to jointly estimate the transmitted symbols. The equations used are as follows:

$$d_1 = h_1^* \cdot r_1 + h_2 \cdot r_2^* \tag{5.146}$$

$$d_2 = h_1^* \cdot r_2 - h_2 \cdot r_1^* \tag{5.147}$$

This can be rewritten in matrix form as follows:

$$\begin{bmatrix} d_1 \\ d_2 \end{bmatrix} = \begin{bmatrix} r_1 & r_2^* \\ r_2 & -r_1^* \end{bmatrix} \cdot \begin{bmatrix} h_1^* \\ h_2 \end{bmatrix} \tag{5.148}$$

After applying the matrix operations, we arrive with the following estimates:

$$\hat{d}_1 = d_1 \cdot \left[|h_1|^2 + |h_2|^2\right] + h_1^* \cdot n_1 + h_2 \cdot n_2^* \tag{5.149}$$

$$\hat{d}_2 = d_2 \cdot \left[|h_1|^2 + |h_2|^2\right] + h_1^* \cdot n_2 - h_2 \cdot n_1^* \tag{5.150}$$

Here we see we have extracted the diversity combining aspect of the channel to improve the receiver's performance.

5.3.2 Closed Loop Transmit Diversity (CLTD)

In this section, we will discuss a CLTD technique. Here the Mobile Station (or User Equipment, UE) calculates the weights to be applied to the Base station Transceiver Station (BTS) (or NodeB) in order to maximize the instantaneous received power at the UE [79–84]. The performance of this CLTD technique is governed by three key issues.

1. *BTS update rate.* How often are the BTS weights updated? For fast time varying channels, we would require a fast update rate in order to accurately track the fading phenomenon so it can be effectively compensated.
2. *Weight quantization.* What information is required by the BTS? Once the UE calculates the weights to be used by the BTS, it must transmit them to the BTS in an efficient manner. Hopefully without loss of important information. This can take on the many forms of phase only, phase and amplitude, progressive updating, and so on.
3. *Weight confidence.* How does the UE confidently know what weights were applied to the BTS? Once the UE has quantized the weights it wants to feed back to the BTS, they are transmitted through a multipath fading channel. So there is a chance that the BTS will apply the incorrect weight, due to errors caused by the channel. At this point the UE can either work under the assumption that the correct weight was applied (even though this maybe false) or it can determine the actual weight applied. In this latter case, there must be a mechanism within the communication system to support this feature.

A block diagram of the CLTD communication system is shown below. Note Fig. 5.72 assumes a single receive antenna to be used by the UE, obviously nothing precludes the UE from using more than 2 antennas.[2] Also, we have drawn a second set of antennas for transmit and receive functionality. In a typical application, an antenna would be used for both transmit and receive.

So for a single receive antenna at the UE, the received signal is given by the following equation:

$$r = d \cdot w_1 \cdot h_1 + d \cdot w_2 \cdot h_2 + \text{noise} \tag{5.151}$$

Note in this transmit diversity example, the same symbol is transmitted on both antennas, the exception being that the antenna weights (w_1 and w_2) are different.

Phase Only Adjustments. We will impose the following constraint: $|w_1|^2 + |w_2|^2 = 1$. Moreover, in an effort to effectively present the CLTD concept, we will further constrain w_1 to equal $1/\sqrt{2}$. This

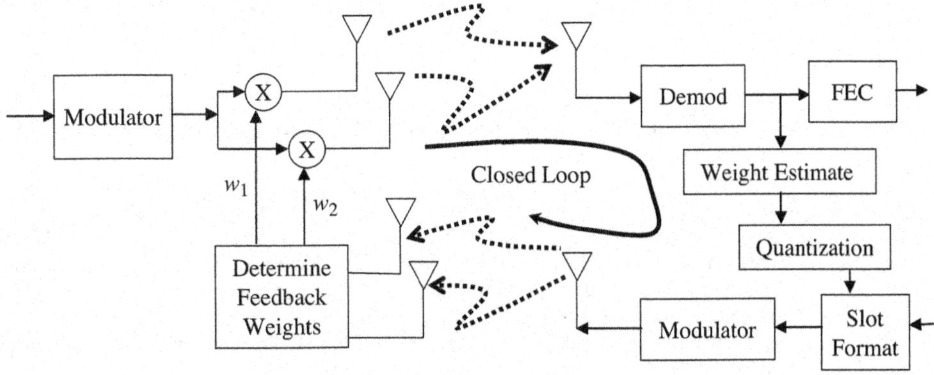

FIGURE 5.72 CLTD system level block diagram.

[2] This architecture is called multiple-input-single-output (MISO).

implies equal amplitude weighting across both antennas; however, the second weight is a complex number and can have phase variations. Let us first present the solution and next we will derive it. The solution is given as

$$w_1 = \frac{1}{\sqrt{2}} \quad \text{and} \quad w_2 = \frac{1}{\sqrt{2}} \cdot \frac{h_1 \cdot h_2^*}{|h_1 \cdot h_2^*|} = \frac{1}{\sqrt{2}} \cdot e^{j(\theta_1 - \theta_2)} \tag{5.152}$$

Assume the received signal is

$$r = d \cdot w_1 \cdot a \cdot e^{j\theta_1} + d \cdot w_2 \cdot b \cdot e^{j\theta_2} \tag{5.153}$$

After applying the transmit weights presented above the received signal is equal to the following:

$$r = d \cdot \frac{1}{\sqrt{2}} \cdot a \cdot e^{j\theta_1} + d \cdot \frac{1}{\sqrt{2}} \cdot e^{j(\theta_1 - \theta_2)} \cdot b \cdot e^{j\theta_2} \tag{5.154}$$

$$r = d \cdot \frac{1}{\sqrt{2}} \cdot \{a \cdot e^{j\theta_1} + b \cdot e^{j\theta_1}\} \tag{5.155}$$

$$r = d \cdot \frac{1}{\sqrt{2}} \cdot e^{j\theta_1} \cdot \{a + b\} \tag{5.156}$$

Here the flat fading, channel responses are $h_1 = a \cdot e^{j\theta_1}$ and $h_2 = b \cdot e^{j\theta_2}$; they were calculated in the channel estimation algorithm in the receiver.

The BTS antenna weights are chosen to maximize the UE's received signal power, P. Assuming an L path channel model we have the following equation that needs to be maximized.

$$P = \underline{w}^* \cdot (H^*H) \cdot \underline{w} \tag{5.157}$$

where

$$\underline{w} = [w_1 \; w_2]; \qquad H = [\underline{h}_1 \; \underline{h}_2]; \qquad \underline{h}_1 = [h_{11} \ldots h_{1L}]^T \tag{5.158}$$

Let us first consider a frequency flat fading channel as the UE would then need to calculate w_1 and w_2 such that the following power is maximized ($L = 1$).

$$P = [w_1 \, w_2] \cdot \begin{bmatrix} h_1^* \\ h_2^* \end{bmatrix} \cdot [h_1 \; h_2] \cdot \begin{bmatrix} w_1 \\ w_2 \end{bmatrix} \tag{5.159}$$

$$P = [w_1 \, w_2] \cdot \begin{bmatrix} |h_1|^2 & h_1^* h_2 \\ h_2^* h_1 & |h_2|^2 \end{bmatrix} \cdot \begin{bmatrix} w_1 \\ w_2 \end{bmatrix} \tag{5.160}$$

After some simple manipulations we come to the final equation shown below for the power of the UE received signal.

$$P = \frac{1}{2}|h_1|^2 + 2\mathrm{Re}\{h_1 h_2^* w_2^*\} + |h_2|^2 \cdot |w_2|^2 \tag{5.161}$$

This is maximized when $w_2 = h_1 h_2^*$. Since w_2 is quantized, the UE can either use the above solution or exhaustively try all the possible options for w_2 into the power equation and then choose the one that maximizes the received signal power.

Now let's suppose there is a frequency selective fading channel such as a two-ray model, then the power equation can be written as follows ($L = 2$).

$$P = [w_1 w_2^*] \cdot \begin{bmatrix} h_{11}^* & h_{12}^* \\ h_{21}^* & h_{22}^* \end{bmatrix} \cdot \begin{bmatrix} h_{11} & h_{21} \\ h_{12} & h_{22} \end{bmatrix} \cdot \begin{bmatrix} w_1 \\ w_2 \end{bmatrix} \tag{5.162}$$

After some simple manipulations we get the following equation that needs to be maximized.

$$P = \frac{1}{2}|h_{11}|^2 + 2\text{Re}\{h_{11}^* h_{21} w_2\} + |w_2 h_{21}|^2 + \frac{1}{2}|h_{12}|^2 + 2\text{Re}\{h_{12}^* h_{22} w_2\} + |w_2 h_{22}|^2 \quad (5.163)$$

Now this consists of two parts: The first part can be maximized when considering only the first multipath while the second part can be maximized when considering only the second multipath. To jointly optimize across both multipaths would possibly sacrifice the performance of each individual path for the sake of the overall multipath channel.

Let us consider the following simplified problem of the channel matrix $R_H = H^*H$.

$$R_H = \begin{bmatrix} a^2 & a^*b \\ ab^* & b^2 \end{bmatrix} \quad (5.164)$$

We know the following equations hold for the determinant and trace of the above matrix.

$$\text{Det}(R_H) = \lambda_1 \cdot \lambda_2 = 0 \quad (5.165)$$

$$\text{Trace}(R_H) = \lambda_1 + \lambda_2 = a^2 + b^2 \quad (5.166)$$

Since the determinant is zero the matrix is singular and hence cannot be inverted, implying the following to hold true when defining the first eigenvalue, $\lambda_2 = 0$. Using the above equations we see the first eigenvalue equals $\lambda_1 = a^2 + b^2$. Applying the ESD, we have the following:

$$R_H \cdot \underline{v}_1 = \lambda_1 \cdot \underline{v}_1 \quad (5.167)$$

The first eigenvalue can be used in the above equation to determine the corresponding eigen vector given below

$$\underline{v}_1 = \begin{bmatrix} \frac{a^*b}{b^2} \\ 1 \end{bmatrix} \cdot K \quad (5.168)$$

where the value of K was chosen to make the eigen vector orthonormal, $K = \sqrt{\frac{b^2}{a^2 + b^2}}$.

Note that the solution that maximizes the power equation is the eigen vector that corresponds to the largest eigenvalue. Using the ESD theorem we can reconstruct the channel matrix as follows:

$$R_H = \lambda_1 \cdot \underline{v}_1 \cdot \underline{v}_1^* + \lambda_2 \cdot \underline{v}_2 \cdot \underline{v}_2^* \quad (5.169)$$

Assuming $\lambda_1 > \lambda_2$, one can simply choose $\underline{v}_1$ as the solution.

5.4 LINK BUDGET DISCUSSION

In this section, we will discuss an example of a link budget of a wireless communication system. The link budget will show where the system power is distributed between the transmitter, channel, and receiver. The link budget consists of the following parameters:

1. Base station transmit power
2. Base station cabling loss
3. Base station antenna gain
4. Path loss
5. Fading margin (large and smaller scale)

6. Interference and noise
7. Receiver interference and noise
8. Receiver antenna gain
9. Receiver KF
10. System data rate

Next, we will provide an example of the link budget where we wish to arrive with a value of SNR or E_b/N_o that the receiver would have to perform. We begin with the following equation for SNR:

$$\text{SNR} = \frac{S_{rx}}{N_o \cdot \text{BW} \cdot \text{NF}} \tag{5.170}$$

where S_{rx} = received signal power, N_o = noise power spectral density, BW = bandwidth, NF = noise figure. We can further elaborate on the above variables as

$$\text{SNR} = \frac{S_{tx} \cdot G_{tx} \cdot G_{rx}}{N_o \cdot \text{BW} \cdot \text{NF} \cdot \text{PL}} \tag{5.171}$$

where we have refined the variable expressing the received signal with S_{tx} = transmit power, G_{rx} of antenna gain of the receiver, G_{tx} = antenna gain of the transmitter and PL = path loss.

The common approach to present and analyze a link budget is through a table as shown in Table 5.5.

TABLE 5.5 Typical Link Budget Parameters

Variable	Unit
(a) Transmit power	dB
(b) BS cabling loss	dB
(c) BS antenna gain	dB
(d) Rx antenna gain	dB
(e) Receiver cabling loss	dB
(f) Thermal noise density	dBm/Hz
(g) Interference density	dBm/Hz
(h) Total $I + N$ density	dBm/Hz
(i) Receiver NF	dB
(j) System BW	dB Hz
(k) Handoff gain	dB
(l) Lognormal margin	dB
(m) Path loss	dB
(n) Building path loss	dB
(o) Various implementations of losses	dB
(p) Receive sensitivity	a − b + c + d − e − k + l − m − n − o
(q) Required $E_b/(I_o + N_o)$	q − h − i − j

A few very useful formulas should be presented; the first is the following very simple relationship.

$$\frac{S}{N} = \frac{E_b}{N_o} \cdot 2 \cdot \frac{1}{\text{PG}} \tag{5.172}$$

Here the processing gain (PG) is the number of chips used to encode the QPSK symbol. The factor of 2 was introduced to convert the symbol energy to bit energy. The second is given as follows, which is essentially a substitution of the above formulas

$$S_{rx} = kT \cdot NF \cdot \frac{E_b}{N_o} \cdot R_b \tag{5.173}$$

These formulas can be used to derive the receiver sensitivity. For example, assume the BER requirements are 1E-3, then we would obtain the required E_b/N_o to meet the requirements. This in turn will be used in the above formula to provide us with the receiver sensitivity.

In Table 5.6 we present an example of link budget for a WCDMA down-link 12.2 kbps voice user. This deployment scenario is for an outdoor UE. The procedure taken below is to estimate the received signal power, then obtain the overall interference plus noise floor of the receiver. With these two parameters, then SNR or $E_c/(I_o + N_o)$ can be calculated. This can be converted to $E_b/(I_o + N_o)$ by utilizing the processing gain and code rate. This value is then used to predict the receiver sensitivity.

We have included a 3-dB rise of interference in the cell. We have also included a variable called various implementation loss. This includes the implementation loss due to impairments plus loss due to the in vehicle loss plus any power control increase.

In Table 5.7 we present an example link budget for a WCDMA down-link 12.2 kbps voice user. This deployment scenario is for an indoor UE. Most of the parameters were kept the same, but we included the loss due to building penetration loss. We have included the loss of the signal traversing the outer wall only. In other words, we didn't place any restriction on the loss due to floor penetration, since this would require us to provide the actual floor the UE is resting on as well as the height of the NodeB antennas [85].

Since this is an indoor deployment scenario we have reduced the various implementation loss due to the in-vehicle loss. And because of the additional loss due to the building penetration we have reduced the propagation path loss to provide a meaningful case.

5.4.1 Areas under Control to Aid Performance Trade-offs

In the previous section, we have presented a reasonably complete link budget of a wireless communication system. Its purpose was to derive the required SNR for the particular set of system values chosen. Once this equivalent E_b/N_o value is derived, we can then use the performance improvement techniques presented in this chapter to achieve the requirements. In this section, we will discuss the particular areas under control in the link budget calculation. In other words, what specific parameters can be adjusted in order to obtain not only a reasonably required SNR, but also a practical communication system.

TABLE 5.6 Link Budget Example for Outdoor Scenario

a	Transmit power	21	dBm	
b	Transmit cable loss	2	dB	
c	Transmit antenna gain	18	dBi	
d	Receiver Antenna Gain	0	dBi	
e	Receiver cable loss	0	dB	
f	Thermal noise density	−174	dBm/Hz	
g	Interference density	−174	dBm/Hz	
h	Total inter + noise density	−170.99	dBm/Hz	
i	Receiver noise figure	7	dB	
j	System BW	65.84331	dB	
k	Hand-off gain	3	dB	
l	Lognormal fading margin	7	dB	
m	Propagation path loss	143	dB	
n	Building penetration loss	0	dB	
o	Various implementation loss	6	dB	
p	**Rx signal power**	**−116**	dBm	a−b+c+d−e+k+l−m−n−o
q	$E_b/(I_o+N_o)$	4.988487	dB	
r	Processing gain	21.0721	dB	
s	**Interference + noise floor**	**−98.1464**	dBm	h+i+j
t	$E_c/(I_o+N_o)$	**−17.8536**	dB	p−s
u	**Sensitivity**	**−118.138**	dBm	h+i+q+v
v	User data rate (12.2 kbps)	40.8636	dB	

TABLE 5.7 Link Budget Example for Indoor Scenario

a	Transmit power	21	dBm	
b	Transmit cable loss	2	dB	
c	Transmit antenna gain	18	dBi	
d	Receiver antenna gain	0	dBi	
e	Receiver cable loss	0	dB	
f	Thermal noise density	−174	dBm/Hz	
g	Interference density	−174	dBm/Hz	
h	Total inter + noise density	−170.99	dBm/Hz	
i	Receiver noise figure	7	dB	
j	System BW	65.84331	dB	
k	Hand-off gain	1	dB	
l	Lognormal fading margin	7	dB	
m	Propagation path loss	134	dB	
n	Building penetration loss	10	dB	
o	Various implementation loss	3	dB	
p	**Rx signal power**	**−116**	dBm	a−b+c+d−e+k−l−m−n−o
q	$E_b/(I_o+N_o)$	4.988487	dB	
r	Processing gain	21.0721	dB	
s	**Interference + noise floor**	**−98.1464**	dBm	h+i+j
t	$\mathbf{E_c/(I_o+N_o)}$	**−17.8536**	dB	p−s
u	**Sensitivity**	**−118.138**	dBm	h+i+q+v
v	User data rate (12.2 kbps)	40.8636	dB	

If the required value of E_b/N_o is low, then the communication system can request the transmit power to be increased. However, assuming all the cells react in a similar fashion, then the interference density values need to be revisited because of the increase in the other cell interference. Moreover, link level simulations need to be rerun to observe the effects on the BLER/TPUT as the interference (both intra- and inter-cell) is increased.

Another potential solution may be to use better performing components (possibly at the expense of cost) that could essentially increase an antenna gain or reduce the desired NF. Here system cost targets must be factored into the system design. Every dB of improvement in the noise figure directly impacts the receiver sensitivity.

Modifying deployment scenarios can also be a valid solution. Here we can restrict the usage to be outdoors only, thus saving on the building penetration loss, not to mention the in-building losses due to the floor attenuation. For this particular case, when the user enters an indoor environment, it hands off to an indoor pico cell, which can better serve the user.

Continuing along the lines of deployment scenarios, we could reduce the cell radius thus having less path required to be compensated for, increasing the overall system performance margin. As you would expect, nothing comes for free; meaning more NodeBs would need to be deployed to cover a particular city area. This involved potentially complicated deployments as well as to increase system cost due to requiring more NodeB to be present for a particular cell.

Generally speaking, performance improvement can be obtained if the NF can be reduced, more powerful error correcting codes can be used, additional transmit power can be used or using smaller cell radii. Moreover, the UE (or NodeB for that matter) can use spatial diversity or other advanced receiver techniques to reduce the required $E_b/(I_o + N_o)$. This reduction can be used in accommodating more users within the cell and/or can be used to increase the user (or cell) data throughput. Lastly, one can use this additional gain to accommodate the use of less costly components for cost reduction reasons. We must add this last option is not the typical application used in the cellular environment.

In any case, an interactive design process is required that interfaces the link budget variables to the link budget simulation to the network level simulations.

APPENDIX 5A

Below we plot the average burst error lengths at the VA output.

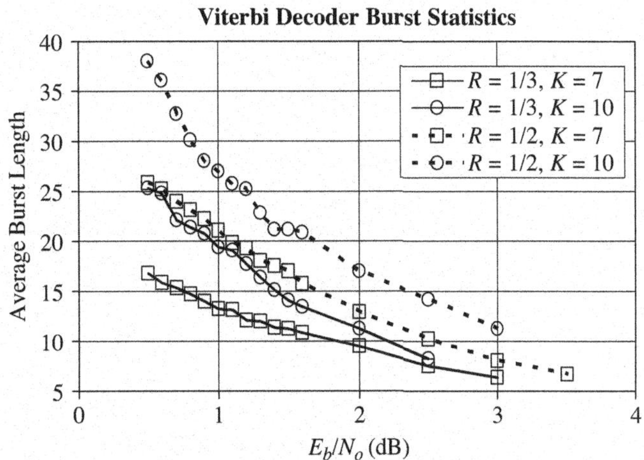

FIGURE 5A.1 Viterbi decoder output burst lengths.

REFERENCES

[1] B. Sklar, *Digital Communications: Fundamental and Applications*, Prentice Hall, 1988, New Jersey.

[2] N. C. Beaulieu and X. Dong, "Level Crossing Rate and Average Fade Duration of MRC and EGC Diversity in Rician Fading," *IEEE Transactions on Communications*, Vol. 51, No. 5, May 2003, pp. 722–726.

[3] S. A. Hanna, "Convolutional Interleaving for Digital Radio Communications," *IEEE ICUPC*, 1993, pp. 443–447.

[4] B. Sklar and F. J. Harris, "The ABCs of Linear Block Codes," *IEEE Signal Processing Magazine*, July 2004, pp. 14–35.

[5] S. Lin and D. J. Costello, Jr., *Error Control Coding: Fundamental and Applications*, Prentice-Hall, 1983, New Jersey.

[6] W. W. Peterson & E. J. Weldon, Jr., *Error-Correcting Codes*, The Massachusetts Institute of Technology Press, 1972.

[7] J. S. Lee & L. E. Miller, *CDMA Systems Engineering Handbook*, Artech House, 1998, Massachusetts.

[8] J. P. Stenbit, "Tables of Generators for Bose-Chadhuri Codes," *IEEE Transactions on Information Theory*, Vol. IT-10, No. 4, Oct. 1964, pp. 390–391.

[9] D. Chase, "A Class of Algorithms for Decoding Block Codes with Channel Measurement Information," *IEEE Transactions on Information Theory*," Vol. IT-18, No. 1, 1972, pp. 170–182.

[10] A. J. Viterbi, "Convolutional Codes and Their Performance in Communication Systems," *IEEE Transactions on Communications Technology*, Vol. COM-19, Oct. 1971, pp. 751–772.

[11] A. J. Viterbi, *CDMA: Principles of Spread Spectrum Communication*, Addison-Wesley, 1995, New York.

[12] G. D. Forney, Jr., "Maximum-Likelihood Sequence Estimation of Digital Sequences in the Presence of Intersymbol Interference," *IEEE Transactions on Information Theory*, Vol. IT-18, No. 3, May 1972, pp. 363–378.

[13] A. J. Viterbi, "A Personal History of the Viterbi Algorithm," *IEEE Signal Processing Magazine*, July 2006, pp. 120–142.

[14] J. K. Omura, "On the Viterbi Decoding Algorithm," *IEEE Transactions on Information Theory*, Vol. IT-15, Jan. 1969, pp. 177–179.

[15] G. D. Forney, Jr., "The Viterbi Algorithm," *Proceedings of the IEEE*, Vol. 61, No. 3, March 1973, pp. 268–278.

[16] H-L. Lou, "Implementing the Viterbi Algorithm," *IEEE Signal Processing Magazine*, Sept. 1995, pp. 42–52.

[17] S-S. Han, J-S. No, Y-C. Jeung, K-O. Kim, Y-B. Shin, S-J. Hahm, and S-B. Lee, "New Soft-Output MLSE Equalization Algorithm for GSM Digital Cellular Systems," *IEEE*, 1995, pp. 919–923.

[18] J. I. Park, S. B. Wicker and H. L. Owen, "Soft Output Equalization for $\pi/4$-DQPSK Mobile Radio," *IEEE Conference on Communications (ICC) 1997*, pp. 1503–1507.

[19] T. Matsumoto and F. Adachi, "BER Analysis of Convolution Coded QDPSK in Digital Mobile Radio," *IEEE Transactions on Vehicular Technology*, Vol. 40, No. 2, May 1991, pp. 435–442.

[20] T. Matsumoto, "Soft Decision Decoding of Block Codes Using Received Signal Envelope in Digital Mobile Radio," *IEEE Journal on Selected Areas in Communications*, Vol. 7, No. 1, Jan. 1989, pp. 107–113.

[21] A. J. Viterbi, "Error Bounds for Convolutional Codes and an Asymptotically Optimum Decoding Algorithm," *IEEE Transactions on Information Theory*, Vol. IT13, No. 2, April 1967, pp. 260–269.

[22] J. A. Heller and I. M. Jacobs, "Viterbi Decoding for Satellite and Space Communications," *IEEE Transactions on Communications Technology*, Vol. COM-19, No. 5, Oct. 1971, pp. 835–848.

[23] J. P. Odenwalder, *Error Control Coding Handbook* (Final Report), Linkabit Corporation, 1976.

[24] Y. Yasuda, K. Kashiki, and Y. Hirata, "High-Rate Punctured Convolutional Codes for Soft Decision Viterbi Decoding," *IEEE Transactions on Communications*, Vol. COM-32, No. 3, March 1984, pp. 315–319.

[25] S. B. Wicker and V. K. Bhargava, *Reed-Solomon Codes and Their Applications*, IEEE Press, 1994, New York.

[26] G. Qun, M. Junfa, and R. Mengtian, "A VLSI Implementation of a Low Complexity Reed-Solomon Encoder and Decoder for CDPD," *International Conference on ASIC*, 2001, pp. 435–439.

[27] R. L. Miller, L. J. Deutsch, and S. A. Butman, "On the Error Statistics of Viterbi Decoding and the Performance of Concatenated Codes," JPL Publication 81-9, 1981.

[28] C. Berrou, A. Glavieus, and P. Thitimajshima, "Near Shannon Limit Error Correcting Coding and Decoding: Turbo Codes," *IEEE International Communications Conference*, 1993, pp. 1064–1070.

[29] B. Vucetic and J. Yuan, *Turbo Codes: Principles and Applications*, Kluwer Academic Publishers, 2000, Massachusetts.

[30] A. Shibutani, H. Suda, and F. Adachi, "Complexity Reduction of Turbo Decoding," *Vehicular Technology Conference*, 1999, pp. 1570–1574.

[31] G. Battail, "A Conceptual Framework for Understanding Turbo Codes," *IEEE Journal on Selected Areas in Communications*, Vol. 16, No. 2, Feb. 1998, pp. 245–254.

[32] E. K. Hall and Stephen G. Wilson, "Design and Analysis of Turbo Codes on Rayleigh Fading Channels," *IEEE Journal on Selected Areas in Communications*, Vol. 16, No. 2, Feb. 1998, pp. 160–174.

[33] J. Hagenauer, E. Offer, and L. Papke, "Iterative Decoding of Binary Block and Convolutional Codes," *IEEE Transactions on Information Theory*, Vol. 42, No. 2, March 1996, pp. 429–445.

[34] C. Berrou and A. Glavieux, "Near Optimum Error Correction Coding and Decoding: Turbo Codes," *IEEE Transactions on Communications*, Vol. 44, No. 10, Oct. 1996, pp. 1261–1271.

[35] W. Oh and K. Cheun, "Adaptive Channel SNR Estimation Algorithm for Turbo Decoder," *IEEE Communications Letters*, Vol. 4, No. 8, Aug, 2000, pp. 255–257.

[36] J. P. Wodward and L. Hanzo, "Comparative Study of Turbo Decoding Techniques: An Overview," *IEEE Transactions on Vehicular Technology*, Vol. 49, No. 6, Nov. 2000, pp. 2208–2233.

[37] B. Sklar, "A Primer on Turbo Code Concepts," *IEEE Communications Magazine*, Dec. 1997, pp. 94–102.

[38] J. Hagenauer and P. Hoeher, "A Viterbi Algorithm with Soft-Decision Outputs and its Application," *IEEE Conference on Global Communications*, 1989, pp. 1680–1686.

[39] S. Benedetto and E. Biglieri, *Principles of Digital Transmission: with Wireless Applications*, Kluwer Academic, 1999, New York.

[40] D. Divsalar and F. Pollara, "Turbo Codes for Deep-Space Communications," TDA Progress Report 42-120, Feb. 1995, pp. 29–39.

[41] B. A. Bjerke, Z. Zvonar, and J. G. Proakis, "Antenna Diversity Combining Schemes for WCDMA Systems in Fading Multipath Channels," *IEEE Transactions on Wireless Communications*, Vol. 3, No. 1, Jan. 2004, pp. 97–106.

[42] M. Ventola, E. Tuomaala, and P. A. Ranta, "Performance of Dual Antenna Diversity Reception in WCDMA Terminals," *IEEE Vehicular Technology Conference*, 2003, pp. 1035–1040.

[43] H. S. Abdel-Ghaffar and S. Pasupathy, "Asymptotical Performance of M-ary and Binary Signals Over Multipath/Multichannel Rayleigh and Rician Fading," *IEEE Transactions on Communications*, Vol. 43, No. 11, Nov. 1995, pp. 2721–2731.

[44] T. T. Tjhung, A. Dong, F. Adachi, and K. H. Tan, "On Diversity Reception of Narrowband 16 STAR-QAM in Fast Rician Fading," *IEEE Transactions on Vehicular Technology*, Vol. 46, No. 4, Nov. 1997, pp. 923–932.

[45] P. Y. Kam, "Bit Error Probabilities of MDPSK Over the Nonselective Rayleigh Fading Channel with Diversity Reception," *IEEE Transactions on Communications*, Vol. 39, No. 2, Feb. 1991, pp. 220–224.

[46] F. Adachi, "Postdetection Optimal Diversity Combiner for DPSK Differential Detection," *IEEE Transactions on Vehicular Technology*, Vol. 42, No. 3, Aug. 1993, pp. 326–337.

[47] H. Kong, T. Eng, and L. B. Milstein, "A Selection Combining Scheme for Rake Receivers," *IEEE International Conference on Universal Personal Communications*, 1995, pp. 426–430.

[48] P. Balaban and J. Salz, "Dual Diversity Combining and Equalization in Digital Cellular Mobile Radio," *IEEE Transactions on Vehicular Technology*, Vol. 40, No. 2, May 1991, pp. 342–354.

[49] P. Balaban and J. Salz, "Optimum Diversity Combining and Equalization in Digital Data Transmission with Applications to Cellular Mobile Radio–Part I: Theoretical Considerations," *IEEE Transactions on Communications*, Vol. 40, No. 5, May 1992, pp. 885–894.

[50] P. Balaban and J. Salz, "Optimum Diversity Combining and Equalization in Digital Data Transmission with Applications to Cellular Mobile Radio–Part II: Numerical Results," *IEEE Transactions on Communications*, Vol. 40, No. 5, May 1992, pp. 895–907.

[51] T. Nihtila, J. Kurjenniemi, M. Lampinen, and T. Ristaniemi, "WCDMA HSDPA Network Performance with Receive Diversity and LMMSE Chip Equalization," *IEEE Conference on Personal, Indoor and Mobile Communications*, 2005, pp. 1245–1249.

[52] J. G. Proakis, "Channel Identification for High Speed Digital Communications," *IEEE Transactions on Automatic Control*, Vol. AC-19, No. 6, Dec. 1974, pp. 916–922.

[53] S. Qureshi, "Adaptive Equalization," *IEEE Communications Magazine*, March 1982, pp. 9–16.

[54] Y. Akaiwa and Y. Nagata, "Highly Efficient Digital Mobile Communications with a Linear Modulation Method," *IEEE Journal on Selected Areas in Communications*, Vol. SAC-5, No. 5, June 1987, pp. 890–895.

[55] W. C. Jakes, *Microwave Mobile Communications*, IEEE Press, 1993, New York.

[56] J. G. Proakis, *Digital Communications*, McGraw-Hill, 1989, New York.

[57] S. Chennakeshu and J. B. Anderson, "Error Rate for Rayleigh Fading Multichannel Reception for MPSK Signals," *IEEE Transactions on Communications*, Vol. 43, No. 2/3/4, Feb./Mar./April 1995, pp. 338–346.

[58] T. Eng, N. Kong, and L. B. Milstein, "Comparison of Diversity Combining Techniques for Rayleigh Fading Channels," *IEEE Transactions on Communications*, Vol. 44, No. 9, Sept. 1996, pp. 1117–1129.

[59] R. A. Monzingo & T. W. Miller, *Introduction to Adaptive Arrays*, John Wiley & Sons, 1980, New York.

[60] S. U. Pillai, *Array Signal Processing*, Springer-Verlag, 1989, New York.

[61] R. G. Vaughan, "On Optimum Combining at the Mobile," *IEEE Transactions on Vehicular Technology*, Vol. 37, No. 4, Nov. 1988, pp. 181–188.

[62] G. E. Bottomley, K. J. Molnar, and S. Chennakeshu, "Interference Cancellation with an Array Processing MLSE Receiver," *IEEE Transactions on Vehicular Technology*, Vol. 48, No. 5, Sept. 1999, pp. 1321–1331.

[63] J. H. Winters, "Optimum Combining in Digital Mobile Radio with Cochannel Interference," *IEEE Journal on Selected Areas in Communications*, Vol. SAC-2, No. 4, July 1984, pp. 528–539.

[64] M. W. Ganz, R. L. Moses, and S. L. Wilson, "Convergence of the SMI and the Diagonally Loaded SMI Algorithms with Weak Interference," *IEEE Transactions on Antennas and Propagation*, Vol. 38, No. 3, March 1990, pp. 394–399.

[65] Y. Wang and H. Scheving, "Adaptive Arrays for High Rate Data Communications," *IEEE Vehicular Technology Conference,* 1998, pp. 1029–1033.

[66] R. L. Cupo et al., "A Four Element Adaptive Antenna Array for IS-136 PCS Base Stations," *IEEE Vehicular Technology Conference*, 1997, pp. 1577–1581.

[67] P. Lancaster and M. Tismenetsky, *The Theory of Matrices*, Academic Press, 1985, California.

[68] E. Lindskog and C. Tidestav, "Reduced Rank Channel Estimation," *IEEE Vehicular Technology Conference*, 1999, pp. 1126–1130.

[69] J. R. Guerci and J. S. Bergin, "Principal Components, Covariance Matrix Tapers and the Subspace Leakage Problem," *IEEE Transactions on Aerospace and Electronic Systems*, Vol. 38, No. 1, Jan. 2002, pp. 152–162.

[70] K. I. Pedersen and P. E. Mogensen, "Performance Comparison of Vector-RAKE Receivers Using Different Combining Schemes and Antenna Array Topologies," *International Journal on Wireless Information Networks*, Vol. 6, No. 3, July 1999, pp. 181–194.

[71] X. Wu and A. M. Haimovich, "Adaptive Arrays for Increased Performance in Mobile Communications," *IEEE Symposium on Personal, Indoor and Mobile Communications (PIMRC) 1995*, pp. 653–657.

[72] J. Boccuzzi, S. U. Pillai, and J. H. Winters, "Adaptive Antenna Arrays Using Subspace Techniques in a Mobile Radio Environment with Flat Fading and CCI," *IEEE Vehicular Technology Conference*, 1999, pp. 50–54.

[73] J. H. Winters, "The Diversity Gain of Transmit Diversity in Wireless Systems with Raleigh Fading," *IEEE Transactions on Vehicular Technology*, Vol. 47, No. 1, Feb. 1998, pp. 119–123.

[74] R. T. Derryberry, S. D. Gray, D. M. Ionescu, G. Mandyam, and B. Raghothaman, "Transmit Diversity in 3G CDMA Systems," *IEEE Communications Magazine*, April 2002, pp. 68–75.

[75] J. Kurjenniemi, J. Leino, Y. Kaipainen, T. Ristaniemi, "Closed Loop Mode 1 Transmit Diversity with High Speed Downlink Packet Access," *Proceedings of International Conference on Communications Technology (ICCT) 2003*, pp. 757–761.

[76] P. Monogioudis et al., "Intelligent Antenna Solutions for UMTS: Algorithms and Simulation Results," *IEEE Communications Magazine*, Oct. 2004, pp. 28–39.

[77] M. Marques da Silva and A. Correia, "Space Time Coding Schemes for 4 or More Antennas," *IEEE Conference on Personal, Indoor and Mobile Communications*, 2002, pp. 85–89.

[78] S. Alamouti, "Space Block Coding: A Simple Transmitter Diversity Technique for Wireless Communications," *IEEE Journal on Selected Areas of Communications*, Vol. 16, Oct. 1998, pp. 1451–1458.

[79] S. Tanaka, T. Ihara, and M. Sawahashi, "Optimum Transmit Antenna Weight Generation Method for Adaptive Antenna Array Transmit Diversity in WCDMA Forward Link," *IEEE Vehicular Technology Conference*, 2001, pp. 2302–2306.

[80] S. Fukumoto, K. Higuchi, M. Sawahashi, and F. Adachi, "Field Experiments on Closed Loop Mode Transmit Diveristy in WCDMA Forward Link," *IEEE Conference on Spread Spectrum Techniques and Applications*, 2000, pp. 433–438.

[81] M. Canales, A. Valdovinos, J. R. Gallego, and F. Gutierrez, "Performance Analysis of Downlink Transmit Diversity System Applied to the UTRA FDD Mode," *IEEE Symposium on Spread Spectrum*, Sept. 2002, pp. 410–414.

[82] A. Hottinen, O. Tirkkonen, and R. Wichman, "Closed-Loop Transmit Diversity Techniques for Multi-element Transceivers," *IEEE Vehicular Technology Conference*, 2000, pp. 70–73.

[83] S. Parkvall et al. "Transmit Diversity in WCDMA: Link and System Level Results," *IEEE Vehicular Technology Conference*, 2000, pp. 864–868.

[84] A. F. Naguib, N. Seshadri, and A. R. Calderbank, "Space Time Coding and Signal Processing for High Data Rate Wireless Communications," *IEEE Signal Processing Magazine*, May 2000, pp. 76–92.

[85] H. L. Bertoni, *Radio Propagation for Modern Wireless Systems*, Prentice Hall, 2000, New Jersey.

CHAPTER 6
RECEIVER DIGITAL SIGNAL PROCESSING

In this chapter we will discuss various receiver digital signal processing functions. We begin with a discussion of equalization methods available assuming a single spatial antenna receiver. Then continue with a presentation of space-time equalization where joint equalization and antenna combining is performed for the purposes of improved performance. Due to an increased interest in the wireless community of operating in the frequency domain, we present an overview of frequency domain equalization (FDE). We then move on to symbol timing recovery (STR), channel quality estimation (CQE), and automatic frequency control (AFC). We conclude the chapter with an overall receiver block diagram to put the above mentioned signal processing algorithms into perspective.

6.1 TEMPORAL EQUALIZATION (EQ)

In channels with memory, such as multipath frequency selective fading channels, the received signal is composed of multiple-delayed versions (or echoes) of the transmitted signal. These delayed versions sum and subtract to result in Inter-Symbol Interference (ISI). This ISI cannot be eliminated by increasing the transmitter power or reducing the receiver noise figure. A method used to remove the ISI is through equalization. Equalizers make use of the "time diversity" of the received signal (as with RAKE receivers in CDMA systems) to extract the multiple-delayed echoes and form a combined symbol. Some commonly used methods are as follows:

Linear equalization
Decision feedback equalization
Maximum likelihood sequential estimation

These methods are shown in Fig. 6.1, where an attempt to provide an outline has been made.

6.1.1 Linear Equalizer

The equalizer structure shown in Fig. 6.2 is a Fractionally Spaced Equalizer (FSE), where the delay elements between the filter coefficients are a fraction of a baud time, T, apart [1]. This type of equalizer has been shown to be more robust against symbol-timing recovery errors. If the delay elements spacing are T seconds apart, then the equalizer is called a Symbol-Spaced Equalizer (SSE) or commonly called T-spaced equalization. The equalizer shown spans for T seconds, depending on the amount of ISI you wish to reduce; the equalizer temporal duration can increase.

The received signal, $r(k)$, enters the adaptive filter (shown below in FIR form) whose responsibility is to remove the ISI present in the signal. The adaptive filter output, $x(k)$, enters a decision device to produce a detected symbol $x'(k)$. There are two modes of operation: the first mode is called

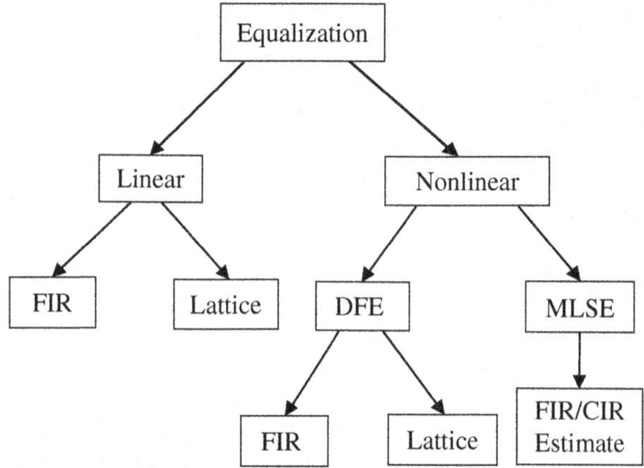

FIGURE 6.1 Outline of equalization methods.

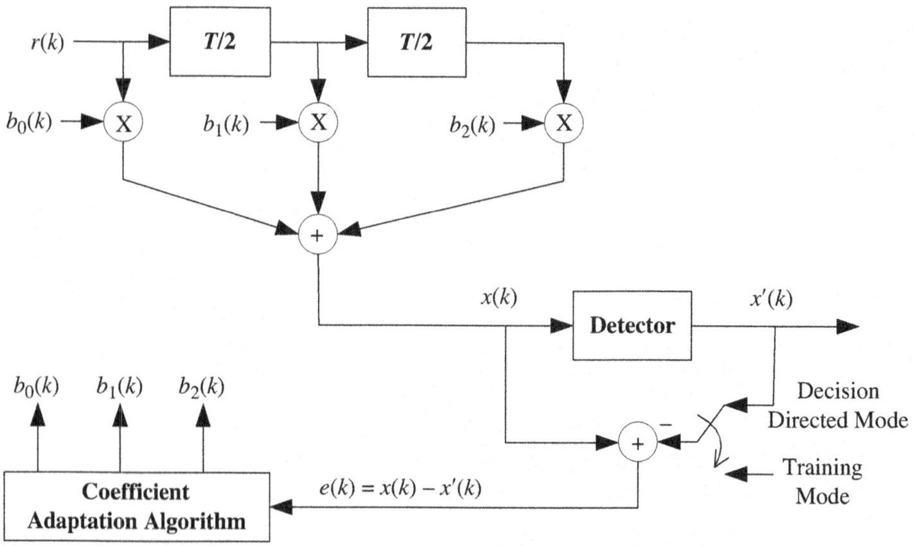

FIGURE 6.2 Linear equalizer structure.

decision directed mode while the second is called the training mode. In most wireless communication systems there is some information known a priori to the receiver such as pilot bits or a pilot channel in the case of WCDMA. This information is typically used to train the coefficients of the adaptive filter. Since the channel is time-varying, the adaptive filter coefficients should also be time-varying in order to track the temporal variations. This is accomplished by the decision-directed mode where detected symbols are used to update the coefficients. The update is accomplished by creating an error signal, $e(k)$. Hence the coefficient update algorithm aims to minimize the error between the adaptive filter and detector outputs.

The received signal is mathematically represented below assuming the channel response has a time duration of N samples.

$$r(k) = \sum_{j=0}^{N-1} h_j(k) \cdot d(k-j) \tag{6.1}$$

Which can be written in vector form as

$$r(k) = \underline{h}^T(k) \cdot \underline{d}(k) \tag{6.2}$$

where k denotes the time index, $\underline{h}$ is the channel vector of size $N \times 1$, and $\underline{d}$ is the desired signal vector of size $N \times 1$.

$$r(k) = [h_0(k) \;\; h_1(k) \;\; \cdots \;\; h_{N-1}(k)] \cdot \begin{bmatrix} d(k) \\ d(k-1) \\ \vdots \\ d(k-N+1) \end{bmatrix} \tag{6.3}$$

Next we can write the equalizer output assuming K taps are used. Here we collect K samples of the input sequence

$$\underline{r}(k) = \begin{bmatrix} h_0 & \cdots & h_{N-1} & 0 & \cdots & 0 \\ 0 & h_0 & \cdots & h_{N-1} & & 0 \\ 0 & & & & & \vdots \\ \vdots & & & \ddots & & 0 \\ 0 & 0 & & & & 0 \\ 0 & \cdots & 0 & h_0 & \cdots & h_{N-1} \end{bmatrix} \cdot \begin{bmatrix} d(k) \\ d(k-1) \\ \\ \vdots \\ \\ d(k-N-K+1) \end{bmatrix} \tag{6.4}$$

$$x(k) = \underline{b}^T(k) \cdot \underline{r}(k) \tag{6.5}$$

$$x(k) = \underline{b}^T(k) \cdot H(k) \cdot \underline{\tilde{d}}(k) \tag{6.6}$$

where the overall channel matrix, H is of size $K \times (K+N)$ and the desired signal vector, $\underline{\tilde{d}}$ is of size $(K+N) \times 1$.

The well-known significant shortcoming of this equalizer is the noise amplifying property in the channels that contain spectral nulls [1]. This problem is exacerbated when using the zero forcing (ZF) criteria and eliminated for the Minimum Mean Squared Error (MMSE). The advantage of this linear equalization technique is that it is very simple to implement and control.

6.1.2 Decision Feedback Equalizer (DFE)

The equalizer structure shown in Fig. 6.3 consists of two filters: a Feed Forward Filter (FFF) and a Feed Back Filter (FBF). The FFF performs similar to the linear equalizer discussed above, it has as its input the received samples. The FBF has as its input the past detected symbols that are feedback. The purpose of the FBF is to remove that part of the ISI from the present estimate caused by the previously detected symbols. Typically, the FFF is in fractionally spaced form and the FBF is T spaced; this combination helps achieve symbol timing recovery insensitivity.

The DFE structure is sometimes referenced as DFE (p, q), where p is the number of FFF taps and q is the number of FBF taps. The DFE also has two modes of operation: decision directed and training. They have the same functions as previously described above. The equalizer output is represented as

$$x(k) = \sum_{j=0}^{N_{FF}-1} b_j(k) \cdot r(k-j) - \sum_{j=0}^{N_{FB}-1} a_j(k) \cdot x'(k-j-1) \tag{6.7}$$

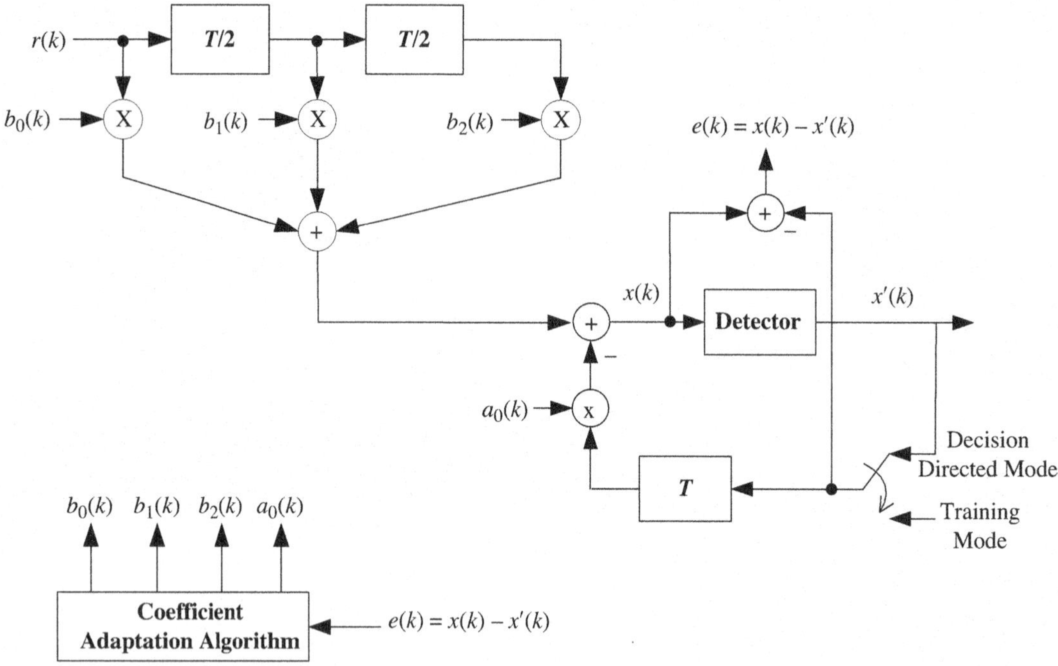

FIGURE 6.3 Decision feedback equalizer structure.

where N_{FF} is the number of feed forward taps and N_{FB} is the number of feed back taps. The DFE output can also be represented in vector notation as

$$x(k) = \underline{b}^T(k) \cdot \underline{r}(k) - \underline{a}^T(k) \cdot \underline{x}'(k-1) \tag{6.8}$$

where $\underline{b}(k)$ is the time varying weight vector of the feed forward section and $\underline{a}(k)$ is the time varying weight vector of the feed back DFE section.

The DFE weights can be optimized in a variety of ways. Approaching this solution from the aspect of creating a MMSE solution by matrix manipulations; the feed forward and feed back weights can be "stacked" in the calculation. This MMSE approach will be discussed in more detail in the following sections. Alternatively each adaptive filter can be separately controlled to arrive at the weight solution.

A block diagram of the DFE equalizer is shown in Fig. 6.3. Using our nomenclature the example structure is called DFE (3, 1). We have chosen to emphasize the FFF with a fractionally spaced structure and the FBF with a symbol spaced structure. As shown in the equalizer outline chart, this type of equalizer is nonlinear in that it contains a feedback component originating from the detector (or decisions device) output.

Bidirectional Equalization. In this section, we will provide some general insight into the training mode of equalization. For time division multiple access (TDMA) systems, a synchronization word is transmitted for a number of reasons—one use is to train the equalizer. Similarly, CDMA systems transmit not only dedicated pilot bits but also pilot channels which serve the same purpose. Since pilot channels are continuously transmitted, the equalizer can essentially be placed into training mode. For sake of this discussion let's assume the following time slot structure (see Fig. 6.4).

The typical procedure is to use a priori information such as pilot bits to train the equalizer and then switch into decision directed mode. The next time slot boundary the mode is switched back into training mode and this procedure starts over again. We call this unidirectional since the adaptive equalizer operation traverses the same direction throughout its operation.

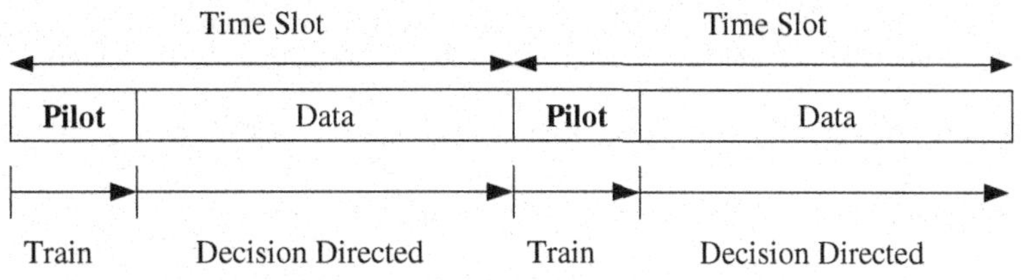

FIGURE 6.4 Unidirectional training.

Depending on the wireless system of interest this a priori information is continually transmitted independent of the number of users in the system. It is important to make total use of the system in order to capitalize on receiver performance. With this assumption the receiver can utilize the pilot information in the adjacent time slots. And then train the equalizer in the reverse direction. This is shown in Fig. 6.5 by the dashed lines. This form of equalization is called bidirectional since the

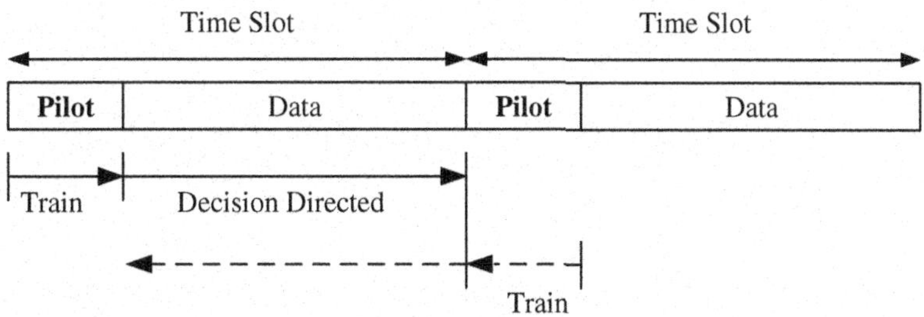

FIGURE 6.5 Bidirectional training.

adaptive weights are being updated from both directions in time. We have drawn the decision directed mode to run for the duration of the data portion of the time slot, it is up to the system designer what to do with the two groups of data available.

One approach to decide when to engage in bidirectional equalization is to use the received signal strength to help. Figure 6.6 shows a receiver operating under the conditions if the received signal

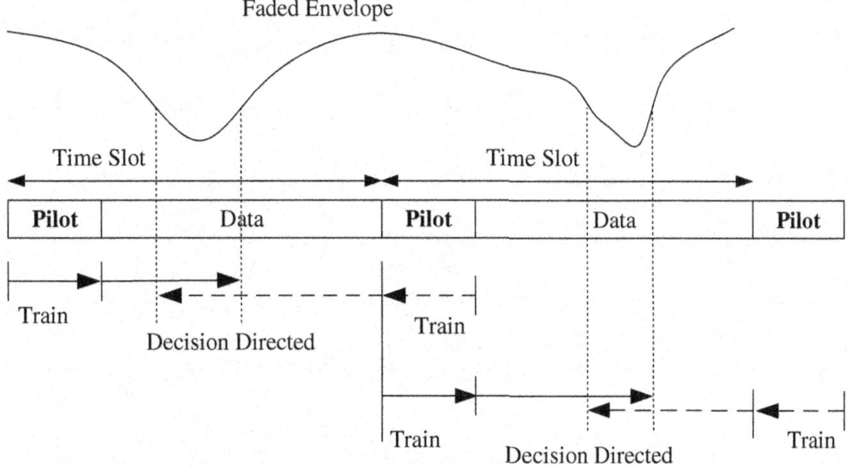

FIGURE 6.6 Bidirectional equalization example.

strength falls below a certain threshold, the adaptive signal processing is halted and the data is stored until the next time slot arrives. When the next pilot bits arrive the training sequence begins, however in the opposite direction. This is shown below for two consecutive time slots using three pilot bit fields. Once again the dashed lines are used to denote the signal processing performed in the reverse direction.

6.1.3 Maximal Likelihood Sequential Estimation (MLSE)

This next form of estimation moves away from the symbol-by-symbol-based detectors presented above; here a sequence of transmitted bits are estimated. In order to proceed, we will assume the received signal is represented as follows:

$$r(k) = \sum_{j=0}^{L} h(j) \cdot s(k - j) + n(k) \tag{6.9}$$

The MLSE makes a decision in favor of the sequence I_k, where $k = 1, 2, \ldots, k$, that maximizes the joint conditional probability density function given as follows:

$$P(r_k, r_{k-1}, \cdots, r_1 | I_k, I_{k-1}, \cdots, I_1) \tag{6.10}$$

Which states the probability that the received sequence, r, is correct given I was the transmitted sequence. Since the noise samples are assumed to be independent and the received signal depends on the most recent L transmitted symbols (due to the ISI of the channel), we have the following relevant equation.

$$P(r_k, r_{k-1}, \cdots, r_1 | I_k, I_{k-1}, \cdots, I_1) = P(r_k | I_k, \cdots, I_{k-L}) \cdot P(r_{k-1}, \cdots, r_1 | I_{k-1}, \cdots, I_1) \tag{6.11}$$

Using this notation we notice only the first term on the RHS needs to be computed for each incoming symbol, r_k. The most often used approach to solve this problem is to use the Viterbi algorithm, which was previously shown to be based on metric calculations. The metric can be iteratively derived making use of the following equality.

$$Ln[P(r_k, r_{k-1}, \cdots, r_1 | I_k, I_{k-1} \cdots, I_1)] = Ln[P(r_k | I_k, \cdots, I_{k-L})] + Ln[P(r_{k-1}, \cdots, r_1 | I_{k-1}, \cdots, I_1)] \tag{6.12}$$

This recursively calculates the joint conditional probability density function, where the first term on the RHS is the incremental metric and the second term is the previously calculated metric. This form of equalization estimates the sequence of transmitted symbols that were most likely (in the maximum likelihood sense) transmitted.

Next we define the incremental metric, $M(k)$, to be used in the above MLSE receiver. We begin with the following likelihood function (written in complex vector notation).

$$p[n(k)] = \frac{1}{\pi^D |R_{nn}(k)|} \cdot e^{-\underline{n}^H(k) \cdot R_{nn}^{-1}(k) \cdot \underline{n}(k)} \tag{6.13}$$

Here H denotes conjugate transpose and the R_{nn} matrix is of size $D \times D$.

After some mathematical manipulations we have the following branch metric.

$$M(k) = \left[r(k) - \sum_{j=0}^{L} h(j) \cdot s(k - j) \right]^H \cdot R_{nn}^{-1}(k) \cdot \left[r(k) - \sum_{j=0}^{L} h(j) \cdot s(k - j) \right] \tag{6.14}$$

We can make some simplifications such as assuming the noise covariance matrix is a diagonal with constant values, then the following metric is formed.

$$M(k) = \left[r(k) - \sum_{j=0}^{L} h(j) \cdot s(k - j) \right]^2 \tag{6.15}$$

Before we pull together the above equations into a form used to produce a receiver block diagram, we would like to point out that certain values need to be estimated within the metric value itself prior to being used in the joint conditional probability equation to determine the sequence of bits that were most likely transmitted. First, the channel response needs to be estimated, we will provide a commonly accepted method below. Second, the transmitted bits need to be exhaustively searched in order to calculate all the metric values. This will be shown below where the iterative metric equation provided earlier is rewritten in another form.

$$J_{k-L}(I_k) = J_{k-L-1}(I_{k-1}) + \left[r(k) - \sum_{j=0}^{L} \hat{h}(j) \cdot \hat{s}(k-j) \right]^2 \tag{6.16}$$

where $J(I_k)$ = accumulated metric using the hypothesis sequence I_k at the kth time instance. $\hat{h}(j)$ is the estimated channel value and $\hat{s}(k)$ is the locally generated modulated symbols (or hypothesis) that is exhaustively searched to find the sequence of symbols that were most likely transmitted.

In an effort to provide some insight into the MLSE receiver, let us consider the following equivalent communication system model provided in Fig. 6.7. Here the modulated symbols enter an FSF channel model whose echoes are spaced T seconds apart and where AWGN is inserted into the signal.

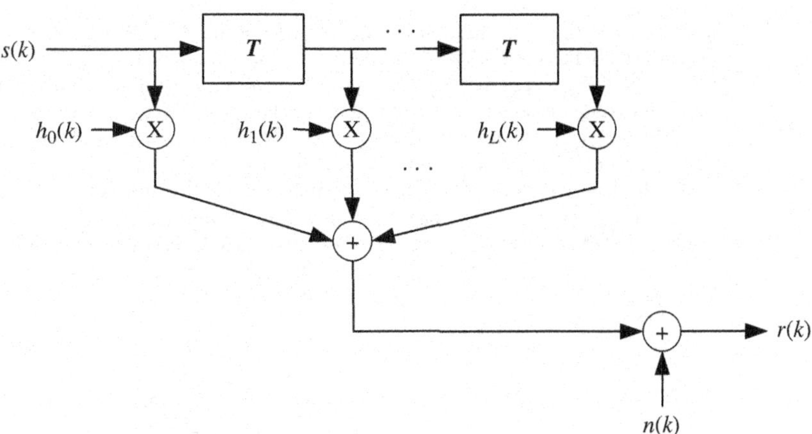

FIGURE 6.7 Equivalent communication model.

It is clear that we can rewrite the received signal to consist of the desired signal plus the interference signal plus noise.

$$r(k) = s(k) \cdot h(0) + \sum_{j=1}^{L} h(j) \cdot s(k-j) + n(k) \tag{6.17}$$

Now we can draw the MLSE receiver structure shown in Fig. 6.8. Notice in the figure we have included a channel estimation block labeled channel impulse estimator. As we previously have shown, the Viterbi algorithm (VA) requires estimates of the channel response in order to compute the incremental metric values. These channel estimates, $\hat{H}(k)$, are derived in the channel response estimator. This estimation requires decisions to have been made so accurate estimates of the channel response can be made. However, reliable decisions cannot occur without accurate channel estimates. Hence we are faced with a dilemma. A favorite solution is to have the VA output tentative decisions in order to start up the channel estimates and then move toward more reliable estimates as time progresses. We have used the variable D to denote the tentative decision delay. Note that typical applications acquire an initial estimate of the channel response to jump start the VA.

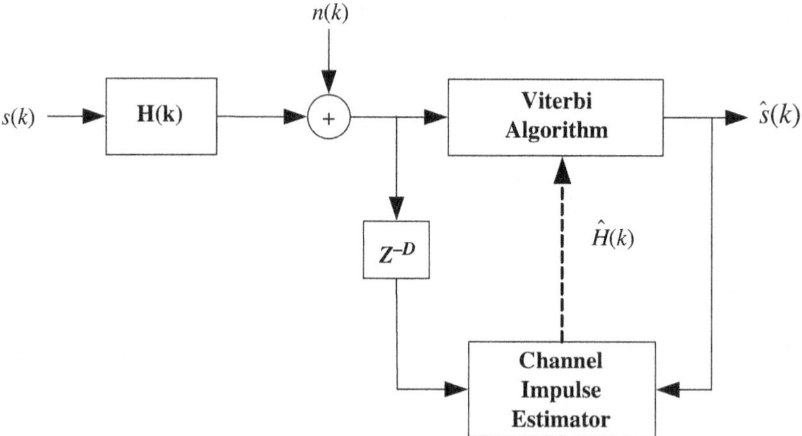

FIGURE 6.8 MLSE receiver structure.

Let's provide a short discussion on the impact of the value D on system performance. If the value D is too small, the symbol decisions will be unreliable and the Channel Impulse Response (CIR) estimator will diverge from the true channel due to error propagation. On one hand, we would prefer D to be small in order to obtain more relevant channel estimates from the channel response estimator; however, since they would be unreliable so will the channel estimate itself. On the other hand, if D is too large, then the CIR estimate will no longer represent the current channel conditions, but the VA decisions will be more reliable. As one can see there are two conditions that should be jointly optimized.

A block diagram for the Channel Response Estimator used to derive the channel estimates to be used by the VA is given in Fig. 6.9. The tentative data decisions and the received signal are inputs to

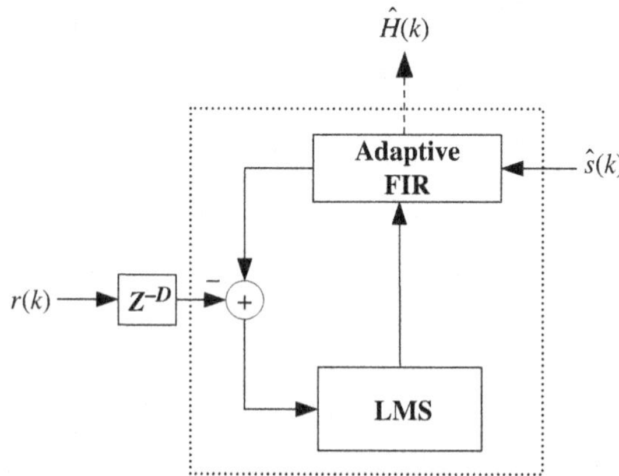

FIGURE 6.9 Channel Response Estimator.

this function. As the VA output decisions are made the channel response is updated according to the Least Mean Square (LMS) algorithm. This will allow us to track time-varying phenomena. The estimates of the channel are output from this function. The channel estimates shown are updated using the LMS algorithm, but nothing precludes the system designer from using other adaptation algorithms (such as RLS, and so on.).

Lastly, we will discuss a few key design items. First, since the VA is used one should be careful of the complexity involved in the metric candidates, since the number of states now depends not only on the number of dominant ISI terms (length of delay spread) but also on the modulation levels of the transmitted signal. The equation used to describe the number of VA states is given below.

$$\text{Number of States} = M^{L-1} \tag{6.18}$$

This equation is similar to that used to describe the number of states in the VA used to decode convolutional codes. In that Forward Error Correction (FEC) application the bits took on a binary form and hence $M = 2$. Also the number of states was dependent on the constraint length of the convolutional code. Recall the constraint length was defined as the number of input bits used to produce an output-coded bit stream. In this MLSE application, the length of the channel ISI (or delay spread) would resemble the constraint length of the convolutional code [2]. The VA implementation of the MLSE is optimum in the sense is it the maximum likelihood (ML) estimator of the "entire" received sequence.

6.1.4 Coefficient Adaptation Techniques

In this section, we will briefly review three adaptation algorithms typically used to update the equalizer coefficients. These algorithms form the basis from which other techniques are based upon [3–5].

Least Mean Square. The first coefficient update technique is commonly used and is called the Least Mean Square. In order to optimize the adaptive filter design, we choose to minimize the mean square value of the estimation error, $e(k)$, using the following cost function.

$$J_{\min} = E\{|e(k)|^2\} = E\{|d(k) - y(k)|^2\} \tag{6.19}$$

where the error signal is defined as follows with $w(k)$ being the adaptive filter with M coefficients.

$$e(k) = d(k) - \sum_{j=0}^{M-1} w(j) \cdot r(k-j) \tag{6.20}$$

The common approach used to solve this problem is to create a gradient vector, set it equal to zero to obtain the steady state solution. This leads to the famous Wiener-Hopf or optimum Wiener solution given as

$$\underline{w}_{\text{opt}} = R^{-1} \cdot \underline{r} \tag{6.21}$$

This solution involves inverting a matrix, which can be complex. A way to avoid this inversion is to use the Steepest Descent[1] method to create an iterative method and provided below for sake of reference

$$\underline{w}(k+1) = \underline{w}(k) + \mu \cdot \{\underline{R}_{rd} - R_{rr} \cdot \underline{w}(k)\} \tag{6.22}$$

This method applies successive corrections in the direction of the negative of the gradient vector to arrive at the optimum weights. Using this equation we can substitute the corresponding correlation values and write the LMS update equation as follows:

$$\underline{w}(k+1) = \underline{w}(k) + \mu \cdot \{\underline{r}(k) \cdot d^*(k) - \underline{r}(k) \cdot \underline{r}^H(k) \cdot \underline{w}(k)\} \tag{6.23}$$

$$\underline{w}(k+1) = \underline{w}(k) + \mu \cdot \underline{r}(k) \cdot e^*(k) \tag{6.24}$$

where $\underline{w}(k)$ is the coefficient weight vector of the adaptive filter at time instance k, μ is the step size, $\underline{r}(k)$ is the equalizer input vector, and $e(k)$ is the error signal representing the difference between the estimate and the true value.

---

[1]Steepest Descent is a method which iteratively searches a multidimensional space.

The step size controls the size of the incremental correction applied to the weight vector. A general rule-of-thumb is that a large step size provides a fast initial convergence, but a large excess noise variation in steady state. While a small step size provides a slow initial convergence, but a small excess noise variation in steady state. Having a variable step size seems to provide a balance of the behaviors.

Additionally, the correction applied to the filter weights at time $k + 1$ is directly proportional to the input vector $\underline{r}(k)$. So when the input level is high the LMS algorithm experiences a gradient noise amplification. A method used to overcome this issue is the Normalized LMS (NLMS). This approach places a constraint to limit the variability of the input signal and results in the following iterative solution.[2]

$$\underline{w}(k + 1) = \underline{w}(k) + \frac{\mu}{\|\underline{r}(k)\|^2} \cdot \underline{r}(k) \cdot e^*(k) \tag{6.25}$$

This can be viewed as LMS with a time-varying step size.

Recursive Least Squares. This second coefficient update technique is called the Recursive Least Squares (RLS). Here we choose to minimize a cost function which is defined as the sum of error squared or least squares. An exponentially weighted Least Squares Cost function is defined below. Where λ is the weighting factor or forgetting factor with constraint $(0 < \lambda < 1)$.

$$J(k) = \sum_{i=1}^{k} \lambda^{k-i} \cdot |e(i)|^2 \tag{6.26}$$

The familiar matrix form solution is

$$\underline{w}(k) = \tilde{R}_{rr}^{-1}(k) \cdot \underline{\tilde{r}}_{rd}(k) \tag{6.27}$$

The recursive updates can be given as

$$\tilde{R}_{rr}(k) = \lambda \cdot \tilde{R}_{rr}(k - 1) + \underline{r}(k) \cdot \underline{r}^*(k) \tag{6.28}$$

$$\underline{\tilde{r}}_{xd}(k) = \lambda \cdot \underline{\tilde{r}}_{xd}(k - 1) + d^*(k) \cdot \underline{r}(k) \tag{6.29}$$

The above correlation matrix and vector are, in fact, the exponentially weighted versions of the traditional block processing solution. After rewriting the cost-function solution in a recursive manner and applying the Matrix Inversion Lemma we arrive with the following set of equations. The weight vector is iteratively calculated as follows:

$$\underline{w}(k) = \underline{w}(k - 1) + \underline{C}(k) \cdot e^*(k) \tag{6.30}$$

where the error signal is defined below

$$e(k) = d(k) - \underline{w}^H(k - 1) \cdot \underline{r}(k) \tag{6.31}$$

The time-varying vector update can be recursively computed using the following two terms.

$$\underline{C}(k) = \frac{\lambda^{-1} \cdot P(k - 1) \cdot \underline{r}(k)}{1 + \lambda^{-1} \cdot \underline{r}^H(k) \cdot P(k - 1) \cdot \underline{r}(k)} \tag{6.32}$$

$$P(k) = \lambda^{-1} \cdot P(k - 1) - \lambda^{-1} \cdot \underline{C}(k) \cdot \underline{r}^H(k) \cdot P(k - 1) \tag{6.33}$$

With this RLS approach we have essentially replaced the inversion of the correlation matrix by scalar division. Many studies have been performed that compares the tracking and steady state performance of not only LMS to RLS, but also their variations as well. We leave this up to the interested reader to consult the references [2][3].

---

[2] $\|\underline{x}\|$ is the Euclidian norm of the vector $\underline{x}$.

Direct Matrix Inversion. This third coefficient update technique is commonly called Direct Matrix Inversion (DMI) and sometimes referred to as Sample Matrix Inversion (SMI). This approach works directly on the optimum filter coefficients by inverting the correlation matrix. Depending on the particular equalizer structure chosen, that is, fractionally spaced or symbol spaced, the correlation matrix can take on familiar forms such as Toeplitz, and so on. These familiar forms can be taken advantage of when inverting the correlation matrix. There are various sources available to the system designer to aid in matrix inversion [7][8].

The optimum weights were given earlier as $\underline{w}_{\text{opt}} = R_{XX}^{-1} \cdot \underline{r}_{Xd}$. In practice, these weights need to be estimated, a typical method is given as follows. Here we have used a single-sided windowing technique to create estimates. If the application allows, this window can be shifted in time (for two-sided windowing) to produce more accurate estimates (i.e., noncausal signal processing).

$$\hat{R}_{XX}(t) = \frac{1}{K} \cdot \sum_{j=0}^{K-1} \underline{x}(t-j) \cdot \underline{x}^*(t-j) \tag{6.34}$$

$$\hat{\underline{r}}_{Xd}(t) = \frac{1}{K} \cdot \sum_{j=0}^{K-1} \underline{x}(t-j) \cdot d^*(t-j) \tag{6.35}$$

Below we will provide a solution to a Hermitian matrix inversion using the famous Eigen Spectra Decomposition (ESD) theorem [7]. Let us start with a covariance matrix defined as follows:

$$R_{XX} = E\{\underline{x} \cdot \underline{x}^*\} \tag{6.36}$$

This matrix is Normal and unitarily diagonalizable. Moreover, the diagonal elements of the diagonalized matrix are indeed the eigenvalues of the matrix itself. In other words, we have the following notation:

$$R_{XX} = U \cdot D \cdot U^* \tag{6.37}$$

where U is the Unitary matrix whose column vectors are the respective eigenvectors of the covariance matrix, D is the diagonal matrix whose elements along the diagonal are the associated eigenvalues. It is well known that the total vector space can be divided into subspaces, specifically signal and nonsignal (i.e., noise). This can be easily written as shown below.

$$R_{XX} = U_S \cdot D_S \cdot U_S^* + U_N \cdot D_N \cdot U_N^* \tag{6.38}$$

where D_S is the diagonal matrix corresponding to the signal subspace and D_N corresponds to the noise subspace. U_S is a matrix representation of the signal subspace where the columns consist of the signal eigenvectors. U_N is a matrix representation of the noise subspace where the columns consist of the noise eigenvectors.

The subspaces are related by the following identity.

$$I = U_S \cdot U_S^* + U_N \cdot U_N^* \tag{6.39}$$

Also the subspaces possess the following property: the signal eigenvectors are orthogonal to the noise subspace, which can be written as

$$U_S^* \cdot U_N = 0 \tag{6.40}$$

We will make use of this property when we discuss the Eigen Canceller in the following chapters. The Eigen Canceller weights become equal to [8]

$$\underline{w}_{EC} = [I - U_S \cdot U_S^*] \cdot \underline{r}_{xd} \tag{6.41}$$

With this information the ESD theorem allows us to write the matrix in the following manner (M is the size of the matrix).

$$R_{XX} = \sum_{i=1}^{M} \lambda_i \cdot \underline{v}_i \cdot \underline{v}_i^* \tag{6.42}$$

where λ_i is the ith eigenvalue of the covariance matrix and $\underline{v}_i$ is the ith eigenvector of R_{XX} associated with the ith eigenvalue.

The proposed solution is to then calculate the eigenvalues and eigenvectors of the matrix so the matrix inversion can be simply written as follows:

$$R_{XX}^{-1} = \sum_{i=1}^{M} \frac{1}{\lambda_i} \cdot \underline{v}_i \cdot \underline{v}_i^* \tag{6.43}$$

Next we can write the optimum weight vector as

$$\underline{w}_{\text{opt}} = \sum_{j=1}^{M} \frac{1}{\lambda_j} \cdot \underline{v}_j \cdot \underline{v}_j^* \cdot \underline{r}_{Xd} \tag{6.44}$$

which can also be written as

$$\underline{w}_{\text{opt}} = U \cdot D^{-1} \cdot U^* \cdot \underline{r}_{xd} \tag{6.45}$$

Now if we assume the cross-correlation vector is in the signal subspace then we can ideally write the optimum weights as

$$\underline{w}_{\text{opt}} = U_S \cdot D_S^{-1} \cdot U_S^* \cdot \underline{r}_{xd} \tag{6.46}$$

Here we have essentially cancelled the noise subspace.

6.2 SPACE-TIME EQUALIZATION (STE)

In the previous sections, we have assumed a single receive antenna was used with the equalization methods presented. In this section, we perform a joint equalizer in both time and across antennas, hence the name space-time equalization. Figure 6.10 presents a space-time equalizer block diagram assuming $M = 2$ antennas, with each antenna supporting a linear equalizer of length K taps. We have chosen to show a fractionally spaced version for sake of discussion. The equalizer weights are jointly derived from information from both antennas [9–15]. We have assumed the square root raised cosine

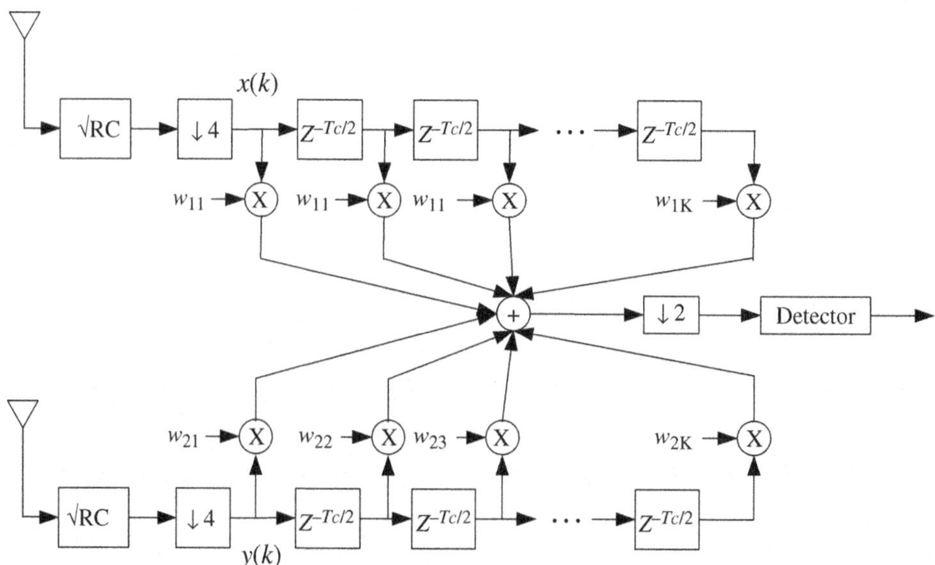

FIGURE 6.10 Fractionally spaced, space-time equalizer architecture with $M = 2$ antennas and K taps on each antenna.

filter is operating with a sampling frequency of eight times the symbol rate. This explains the decimation by a factor of four to produce half-symbol rate information. Once the equalization and combining operations have completed, the STE output is decimated by a factor of two to enter symbols to the detector chosen. This assumes the detector is a symbol rate-based detector otherwise the decimation operation is not needed.

After collecting K consecutive samples on both antennas we can create the following column vectors.

$$\underline{x}(k) = \begin{bmatrix} x(k) \\ x(k - T/2) \\ \vdots \\ x(k - (K - 1)T/2) \end{bmatrix} \qquad \underline{y}(k) = \begin{bmatrix} y(k) \\ y(k - T/2) \\ \vdots \\ y(k - (K - 1)T/2) \end{bmatrix} \tag{6.47}$$

Recall the MMSE weight calculation involves a covariance matrix and a correlation vector. The covariance matrix can be obtained by stacking the received samples from both antennas into a single column vector which results in the following $MK \times MK$ square matrix.

$$R = E\left\{\begin{bmatrix} \underline{x}(k) \\ \underline{y}(k) \end{bmatrix} \cdot \begin{bmatrix} \underline{x}(k) \\ \underline{y}(k) \end{bmatrix}^*\right\} \tag{6.48}$$

which is also written in block matrix form assuming $M = 2$ receive antennas.

$$R = \begin{bmatrix} R_{xx} & R_{xy} \\ R_{xy}^* & R_{yy} \end{bmatrix} \tag{6.49}$$

We will show in the following sections that the covariance matrix can be written in the simplest terms below.

$$R = H + \sigma_n^2 I \tag{6.50}$$

where H is the covariance matrix of the overall channel response and σ_n^2 is the noise power. Thus the MMSE-based weights are written as

$$\underline{w}_{\text{MMSE}} = R^{-1} \cdot \underline{r}_{xd} \tag{6.51}$$

where $\underline{r}_{xd}$ is the cross-correlation vector between the received signal and the reference (desired) signal, denoted as d.

In the sections that follow, we vary the channel conditions gradually creating a more and more complex covariance matrix, along the way we provide insight into the components that make up the covariance matrix and correlation vector. We will emphasize the DMI method to calculate the STE weights. More specifically we will utilize the ESD of the covariance matrix to provide insight into the STE weights. This will be accomplished using the following spatio-temporal signal processing receiver, shown with $M = 2$ antennas (see Fig. 6.11).

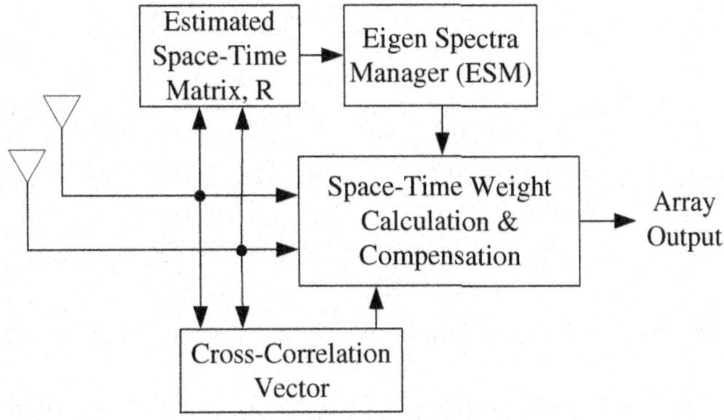

FIGURE 6.11 Unified eigenbased, spatio-temporal signal processing block diagram.

The proposed receiver continually estimates the eigenspectra of the received signal and uses a priori information (obtained through extensive simulation and field trial campaigns) to adjust the STE weights. Since the wireless environment is time varying (i.e., delay spread, interference, and so on) the STE weights should also adapt to the changing environment. The ESM collects the eigenspectrum, operates on the subspaces and creates a structured covariance matrix. The term structured is used here to emphasize that the estimated matrix is manipulated to contain the desired properties.

In the following sections, we will present certain statistics of the space-time equalizer weights. Specifically, the eigenspectra will be presented for a given channel condition. These results will show that reduced rank signal processing techniques can perform better than the full rank techniques, depending on the channel conditions.

6.2.1 Flat Fading Environment

In this section, we will derive the STE weights and present some simulation results. In doing so, we hope to provide some insight into the effects of the environment on the receiver. This section will consider a flat fading channel without the presence of interference. The signal at each of the two receive antennas is given below, assuming independent fading on each antenna and $s(k)$ representing the desired signal.

$$x(k) = s(k) \cdot h_x(k) + n_x(k) \tag{6.52}$$

$$y(k) = s(k) \cdot h_y(k) + n_y(k) \tag{6.53}$$

The signal-to noise-ratio on each antenna is given as

$$\text{SNR} = \frac{E\{|s(k) \cdot h_x(k)|^2\}}{E\{|n_x(k)|^2\}} = \frac{\sigma_s^2}{\sigma_n^2} \cdot E\{|h_x(k)|^2\} = \frac{\sigma_s^2}{\sigma_n^2} \cdot E\{|h_y(k)|^2\} \tag{6.54}$$

In order to provide some insight, let us fix the dimension of the problem such that there are two receive antennas, $M = 2$, and there are three taps in the linear equalizer, $K = 3$. With this we can write the covariance block matrix as

$$R_{xx} = \begin{bmatrix} \sigma_s^2 |h_x(k)|^2 & R_d\left(\frac{T}{2}\right) \cdot h_x(k) \cdot \overline{h}_x\left(k - \frac{T}{2}\right) & R_d(T) \cdot h_x(k) \cdot \overline{h}_x(k - T) \\ \overline{R}_d\left(\frac{T}{2}\right) \cdot \overline{h}_x(k) \cdot h_x\left(k - \frac{T}{2}\right) & \sigma_s^2 \left|h_x\left(k - \frac{T}{2}\right)\right|^2 & R_d\left(\frac{T}{2}\right) \cdot h_x\left(k - \frac{T}{2}\right) \cdot \overline{h}_x(k - T) \\ \overline{R}_d(T) \cdot \overline{h}_x(k) \cdot h_x(k - T) & \overline{R}_d\left(\frac{T}{2}\right) \cdot \overline{h}_x\left(k - \frac{T}{2}\right) \cdot h_x(k - T) & \sigma_s^2 |h_x(k - T)|^2 \end{bmatrix} + \sigma_n^2 I \tag{6.55}$$

where $R_d(k)$ represents the autocorrelation of the pulse-shaped desired signal. And the cross-covariance block matrix as

$$R_{xy} = \begin{bmatrix} \sigma_s^2 h_x(k) \cdot \overline{h}_y(k) & R_d\left(\frac{T}{2}\right) \cdot h_x(k) \cdot \overline{h}_y\left(k - \frac{T}{2}\right) & R_d(T) \cdot h_x(k) \cdot \overline{h}_y(k - T) \\ \overline{R}_d\left(\frac{T}{2}\right) \cdot h_x\left(k - \frac{T}{2}\right) \cdot \overline{h}_y(k) & \sigma_s^2 h_x\left(t - \frac{T}{2}\right) \cdot \overline{h}_y\left(k - \frac{T}{2}\right) & R_d\left(\frac{T}{2}\right) \cdot h_x\left(k - \frac{T}{2}\right) \cdot \overline{h}_y(k - T) \\ \overline{R}_d(T) \cdot h_x(k - T) \cdot \overline{h}_y(k) & \overline{R}_d\left(\frac{T}{2}\right) \cdot h_x(k - T) \cdot \overline{h}_y\left(k - \frac{T}{2}\right) & \sigma_s^2 h_x(k - T) \cdot \overline{h}_y(k - T) \end{bmatrix} \tag{6.56}$$

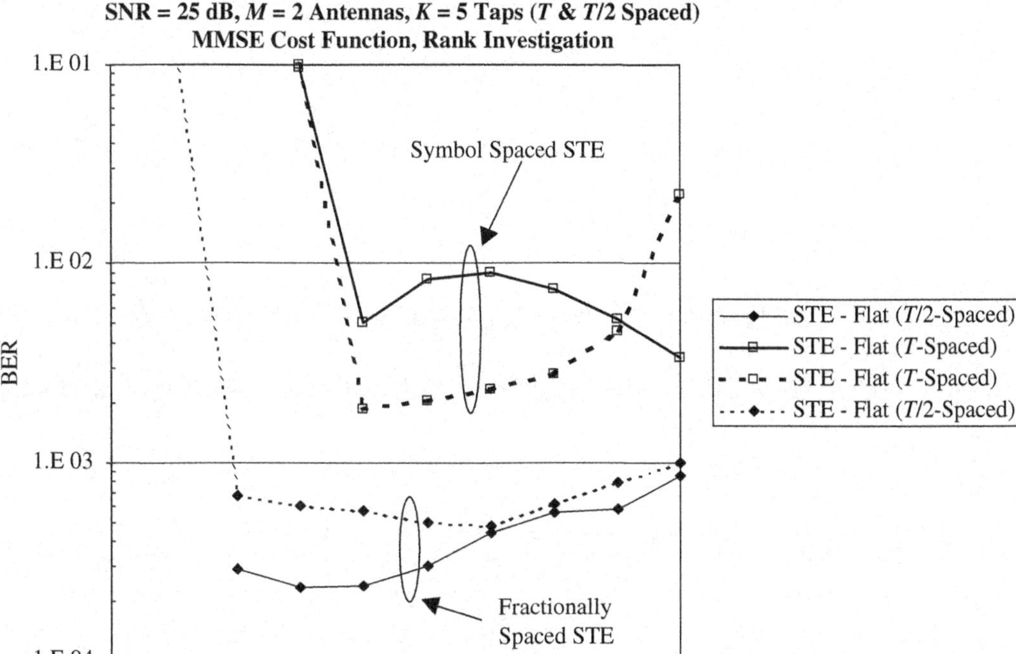

FIGURE 6.12 $M = 2$ antennas and $K = 5$ taps STE performance for various estimation window size lengths.

In Fig. 6.12, we plot the symbol and fractionally spaced STE ($K = 5$ taps) performance versus the rank of the covariance matrix or the signal subspace. First, we see that the fractionally spaced STE performed better than the symbol spaced STE. Next we see that the rank of the symbol-spaced STE covariance matrix is equal to 5, while the rank of the fractionally spaced covariance matrix is equal to 3. This is due to the fact as the taps become closer together in time, the matrix becomes more coherent, and the rank now becomes dependent on the channel conditions rather than the STE architecture chosen. It can be easily shown that the rank for the symbol-spaced covariance matrix is equal to the number of taps on each antenna. However; for the fractionally spaced STE, the overall channel response dictates the covariance matrix rank, which in this case is 3 due to the overall pulse-shaping and flat-fading channel model investigated.

We have decided to plot the BER performance versus the covariance matrix rank for the purposes of studying the possibility of reduced rank signal processing. The curves show the covariance matrix rank is less than full and hence reduced rank opportunities exist.

6.2.2 Frequency Selective Fading Environment

In this next section, we introduce a two-ray channel model to address the presence of delay spread. The received signals at both antennas are given as

$$x(k) = s(k) \cdot h_{1x}(k) + s(k - T) \cdot h_{2x}(k - T) + n_x(k) \tag{6.57}$$

$$y(k) = s(k) \cdot h_{1y}(k) + s(k - T) \cdot h_{2y}(k - T) + n_y(k) \tag{6.58}$$

With this channel environment, we can write the covariance block matrix as follows:

$$R_{xx} = \begin{bmatrix} R_d(0)|h_1(k)|^2 & R_d\left(\frac{T}{2}\right)h_1(k)\overline{h}_1\left(k-\frac{T}{2}\right) & R_d(T)h_1(k)\overline{h}_1(k-T) \\ R_d\left(-\frac{T}{2}\right)\overline{h}_1(k)h_1\left(k-\frac{T}{2}\right) & R_d(0)\left|h_1\left(k-\frac{T}{2}\right)\right|^2 & R_d\left(\frac{T}{2}\right)h_1\left(k-\frac{T}{2}\right)\overline{h}_1(k-T) \\ R_d(-T)\overline{h}_1(k)h_1(k-T) & R_d\left(-\frac{T}{2}\right)\overline{h}_1\left(k-\frac{T}{2}\right)h_1(k-T) & R_d(0)\left|h_1(k-T)\right|^2 \end{bmatrix}$$

$$+ \begin{bmatrix} R_d(0)|h_2(k)|^2 & R_d\left(\frac{T}{2}\right)h_2(k-T)\overline{h}_2\left(k-\frac{3T}{2}\right) & R_d(T)h_2(k-T)\overline{h}_2(k-2T) \\ R_d\left(-\frac{T}{2}\right)\overline{h}_2(k-T)h_2\left(k-\frac{3T}{2}\right) & R_d(0)\left|h_2\left(k-\frac{3T}{2}\right)\right|^2 & R_d\left(\frac{T}{2}\right)h_2\left(k-\frac{3T}{2}\right)\overline{h}_2(k-2T) \\ R_d(-T)\overline{h}_2(k-T)h_2(k-2T) & R_d\left(-\frac{T}{2}\right)\overline{h}_2\left(k-\frac{3T}{2}\right)h_2(k-2T) & R_d(0)|h_2(k-2T)|^2 \end{bmatrix}$$

$$+ \begin{bmatrix} R_d(T)h_1(k)\overline{h}_2(k-T) & R_d\left(\frac{3T}{2}\right)h_1(k)\overline{h}_2\left(k-\frac{3T}{2}\right) & R_d(2T)h_1(k)\overline{h}_2(k-2T) \\ R_d\left(-\frac{3T}{2}\right)\overline{h}_1(k)h_2\left(k-\frac{3T}{2}\right) & R_d(T)h_1\left(k-\frac{T}{2}\right)\overline{h}_2\left(k-\frac{3T}{2}\right) & R_d\left(-\frac{3T}{2}\right)h_1\left(k-\frac{T}{2}\right)\overline{h}_2(k-2T) \\ R_d(-2T)\overline{h}_1(k)h_2(k-2T) & R_d\left(-\frac{3T}{2}\right)\overline{h}_1\left(k-\frac{T}{2}\right)h_2(k-2T) & R_d(T)h_1(k-T)\overline{h}_2(k-2T) \end{bmatrix}$$

$$+ \begin{bmatrix} R_d(-T)\overline{h}_1(k)h_2(k-T) & R_d\left(-\frac{T}{2}\right)h_2(k-T)\overline{h}_1\left(k-\frac{T}{2}\right) & R_d(0)h_2(k-T)\overline{h}_1(k-T) \\ R_d\left(-\frac{T}{2}\right)\overline{h}_2(k-T)h_1\left(k-\frac{T}{2}\right) & R_d(-T)h_2\left(k-\frac{3T}{2}\right)\overline{h}_1\left(k-\frac{T}{2}\right) & R_d\left(-\frac{T}{2}\right)h_2\left(k-\frac{3T}{2}\right)\overline{h}_1(k-T) \\ R_d(0)\overline{h}_2(k-T)h_1(k-T) & R_d\left(-\frac{T}{2}\right)\overline{h}_2\left(k-\frac{3T}{2}\right)h_1(k-T) & R_d(-T)h_2(k-2T)\overline{h}_1(k-T) \end{bmatrix} \quad (6.59)$$

As one can see this block covariance matrix can be separated into four submatrices. The first corresponds to the first arriving ray, the second submatrix corresponds to the second arriving ray, and the third and fourth correspond to a mixture of both. They have been written in the absence of noise. The cross-covariance block matrix can be written as

$$R_{xy} = \begin{bmatrix} R_d(0)h_{x1}(k)\overline{h}_{y1}(k) & R_d\left(-\frac{T}{2}\right)h_{x1}(k)\overline{h}_{y1}\left(k-\frac{T}{2}\right) & R_d(T)h_{x1}(k)\overline{h}_{y1}(k-T) \\ R_d\left(-\frac{T}{2}\right)h_{x1}\left(k-\frac{T}{2}\right)\overline{h}_{y1}(k) & R_d(0)h_{x1}\left(k-\frac{T}{2}\right)\overline{h}_{y1}\left(k-\frac{T}{2}\right) & R_d\left(\frac{T}{2}\right)h_{x1}\left(k-\frac{T}{2}\right)\overline{h}_{y1}(k-T) \\ R_d(-T)h_{x1}(k-T)\overline{h}_{y1}(k) & R_d\left(-\frac{T}{2}\right)h_{x1}(k-T)\overline{h}_{y1}\left(k-\frac{T}{2}\right) & R_d(0)h_{x1}(k-T)\overline{h}_{y1}(k-T) \end{bmatrix}$$

$$+ \begin{bmatrix} R_d(0)h_{x2}(k-T)\overline{h}_{y2}(k-T) & R_d\left(\frac{T}{2}\right)h_{x2}(k-T)\overline{h}_{y2}\left(k-\frac{3T}{2}\right) & R_d(T)h_{x2}(k-T)\overline{h}_{y2}(k-2T) \\ R_d\left(-\frac{T}{2}\right)h_{x2}\left(k-\frac{3T}{2}\right)\overline{h}_{y2}(k-T) & R_d(0)h_{x2}\left(k-\frac{3T}{2}\right)\overline{h}_{y2}\left(k-\frac{3T}{2}\right) & R_d\left(\frac{T}{2}\right)h_{x2}\left(k-\frac{3T}{2}\right)\overline{h}_{y2}(k-2T) \\ R_d(-T)h_{x2}(k-2T)\overline{h}_{y2}(k-T) & R_d\left(-\frac{T}{2}\right)h_{x2}(k-2T)\overline{h}_{y2}\left(k-\frac{3T}{2}\right) & R_d(0)h_{x2}(k-2T)\overline{h}_{y2}(k-2T) \end{bmatrix}$$

$$+\begin{bmatrix} R_d(T)h_{x1}(k)\bar{h}_{y2}(k-T) & R_d\left(\frac{3T}{2}\right)h_{x1}(k)\bar{h}_{y2}\left(k-\frac{3T}{2}\right) & R_d(2T)h_{x1}(k)\bar{h}_{y2}(k-2T) \\ R_d\left(\frac{T}{2}\right)h_{x1}\left(k-\frac{T}{2}\right)\bar{h}_{y2}(k-T) & R_d(T)h_{x1}\left(k-\frac{T}{2}\right)\bar{h}_{y2}\left(k-\frac{3T}{2}\right) & R_d\left(\frac{3T}{2}\right)h_{x1}\left(k-\frac{T}{2}\right)\bar{h}_{y2}(k-2T) \\ R_d(0)h_{x1}(k-T)\bar{h}_{y2}(k-T) & R_d\left(\frac{T}{2}\right)h_{x1}(k-T)\bar{h}_{y2}\left(k-\frac{3T}{2}\right) & R_d(T)h_{x1}(k-T)\bar{h}_{y2}(k-2T) \end{bmatrix}$$

$$+\begin{bmatrix} R_d(-T)h_{x2}(k-T)\bar{h}_{y1}(k) & R_d\left(-\frac{T}{2}\right)h_{x2}(k-T)\bar{h}_{y1}\left(k-\frac{T}{2}\right) & R_d(0)h_{x2}(k-T)\bar{h}_{y1}(k-T) \\ R_d\left(-\frac{3T}{2}\right)h_{x2}\left(k-\frac{3T}{2}\right)\bar{h}_{y1}(k) & R_d(-T)h_{x2}\left(k-\frac{3T}{2}\right)\bar{h}_{y1}\left(k-\frac{T}{2}\right) & R_d\left(-\frac{T}{2}\right)h_{x2}\left(k-\frac{3T}{2}\right)\bar{h}_{y1}(k-T) \\ R_d(-2T)h_{x2}(k-2T)\bar{h}_{y1}(k) & R_d\left(-\frac{3T}{2}\right)h_{x2}(k-2T)\bar{h}_{y1}\left(k-\frac{T}{2}\right) & R_d(-T)h_{x2}(k-2T)\bar{h}_{y1}(k-T) \end{bmatrix} \tag{6.60}$$

With the above defined block matrices we can see that the MMSE array weight vector can be written as

$$\underline{w}_{\text{MMSE}} = [H_1 + H_2 + \sigma_n^2 I]^{-1} \cdot \underline{r}_d \tag{6.61}$$

$$\underline{w}_{\text{MMSE}} = [H_{1A} + H_{1B} + H_{1C} + H_{1D} + \sigma_n^2 I]^{-1} \cdot \underline{r}_d \tag{6.62}$$

where H_{1A} is the channel matrix for the first arriving ray, H_{1B} is the channel matrix for the second arriving ray, H_{1C} and H_{1D} are the channel matrices involving both the first and second arriving rays.

In Fig. 6.13 we plot the BER performance of a $K = 3$ tap fractionally spaced STE using $M = 2$ antennas. The results are for the flat-fading and frequency-selective two-ray channel models. For the

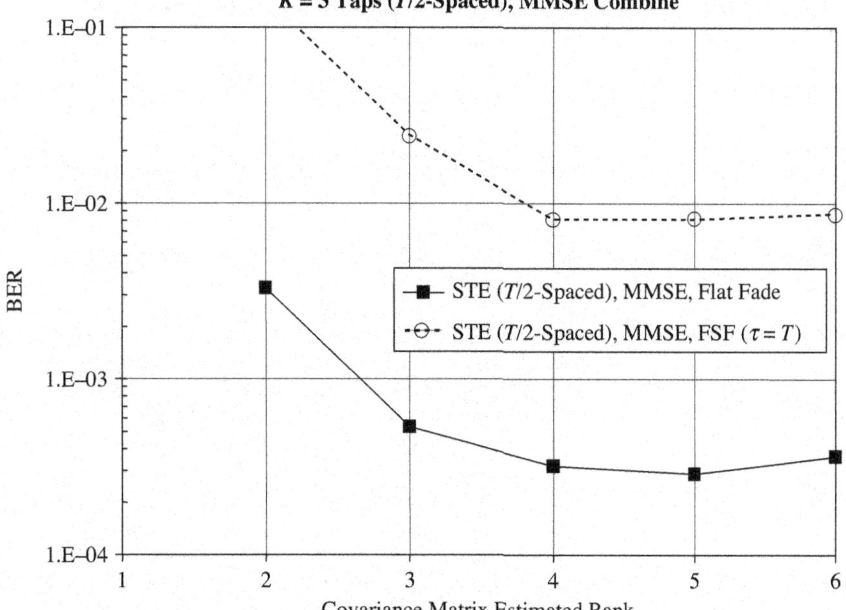

FIGURE 6.13 Fractionally spaced STE matrix rank investigation.

flat-fading channel the covariance matrix rank is shown to be 3 and increased to 4 when a second ray is inserted with significant delay.

6.2.3 Frequency Selective Fading + CCI Environment

In this section, we consider a two-ray model plus interference that contains delay spread as well. This is modeled as follows with $i(k)$ representing the interfering signal.

$$x(k) = s(k)h_{x1}(k) + s(k - T)h_{x2}(k - T) + i(k)h_{ix1}(k) + i(k - T)h_{ix2}(k - T) + n_x(k) \quad (6.63)$$

$$y(k) = s(k)h_{y1}(k) + s(k - T)h_{y2}(k - T) + i(k)h_{iy1}(k) + i(k - T)h_{iy2}(k - T) + n_y(k) \quad (6.64)$$

Given the results of the previous sections we can immediately write down the MMSE array weights given as

$$\underline{w}_{\text{MMSE}} = [H_1 + H_2 + H_{I1} + H_{I2} + \sigma_n^2 I]^{-1} \cdot \underline{r}_d \quad (6.65)$$

which can also be written as

$$\underline{w}_{\text{MMSE}} = [H_{1A} + H_{1B} + H_{1C} + H_{1D} + H_{IA} + H_{IB} + H_{IC} + H_{ID} + \sigma_n^2 I]^{-1} \cdot \underline{r}_d \quad (6.66)$$

In looking at this array weight we have the following comments to make: H_{1A} is the desired channel matrix of the first arriving ray, H_{1B} is the desired channel matrix of the second arriving ray, H_{1C} and H_{1D} are the cross terms relating to the first and second arriving rays of the desired signal. Similarly, H_{IA} is the interfering channel matrix of the first arriving ray, H_{IB} is the interfering channel matrix of the second arriving ray, and H_{IC} and H_{ID} are the cross terms relating the first and second arriving rays of the interfering signal.

Next we plot the BER versus the covariance matrix rank for three channel conditions: flat fading, frequency selective fading with a time delay ($\tau = T/2$), and frequency selective fading with a time delay ($\tau = T$). One can see at the delay spread increases the rank of the covariance matrix also increases. Also note in the flat-fading channel we can see that the addition of the interference increased the rank by 1 (see Fig. 6.14).

6.2.4 Covariance Matrix Eigen Spectra Properties

In this section, we will present some measured statistics of eigenvalues of the space-time equalizer given above. The STE architecture was a fractionally spaced, $T/2$-spaced STE using $K = 5$ taps and $M = 2$ receive antennas. First, we plot the probability density function (PDF) of the ten eigenvalues below (see Fig. 6.15). The channel model considered was flat fading plus CCI interference. As shown in the previous sections, the presence of CCI will increase the covariance matrix rank by 1. Thus bringing the total number of dominant eigenvalues to 4.

The covariance matrix rank is more visible in the plot of the eigenvalue CDF. Here we see the four dominant eigenvalues (see Fig. 6.16).

We wish to conclude this section by stating that the covariance matrix vector space can be divided into subspaces: a signal subspace and noise subspace.

$$R^{-1} = \sum_{j=1}^{N_s} \frac{1}{\lambda_j + \sigma_n^2} \cdot \underline{v}_j \cdot \underline{v}_j^* + \sum_{j=N_s+1}^{MK} \frac{1}{\sigma_n^2} \cdot \underline{v}_j \cdot \underline{v}_j^* \quad (6.67)$$

Ideally when the cross-correlation vector is projected on to the inverted covariance matrix, the noise subspace contribution is zero. However, practically speaking there is residual estimation error thus causing performance degradation. This observation leads us to the Unified eigenbased, spatio-temporal receiver presented earlier which continually calculated the eigen spectra of the received signal.

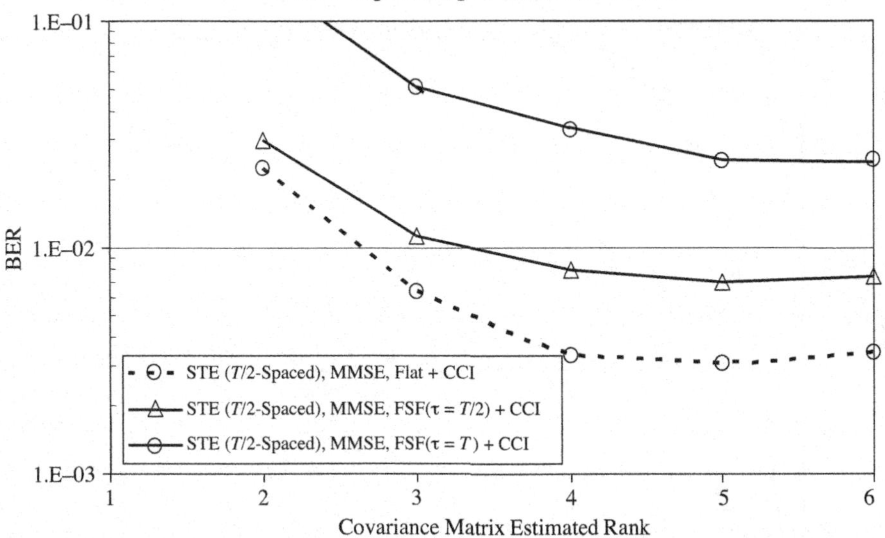

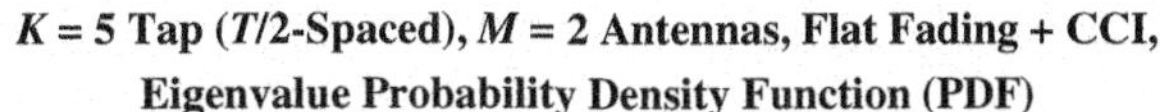

FIGURE 6.14 STE covariance matrix rank investigation.

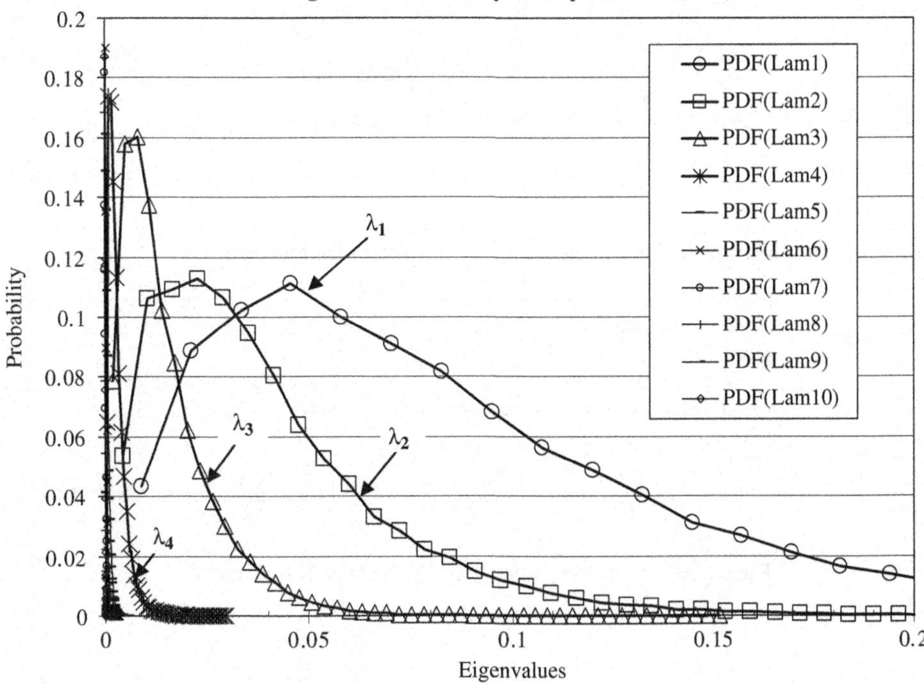

FIGURE 6.15 STE Eigenvalue PDF plot.

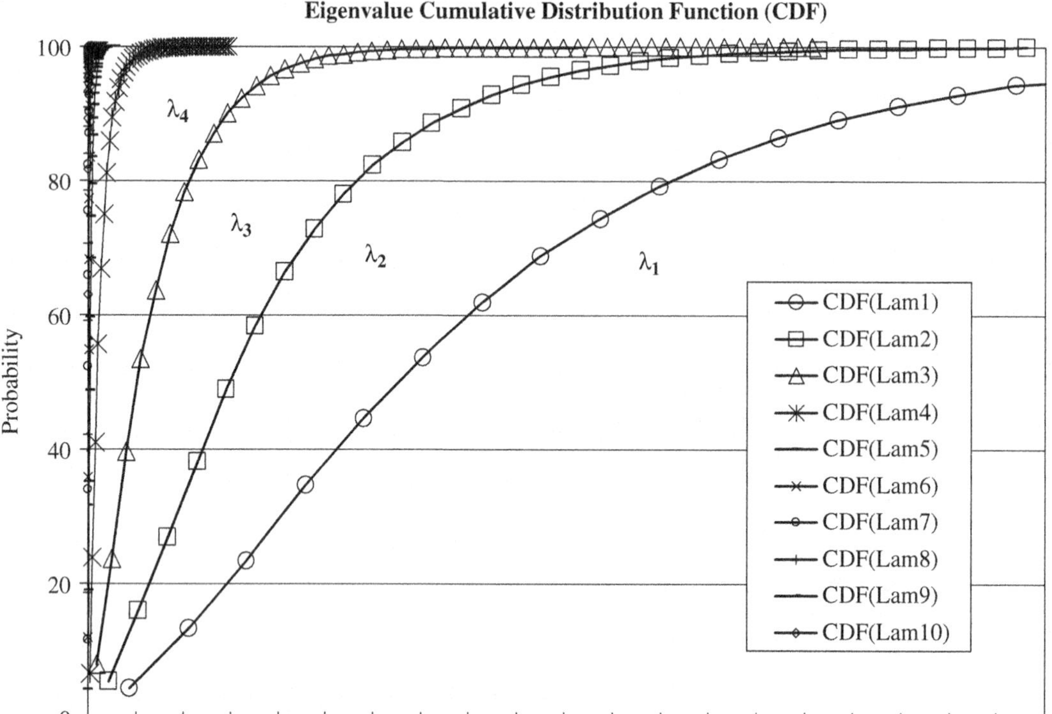

FIGURE 6.16 STE eigenvalue CDF plot (shown in terms of percentages).

6.2.5 Linear Minimum Mean Square Error (LMMSE) Discussion

In this section, we will continue along the lines of using a linear equalizer structure on each antenna as well as the MMSE cost function. Below we provide an alternative approach to performing a joint space-time equalizer, assuming M receiver antennas [16][17].

We begin by writing the received signal on the first antenna which consists of the desired signal with L ISI terms, C interfering signals with P ISI terms and noise.

$$r_1(t) = \sum_{j=0}^{L-1} h_{1,j} \cdot d(t-j) + \sum_{i=0}^{C-1} \sum_{j=0}^{P-1} h_{1,j}^{(i)} \cdot d_i(t-j) + n_1(t) \tag{6.68}$$

The received signal vector notation (of size $M \times 1$) is given as

$$\underline{r}(t) = H(t) \cdot \underline{d}(t) + \sum_{i=0}^{C-1} H_i(t) \cdot \underline{d}_i(t) + \underline{n}(t) \tag{6.69}$$

The desired signal's channel matrix (of size $M \times L$) can be written as (we have dropped the index over time for simplicity). The interfering signal's channel matrix can be similarly written.

$$H = \begin{bmatrix} h_{1,0} & h_{1,1} \cdots & h_{1,L-1} \\ & \vdots & \vdots \\ h_{M,0} & h_{M,1} \cdots & h_{M,L-1} \end{bmatrix} \tag{6.70}$$

$$H = [\underline{h}_0 \ \underline{h}_1 \ \cdots \ \underline{h}_{L-1}]$$

And the desired signal vector (of size $L \times 1$) is given as

$$\underline{d}(t) = \begin{bmatrix} d(t) \\ d(t-1) \\ \vdots \\ d(t-L+1) \end{bmatrix} \tag{6.71}$$

Let us collect $K + 1$ consecutive samples so we can write the received signal vector (of size $M^*(K + 1) \times 1$).

$$\underline{\tilde{r}}(t) = [\underline{r}^T(t) \quad \underline{r}^T(t-1) \quad \cdots \quad \underline{r}^T(t-K)]^T \tag{6.72}$$

Then the overall channel matrix (of size $M^*(K + 1) \times (L + K)$) can be written in a very nicely compacted form.

$$\tilde{H} = \begin{bmatrix} \underline{h}_0 & \underline{h}_1 & \cdots & \underline{h}_{L-1} & 0 & 0 \\ 0 & \underline{h}_0 & \underline{h}_1 & \cdots & \underline{h}_{L-1} & \cdots \\ \vdots & & & \ddots & & \vdots \\ 0 & \cdots & \underline{h}_0 & \underline{h}_1 & \cdots & \underline{h}_{L-1} \end{bmatrix} \tag{6.73}$$

And the overall desired signal vector of size $(L + K) \times 1$ can be written as

$$\underline{\tilde{d}}(t) = [d(t) \quad d(t-1) \ \cdots \ d(t-L-K+1]^T \tag{6.74}$$

Using the above new definitions we can write the received signal as

$$\underline{\tilde{r}}(t) = \tilde{H}(t) \cdot \underline{\tilde{d}}(t) + \sum_{i=0}^{C-1} \tilde{H}_i(t) \cdot \underline{\tilde{d}}_i(t) + \underline{\tilde{n}}(t) \tag{6.75}$$

The MMSE-based STE weights are created with the estimate of the covariance matrix and correlation vectors provided below.

$$R_{\tilde{r}\tilde{r}}(t) = E\left\{\underline{\tilde{r}}(t) \cdot \underline{\tilde{r}}^*(t)\right\} \tag{6.76}$$

$$R_{\tilde{r}\tilde{r}}(t) = \tilde{H}(t) \cdot R_{\tilde{d}\tilde{d}}(t) \cdot \tilde{H}^*(t) + \sum_{i=0}^{C-1} \tilde{H}_i(t) \cdot R_{\tilde{d}_i\tilde{d}_i}(t) \cdot \tilde{H}_i^*(t) + \sigma_n^2 I \tag{6.77}$$

Now if we are allowed to invoke a very popular assumption that the received signal is only correlated with itself and that the interfering signals all have the same property then the signal covariance matrices can be approximated by diagonal matrices thus simplifying the result as follows.

$$R_{\tilde{r}\tilde{r}}(t) = \sigma_s^2 \cdot \tilde{H}(t) \cdot \tilde{H}^*(t) + \sum_{i=0}^{C-1} \sigma_i^2 \cdot \tilde{H}_i(t) \cdot \tilde{H}_i^*(t) + \sigma_n^2 I \tag{6.78}$$

Similarly, the correlation vector can be written as

$$\underline{r}_{\tilde{r}d}(t) = E\{\underline{\tilde{r}}(t) \cdot d^*(t - D)\} \tag{6.79}$$

$$\underline{r}_{\tilde{r}d}(t) = \sigma_s^2 \cdot \tilde{H}(t) \cdot O_D \tag{6.80}$$

where O_D is the operator which extracts the Dth column from the preceding matrix. The LMMSE-based weights can be written as follows.

$$\underline{w}(t) = \left[\sigma_s^2 \cdot \tilde{H}(t) \cdot \tilde{H}^*(t) + \sum_{i=0}^{C-1} \sigma_i^2 \cdot \tilde{H}_i(t) \cdot \tilde{H}_i^*(t) + \sigma_n^2 I\right]^{-1} \cdot \sigma_s^2 \cdot \tilde{H}(t) \cdot O_D \tag{6.81}$$

Note that the interference cancellation ability of this receiver is strictly related to including the interference statistics into the calculation of the weights.

As we have discussed in the previous sections inverting the above matrix can be tedious and complex, and alternative solutions exist. We will give a solution based on the LMS algorithm. The equalizer weights are updated according to the following rule:

$$\begin{bmatrix} \underline{w}_1(k+1) \\ \vdots \\ \underline{w}_M(k+1) \end{bmatrix} = \begin{bmatrix} \underline{w}_1(k) \\ \vdots \\ \underline{w}_M(k) \end{bmatrix} + \mu \cdot \begin{bmatrix} \sigma_d^2 \underline{h}_1(k) - \underline{r}_1(k) \cdot d^*(k) \\ \vdots \\ \sigma_d^2 \underline{h}_M(k) - \underline{r}_M(k) \cdot d^*(k) \end{bmatrix} \tag{6.82}$$

Here we see the update error vector has been slightly modified from previous discussions to include the update dependent on the channel estimate. The original intent was to use this technique without the use of a training sequence but nothing precludes the system designer form using a training sequence to improve the channel estimates.

6.2.6 Maximum Likelihood Estimation (MLE)

In this section, we will present the STE solution from the ML perspective. The discussion begins with the familiar ML metric and then shows how this metric changes as the channel conditions vary. Assume the received signal on antenna #1 is

$$r_1(k) = \sum_{j=0}^{J-1} h_1(j) \cdot s(k-j) + \sum_{i=0}^{I-1} \sum_{m=1}^{M} h_{1,m}(i) \cdot s_m(k-i) + n_1(k) \tag{6.83}$$

where we have generally assumed the desired signal contains delay spread components on the order of J, and there are M interfering signals with each having I delay spread components.

In order for us to gain some insight into the ML relationship, let us assume the desired signal experiences flat fading as well as the interfering signals. We then have the following representation of the received signal given in vector notation (assuming D receive antennas).

$$\underline{r}(k) = \underline{h}(k) \cdot s(k) + \sum_{m=1}^{M} \underline{h}_m(k) \cdot s_m(k) + \underline{n}(k) \tag{6.84}$$

This allows us to write down the received signal's covariance matrix as

$$R_{rr}(k) = \sigma_s^2 \cdot \underline{h}(k) \cdot \underline{h}^*(k) + \sum_{m=1}^{M} \sigma_m^2 \cdot \underline{h}_m(k) \cdot \underline{h}_m^*(k) + \sigma_n^2 I \tag{6.85}$$

With these above definitions we can write down the ML metric to be used in the MLE procedure and this is given as

$$M(k) = \left[\underline{r}(k) - \underline{\hat{h}}(k) \cdot s(k)\right]^* \cdot R_{I+n}^{-1}(k) \cdot \left[\underline{r}(k) - \underline{\hat{h}}(k) \cdot s(k)\right] \tag{6.86}$$

We used $\underline{\hat{h}}(k)$ to denote the estimate of the channel response. If we temporarily assume the interference can be modeled as either white or is nonexistent then the interference plus noise covariance matrix will be a diagonal and greatly simplify the metric calculation. In fact, if we further assume that each of the receive antennas has the same noise power then we can further simplify the metric. We will utilize the VA to perform the MLE where each of the D antennas will be utilized sequentially as follows.

$$M(k) = \sum_{j=1}^{D} \left| r_j(k) - \hat{h}_j(k) \cdot s_h(k) \right|^2 \tag{6.87}$$

This can be easily extended to the case when the desired signal has delay spread components.

$$M(k) = \sum_{j=1}^{D} \left| r_j(k) - \sum_{i=0}^{J-1} \hat{h}_j(i) \cdot s_h(k-i) \right|^2 \tag{6.88}$$

In Fig. 6.17 we provide an overall block diagram of a system where the desired signal encounters a two-ray channel. The channel has a single interferer experiencing flat fading. The channel estimates are obtained as previously discussed where q data symbols are consumed before an output symbol is generated. The MLSE generates various hypothesis signals, $s_h(k)$, to use in calculating the metrics.

A simple block diagram is provided in Fig. 6.18, although it is shown for the flat fading scenario, it can easily be extended to the frequency selective fading scenario. Each antenna has an independent channel estimate and the MLSE hypothesis is applied to each antenna.

One can also show that minimizing the above metric, $M(k)$, is equivalent to maximizing the following relationship [18–20].

$$L(k) = \mathrm{Re}\{\underline{\hat{h}}^*(k) \cdot s_h^*(k) \cdot \underline{r}(k)\} \tag{6.89}$$

This can be represented with the following block diagram which after careful inspection is equivalent to MRC combining (see Fig. 6.19). This is expected since we assumed no interference was present and the desired signal experienced flat fading.

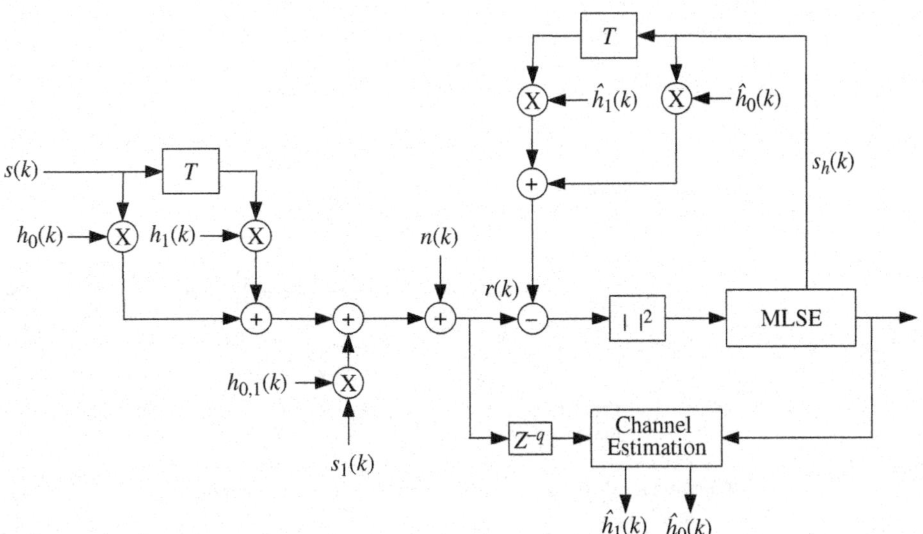

FIGURE 6.17 Overall system block diagram using an MLSE-based receiver.

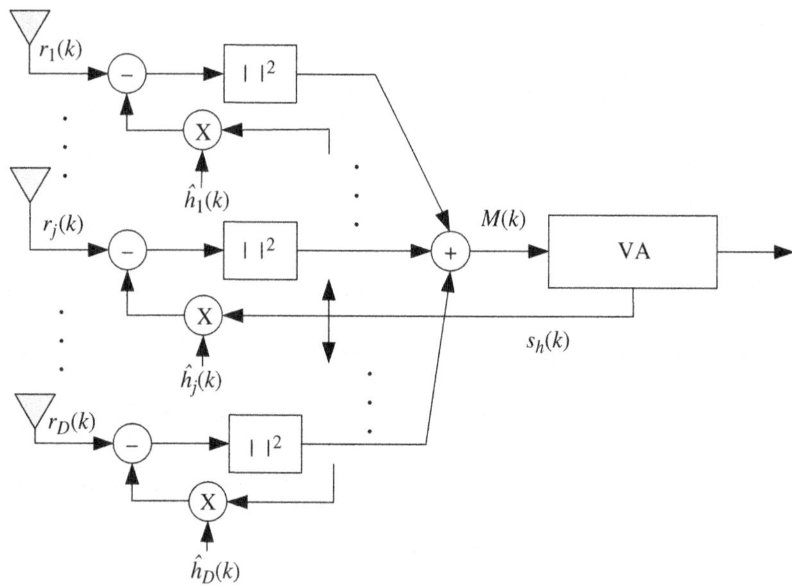

FIGURE 6.18 Simplified block diagram of an MLSE-based receiver.

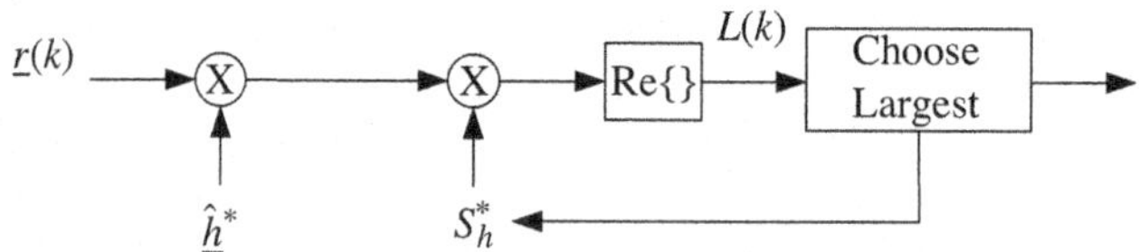

FIGURE 6.19 ML equivalent representation of maximal ratio combining.

Now let's return to the presence of interference which is not white, then the previously provided ML delay-spread metric should be minimized. Alternatively, it is easy to show that the above metric is equivalent to maximizing the following relationship.

$$L(k) = \mathrm{Re}\{\underline{\hat{h}}^*(k) \cdot R_{I+n}^{-1}(k) \cdot \underline{r}(k) \cdot s_h^*(k)\} \tag{6.90}$$

Which can be alternatively shown next (see Fig. 6.20).

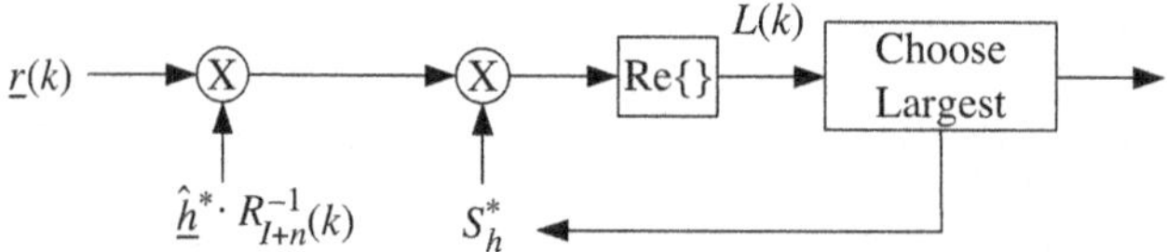

FIGURE 6.20 ML equivalent representation of optimum combining using the MSINR-based weights.

What this result shows us is that the ML antenna-combining technique discussed herein is equivalent to the adaptive antenna array using the MSINR cost function-based weights, $\underline{w}_{MSINR} = R_{I+n}^{-1} \cdot \underline{h}$. This tells us that the MLE has interference suppression capabilities provided the metric value is correctly computed. In other words, the interference statistics are included in the metric calculation.

A unified MLSE block diagram is shown in Fig. 6.21. The received signals are grouped together to form a vector $\underline{r}(k)$. This received signal vector is subtracted by using the hypothesis signals. The metric is then computed using the inverse of the interference plus noise covariance matrix.

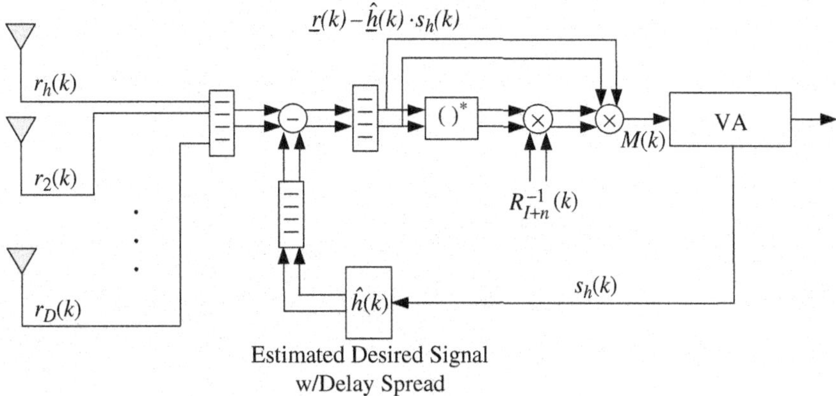

FIGURE 6.21 Unified MLSE-based receiver architecture.

Let us now discuss methods to include interference statistics in the ML metric calculation. In the method presented above the interference is considered in the interference plus noise covariance matrix. This is usually obtained by subtracting the estimated desired signal from the received signal. Hence the accuracy of the R_{I+n} matrix depends on how accurate the desired signal can be estimated. Alternatively, the R_{I+n} matrix inversion can be avoided if the M interfering signals with I multipaths are estimated and also removed from the received signal. The very significant drawback of this second approach is that the number of states in the VA increases and the metric now also becomes dependent on the accuracy of estimating the interference. The metric where not only the desired channel is estimated, but also the interfering channel as well is given below. Note for this case, hypothesis signals for both the desired and interfering signals are required.

$$M_I(k) = \sum_{j=1}^{D} \left| r_j(k) - \sum_{i=0}^{J-1} \hat{h}_j(i) \cdot s_h(k-i) - \sum_{p=0}^{I-1} \sum_{m=1}^{M} \hat{h}_{j,m}(p) \cdot s_{h,m}(k-p) \right|^2 \tag{6.91}$$

One last comment to make here is that we have assumed symbol-sampled data in the metric. This can be extended to using fractionally spaced MLE processing which can have additional benefits. In this case, the metrics are computed for each of the timing phase instances of the over-sampled signal and then the respective metrics are calculated and then combined across antennas and timing instances.

6.3 FREQUENCY DOMAIN EQUALIZATION

In this next section, we will provide an overview of frequency domain equalization (FDE) applied to single carrier (SC) systems [21]. We have presented the time-domain equalization methods since they are effective in combating ISI and suppressing interference. Some of the motivation in moving our attention to the frequency domain are that OFDMA-based systems are proposed for the next generation of wireless communications, hence requiring frequency domain signal processing. Also as the user data rate increases the complexity of the equalizer grows since more symbols now span across the same channel delay spread. Also if higher-order modulation is deemed as a viable solution then the equalizer complexity also grows [22]. One can also suggest other reasons, in any case sometimes it makes sense to move the signal processing into the frequency domain; we will discuss this shortly [23–25].

6.3.1 MMSE Based

A general SC-FDE-based receiver is shown in Fig. 6.22 for the case of M antennas. First the received signal is transformed to the frequency domain through the use of a discrete Fourier transform (DFT), next the equalizer weights are applied by multiplication. The combined signal is transformed back into the time domain using an Inverse DFT (I-DFT) operation. The received signals are represented as r_j where $j = 1$ to M.

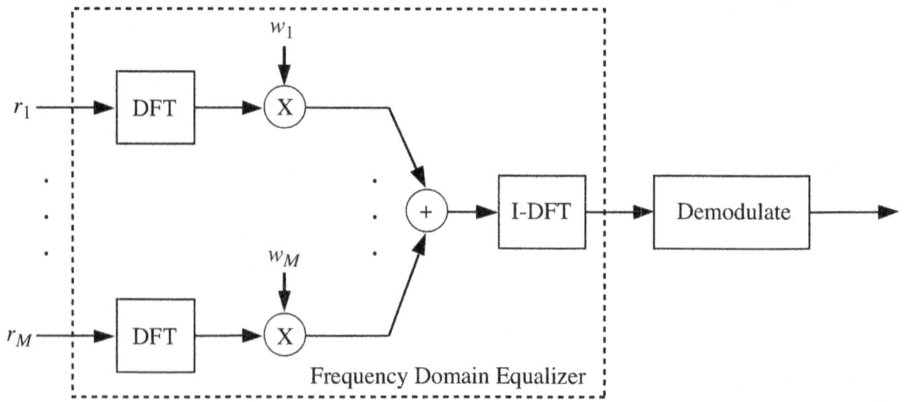

FIGURE 6.22 Single-carrier frequency domain equalizer receiver.

When determining the size of the DFT, one should consider the size to be enough to represent the inverse of the channel response in the time domain.

The well-known weights used to perform the combining are defined below assuming the MMSE cost function.

$$w_j(k) = \frac{\left(\frac{\sigma_d}{\sigma_j}\right)^2 \cdot H_j^*(k)}{\sum_{i=1}^{M}\left(\frac{\sigma_d}{\sigma_i}\right)^2 \cdot |H_i(k)|^2 + 1} \qquad (j = 1, 2, \ldots, M) \tag{6.92}$$

Above σ_d^2 is the desired signal's power, σ_j^2 is the noise power present on the jth antenna. Allow us to assume $M = 2$ in order to provide some insight into the above general equation. We then have the following two FDE weights [24].

$$w_1(k) = \frac{H_1^*(k)}{|H_1(k)|^2 + \left(\frac{\sigma_1}{\sigma_2}\right)^2 \cdot |H_2(k)|^2 + \left(\frac{\sigma_1}{\sigma_d}\right)^2} \tag{6.93}$$

$$w_2(k) = \frac{H_2^*(k)}{\left(\frac{\sigma_2}{\sigma_1}\right)^2 \cdot |H_1(k)|^2 + |H_2(k)|^2 + \left(\frac{\sigma_2}{\sigma_d}\right)^2} \tag{6.94}$$

One can further assume the noise power is the same for all receive antennas, then the denominator becomes the same for both weights and can be applied after the summing operation shown in Fig. 6.22. In either case we can see the denominator contains a term inversely proportional to the SNR.

One last point to make here is that this type of equalization was applied to High Speed Downlink Packet Access (HSDPA) in [23] and shown to produce better performance than the conventional RAKE receiver. However, this was accomplished by making a slight modification into the WCDMA standard by inserting a guard interval or cyclic prefix. Nevertheless this continues to prove to be a viable alternative solution to the time-domain approaches.

6.4 SYMBOL TIMING RECOVERY

This next section addresses the need for time synchronization. This occurs when there is a difference in the sampling clock frequencies between the received signal and locally generated signal. Also time synchronization is required when the receiver clock sampling is not synchronized with the received data symbols [4][26–33].

We will base our discussion on the following block diagram and assume the sampling clock to operate at twice the symbol rate (see Fig. 6.23). The complex samples are defined in terms of the even and odd sampling, $r_e(k)$ and $r_o(k)$, respectively.

$$r_e(k) = r(t)|_{t=\tau+nT} \tag{6.95}$$

$$r_o(k) = r(t)|_{t=\tau+nT+\frac{T}{2}} \tag{6.96}$$

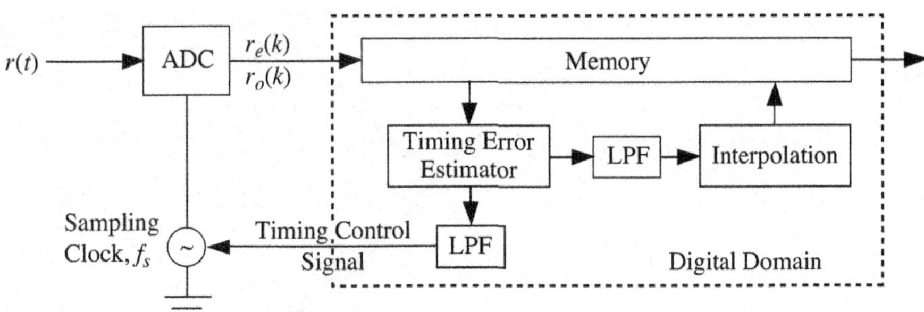

FIGURE 6.23 Timing recovery functional block diagram.

The complex, oversampled samples are temporarily stored in memory so the timing error estimator can estimate the sampling phase offset between the received signal and the locally generated timing. Once this timing error has been determined this can be feedback to the analog domain instructing the oscillator to either speed up or slow down in order for the samples to have proper sampling. If a timing offset exists then the receiver may want to use this for the present samples waiting inside the memory. This would be accomplished through a feed forward path using an interpolator. Whether the timing correction is feed back or feed forward, an LPF was inserted to reduce the effects of noise and jitter.

In order to provide more details about the optimum sampling point we show an eye diagram in Fig. 6.24 and highlight the possible sampling instances. This diagram assumes an overall raised cosine response with a roll-off factor, $\alpha = 0.22$. The optimal sampling point is indeed the center of the eye diagram. This sampling time instance is optimal since there is zero ISI at this instance.

Next the roll-off factor was increased to $\alpha = 0.9$ to use as an example. The overall eye diagram can vary from one system to another and depending on the roll-off factor of the raised cosine filter, timing requirements may differ (see Fig. 6.25).

In the subsections that follow we will briefly present some timing error estimators that can be used in a variety of places. The purpose is to introduce the estimator categories to the system designer.

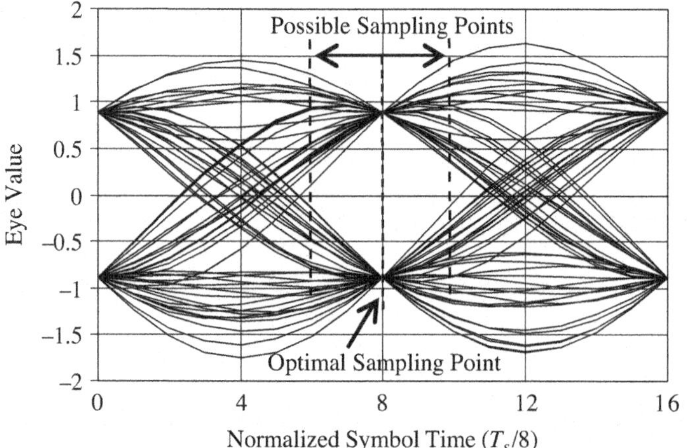

FIGURE 6.24 QPSK eye diagram with roll-off factor = 0.22.

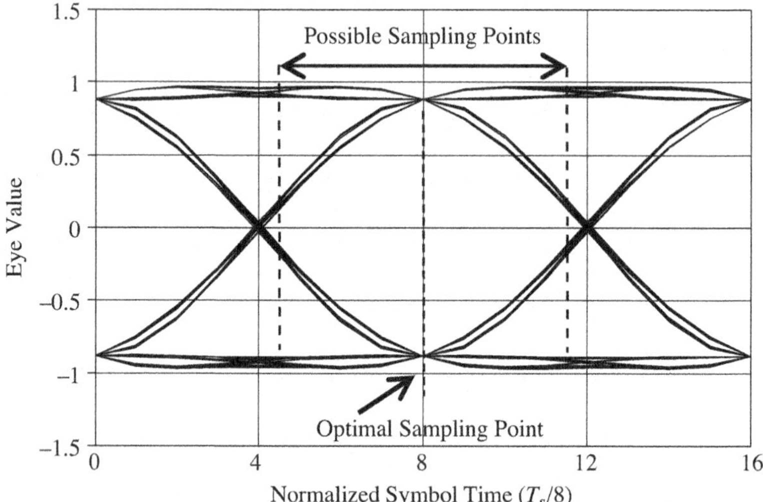

FIGURE 6.25 QPSK eye diagram with roll-off factor = 0.9.

6.4.1 Data Aided

This first form of timing error estimation is termed data aided since the output of the decision device is used to determine the timing adjustment. The basic premise of operation is that the samples entering the decision device are compared to the samples output by the decision device and the timing instance is varied in order to minimize this difference. This assumes a timing instance was chosen prior to performing these operations. A simplified block diagram of a data-aided technique is provided in Fig. 6.26.

The symbol timing estimator is mathematically represented as

$$e(k) = \text{Re}\{r(k-1) \cdot a^*(k) - r(k) \cdot a^*(k-1)\} \tag{6.97}$$

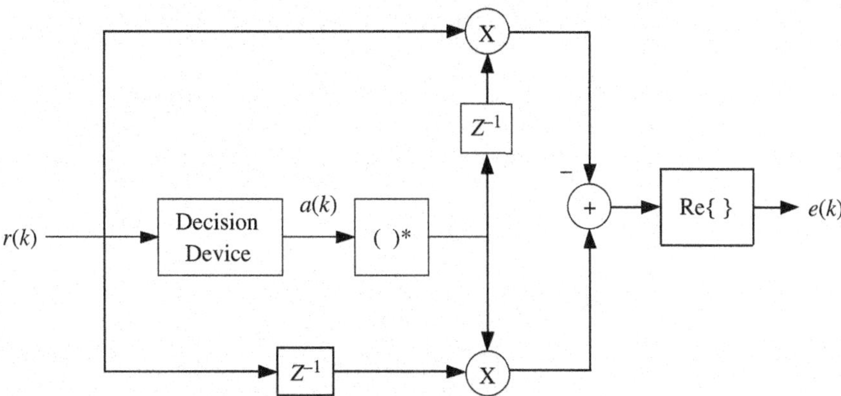

FIGURE 6.26 Data-aided timing error estimator.

Continuing with the assumption that the data symbols are complex valued we have

$$e(k) = (-r_I(k) \cdot a_I(k-1) + r_I(k-1) \cdot a_I(k)) + (-r_Q(k) \cdot a_Q(k-1) + r_Q(k-1) \cdot a_Q(k)) \quad (6.98)$$

Depending on the channel conditions and wireless application the received symbols can be severely distorted. In this case, the above algorithm degrades due to possible asymmetric behavior.

We would like also to point out that there are timing error estimators that work directly on the received signal using non-decision-directed principles. A simple example is given below where we have decided to show error estimators involving differentiation.

$$e(k) = r_I\left(k - \frac{1}{2}\right) \cdot [r_I(k) - r_I(k-1)] + r_Q\left(k - \frac{1}{2}\right) \cdot [r_Q(k) - r_Q(k-1)] \quad (6.99)$$

When considering the quantity in brackets to represent differences, the error is similarly expressed below.

$$e(k) = r_I\left(k - \frac{1}{2}\right) \cdot \frac{dr_I}{dk} + r_Q\left(k - \frac{1}{2}\right) \cdot \frac{dr_Q}{dk} \quad (6.100)$$

The impulse response of an example digital differentiator [4] is given below

$$h_{\text{diff}}(kT_s) = \begin{cases} 0, & k = 0 \\ \frac{1}{kT_s}(-1)^k, & \text{otherwise} \end{cases} \quad (6.101)$$

6.4.2 Maximum Likelihood (ML) Based

In this timing error estimation technique the timing instance that maximizes a likelihood function is chosen. We begin by defining the likelihood function as

$$\Lambda(\hat{\tau}) = \sum_{k=0}^{L-1} a(k) \cdot r(kT + \hat{\tau}) \quad (6.102)$$

where $a(k)$ are the detected data symbols, $r(kT)$ is the received, oversampled signal but subsampled at a rate of $1/T$, k is used to denote the time index, $\hat{\tau}$ is the timing offset error to be estimated, and L is the duration of the comparison.

Hence the timing offset $\hat{\tau}$, is varied so as to maximize the above function over the L symbol observation window. A block diagram showing these operations is shown in Fig. 6.27. Here a timing instance, τ_j, is chosen and its respective samples enter the decision device so that the output can be used in the comparison. The likelihood value is computed and then the next timing instance, τ_{j+1}, is chosen and a new likelihood value is computed. This procedure continues for as many times as the symbol was oversampled. The best choice, that is, the sample timing with the largest likelihood value, is chosen and used for the remainder of the signal processing functions downstream.

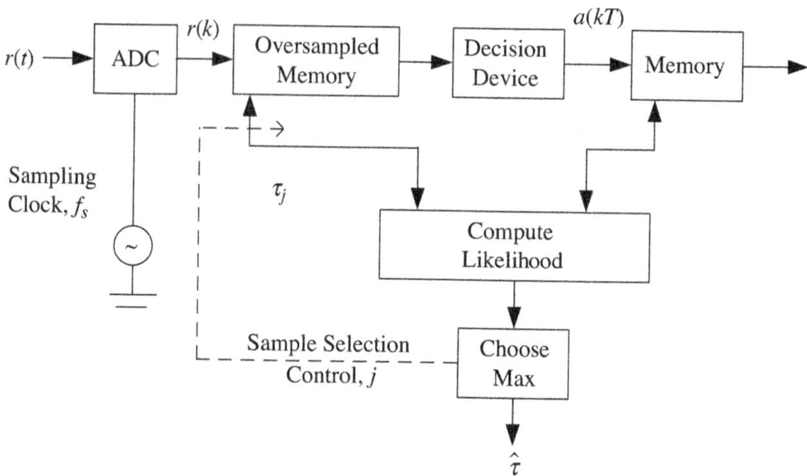

FIGURE 6.27 Maximum likelihood timing error estimator.

It should be pointed out that an oversampled signal is used above, as long as the received signal is sampled at twice the baud rate, then an interpolator can be used to create the further oversampled signal to be used in the ML calculation.

Since timing drifts can occur due to uncertainties in the transmitter, channel, and receiver, it is recommended that this estimation be repeated periodically to track such drifts. These estimation updates can occur at frame, time slot, or subtime slot boundaries. The update rate depends on the expected worse case timing drift. This technique can be performed blindly (directly on samples) or a priori information such as pilot bits can be used to improve the accuracy of the estimate.

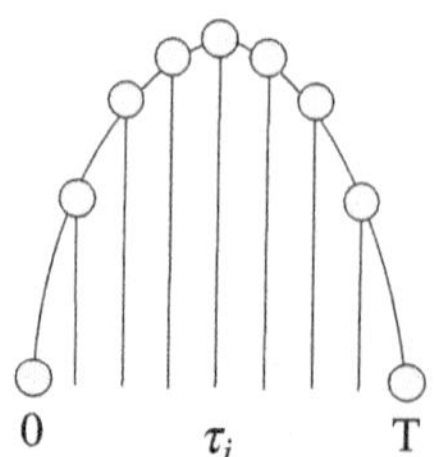

FIGURE 6.28 Oversampled symbol diagram.

Returning our attention to the above ML timing error estimator block diagram, we essentially have the following oversampled eye diagram and wish to choose the best timing instance (see Fig. 6.28). Below we provide an example where the received symbol is oversampled by a factor of 8.

As discussed above, τ_j ($j = 0, 1, \ldots, 7$) is used in the comparison. One wishes to have a sampled signal sufficiently oversampled in order to meet the performance requirements.[3] The disadvantage of this technique is that the sampling clocks need to be rather high and the memory storage requirements can be excessive. An alternative approach is to sample the received signal at a rate equal to the Nyquist rate and then perform the interpolation operations as needed. This will result in the minimal storage of intermediate values.

[3] In a section that follows, we plot the performance as a function of timing offset. The system designer will allocate a certain amount of degradation allowed in order to meet the overall required performance link budget.

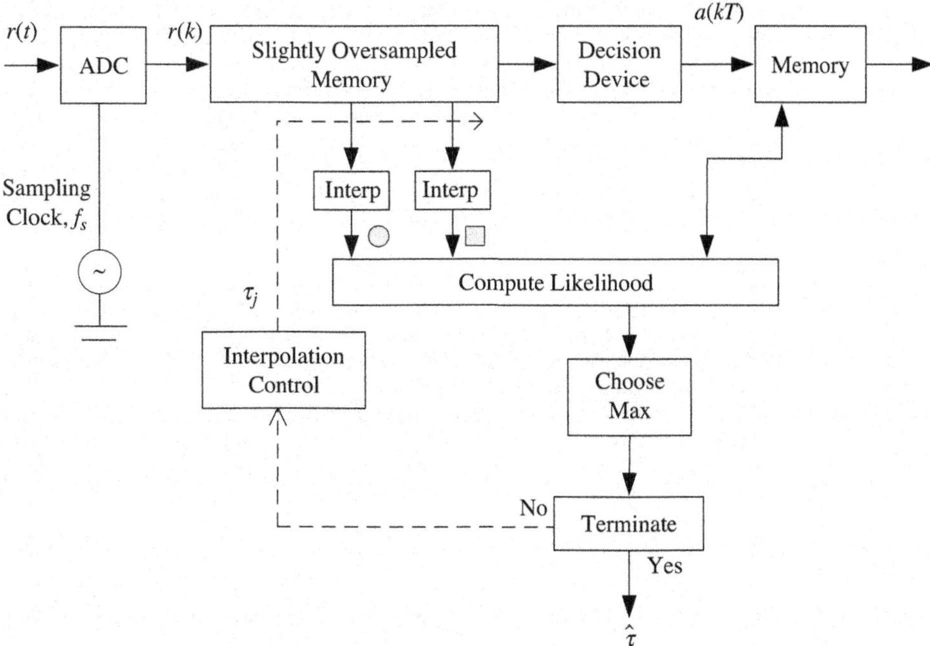

FIGURE 6.29 An iterative ML-based timing recovery method using a free-running clock.

We will discuss this alternative technique in the context of iterative timing-error estimation. Here the available samples are first tested against the ML criteria. After this test, one timing instance is chosen and we interpolate to either side of the temporary estimate to provide new timing offset error candidates. Each of these new samples are tested against the ML criteria and upon selection of a new temporary estimate, two newer samples are interpolated on either side and this iterative procedure repeats until a certain criteria is met. For example, we can state when the temporal distance between the updated timing error estimates is less than $T/16$, then the procedure terminates. A block diagram depicting this procedure is shown in Fig. 6.29.

We have used "circles" and "boxes" to denote the time instances used in the interpolation. The timing recovery begins with the circles and then updates the timing to produce the boxes. This procedure is shown in Fig. 6.30 using the eye diagram.

In this figure, we show the iterative procedure operating on the eye diagram directly. Let us assume the initial conditions are valid at time, $t = 0$ with $T/2$ sampling creating two samples, A_0 and B_0.

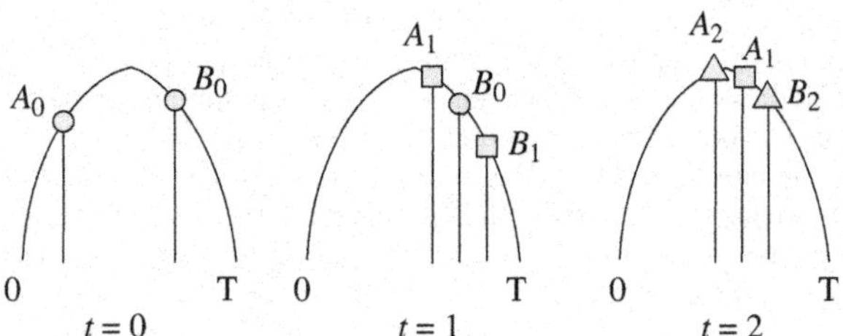

FIGURE 6.30 Eye diagram representation at each iteration of the ML-based timing recovery method.

The ML metric will select B_0 as the better choice of the two. Next we interpolate near this timing phase creating A_1 and B_1. The interpolation distance is a design parameter and chosen to be $T/8$ for this example. As discussed above, the desired resolution should be obtained through system simulations. The updated ML metric is recomputed and timing-phase instance A_1 is chosen at $t = 1$. Next another iteration is performed near A_1 to create A_2 and B_2. Once again a new ML metric is computed and A_2 is chosen. As we can see each iteration will bring the estimated timing closer and closer to the desired point. The number of iterations required to accomplish this task is dependent on the channel conditions, interpolation polynomial used, and desired timing offset requirement.

6.4.3 DFT Based

The next timing error estimation method is based on calculating the Fourier transform of a signal to create a useful spectral component. The timing error is related to the phase of this spectral component. The method used to perform the transformation can either be discrete Fourier transform (DFT) or fast Fourier transform (FFT) based [26].

We can continue to assume the clock controlling the Analog to Digital Convertor (ADC) is free running as shown in the block diagrams above. The timing error is generated by first squaring the received signal and then the DFT is computed using LN samples, where N is the oversampling factor. The block diagram for this technique is given in Fig. 6.31.

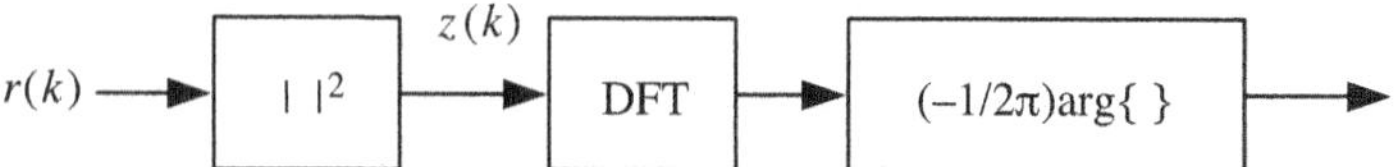

FIGURE 6.31 DFT-based timing error estimator.

The timing error estimator is mathematically represented as

$$\hat{\tau} = \frac{1}{2\pi} \cdot \tan^{-1}\left[\frac{\left(\sum_{k=mLN}^{(m+1)LN-1} z[k] \cdot \sin\left[\frac{2\pi k}{N}\right]\right)}{\left(\sum_{k=mLN}^{(m+1)LN-1} z[k] \cdot \cos\left[\frac{2\pi k}{N}\right]\right)}\right] \tag{6.103}$$

where N is the number of samples per symbol, L is the length of the symbols used in the average, and $z(k)$ is the sampled squared value of the received signal. Hence for each group of samples, a new timing error estimate is obtained and used for timing correction/compensation.

6.4.4 Delay Locked Loop (DLL)

Next we will discuss a technique called DLL which is typically used in spread spectrum applications for chip-time synchronization. More details will be provided in the next chapter where we address CDMA; however, we felt it was important to provide a preview in this related section. The block diagram shows there are two parallel correlators, one for the early timing and one for the late timing (see Fig. 6.32). We used $c(t - \tau - \Delta)$ to denote the chip sequence appropriately delayed. This technique assumes the received eye diagram is reasonably symmetrical thus forcing us to say the "on time" sample timing instance is directly in the center of the "early" and "late" timing instances.

In the spread spectrum application, the early and late correlators exploit the autocorrelation properties of the PN sequence used. The time separation between the early and late correlators is given by 2Δ. The autocorrelation is evaluated at early and late lags and then squared to remove the

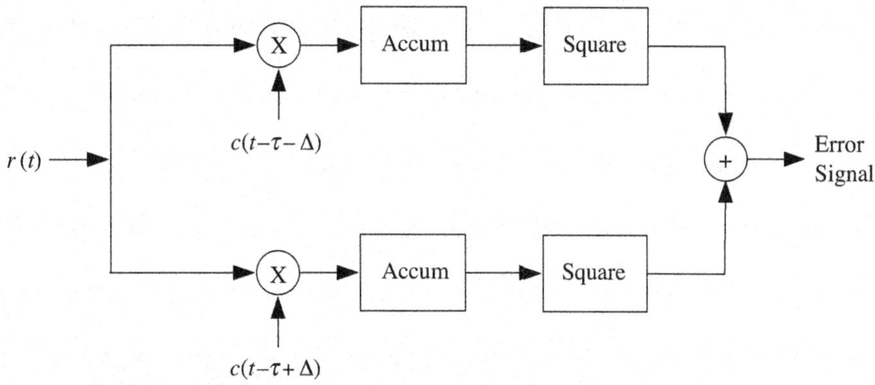

FIGURE 6.32 Simplified DLL timing error estimator based on noncoherent signal processing.

influence of the channel response. Lastly, the squared values are subtracted from each other to create an error signal.

$$\text{Error Signal} = \left|\sum r(t)\cdot c(t-\tau-\Delta)\right|^2 - \left|\sum r(t)\cdot c(t-\tau+\Delta)\right|^2 \qquad (6.104)$$

6.4.5 BER Performance Degradation

In this section, we will attempt to quantify the performance degradation due to timing offsets in the receiver signal processing. We hope the following figures would encourage system designers to fully characterize their detectors in an effort to create a specification for timing jitter allowed for a certain tolerable performance degradation.

Figure 6.33 clearly shows as the timing offset increases the performance degradation exponentially increases. The curves reflect using a differential detector for $\pi/4$-DQPSK modulation. The symbol

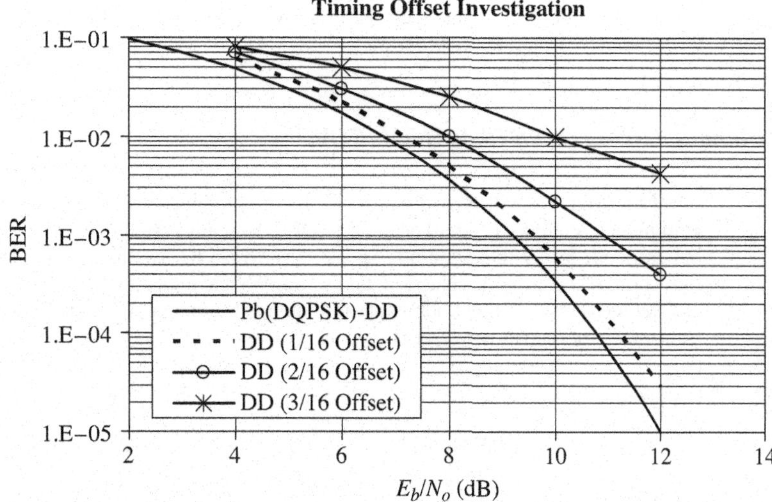

FIGURE 6.33 Differential detection performance degradation as a function of timing offset.

stream entering the differential detector is offset in time by a fixed amount as indicated above and maintained throughput the simulation run.

Next we plot the performance degradation for the various noncoherent detectors presented earlier: differential detection, 1NEC, 2NEC, and MLDD. Generally speaking, increasing the time offset degrades performance. Moreover, the performance of the detectors with the larger observation window is still better than that of the conventional differential detector for small timing errors. These curves were plotted assuming an E_b/N_o equal to 8 dB (see Fig. 6.34).

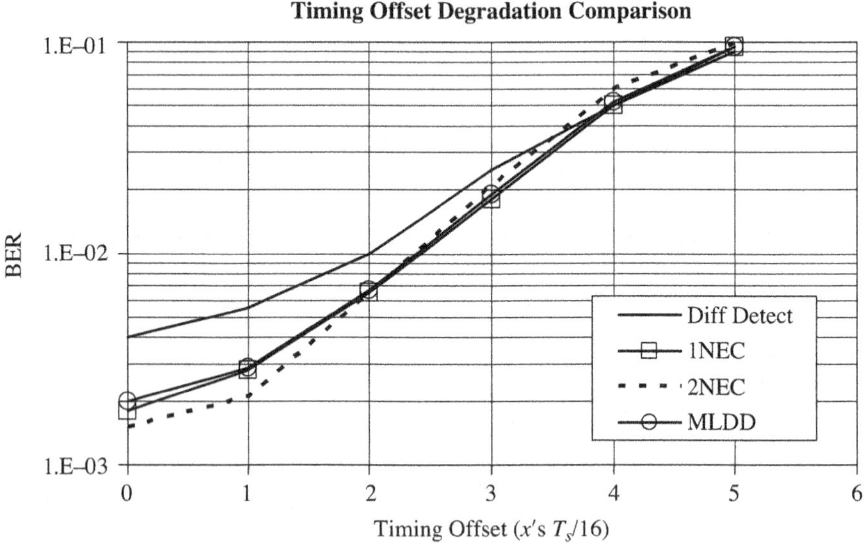

FIGURE 6.34 Timing offset degradation comparison for various detection techniques.

The detection techniques with the larger observation window degrade at a faster rate than those with a smaller observation window.

6.5 CHANNEL QUALITY ESTIMATION (CQE)

In this section, we will provide an overview of some techniques used to estimate the quality of the communication link, they will be called CQE. The many actual uses of the CQE will not be discussed, but suffice it to say that generally speaking they can be used for performance improvement, power reduction, receiver mode selection, and so on. We leave this up to the system designer to be creative.

The methods discussed below are not intended to be an exhaustive listing, but to form a foundation and spark innovative designs.

6.5.1 Pilot Error Rate

This first quality estimation technique is based on the assumption that certain bit fields (e.g., pilot bits) will be known to the receiver; similar assumptions were made earlier on the topic of training the equalizer. In this case, we will use this a priori knowledge to calculate the Pilot Error Rate (PER) of the communication link. We will provide a time slot structure for sake of discussion (see Fig. 6.35).

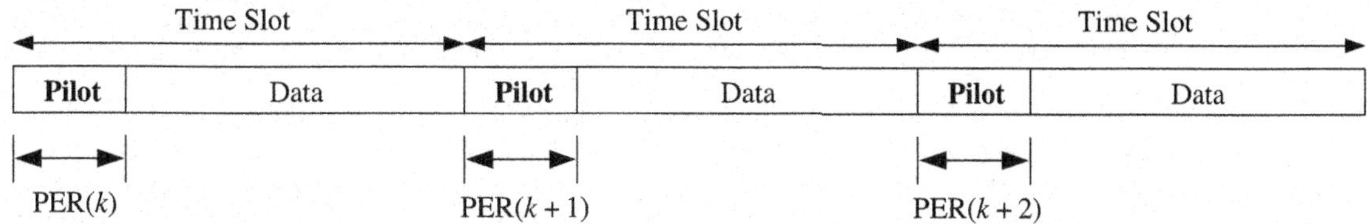

FIGURE 6.35 General time slot structure.

Some of the disadvantages of this technique is that it requires time slot synchronization. This may not be an issue for channels actively being used in the communication link but may be an issue if the receiver is instructed to change frequency bands to measure its quality indication in a relatively short period of time. Also depending on the operating point of the communication system, available information, channel conditions, and so on, many slots may be required to provide statistical confidence in the measured quantity.

Here a simplified receiver block diagram that calculates the pilot errors over the current time slot and then averages this value using the previously measured values is shown in Fig. 6.36.[4]

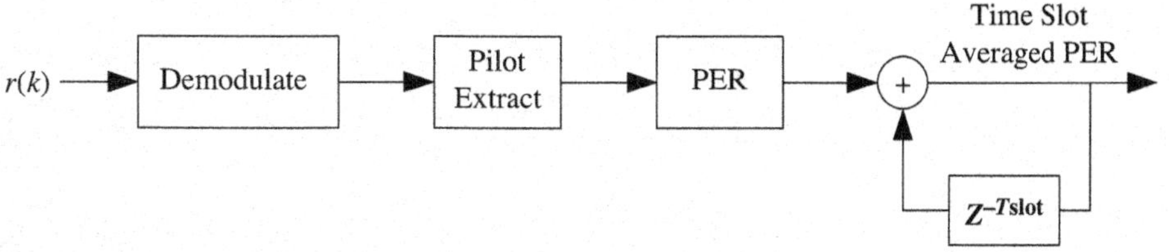

FIGURE 6.36 PER receiver block diagram.

6.5.2 Nonredundant Error Correction Based

This second quality estimation technique is based on the Nonredundant Error Correction (NEC) demodulation technique presented in the early chapters. Recall the NEC demodulator calculates syndromes in order to determine if an error is present. This CQE counts the number of times (or frequency) the error signal is enabled. A simplified block diagram of the NEC-based CQE is given in Fig. 6.37.

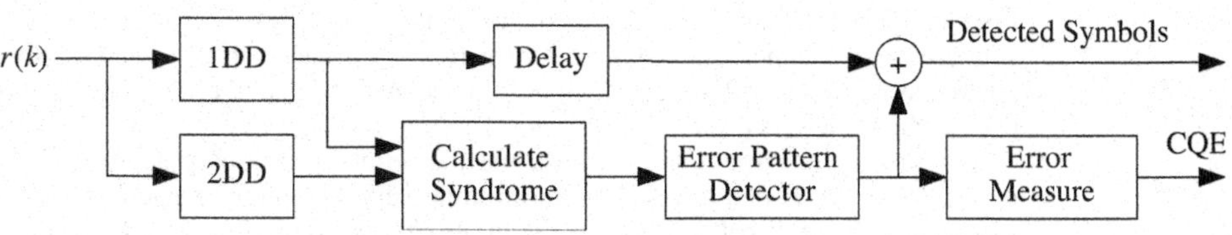

FIGURE 6.37 Block diagram of an NEC-based CQE.

[4] United States Patent # 5,159,282.

In Fig. 6.37, we show the basic elements of the NEC demodulator. The CQE output comes from the block labeled error measure. Here, the error signal is counted to determine an estimate of the symbol error rate and then produces the CQE output.

Since the CQE is now based on all the symbols in the time slot, a priori information is not needed. Also the receiver doesn't need to have time slot synchronization before calculating the CQE. This is a useful property to exploit, especially if the receiver is asked to measure the quality of another link; in this case it would immediately be able to perform this CQE task. Lastly, since the entire slot is used in the calculation it is conceivable that fewer slots are needed to achieve the desired confidence level in the measured quantity, when compared to techniques involving only the pilot bit field.

In Fig. 6.38 we plot the estimated error rate using the 1NEC demodulator receiver in an AWGN channel. The CQE results follow reasonably well when compared to the conventional differential detector.

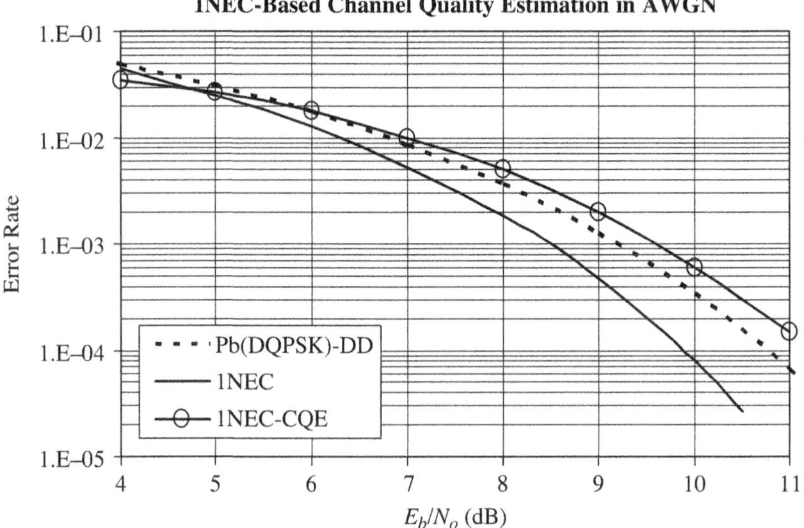

FIGURE 6.38 1NEC-based channel quality estimation performance.

6.5.3 Channel Impulse Response Based

This third quality estimation technique is based on the receiver's knowledge of the Power Delay Profile (PDP) or Channel Impulse Response (CIR). Regardless of the multiple access scheme used, some form of channel delay spread information is estimated in the receiver. For example, in DS-CDMA systems the PDP is continually estimated and updated in order to find multipaths that can be used in the RAKE receiver.

For sake of discussion we will assume the wireless multipath channel consists of a two-ray model. Where $h_1(t)$ and $h_2(t)$ represent the time-varying channel response of the first arriving and second arriving multipaths, respectively. In the case of "*K*" multipath rays we have provided earlier an equation to determine the rms delay spread of the channel that can also be used. Continuing along with the simplified example, we can take the ratio of the relative channel powers to determine if the channel is very frequency selective (i.e., both have large average power) or somewhat frequency selective (i.e., large difference in average powers) or nonfrequency selective (i.e., single large average power).

As an example, we may want to enable another mode (say mode #2) in the receiver if the delay spread is significant otherwise continue with mode #1. This can be based on the following comparison:

$$\frac{|h_1|^2}{|h_2|^2} \geq \text{Threshold} \quad \text{or} \quad \frac{|h_2|^2}{|h_1|^2} \geq \text{Threshold} \tag{6.105}$$

The benefit of this technique is that time slot synchronization may not be needed depending on the method used to derive the channel estimates. Also we can make use of the information that is already available in the receiver so additional signal processing may not be required.

6.5.4 Differential-Based Statistics

This fourth quality estimation technique is based on the differential detector's output signal. Here we will use the difference between adjacent time samples to determine if there is significant delay spread or interference present in the channel. First we will rewrite the differential detector equations using the Cartesian coordinate system.

$$X_k = I_k \cdot I_{k-1} + Q_k \cdot Q_{k-1} \tag{6.106}$$

$$Y_k = Q_k \cdot I_{k-1} - Q_{k-1} \cdot I_k \tag{6.107}$$

We can clearly see if the expectation is taken on the real output signal, X_k, this resembles the sum of the autocorrelations of the I- and Q-channels. Similarly, the expectation on the imaginary output signal, Y_k, resembles the difference of the cross-correlations of the I- and Q-channels.

Consider the real output, this is a measure of the similarity between the present symbol at time instant k and the previous symbol at time instant $k-1$. Assuming an overall raised cosine pulse shaping filter is employed in the system, then this is a measure of how much delay spread is present in the channel. This output value should be averaged due to its discrete nature and calibrated against various channel conditions. Let us assume a two-ray, frequency-selective fading channel where the separation in time between the two rays is equal to τ sec. Then we can write the differential detector output as $d(t) = X(t) + j\,Y(t)$.

$$d(t) = r(t) \cdot r^*(t - T) \tag{6.108}$$

$$d(t) = \{h_1(t) \cdot s(t) + h_2(t - \tau) \cdot s(t - \tau)\}$$
$$\{h_1(t - T) \cdot s(t - T) + h_2(t - \tau - T) \cdot s(t - \tau - T)\}^* \tag{6.109}$$

Taking the expectation operation reveals the following which can be used to determine the presence of delay spread ($\phi(t)$ that is the differential modulation symbol).

$$E\{d(t)\} = R_{h_1}(T) \cdot e^{j\phi(t)} + R_{h_2}(T) \cdot e^{j\phi(t-\tau)} \tag{6.110}$$

This consists of the sum of the autocorrelations of the channel response weighted by the modulation.

6.5.5 Eye Opening Statistics

This fifth quality estimation technique is based on the symbol eye opening at the output of the demodulator. It is well known that the received signal's eye diagram is indicative of the quality of the received signal and in turn the quality of the channel. In this CQE we will assume the normal operation is to have a received signal with an opened eye diagram. Hence the block diagram shown in Fig. 6.39 is used to convey the idea of using the averaged eye opening power to provide a CQE.[5]

[5] United States Patent #5,159,282.

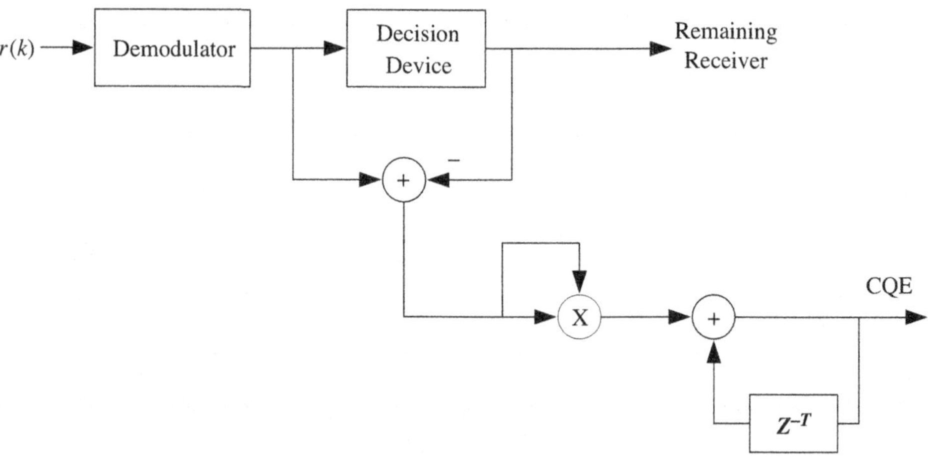

FIGURE 6.39 Eye opening CQE block diagram.

The above block diagram produces a CQE that is inversely proportional to the received eye opening. Consider the received signal to have much noise, interference, and ISI then the eye diagram that would be considered to have a small opening, if not closed. This would create a large error signal which is the difference between the decision device output and its input. In turn, this will create an even larger error power signal. On the other hand, if the received eye opening is large, then the difference between this signal and the output of the decision device is small, and so will be its power. Hence we have a CQE value that is inversely proportional to the eye opening.

In these subsections we have presented a few CQE techniques that can be used to generally provide a qualitative and quantitative measurement of the channel.

6.6 AUTOMATIC FREQUENCY CONTROL

When performing any spectral conversion operation, frequency offsets exist due to component tolerances, temperature changes, estimation error, device aging, and so on. This frequency offset is defined as the offset in frequency between the carrier frequency of the received signal and the locally generated frequency. Let us recall the following functionality of the quadrature demodulator shown in Fig. 6.40.

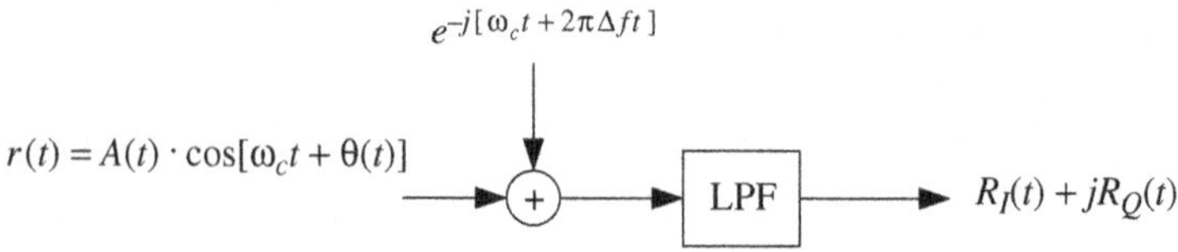

FIGURE 6.40 Quadrature demodulation operation.

The receiver will estimate this residual offset in frequency (Δf) and correct it by performing frequency compensation or what is commonly called Automatic Frequency Control (AFC) [34–37]. This frequency offset should be diminished since its presence will degrade overall system performance. With the help of the above block diagram we can represent the offset mathematically in the following manner

$$\begin{bmatrix} R_I(t) \\ R_Q(t) \end{bmatrix} = \begin{bmatrix} \cos[\theta(t)] & \sin[\theta(t)] \\ \sin[\theta(t)] & -\cos[\theta(t)] \end{bmatrix} \cdot \begin{bmatrix} \cos[2\pi\Delta f t] \\ \sin[2\pi\Delta f t] \end{bmatrix} \tag{6.111}$$

6.6.1 Typical Receiver Architecture

In order to facilitate a technical discussion around this topic we need to introduce a receiver architecture to be used as a baseline. In Fig. 6.41 we show a simplified block diagram that emphasizes the frequency compensation functionality.

This diagram shows the receiver performing the spectral down conversion operation through the use of the quadrature demodulator. The baseband section will estimate the frequency offset and then send a correction signal to the analog front end of the receiver. This particular architecture performs the spectral down conversion in the analog domain; if the application pushes this operation closer toward the antenna, then an analog correction may not be needed. In any event this frequency control loop is continually running since the offset can vary over temperature and time as discussed above.

6.6.2 ARCTAN Based

In this section, we will provide a frequency offset estimation technique using differential detection and $\pi/4$-DQPSK modulation scheme. Recall the differential detector's output constellation consists of four states since there are only four possible transitions allowed. Also the presence of any constant frequency offsets prior to the differential detector will show up as constant phase offsets at the output of the detector. The amount of phase offset is directly proportional to the amount of frequency offset present on the received signal. Moreover, the direction of the phase offset depends on if the offset is a positive value or a negative value. Hence the differential detector output can be used as an indication of the amount of frequency offset present on the received signal (see Fig. 6.42).

The AFC loop will generally consist of three components: the phase detector, Voltage Control Oscillator (VCO), and loop filter. An example of such components will be provided next.

With respect to the above AFC block diagram, let us represent the differential detector output as (assuming the compensation loop is open)

$$d(t) = \{h(t) \cdot e^{j[\theta(t)+\Delta\omega t]}\} \cdot \{h(t - T) \cdot e^{j[\theta(t-T)+\Delta\omega(t-T)]}\}^* \tag{6.112}$$

where T = the time separation between samples compared and $\Delta\omega$ = residual frequency offset present on the signal.

$$d(t) = h(t) \cdot h^*(t - T) \cdot e^{j[\theta(t)-\theta(t-T)+\Delta\omega T]} \tag{6.113}$$

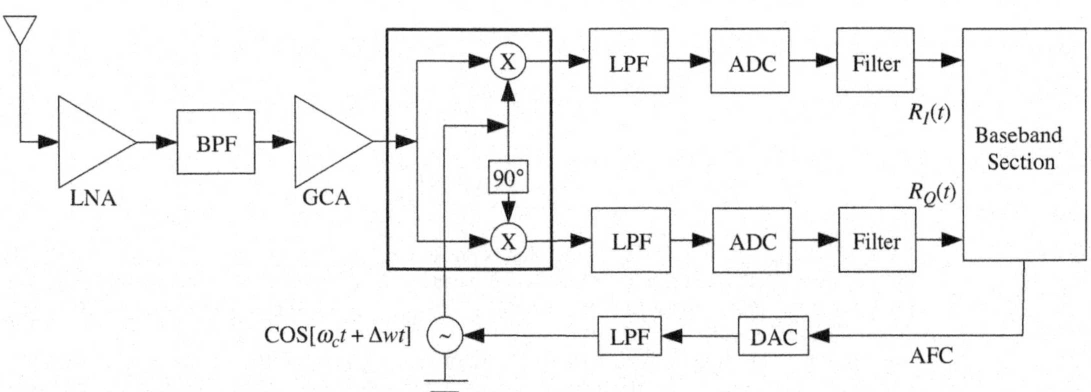

FIGURE 6.41 Typical receiver architecture with emphasis on frequency control.[6]

[6] LNA = Low Noise Amplifier, BPF = Band Pass Filter, and GCA = Gain Control Amplifier.

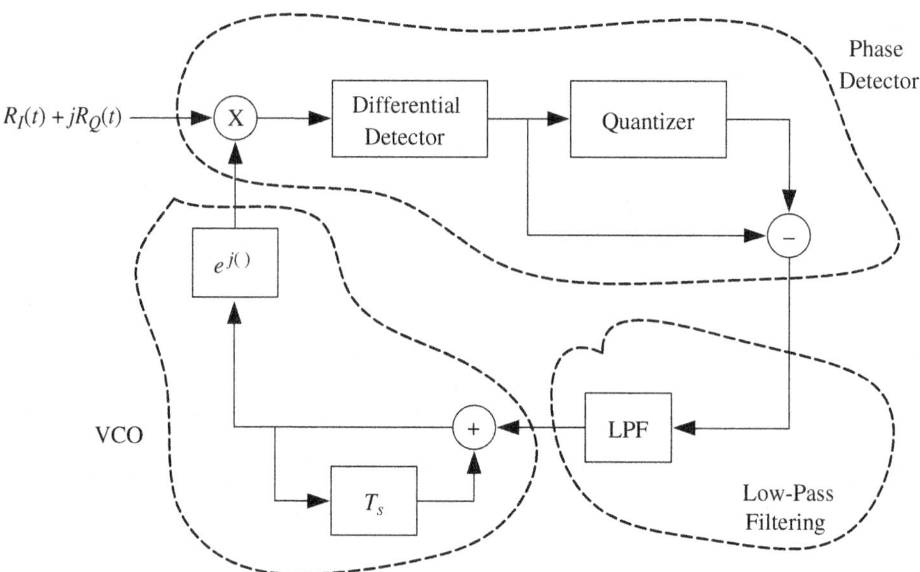

FIGURE 6.42 π/4-DQPSK frequency estimation and compensation block diagram.

Assuming differential encoding this can be represented as

$$d(t) = h(t) \cdot h^*(t - T) \cdot e^{j[\phi(t) + \Delta\omega T]} \tag{6.114}$$

We discuss two options: first without differential encoding modulation and second with differential encoding. In the absence of modulation the expectation of the differential detector output produces the channel autocorrelation function.

$$E\{d(t)\} = R_h(T) = \sigma_h^2 \cdot J_o(\omega_d T) \cdot e^{j\Delta\omega T} \tag{6.115}$$

where ω_d= the maximum Doppler spread, σ_h^2= fading channel power. The frequency error can then be easily determined as

$$\Delta\omega = \frac{1}{T} \cdot \arg\{R_h(T)\} \tag{6.116}$$

$$\Delta\omega = \frac{1}{T} \cdot \text{ATAN}\left(\frac{d_Q(t)}{d_I(t)}\right) \tag{6.117}$$

The second option is when the modulation is present and is discussed next. Below we present an AFC technique applied to the π/4-DQPSK, which can be extended to other modulation schemes. We will also provide approximations to the error estimator. The first technique will be called AFC-One and is shown in Fig. 6.43.

In the above AFC algorithm we have made use of the differential detector. The differential detector's phase is denoted as θ_2 and the quantized symbols phase is denoted as θ_1. This was accomplished through the help of an ATAN function which can be implemented with Taylor series expansions or LUT. The error signal, $e(t)$, is obtained by the difference of the two estimated phases. The estimated error signal enters the loop filter and then goes to the VCO. The design of the loop filter is important since it will define the loop noise bandwidth, control convergence time, and affect system performance.

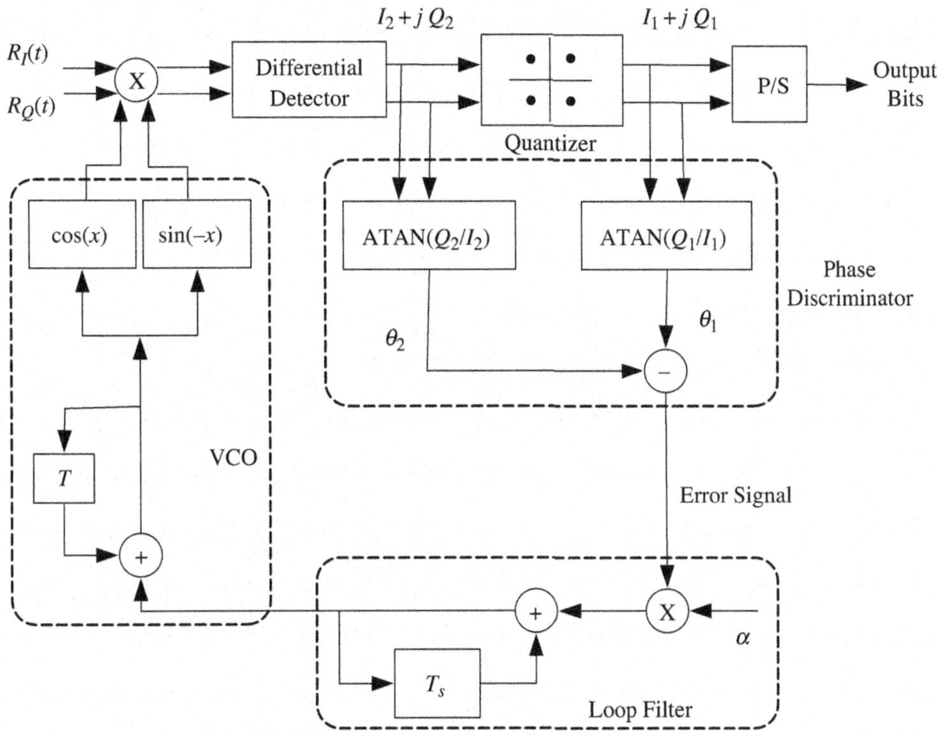

FIGURE 6.43 AFC-One block diagram.

The corresponding error signal for this AFC-One algorithm can be written as

$$e = \text{ATAN}\left[\frac{I_1 \cdot Q_2 - I_2 \cdot Q_1}{I_1 \cdot I_2 + Q_1 \cdot Q_2}\right] \tag{6.118}$$

An alternative way to view this is through the Cartesian coordinate system, where we have plotted the received symbol (shown by a box) along with the quantized version (shown by a circle). Assume the data symbol, $1 + j\,1$, was transmitted and thus we have the following event described in Fig. 6.44. Here a positive frequency offset is considered producing the error.

We notice in the absence of a frequency offset; the error signal is completely projected onto the real axis. In the presence of a frequency offset the error signal has a projection onto the imaginary axis, which can be used to indicate a frequency offset. This leads us to present the AFC-Two algorithm shown in Fig. 6.45.

The corresponding error signal is written as follows which is accurate for small frequency offsets.

$$e = I_1 \cdot Q_2 - I_2 \cdot Q_1 \tag{6.119}$$

Note this error signal is basically the small angle approximation of the error signal used for the AFC-One algorithm presented earlier.

We will also present two alternative forms of representing the phase discriminator for the AFC-Two algorithm (see Fig. 6.46). These forms are intended to provide insight into the mechanisms used to create the error signal.

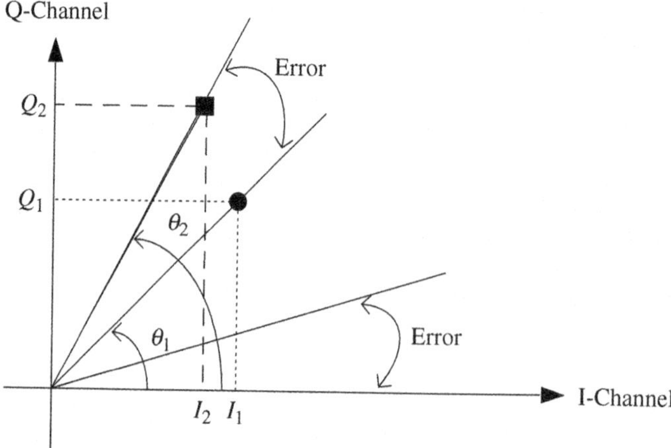

FIGURE 6.44 The effect of frequency offsets on detected constellation states.

In fact, there are quite a few options available to the system designer, the reader is encouraged to refer to [38] for further discussions and techniques.

Next we will discuss another form of the AFC without using the differential detector in the phase discriminator operation. In some publications this has been called Digital Tan lock Loop (DTL) [39].

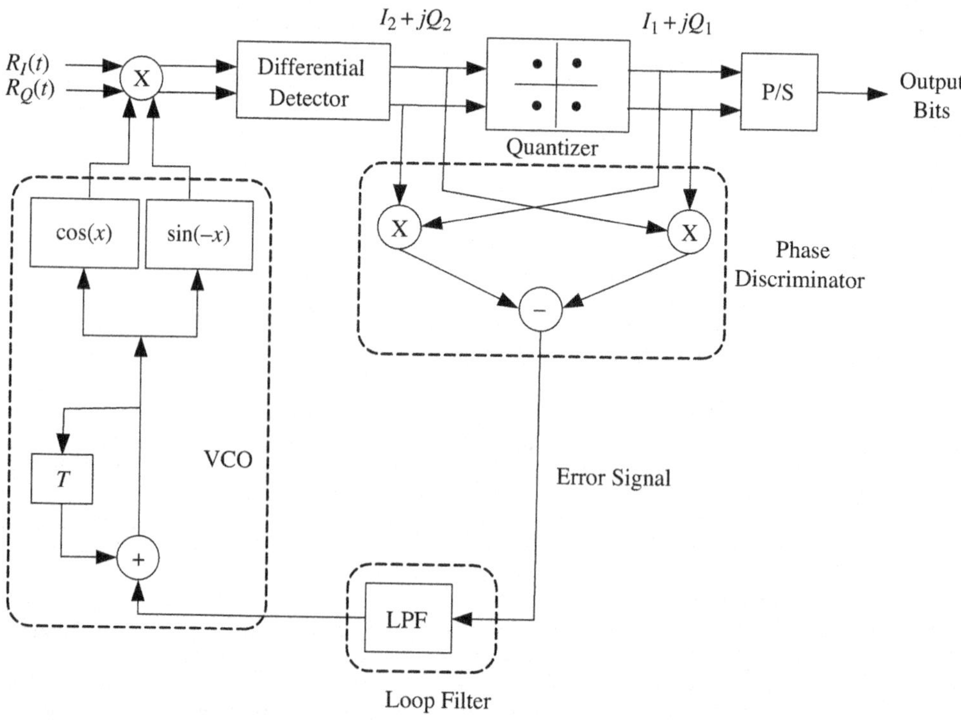

FIGURE 6.45 AFC-Two block diagram.

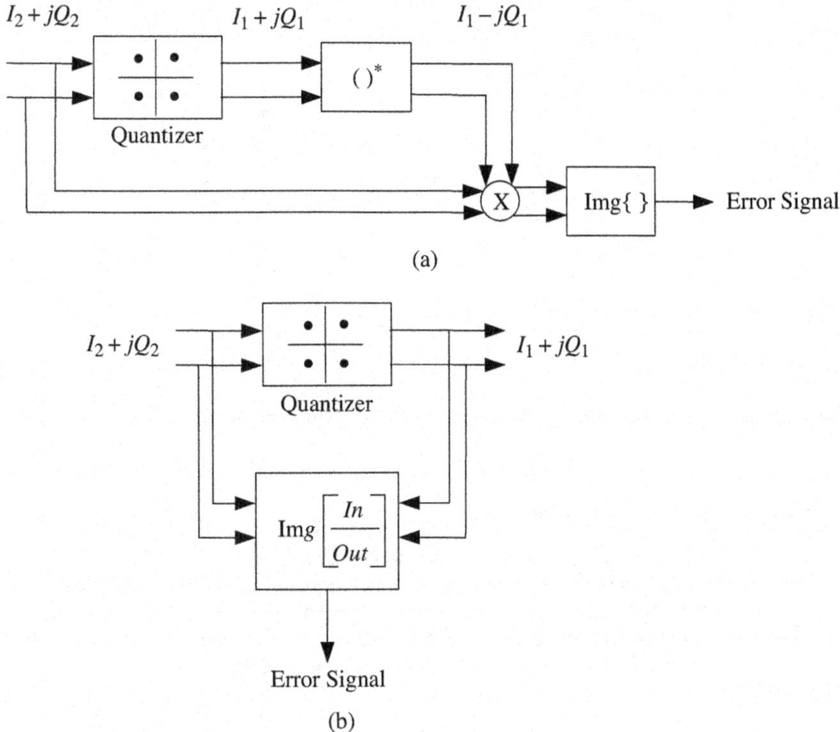

FIGURE 6.46 Alternative forms of the AFC-Two phase discriminator.

The basic premise of the loop is the same, with the difference coming from the method used in the phase discriminator. For $\pi/4$-DQPSK modulation, the error signal is expressed as

$$x(t) = [I(t) + jQ(t)]^8 = [e^{j(\theta(t)+\Delta\omega t)}]^8 \tag{6.120}$$

$$x(t) = \cos[8\theta(t) + 8\Delta\omega t] + j\sin[8\theta(t) + 8\Delta\omega t] \tag{6.121}$$

Since $\theta(t) = k \cdot {}^{\pi}/_{4}$ where $k = 0, 1, \ldots, 7$ then $8\,\theta(t) = k \,.\, 2\pi$, which places all 8 phase states onto a single state on the real axis. By raising the modulated symbol stream to the power of 8, we have essentially removed the modulation. Similarly, QPSK can be raised to the power of 4 to remove the modulation present on the received signal (see Fig. 6.47).

A point worth remembering is when the modulation is raised to a power of say K, then the phase change has increased by the same amount. This phase amplification should be removed when estimating absolute terms. Hence the error signal can be derived as

$$e = \frac{1}{8} \cdot \operatorname{Im} g\{[I(t) + jQ(t)]^8\} \tag{6.122}$$

The interested reader should consult [38] for alternative forms of frequency error discrimination techniques, such as "Differentiator AFC" and "Cross Product AFC," respectively.

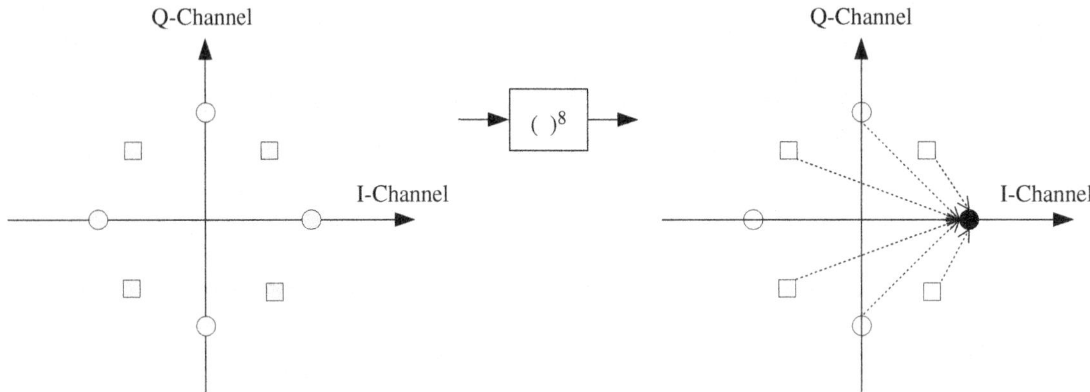

FIGURE 6.47 Phase state transitions due to raising $\pi/4$-DQPSK to the power of 8.

6.6.3 BER Performance Degradation

In this section, we will present the BER performance of the 1NEC, 2NEC, MLDD, and DD detectors previously described for various residual frequency offsets. In Fig. 6.48 we start with the performance of the 2NEC as the frequency offset is varied from 200 to 800 Hz. The results clearly show as the residual frequency offset increases performance degradation increases. Even for a small frequency offset of 200 Hz, a fraction of a dB degradation exists.

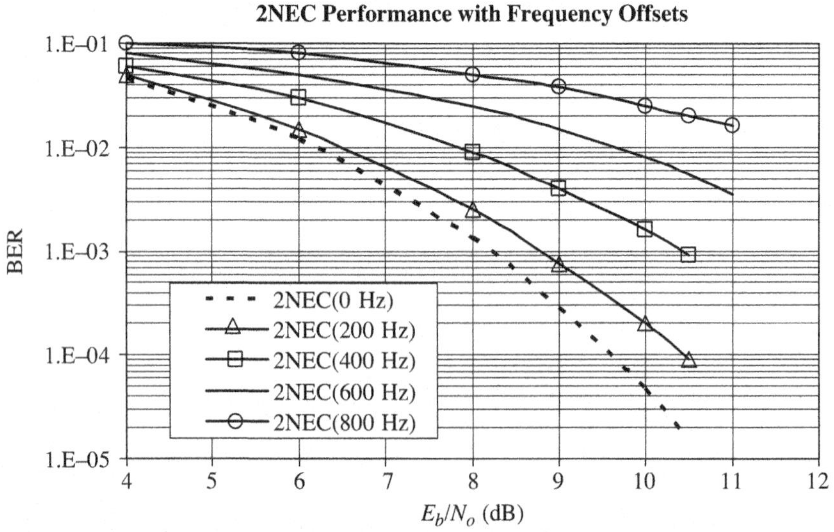

FIGURE 6.48 The BER performance of the 2NEC receiver for various frequency offsets.

This system used the $\pi/4$-DQPSK modulation in the AWGN channel. The symbol rate used in the simulation was set to 24.3 ksps, which for a frequency offset of 400 Hz produced a normalized frequency offset of $f_{off} \cdot T_s = 0.016461$.

Next we wanted to show the performance degradation of the earlier described detectors as the frequency offset is increased. Here we see for the zero offset case, widening the observation window when making a decision improves overall performance. However, when there is frequency offset present the behavior is different. First we compare the performance of 1NEC or MLDD to the differential detector, we notice that there is gain in widening the observation window as long as the residual frequency offset is less than approximately 600 Hz, for this particular system simulation. Moreover, when comparing the differential detector to the 2NEC, the residual offsets should be less than 300 Hz. The message we wish the reader to take away from Fig. 6.49 is that when widening the observation window, the receiver is more susceptible to frequency offsets. This is evident by the rapid degradation exhibited by the respective receivers.

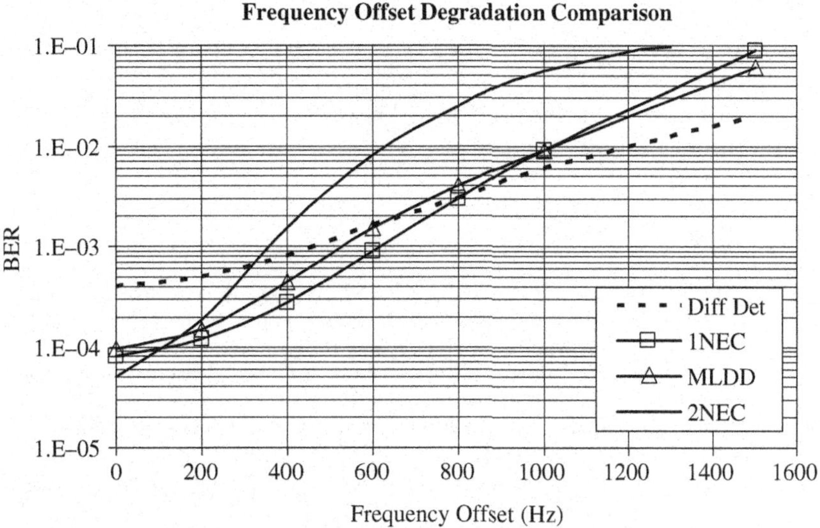

FIGURE 6.49 BER Performance degradation comparison for various frequency offsets.

6.7 OVERALL RECEIVER BLOCK DIAGRAM

In this section, we will build upon the above described functions and present a receiver block diagram. The diagram will include more of the RF and baseband interface signals.

6.7.1 Direct Conversion Architecture (ZIF)

In this section the direction conversion receiver will be presented and some details highlighted. This type of receiver architecture does not create an intermediate frequency (IF), the received carrier signal is directly converted to baseband (see Fig. 6.50). This type of spectral down conversion is sometimes called a Zero IF (ZIF) receiver.

We have intentionally denoted the baseband section as a large box. This section will perform the signal processing techniques discussed so far in this book. Specifically the algorithms include: symbol timing recovery, demodulation, FEC, equalization, frequency offset estimation/compensation, DC offset compensation, automatic gain control, and so on. We have chosen to emphasize the analog front end (AFE) control signals such as AGC, AFC, and DC Offset. Since an ADC is placed in the data path the received signal will have a fixed dynamic range. A control signal is used to adjust the

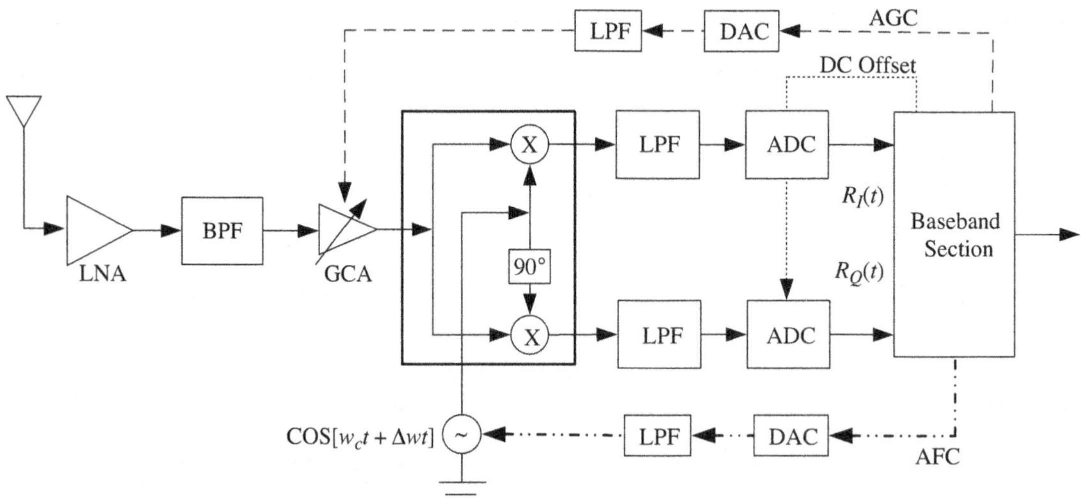

FIGURE 6.50 Direct conversion receiver architecture.

Gain Control Amplifier (GCA) in the RF front end to keep the signal within the operating dynamic range. This correction signal will be used to compensate for path loss and time varying phenomenon such as shadowing [40].

The frequency offset control signal (i.e., AFC) will be sent to the AFE to compensate for residual frequency offsets due to tolerances in the crystal oscillators used in the design. In addition to the fact that large frequency offsets degrade performance, wireless standards dictate certain frequency uncertainties be met. For example, in the WCDMA standard a certain offset uncertainty should be achieved (i.e., 0.1 ppm for the UE) before transmission on the uplink can occur. Lastly, analog components tend to have DC drifting issues which basically shifts the information bearing signal. However, the baseband signal processing algorithms typically operate under the impression that the origin is located at (0, 0). Moving the origin to another location will degrade performance unless this DC offset is estimated and removed [41].

The quadrature demodulation requires the use of the LPF to filter out the spectral components at twice the input frequency. Recalling our earlier discussion on pulse shaping, this LPF can, in fact, have a response that this is matched to the transmitted waveform.

6.7.2 Software Defined Radio (SDR) Evolution

In this section, we will present some comments on the notation of having a radio (MS or UE) be completely programmable through software changes [42]. The present day approach is to optimize the design with respect to cost, size, power, and overall performance. Much effort is placed in performing functionality partitioning of the signal processing algorithms to achieve a competitive consumer product.

The current state of the art in digital signal processors (DSPs) preclude performing all the baseband tasks within the DSP. Until such a path exists various options are available such as using multiple processors and/or FPGA fabric [42]. The SDR Forum [43] defines five tiers of solutions, the reader should consult the variety of references for further reading. The most likely path to take form is to have a solution possible to support a variety of predefined standards. This joint optimization across all standards may be possible and still allow for software programmability across the same platform.

The most likely multistandard terminal will consists of GSM, WCDMA/LTE, Bluetooth, and Wireless LAN. Generally speaking, the radio access technologies can be divided into short range communications (i.e., Bluetooth) to enable hands-free voice calls, remote keyboards and mice, allow FM music, as well as a variety of other applications. Also, data transfer (i.e., WiFi) would be used by portable computing devices such as PDAs, Smart Phones, laptops, and so on. Lastly, the cellular umbrella is essential to accommodate mobility, not to mention roaming where we have specifically mentioned GSM/WCDMA/LTE. Lastly, support of location based services will be required (i.e., GPS, and so on).

Relative design issues are discussed in [40] and [41] for Reconfigurable Multistandard Terminals.

As mentioned in the references used in this section, creating a multistandard terminal is challenging in and of itself. Placing additional constraints such as being cost-competitive, size and power aggressive designs make this task even more difficult. A simple argument is that different standards have differing levels of requirements; creating a device that assumes the most challenging requirements for all applications will essentially overdesign the terminal and potentially lose certain design goals.

In Fig. 6.51 we provide a high-level block diagram highlighting certain features to be considered when designing SDR applications. We have chosen to split the design space into three areas: Spectral Conversion (RF), Baseband—Hardware and Software.

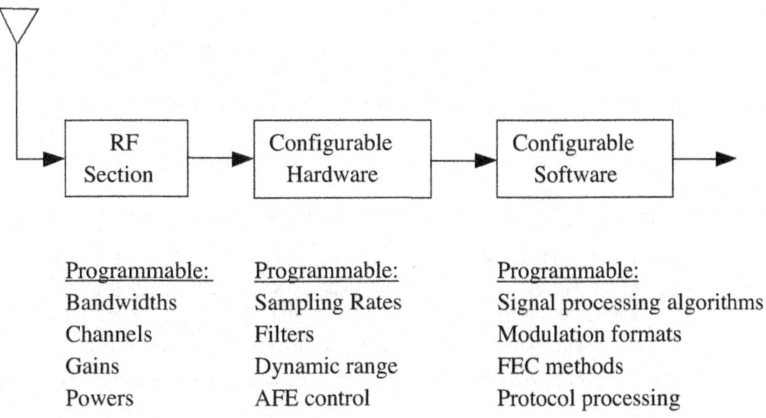

FIGURE 6.51 Software defined radio features.

Some of the reasons for the segmentation are given in the block diagram. We know that the multistandard terminal would need to operate in different frequency bands. These bands will not only differ within the geography of use, but also from one country to another. These bands will have different bandwidths and operating specifications (i.e., FCC, and so on) that need to be met.

The baseband would need to have a modem to support various modulation schemes and widely varying data rates. The dynamic range and quantization effects should be carefully considered when jointly optimizing the multistandard terminal.

Lastly, the software would need to perform demanding signal processing functions for the various modulation schemes and at the widely varying data rates. We have focused our attention to radio access technologies that cover Personal Area Network (PAN) to Wide Area Network (WAN). This wide variation will be visible in the varying channel conditions the user experiences when the terminal is utilized. Since the channel conditions dictate the complexity of the receiver signal processing, the terminal will have differing levels of signal processing demands.

In Fig. 6.52 we show a hierarchical diagram of the software architecture. We begin with the reconfigurable hardware that has been optimized to handle concurrent and separate wireless standards. This optimized reconfigurable hardware can potentially have functionality partitioning to accommodate power restrictions. The drivers will essentially control the functionality partitioning. Since the software designer would strive to reuse as much software as reasonably possible, middleware is inserted to act as

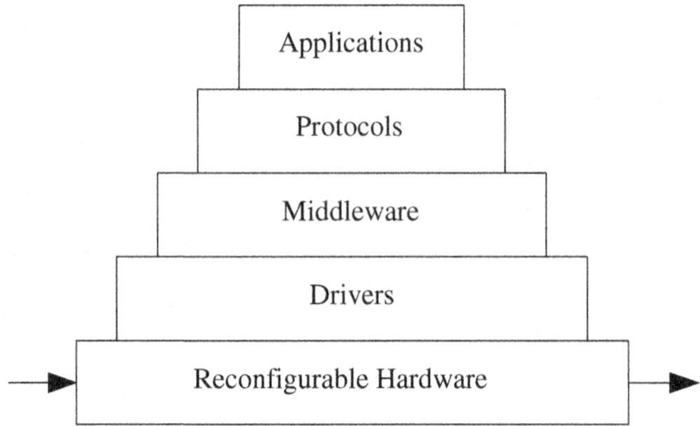

FIGURE 6.52 Exemplary hierarchical architecture.

a liaison between the specific drivers and the wireless protocols that allow the terminal to communicate to the base station (BS) or access point or another node. Lastly, various applications will be used to provide the feature rich terminals desired by various consumers. Since concurrent applications will become more and more mainstream the ability to handle multiple processor threads is essential.

REFERENCES

[1] J. G. Proakis, *Digital Communications*, McGraw Hill, 1989, New York.

[2] G. D. Forney, Jr., "Maximum-Likelihood Sequence Estimation of Digital Sequences in the Presence of Intersymbol Interference," *IEEE Transactions on Information Theory*, Vol. IT-18, No. 3, May 1972, pp. 363–378.

[3] S. Haykin, *Adaptive Filter Theory*, Prentice Hall, 1996, New Jersey.

[4] H. Meyr, M. Moeneclaey, and S. A. Fechtel, *Digital Communication Receivers*, J. Wiley & Sons, 1998, New York.

[5] S. U. H. Qureshi, "Adaptive Equalization," *Proceedings of the IEEE*, Vol. 73, Sept. 1985, pp. 1349–1387.

[6] G. H. Golub and C. F. Van Loan, *Matrix Computations*, Johns Hopkins University Press, 1989, Maryland.

[7] P. Lancaster and M. Tismenetsky, *The Theory of Matrices*, Academic Press, 1985, California.

[8] X. Wu and A. M. Haimovich, "Adaptive Arrays for Increased Performance in Mobile Communications," in: *IEEE Personal, Indoor and Mobile Radio Communications Conference*, 1995, pp. 653–657.

[9] T. Boros, G. G. Raleigh, and M. A. Pollack, "Adaptive Space-Time Equalization for Rapidly Fading Communications Channels," in: *IEEE Global Telecommunications Conference*, 1996, pp. 984–989.

[10] Y. Li, J. H. Winters, and N. R. Sollenberger, "Parameter Tracking of STE for IS-136 TDMA Systems with Rapid Dispersive Fading and Co-Channel Interference," *IEEE Transactions on Vehicular Technology*, Vol. 48, No. 4, Dec. 1999, pp. 3381–3391.

[11] A. J. Paulraj and C. B. Papadias, "Space-Time Processing for Wireless Communications," *IEEE Signal Processing Magazine*, 1997, pp. 49–83.

[12] F. Choy and M. Chernikov, "Combinations of Adaptive Antennas and Adaptive Equalizers for Mobile Communications," in: *IEEE TENCON*, 1997, pp. 497–500.

[13] A. A. Mostafa, "Single Antenna Interference Cancellation (SAIC) Method in GSM Network," in: *IEEE Vehicular Technology Conference*, 2004, pp. 3748–3752.

[14] P. Balaban and J. Salz, "Optimum Diversity Combining and Equalization in Digital Data Transmission with Applications to Cellular Mobile Radio—Part I: Theoretical Considerations," *IEEE Transactions on Communications*, Vol. 40, No. 5, May 1992, pp. 885–894.

[15] P. Balaban and J. Salz, "Optimum Diversity Combining and Equalization in Digital Data Transmission with Applications to Cellular Mobile Radio—Part II: Numerical Results," *IEEE Transactions on Communications,* Vol. 40, No. 5, May 1992, pp. 895–907.

[16] T. Nihtila, J. Kurjenniemi, M. Lampinen, and T. Ristaniemi, "WCDMA HSDPA Network Performance with Receive Diversity and LMMSE Chip Equalization," in: *IEEE PIMRC*, 2005, pp. 1245–1249.

[17] M. J. Heikkila, P. Komulainen, and J. Lilleberg, "Interference Suppression in CDMA Downlink through Adaptive Channel Equalization," in: *IEEE Vehicular Technology Conference,* 1999, pp. 978–982.

[18] G. E. Bottomley and K. Jamal, "Adaptive Arrays and MLSE Equalization," in: *IEEE Vehicular Technology Conference,* 1995, pp. 50–54.

[19] G. E. Bottomley, K. J. Molnar, and S. Chennakeshu, "Interference Cancellation with an Array Processing MLSE Receiver," *IEEE Transactions on Vehicular Technology,* Vol. 48, No. 5, Sept. 1999, pp. 1321–1331.

[20] H. Yoshino, K. Fukawa, and H. Suzuki, "Interference Canceling Equalizer (ICE) for Mobile Radio Communication," *IEEE Transactions on Vehicular Technology,* Vol. 46, No. 4, Nov. 1997, pp. 849–861.

[21] M. V. Clark, "Adaptive Frequency-Domain Equalization and Diversity Combining for Broadband Wireless Communications," *IEEE Journal on Selected areas of Communications,* Vol. 16, No. 8, Oct. 1998, pp. 1385–1395.

[22] D. Falconer et al. "Frequency Domain Equalization for Single-Carrier Broadband Wireless Systems," *IEEE Communications Magazine,* April 2002, pp. 58–66.

[23] K. Ishihara, K. Takeda, and F. Adachi, "Pilot-Assisted Channel Estimation for Frequency-Domain Equalization of DSSS Signals," in: *IEEE Vehicular Technology Conference,* 2005, pp. 783–787.

[24] D. L. Iacono et al. "Serial Block Processing for Multi-Code WCDMA Frequency Domain Equalization," *IEEE Communications Society, WCNC,* 2005, pp. 164–170.

[25] K. Takeda and F. Adachi, "Inter-Chip Interference Cancellation for DS-CDMA with Frequency-Domain Equalization," in: *IEEE Vehicular Technology Conference*, 2004, pp. 2316–2320.

[26] M. Oerder and H. Meyr, "Digital Filter and Square Timing Recovery," *IEEE Transactions Communications,* Vol. COM-36, No. 5, May 1988, pp. 605–612.

[27] K. H. Mueller and M. Muller, "Timing Recovery in Digital Synchronous Data Receiver," *IEEE Transactions Communications,* Vol. COM-24, No. 5, May 1976, pp. 516–531.

[28] F. Gardner, "A BPSK/QPSK Timing Error Detector for Sampled Receivers," *IEEE Transactions Communications,* Vol. COM-34, May 1986, pp. 423–429.

[29] G. Bolding and W. G. Cowley, "A Computationally Efficient Method of Timing and Phase Estimation in TDMA Systems Using a Preamble Sequence," *International Journal of Satellite Communications,* Vol. 13, 1995, pp. 441–452.

[30] F. M. Gardner, "Interpolation in Digital Modems Part I: Fundamentals," *IEEE Transactions on Communications,* Vol. 41, No. 3, March 1993, pp. 501–507.

[31] L. Erup, F. M. Gardner, and R. A. Harris, "Interpolation in Digital Modems Part II: Implementation and Performance," *IEEE Transactions on Communications,* Vol. 41, No. 6, June 1993, pp. 998–1008.

[32] J. Armstrong, "Symbol Synchronization Using Baud Rate Sampling and Data Sequence Dependent Signal Processing," *IEEE Transactions on Communications,* Vol. 39, No. 1, Jan. 1991, pp. 127–132.

[33] W. G. Cowley and L. P. Sabel, "The Performance of Two Symbol Timing Recovery Algorithms for PSK Demodulators," *IEEE Transactions on Communications,* Vol. 42, No. 6, June 1994, pp. 2345–2355.

[34] C. R. Cahn, "Improving Frequency Acquisition of a Costas Loop," *IEEE Transactions on Communications,* Vol. COM-25, No. 12, Dec. 1977, pp. 1453–1459.

[35] U. Fawer, "A Coherent Spread-Spectrum Diversity Receiver with AFC for Multipath Fading Channels," *IEEE Transactions on Communications,* Vol. 42, No. 2/3/4, Feb./March/April 1994, pp. 1300–1322.

[36] M. Ikura, K. Ohno, and F. Adachi, "Baseband Processing Frequency-Drift-Compensation for QDPSK Signal Transmission," *Electronics Letter,* Vol. 27, No. 17, Aug. 1991, pp. 1521–1523.

[37] J. C. I. Chuang and N. R. Sollenberger, "Burst Coherent Demodulation with Combined Symbol Timing, Frequency Offset Estimation and Diversity Selection," *IEEE Transactions on Communications,* Vol. 39, No. 7, July 1991, pp. 1157–1164.

[38] F. D. Natali, "AFC Tracking Algorithms," *IEEE Transactions on Communications,* Vol. COM-32, No. 8, Aug. 1984, pp. 935–947.

[39] C. Wei and W. Chen, "Digital Tanlock Loop for Tracking π/4-DQPSK Signals in Digital Cellular Radio," *IEEE Transactions on Vehicular Technology,* Vol. 43, No. 3, Aug. 1994, pp. 474–479.

[40] A. Baschirotto et al. "Baseband Analog Front-End and Digital Back-End for Reconfigurable Multi-Standard Terminals," *IEEE Circuits and Systems Magazine,* Jan.–March 2006, pp. 8–28.

[41] F. Agnelli et al. "Wireless Multi-Standard Terminals: System Analysis and Design of a Reconfigurable RF Front-End," *IEEE Circuits and Systems Magazine,* Jan.–March 2006, pp. 38–58.

[43] www.sdrforum.org

[42] J. Mitola, "The Software Radio Architecture," *IEEE Communications Magazine,* Vol. 33, No. 5, May 1995, pp. 26–38.

CHAPTER 7
3G WIDEBAND CDMA

In this chapter, we will discuss a spread spectrum multiple-access technique used for the third generation (3G) Digital Cellular System called wideband CDMA, also referred to as WCDMA. We begin with some examples of the 2G spread spectrum cellular system called IS-95. The purpose is to introduce key topics and issues surrounding CDMA systems in general. We move on to a discussion of Pseudorandom Noise (PN) code generation techniques as well as their properties. A short comparison will be made and applications will be discussed. Next, the RAKE receiver will be presented and used as our reference receiver. At this point, we will have developed a baseline understanding to prepare us for the rest of the chapter, where we discuss the WCDMA standard. We expand this discussion to include high-speed downlink packet access (HSDPA) and high-speed uplink packet access (HSUPA) features of the 3GPP standard.

7.1 INTRODUCTION

In a spread spectrum system, the narrowband information signal is transmitted over a larger bandwidth to overcome the multipath fading channel distortion discussed earlier in Chap. 3. In Fig. 7.1, we provide an overview of two types of spread spectrum techniques: Direct Sequence (DS) and Frequency Hopping (FH). This multiple-access scheme is used when all users have a PN sequence code that is ideally uncorrelated with that assigned to any other user occupying the same channel [1].

Next we provide some general comments for each of the two techniques discussed above. First, the DS technique is greatly affected by the near-far problem, which, we will see, can be solved by the use of power control. Transmitting chips instead of symbols can lead to good multipath performance through the use of a RAKE receiver. This performance can be achieved by using PN codes with excellent correlation properties in a frequency selective fading (FSF) environment.

This technique, though, suffers from multiple access interference (MAI). The source of this interference is the addition of DS users operating in the same frequency band. The best and most often used analogy is the cocktail party effect. The party starts with two people having a conversation; as more couples engage in conversation of their own, using a different language, the background noise or sound begins to increase. This noise can cause the original two conversants to speak louder and, in turn, cause the other attendees to speak louder themselves. Caution must be exercised here, since instability (i.e., unintelligible speech) can be reached. The FH technique is less affected by the near-far problem, but requires a more complex frequency synthesizer so that the information signal can "hop" around the wide bandwidth available for transmission.

In an FSF channel, certain frequency bands experience spectral "nulls" while others don't. Hopping through the entire bandwidth will reduce the possibility that your transmitted signal is in a fade. Lastly, depending on the amount of time, the signal spends at a particular frequency channel; there could be a coherent-detection (i.e., carrier phase recovery) acquisition time issue.

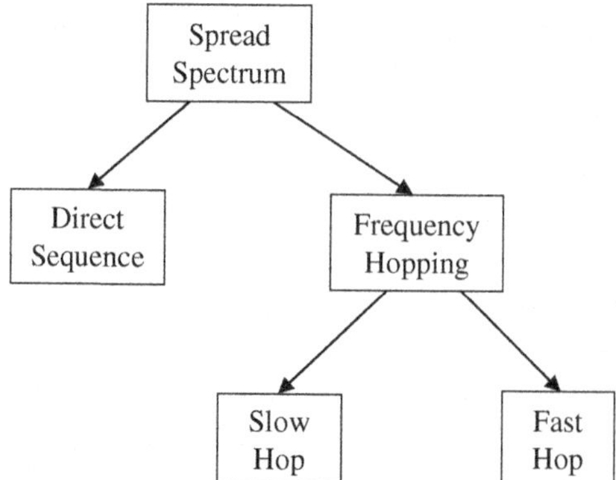

FIGURE 7.1 Spread spectrum categories.

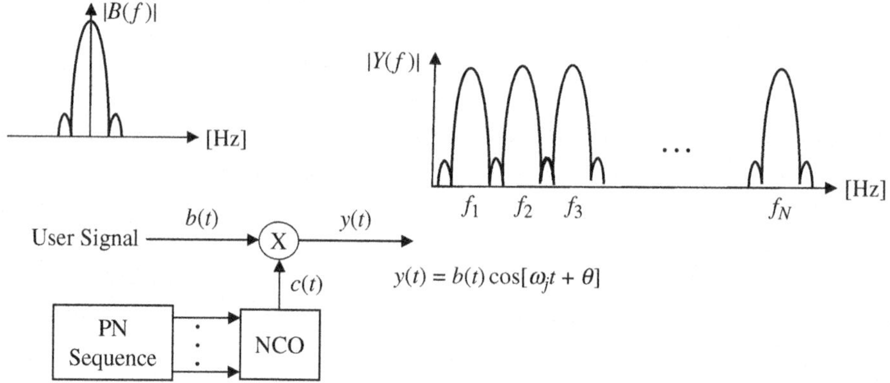

FIGURE 7.2 Frequency hopping spread spectrum transmitter block diagram.

Let's discuss the techniques in more detail. We will begin with the FH method (see Fig. 7.2). This spread spectrum technique uses N frequency channels to transmit the data signal. The N channels are "hopped" in a predetermined, but appearing in a random manner. Assume we have N channels with an information bandwidth of $R_b = 10$ kHz. The transmit bandwidth now becomes spread to $N \times 10$ kHz.

The information-bearing signal is denoted as $b(t)$, which is then spectrally up-converted by the carrier signal $c(t)$. The PN sequence will select one of the N channels in a random fashion. There are two classifications that depend on the hopping rate f_h. The first is fast hopping, where we hop more than one channel for the duration of a symbol ($f_h > R_b$). The second is slow hopping, where we hop on a channel for the duration of many symbols ($f_h < R_b$). Since the WCDMA standard is based on the DS technique, we will end our discussion on FH and emphasize the DS method next.

The DS technique encodes each bit of information with many chips. Thus, the frequency content of the signal has changed in order to take advantage of the wireless propagation channel. Let us consider a bit stream with $R_b = 10$ kbps and say each bit is coded with 100 chips; thus the chip rate is $R_c = 1$ Mcps. A simple transmitter block diagram is shown in Fig. 7.3.

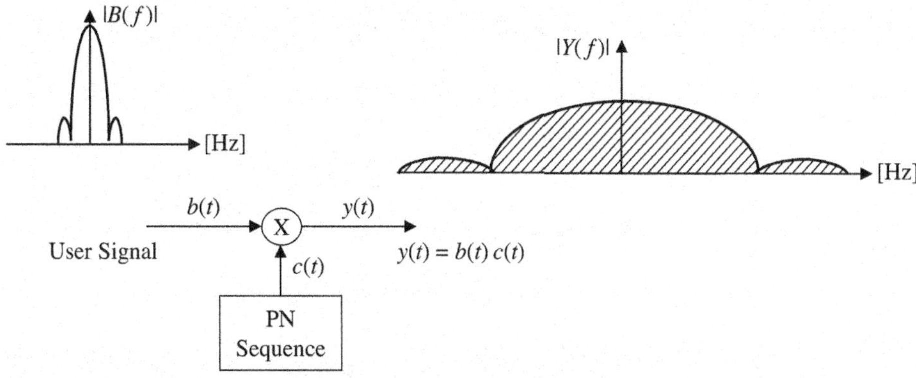

FIGURE 7.3 Direct sequence spread spectrum transmitter block diagram.

In the block diagram of the DS transmitter in Fig. 7.3, the narrowband user information signal $b(t)$ is spectrally spread to become $y(t)$, using the PN code sequence $c(t)$. This spread spectrum technique is measured by what is called the *processing gain* or *PG*, which is defined as[1]

$$\mathrm{PG} = 10\log\left[\frac{R_c}{R_b}\right] = 10\log\left[\frac{T_b}{T_c}\right] \tag{7.1}$$

For the example given above, we have the following value:

$$\mathrm{PG} = 10\log\left[\frac{1\,\mathrm{MHz}}{10\,\mathrm{kHz}}\right] = 20\ \mathrm{dB} \tag{7.2}$$

When we address the RAKE receiver, along with the frequency-selective fading channel, we will discuss how the PG quantity is used as a measurement of interference suppression capability. It is also common to refer to the number of chips used to encode a modulation symbol as the spreading factor (SF).

7.2 RAKE RECEIVER PRINCIPLE

Let us consider a simple BPSK-based CDMA system. The transmitter example is given in Fig. 7.4, where the input data bits $b(t)$ are multiplied by a chip sequence $c(t)$, in order to produce the spectrally spread signal $y(t)$. This spread signal is then spectrally up-converted to its allowed frequency band centered at a carrier frequency denoted by f_c. The block diagram in Fig. 7.4 highlights the two operations performed at the transmitter, namely, spreading and spectral up-conversion.

We can immediately draw the receiver architecture with the assumption of ideal carrier recovery, chip-time synchronization, and the knowledge of the actual PN chip sequence used at the transmitter. The receiver block diagram shown in Fig. 7.5 highlights the two operations performed at the receiver, namely, spectral down conversion and despreading.

The spectral down-conversion operation involves the multiplication of a sinusoid followed by the low-pass filter (LPF) to remove the unwanted spectral product. This baseband (ZIF) signal becomes despread by first multiplying by the PN chip sequence $c(t)$ and then accumulating the chips across the SF interval. In this example, the SF interval is equivalent to the bit time duration.

---

[1] Chip rate = $R_c = 1/T_c$ and the bit rate = $R_b = 1/T_b$.

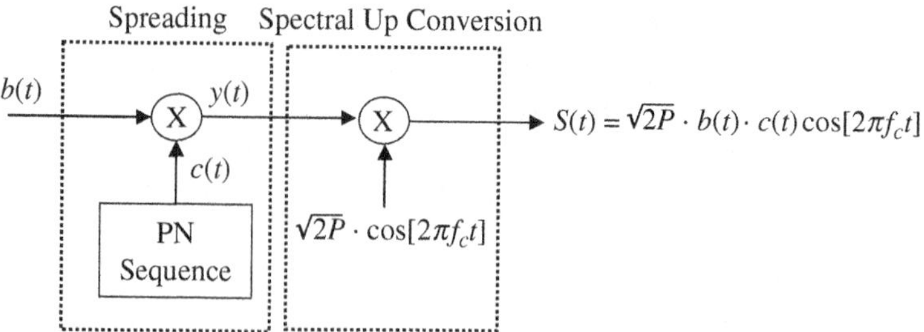

FIGURE 7.4 BPSK-based CDMA transmitter block diagram.

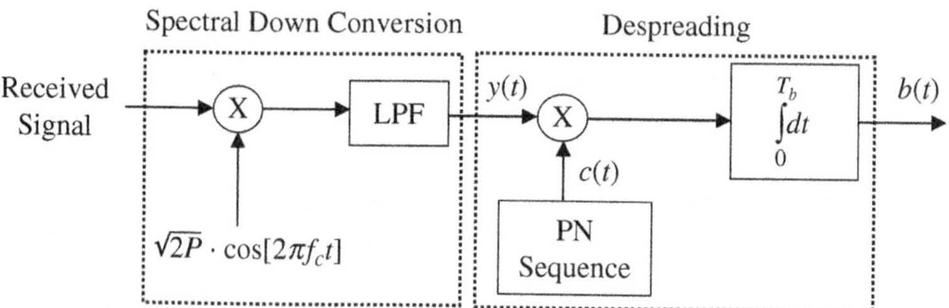

FIGURE 7.5 BPSK-based CDMA receiver block diagram.

In the above example, we assumed ideal coherent detection and perfect time synchronization of the PN sequences. More specifically, we assumed the locally generated oscillator used in the spectral down-conversion operation was exactly in phase and of the same frequency with the received signal. A more realistic analysis is as follows: Assuming the channel introduces a propagation time delay of τ sec. Figure 7.6 can be used to express PN code timing offsets.

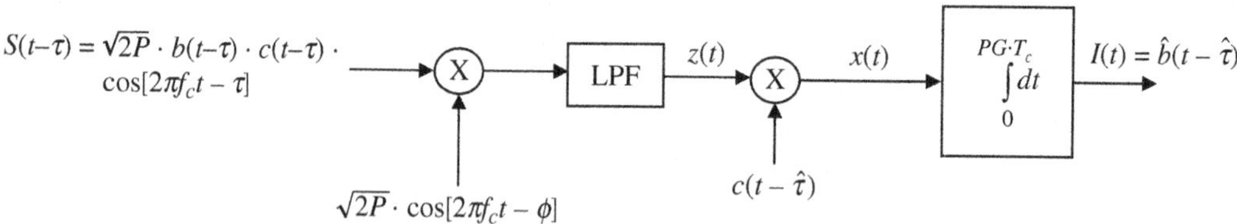

FIGURE 7.6 BPSK-based CDMA receiver.

The first operation performed in the above receiver is to spectrally down convert the received signal $s(t - \tau)$ to the baseband $z(t)$, which is given below with the assumption of ideal carrier recovery.

$$z(t) = P \cdot b(t - \tau) \cdot c(t - \tau) \tag{7.3}$$

Once this signal is multiplied by the locally generated PN sequence, appropriately shifted in time to account for the propagation time delay, we get the following intermediate value, assuming $\hat{\tau}$ is an estimate of the propagation time delay τ:

$$x(t) = P \cdot b(t - \tau) \cdot c(t - \tau) \cdot c(t - \hat{\tau}) \tag{7.4}$$

If $\tau = \hat{\tau}$, there is no timing offset and perfect synchronization occurs. In this case, we have the despreader output signal represented as

$$I(t) = P \cdot \hat{b}(t - \tau) \cdot \text{PG} \tag{7.5}$$

assuming the autocorrelation of the PN sequence has been set to the PG value. A point worth mentioning here is if the time delay τ between two arriving multipath rays is less than the chip time T_c, then the received signals are difficult to be resolved due to the interchip interference (ICI). This is commonly called interpath interference (IPI). On the other hand, if $\tau > T_c$, we can resolve the multipaths with the conventional RAKE receiver. Since a particular chip now arrives at distinctly different times, a time diversity gain can be observed.

Let us take a step to the side to try to describe the transmit and receive operations required in the CDMA system. In Fig. 7.7, we show the baseband signal is spread to a larger BW to be transmitted. This signal, along with other spread users, and narrowband interference are summed in the Mobile Station (MS) antenna. At this point, the desired signal is extracted by performing the despreading operation to create the narrowband signal at the receiver output.

What is interesting in Fig. 7.7 is that any narrowband interference present in the received signal becomes spread in the despreading operation in the receiver. The desired signal is extracted or despread because the correct PN code and timing offset was used during the despreading operations. Lastly, other CDMA users entering the receiver remain spread because they have different PN codes. Hence the receiver relies on autocorrelation and cross-correlation properties of the PN sequence for this CDMA system to operate properly. The despreader output not only has the desired signal, but also has other wideband interference and noise. In the block diagram of Fig. 7.7, we inserted an LPF after despreading, in order to further filter the wideband interference and noise for illustrative purposes.

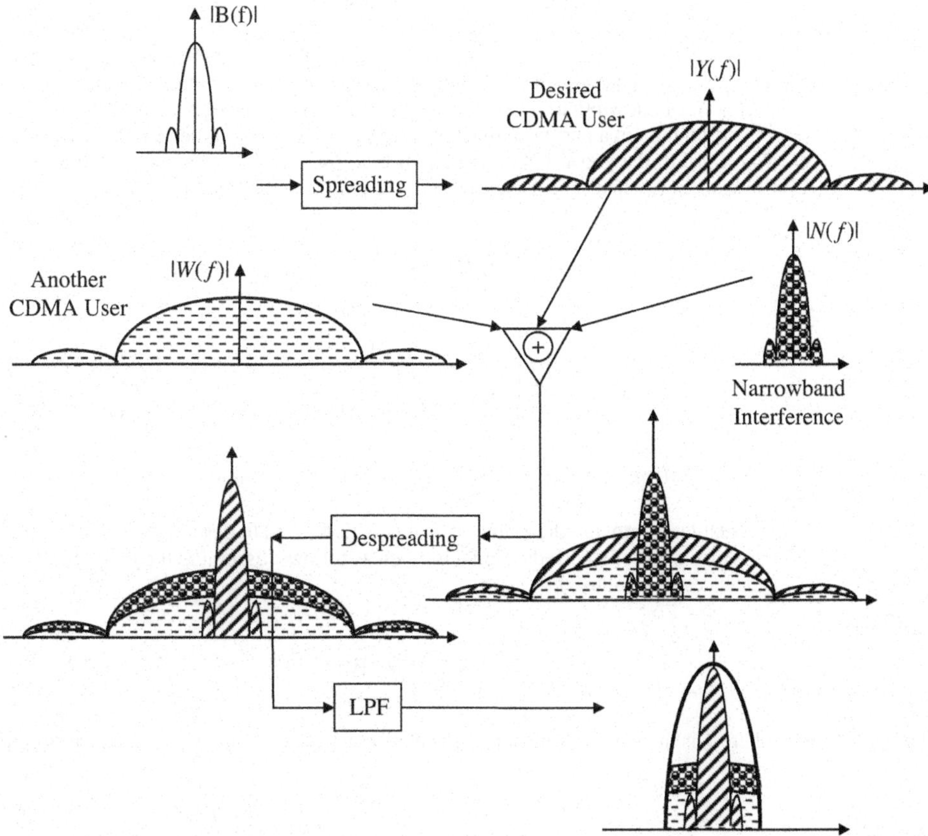

FIGURE 7.7 DS-CDMA system principles.

Let us briefly describe this PN sequence generator. The output chip sequence is usually generated by linear maximal–length sequence generators. These sequences are easy to generate and, depending on the generator polynomial, have very interesting properties. We will discuss the properties of these polynomials, in more detail, later in this chapter; suffice it to say that the following three properties are desirable [1–4]:

1. **Looks like random data.** There should be a balance of 1s and 0s, where the period is defined as $T = 2^N - 1$, where N is the number of shift registers or order of the polynomial.
2. **Cyclic addition.** If one were to cyclically shift the PN sequence by X shifts, it will produce another code word which, when added to a code word of Y shifts, results in still a third code word, Z.
3. **Good correlation properties.** Both autocorrelation and cross-correlation are important if one wants to confidently decide whether this is your PN code word or not.

For example, consider this simple PN generator with the polynomial given as $g(x) = 1 + x^3 + x^4$. The linear feedback shift register (LFSR) block diagram is shown in Fig. 7.8.

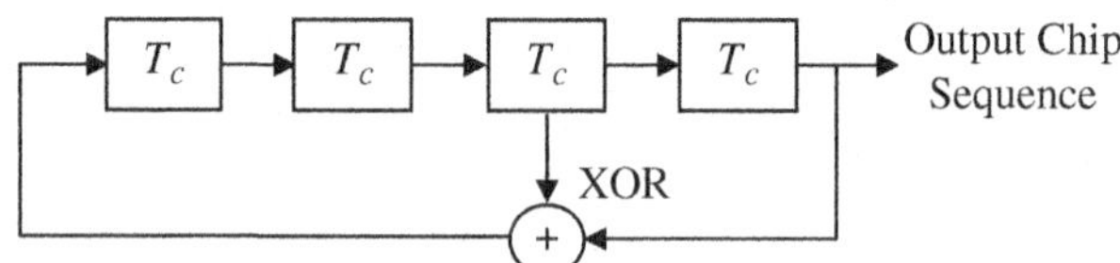

FIGURE 7.8 LFSR example of PN code generation.

In this PN code generator, the state is updated by the combination of shifting the contents of the registers at the chip rate and exclusive ORing certain register outputs to form the feedback signal.

It is well known that the autocorrelation of a square pulse is a triangular waveform [5]. Also, the autocorrelation of a periodic signal is periodic; these two comments explain the autocorrelation function $R_{cc}(\tau)$ of the above PN sequence. In Fig. 7.9, we plot the autocorrelation function of the above PN sequence, where T is used to denote the period of the PN sequence.

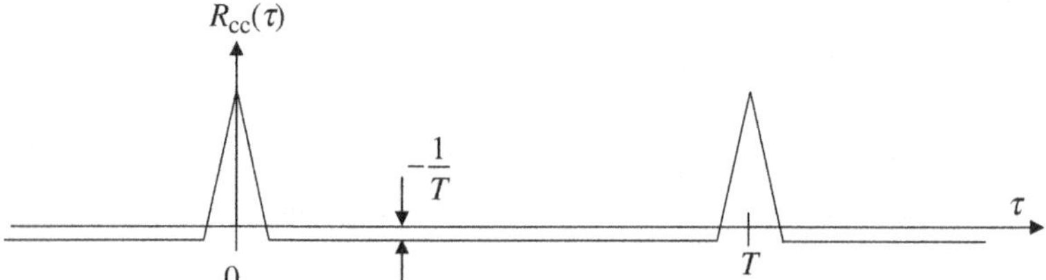

FIGURE 7.9 Autocorrelation function of the PN sequence.

Next let us provide the mathematical equations used to describe the autocorrelation function. The autocorrelation of a periodic signal is given as follows (assuming the time difference between the PN sequences is equal to τ):

$$R_{xx}(\tau) = \frac{1}{T}\int_{-\frac{T}{2}}^{\frac{T}{2}} x(t) \cdot x(t + \tau)d\tau \qquad (-\infty < \tau < \infty) \tag{7.6}$$

where T is the period of the signal. The average power of the received signal is denoted as

$$R_{xx}(0) = \frac{1}{T}\int_{-\frac{T}{2}}^{\frac{T}{2}} x^2(t)\,dt \tag{7.7}$$

Normalizing the above autocorrelation by the average signal power gives the following:

$$R_x(\tau) = \frac{1}{R_{xx}(0)} \cdot \frac{1}{T} \int_{-\frac{T}{2}}^{\frac{T}{2}} x(t) \cdot x(t + \tau)\, d\tau \tag{7.8}$$

This can also be described by the following formula:

$$R_x(\tau) = \frac{N_{\text{Agrees}} - N_{\text{Disagrees}}}{T} \tag{7.9}$$

Here we count the number of chip positions where the values agree or have the same sign, and we subtract from this the number of chip positions where the values are not the same or disagree. This is used to explain the value of $R_{cc}(\tau)$ being equal to $-1/_{15}$ for the previous example. In this case, the all-zero PN sequence is not a valid sequence which can be easily confirmed if one sets the state of the above PN generator to the all-zero state. Also, the autocorrelation waveform drawn above assumes no pulse-shaping filter was used in the comparison. In reality, an LPF needs to be used, and this will result in widening the width as well as changing the general shape of the autocorrelation function to resemble the windowed sinc(x) function.

7.3 2G IS-95 CDMA DISCUSSION

CDMA is a multiple-access technique, where multiple users can share the same time and frequency domains while remaining distinct in the code domain (e.g., applying the orthogonality principle). Let us consider the case when we have two users occupying the same frequency channel and transmitting at the same time. For TDMA systems, this can be a nightmare! For our CDMA system, each user has a different PN code sequence. The users will be able to become separable with the use of the statistical properties of the PN code. This can be accomplished in two ways:

1. Use physically different PN sequence-generator polynomials. Here we must assure the cross-correlation between $C_1(t)$ and $C_2(t)$ is minimal or 0 for the orthogonal case, in order to reduce the MAI.
2. Use the same PN generator polynomials, but have the two users start spreading at different points in time of the chip sequence period. We rely on good autocorrelation properties of the PN sequence generator to reduce "multiple user interference."

In fact, we can quickly draw the analogy of the above two ways to how the North American CDMA system (IS-95) operates [6]. Specifically, the first way parallels the downlink spreading and the second way parallels the uplink spreading. Let us discuss this in more detail in the following subsections.

7.3.1 Downlink Spreading Example

Let us consider a simple example of two users, with their information signals given by $b_1(t)$ and $b_2(t)$. The base station (BS) transmitter looks like that in Fig. 7.10, where different PN sequences, $C_1(t)$ and $C_2(t)$, have been used for each user.

The transmitted signal is represented as

$$S(t) = \sqrt{2P} \cdot \left[\sum_{i=1}^{2} b_i(t) \cdot C_i(t) \right] \cdot COS\,[2\pi f_c t] \tag{7.10}$$

We will assume we have a frequency flat channel with AWGN, so the received signal becomes

$$X(t) = h(t) \cdot S(t - \tau) + N(t) \tag{7.11}$$

where $h(t)$ = fading channel variations, τ = propagation time delay, and $N(t)$ = Gaussian noise with zero mean.

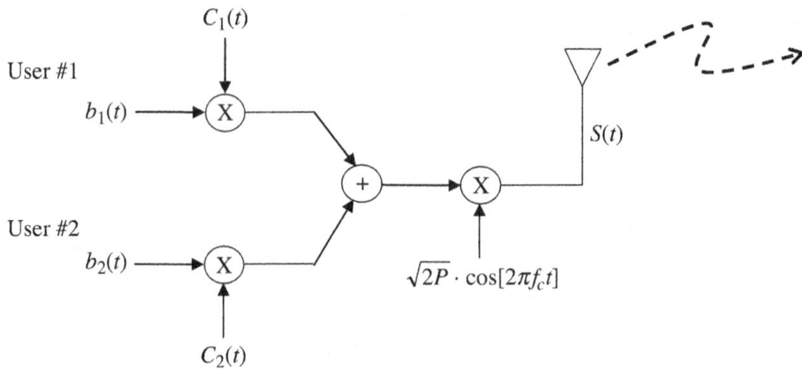

FIGURE 7.10 Downlink spreading example.

The mobile station received signal is represented as

$$X_j(t) = h_j(t)\left\{\sum_{i=1}^{2} C_i(t - \tau_j) \cdot b_i(t - \tau_j)\right\} \cdot \cos[2\pi f_c(t - \tau_j)] + N_j(t) \tag{7.12}$$

which can be rewritten as (for $j = 1, 2$)

$$X_j(t) = h_j(t)\left\{\sum_{i=1}^{2} C_i(t - \tau_j) \cdot b_i(t - \tau_j)\right\} \cdot \cos[2\pi f_c t - \phi_j] + N_j(t) \tag{7.13}$$

Here we have used the index j to denote the user of interest number and the index i for the number of users transmitted. Since the receiving MS will be located in physically different locations within the cell, the fading and propagation time delays are assumed to be different. In Fig. 7.11, we present the receivers for each user.

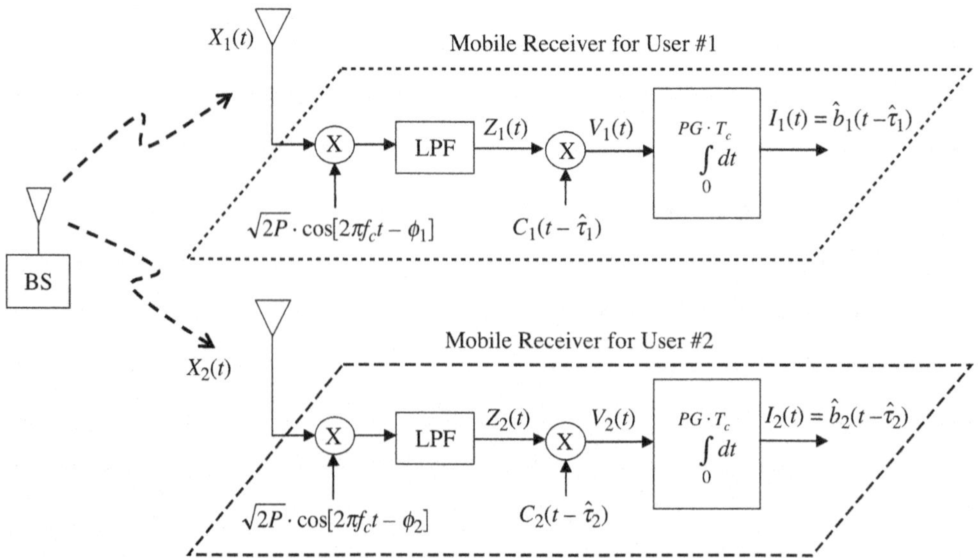

FIGURE 7.11 MS example receiver operations for the downlink spreading scenario.

If we consider User 1 only, we can write the following equations, assuming ideal carrier recovery (coherent detection). Although the above block diagrams imply carrier recovery should occur prior to despreading, in practice, it is not necessary to perform spectral down conversion with both phase and frequency coherence. By this, we mean a "sloppy" spectral down-conversion operation can occur where the residual frequency offsets will be compensated by the automatic frequency control (AFC) algorithm. Similarly, the residual phase/amplitude offsets will be compensated by the channel estimation (CE) algorithm. For this discussion, let us work with the following equation for User 1:

$$Z_1(t) = P \cdot [b_1(t - \tau_1) \cdot C_1(t - \tau_1) + b_2(t - \tau_1) \cdot C_2(t - \tau_1)] \tag{7.14}$$

After the multiplication of the locally generated PN sequence for User 1, we have an expression for the first part of despreading operation, which is given below:

$$V_1(t) = P \cdot [b_1(t - \tau_1) \cdot C_1(t - \tau_1) \cdot C_1(t - \hat{\tau}_1) + b_2(t - \tau_1) \cdot C_2(t - \tau_1) \cdot C_1(t - \hat{\tau}_1)] \tag{7.15}$$

The second part of the despreading operation occurs after the integration over the symbol or bit time (T_b), resulting in the following despreader output signal (assuming perfect time synchronization, $\tau_1 = \hat{\tau}_1$):

$$I_1(t) = P \cdot \int_0^{T_b} b_1(t - \tau_1) \cdot C_1^2(t - \tau_1)\, dt + P \cdot \int_0^{T_b} b_2(t - \tau_1) \cdot C_2(t - \tau_1) \cdot C_1(t - \tau_1)\, dt \tag{7.16}$$

Since the BPSK symbol is constant over the duration of the integration, it can be moved outside the integration. We have assumed the absence of a channel.

$$I_1(t) = P \cdot b_1(t - \tau_1) \cdot \int_0^{T_b} C_1^2(t - \tau_1)\, dt + P \cdot b_2(t - \tau_1) \cdot \int_0^{T_b} C_2(t - \tau_1) \cdot C_1(t - \tau_1)\, dt \tag{7.17}$$

What we notice is that the first integration on the RHS is the autocorrelation function evaluated at a lag of zero. This was due to an earlier assumption of perfect time synchronization. The second summation on the RHS is the cross-correlation of the two PN sequences evaluated at a lag of zero. Hence we see the importance of the properties of the PN codes used in the system. The resulting signal contains the desired plus interference plus noise components:

$$I_1(t) = P \cdot b_1(t - \tau_1) \cdot \mathrm{PG} + b_2(t - \tau_1) \cdot R_{C_1C_2}(0) + \text{noise terms} \tag{7.18}$$

Notice, ideally, the second term would be zero if the PN codes were chosen to satisfy the orthogonality principle.

7.3.2 Uplink Spreading Example

Let us consider the case when each user had a shifted version of the same PN sequence. For sake of discussion, we have chosen User 2 to be spread with $C_1(t)$ but shifted in time by half the PN code period (see Fig. 7.12). A major point to consider here is that this time difference should be larger than the expected maximum delay spread of the channel. This restriction is used in order to help the BS to not incorrectly add multipaths from different users within the RAKE.

The uplink transmitted signals are given as

$$S_1(t) = \sqrt{2P} \cdot b_1(t) \cdot C_1(t) \cdot \cos[2\pi f_c t] \tag{7.19}$$

$$S_2(t) = \sqrt{2P} \cdot b_2(t) \cdot C_1\left(t - \frac{T}{2}\right) \cdot \cos[2\pi f_c t] \tag{7.20}$$

The BS receiver will see the summation of these two signals, respectively scaled and shifted in time due to the propagation channel conditions.

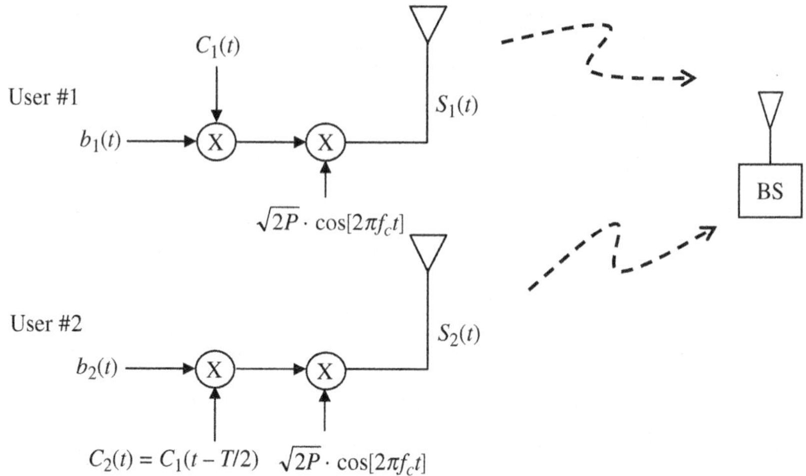

FIGURE 7.12 Uplink spreading example.

If we continue along the lines that we have a frequency flat fading channel, then the receiver architecture in Fig. 7.13 exists for both users at the BS.

Considering User 1 only, we have the following equation where we have included a term to represent the phase offset ϕ_e between the two received signals carrier frequencies:

$$Z_1(t) = P \cdot b_1(t - \tau_1) \cdot C_1(t - \tau_1) + P \cdot b_2(t - \tau_1) \cdot C_1\left(t - \tau_1 - \frac{T}{2}\right) \cdot \cos[\phi_e] \tag{7.21}$$

This phase offset arises from the fact that ideal coherent detection was applied to User 1 in this example and not to User 2.

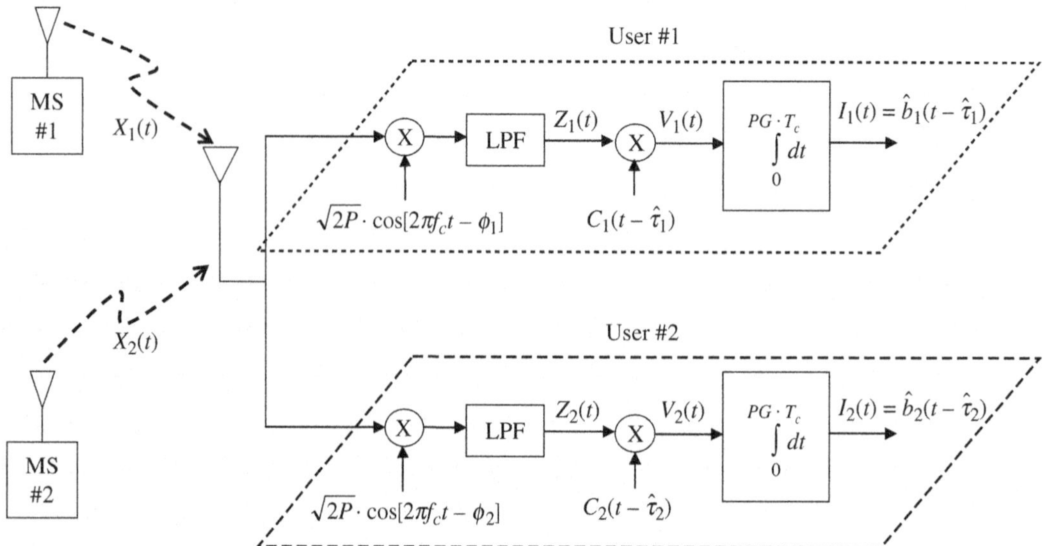

FIGURE 7.13 Uplink receiver structure for the uplink spreading example.

In the first part of the despreader, we have (omitting the AWGN terms)

$$V_1(t) = P \cdot b_1(t - \tau_1) \cdot C_1(t - \tau_1) \cdot C_1(t - \hat{\tau}_1) + P \cdot b_2(t - \tau_1) \cdot C_1\left(t - \tau_1 - \frac{T}{2}\right) \cdot C_1(t - \hat{\tau}_1) \cdot \cos[\phi_e] \tag{7.22}$$

The second part of the descrambling is given after integration.

$$I_1(t) = P \cdot \int_0^{T_b} b_1(t - \tau_1) \cdot C_1(t - \tau_1) \cdot C_1(t - \hat{\tau}_1)\,dt + P \cdot \int_0^{T_b} b_2(t - \tau_1) \cdot C_1\left(t - \tau_1 - \frac{T}{2}\right) \cdot C_1(t - \hat{\tau}_1) \cdot \cos[\phi_e]\,dt \tag{7.23}$$

Using the same approach as in the previous subsection allows us to write the following:

$$I_1(t) = P \cdot b_1(t - \tau_1) \cdot \int_0^{T_b} C_1^2(t - \tau_1)\,dt + P \cdot b_2(t - \tau_1) \cdot \cos[\phi_e] \cdot \int_0^{T_b} C_1\left(t - \tau_1 - \frac{T}{2}\right) \cdot C_1(t - \tau_1)\,dt \tag{7.24}$$

The first integration on the RHS represents the autocorrelation of the PN sequence evaluated at a lag of zero. The second integration on the RHS also represents the autocorrelation, except that this is for a lag of $T/2$, half the PN code period.

$$I_1(t) = P \cdot b_1(t - \tau_1) \cdot \mathrm{PG} + P \cdot b_2(t - \tau_1)\cos[\phi_e] \cdot R_{cc}\left(\frac{T}{2}\right) + \text{noise terms} \tag{7.25}$$

The interference component of the second summation can be minimized by proper choice of the PN codes with excellent autocorrelation properties. Hence the signal contains the desired signal plus interference plus noise components.

Let us take a moment to provide a brief summary of what has been discussed thus far. Both examples showed how the desired signal can be extracted. The first example showed the interference components to be dependent on the cross-correlation properties, while the second example stressed the autocorrelation properties of the PN sequence. This choice of the PN sequence with proper statistical properties is essential. Recall these equations have been derived assuming a frequency flat fading channel with AWGN.

7.4 RAKE FINGER ARCHITECTURE AND PERFORMANCE

In the previous subsections, we have considered an example of downlink and uplink spreading. These examples assumed a frequency flat fading channel; hence the receiver simply needed to calculate the propagation time delay $\hat{\tau}$. In this subsection, we will discuss the FSF channel and its effects on the receiver architecture. For sake of discussion, we have decided to choose the downlink spreading example provided earlier. If we have a two-ray channel, then the received signal becomes

$$X(t) = h_1(t) \cdot S(t - \tau_1) + h_2(t) \cdot S(t - \tau_2) + N(t) \tag{7.26}$$

We can generalize this to the following:

$$X_j(t) = \sum_{k=1}^{2} h_{j,k}(t) \cdot \left\{\sum_{i=1}^{2} C_i(t - \tau_{j,k}) \cdot b_i(t - \tau_{j,k})\right\} \cdot \cos[2\pi f_c(t - \tau_{j,k})] + N_j(t) \tag{7.27}$$

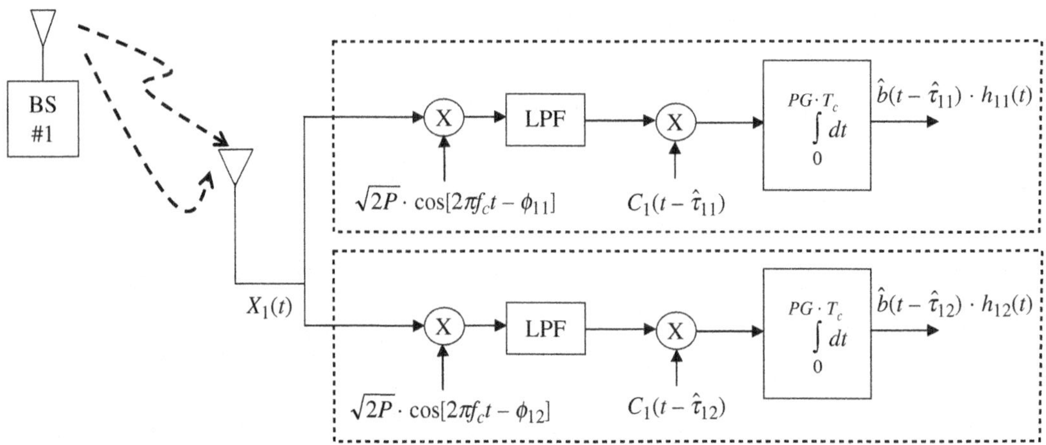

FIGURE 7.14 Two-finger RAKE receiver for User 1.

where the index j is used to denote the receiver user of interest number, the index k is used to denote the multipath number, and the index i is used to denote the number of transmit users present. Here we see each user receives multiple signals or replicas of the same information that has traversed different propagation paths.

Generally speaking, a RAKE receiver has individual despreading functions called fingers [3]. One typically assigns a finger to each arriving multipath. An example of a two-finger RAKE for User 1 is shown in Fig. 7.14.

Above we have decided to have two entirely separate receiver chains to start a discussion on spectral down conversion. In practice, a single spectral down-conversion operation is needed for all the arriving multipaths. In doing so, a phase offset term is introduced since a single phase reference is used for all the arriving multipaths. In fact, this term can be lumped together with the channel $h_{j,k}(t)$ to produce an overall channel response.

If we remove the separate spectral down converters, then we have the simplified block diagram in Fig. 7.15.

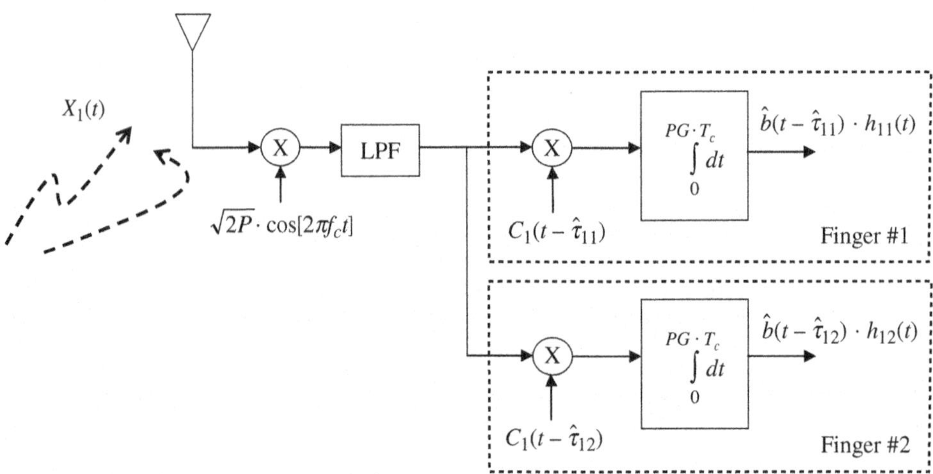

FIGURE 7.15 Two-finger RAKE receiver block diagram.

In this figure, the first finger is demodulating the path arriving at $t = \tau_{11}$ while the second finger is demodulating the path arriving at $t = \tau_{12}$. We see the same information is present at both outputs, except they are time skewed and independently Rayleigh faded. So the next function required is to introduce the CE algorithm. The CE independently estimates the channel response for each arriving ray and then performs CD (for this example). Depending on the relative time difference of the two rays and its relation to the PG or SF, a time-deskewing buffer is needed to time align the despread symbols prior to being added or combined to create a single, received RAKE combined output symbol.

Whether we study the uplink or the downlink, there will be some known pilot bits/symbols that the receiver can use to estimate the channel response. If the pilot bits are time multiplexed, then some sort of demultiplexing operation is needed to extract the pilot bits. If the pilot symbols are code multiplexed, then a parallel despreader (at the same time of arrival) is needed to determine the symbol. These individual symbols are combined in order to produce an aggregate symbol. The actual combining method shown in Fig. 7.16 follows the Maximal Ratio Combining (MRC) approach presented in Chap. 5. For sake of completeness, we should mention that any of the previously mentioned techniques used for receiver diversity can now be applied in the RAKE finger combining operation.

As shown in Fig. 7.16, a RAKE receiver consists of multiple parallel demodulators, who individually track each of the arriving replicas (or delay spread echos). This is a form of time diversity. One common technique is to perform combining of the individual demodulated signals. In Fig. 7.17, we show a simple high-level block diagram of the RAKE receiver for illustrative purposes.

In Fig. 7.17, we have shared a single, spectral down-conversion receiver, called the RF section. Also we have placed the ADC directly before the RAKE to emphasize the RAKE operations are typically performed in the digital domain. A finger can only despread multipaths that it is aware of, which is accomplished by the use of an Echo Profile Manager (EPM) or a searcher. The EPM finds valid multipaths by maintaining an up-to-date power delay profile $P(\tau)$. The EPM selects a particular arriving multipath and assigns a finger to demodulate the information present on that arriving multipath. In an FSF channel, many multipaths can be found, and this could result in many fingers being used in the demodulation process.

A high layer of control over the assignment and de-assignment of multipaths is the role of the finger management. Although not shown, it is responsible for the handling of the birth/death scenarios discussed in the channel modeling section of Chap. 3. Another role of the EPM is to perform course time synchronization, say within $T_c/2$. Once this multipath is assigned to a finger, then fine time tracking (i.e., DLL) will be activated to track moving propagation environments. One of the roles of the finger manager is to always have at least one finger actively demodulating a multipath, so there are no time gaps of symbols at the RAKE combiner output.

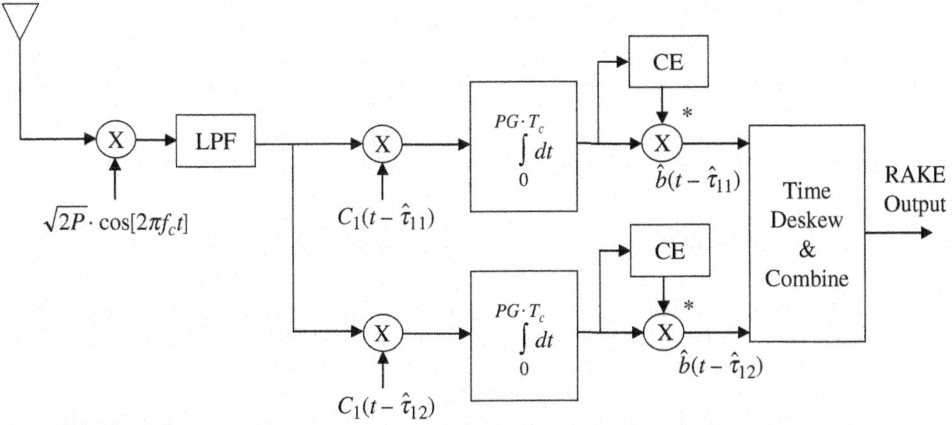

FIGURE 7.16 Two-finger RAKE receiver with CE block diagram.

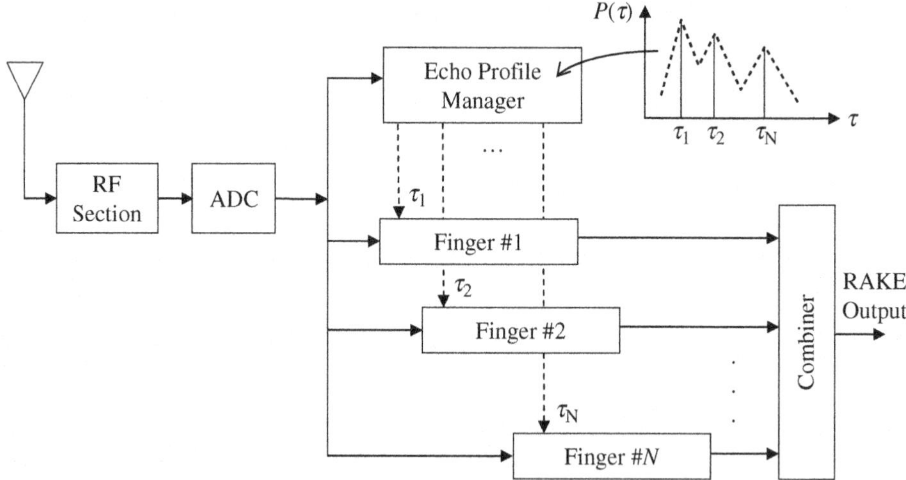

FIGURE 7.17 RAKE high-level block diagram.

Lastly, this RAKE receiver provides BER performance improvements due to the time diversity of the received symbols. An example of the performance improvement is given in Fig. 7.18 for a RAKE having 1, 2, and 3 fingers in a fast fading environment with a Doppler spread = 300 Hz, assuming a convolutional Forward Error Correction (FEC) code.

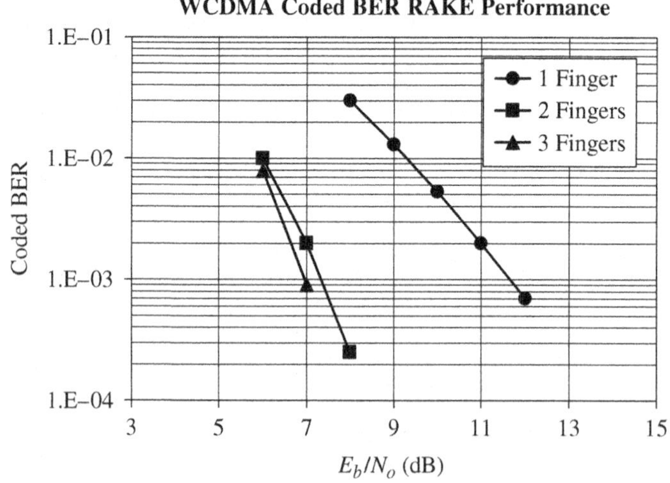

FIGURE 7.18 RAKE receiver BER performance results.

The results were obtained as follows: The one-finger RAKE performance curve assumed a flat fading channel model. The two- and three-finger RAKE curves assume an FSF channel model with two and three multipaths, respectively. All multipaths had equal average power. Here, one can see the tremendous benefit in having an FSF channel with DS-CDMA. An important piece of information to have is that the SF of the physical channel was set to 128, a relatively large number. This large PG is used to suppress interference from other arriving multipaths in the RAKE finger.

7.4.1 General RAKE

In this section, we will present a more general RAKE architecture in detail. Let us start with one of the early proposals for the RAKE receiver shown in Fig. 7.19 [3].

The time separation between the PN code correlations is Δ sec. Recall from our earlier discussion on the RAKE: We placed fingers at each significant time-of-arrival multipath. There was no mention of preserving a distance constraint between each finger, as shown in Fig. 7.19.

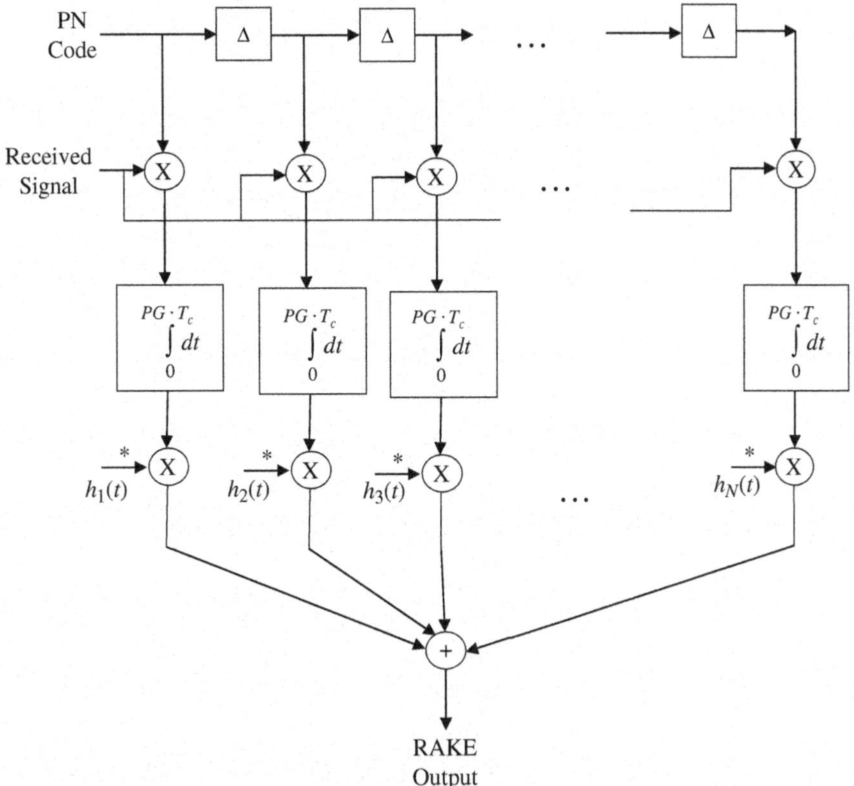

FIGURE 7.19 General RAKE receiver block diagram using the MRC method.

In [7], a generalized RAKE receiver called GRAKE was presented, and we have provided a block diagram in Fig. 7.20. This is a generalized version of the RAKE diagram shown in Fig. 7.19. The only difference being the combining weight calculation.

Note that with this technique, a group of N fingers can be placed contiguously in time, in order to despread multipaths arriving within a group of N multipaths.

The RAKE output is given as $I = \underline{w}^* \cdot \underline{x}$. In the GRAKE, the weight is calculated using the following (assuming the MSINR cost function is used):

$$\underline{w} = R_{I+N}^{-1} \cdot \underline{r}_{xd} \tag{7.28}$$

where R_{I+N} is the interference plus noise covariance matrix and $\underline{r}_{xd}$ is the cross-correlation vector between the received signal and the desired or reference signal. It is common to define R to be the covariance matrix of the received signal as the sum of the desired, interference and noise covariance

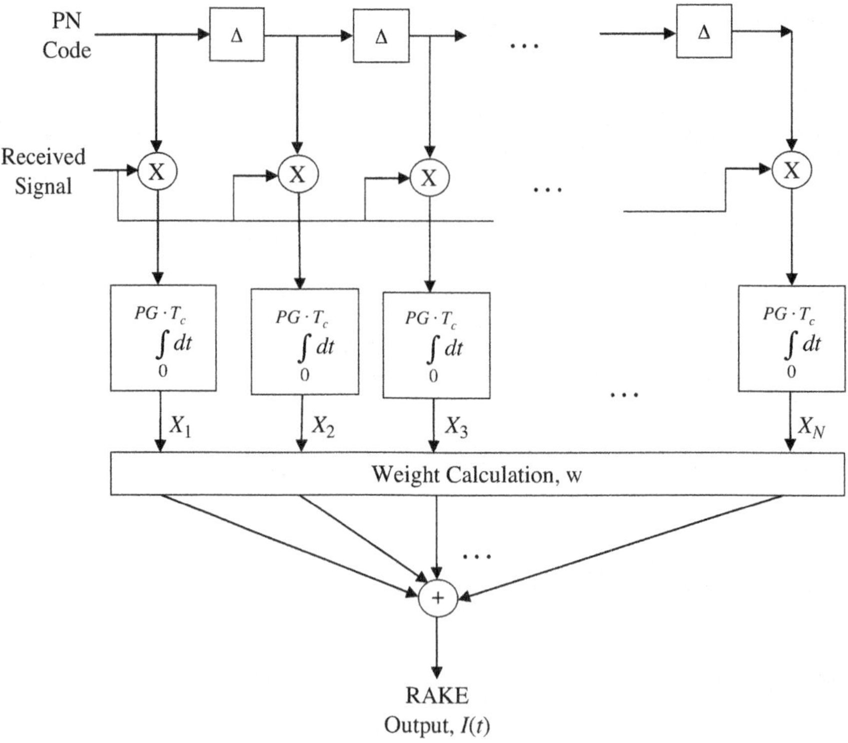

FIGURE 7.20 GRAKE architecture.

matrices, as in Eq. (7.29). Here we have assumed the desired signal, interfering signals, and noise are all uncorrelated from each other.

$$R = R_D + R_I + R_N \tag{7.29}$$

Suffice it to say that various combining or weight calculation options exist [8]. Let us briefly apply the eigenanalysis to this receiver by providing the Eigen Canceler (EC) combining weights. First, recall this interference plus noise covariance matrix can be written as [9]

$$R_{I+N} = U_I \cdot D_I \cdot U_I^* + U_N \cdot D_N \cdot U_N^* \tag{7.30}$$

The EC-based RAKE combiner is then expressed as

$$\underline{w} = (I - U_I \cdot U_I^*) \cdot \underline{r}_{xd} \tag{7.31}$$

7.4.2 Predespread and Postdespread Combining

In this section, we will briefly mention a couple of options that are available to perform the multipath combining in the RAKE functionality. The previous subsection had the despreading operations occurring first and then the time-deskewing and combining operations being performed at the despread symbol level. One can move the deskewing and combining into the chip level, prior to the actual despreading operations. In this case, a simple block diagram is shown in Fig. 7.21.

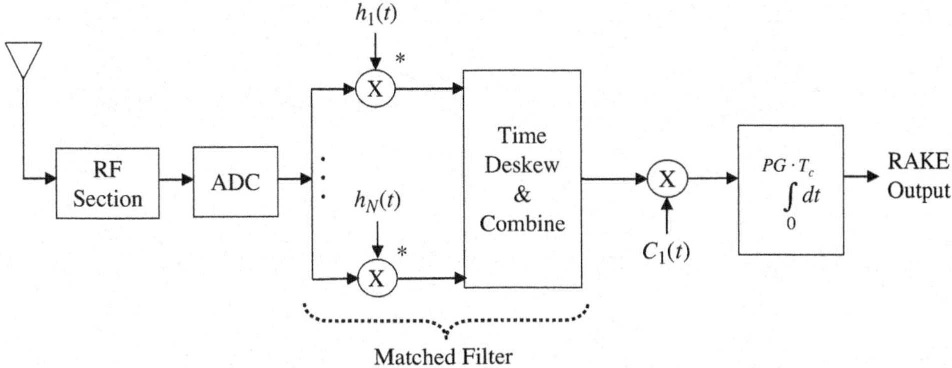

FIGURE 7.21 Chip-level RAKE receiver.

If we have to combine at the chip level, then knowledge of the arrival and interarrival times of the multipaths is necessary. This information is obtained from the multipath-searcher function and is used in the time-deskewing operation. In addition to knowing the arrival time information, the channel response is required before coherent combining can take place. As a simple example, consider a 2-ray channel model, where the separation between the 2 rays is 4 chips. A very simple implementation would be to delay the first arriving multipath by 4-chip times and combine this sequence with a nondelayed sequence (which represents the second ray). In this architecture, one needs to know the maximum expected delay spread to be used in the matched filter. The benefit in using this method is that a single despreader function is required to extract the desired signal. This benefit becomes more useful as the number of fingers supported in the RAKE receiver increases.

7.5 PN CODE PROPERTIES

In this section, we will begin by presenting some properties of PN codes and then discuss the maximal length sequences, Gold codes, and orthogonal codes. Some preliminary information includes a discussion on Galois fields (GFs). A field is a set of elements in which we can do addition, subtraction, multiplication, and division without ever leaving the set. When this field contains a finite number of elements, it is a finite field, also known as Galois field [10]. Here we list some general statements.

- GF(2^n) means there exists a Galois field of 2^n elements.
- A polynomial with coefficients from a binary field, GF(2), is called a binary polynomial.
- A binary polynomial $p(x)$ of degree m is said to be irreducible if it is not divisible by any binary polynomial of degrees less than m and greater than 0.
- An irreducible polynomial $p(x)$ of degree m is said to be a primitive if the smallest positive integer n for which $p(x)$ divides $x^n + 1$ is $n = 2^m - 1$. For example, $p(x) = 1 + x + x^4$ is a primitive polynomial because the smallest integer for which $1 + x + x^4$ divides $x^n + 1$ is $n = 2^4 - 1 = 15$.

We will present two approaches used to generate PN code sequences: namely, the Fibonacci and Galois approaches. The relationship between these two approaches will be provided in order to gain insight into the PN codes [4].

General shift register implementations of PN sequences are given by $C_i \in$ GF(2). The first approach is called the Fibonacci approach and is given in Fig. 7.22.

The polynomial for this implementation is given as (where x represents a shift in time)

$$p(x) = 1 + C_1x + C_2x^2 + \cdots + C_{n-1}x^{n-1} + x^n \tag{7.32}$$

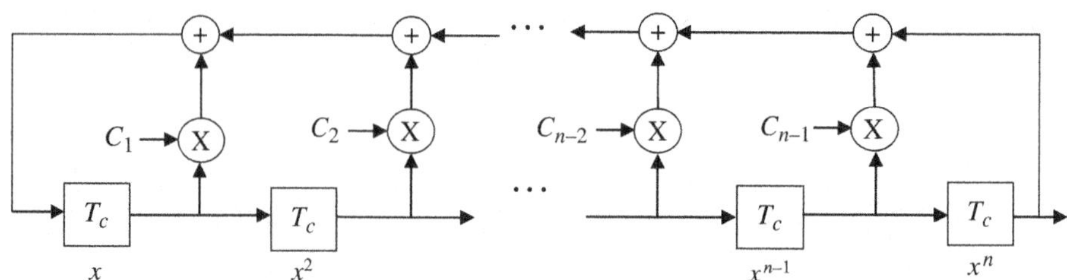

FIGURE 7.22 Fibonacci-type PN generator.

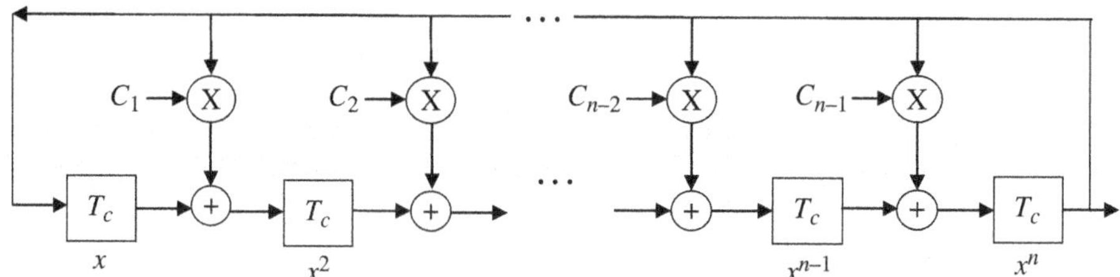

FIGURE 7.23 Galois-type PN generator.

The second approach is called the Galois approach and is given in Fig. 7.23.

The polynomial for this implementation is given as (where x represents a shift in time)

$$q(x) = 1 + C_1x + C_2x^2 + \cdots + C_{n-1}x^{n-1} + x^n \tag{7.33}$$

where the following relationship holds true between the two approaches: The value of n is used to represent the maximum number of shift registers used in the generation of the PN code.

$$q(x) = x^n \cdot p(x^{-1}) \tag{7.34}$$

The PN sequence is controlled by the state of the shift registers, the number of shift registers, and the feedback connections. The latter two are described by the primitive polynomials. In Table 7.1, we show some primitive polynomials.

To generate the PN sequence, the shift registers must be initialized or "loaded" with the first n chips of the sequence, otherwise a shift or "delay" will be introduced into the output PN sequence. This implies if one desires a certain shift in the PN sequence, then a proper initialization vector should be loaded into the PN sequence generators shown in Figs. 7.22 and 7.23. Let's discuss time shifts in more detail with a working example.

Consider the simple third-order ($m = 3$) case, with the primitive polynomial given as $p(x) = 1 + x + x^3$. The Fibonacci-type generator is given in Fig. 7.24.

We wish to analyze the generated PN sequence by first discussing the state of the PN generator. The state of the PN generator is defined by the contents of the shift registers. In Fig. 7.25, we show the PN state as a function of time. The initial state of the PN generator is set to 100.

Note that shifts in sequences can be obtained and controlled by initial loading of the shift registers. Also, the use of "masks" allows us to obtain a desired shift without changing the loaded values. This is accomplished by combining certain outputs of the shift registers.

Earlier, we have shown the EPM's role is to find valid multipaths. Sometimes, in searching for multipath candidates, there is a need to test the present time hypothesis against phase offsets of the

TABLE 7.1 GF(2) Primitive Polynomials

Degree	Primitive polynomials
2	$x^2 + x + 1$
3	$x^3 + x + 1$
4	$x^4 + x + 1$
5	$x^5 + x + 1$
6	$x^6 + x + 1$
7	$x^7 + x^3 + 1$
8	$x^8 + x^4 + x^3 + x^2 + 1$
9	$x^9 + x^4 + 1$
10	$x^{10} + x^3 + 1$
11	$x^{11} + x^2 + 1$
12	$x^{12} + x^6 + x^4 + x + 1$
13	$x^{13} + x^4 + x^3 + x + 1$
14	$x^{14} + x^{10} + x^6 + x + 1$
15	$x^{15} + x + 1$
16	$x^{16} + x^{12} + x^3 + x + 1$
17	$x^{17} + x^3 + 1$
18	$x^{18} + x^7 + 1$
19	$x^{19} + x^5 + x^2 + x + 1$
20	$x^{20} + x^3 + 1$
21	$x^{21} + x^2 + 1$
22	$x^{22} + x + 1$
23	$x^{23} + x^5 + 1$
24	$x^{24} + x^7 + x^2 + x + 1$
25	$x^{25} + x^3 + 1$
26	$x^{26} + x^6 + x^2 + x + 1$
27	$x^{27} + x^5 + x^2 + x + 1$
28	$x^{28} + x^3 + 1$
29	$x^{29} + x^2 + 1$
30	$x^{30} + x^{23} + x^2 + x + 1$

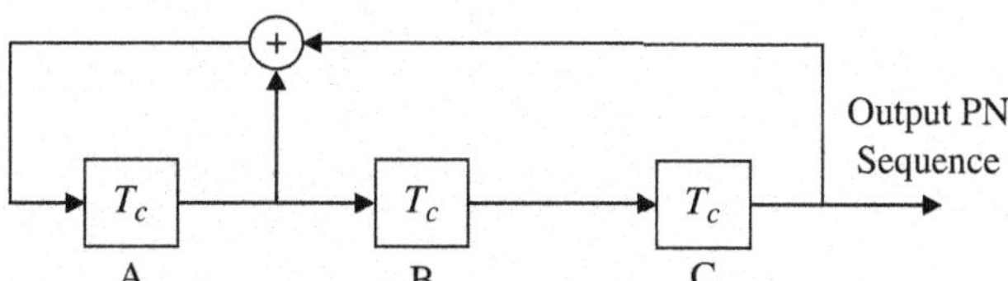

FIGURE 7.24 Simple PN sequence generator block diagram.

PN sequence. Also, some CDMA systems separate users by phase shifts provided to the MS by the higher layers. It is for this reason that we focus our next discussion on shifting PN sequences. We will discuss two ways to shift a PN output chip sequence [11].

Method	Tool
Initial loading	Polynomial = $a_0 + a_1x + a_2x^2$
Masks	Polynomial = $m_0 + m_1x + m_2x^2$

Both of these methods are possible with the help of Fig. 7.26. The initial contents of the PN generator are shown in the shift registers. The mask is shown as multiplier coefficients for the PN states.

Chip-Time Index	A	B	C
0	1	0	0
1	1	1	0
2	1	1	1
3	0	1	1
4	1	0	1
5	0	1	0
6	0	0	1
7	1	0	0
8	1	1	0
9	1	1	1
10	0	1	1
11	1	0	1

FIGURE 7.25 PN generator states for $p(x) = 1 + x + x^3$.

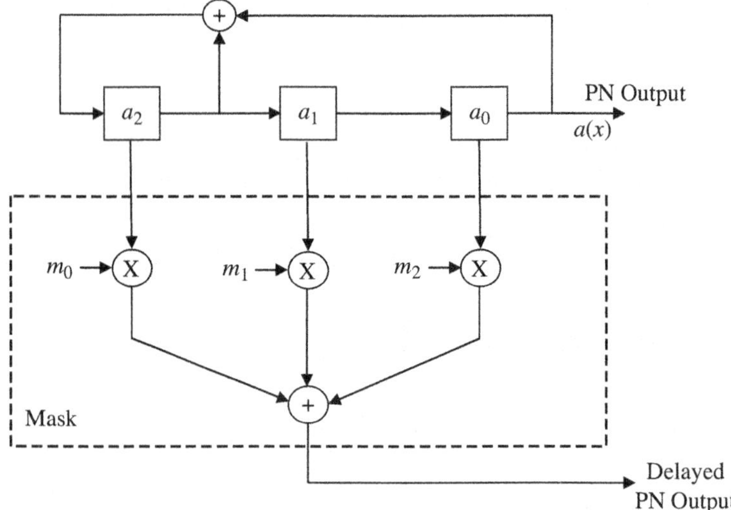

FIGURE 7.26 Shifting the PN sequence output sequence.

Consider the initial loading approach first. If we assume the initial contents of the PN generator are 100, this corresponds to a polynomial of x^2 from the polynomial defined above. Next we will multiply this initial contents polynomial by the PN code polynomial and maintain the variables with degree less than m, in this case $m = 3$.

$$d(x) = [x^2 \cdot p(x)]_{\text{Degree}<3} \tag{7.35}$$

$$d(x) = [x^2 \cdot (1 + x + x^3)]_{\text{Degree}<3} \tag{7.36}$$

$$d(x) = x^2 \tag{7.37}$$

Hence for this example, this corresponds to a delay or shift of 2 chips in the output PN sequence.

Now that was a rather simple example, let's consider initial contents equal to 001, which corresponds to a polynomial of 1. Hence, we go through the same procedure as shown above to generate $d(x)$.

$$d(x) = [1 \cdot p(x)]_{\text{Degree}<3} \tag{7.38}$$

$$d(x) = [1 \cdot (1 + x + x^3)]_{\text{Degree}<3} \tag{7.39}$$

$$d(x) = 1 + x \tag{7.40}$$

In order to determine what offset this corresponds to, we need to perform the following step of dividing by the PN code polynomial:

$$a(x) = \frac{d(x)}{p(x)} = \frac{1 + x}{1 + x + x^3} \tag{7.41a}$$

$$a(x) = 1 + x^3 + x^4 + x^5 + x^7 + x^{10} + \cdots \tag{7.41b}$$

$$a(x) = 10011101001\ldots \tag{7.42}$$

After careful inspection of the tabulated PN contents (provided earlier in Fig. 7.25), we notice this corresponds to a shift of 6 chips at the output.

The second approach we will discuss is called masking. Let us consider the block diagram redrawn in Fig. 7.27 with the mathematical expressions for the PN output and delayed PN output sequences.

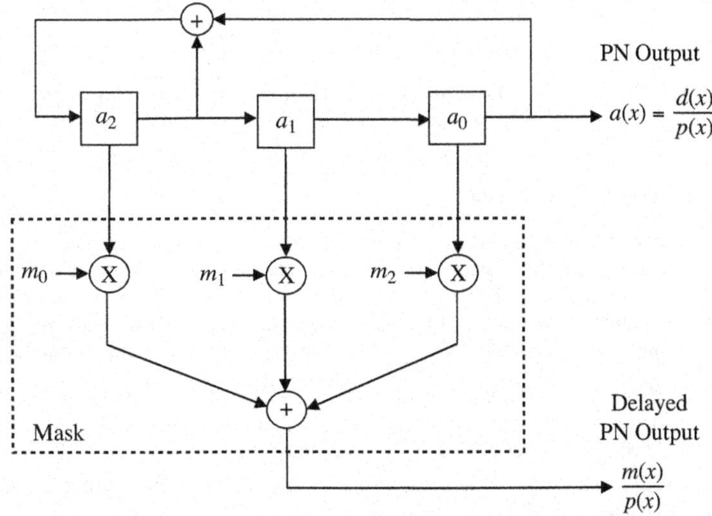

FIGURE 7.27 Masking technique applied to the PN sequence generator.

Using the above definitions, we have the following variables:

$$\begin{aligned} p(x) &= 1 + x + x^3 \\ a(x) &= [100] = x^2 \\ m(x) &= [110] = 1 + x \\ d(x) &= x^2 \end{aligned} \tag{7.43}$$

Note that in Fig. 7.27, the PN output and delayed PN output sequences can be derived by division of two polynomials. In the top case, the PN sequence is delayed by the initial contents of the shift registers, which is used to generate the $d(x)$ polynomial. The bottom case can delay the output PN sequence by dividing the mask polynomial by the code polynomial.

Note the mask produces a delay at the output, regardless of the initial loading of the PN shift registers. This can best be explained as follows: The PN sequence is periodic, and so we represent this behavior having a circular definition (see Fig. 7.28). The PN sequence starts at $t = 0$ and ends at $t = T - 1$, representing the PN sequence period of T chips. Each state on the circle represents an output chip from the generator.

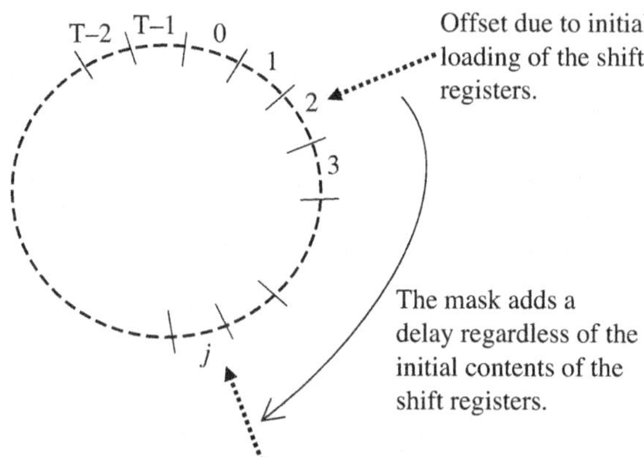

FIGURE 7.28 PN sequence represented as a circular state diagram.

7.5.1 Maximal Length Sequences

Maximal length sequences are sometimes referred to as m-sequences. Assume the generator polynomial is a primitive polynomial of degree L. The maximum number of different states of the linear feedback shift register (LFSR) generating the sequence is $N = 2^L - 1$, where N is defined as the period of the sequence, thus obtaining the maximum possible repetition sequence. Sequences that exhibit this property are called maximal length sequences. Some interesting properties are provided next.

1. *Balance:* The number of "1s" and the number of "0s" should be approximately the same. Let N_1 denote the number of times "1" occurs in a single period, and let N_0 be the number of times "0" occurs in one period. Then we have the relationship $N_1 = N_0 + 1$.
2. *Shift and Add:* If the m-sequence is added to its time-shifted version, the result is the same m-sequence with a new shift.
3. *Correlations:* A measure of similarity or dissimilarity between the m-sequences of interest.

A goal of WCDMA system design is to find a set of spreading codes such that as many users as possible can utilize the same bandwidth with as little multiple access interference (MAI). In order to achieve such a goal, we are forced to study the autocorrelation and cross-correlation properties of PN sequences.

Consider the following autocorrelation of the m-sequence:

$$R_{cc}(k) = \sum_{n=1}^{N} C_{k+n} \cdot C_n = \begin{cases} N; & k = 0 \\ 1; & k \neq 0 \end{cases} \tag{7.44}$$

Similarly, the cross-correlation of the m-sequence is

$$R_{c_1c_2}(k) = \sum_{n=1}^{N} C_{k+n}^{(1)} \cdot C_n^{(2)} \tag{7.45}$$

In Fig. 7.29, we plot the results of the autocorrelation and cross-correlation for a few chosen m-sequences. The sequences used a primitive polynomial of degree $L = 5$. From this particular example, it is clear that the autocorrelation function has values that vary and depend on the lag of interest.

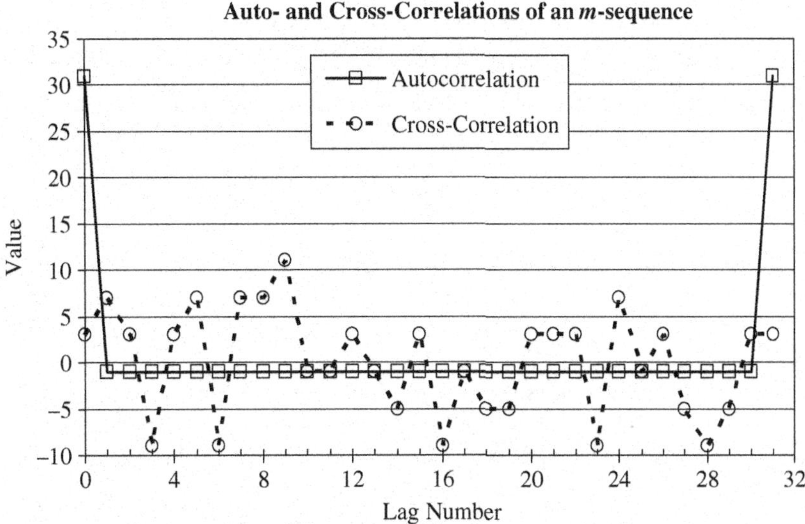

FIGURE 7.29 Auto- and cross-correlation plots for an m-sequence of degree = 5.

7.5.2 Gold Codes

In this section, we will discuss Gold codes (1967) and their properties [12]. Gold codes obey the following theorem: Sequences generated by combining binary m-sequences, $a(x)$ and $b(x)$, give cross-correlation peaks that are not greater than the minimum possible cross-correlation peaks between any pair of maximal length sequences. This is a tremendous property to have since having a low cross-correlation function can directly relate to having low MAI as new users begin to use the available spectrum. A general block diagram showing the method used to generate Gold codes is provided in Fig. 7.30.

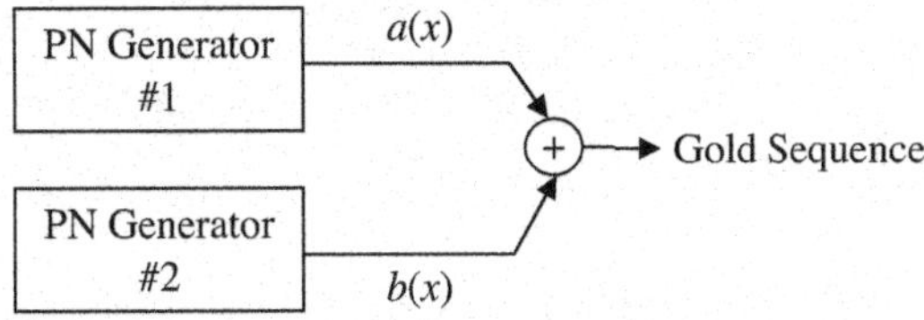

FIGURE 7.30 General Gold sequence generator block diagram.

In Fig 7.30, the binary m-sequences are sometimes referred to as "preferred pairs." Gold sequences are not maximal sequences. Hence, the autocorrelation function is not two valued; it is three valued.

Given $a(x)$ is an m-sequence of length $N = 2^L - 1$, $b(x)$ can be either an m-sequence or decimated version from $a(x)$, say $a(rx)$. For this special case of $r = 2^{[(L+2)/2]} + 1$, assuming L = even, we have the following autocorrelation function:

$$R_{cc}(k) = \sum_{n=1}^{2^L-1} C_{n+k} \cdot C_n = \begin{cases} -1 \\ -1 + 2^{[(L+2)/2]} \\ -1 - 2^{[(L+2)/2]} \end{cases} \tag{7.46}$$

What this does is to provide a set of sequences of size $2^L + 1$. This value was obtained as follows: one for each different shift of $b(x)$ plus the two original sequences used to generate the Gold code. This number of sequences is larger than m-sequences.

Now the Gold codes generated based on primitive polynomials satisfy the following cross-correlation function:

$$|R_{C,t}(k)| \leq \begin{cases} 2^{(N+1)/2} + 1; & N = \text{odd} \\ 2^{(N+2)/2} + 1; & N = \text{even} \end{cases} \tag{7.47}$$

where the code number is given as

$$t = \begin{cases} 2^{(N+1)/2} + 1; & N = \text{odd} \\ 2^{(N+2)/2} + 1; & N = \text{even} \end{cases} \tag{7.48}$$

We want to take some time to present another PN sequence called Kasami codes [13]. These codes have very low cross-correlation properties. The procedure is similar to generating Gold codes; however, this results in a small set of Kasami codes, denoted as $M = 2^{L/2}$ sequences of period equal to $N = 2^L - 1$.

Let us consider $a(x)$ to be an m-sequence of period $N = 2^L - 1$, and $b(x) = a(rx)$ equal the m-sequence $a(x)$ decimated by $r = 2^{L/2} + 1$ (period $= 2^{L/2} - 1$). The Kasami codes are equal to $a(x) + a(rx)$. The cross-correlation property is [with $2^{L/2}$ (2 cyclic shifts of $a(rx)$]

$$R_{cc}(k) = \begin{cases} -1 \\ 2^{L/2} - 1 \\ -2^{L/2} - 1 \end{cases} \tag{7.49}$$

Kasami codes are optimal in the sense they achieve the Welch bound by minimizing the maximum absolute correlation parameters, $P_{\max}$, between any pair of binary sequences of period n in a set of m-sequences

$$P_{\max} \geq n\sqrt{\frac{M-1}{Mn-1}} \tag{7.50}$$

For large values of M and n, the Welch bound can be approximated as $P_{\max} \geq \sqrt{n}$. One of the shortcomings of the Kasami codes discussed so far is the limited number of possible code sequences that can be generated. This essentially translates to limited number of possible users that can simultaneously communicate with each other.

Next we show how a very large set of Kasami codes can be generated. Given the following participating m-sequences:

$a(x) = m$-sequence

$a(rx) = m$-sequence $a(x)$ decimated by $r = 2^{n/2} + 1$

$a(sx) = m$-sequence $a(x)$ decimated by $s = 2^{(n+2)/2} + 1$

The very large set of Kasami codes can be generated as

$$\text{Very large Kasami} = a(x) + a(rx) + a(sx) \tag{7.51}$$

Here the number of sequences is $M = 2^{3n/2}$, and the autocorrelation function now takes on three values.

We will leave this subsection by stating that m-sequences have a period equal to the maximum length and reasonable cross-correlation properties. These properties can be improved by using Gold codes without sacrificing the number of available codes. Furthermore, the cross-correlation properties can be improved using Kasami codes at the expanse of the limited availability of codes.

7.5.3 Orthogonal Codes

In this section, we will provide an example of an orthogonal code which has the property that the cross-correlation of any two codes produces a result equal to 0. We will discuss the Walsh codes or Hadamard sequences. Recall the orthogonality property:

$$\int_0^T C_j(t) \cdot C_k(t)\, dt = \begin{cases} 0; & j \neq k \\ T; & j = k \end{cases} \tag{7.52}$$

What this equation states is that the chip sequence is only correlated with itself and not with any other. An alternative viewpoint is that the projection of one code onto another is 0. These sequences can be generated using the Hadamard matrices.

$$H_2 = \begin{bmatrix} 1 & 1 \\ 1 & -1 \end{bmatrix} \quad \text{or} \quad \begin{bmatrix} 0 & 0 \\ 0 & 1 \end{bmatrix} \tag{7.53}$$

Higher-order code words can be generated using the following relationship:

$$H_{2N} = \begin{bmatrix} H_N & H_N \\ H_N & \overline{H_N} \end{bmatrix} \tag{7.54}$$

For example, a code of length $N = 4$ can be generated as

$$H_4 = H_2 \times H_2 \tag{7.55}$$

$$H_4 = \begin{bmatrix} 0 & 0 \\ 0 & 1 \end{bmatrix} \times \begin{bmatrix} 0 & 0 \\ 0 & 1 \end{bmatrix} \tag{7.56}$$

$$H_4 = \begin{bmatrix} 0 & 0 & 0 & 0 \\ 0 & 1 & 0 & 1 \\ 0 & 0 & 1 & 1 \\ 0 & 1 & 1 & 0 \end{bmatrix} = \begin{bmatrix} w_1 \\ w_2 \\ w_3 \\ w_4 \end{bmatrix} \tag{7.57}$$

This technique generates a square matrix of code words.

Here we generate a 4 × 4 matrix where each row is a code word w_j. If we compare w_2 to w_4, we notice the number of positions in which the data agree is 2; also, the number of positions in which the data disagree is also 2. Recall that, earlier, we presented the autocorrelation as

$$R = \frac{N_{\text{Agree}} - N_{\text{Disagree}}}{T} = \frac{2 - 2}{4} = 0 \tag{7.58}$$

The WCDMA standard uses variable-length orthogonal codes which are called Orthogonal Variable Spreading Factor (OVSF) codes. These OVSF codes are essentially equivalent to the Hadamard code described above.

Now let us provide a brief summary to what has been presented thus far.

Autocorrelation:
- Periodic with $N = 2^L - 1$.
- m-sequences are not orthogonal.
- m-sequences have 2-level property.

Cross-correlation:
- Periodic with $N = 2^L - 1$.
- Provides increase in MAI in DS-CDMA.
- Cross-correlation peaks for certain pairs of m-sequences are larger than those of other pairs. Solution is to use Gold codes or Kasami codes.

7.6 WCDMA PHYSICAL LAYER OVERVIEW

In this section, we provide an overview of the Universal Mobile Telecommunication System (UMTS) cellular system [14–16]. We begin by describing the elements inside the UMTS architecture as shown in Fig. 7.31. In the 2G cellular system, the terminals were called Mobile Stations (MS), and in the 3G cellular system, they are called User Equipment (UE). Also, the numerology for the BS and BSC have changed to NodeB and Radio Network Controller (RNC), respectively. Each RNC can have multiple

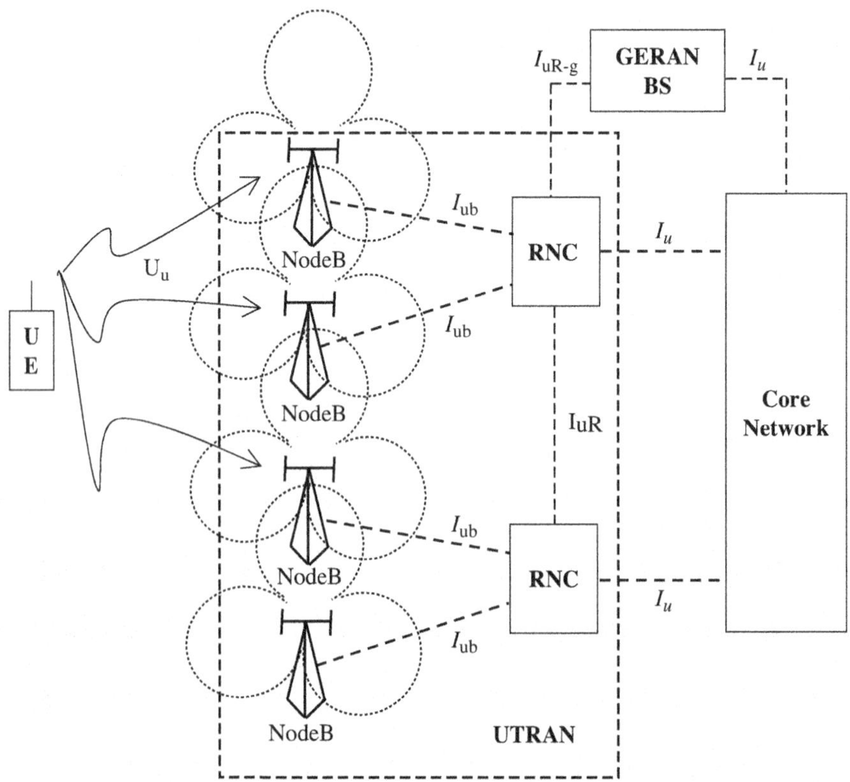

FIGURE 7.31 UMTS architecture overview.

NodeBs; the communication mechanism between them is labeled I_{ub}. Moreover, RNCs can communicate over an interface labeled I_{uR}, which is important if soft handoff is supported. The RNCs communicate to the UMTS core network via the I_u interface.

In the UMTS architecture diagram in Fig. 7.31, we have a single UE communicating to three NodeBs. For sake of discussion, we have assumed each NodeB operates at a single RF carrier and has three sectors. These sectors are highlighted by the circular antenna patterns provided. The traffic sent to the UE is sent via the NodeBs shown; one will be called the serving NodeB. As the UE travels toward the bottom of the page, NodeBs closer to the UE will start to communicate while others further away begin to cease communication. This procedure is called handoff. We have also shown the interface to the GSM network, which will be discussed shortly.

Now that we have presented the architecture, the next step is to provide an overview of the software control architecture, in terms of functionality partitioning [17]. Our goal is to show which element performs what tasks in the overall functional hierarchy. The diagram in Fig. 7.32 is drawn with

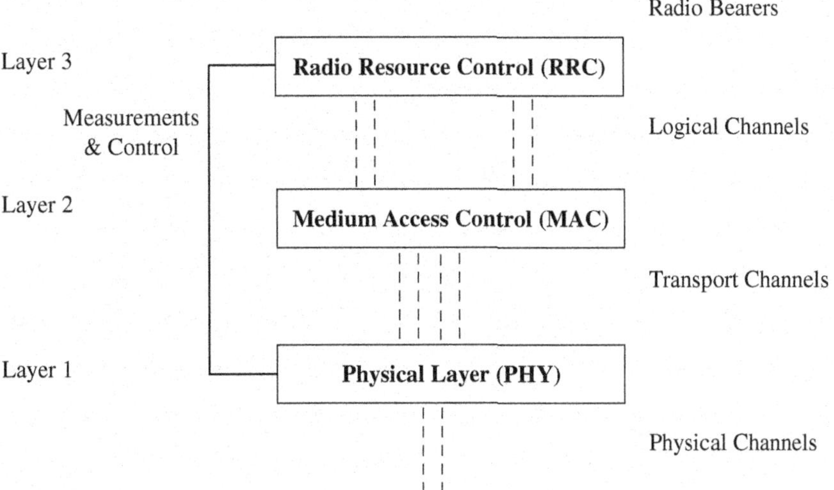

FIGURE 7.32 Radio interface protocol architecture.

the OSI layers in mind and represents the radio interface protocol architecture. We have shown the first three layers of the architecture. The PHY layer represents the WCDMA physical channel links. The MAC (medium access control) is the mechanism that manages the physical resources. Lastly, the RRC (radio resource control) is responsible for deciding on the radio resources to be used.

The communication between layers is conveyed through channels. A logical channel is defined by the type of information transferred. The physical channel is defined by the physical resources occupied. This is either a code, frequency, or relative I/Q phase. A transport channel is defined by how the information is transferred over the radio interface.

The UMTS Terrestrial Radio Access Network (UTRAN) performs the following functions: Broadcasting system information, admission and congestion control, mobility handling, radio resource management, ciphering and deciphering, multimedia broadcasting multicast services (MBMS) provisioning, allocation and de-allocation of radio bearers, connection setup and release, and so forth.

The GSM/EDGE Radio Access Network (GERAN) reference architecture is given in Fig. 7.33, for sake of reference. The MS is communicating to two sectors of different BTSs and through the use of the $I_{uR\text{-}g}$ can interface to the RNC.

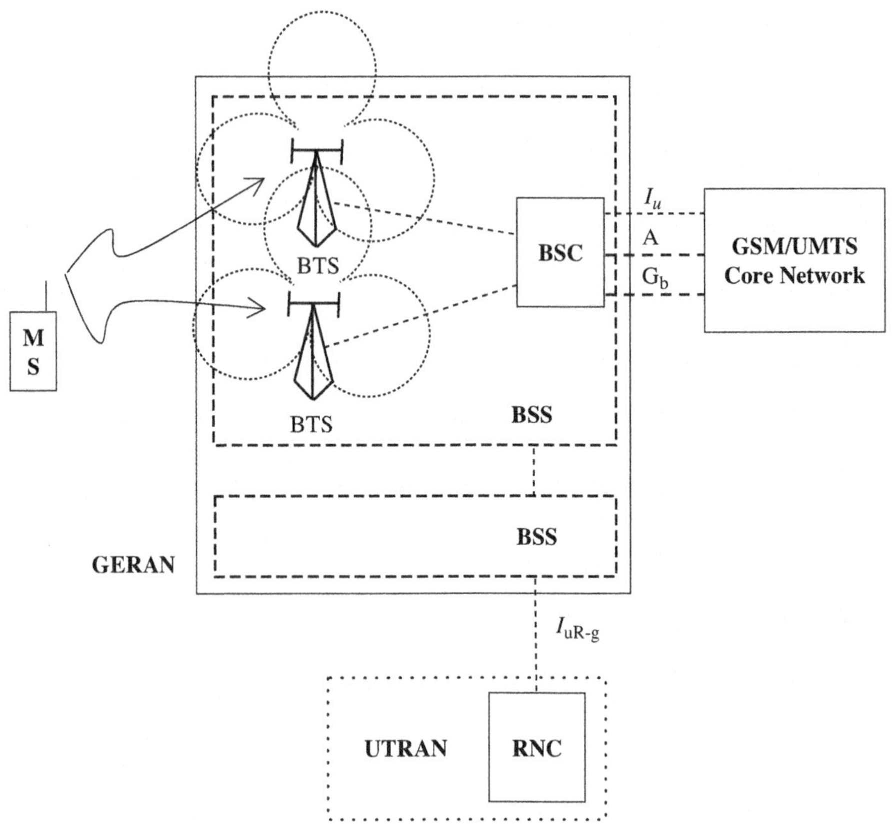

FIGURE 7.33 GSM/EDGE reference architecture.

7.6.1 Dedicated Channel Time Slot Structure

In this subsection, we will briefly present the frame and time slot structures for the uplink and downlink channels. First we provide a summary of the mapping of transport channels onto physical channels. The acronyms will be defined in the latter sections, where we will discuss each channel separately. First the uplink channels are presented, followed by the downlink channels (see Fig. 7.34).

The FDD mode chip rate is fixed at 3.84 Mcps. The physical channels are carried in the form of the following resources:

Radio frame	Equals 10 msec in duration Consists of 15 time slots Consists of 5 subframes Consists of 38,400 chips
Time slot	Equals 10/15 msec (or 0.667 msec) in duration Consists of 2560 chips
Subframe	Equals 2 msec in duration Consists of 3 time slots Consists of 7680 chips

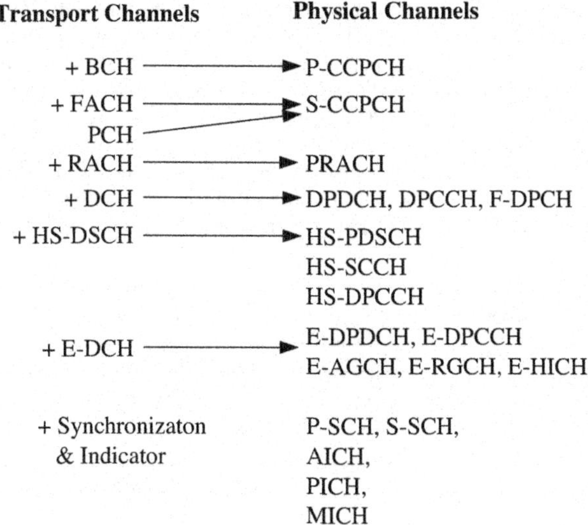

FIGURE 7.34 Transport channel mapping.

Uplink Discussion. There are five types of uplink dedicated physical channels (DPCHs): DPDCH, DPCCH, E-DPDCH, E-DPCCH, and HS-DPCCH. The dedicated physical data channel (DPDCH) is used to carry the DCH transport channel. The DPCCH is used to carry the Layer 1 control information such as transmit power control (TPC) command, transport format combination indicator (TFCI), and feedback indicator (FBI) bits.

The uplink frame and time slot structure is given in Fig. 7.35 for the first two uplink DPCHs mentioned above, namely, DPDCH and DPCCH.

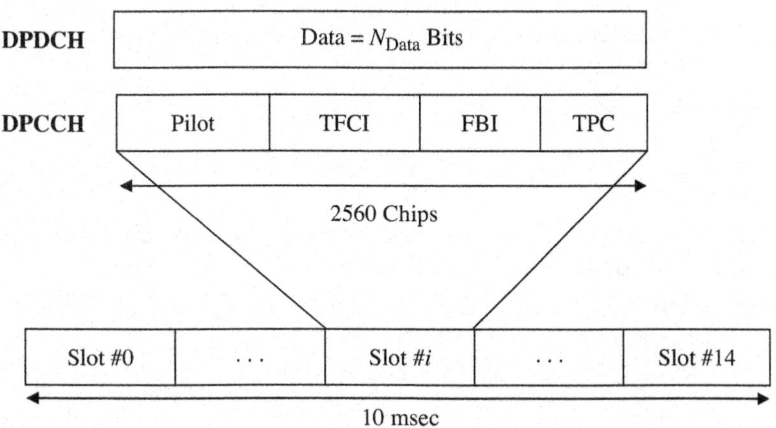

FIGURE 7.35 Uplink DPDCH and DPCCH time slot and frame structure.

The SF of DPCCH is constant and equal to 256 while the SF of DPDCH varies from 256 down to 4, depending on the user data rate selected. The duration of the bit fields are clearly provided in the 3GPP Technical Specification (TS) 25.211 and will not be repeated here [18]. The pilot bit field contains a known sequence of bits that changes in each time slot and repeats every frame. These pilot bits can be used to aid CE for coherent detection, assist in beam forming, aid SIR estimation, perform frames synchronization, and so forth. The TFCI bits inform the receiver about the instantaneous transport format combination of the transport channels mapped to the simultaneously transmitted DPDCH radio frame.

The FBI bits are used to support feedback information required from the UE for such application as CLTD. Lastly, the TPC bits are used to convey power control commands to adjust the downlink NodeB transmit power. Since these TPC bits are all of the same polarity (i.e., either all 1s or all 0s), it is conceivable that they also can be used to aid the CE or other algorithms, if so required.

The frame and time slot structure for the next two uplink DPCHs is given in Fig. 7.36 for E-DPDCH and E-DPCCH. These channels correspond to the enhanced DPCH support.

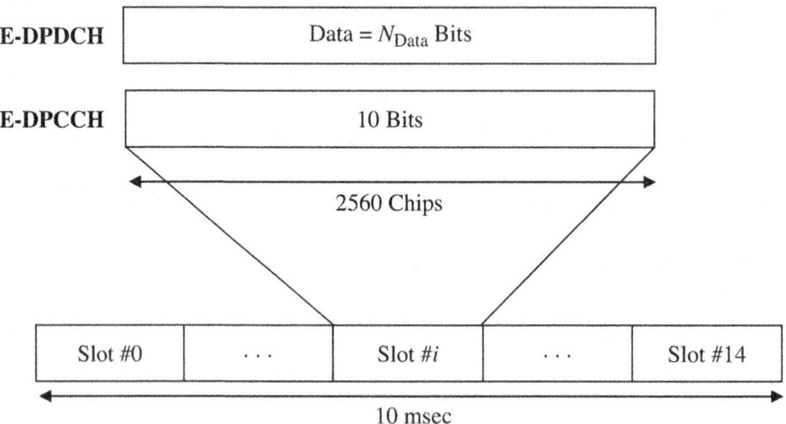

FIGURE 7.36 Uplink E-DPDCH and E-DPCCH time and frame structure.

The SF of E-DPCCH is fixed at 256 while the SF for E-DPDCH can vary from 256 down to 2. E-DPDCH is used to carry the E-DCH transport channel, and the E-DPCCH is used to carry the control information. The enhanced DPCH channel will be used to send uplink packet data, for HSUPA services. Based on certain downlink measurements, the UE will send appropriate packets on the uplink E-DPCH channels. The number of channels will be discussed in the HSUPA section later in this chapter.

The last uplink DPCH to be discussed is HS-DPCCH; its subframe and time slot structure is provided in Fig. 7.37.

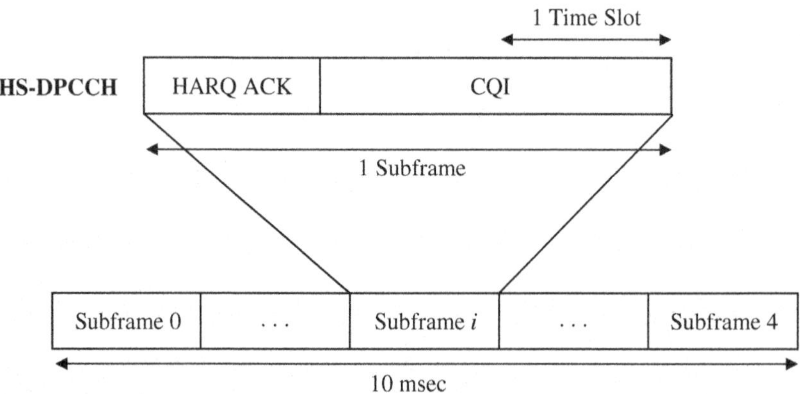

FIGURE 7.37 Uplink HS-DPCCH time slot and subframe structure.

The SF for HS-DPCCH is fixed at 256 and has a single slot format. The HS-DPCCH carries the uplink feedback signaling related to the downlink transmission of HS-DSCH, supporting HSDPA services. The uplink feedback signaling consists of Hybrid Automatic Repeat Request (HARQ) acknowledgment (ACK/NACK) and the Channel Quality Indication (CQI). The first part of the subframe consists of the acknowledgment while the remaining consists of the CQI information.

The acknowledgment is used by the NodeB as a response to the demodulation of the downlink HS-DSCH packet transmitted. If the HS-DSCH CRC decode is successful, then an ACK is inserted. If the HS-DSCH CRC decode is unsuccessful, then a NACK is inserted. The CQI is a time-varying indication of the channel the UE observes. The mapping of this quality to an indicator is accomplished by sending the NodeB an index into a channel quality table indicating the best possible combinations of modulation scheme, transport block size, and code rate the UE can receive at that particular time instant. We will discuss this operation in more detail in the HSDPA section later in this chapter.

Five uplink physical channels were presented. The DPDCH and DPCCH are time slot aligned along with the E-DPDCH and E-DPCCH channels. The HS-DPCCH channel is not necessarily time aligned with these channels due to the uplink and downlink timing relationship required to support HSDPA services.

Downlink Discussion. There are five types of downlink DPCHs: DPCH, F-DPCH, E-RGCH, E-AGCH, and E-HICH. The downlink frame and time slot structure for the DPCH is given in Fig. 7.38. The DPCH consists of DPDCH and DPCCH time multiplexed into the physical channel.

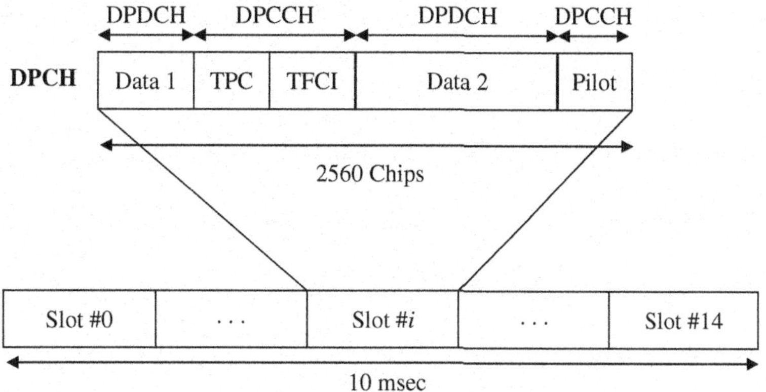

FIGURE 7.38 Downlink DPDCH and DPCCH time slot and frame structure.

The SF can vary from 512 down to 4. Since the total number of chips transmitted in a time slot is fixed, the variable SF translates into having variable number of bits in each time slot field. The time slot formats are well defined in 3GPP Technical Specification 25.211 [18]. The fields are defined to have the same functionality as described in the previous subsection for the uplink. The major difference is on the uplink, where the DPDCH and DPCCH channels are code multiplexed, while they are time multiplexed on the downlink.

There are two types of DPCH channels: those that do not include the TFCI field, for fixed rate services, and those that include the TFCI field, for several simultaneous services. The UTRAN determines if the TFCI field should be included.

The downlink TPC bit field duration can be 2, 4, or 8 bits, depending on the SF used. Moreover, the possibility to have 16 bits exists when frames are affected by compressed mode (CM) operation, specifically when SF reduction is used.

The E-DCH Relative Grant Channel (E-RGCH) time slot and frame structure is given in Fig. 7.39. The SF is fixed at 128 and is used to carry the E-DCH relative grants.

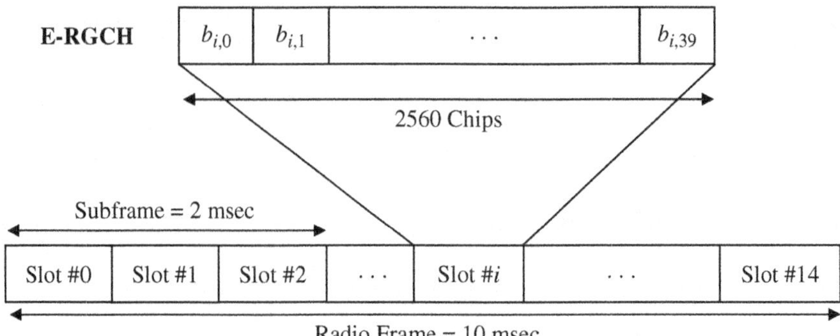

FIGURE 7.39 Downlink E-RGCH time slot and frame structure.

The relative grant is transmitted using either 3, 12, or 15 consecutive time slots—each slot consists of 40 ternary values. The 3– and 12–time slot durations are used on E-RGCH in cells which have E-RGCH and E-DCH in the serving radio link set. The 3– and 12–time slot durations correspond to the E-DCH TTI of 2 and 10 msec, respectively. The 15–time slot duration is used on E-RGCH in cells which have E-RGCH not in the E-DCH serving radio link set. The 40 ternary values are generated by multiplying an orthogonal signature sequence of length 40 by the relative grant value a. The relative grant value has two definitions: $a \in \{-1, 0, +1\}$ in a serving E-DCH radio link set and $a \in \{0, -1\}$ in a radio link not belonging to the serving E-DCH radio link set. Lastly, the orthogonal signature sequence to be used is provided by the higher layers. E-AGCH will be discussed later.

The last downlink DPCH is the fractional DPCH (F-DPCH). The time slot and frame structure is shown in Fig. 7.40. The SF is fixed at 256 and is used to carry control information for Layer 1,

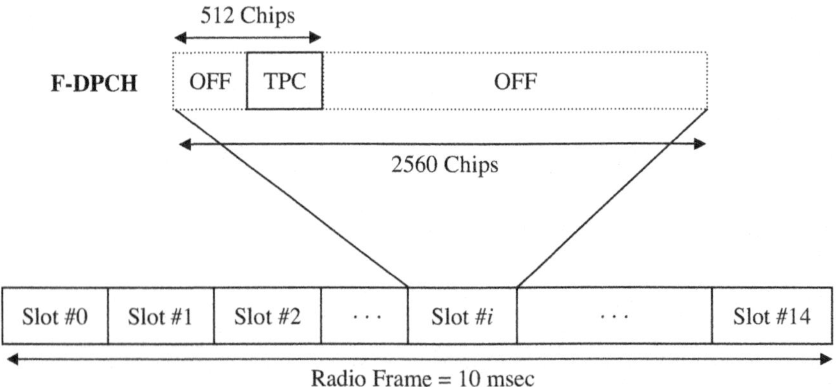

FIGURE 7.40 Downlink F-DPCH time slot and frame structure.

specifically the TPC commands. We will leave the E-HICH channel discussion until the HSUPA services are discussed.

F-DPCH is a special case of the downlink DPCCH channel; it contains only the uplink power control commands. This channel removes the need to have a downlink DPDCH channel, if not necessary. A typical scenario is to use the F-DPCH channel with traffic sent over the HS-DSCH channel. One such scenario can be the VoIP application. All downlink data/voice will be sent to the UE via the HS-DSCH with F-DPCH present to only signal power control commands. The uplink channels remain the same as discussed in the previous sections, namely, DPDCH/DPCCH/HS-DPCCH. The last point to mention here is by dramatically reducing the downlink signal, for example, no DPDCH transmission, and a fraction of the DPCCH by using F-DPCH, the downlink interference (both within the cell and outside the cell) is reduced. Later, we will discuss other solutions that reduce the uplink noise rise from UE transmissions.

Due to the time slot structure, it is conceivable that multiple users can have their F-DPCH channels time multiplexed onto a single physical channel where each user has a multiple of 256-chip time offset. This time-multiplexed scenario can also be used to save on the usage of channelization codes.

7.6.2 Spreading and Scrambling Procedure

In this section, we will discuss the spreading and scrambling operations used in WCDMA physical layer for both the uplink and downlink communication.

Downlink Spreading and Scrambling. The downlink operations can be summarized with Fig. 7.41, where only a single physical channel is shown. The bits to be transmitted are first converted to symbols,

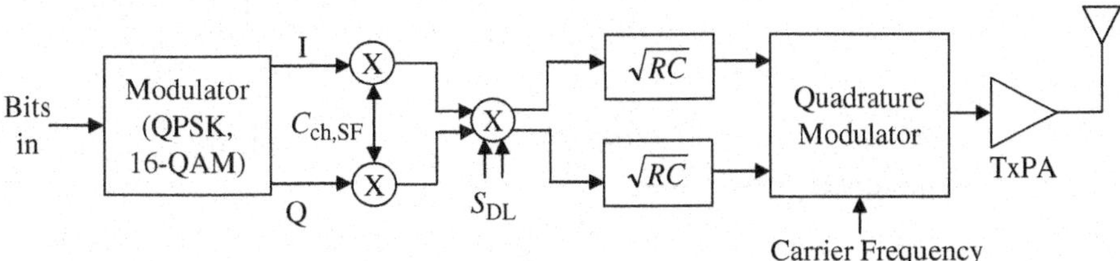

FIGURE 7.41 Downlink spreading and scrambling.

depending on the modulation scheme used. The modulation symbols are then spread by a channelization code; here, the OVSF codes are used. This is represented as $C_{\mathrm{CH,SF}}$, where the subscript CH stands for channelization and SF stands for spreading factor. Next the complex spread waveform is scrambled by a Gold code. This is represented by S_{DL}, where the double lines indicate a complex multiplication. The chips then become filtered and enter the quadrature modulator for frequency translation.

The modulator maps the even- and odd-numbered bits to the I- and Q-channels, respectively. The channelization code is time aligned with the symbol boundary. The scrambling code is time aligned with the P-CCPCH frame. For other channels not frame aligned with P-CCPCH, they will not have the scrambling code time aligned with their physical frames. The pulse-shaping filter used is the square root raised cosine (SRC) with a roll-off factor $\alpha = 0.22$.

The channelization codes are called OVSF and preserve the orthogonality constraint between a user's different physical channels. They are typically shown using the code tree, in Fig. 7.42, with the following notation: $C_{\mathrm{CH,SF,k}}$, where CH is a channelization code, SF is the spreading factor, and k is the index into the possible codes.

The OVSF codes are generated similar to the well-known Walsh codes defined in the previous sections and rewritten below for sake of reference for the SF = 4 case:

$$H_4 = \begin{bmatrix} 1 & 1 & 1 & 1 \\ 1 & -1 & 1 & -1 \\ 1 & 1 & -1 & -1 \\ 1 & -1 & -1 & 1 \end{bmatrix} \tag{7.59}$$

Here you can see the row numbers of the generated Walsh codes are not the same as the OVSF codes; however, the row vectors are equivalent after rearranging them.

The channelization code for P-CPICH is $C_{\mathrm{CH,256,0}}$ and the channelization code for P-CCPCH is fixed to $C_{\mathrm{CH,256,1}}$. The channelization codes for all other physical channels are determined by the UTRAN. In each cell, the E-RGCH and E-HICH assigned to a UE have the same channelization code.

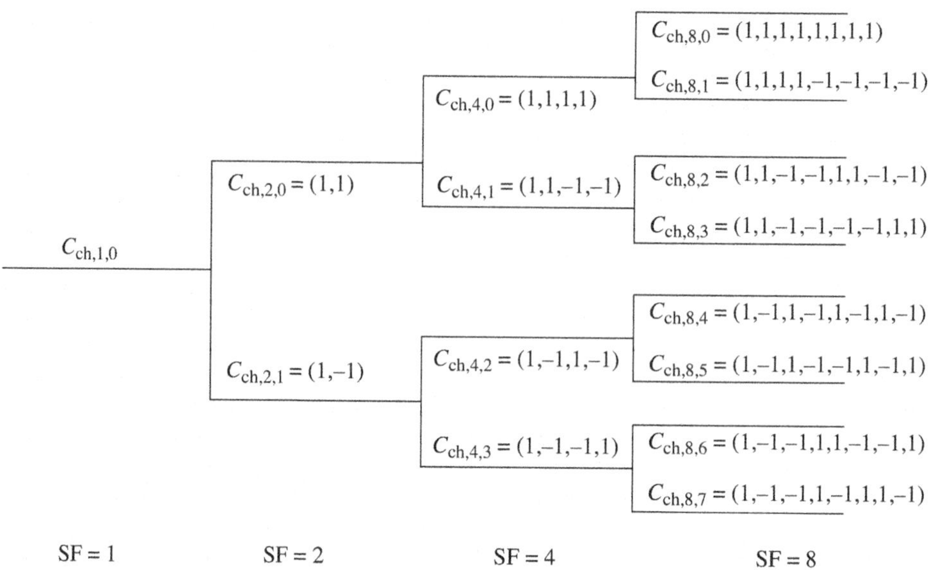

FIGURE 7.42 OVSF code tree.

There are a total of $2^{18} - 1 = 262{,}143$ scrambling codes; however, not all are used. The scrambling codes are divided into 512 sets; each set consists of a primary scrambling code and 15 secondary scrambling codes. Hence the total number of scrambling codes = 8192, not including left-alternative and right-alternative scrambling codes used in CM frames.

The set of primary scrambling codes (512 of them) is further divided into 64 scrambling code groups, each consisting of 8 primary scrambling codes. Each cell is given one and only one primary scrambling code. The P-CCPCH, P-CPICH, PICH, MICH, AICH, and S-CCPCH carrying PCH are always transmitted using the primary scrambling code. In the case of HS-DSCH, the HS-PDSCH and HS-SCCH that a single UE may receive under a single scrambling code may be either primary or secondary. The F-DPCH, E-RGCH, E-HICH, and E-AGCH assigned to a UE are configured with the same scrambling code as the assigned phase reference (either primary or secondary CPICH). The UE can be configured to at most support two simultaneous scrambling codes.

The scrambling codes are Gold codes using 2 binary m-sequences generated by 2 generator polynomials of degree 18. The scrambling codes are repeated every 10 msec. The two polynomials are $x_1 = 1 + x^7 + x^{18}$ and $x_2 = 1 + x^5 + x^7 + x^{10} + x^{18}$. The block diagram of the code generator for S_{DL} is shown in Fig. 7.43.

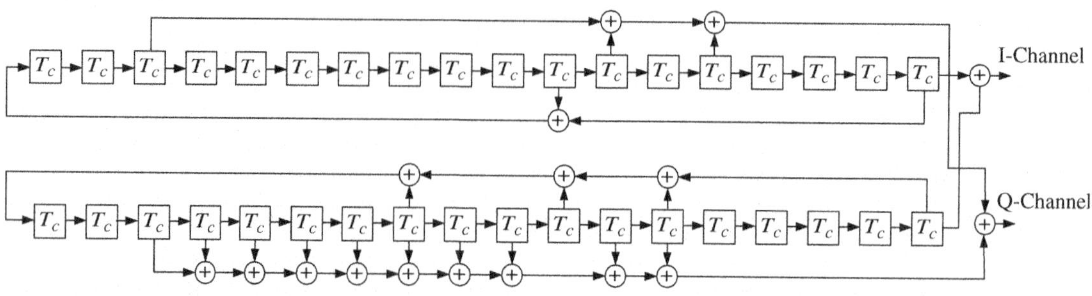

FIGURE 7.43 Downlink PN scrambling code generator.

Hence the complex scrambling code has a real part, I-channel, containing the unshifted (in time) Gold code and the imaginary part, Q-channel, that contains the shifted (in time) Gold code by 131,072 chips. This value corresponds to half the period of the 18th-order *m*-sequence.

Uplink Spreading and Scrambling. The uplink spreading and scrambling operations can best be described with the help of Fig. 7.44 for DPCH channels.

In this figure, we see that multiple DPDCH channels are possible, up to six to be exact. They are alternatively placed between the I- and Q-channels. The DPCCH is always placed on the Q-channel. The HS-DPCCH is alternatively placed between the I- and Q-channels, depending on the number of multiple channels used. The E-DPDCH is also alternatively placed between the I- and Q-channels. Lastly, the E-DPCCH is placed on the I-channel.

The relative powers between the channels are controlled by the β multipliers. The gains, β values, are signaled by the higher layers. The DPCCH is always spread by channelization code $C_{CH,256,0}$. When only one DPDCH is to be transmitted, DPDCH shall be spread by $C_{CH,SF,k}$, where SF is the spreading factor of the DPDCH channel and k is set to SF/4.

The uplink scrambling codes are assigned by higher layers. The scrambling codes are generated by modulo-2 sum of 2 binary *m*-sequences generated by 2 generator polynomials of degree 25. The polynomials are $x_1 = 1 + x^3 + x^{25}$ and $x_2 = 1 + x + x^2 + x^3 + x^{25}$. A block diagram of uplink scrambling code generator is given in Fig. 7.45.

The uplink scrambling code is a complex-valued sequence where the imaginary part C_2 is a time-shifted version (by 16,777,232 chips) of the real part C_1.

The next part of the scrambling code generation is to create the Hybrid PSK (HPSK) modulation [19]. Here the number of 180-degree phase variations is kept to a minimum and is accomplished as shown in Fig. 7.46.

The above functionality was introduced to lower the uplink peak-to-average power ratio (PAPR). After scrambling, the uplink waveform is said to be HPSK modulated. Essentially, the number of times the 180-degree phase change is encountered has been reduced, thus having the desirable side effect of lowering the out-of-band emissions. In Fig. 7.46, the letter n is used to denote the chip number index.

7.6.3 Downlink Common Channels

In this section, we will present the downlink common physical channels, specifically their time slot and frame structures. We also provide a summary of the relative timing relationships between the physical channels.

Common Pilot Channel (CPICH). The CPICH time slot and frame structure is given in Fig. 7.47. The SF is fixed at 256 and carries a predefined pilot symbol sequence. Since QPSK modulation is used for CPICH, a constant symbol of $A = 1 + j\,1$ is transmitted continuously.

The CPICH channel also supports transmit diversity. If transmit diversity is used on any downlink channel in the cell, CPICH shall be transmitted on both antennas using the same channelization and scrambling code. In order to help the UE in estimating the channel response from each NodeB transmit antenna, a different symbol sequence is transmitted on antenna 2.

The CPICH symbol sequences transmitted on each of the two transmit antennas is shown in Fig. 7.48, where we have defined the symbol A for sake of convenience. This symbol has an SF = 256 chips.

In viewing the above CPICH pilot symbol patterns, we notice the pilot patterns are orthogonal over blocks of symbols that comprise multiples of 512 chips or groups of 2 symbols. Note the pattern on the second antenna follows this order: A, −A, −A, A, which is repeated throughout the frame. This four-symbol repetition is violated at the end of the frame. At the frame boundary, orthogonality is maintained by viewing groups of four symbols.

There are two types of CPICHs: primary and secondary. Below we list some of their characteristics:

Primary CPICH (P-CPICH):

- Same channelization code is always used for the P-CPICH across all the cells.
- It is scrambled by the primary scrambling code.

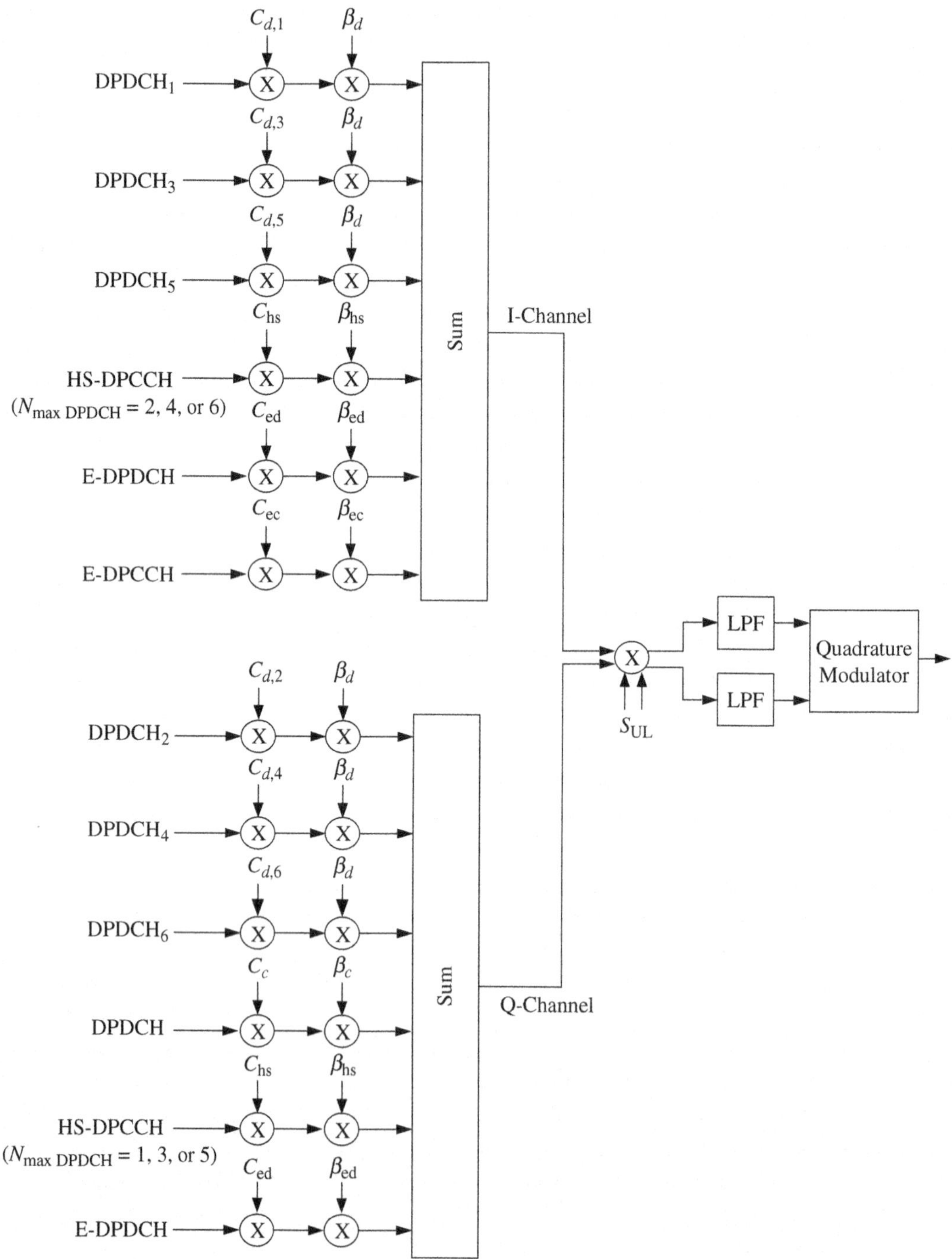

FIGURE 7.44 Uplink scrambling and spreading overview.

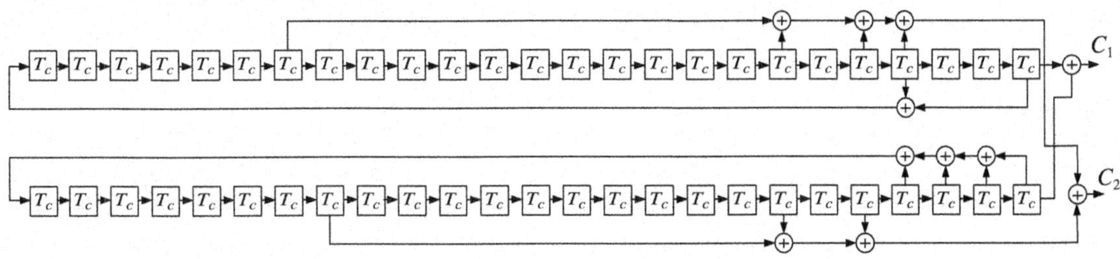

FIGURE 7.45 Uplink PN scrambling code generator.

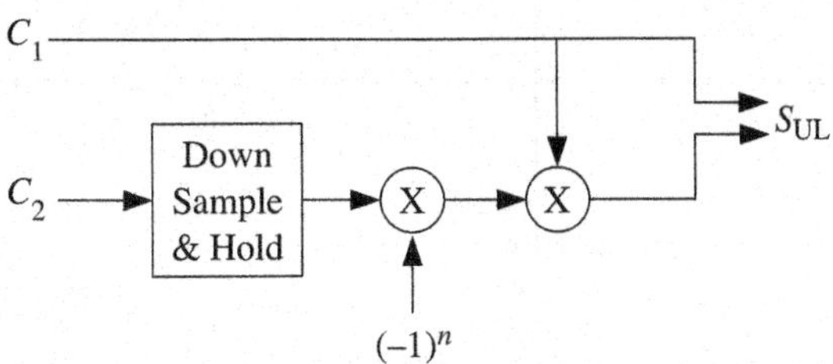

FIGURE 7.46 Uplink scrambling code generator for HPSK modulation.

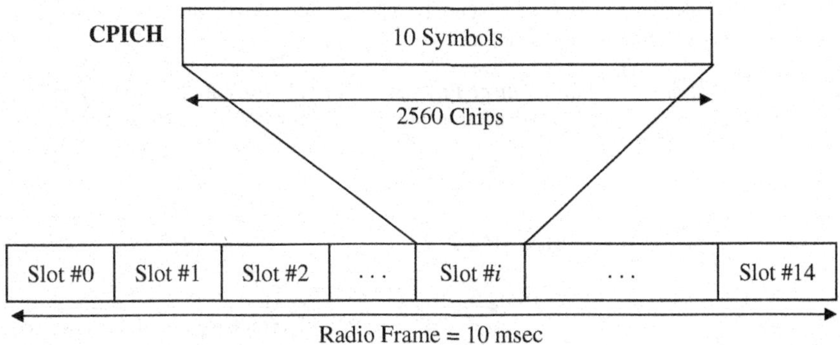

FIGURE 7.47 Downlink CPICH time slot and frame structure.

- There is only one P-CPICH per cell.
- It is broadcast over the entire cell.
- It is the default channel to be used as a phase reference for coherent demodulation in the UE receiver.

Secondary CPICH (S-CPICH):

- An arbitrary channelization code is used for S-CPICH.
- It can be scrambled with either the primary or a secondary scrambling code.
- There can be greater than or equal to zero S-CPICH per cell.
- It can be transmitted over the entire cell or part of the cell.
- It can be used as a phase reference for coherent demodulation, if notified by the higher layers.

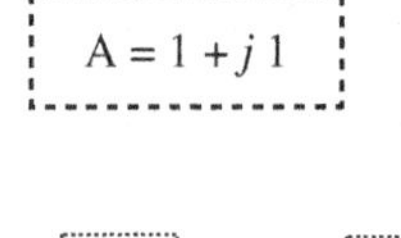

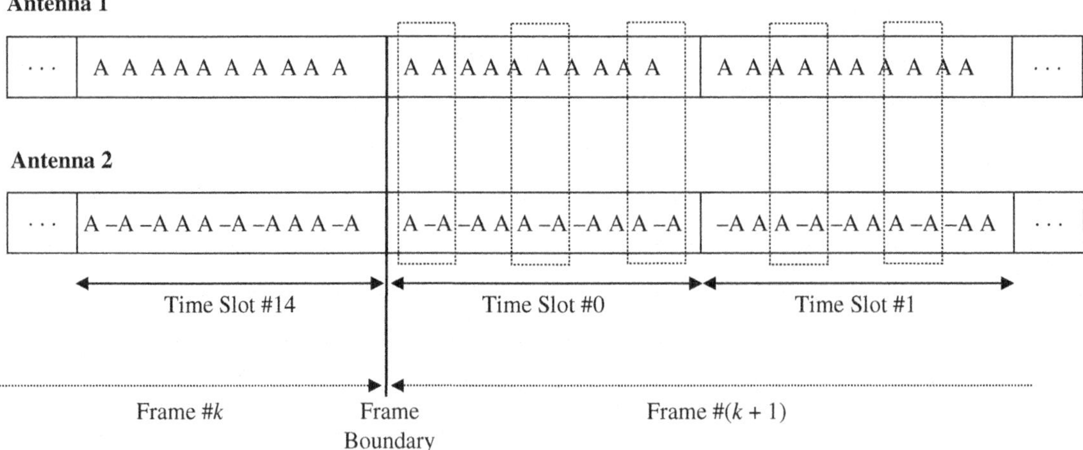

FIGURE 7.48 CPICH transmitted symbols when transmit diversity is used.

Note adding an S-CPICH using the secondary scrambling code as well as having the UE DPCH spread with the secondary scrambling code will increase the MAI due to increase in cross-correlation between the used scrambling codes.

Primary Common Control Physical Channel (P-CCPCH). The P-CCPCH physical channel is used to carry the broadcast transport channel (BCH). The time slot and frame structure is presented in Fig. 7.49.

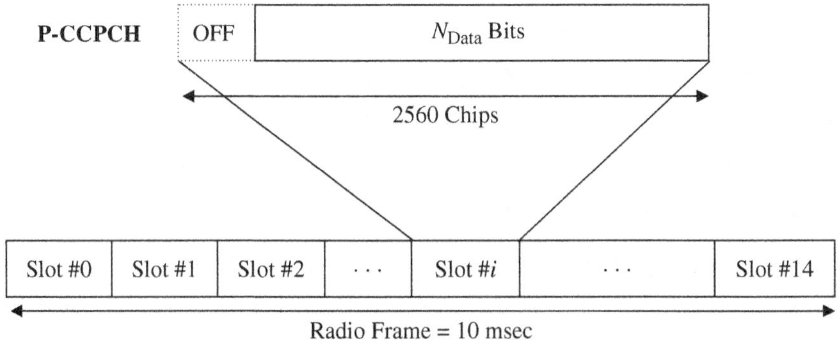

FIGURE 7.49 Downlink P-CCPCH time slot and frame structure.

The P-CCPCH is spread with a constant SF = 256. The P-CCPCH is not transmitted during the first 256 chips of every time slot. Instead, the primary/secondary synchronization channels (P/S-SCHs) are transmitted during this time. This allows the P-SCH and S-SCH to be transmitted with minimal interference. The P-CCPCH supports open-loop transmit diversity only since this is a common channel used by all UE within the cell. Because of the time slot structure, special operations are used to support Space-Time Transmit Diversity (STTD). These include STTD decoding across time slot boundaries as well as no encoding at the end of a radio frame. The P-CPICH is used as a phase reference for demodulation of the P-CCPCH.

The UE has two methods to determine if STTD encoding is used on the P-CCPCH: higher-layer signaling and modulation on SCH. Higher layers signal if STTD is used or not, so the UE can rely on the assistance of the information obtained from the protocol stack. In addition, the SCHs (both the P-SCH and S-SCH) are modulated by a diversity indicator bit that can also be used to determine if STTD encoding is used on P-CCPCH in that cell. Suffice it to say that the WCDMA standard does not preclude the UE from using both methods to determine if transmit diversity is used.

Secondary Common Control Physical Channel (S-CCPCH). The S-CCPCH is used to carry the FACH and PCH. There are two types of S-CCPCH: those with the TFCI bit field and those without the TFCI bit field. The FACH and PCH can be mapped to the same or separate S-CCPCH. As a result, the S-CCPCH has variable transport format combinations, which requires the use of the TFCI bits to indicate this behavior. When comparing this channel to the P-CCPCH, we notice the P-CCPCH does not have TFCI bits included, which means a single predefined transport format combination exists.

The S-CCPCH time slot and frame structure is provided in Fig. 7.50.

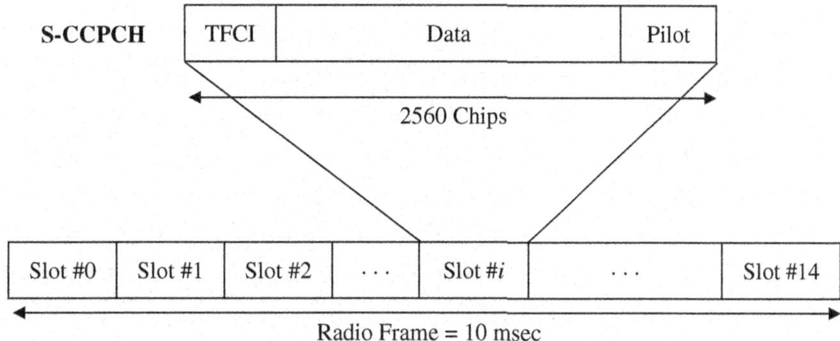

FIGURE 7.50 Downlink S-CCPCH time slot and frame structure.

The SF can vary from 256 down to 4. Note for those scenarios where TFCI is not required; then DTX (discontinuous transmission) will be used in those bit field positions.

Primary/Secondary Synchronization Channel (P/S-SCH). The SCH is used by the UE to search for active and neighboring cells, more commonly called cell search. The SCH consists of two subchannels, the P-SCH and S-SCH. The SCH is transmitted in each time slot with the following time slot and frame structure.

The P-SCH is a modulated Galois code of length 256 chips. The primary synchronization code is given by the variable C_p shown in Fig. 7.51. Notice it is transmitted in every time slot and is the same for every cell in the system.

The S-SCH is also a modulated Galois code of length 256 chips, but differs from the P-SCH in that it is different in every time slot, but repeats every frame. The secondary synchronization code is given by $C_s^{i,k}$ shown above, where $i \in \{0, 1, \ldots, 63\}$ and represents the scrambling code group number and $k \in \{0, 1, \ldots, 14\}$ represents the time slot number. Each C_s is chosen from a set of 16 different codes of length 256 chips. The sequence of the C_s on the S-SCH indicates in which code group the cell's primary scrambling code belongs.

As discussed above, the P-SCH and S-SCH channels are modulated by a bit labeled a, which is used to indicate the presence or absence of STTD encoding on the P-CCPCH. A very simple encoding is used. If $a = 1$, then STTD is used; else if $a = -1$, then STTD is not used on P-CCPCH.

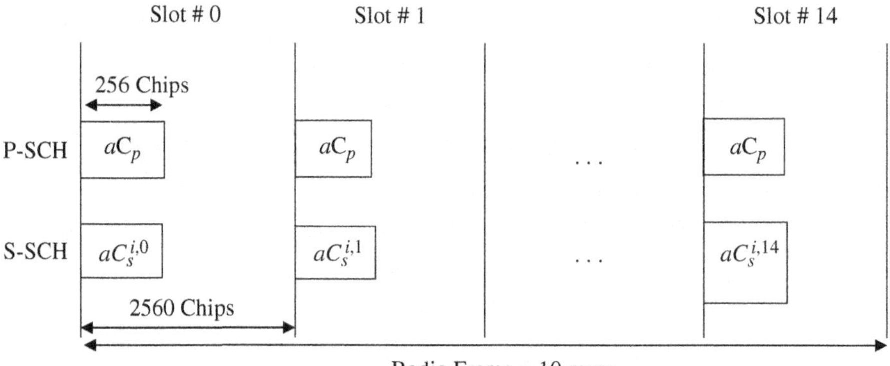

FIGURE 7.51 Downlink P-SCH and S-SCH time slot and frame structure.

Now a special case of transmit diversity is applied to the SCH, which is called Time Switched Transmit Diversity (TSTD). Here the transmission of the SCH is switched between the NodeB antenna 1 and antenna 2, in order to capitalize on usage of spatial diversity. In the even-numbered time slots, the SCHs are transmitted on antenna 1, and in the odd-numbered time slots, the SCHs are transmitted on antenna 2. Hence you have this very simple spatial hopping pattern defined. Here the SCH alternatively samples the channel response from each of the NodeB transmit antennas.

The use of the P-SCH and S-SCH channels will be discussed in latter sections. Suffice it to say that the UE will use the SCH channels to synchronize to the cell and obtain the cell's primary scrambling code. Time slot synchronization is obtained from P-SCH. Frame synchronization and code group identification is obtained for the S-SCH channel.

Acquisition Indicator Channel (AICH). The AICH is used to carry the acquisition indicators (AIs). The AIs correspond to signatures transmitted on the uplink PRACH channel. The time slot (or access slot, AS, terminology used for this channel) and frame structures are presented in Fig. 7.52.

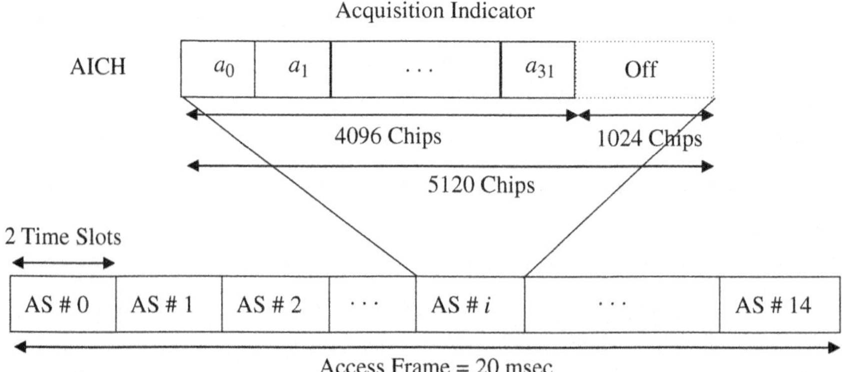

FIGURE 7.52 Downlink AICH AS and access frame structure.

The AICH consists of 15 ASs where each AS has a duration of 5120 chips (or 2 time slots). The AS consists of 2 parts: the AI and a DTX part. The AI part comprises 32 real-valued signals ($a_0, a_1, \ldots, a_{31}$) and collectively form a duration of 4096 chips. The DTX part has a duration of 1024 chips. The SF of AICH channel is fixed at 256. Also notice the access frame has a time duration of 20 msec, which is equal to 2 radio frames.

The AI real values are generated as follows:

$$a_j = \sum_{s=0}^{15} AI_s \cdot b_{s,j} \qquad j \in \{0, \ldots, 31\} \qquad s \in \{0, \ldots, 15\} \tag{7.60}$$

The AIs can take on values +1, 0, and −1. A table exists within the 3GPP Technical Specification 25.211 that provides the values for the signature patterns $b_{s,j}$ [18].

Paging Indicator Channel (PICH). The PICH channel is used to carry paging indicators. The PICH is always associated with an S-CCPCH which is carrying the PCH transport channel. The frame structure is provided in Fig. 7.53.

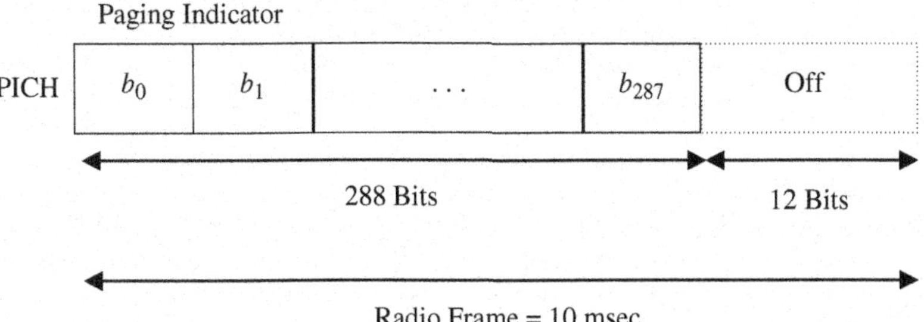

FIGURE 7.53 Downlink PICH radio frame structure.

The entire PICH frame consists of 300 bits ($b_0, b_1, \ldots, b_{299}$), but only the first 288 bits are transmitted and used. We can say the PICH frame consists of two parts: paging indicator and DTX.

In each PICH frame, the number of paging indicators transmitted is N_p. N_p can take on the following values: 16, 36, 72, and 144. Which paging indicators to view, say P_q, is computed in the UE with the help of the following variables: Paging indicator (PI) is calculated by the higher layers, the System Frame Number (SFN) of the P-CCPCH radio frame that overlaps with the start of the PICH frame, and lastly the number of paging indicators in the frame, N_p. All of these values are used to determine where in the frame to view the paging indicator bits. If the paging indicator was set to "1," then the UE associated with this paging indicator should read the next frame of S-CCPCH; otherwise, they should continue to sleep and wake up at the next scheduled time interval.

During idle mode operation, the UE will periodically wake up and demodulate the PICH bits. This discontinuous reception is called a DRX cycle and takes on values of 5.12, 2.56, 1.28, 0.64, 0.32, 0.16, and 0.08 sec. Obviously, the larger the DRX cycle, the better your standby time will be for the UE. However, the price to pay is the ability to control the time drifting of the locally generated timing (time and frequency), as well as the potential of a service region change, if the UE is mobile.

Overall Physical Channel Timing Relationship. In Fig. 7.54, we present various timing diagrams for the physical channels discussed above. A few points should be mentioned, prior to our discussion. First, the P-CCPCH is used as a timing reference for all physical channels for the downlink, and since the uplink has a fixed relationship to the downlink, the P-CCPCH indirectly acts as a timing reference for the uplink. Second, the P-CCPCH, P-CPICH, S-CPICH, P-SCH, S-SCH, and HS-SCCH have identical frame timing. In Fig. 7.54, we present the frame timing relationship for some of the downlink physical channels taken from 3GPP [18].

The S-CCPCH frame timing may be different for different S-CCPCHs transmitted. The offset from the P-CCPCH frame timing is measured in multiples of 256 chips: $\tau_{S\text{-CCPCH},k} = T_k \cdot 256$ chips, with $T_k \in \{0, 1, \ldots, 149\}$. The PICH frame timing is $\tau_{PICH} = 7680$ chips, prior to the corresponding

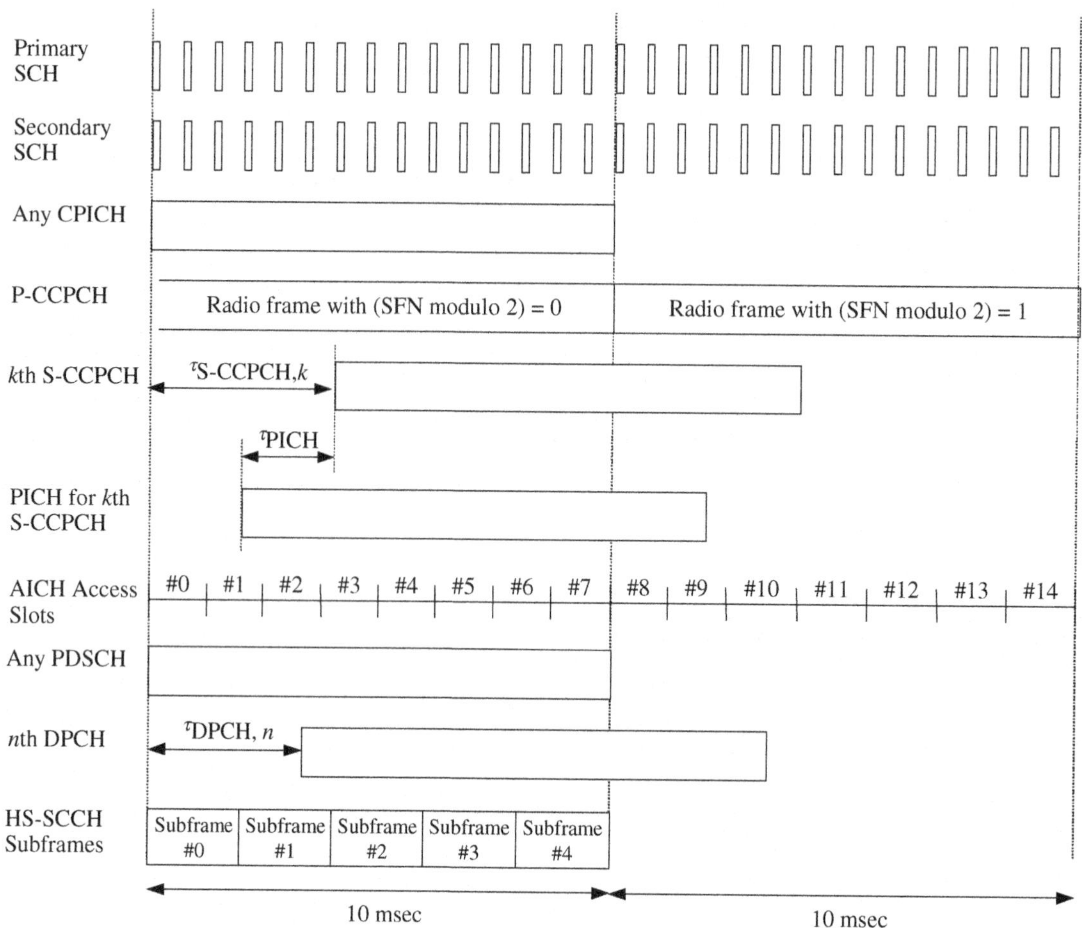

FIGURE 7.54 Downlink physical channel frame timing relationship.

S-CCPCH frame timing. Figure 7.54 can be somewhat misleading; let's try to clear it up. The paging indicator is sent in a PICH frame that corresponds to the PCH on the S-CCPCH which has a frame timing starting τ_{PICH} chips after the transmitted PICH frame. This discussion on timing is also applicable to the MBMS indication channel (MICH) and S-CCPCH relationship, which will be addressed later.

The AICH access slot #0 (AS #0) starts the same time as the P-CCPCH frame that has SFN modulo-2 = 0. The downlink AICH access slots have a length of 5120 chips. The uplink PRACH access slots are also of length 5120 chips. The uplink AS #*k* is transmitted from the UE $\tau_{p\text{-}a}$ chips before the reception of the downlink AS #*k*. The timing relationship between AICH and PRACH is given in Fig. 7.55 (taken from 3GPP).

The variables shown in Fig. 7.55 are defined below:

$\tau_{p\text{-}a}$ = time between transmission of the uplink RACH preamble and downlink AICH

$\tau_{p\text{-}p}$ = time between consecutive transmission of the uplink RACH preambles

$\tau_{p\text{-}m}$ = time between transmission of uplink RACH preamble and RACH message

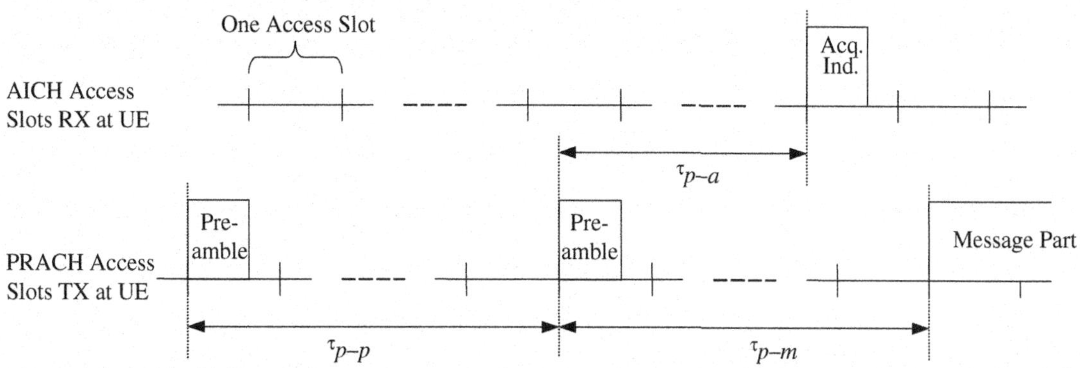

FIGURE 7.55 AICH and PRACH timing relationship.

The distance between the uplink PRACH preamble transmissions is defined by $\tau_{\text{p-p}}$ and controlled by the AICH transmission timing (ATT) parameter, which is signaled from the higher layers to the UE. The following values are derived from this parameter:

If (ATT = 0) then

$$\tau_{\text{p-p,min}} = 15{,}360 \text{ chips (3 ASs)}$$

$$\tau_{\text{p-a}} = 7680 \text{ chips}$$

$$\tau_{\text{p-m}} = 15{,}360 \text{ chips (3 ASs)}$$

If (ATT = 1) then

$$\tau_{\text{p-p,min}} = 20{,}480 \text{ chips (4 ASs)}$$

$$\tau_{\text{p-a}} = 12{,}800 \text{ chips}$$

$$\tau_{\text{p-m}} = 20{,}480 \text{ chips (4 ASs)}$$

Note that the time available for the UE to reliably detect the AI on the AICH and prepare the uplink transmission for the PRACH message part is denoted as D and defined as

$$D = \tau_{\text{p-m}} - \tau_{\text{p-a}} - 4096 \quad \text{(chips)} \tag{7.61}$$

The DPCH timing may be different for different DPCHs. The frame offset from P-CCPCH is a multiple of 256 chips, $\tau_{\text{DPCH},n} = T_n \cdot 256$ chips, where $T_n \in \{0, 1, \ldots, 149\}$. Similar comments can be made for the F-DPCH channels.

As discussed above, the P-CCPCH indirectly acts as a timing reference for the uplink because of the constant offset between the uplink and the downlink. At the UE side, the uplink DPCCH/DPDCH frame transmission takes place approximately 1024 chips after the reception of the first detected path (in time of arrival) of the corresponding downlink DPCCH/DPDCH or F-DPCH frame.

Cell Geometry Definition (G). In the WCDMA 3G digital cellular standard, all downlink physical channels are allocated a fraction of the total available downlink power. The total transmitted power I_{or} consists of contributions from common, shared, and dedicated channels. For simulation and controlled testing purposes, the total transmit power is normalized to unity (or a constant), and if the sum of all the common plus shared plus dedicated users is less than the normalized value, then the other channel noise (OCN) is inserted in order to maintain the total transmit power to be constant. Figure 7.56 will assist in the variable definitions.

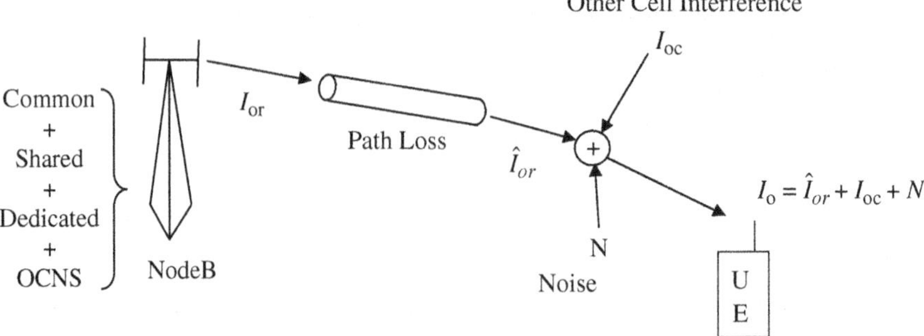

FIGURE 7.56 Definition of NodeB power allocation.

The fractional power allocations for the downlink are commonly defined by the ratio of the chip energy to the total transmit power, E_c/I_{or}, which can be written as

$$\left(\frac{E_c}{I_{or}}\right)_{\text{Common}} + \left(\frac{E_c}{I_{or}}\right)_{\text{Shared}} + \left(\frac{E_c}{I_{or}}\right)_{\text{Dedicated}} + \left(\frac{E_c}{I_{or}}\right)_{\text{OCNS}} = 1 \qquad (7.62)$$

Next we provide a brief comment about the usage of the above equation in actual deployments. First, the total power available is related to the transmission PA's capability (minus a certain back-off); second, OCN is not intentionally inserted since it will increase both the intra- and intercell interference. Lastly, due to power control, addition and deletion of UEs within the cell and network adjustments, the power allocations of the common, shared, and dedicated channels vary during operation. The NodeB will allocate power to the respective channels, as the need arises, in order to provide service to the UE within that cell.

Based on Fig. 7.56, the total power received by the UE consists of $\hat{I}_{or} + I_{oc} + N$. If we assume (for sake of this discussion) that there is no path loss, then we can define a geometry variable G that provides a measure of SNR as follows (neglecting thermal noise):

$$G = \frac{I_{or}}{I_{oc}} \qquad (7.63)$$

This value also gives an indication of the distance from the NodeB. Very simply, large geometries imply the UE is close to the NodeB, whereas small geometries imply the UE is close to the cell edge. In the following sections, we will use the geometry (or G factor) extensively in our discussions on system performance.

Depending on cell deployment scenarios (i.e., number of NodeBs, adjacent RATs, etc.), surrounding environments (i.e., suburban, urban, and rural), overall system usage (i.e., number of users, throughput, etc.), as well as UE distance to the NodeB, the geometry value can vary considerably. It is common practice to capture the randomness of the geometry by plotting the CDF. In Fig. 7.57, we plot the CDF of the geometry for a scenario in downtown London [20].

We have defined two variables that will be used extensively in evaluating system performance, namely, E_c/I_{or} and I_{or}/I_{oc}. The first is used to determine what fraction of the total NodeB transmit power is required to meet a certain performance target. The second is used to control the amount of interference seen by the UE, by introducing a dependency on distance separated from the NodeB. We will use these variables plus others to help in the system performance study.

We will consider two values from the downtown London geometry values, specifically $I_{or}/I_{oc} = 10$ and -5 dB. These geometry values correspond to approximately 97% and 5% occurrence. These

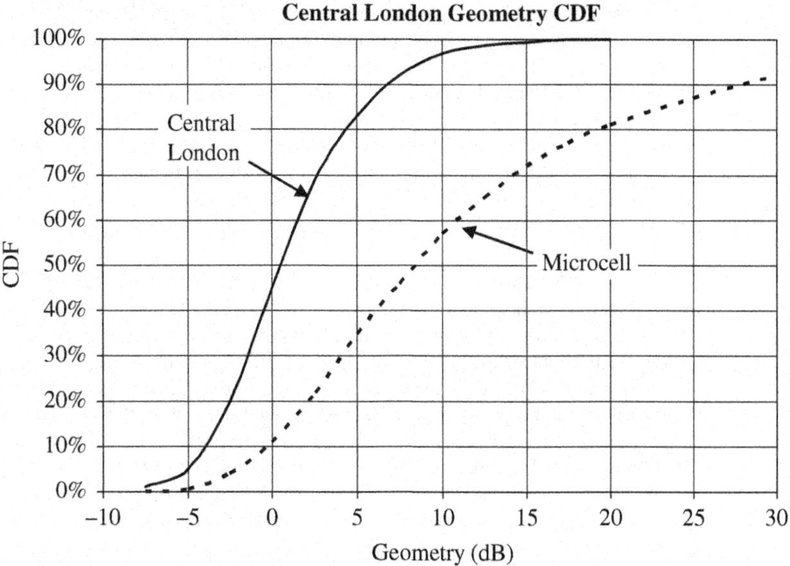

FIGURE 7.57 Geometry values from downtown London.

values were chosen to represent the UE location near the NodeB and at the cell edge, respectively. Generally speaking, the E_b/N_o equation assuming QPSK modulation is given below.

$$\frac{E_b}{N_o} = \frac{E_c}{I_{or}} \cdot \frac{I_{or}}{I_{oc}} \cdot \mathrm{SF} \cdot \frac{1}{2} \tag{7.64}$$

The assumed power allocation for P-CCPCH is $E_c/I_{or} = -12$ dB. The SNR per bit of the P-CCPCH is given as follows

$$\begin{aligned} \left(\frac{E_b}{N_o}\right)_1 &= -12\text{ dB} + 10\text{ dB} + 10\log(256) - 3\text{ dB} \\ \left(\frac{E_b}{N_o}\right)_1 &= 19.08\text{ dB} \end{aligned} \tag{7.65}$$

Similarly, we have

$$\begin{aligned} \left(\frac{E_b}{N_o}\right)_2 &= -12\text{ dB} - 5\text{ dB} + 10\log(256) - 3\text{ dB} \\ \left(\frac{E_b}{N_o}\right)_2 &= 4.08\text{ dB} \end{aligned} \tag{7.66}$$

We can clearly see the difference in E_b/N_o and expect significant difference in BER, depending on the channel conditions.

7.6.4 Searching Operation

In this section, we will describe two searching operations, namely, the cell search procedure and the multipath searcher. Since this is a spread spectrum system, the receiver is anticipating various echos of the transmitted chip sequence, also called delay spread. Hence the UE will need to not only find (or search for) the nearest neighbor cell, but also find as many "good" multipaths as possible, in order to keep the RAKE receiver satisfied. A "satisfied" RAKE is a term used to describe a scenario where the RAKE is actively demodulating at least one valid multipath signal.

The first searching operation is cell search, where the UE needs to find the nearest NodeB. This is accomplished by the use of the following physical channels: P-SCH, S-SCH, P-CPICH, and P-CCPCH.

Once a NodeB has been found and the necessary broadcast system information has been read from the P-CCPCH, the UE would then need to continue to find multipaths that were spread with the same modulation symbols. This is commonly called multipath searching.

Cell Search Procedure. One major difference between the CDMA2000 cellular system [6] and the WCDMA system is cell site synchronization. In CDMA2000, each BS has a GPS receiver, in order to maintain frame timing between them. In this case, the UE needs to find a single BS first and can then find its neighbors simply by looking to the "left" and to the "right" of the present frame synchronization. Now in the WCDMA system, the cell sites are asynchronous, meaning the frame timing among the neighboring cells is not strictly time aligned. However, the standard allows for downlink timing adjustments, so handoffs between NodeBs have timing uncertainties that are within a reasonable range. In this case, the UTRAN relies on measurements made by various UE to update NodeB system timing.

As mentioned above on the downlink, 8192 scrambling codes are used, and if left to a brute force method to determine the scrambling code used by the closest cell, the cell search time would be excessive. Hence these codes are split up into groups; there are 64 scrambling code groups, where each group consists of 8 primary scrambling codes. It would be beneficial to reduce the search candidates, in order to speed up the cell search time.

In an effort to provide a strategy for the UE to be able to efficiently search the nearest cell's primary scrambling code, the three-step method was introduced [21–23]. The first step is aimed at determining the time slot synchronization of the strongest (possibly nearest) NodeB. This is accomplished by detecting the P-SCH, since it is a Golay sequence repeated every time slot. The same P-SCH code is used for all NodeBs.

Once time slot synchronization is obtained, we can proceed to Step 2. Here frame synchronization is achieved by detecting the S-SCH. Recall the S-SCH transmits a different synch code every time slot which can be used to determine where the UE is with respect to the frame boundary. The SCH codes are constructed such that their cyclic shifts are unique. By this, we mean a nonzero cyclic shift less than 15 of any of the SCH codes is not the same as some cyclic shift of any other SCH codes. The sequence of the S-SCH codes is used to encode the 64 scrambling code groups mentioned earlier. The 3GPP standard supplies code group tables to the UE for it to use in determining frame synchronization. What this means is, in Step 2, we have not only determined the frame boundary timing, but also the code group the NodeB belongs to.

In Step 3, we aim to determine which one of the 8 primary scrambling codes is being used by the NodeB presently tracked. This is usually accomplished by trying to demodulate the CPICH with the locally generated primary scrambling code candidates. The one with the largest accumulated (coherent and/or noncoherent) energy is assumed to be the valid one. At this point, the three-step method is complete, and we have obtained synchronization with a particular NodeB of which we have obtained its primary scrambling code.

Given the primary scrambling code for a particular NodeB, the next logical phase is one of verification, where the UE attempts to demodulate the BCH transport channel on the P-CCPCH. If the CRC check passes, then it is with high degree of confidence that the primary scrambling code found is indeed the code used by the NodeB of interest. Otherwise, other candidates should be used. In Fig. 7.58 is a flow chart providing these details.

Next we wish to provide some cell search time statistics obtained through a link level simulation. Some of the details of the simulation environment parameters are listed below:

Number of NodeBs = 20

NodeB distance = 2 km

Channel conditions = vehicular

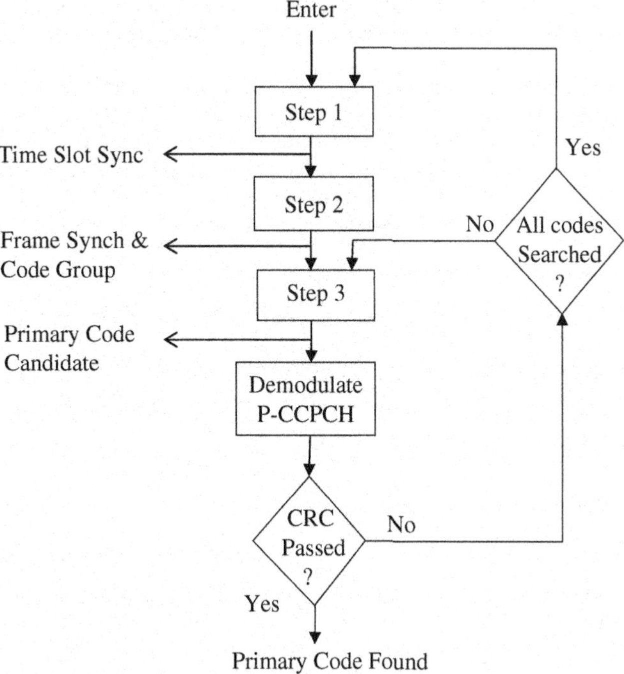

FIGURE 7.58 Cell search control algorithm.

The CDF of the cell search time is provided in Fig. 7.59, where three test cases were considered. The test cases range from good channel conditions to bad channel conditions (i.e., test 1 through test 3).

Figure 7.59 shows us possible acquisition times using the three-step method; hence it is a powerful strategy. A few of the potential benefits in deploying asynchronous cells are the choice of flexible

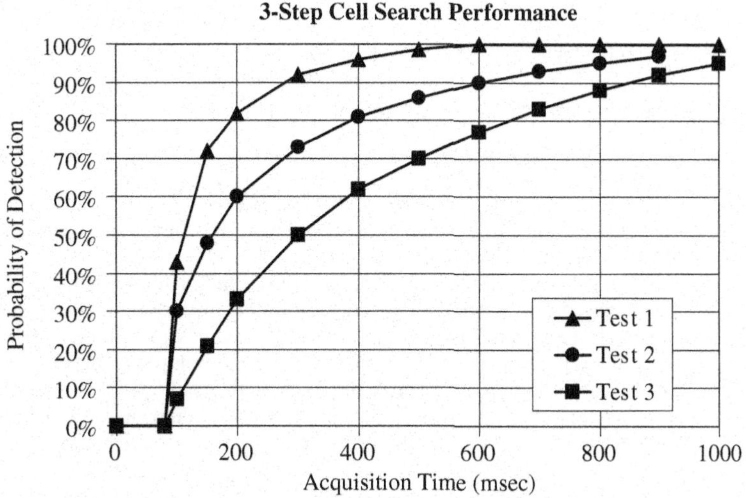

FIGURE 7.59 Cell search time statistics for a specific set of system simulation parameters.

system upgrades without the need for extensive code replanning. Also easy indoor deployments are possible since there is no need to receive GPS signals in underground places.

Multipath Search Procedure. Next we describe the searching operations that not only the UE but also the NodeB must perform, in order to obtain valid multipaths to be used in the RAKE receiver. Multipath searching is sometimes referred to as PN acquisition in that we are trying to acquire time synchronization between the locally generated PN and the received PN arriving at random propagation time intervals. Essentially, the multipath searcher finds valid multipaths or candidate multipaths that become processed to determine if they can be used in the RAKE receiver. The multipath searcher builds up a power delay profile information and keeps it up-to-date since there exists the multipath birth/death phenomenon [4, 24, 25].

There are a few parameters that affect the searcher performance; they are listed below

- Threshold setting (τ)
- Correlation time or accumulation length (NT_c)
- Number of tests or dwells per time hypothesis
- Coherent and noncoherent integration
- Searching strategy (parallel versus serial)
- Searcher window size
- Time hypothesis increment

Before we discuss the parameters, let us begin with the block diagram of a simple serial searching operation shown in Fig. 7.60. The received chip sequence is correlated against a locally generated

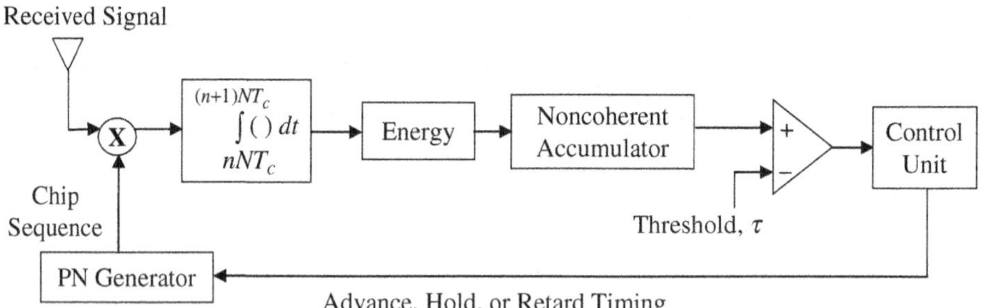

FIGURE 7.60 General block diagram of serial multipath searching.

chip sequence over a certain period of time after which the accumulated energy (coherently and/or noncoherently accumulated) is compared against a threshold to determine if there is a multipath present at this tested time offset or not.

For a particular timing instance, if there is a multipath present, then the despread symbol energy would be substantial and the PN generator timing info is sent to the RAKE finger to perform demodulation and then added to the combiner. Now, if there is no multipath or the timing difference between the received chip sequence and the locally generated chip sequence is large, then the despread energy will be small and hopefully below the decision threshold. If there is no multipath, then we are correlating noise against the PN sequence; so the results should have low accumulated energy. On the other hand, if there are interfering multipaths present, then we are calculating the cross-correlation of the received chip sequence and the locally generated chip sequence. Lastly, if a timing difference exists, then we are calculating the autocorrelation of the two chip sequences. In Fig. 7.61, we present a PN acquisition flow chart.

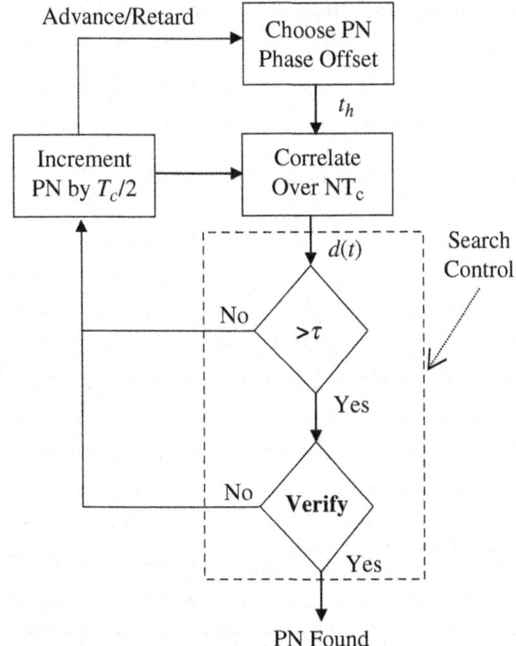

FIGURE 7.61 Multipath searcher control flow chart.

The accumulated symbol energy $d(t)$, obtained by calculating the autocorrelation signal at a time hypothesis of t, is compared to some threshold. This threshold is high enough to reduce the false alarm probability, but also low enough to increase the probability of detection. Each timing hypothesis is compared to a threshold; if the test fails, we should move on to the next timing offset and start a new comparison. However, if the test succeeds, a larger dwell time KT_c ($K > N$) can be used and, similarly, a larger threshold value can be used. Here we will define dwell to consist of an accumulation length and threshold comparison (test). A block diagram of a two-dwell serial search operation is shown in Fig. 7.62 for $\tau_2 > \tau_1$.

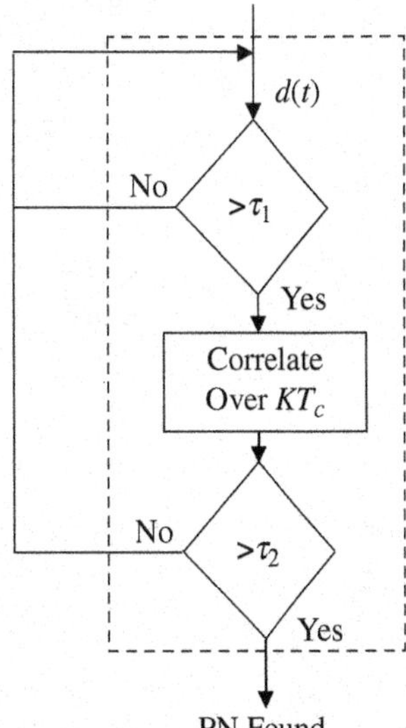

FIGURE 7.62 Multidwell searcher control.

In the above searcher control flow chart, the accumulated energy is first compared against the threshold τ_1. If this test succeeds, then a potentially longer accumulation length can be calculated and compared to a threshold τ_2. At this point, if the second test passes, then there is a higher degree of confidence that the multipath tested is indeed valid.

Instead of making a hard decision when comparing the energy to a threshold, we can perform the following control strategy:

- If the accumulated energy is much greater than or much smaller than the thresholds, then we can be sure of a decision for that particular time hypothesis.
- Now if the accumulated energy is close, either above or below the threshold, it may be better to continue the integration in order to improve the likelihood of the correct decision.

Let us consider the following example:

Case A:	$d(t) > \tau + D$	Go to verification and declare PN is present.
Case B:	$d(t) < \tau - D$	Declare PN absent and increase PN time offset.
Case C:	$\tau - D < d(t) < \tau + D$	Increase integration length and compare again.

where D = decision uncertainty region (see Fig. 7.63).

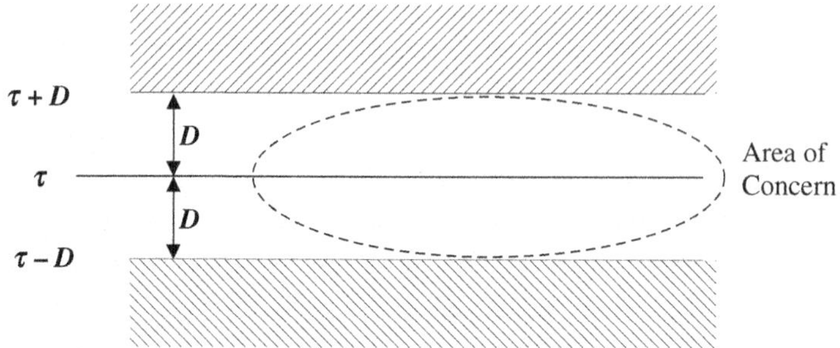

FIGURE 7.63 Searcher thresholds using soft decisions.

Mean Multipath Acquisition Time (MMAT) is a figure of merit associated with any searching operation. One way to reduce this time is to perform more operations or comparisons in parallel (at the same time). In fact, the more parallel correlations that can be performed against different time hypotheses, the more likely we are to find valid multipaths sooner. In Fig. 7.64, we show the periodic

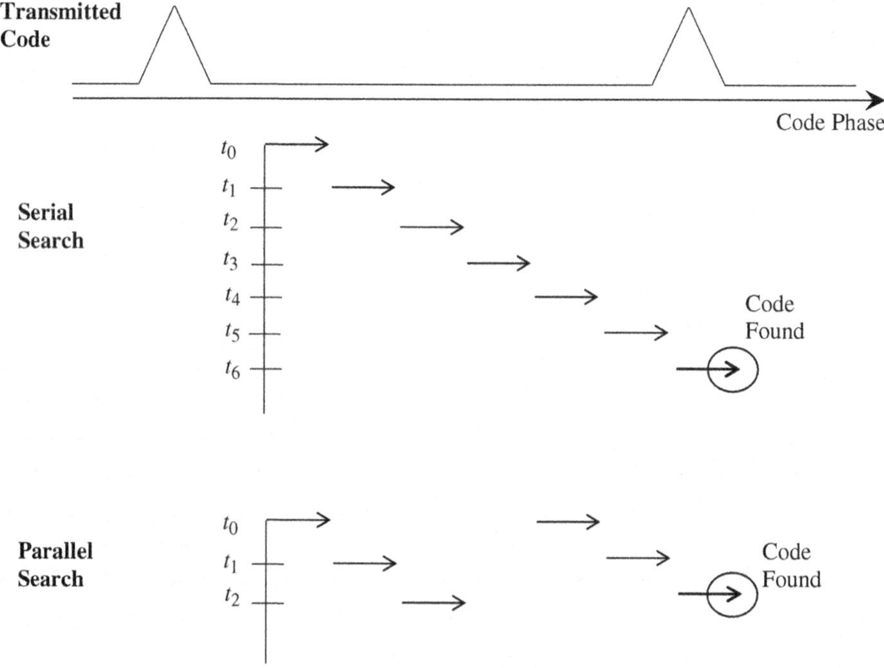

FIGURE 7.64 Serial versus parallel PN code searching.

autocorrelation function that the receiver is trying to determine, assuming a single arriving multipath. First, the serial search operation is shown where each time hypothesis is tested sequentially until the accumulated energy satisfies the above-mentioned test conditions. Next, we show the parallel search operation where two parallel correlators are simultaneously operated. This simple example is meant to emphasize the importance of parallel searching to reduce the MMAT.

A last comment on threshold is made next. Proper choice of the thresholds should be made to achieve a target probability of detection (P_D) and probability of false alarm (P_{FA}). A typical application uses a PN increment of $T_c/2$; however, the system designer can vary this parameter, depending on the receiver architecture and desired performance.

Once the PN code phase is found, the acquisition algorithm searches "around" the found offset to find other multipath components. Looking only at the PDF of the propagation delay spread measurements, say 20 μsec, the search window may become normalized by the chip time

$$\text{Search window} = \frac{\text{Delayspread}}{T_c} = \frac{20\ \mu\text{sec}}{260\ \text{nsec}} < 80\ \text{chips} \tag{7.67}$$

Another factor that affects window size is soft handoff. NodeBs will have a certain frame timing offset that should also be considered in the searching operation. Also the operating environment, whether it is indoors, pedestrian, or outdoors, and lastly the distance between the NodeBs also affects the delay spread present in the wireless channel. We have discussed the strong correlation between the delay spread and distance in Chap. 3.

7.6.5 RAKE Finger Signal Processing

In this subsection, we will present some of the signal processing functions performed within the RAKE receiver. We will begin with channel estimation (CE) which is used to perform coherent demodulation for each arriving multipath. A few fine–time tracking algorithms will be discussed that are performed within each finger. These time tracking mechanisms are used to track variations in the multipath arrival times. Next, power control will be reviewed to introduce power control loops. Both the inner- and outer-loop power control will be discussed and an example SIR estimation algorithm is presented.

Each finger is assigned a multipath maintained by the Echo Profile Manager (EPM). The Analog Front End (AFE) control is an essential part of the baseband signal processing; we have chosen to present the Frequency Lock Loop (FLL). The corresponding AFC algorithm is shown for example purposes only. Lastly, a simple TFCI Reed-Muller decoder is presented.

CE Algorithms. Here we will present some CE algorithms that can be used by each finger in the RAKE. Since the uplink and the downlink physical channels of the WCDMA standard are different, we will present a receiver that can be used on the downlink (UE receiver) and another receiver to be used on the uplink (NodeB receiver). First, the MRC RAKE combining algorithm is discussed, and then, the optimal combining (OC) algorithm is presented.

MRC Combining. Let us provide a finger block diagram for the UE receiver. When an FSF environment is encountered, multiple fingers are used to combine the received symbols. The single finger is shown in Fig. 7.65, assuming the DPCH is using the primary scrambling code and the P-CPICH is used as the phase reference. The downlink received signal is given as

$$r(t) = \{p_{\text{CPICH}}(t) \cdot S_{cr}(t) \cdot C_{\text{CPICH}}(t) + d(t) \cdot S_{cr}(t) \cdot C_{\text{DPCH}}(t)\} \cdot h(t) + n(t) \tag{7.68}$$

The pilot despreader output is given as

$$\begin{aligned} x(t)\big|_{t=kT_c} = x(k) = {} & \sum_{256} |S_{cr}(k)|^2 \cdot |C_{\text{CPICH}}(k)|^2 \cdot p_{\text{CPICH}}(k) \cdot h(k) \\ & + \sum_{256} |S_{cr}(k)|^2 \cdot C_{\text{CPICH}}(k) \cdot C_{\text{DPCH}}(k) \cdot d(k) \cdot h(k) + \tilde{n}(k) \end{aligned} \tag{7.69}$$

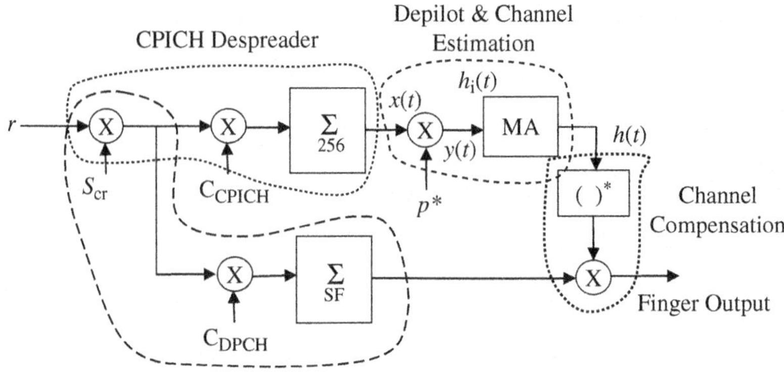

FIGURE 7.65 A RAKE finger-block diagram using CPICH-based CE.

Forcing orthogonality between CPICH and DPCH, we have the following simplification:

$$x(k) = \text{PG} \cdot p_{\text{CPICH}}(k) \cdot h(k) + \tilde{n}(k) \tag{7.70}$$

The signal must now be depiloted; since we force $|p_{\text{CPICH}}(k)|^2 = 1$, this is trivial. Moreover, we assumed the channel did not change over the despreading operation.

$$y(k) = \text{PG} \cdot h(k) + \tilde{n}(k) \tag{7.71}$$

The CPICH is not only despread but also depiloted in order to have the true instantaneous channel estimate. In this case, $p(t) = 1 + j\,1$ and is continuously transmitted. The instantaneous channel estimates $h_i(t)$ are averaged, using a finite-length moving average (MA) filter, to produce a more reliable estimate $h(t)$ by reducing the effects of noise. This channel estimate is then used to derotate the despread DPCH symbols by the channel compensation block. In the block diagram in Fig. 7.65, we have assumed we have ideal timing synchronization.

As discussed earlier in this chapter, on the downlink, the pilot symbols and data symbols are code multiplexed so that the UE performs parallel despreading of the two physical channels. On the uplink, the NodeB receiver performs slightly different processing. In this case, the pilot bits are sent on the Q-channel while the data bits are sent on the I-channel. In other words, they are quadrature multiplexed.

The block diagram in Fig. 7.65 represented a single finger in the RAKE; these operations will be repeated for each finger since it is well known that differing multipaths have independent fading.

Let us shift our attention to the uplink for now. The NodeB received signal in a flat fading channel is given as

$$r(t) = \{d(t) \cdot C_{\text{DCH}}(t) + j \cdot p(t) \cdot C_{\text{CCH}}(t)\} \cdot S_{cr}(t) \cdot h(t) + n(t) \tag{7.72}$$

After multiplying and despreading by the control channel (CCH) OVSF code, we have

$$r_1(t) = h(t) \cdot j \cdot p(t) \cdot \sum_{k=1}^{256} |C_{\text{CCH}}(t_k)|^2 \cdot |S_{cr}(t_k)|^2 + \tilde{n}(t) \tag{7.73}$$

where the DCH has been removed by forcing orthogonality between DCH and CCH. Since the pilot bits vary with the time slot, we next depilot the despread symbol sequence, which gives

$$\tilde{r}_1(k) = \sum_{k=1}^{Np} r_1(k) \cdot p^*(k) \tag{7.74}$$

$$\tilde{r}_1(k) = j \cdot \text{PG} \cdot N_p \cdot h(k) + \hat{n}(k) \tag{7.75}$$

Next, an MA filter is inserted to reduce the effects of noise. The output of the despreader for the DCH channel is mathematically represented as follows:

$$r_2(k) = \sum_{t=1}^{SF} |S_{cr}(k)|^2 \cdot |C_{DCH}(k)|^2 \cdot d(k) \cdot h(k)$$
$$+ j\sum_{t=1}^{SF} |S_{cr}(k)|^2 \cdot p(k) \cdot C_{CCH}(k) \cdot C_{DCH}(k) \cdot h(k) + n(k) \tag{7.76}$$

Once again forcing the DCH and CCH codes to be orthogonal, the output is given as

$$r_2(k) = \mathrm{PG} \cdot d(k) \cdot h(k) + \tilde{n}(k) \tag{7.77}$$

If we neglect the noise terms for the moment and use the previously defined, depiloted output signal, we can then perform channel compensation in the following manner:

$$z(k) = r_2(k) \cdot \tilde{r}_1^*(k) \tag{7.78}$$

$$z(k) = \{\mathrm{PG} \cdot d(k) \cdot h(k)\}\{-j \cdot \mathrm{PG} \cdot N_p \cdot h^*(k)\} \tag{7.79}$$

$$z(k) = -j \cdot K \cdot |h(k)|^2 \cdot d(k) \tag{7.80}$$

Here k is a constant over the observed time interval. Since the estimated channel will slightly differ from the actual channel, the compensated symbols $z(k)$ will be complex valued. Hence the imaginary part of the complex signal is used for DCH processing. The projection onto the real axis is due to the estimation error of the channel estimate (see Fig. 7.66).

The uplink modulation scheme is BPSK with HPSK spreading. The CCH contains not only the pilot bits but also other Layer 1 control information, such as TPC and TFCI bits. Hence the pilot bits must be extracted from the received bit stream through the pilot-gating function. In this case, using pilot bits from adjacent time slots can improve system performance. We will quantify this improvement when the BER performance results are presented later.

As stated earlier, in the above uplink and downlink finger-block diagrams, we have assumed fine time tracking is performed elsewhere and relevant timing info is passed to these fingers. One of the most important functions in the RAKE finger is CE. As one would expect, various options exist to the system designer. What has been presented thus far is averaging a combination of the present and past instantaneous channel estimates. This type of signal processing is called causal estimation. Next we provide an alternative estimator where future estimates are additionally used, at the expense of delay through the finger output.

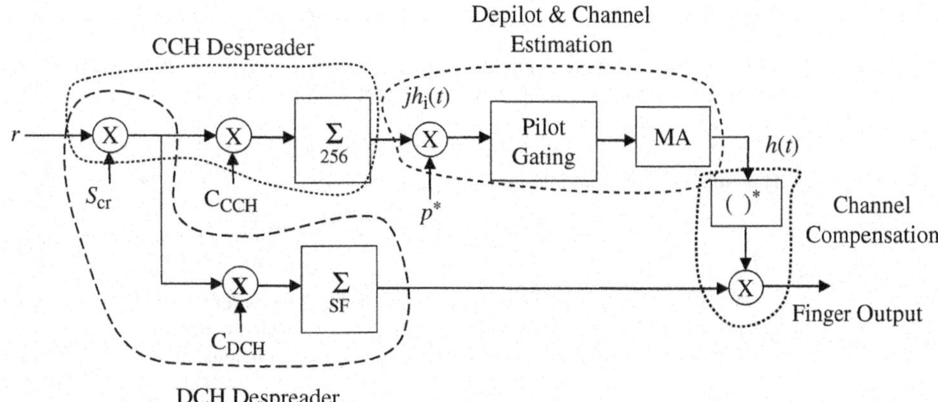

FIGURE 7.66 WCDMA uplink RAKE finger-block diagram using time division–multiplexed dedicated pilot bits.

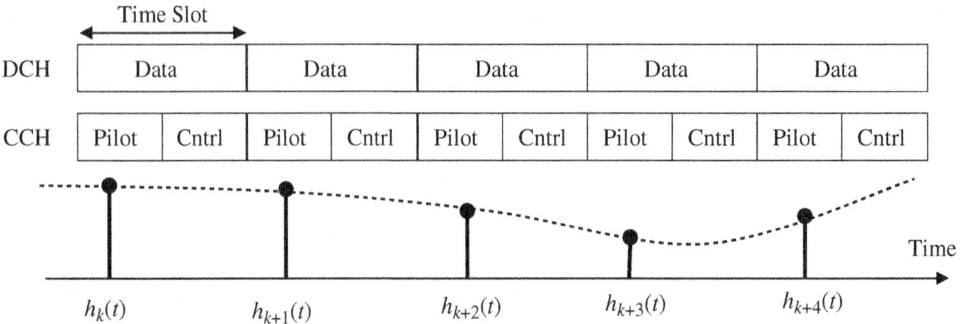

FIGURE 7.67 Uplink noncausal CE overview.

Recall, on the uplink, we have the quadrature-multiplexed physical channels shown in Fig. 7.67. In Fig. 7.67, 5 time slots are used to provide for an estimate of the channel to be used in the $k + 2$ time slot (the center of the 5 time slot) window. What this means is that the DCH and CCH must be stored in memory (e.g., FIFO), so channel estimates can be formed. Then the channel compensation block traverses back into time to apply the channel estimate in the middle time slot. Traversing back into time implies these samples are available to be operated on, so some buffering is required. It is the addition of this buffer that introduces the time delay discussed above. This function must be performed for each finger actively used in the RAKE receiver.

On the downlink, the P-CPICH is commonly used as the phase reference for coherent demodulation. The pilot channel time slot and frame structure are continuous in time. However, on the uplink, there is no pilot channel, simply time-multiplexed pilot bits. Since Layer 1 control information is also time multiplexed on this channel, the pilot sequence and hence the channel estimate is discontinuous in time. This is indicated in Fig. 7.67, where we have provided a time slot rate channel estimate labeled $h_k(t)$, where k = time slot number. A commonly used technique to overcome this shortcoming is to use interpolation. Below we provide a simple example using the Lagrange Interpolation technique [26]. The interpolation coefficients are given as

$$L_j(x) = \prod_{\substack{k=0 \\ k\neq j}}^{n} \left(\frac{x - x_k}{x_j - x_k}\right) \qquad (j = 0, 1, \ldots, n) \tag{7.81}$$

The interpolating polynomial with $n = 2$ can be constructed as

$$p_2(x) = L_0(x) \cdot y_0 + L_1(x) \cdot y_1 + L_2(x) \cdot y_2 \tag{7.82}$$

Given the following, sample points are available to be used by the interpolator. These sample points are, in fact, the channel estimates in the pilot bit field (i.e., $h_k(t)$, $h_{k+1}(t)$, etc.):

$$(x_0, y_0) \qquad (x_1, y_1) \qquad (x_2, y_2)$$

Then the interpolating polynomial is given as

$$p_2(x) = y_0\frac{(x - x_1)(x - x_2)}{(x_0 - x_1)(x_0 - x_2)} + y_1\frac{(x - x_0)(x - x_2)}{(x_1 - x_0)(x_1 - x_2)} + y_2\frac{(x - x_0)(x - x_1)}{(x_2 - x_0)(x_2 - x_1)} \tag{7.83}$$

Let us assume the simple example given in Fig. 7.68, where 3 time slots are drawn. Each time slot has 10 bits in total, and we assume the first 5 bits in each slot are the pilot bits.

Each pilot bit field will allow us to obtain an instantaneous channel estimate. The group of these estimates is averaged in order to produce a more reliable channel estimate. This averaged estimate is indicative of the channel at the center of the pilot bit fields highlighted by the ovals and denoted as $h_1(3)$, $h_2(13)$, and $h_3(23)$. These data points will become our reference.

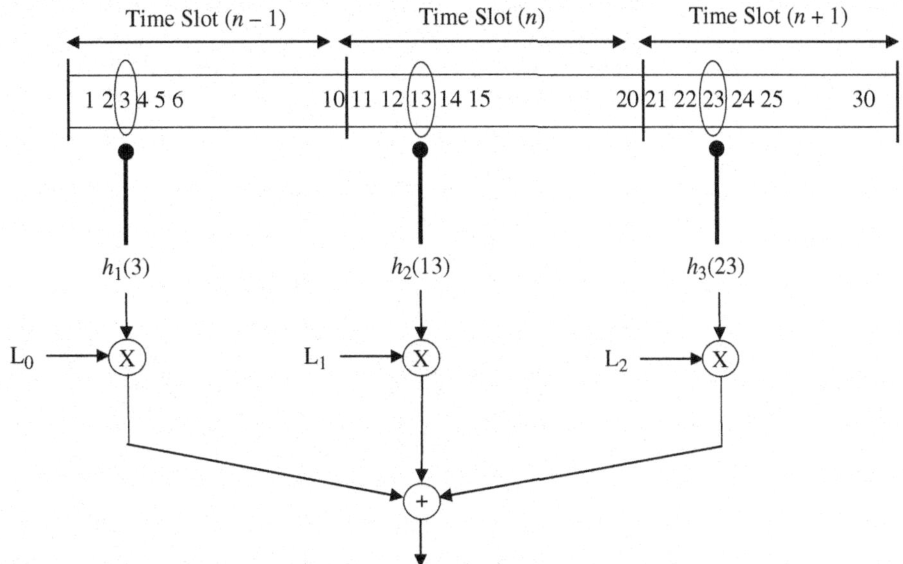

FIGURE 7.68 Channel estimation using Lagrange Interpolation.

Using the three sets of points available to us, namely, $(3, h_1)$, $(13, h_2)$, and $(23, h_3)$, one can easily show the interpolation coefficients are given as follows, where x is used to indicate time in terms of bit duration increments:

$$\begin{aligned} L_0(x) &= \frac{1}{200}(x^2 - 36x + 299) \\ L_1(x) &= \frac{-1}{100}(x^2 - 26x + 69) \\ L_2(x) &= \frac{1}{200}(x^2 - 16x + 39) \end{aligned} \tag{7.84}$$

If we wish to determine the interpolated channel estimate at bit time $x = 17$, then we have the following operation to perform:

$$p_2(17) = -0.12 \cdot h_1 + 0.84 \cdot h_2 + 0.28 \cdot h_3 \tag{7.85}$$

In a fast fading environment, the channel will change more rapidly than a slow fading environment. In an effort to track these rapid changes, the interpolation coefficients should be time varying. This way the channel estimate is updated during the bit times between the pilot fields. Other types of interpolation techniques exist such as linear, and cubic splines [27]. Depending on the pilot field separation, the number of pilot bits, the SNR of the pilot bits, fading channel parameters, and so forth, some techniques will work better than others.

A general NodeB receiver block diagram is given in Fig. 7.69 for a RAKE. We have assumed the multipath searcher has found K distinct paths and sent their respective timing information to each finger. Each finger will despread the CCH, extract the pilot bits, and perform some averaging across multiple time slots. We have used the well-known Weighted Multi Slot Averaging (WMSA) approach [28]. Now since we make use of not only the present and past estimates but also the future estimates, the data channel must have some memory (e.g., FIFO) to prevent loss of information. This FIFO essentially allows the channel compensation to go back into time to perform channel compensation.

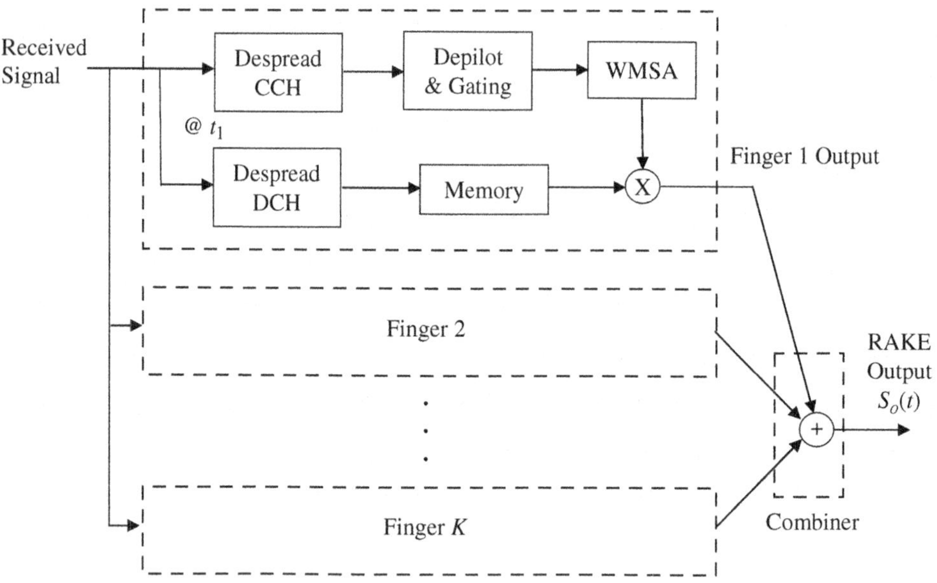

FIGURE 7.69 RAKE block diagram using K fingers.

If we look at the RAKE output signal, we can observe the following (assuming K distinct mutipaths are being demodulated):

$$S_o(t) = d(t) \cdot \sum_{j=1}^{K} |h_j(t)|^2 \tag{7.86}$$

which is essentially performing the MRC function, similar to what was presented when we discussed receive antenna diversity techniques. What Eq. (7.86) shows us is that performance improvements can be achieved as more multipaths with significant energy are actively demodulated. This performance improvement is a direct result of exploiting the time diversity present in the wireless channel.

In Fig. 7.70, we present some uncoded BER simulation results for the downlink. The performance curves are plotted against E_b/N_o, which is derived below, assuming QPSK modulation (generally speaking).

$$\frac{E_c}{N_o} = \frac{S}{N} = \frac{E_b}{N_o} \cdot \frac{1}{\text{PG}} \cdot 2 \tag{7.87}$$

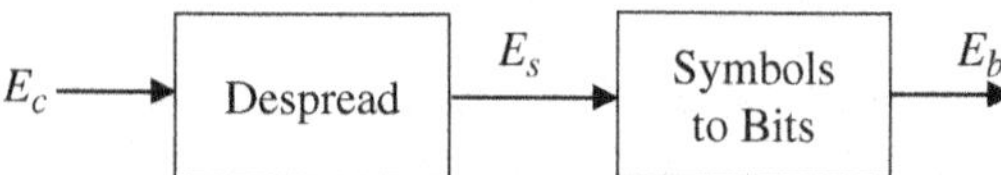

FIGURE 7.70 Receiver block diagram showing signal energies.

We can provide a more accurate definition in the sense of relating this to what was previously defined earlier for the figure of merit. On the downlink, this is given as

$$\frac{E_b}{N_o} = \frac{E_c}{I_{or}} \cdot \frac{I_{or}}{I_{oc}} \cdot \text{PG} \cdot \frac{1}{2} \tag{7.88}$$

where $\frac{E_c}{I_{or}}$ = fraction of the total power allocated to the DPCH channel of interest, $\frac{I_{or}}{I_{oc}}$ = form of SNR or SIR using the total (common plus dedicated channels) for the transmitted power value, PG = processing gain of the channel of interest, and the factor of $^1/_2$ is for conversion from the symbol-to-bits factor.

The BER performance results assume the CPICH transmit power is 10% of the total transmitted power. Recall the following closed-form BER expressions were presented previously and are rewritten and slightly modified for sake of reference [1]:

AWGN:

$$P_b\left(\frac{E_b}{N_o}\right) = Q\left(\sqrt{\frac{2 \cdot E_b}{N_o}}\right) \tag{7.89}$$

Flat fading:

$$P_b\left(\frac{E_b}{N_o}\right) = \frac{1}{2}\left\{1 - \frac{\mu}{\sqrt{2 - \mu^2}}\right\}; \qquad \mu = \sqrt{\frac{\dfrac{2 \cdot E_b}{N_o}}{1 + \dfrac{2 \cdot E_b}{N_o}}} \tag{7.90}$$

Two-ray, equal power fading:

$$P_b\left(\frac{E_b}{N_o}\right) = \frac{1}{2}\left\{1 - \frac{\beta}{\sqrt{2 - \beta^2}} \cdot \left[1 + \frac{1 - \beta^2}{2 - \beta^2}\right]\right\} \qquad \beta = \sqrt{\frac{\dfrac{E_b}{N_o}}{1 + \dfrac{E_b}{N_o}}} \tag{7.91}$$

Two CE algorithms were compared to each other under various channel conditions. The first technique used the CPICH channel (common pilot, CP), where an MA filter of 5 symbols in length was used and the channel estimate was updated every 256 chips. Recall the CPICH is transmitted in parallel, so the estimate can be updated faster. The second technique used the dedicated pilot (DP) bits of DPCCH, where the channel estimate from time slot k is used for the duration of time slot $k + 1$. Since we need to use gated pilot bits to estimate the channel, this technique assumes the channel is relatively constant over the time slot. The user data rate is 384 kbps, which corresponds to an SF of 8. The particular time slot format chosen had eight pilot symbols in the DPCCH. Both of these techniques are based on causal signal processing. Fig. 7.71 represents the uncoded BER versus E_b/N_o for a flat fading channel with a Doppler spread of 20 Hz.

In Fig. 7.71, we compare the first technique of using five common pilot symbols, denoted as 5CP, to the second technique of using eight dedicated pilot bits, denoted as 8DP. As we can see, for this slowly fading (Doppler spread = 20 Hz) frequency flat channel, the first technique performs very close to the theoretical predictions, while the second technique performs worse. The two-ray theoretical curve has been inserted into the figure for reference. In fact, both techniques provide an acceptable operating characteristic in this slowly time-varying channel.

Next, the maximum Doppler spread was increased to 200 Hz (see Fig. 7.72) and the results are significantly worse than the earlier case. It is clear assuming the channel is time invariant within the time slot is not accurate. (Refer to Chap. 3, where the LCR and AFTD were presented.) This can be seen with the results from the second technique. The first technique has a CE that is updated more often and produces better results. It is well known that the best solution for this channel condition is to use noncausal signal processing with time-varying interpolation coefficients.

Figures 7.73 and 7.74 plot the BER performance for a two-ray FSF channel with Doppler spreads of 20 and 200 Hz. The time separation between the two multipath rays is set to a few chip times and the average power of each ray is equal. Below we will see the performance of the first technique to be slightly better than that of the second technique. Another component that we must be aware of is

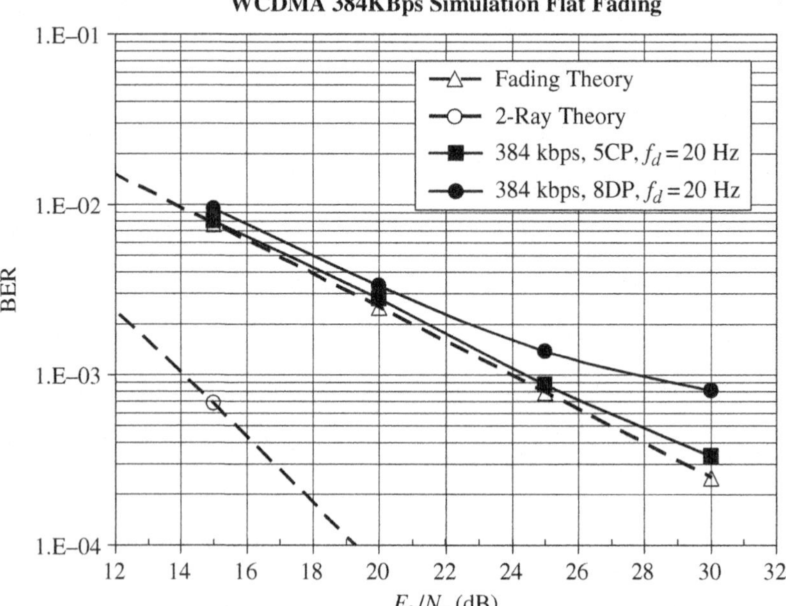

FIGURE 7.71 Flat-fading BER performance with Doppler spread (20 Hz).

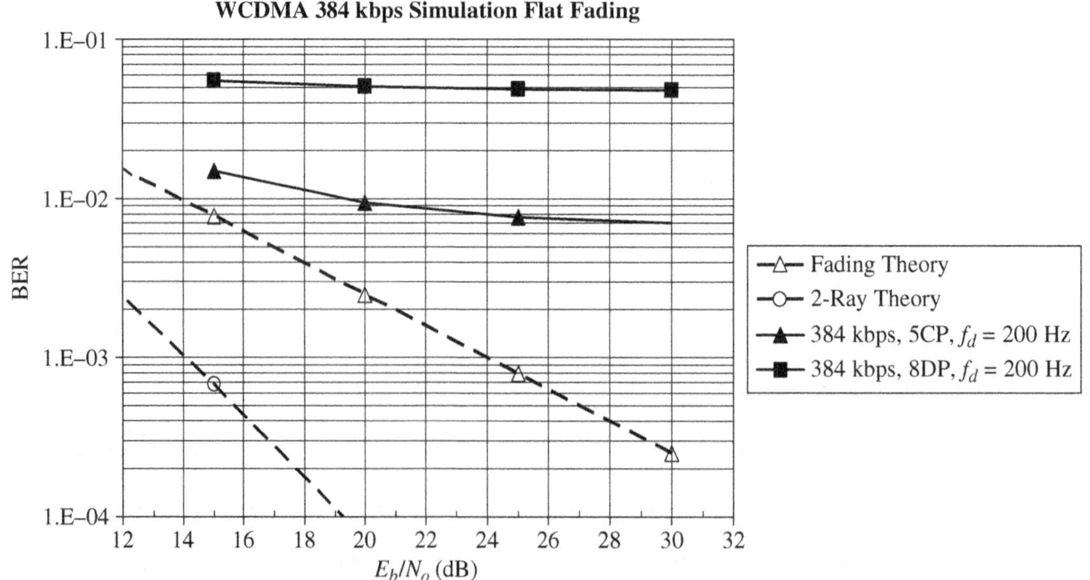

FIGURE 7.72 Flat-fading BER performance with Doppler spread (200 Hz).

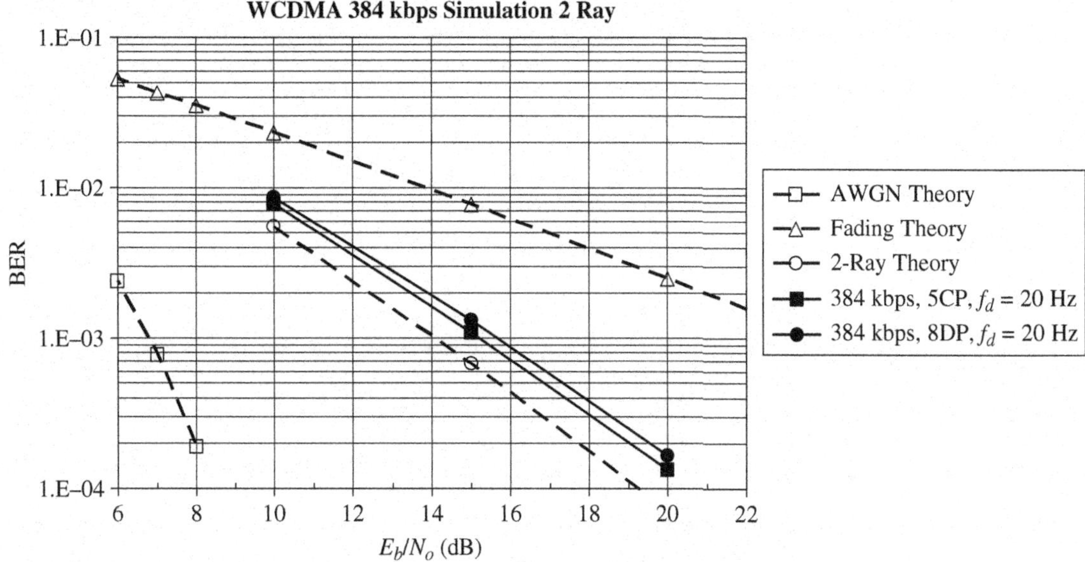

FIGURE 7.73 Two-ray frequency-selective fading BER performance with Doppler spread (20 Hz).

that the insertion of the second ray leads to interpath interference (IPI) that is not negligible for the low SF of interest. We will discuss the impact of IPI on the RAKE output in the latter sections.

The maximum Doppler spread was increased to 200 Hz, and once again, we can clearly see the degradation of the second technique (DP) to be larger than the first technique (CP). We have also plotted the performance of the first technique, where the maximum Doppler spread was further increased to 300 Hz.

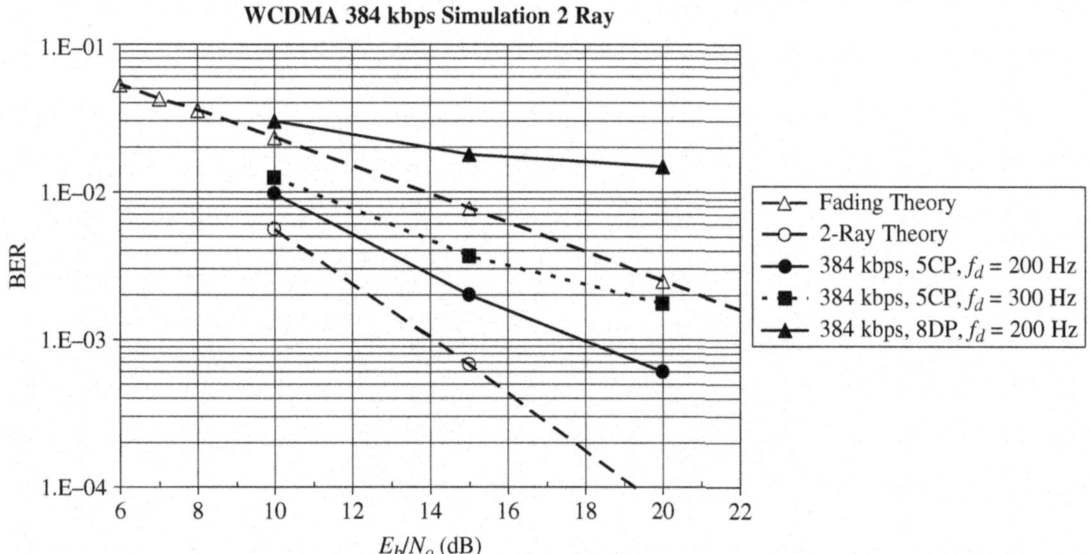

FIGURE 7.74 Two-ray frequency-selective fading BER performance with Doppler spread (200 Hz).

To summarize the observations, we begin by restating that two causal CE techniques were compared. Generally speaking, these techniques seem to break down at high Doppler spreads. Moreover, assuming a time-invariant channel in a fast-fading channel is incorrect, but acceptable at the very low Doppler spread values. Due to the low SF used, the effects of IPI are more pronounced, and thus, using the dedicated pilot bits to aid CE can suffer in quality due to this increased interference.

Now for the uplink, the situation is slightly different; here the modulation scheme is BPSK, and the DCH and CCH are quadrature multiplexed. The 3GPP standard definition applies as follows (assuming a voice service application):

$$\frac{E_b}{N_o} = \frac{E_c}{N_o} \cdot \frac{3840 \text{ kcps}}{12.2 \text{ kbps}} \tag{7.92}$$

If we consider the relative power offset (pilot to traffic channel) defined as $\Delta = \frac{\text{CCH}}{\text{DCH}}$, then we can write down the E_b/N_o of the data channel as

$$\left(\frac{E_b}{N_o}\right)_{\text{DCH}} = \left(\frac{E_d}{N_o}\right)_{\text{dB}} = \left(\frac{E_c}{N_o}\right)_{\text{dB}} - 10 \cdot \log(1 + \Delta) + 10 \cdot \log(\text{PG}_{\text{DCH}}) \tag{7.93}$$

Then the CCH E_b/N_o is given as

$$\left(\frac{E_b}{N_o}\right)_{\text{CCH}} = \left(\frac{E_P}{N_o}\right)_{\text{dB}} = \left(\frac{E_c}{N_o}\right) - 10 \cdot \log\left(1 + \frac{1}{\Delta}\right) + 10 \cdot \log(\text{PG}_{256}) \tag{7.94}$$

Hence, when evaluating the error rate performance on either the DCH or CCH channels, the above equations will help determine the respective SNR per bit value. These equations take into account not only the PG but also their respective powers.

Optimal Combining (OC). Let us, first, redraw the above RAKE finger receiver in the simplified form as in Fig. 7.75.

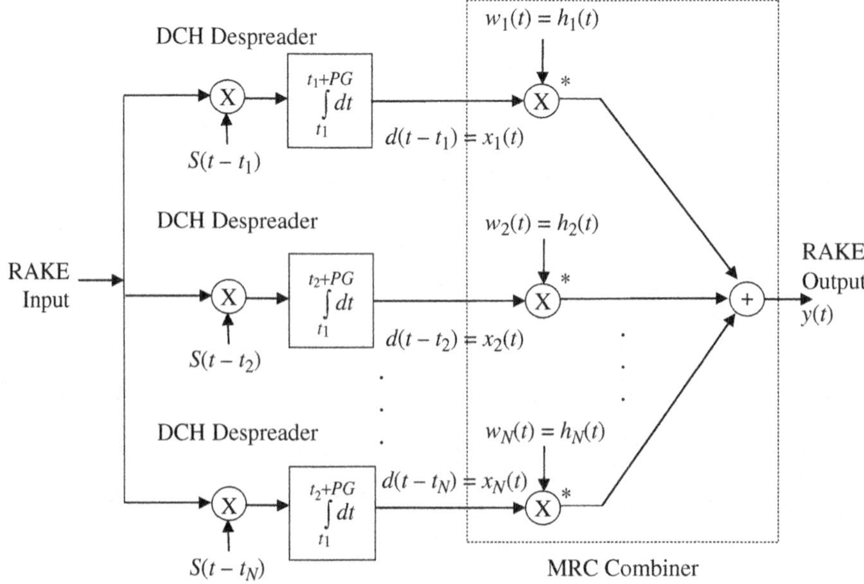

FIGURE 7.75 Generalization of a RAKE combiner.

The question we wish to answer in this section is "Can alternative combining techniques be used in the RAKE?" The answer is an overwhelming YES. If we treat the combining function as a multi-sensor problem, then whatever was discussed in the receiver spatial diversity section in Chap. 5 can be applied in the RAKE receiver (i.e., SS, EGC, MRC, etc.). We wish to focus on what is called the OC technique [8, 29].

The MRC combiner performs the following operations:

$$y(t) = \underline{w}^*(t) \cdot \underline{x}(t) \tag{7.95}$$

In this case, the weights are the channel estimates themselves [$\underline{w}_{\text{MRC}}(t) = \underline{h}(t)$]. We could apply the MMSE-based cost function and arrive with the following combining weight equation:

$$\underline{w}_{\text{MMSE}} = \hat{R}_{xx}^{-1} \cdot \underline{\hat{r}}_{xd} \tag{7.96}$$

Where $\hat{R}_{xx}$ = the estimated covariance matrix of the received and despread data symbols and $\underline{\hat{r}}_{xd}$ = estimated cross-correlation vector between the received, despread data symbols, and the desired signals.

Recall MRC is a special case of OC in that MRC assumes the covariance matrix to be a diagonal. In fact, depending on the SF of the desired physical channel, the number of multicodes, the channel model, and the amount of interference in the system, this can be an inaccurate assumption. In other words, the IPI should be taken into account in the receiver.

This approach is also called generalized RAKE or GRAKE for short [7]. In the first public appearance of this technique, the time delay between adjacent fingers was not specified and the interference plus noise matrix was used in the weight calculation. Applying the same principles provided in [3] can result in performance improvement.

The optimal combiner includes the IPI statistics in the finger weight calculation. The effects of the IPI become more apparent as the SF of the desired physical channel decreases. Let us present the derivation for the statistical signal properties of the RAKE finger output, assuming QPSK modulation. This will give us insight into the signal characteristics. For the sake of illustrative purposes, we will assume a two-ray channel model with an arbitrary time delay separation between the arriving two multipaths.

The first RAKE finger output is assumed to be tracking the first arriving ray and is given as follows:

$$x_1 = h_1 \cdot d_k \cdot \sum_{i=1}^{\text{SF}} C_i \cdot C_i^* + h_2 \cdot d_{k-1} \cdot C_{\text{SF}} \cdot C_1^* + h_2 \cdot d_k \sum_{i=1}^{\text{SF}-1} C_i \cdot C_{i+1}^* \tag{7.97}$$

The second RAKE finger output is written as

$$x_2 = h_2 \cdot d_k \cdot \sum_{i=1}^{\text{SF}} C_i \cdot C_i^* + h_1 \cdot d_{k+1} \cdot C_1 \cdot C_{\text{SF}}^* + h_1 \cdot d_k \cdot \sum_{i=1}^{\text{SF}-1} C_{i+1} \cdot C_i^* \tag{7.98}$$

In Eqs. (7.97) and (7.98), we used h_1 and h_2 to represent the channel response of the first and second arriving rays. C_i is the ith chip of the complex-valued PN sequence, and d_k is the QPSK symbol at the time instance k. We have assumed that the channel impulse is not changing during the symbol time, so the channel behaves time invariant during the despreading operation. Also we initially assumed the delay spread is within the SF of the QPSK symbol. Note that depending on the channel model of interest (i.e., outdoors and indoors), this may or may not be a valid assumption.

Then we can write the RAKE output signal as follows:

$$\begin{aligned} y = {} & \{|h_1|^2 + |h_2|^2\} \cdot d_k \cdot \sum_{i=1}^{\text{SF}} C_i \cdot C_i^* + h_2 \cdot h_1^* \cdot d_{k-1} \cdot C_{\text{SF}} \cdot C_1^* \\ & + h_1 \cdot h_2^* \cdot d_{k+1} \cdot C_1 \cdot C_{\text{SF}}^* + 2 \cdot d_k \cdot \text{Re}\left\{ h_2 \cdot h_1^* \cdot \sum_{i=1}^{\text{SF}-1} C_i \cdot C_{i+1}^* \right\} \end{aligned} \tag{7.99}$$

The RAKE output consists of four components:

1. First is the desired signal output.
2. Second is the interference from the past symbol.
3. Third is the interference from the future symbol.
4. Fourth is the interference due to itself.

It can be easily seen that the presence of delay spread creates interference from either side of the desired symbol, resembling a three-tap channel impulse response whose time duration spans two symbol times. Hence the IPI takes on the form of ISI and can be compensated through the use of an adaptive equalizer.

As the SF decreases, the delay spread will become larger than the symbol duration. With this in mind, if we increase the delay spread by a symbol time, then we can get the following modified RAKE finger output signals: The first finger output is

$$x_1 = h_1 \cdot d_k \sum_{i=1}^{\mathrm{SF}} C_i \cdot C_i^* + h_2 \cdot d_{k-2} \cdot C_{\mathrm{SF}} \cdot C_1^* + h_2 \cdot d_{k-1} \cdot \sum_{i=1}^{\mathrm{SF}-1} C_i \cdot C_{i+1}^* \tag{7.100}$$

and the second finger output is

$$x_2 = h_2 \cdot d_k \cdot \sum_{i=1}^{\mathrm{SF}} C_i \cdot C_i^* + h_1 \cdot d_{k+2} \cdot C_1 \cdot C_{\mathrm{SF}}^* + h_1 \cdot d_{k+1} \cdot \sum_{i=1}^{\mathrm{SF}-1} C_{i+1} \cdot C_i^* \tag{7.101}$$

Then the RAKE output is written as

$$\begin{aligned} y = {} & \{|h_1|^2 + |h_2|^2\} \cdot d_k \cdot \sum_{i=1}^{\mathrm{SF}} C_i \cdot C_i^* + h_2 \cdot h_1^* \cdot d_{k-2} \cdot C_{\mathrm{SF}} \cdot C_1^* \\ & + h_2 \cdot h_1^* \cdot d_{k-1} \cdot \sum_{i=1}^{\mathrm{SF}-1} C_i \cdot C_{i+1}^* + h_1 \cdot h_2^* \cdot d_{k+1} \cdot \sum_{i=1}^{\mathrm{SF}-1} C_{i+1} \cdot C_i^* \\ & + h_1 \cdot h_2^* \cdot d_{k+2} \cdot C_1 \cdot C_{\mathrm{SF}}^* \end{aligned} \tag{7.102}$$

Here we see the presence of a large delay spread that introduces additional terms at the RAKE output. It now appears that the impulse response duration spans four symbol times.

Next we would like to present a similar analysis for the WCDMA uplink physical channel [30]. Since the SF is different on the CCH and DCH, regions of the DPCCH symbol must be separated. We have split the symbol into three regions:

1. First DPDCH symbol on DPCCH boundary
2. Middle DPDCH symbols
3. Last DPDCH symbol on the DPCCH boundary

Before we begin, lets present the notation used for the data and control as (assuming a SF = 64).

Here $d_{I_{j-1,4}}$ represents the $(j-1)$th DPCCH bit time and the 4th DPDCH bit on that time interval. This particular symbol is circled in Fig. 7.76.

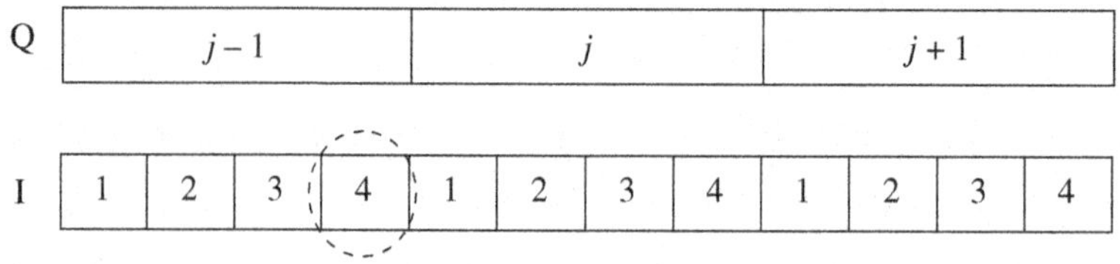

Q	$j-1$				j				$j+1$			
I	1	2	3	4	1	2	3	4	1	2	3	4

FIGURE 7.76 DPDCH and DPCCH symbol timing relationship.

REGION 1. Let us consider the first region where we again assumed a two-ray channel, with a time delay separation of a chip. For the voice application, there are 4 DPDCH bits per 1 DPCCH bit. The first RAKE finger output becomes

$$
\begin{aligned}
x_1 = {} & h_1 \cdot d_{I_{j,1}} \cdot \sum_{i=1}^{\mathrm{SF}} |C_{I_i}|^2 \cdot |C_i|^2 \\
& + h_2 \cdot \{ d_{I_{j-1,4}} \cdot C_{I_{\mathrm{SF}}} \cdot C_{\mathrm{SF}} + jd_{Q_{j-1}} \cdot C_{Q_{256}} \cdot C_{256} \} \cdot C_{I_i} \cdot C_1^* \\
& + h_2 \cdot \left\{ d_{I_{j,1}} \cdot \sum_{i=1}^{\mathrm{SF}-1} C_{I_i} \cdot C_i + jd_{Q_j} \cdot \sum_{i=1}^{\mathrm{SF}-1} C_{Q_i} \cdot C_i \right\} \cdot C_{I_{i+1}} \cdot C_{i+1}^*
\end{aligned}
\tag{7.103}
$$

The three components are better described as consisting of the desired signal, interference from the previous symbol due to second arriving ray, and interference from the present symbol due to the second arriving ray.

The second RAKE finger output becomes

$$
\begin{aligned}
x_2 = {} & h_2 \cdot \sum_{i=1}^{\mathrm{SF}} d_{I_{j,1}} \cdot |C_{I_i}|^2 \cdot |C_i|^2 \\
& + h_1 \cdot \left\{ d_{I_{j,1}} \cdot \sum_{i=1}^{\mathrm{SF}-1} C_{I_{i+1}} \cdot C_{i+1} + jd_{Q_j} \cdot \sum_{i=1}^{\mathrm{SF}-1} C_{Q_{i+1}} \cdot C_{i+1} \right\} \cdot C_{I_i} \cdot C_i^* \\
& + h_1 \cdot d_{I_{j,2}} \cdot C_{I_1} \cdot C_1 \cdot C_{I_{\mathrm{SF}}} \cdot C_{\mathrm{SF}}^*
\end{aligned}
\tag{7.104}
$$

The three components of this second output are described as: The first part represented the desired signal, the second part represented interference from the present symbol due to the first arriving ray, and the last term is the interference from the future symbol due to the first arriving ray.

REGION 2. Next we will discuss the middle region, specifically the middle (kth) DPDCH symbol of the DPCCH boundary. Now we detect DPDCH symbols inside the DPCCH symbol, so the first finger output is

$$
\begin{aligned}
x_1 = {} & h_1 \cdot \sum_{i=1}^{\mathrm{SF}} d_{I_{j,k}} \cdot |C_{I_i}|^2 \cdot |C_i|^2 + h_2 \cdot d_{I_{j,k-1}} \cdot C_{I_{\mathrm{SF}}} \cdot C_{\mathrm{SF}} \cdot C_1^* \\
& + h_2 \cdot \left\{ d_{I_{j,k}} \cdot \sum_{i=1}^{\mathrm{SF}-1} C_{I_i} \cdot C_i + jd_{Q_j} \cdot \sum_{i=0}^{\mathrm{SF}-1} C_{Q_{(k-1)\mathrm{SF}+i}} \cdot C_{(k-1)\mathrm{SF}+i} \right\}
\end{aligned}
\tag{7.105}
$$

The second finger output is

$$
\begin{aligned}
x_2 = {} & h_2 \cdot d_{I_{j,k}} \cdot \sum_{i=1}^{\mathrm{SF}} |C_{I_i}|^2 \cdot |C_i|^2 + h_1 \cdot d_{I_{j,k}} \sum_{i=1}^{\mathrm{SF}-1} C_{I_{i+1}} \cdot C_{i+1} \cdot C_{I_i} \cdot C_i^* \\
& + h_1 \cdot d_{I_{j,k+1}} \cdot C_{I_1} \cdot C_1 \cdot C_{I_{\mathrm{SF}}} \cdot C_{\mathrm{SF}}^* \\
& + jh_1 \cdot d_{Q_i} \cdot \sum_{i=1}^{\mathrm{SF}-1} C_{Q_{(k-1)\mathrm{SF}+1+i}} \cdot C_{(k-1)\mathrm{SF}+1+i} \cdot C_{I_i} \cdot C_i^*
\end{aligned}
\tag{7.106}
$$

REGION 3. The last region is the last DPDCH symbol on the DPCCH boundary. The results are now presented for demodulating the last DPDCH symbol on the DPCCH boundary. The first RAKE finger output is given as (assuming $k = 4$)

$$x_1 = h_1\left\{d_{I_{j,4}} \cdot \sum_{i=1}^{SF} C_{I_i} \cdot C_i + d_{Q_j} \cdot \sum_{i=1}^{SF} C_{Q_{(SF-1)SF+i}} \cdot C_{(k-1)SF+i}\right\} \cdot C_{I_i} \cdot C_i^*$$
$$+ h_2 \cdot d_{I_{j,k-1}} \cdot C_{I_{SF}} \cdot C_{SF} \cdot C_{I_1} \cdot C_1^*$$
$$+ h_2 \cdot \left\{d_{I_{j,k}} \cdot \sum_{i=1}^{SF-1} C_{I_i} \cdot C_i \cdot C_{I_{i+1}} \cdot C_{i+1}^* + jd_{Q_j} \cdot \sum_{i=0}^{SF-1} C_{Q_{(k-1)SF+i}} \cdot C_{(k-1)SF+i} \cdot C_{I_{i+1}} \cdot C_{i+1}^*\right\} \quad (7.107)$$

The second RAKE finger output is

$$x_2 = h_2 \cdot \left\{d_{I_{j,k}} \cdot \sum_{i=1}^{SF} C_{I_i} \cdot C_i + jd_{Q_j} \cdot \sum_{i=1}^{SF} C_{Q_{(k-1)SF+i}} \cdot C_{(k-1)SF+i}\right\} \cdot C_{I_i} \cdot C_i^*$$
$$+ h_1 \cdot d_{I_{j,k}} \cdot \sum_{i=1}^{SF-1} C_{I_{i+1}} \cdot C_{i+1} \cdot C_{I_i} \cdot C_i + h_1 \cdot d_{I_{j+1,1}} \cdot C_{I_1} \cdot C_1 \cdot C_{I_{SF}} \cdot C_{SF}^*$$
$$+ jh_1 \cdot d_{Q_j} \cdot \sum_{i=1}^{SF-1} C_{Q_{(k-1)SF+1+i}} \cdot C_{(k-1)SF+1+i} \cdot C_{I_1} \cdot C_i^*$$
$$+ jh_1 \cdot d_{Q_{j+1}} \cdot C_{Q_1} \cdot C_1 \cdot C_{I_{SF}} \cdot C_{SF}^* \quad (7.108)$$

We have presented the RAKE finger output signals as the receiver despreads the CCH and DCH channels. What is very interesting is that the finger output statistics depend on which DCH symbol we are despreading at the particular moment. Hence, having an adaptive signal processing algorithm that makes use of this a priori information should produce favorable results.

Lastly, we provide the RAKE combiner output for the first DPDCH symbol at the DPCCH boundary (i.e., region 1), using the following definitions of the PN code sequence product:

$$C_{A_i} \equiv C_{I_1} \cdot C_i$$
$$C_{B_i} \equiv C_{Q_i} \cdot C_i \quad (7.109)$$

The RAKE receiver output y is given by the following:

$$y = \{|h_1|^2 + |h_2|^2\} \cdot d_{I_{j,1}} \cdot \sum_{i=1}^{SF} |C_{A_i}|^2$$
$$+ h_2 \cdot h_1^* \cdot \left\{d_{I_{j,1}} \sum_{i=1}^{SF-1} C_{A_i} + jd_{Q_j} \sum_{i=1}^{SF-1} C_{B_i}\right\} \cdot C_{A_{i+1}}^*$$
$$+ h_1 \cdot h_2^* \cdot \left\{d_{I_{j,1}} \cdot \sum_{i=1}^{SF-1} C_{A_{i+1}} + jd_{Q_j} \cdot \sum_{i=1}^{SF-1} C_{B_{i+1}}\right\} \cdot C_{A_i}^*$$
$$+ h_2 \cdot h_1^* \cdot \{d_{I_{j-1,4}} \cdot C_{A_{SF}} + jd_{Q_{j-1}} \cdot C_{B_{256}}\} \cdot C_{A_1}^*$$
$$+ h_1 \cdot h_2^* \cdot \{d_{I_{j,2}} \cdot C_{A_1} + jd_{Q_j} C_{B_{SF+1}}\} \cdot C_{A_{SF}}^* \quad (7.110)$$

This equation shows how an FSF channel can cause the RAKE output to contain ISI from past and future symbols, not to mention interference from the same symbol time. As previously stated, the effects of IPI on the RAKE output are more visible for the lower values of SF, say 8. Note that the ISI terms presented earlier also contain components from both the I- and Q-channels, which we will call the Inter-Channel Interface (ICI).

The solution of OC creates RAKE finger weight vectors based on the covariance matrix of the despread symbols. This matrix contains the ISI and ICI terms presented above and would create more accurate finger weights, especially in the low-SF application. A previous solution was to add an equalizer at the RAKE output to decrease the detrimental effects of ISI. A simplified block diagram is shown in Fig. 7.77.

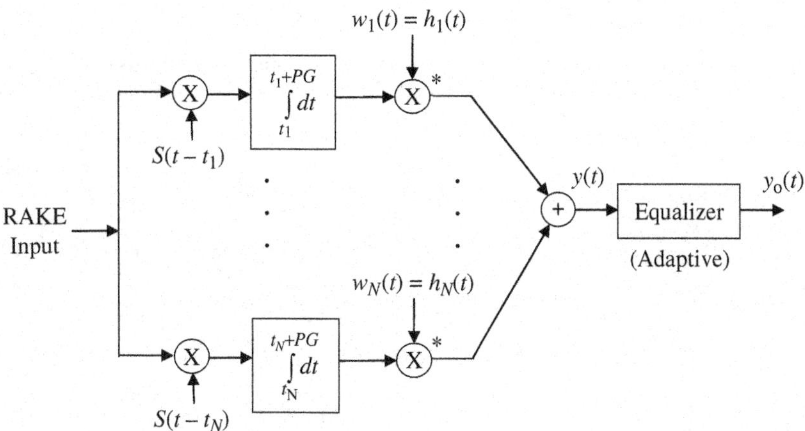

FIGURE 7.77 Modified RAKE receiver block diagram.

At the RAKE output, there are certain bit fields (i.e., pilot bits) that can be used to train the equalizer. Also a RAKE, actively demodulating a few multipaths has an output symbol sequence that can be reliably used in a decision-directed mode to update the equalizer taps for the time-varying channel conditions. On the downlink, dedicated pilot bits and the CPICH are also present to aid in the equalizer weight update.

RAKE Finger Memory Requirements. In this section, we will discuss the memory requirements to support CE used for coherent demodulation. As shown earlier, on the uplink, the dedicated pilot bits are time multiplexed within a time slot. It is well known, as shown with WMSA, that using channel estimates of not only the present and past but also the future can significantly improve performance. Here we briefly compare two RAKE finger architectures and discuss the advantages and disadvantages of both approaches.

The first approach is based on storing despread symbols; we will call this symbol level (SL) buffering. A simplified block diagram of a RAKE finger is given in Fig. 7.78.

Let us assume we need to store 2 time slots of symbols in this buffer, which is equivalent to (2×2560)/SF complex symbols. If we let the number of fingers equal N_{fing} and assume the I/Q

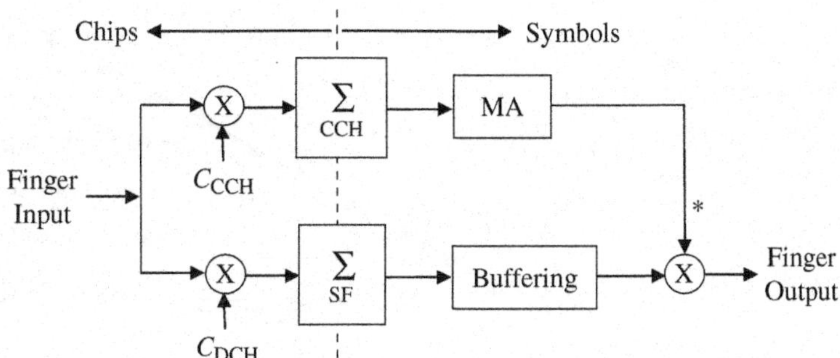

FIGURE 7.78 RAKE finger architecture based on symbol level buffering.

samples require 8 bits of finite precision representation, then the total number of bytes required is given as

$$\mathrm{SL} = \frac{\left\{\left\lceil \frac{2*2560 \text{ chips}}{\mathrm{SF}} \right\rceil *2\mathrm{I/Q}*8 \text{ bits}\right\}}{8} *N_{\text{fing}} \quad \text{[bytes]} \tag{7.111}$$

The second approach is based on storing predespread oversampled chips; we will call this chip level (CL) buffering. A simplified block diagram of a RAKE finger is given in Fig. 7.79.

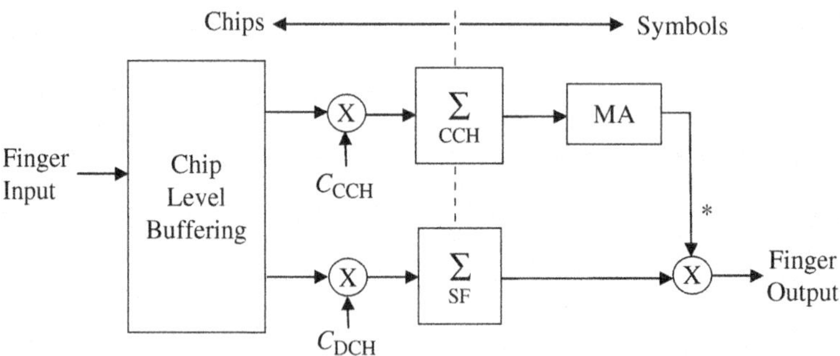

FIGURE 7.79 RAKE finger architecture based on chip-level buffering.

The advantage of this architecture is that the memory doesn't need to be repeated for each finger. Let us still assume we need to store 2 time slots where each chip is oversampled 4 times. We will assume the I/Q samples require 5 bits of finite precision representation. Then the total number of bytes is given as

$$\mathrm{CL} = \frac{2 \cdot 2560 \text{ chips} \cdot 4 \text{ times} \cdot 5 \text{ bits} \cdot 2\mathrm{I/Q}}{8} \quad \text{[bytes]} \tag{7.112}$$

If we assume the number of fingers is a design parameter, then we can plot the memory requirements comparing the chip-level buffer to the symbol-level buffer.

In Fig. 7.80, we plot the symbol-level buffering requirements for voice (SF = 64) and 384-kbps data service (SF = 4). Notice a crossover region occurs when the number of fingers is greater than 10 (assuming the above finger architecture parameters).

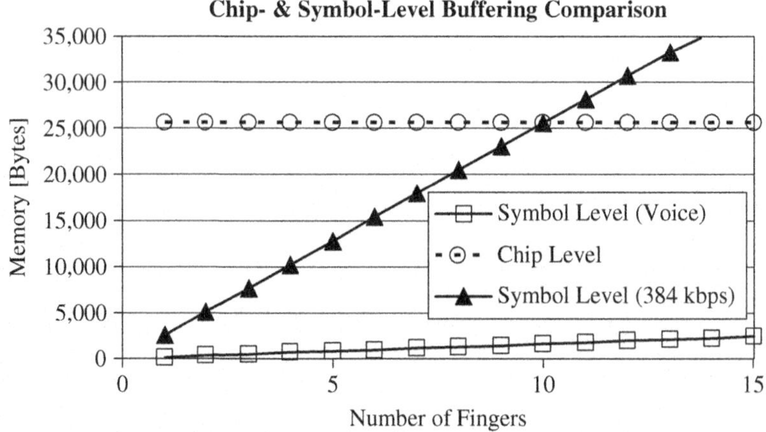

FIGURE 7.80 Chip- and symbol-level comparison.

Typically, a RAKE receiver is implemented with the worse-case scenario in mind, namely, the maximum number of fingers supported are all used. If the modem was solely designed for voice applications, then using the chip-level buffer would significantly increase the memory requirements over the symbol-level buffer. However, if one expects a 16-finger RAKE actively demodulating 384-kbps data services, then the chip-level buffer makes more sense. It is not the intention of this section to suggest an architectural implementation, rather to inform the reader of choices that are available.

Fine Time Tracking. Once a multipath is found and assigned to a finger, certain time offset exists due to the course resolution of the multipath searcher. Hence the finger must perform fine time code tracking to not only reduce this potentially large initial time offset, but also to track time due to movement and tolerances in the transmit and receive chip clocks. In this section, two fine–time tracking methods will be discussed.

Delay Locked Loop (DLL) Time Tracking. The first technique used for fine time tracking is the DLL. This technique assumes the autocorrelation function of the received chip sequence is symmetrical. The DLL has two parallel correlators: the first tracking the early time offset and the second tracking the late time offset. The time corresponding to the center of the earlier and late offset is the desired timing instance to be used by the finger despreader. This implies the received chip sequence is oversampled to allow for tracking time offsets.

In Fig. 7.81, we present an oversampled autocorrelation function to demonstrate the principle. The signal is oversampled by a factor of 4; alternatively stated, the time separation between time samples is $T_c/4$. This first figure shows an example where the correct receiver time sampling is performed. We have separated the early and late correlators by T_c chips apart.

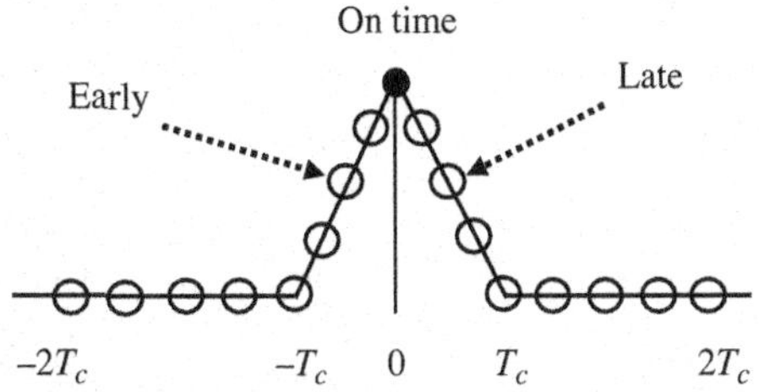

FIGURE 7.81 Optimal chip-time sampling example.

In Fig. 7.82, we show a time offset of $T_c/4$ between the received signal and the locally generated chip sequence.

In this example, the late correlator output energy will be larger than that of the early output, and this will indicate that we are sampling too early or too fast. So a time correction must be applied to move the on-time correlator to the peak position. The same line of reasoning can be applied to the opposite case. In either case, this technique does require an oversampled chip signal to operate properly.

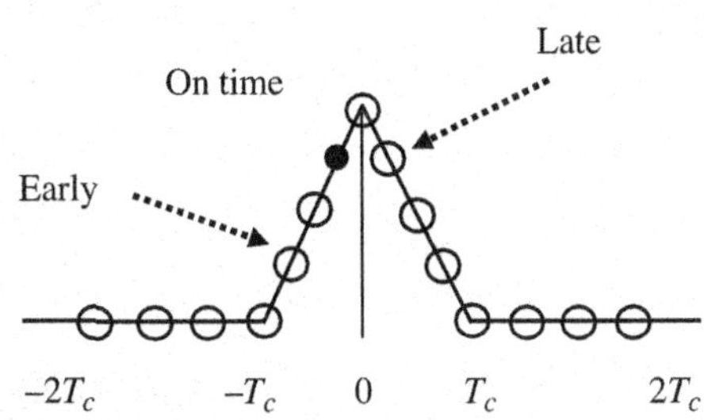

FIGURE 7.82 Chip-timing offset example.

A block diagram of the noncoherent-based DLL time-tracking synchronizer is shown in Fig. 7.83. The received signal is despread by two parallel correlators. The length of the despreading is dependent on the signal that is used for tracking. If the DPCCH was used, then depiloting would also be required to further coherent accumulation. In the block diagram in Fig. 7.83, we have assumed to be tracking the CPICH physical channel. In this case, we can improve noise immunity by integrating longer; however, we would be susceptible to degradation due to larger frequency offsets and time-varying channels.

The RAKE finger block diagram in Fig. 7.83 essentially has two sections: time tracking and data demodulation. The upper part consists of the time-tracking functions while the lower part shows the data-despreading/demodulating functions.

A tremendous amount of literature exists about the analysis of the DLL; we refer them to the interested reader [4, 24, 31, 32]. However, we will provide the baseline mathematical description for illustrative purposes. We assume the received signal encountered a flat fading channel with time delay τ. In the case of a frequency-selective fading channel, each of the distinct arriving multipaths will have

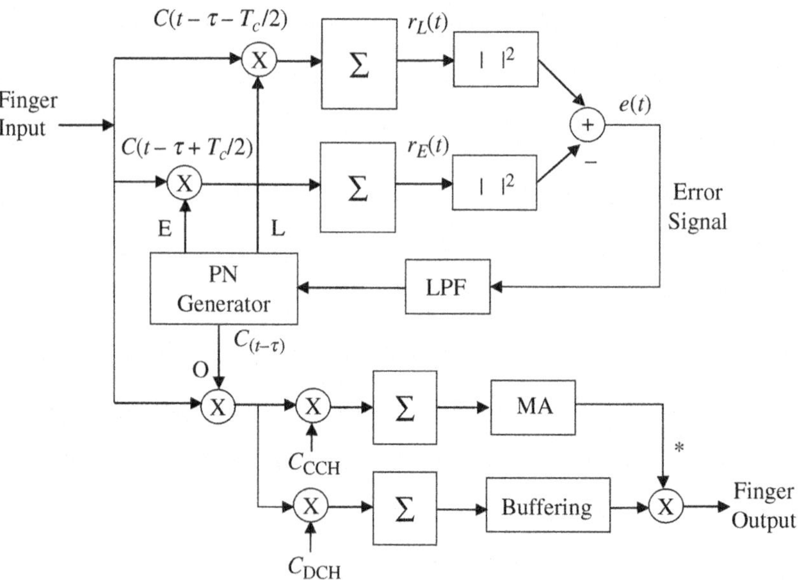

FIGURE 7.83 Noncoherent DLL RAKE finger-block diagram.

a DLL tracking them individually. In this case, there will be autocorrelation contributions among all the paths. The received signal is given as

$$r(t) = d(t - \tau) \cdot c(t - \tau) \cdot h(t - \tau) + n(t) \tag{7.113}$$

where $d(t)$ is the information symbol, $c(t)$ is the concatenated scrambling plus spreading code, $h(t)$ is the channel response, and $n(t)$ is the AWGN. The late correlator output signal is given as

$$r_L(k) = r_L(t)\Big|_{t=k\frac{T_c}{4}} = \sum_k r(k) \cdot c\left(k - \tau - \frac{T_c}{2}\right) + \text{noise terms} \tag{7.114}$$

$$r_L(k) = d(k - \tau) \cdot h(k - \tau) \cdot \sum_k c(k - \tau) \cdot c\left(k - \tau - \frac{T_c}{2}\right) + \text{noise terms} \tag{7.115}$$

$$r_L(k) = d(k - \tau) \cdot h(k - \tau) \cdot R_{cc}\left(\frac{T_c}{2}\right) + \text{noise terms} \tag{7.116}$$

Since it is commonly assumed that the pilot channel is used to perform fine time tracking, then $d(k)$ is either known to be depiloted or constant. In either case, it can be removed from Eq. (7.116), so we now have

$$r_L(k) = h(k - \tau) \cdot R_{cc}\left(\frac{T_c}{2}\right) + \text{noise terms} \tag{7.117}$$

Hence the late correlator output consists of a shifted autocorrelation function weighed by the channel response. Similarly, we can write down the early correlation output signal as

$$r_E(k) = \sum_k r(k) \cdot c\left(k - \tau + \frac{T_c}{2}\right) + \text{noise terms} \tag{7.118}$$

$$r_E(k) = h(k - \tau) \cdot R_{cc}\left(-\frac{T_c}{2}\right) + \text{noise terms} \tag{7.119}$$

As shown above, we wish to subtract these two correlator outputs in order to determine if the optimal sampling point is the one presently being used or not. If we generate this error signal, we get the following (neglecting the noise terms):

$$e(t) = r_L(k) - r_E(k) \tag{7.120}$$

$$e(t) = h(k-\tau)\cdot R_{cc}\left(\frac{T_c}{2}\right) - h(k-\tau)\cdot R_{cc}\left(-\frac{T_c}{2}\right) \tag{7.121}$$

$$e(t) = h(k-\tau)\cdot \left\{R_{cc}\left(\frac{T_c}{2}\right) - R_{cc}\left(-\frac{T_c}{2}\right)\right\} \tag{7.122}$$

In an AWGN channel, then ideally, $h(k)$ would be a constant and not influence the error signal over time. However, for time-varying signals, the channel estimate influence should be removed. The common technique is to use a noncoherent DLL, where the late and early powers are determined to obtain the error signal. In this case, we have the following error signal, neglecting noise cross terms as

$$e_{\mathrm{NC}}(t) = |h(k-\tau)|^2\cdot \left\{\left|R_{cc}\left(\frac{T_c}{2}\right)\right|^2 - \left|R_{cc}\left(-\frac{T_c}{2}\right)\right|^2\right\} \tag{7.123}$$

The channel power can be removed or normalized by obtaining the channel power from the on-time correlator output. The LPF can be replaced by an accumulator with appropriate decisions functions.

In Fig. 7.83, the top correlator is demodulating the late timing instance, while the lower correlator is demodulating the early timing instance. The despread waveforms are squared to obtain the symbol energies or powers to be used to calculate the error signal. The error signal computes the late minus early correlation powers, which then become low-pass filtered to reduce the results of smaller integration output values, if used. This low-pass filtered error signal will be used to either speed up or slow down the PN generator clock.

Alternatively, a coherent-detection DLL can be used to perform fine time tracking [33]. In this case, the modified error signal is given in Fig. 7.84.

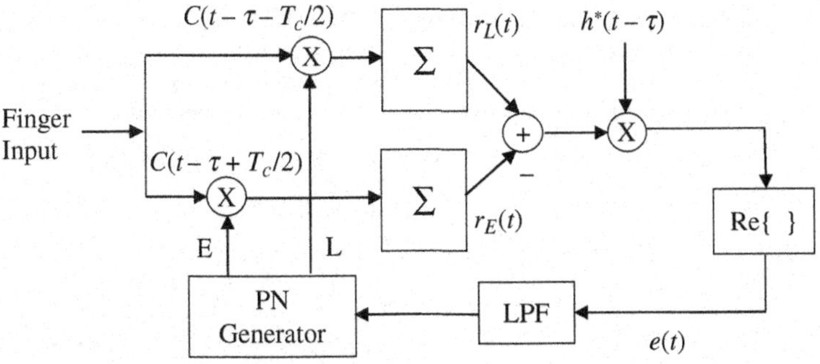

FIGURE 7.84 Coherent DLL block diagram.

Here the estimate of the channel, obtained possibly from the on-time correlator output, is used to derotate the error signal. Lastly, the projection onto the real axis is used as the error signal. The coherent-based DLL error signal is given as follows:

$$e_c(t) = \mathrm{Re}\left\{|h(t-\tau)|^2\cdot \left[R_{cc}\left(\frac{T_c}{2}\right) - R_{cc}\left(-\frac{T_c}{2}\right)\right]\right\} \tag{7.124}$$

Let us mention a few parameters that affect the performance of the DLL:

a. Coherent DLL versus noncoherent DLL.
b. Early-late separation. A larger separation would have a less–steep error characteristics curve than a smaller separation.
c. Pulse shaping used.
d. Update rate. How often does one update the time phase adjustment?
e. Oversampling factor used in the fine time tracking.
f. Quality of the channel estimate.

Tau-Dithered Loop (TDL) Time Tracking. In this section, we will present the second technique used to perform fine time tracking. A simplified block diagram is shown in Fig. 7.85.

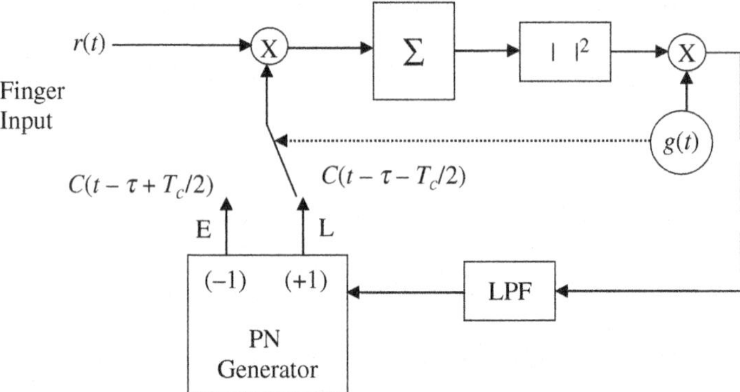

FIGURE 7.85 TDL block diagram.

This is a single correlator realization of the DLL (which, as shown above, uses two parallel correlators). The received signal is alternatively correlated with the early and late PN codes. Alternatively stated, this TDL can be viewed as a DLL with early and late correlations time sharing a single correlator. The dithering signal is denoted as $g(t)$ and shown in Fig. 7.86.

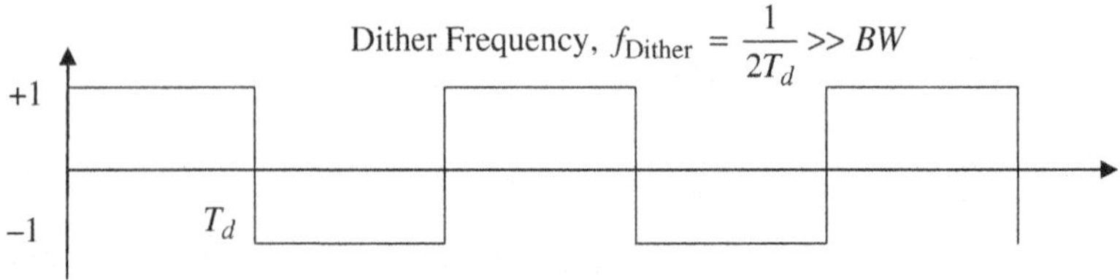

FIGURE 7.86 Dither control signal example.

The dithering signal controls which of the two correlators, early or late, are to be used. The actual subtracting functions are performed with the help of the inverter and the LPF.

Inner/Outer-Loop Power Control. In this section, we will present a mechanism that adjusts the transmission power inversely proportional to the channel fading characteristics. This is called transmit power control (TPC). We wish to increase system capacity by maximizing the number of simultaneous users we can accommodate in the WCDMA system. If each UE transmit power is adjusted so that the SIR level at the NodeB is at minimal acceptable level (i.e., QoS), then capacity can be increased.

We will discuss three forms of power control, namely,

Open loop

Closed loop inner

Closed loop outer

The above forms of power control will be discussed with the uplink and downlink frame timing relationship shown in Fig. 7.87. The uplink is transmitted T_o = 1024 chips behind the downlink time slot.

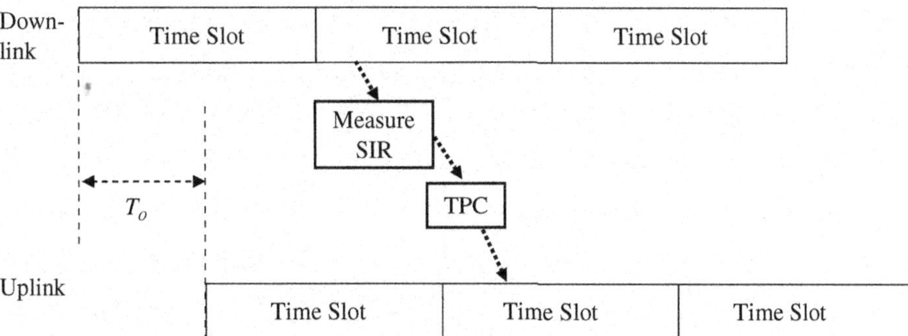

FIGURE 7.87 Uplink and downlink time slot relationship with power control timing emphasis.

The basic operations consist of measuring the received SIR and comparing this filtered value to a target $\text{SIR}_{\text{Target}}$. The outcome of the comparison will be used to generate TPC commands to be inserted into the transmit bit stream.

The first form of power control is called open loop. Here the measured received signal power from the NodeB is used to determine the transmit power of the UE. In an effort to help the UE determine a reliable initial transmit power, an equation is used by the UE. We have simplified the equation, and it is given below.

$$P_{\text{UE}} = P_{\text{CPICH}} + I_{\text{BTS}} + K \tag{7.125}$$

where P_{UE} = UE transmit power; P_{CPICH} = path loss of CPICH channel; I_{BTS} = interference power seen by the NodeB; and K = constant value (set by higher layers).

The open-loop correction provides a quick response to the mobile environment. The open-loop correction is generated as in Fig. 7.88, where $r(t)$ is the received signal.

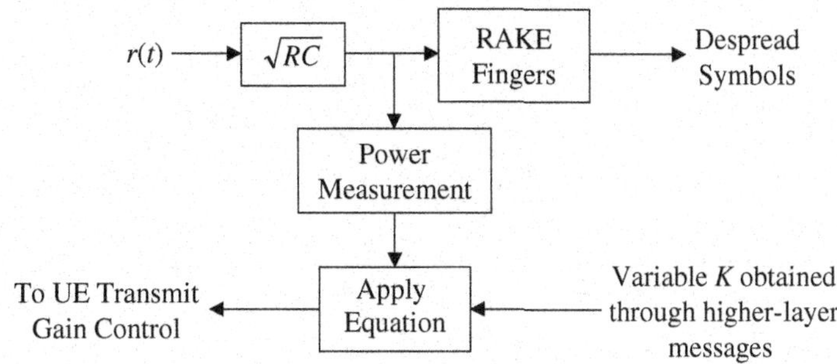

FIGURE 7.88 Open-loop power control measurement flow.

This is used when a UE is powered ON, extracts the cell system information from the BCH transport channel, and then needs to transmit a message such as PRACH.

The second form of power control is the closed inner loop. Two inner loops exist, one for the UE and one for the NodeB, which respectively measure their received signal power and determine the correction. For example, in an effort to refine the open-loop estimate, the NodeB must measure the UE's transmit power and then feed back a power control command to the UE to increase or decrease its transmit power. The rate at which the correction is applied for WCDMA is 1500 Hz, with step sizes of ± 1 or ± 2 dB. A high-level block diagram of the two inner-loop controls is shown in Fig. 7.89.

The loop designated by the bold solid lines is called the downlink power control, since the UE measures the downlink signal and determines the adjustment on the NodeB transmission. Here the RAKE output signal enters an SIR estimator. This downlink SIR estimate is converted to a TPC command that is multiplexed into the uplink transmit data stream. The loop designated by the dashed line is called uplink power control, where the NodeB measures the uplink signal and then sends a correction to the UE to adjust its transmit power, if necessary.

Let us discuss the UE side with some more details. This is accomplished through the use of Fig. 7.90. As shown in Fig. 7.90, the received SIR value is compared to an SIR_{REF} (or target value); we use the following rules of operation:

If $SIR_{EST} > SIR_{REF}$, then decrease transmit power by ΔdB.

If $SIR_{EST} < SIR_{REF}$, then increase transmit power by ΔdB.

The closed-loop inner power control is used to track the channel variations. This is possible since a TPC command is inserted every time slot, which translates to a rate of 1.5 kHz.

In the block diagram in Fig. 7.90, the received SIR is measured (SIR_{EST}) and compared to a reference value (determined by channel conditions and user data rate, QoS, etc.). This is shown by the dashed line. If the SIR_{REF} is lower than SIR_{EST}, then the UE instructs the NodeB to decrease the downlink transmit power. Conversely, the opposite holds true. The UE also extracts the downlink TPC commands within the Layer 1 control messages and uses this to increase/decrease the uplink transmit power. This is shown by the solid line and is adjusted once per time slot. Later in this section, we will provide an example SIR estimation algorithm.

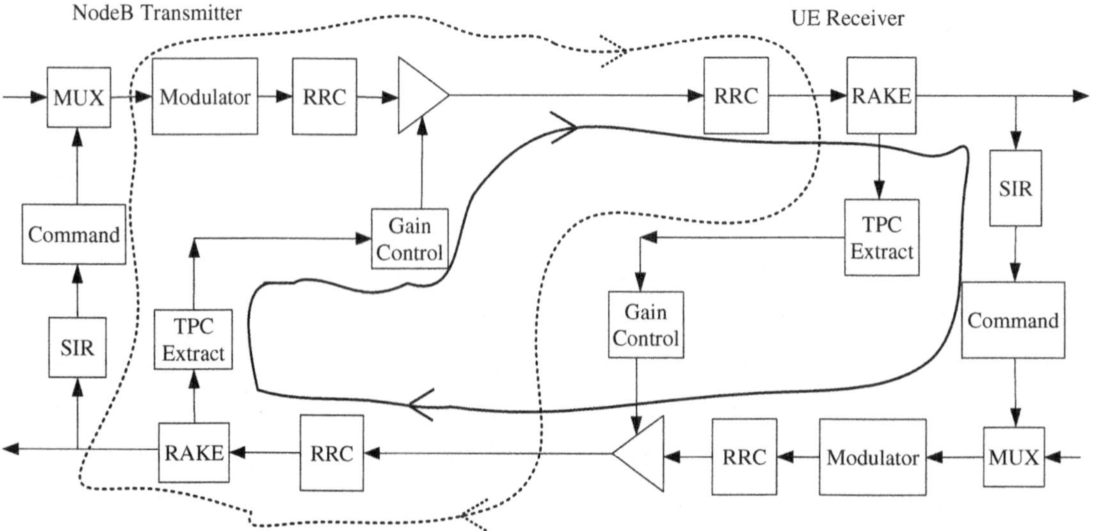

FIGURE 7.89 Uplink and downlink inner-loop power control.

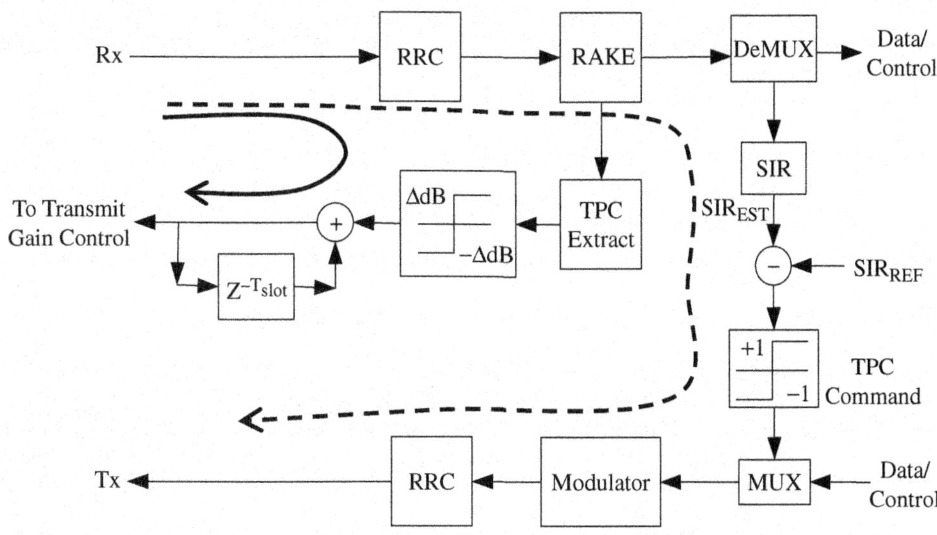

FIGURE 7.90 UE side of power-control loops.

The solid loop shown in Fig. 7.90 corresponds to the uplink power control. The dashed loop corresponds to the downlink power control. These two loops work simultaneously to combat multipath fading.

The third form of power control is the closed outer loop. In order to manage the power control mechanism, an outer loop is inserted that adjusts the target or reference SIR value. Here the receiver calculates the Frame Error Rate (FER) and converts it to a reference value in decibel (dB). Moreover, depending on the system capacity, QoS, and so forth, the higher layers can also adjust this reference value [34, 35]. A simplified block diagram is shown in Fig. 7.91.

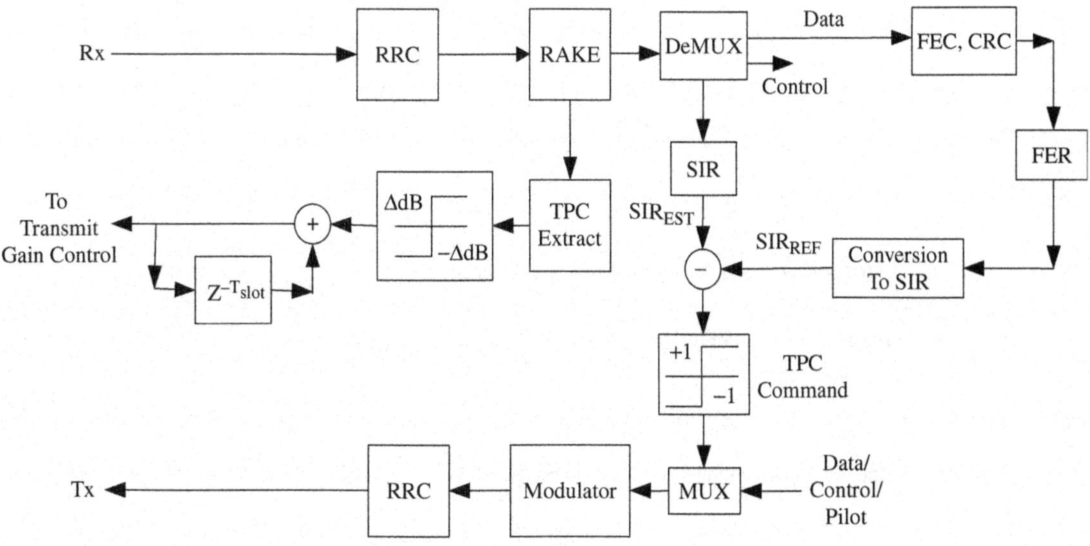

FIGURE 7.91 Outer-loop power control.

Next we will explain an outer-loop adjustment based solely on FER or BLER measurements. The outer loop will adjust the SIR target based on the BLER derived from the CRC check every TTI. The change in SIR is explained as

$$\Delta \text{SIR}_{\text{Target}} = \Delta \text{SIR}_{\text{Up}} \cdot \text{BLER} - \Delta \text{SIR}_{\text{Down}} \cdot (1 - \text{BLER}) \tag{7.126}$$

In other words, the SIR target is increased when a BLER increases, and it decreases with correctly decoded blocks. Some values to use for the $\Delta\text{SIR}_{\text{Up}}$ and $\Delta\text{SIR}_{\text{Down}}$ adjustments are given in Ref. [36]. In steady state, the measured BLER is equal to the target BLER, so we have

$$0 = \Delta \text{SIR}_{\text{Up}} \cdot \text{BLER} - \Delta \text{SIR}_{\text{Down}} \cdot (1 - \text{BLER}) \tag{7.127}$$

$$\frac{\Delta \text{SIR}_{\text{Up}}}{\Delta \text{SIR}_{\text{Down}}} = \frac{1 - \text{BLER}}{\text{BLER}} \tag{7.128}$$

Next we will present an example of an SIR estimator algorithm [37]. The SIR estimator guidelines can resemble the following: The I estimate time period must be long enough to estimate the ensemble average of I. The S estimate time period should be short so as not to average out the multipath small-scale fading effects.

Below we provide an example of an SIR estimation algorithm applied to the uplink NodeB receiver. The SIR measurement is performed on the DPCCH channel and uses the following definition (as defined within the 3GPP standard):

$$\text{SIR} = \frac{\text{RSCP}}{\text{ISCP}} \cdot \text{SF} \tag{7.129}$$

where RSCP is the received signal code power and ISCP is the interfering signal code power. We use Fig. 7.92 to calculate the signal energy, assuming an L-path channel impulse response and a pilot sequence given by $x(t)$, having a duration of N bits.

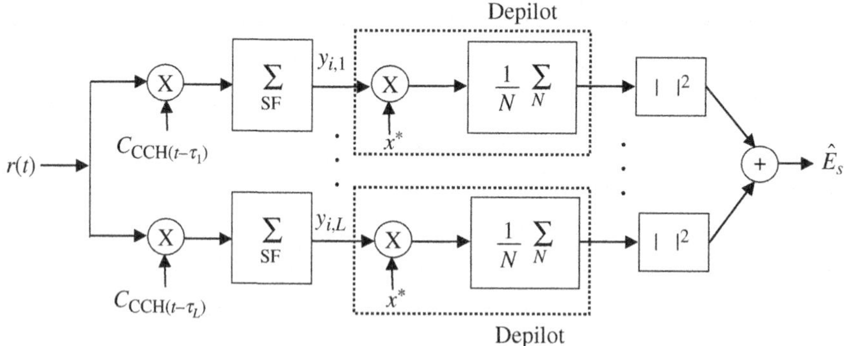

FIGURE 7.92 Signal energy estimator block diagram.

Our goal is to derive the following SIR estimate:

$$\text{SIR} = \frac{\hat{E}_s}{\hat{I}_s} \tag{7.130}$$

Assume the L-path channel model is given as follows:

$$h(t) = \sum_{l=1}^{L} c_l \cdot \partial(t - \tau_l) \tag{7.131}$$

The transmitted signal is given as (assuming $A_d^2 + A_p^2 = 1$)

$$s(t) = A_d \cdot C_{\text{DCH}} \cdot d + jA_p \cdot C_{\text{CCH}} \cdot x \tag{7.132}$$

where A_d is the data amplitude, C_{DCH} is the DCH OVSF code, d is the data bits, A_p is the pilot or control amplitude, C_{CCH} is the CCH OVSF, and x = control bit.

The received signal is given by

$$r(t) = \sqrt{E_c} \cdot A_d \cdot \sum_{l=1}^{L} c_l \cdot C_{\text{DCH}}(t - \tau_l) \cdot d(t - \tau_l) + j\sqrt{E_c} \cdot A_p \cdot \sum_{l=1}^{L} C_{\text{CCH}}(t - \tau_l) \cdot x(t - \tau_l) \cdot c_l \tag{7.133}$$

The received signal energy is estimated as

$$\hat{E}_s = \sum_{l=1}^{L} \left| \frac{1}{N} \sum_{i=1}^{N} y_{i,l} \cdot x_i^* \right|^2 \tag{7.134}$$

where we have defined the despreader output signal as

$$y_{i,l} = A_p \cdot N_p \cdot c_l \cdot \sqrt{E_c} \cdot x(t - \tau_l) \tag{7.135}$$

The ensemble average can be shown to equal

$$E\{\hat{E}_s\} = E\left\{ \sum_{l=1}^{L} \left| \frac{1}{N} \sum_{i=1}^{N} A_p \cdot N_p \cdot c_l \cdot \sqrt{E_c} + \text{noise} \right|^2 \right\} \tag{7.136}$$

$$E\{\hat{E}_s\} = A_p^2 \cdot N_p^2 \cdot E_c \cdot \sum_{l=1}^{L} \left| c_l \right|^2 + \frac{L \cdot I_s}{N} \tag{7.137}$$

Notice the second term on the RHS is a bias term. Now we may write the equation for the signal energy, assuming the bias term has been corrected.

$$\tilde{E}_s = \hat{E}_s - L\frac{I_s}{N} \tag{7.138}$$

$$\tilde{E}_s = \sum_{l=1}^{L} \left| \frac{1}{N} \sum_{i=1}^{N} y_{i,l} \cdot x_i^* \right| - L\frac{I_s}{N} \tag{7.139}$$

The denominator of the SIR can be calculated next. Here we assume the channel does not vary over two adjacent symbols and arrive with the following estimate of the interference:

$$\hat{I}_s = \frac{1}{2(N-1)L} \sum_{l=1}^{L} \sum_{i=1}^{N-1} [y_{i,l} \cdot x_i^* - y_{i+1,l} \cdot x_{i+1}^*]^2 \tag{7.140}$$

Since this estimate is not biased, the unbiased SIR estimates can be written below.

$$\text{SIR} = \frac{\tilde{E}_s}{\hat{I}_s} \tag{7.141}$$

Next we plot the SIR estimation simulation results for this SIR estimator in an AWGN channel (see Fig. 7.93) and fading channel (see Fig 7.94). An LPF (IIR based) was used on the interference estimate. The SIR error is defined as the difference between the estimated values and the true SIR values.

Here we see the SIR error of less than 0.75 dB can be achieved for 97% of the time and the SIR error of 1.5 dB can be achieved for 95% of the time for the flat fading channel considered. Since the channel is time varying, the SIR estimator will have a short-term estimate that varies as well.

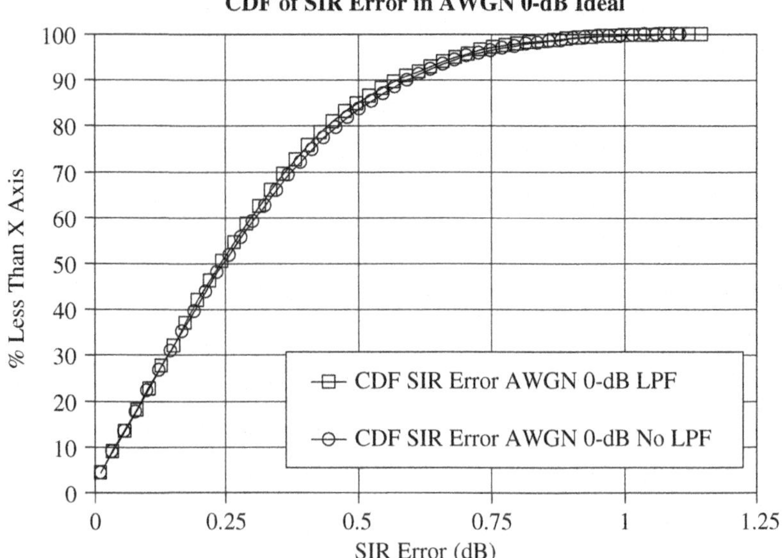

FIGURE 7.93 AWGN SIR estimation results.

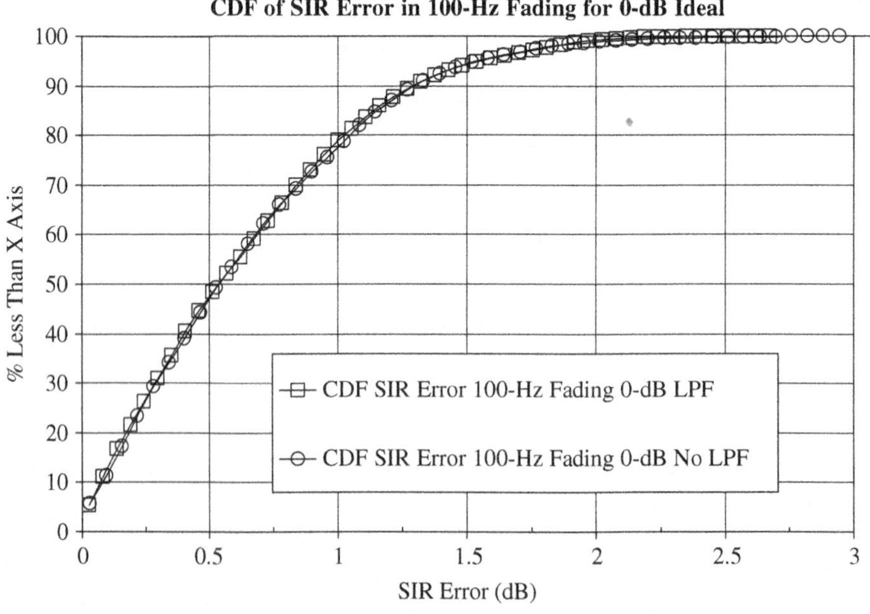

FIGURE 7.94 Frequency flat fading (f_d = 100 Hz) AWGN SIR estimation results.

As discussed above, the receiver (NodeB or UE) estimates the received SIR; then this estimate is compared to the SIR target. If the measured SIR is lower than the target, then the NodeB or UE transmits a command to increase the transmission power of the UE or NodeB. This closed-loop power control mechanism is used to mitigate the effects of multipath fading and thus improve overall system performance.

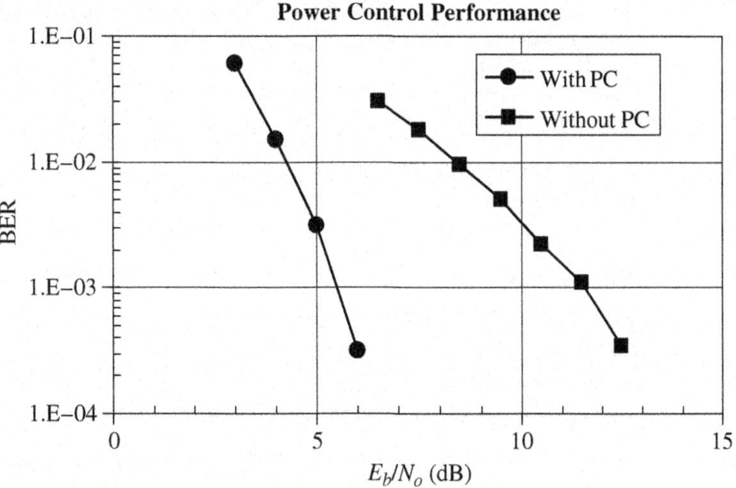

FIGURE 7.95 Power control BER performance comparison.

In Fig. 7.95, we plot the BER versus E_b/N_o, with and without power control enabled.

Echo Profile Manager (EPM). In this section, we present a function that manages the power delay profile determined by the searcher, along with the multipaths that are actively being combined in the RAKE. The list of multipaths that are actively being tracked in the RAKE is called the active list. This overall operation is called the EPM. A block diagram that describes the overall function is presented in Fig. 7.96.

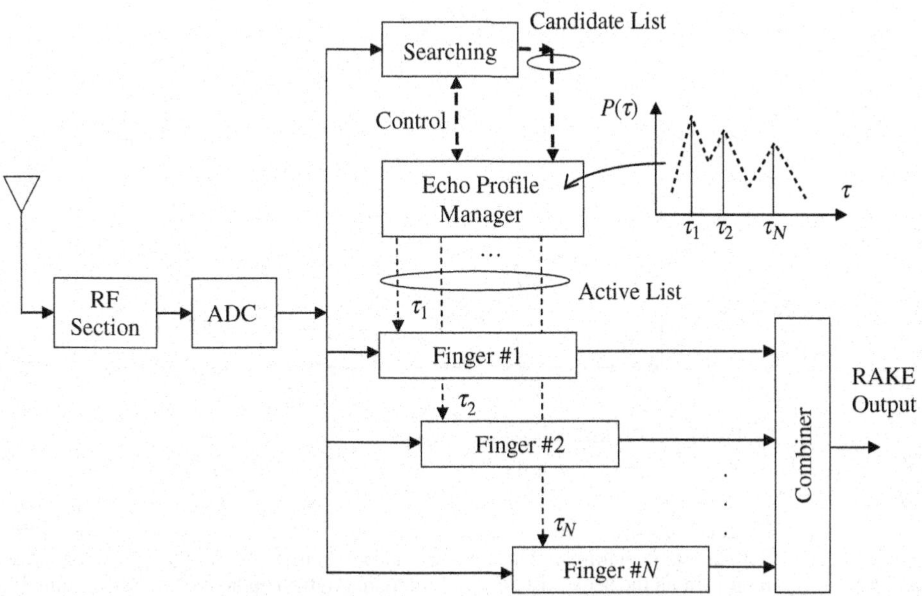

FIGURE 7.96 EPM and RAKE receiver block diagram.

Here the searcher finds N_{MP}-valid MPs within a chosen window size. These multipaths are valid because their energies after coherent and/or noncoherent integration have exceeded the hypothesis criteria. These multipaths are considered candidates to possibly be chosen to be used in the RAKE. The actively demodulated multipaths are being tracked in the RAKE receiver. Hence the EPM compares the candidate list of multipaths to what is actively being used to determine if a better multipath exists that can either be added or swapped into the active list. The finger management is continually operating due to the birth/death phenomenon of the multipaths.

The EPM will send control information to the searcher such as

- Enable/disable functional search
- Window size to look for multipaths
- Thresholds to compare against
- Integration length (coherent and noncoherent)
- Number of dwells (verification method)
- Time-hypothesis increment (say $T_c/2$ or T_c or other)
- Scrambling/channelization codes to be used

A simplified version of the RAKE finger management is provided in Fig. 7.97. If the accumulated energy E_j is above a threshold, then we will verify this by using longer statistics to confirm. This

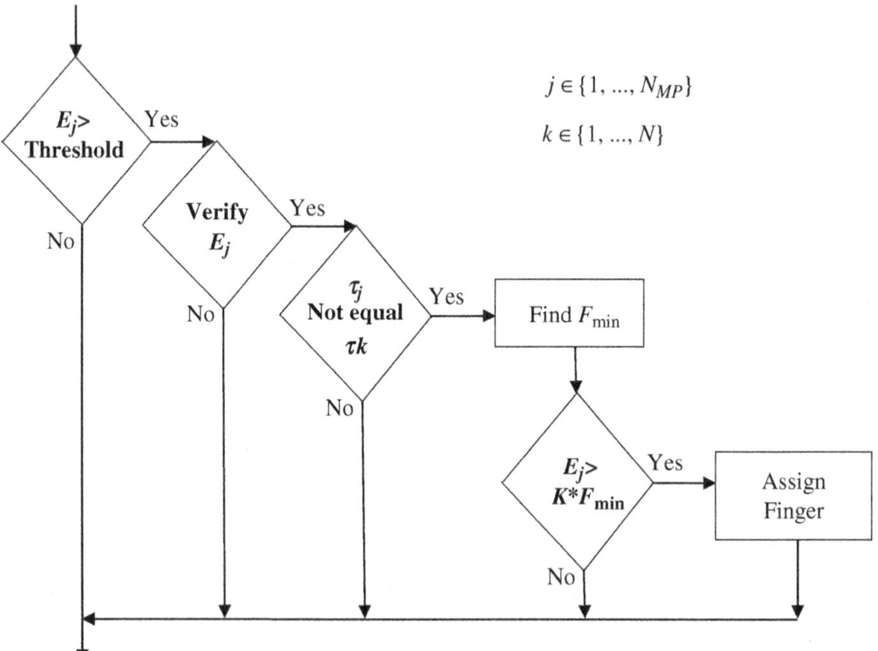

FIGURE 7.97 RAKE finger-manager control.

candidate list (τ_j) is compared to the active list (τ_k), so actively tracked multipaths are not swapped with the same multipath. Once the active list is removed from the candidate list, the smallest energy multipath on the active list (F_{min}) is compared to the candidate list to determine if a new assignment into the RAKE is necessary. We have included a small margin, labeled K, so as to prevent adding multipaths with essentially the same average power as what is being actively combined.

Frequency Lock Loop (FLL). Let us consider the downlink in this AFC example. The NodeB and UE will use oscillators to assist in the frequency up-conversions and down-conversions of the WCDMA signal. These oscillators will have some uncertainty of frequency error associated with it. As shown earlier, frequency offsets rotate the constellation with an angular frequency proportional to the residual frequency offset present on the signal. We will present a technique called Frequency Lock Loop (FLL), which is used to estimate the frequency offset present in the received signal (Δf) and then apply a correction signal in an effort to reduce the frequency offset.

The FLL block diagram emphasizes a single RAKE finger. The residual frequency offset is intentionally introduced in the analysis. A square root raised cosine is assumed in the receive path, followed by despreading and then depiloting. At this point, the modulation has been removed and an intersymbol phase difference can be used and accumulated over N pilot symbols.

In the block diagram in Fig. 7.98, the received signal is represented as

$$r(t) = S_{cr}(t) \cdot p(t) \cdot h(t) \tag{7.142}$$

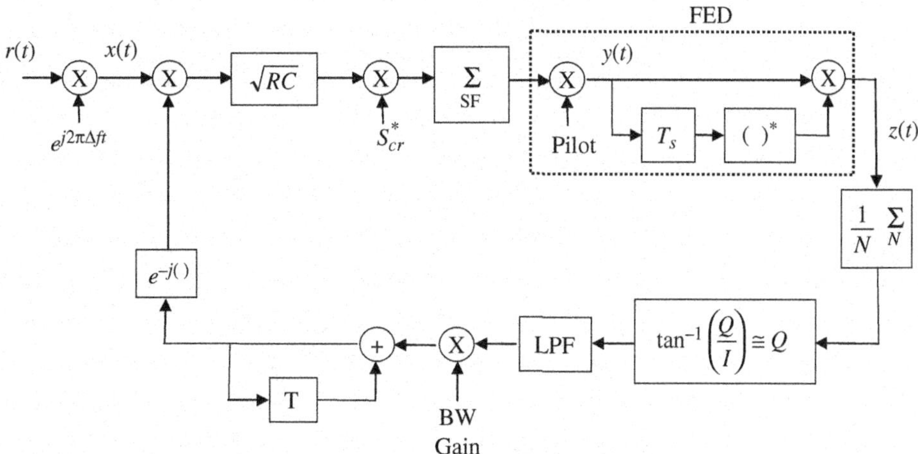

FIGURE 7.98 FLL block diagram.

Let's assume the spectral down-conversion operation performed by the quadrature demodulator has a residual frequency offset equal to Δf Hz. The residual frequency offset manifests itself in the following manner:

$$x(t) = S_{cr}(t) \cdot p(t) \cdot h(t) \cdot e^{j2\pi\Delta ft} \tag{7.143}$$

After despreading and depiloting, we have

$$y(t) = |S_{cr}(t)|^2 \cdot |p(t)|^2 \cdot h(t) \cdot e^{j2\pi\Delta ft} \tag{7.144}$$

This signal enters the Frequency Error Discriminator (FED) to produce

$$z(t) = h(t) \cdot h^*(t - T_s) \cdot e^{j2\pi\Delta ft} \cdot e^{-j2\pi\Delta f(t-T_s)} \tag{7.145}$$

$$z(t) \cong |h(t)|^2 \cdot e^{j2\pi\Delta fT_s} \tag{7.146}$$

where the frequency offset estimated is equal to $\Delta f = \theta/2\pi T_s$. The diagram in Fig. 7.98 shows a one-finger RAKE receiver. In practice, the RAKE will have many fingers actively demodulating and combining multipaths. Hence the FED output $z(t)$ should be summed across the pilot bits of interest, averaged across all fingers, and averaged across all receive antennas to create a reliable estimate of the received residual frequency offset.

Large residual frequency offsets are extremely detrimental to the overall system performance, if left uncompensated. Large frequency offsets prevent extended accumulation lengths, degrade CE, and can diminish system capacity. The NodeB requirement is 0.05 ppm, and the UE requirement is 0.1 ppm.

TFCI Decoding. In this section, we will briefly review the generation of the TFCI bits. The transport channel demultiplexor, in the receiver, will use the information to extract parameters to aid it in decoding the received physical channel. An example of what info is conveyed by these control bits is the number of transport channels, the size of the transport block, power ratios, and so forth.

We begin by presenting the generation or encoding procedure of the TFCI bits. We briefly discuss the mapping of these bits to the physical channel. Next we present the TFCI decoder and discuss its interaction with the physical channel. Lastly, we present a few suggestions on how to improve decoding performance and possibly reduce implementation complexity [38].

TFCI Encoder Details. The encoding bit stream is based on a second-order Reed-Mueller code. This code can be represented as a combination of two vector spaces. The first vector space comprises of a set of orthogonal vectors, while the second comprises of mask vectors (not orthogonal to the first vector space).

There are a total of 1024 combinations of the transport format, and thus, we only require the TFC to be 10 bits in length. This is referred to as the TFC index. This index will essentially choose a certain combination of the above-mentioned vector spaces to create a single TFCI code word of length 32 bits. A graphical representation of this is shown in Fig. 7.99.

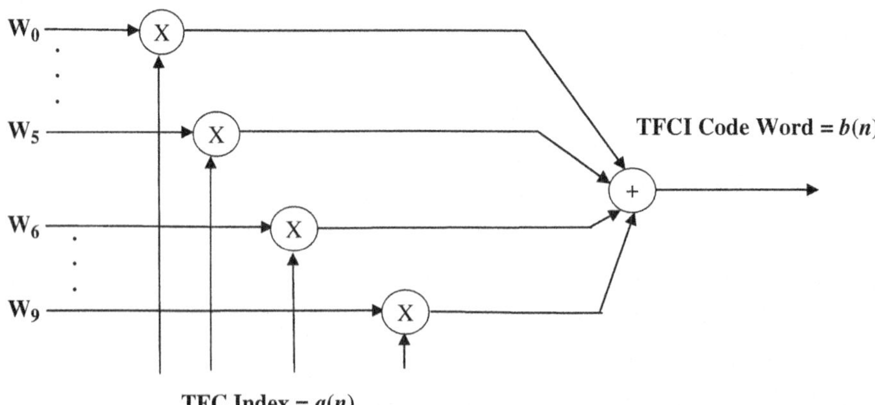

FIGURE 7.99 Graphical representation of TFCI code-word generation.

The block diagram in Fig. 7.99 can be mathematically represented as follows:

$$b(n) = \left[\sum_{j=0}^{9} a(j) \cdot w_j(n) \right]_{\text{Modulo 2}} ; \qquad (n = 0, \ldots, 31) \tag{7.147}$$

Alternatively, this can be written as (given $\underline{a}$ and $\underline{b}$ are row vectors and $\underline{w}$ are column vectors)

$$\underline{b} = \underline{a} \cdot W$$

$$\underline{b} = \underline{a} \cdot \begin{bmatrix} \underline{w}_0^T \\ \underline{w}_1^T \\ \cdots \\ \underline{w}_9^T \end{bmatrix} \tag{7.148}$$

Let us provide some insight into the generated TFCI code word based on the TFC index value. Even though any of the individual vectors can be enabled or disabled by a multiplication of 1 or 0, respectively, the resulting code word is a modulo-2 sum of all the enabled vectors.

Since we have a maximum of 15 time slots in a frame, we will typically receive 30 bits (assuming 2 bits per time slot). The last 2 bits are punctured in the transmission. In the case of compressed mode, the code word will be transmitted more than once per frame and thus the individual code words are combined prior to entering the decoder. So care must be taken in considering possible scenarios when designing the TFCI decoder.

TFCI Decoder Details. In this section, we will present the details of the TFCI decoder. As previously discussed, the code word consists of a modulo-2 summed combination of the full vector space. The receiver must perform some sort of joint detection of the individual basis vectors.

The basic decoding principle is as follows: Since there are 16 mask vectors and 64 orthogonal vectors, we will first premultiply the received signal by each of the combinations of the enabled mask vectors. Then perform an exhaustive correlation search over the 64 orthogonal vectors. Since the orthogonal vectors can be treated as OVSF or Walsh vectors, a fast Hadamard transform (FHT) is a computationally efficient method for implementation. For our purposes, we have chosen to implement an exhaustive searching operation; the reason will become apparent in the later sections.

The above operations are repeated 16 times, one for each mask vector combination. The premultiplication operation is needed because the mask vectors are not orthogonal and we wouldn't be able to use the FHT principle as it applies.

A block diagram of the decoder is shown in Fig. 7.100.

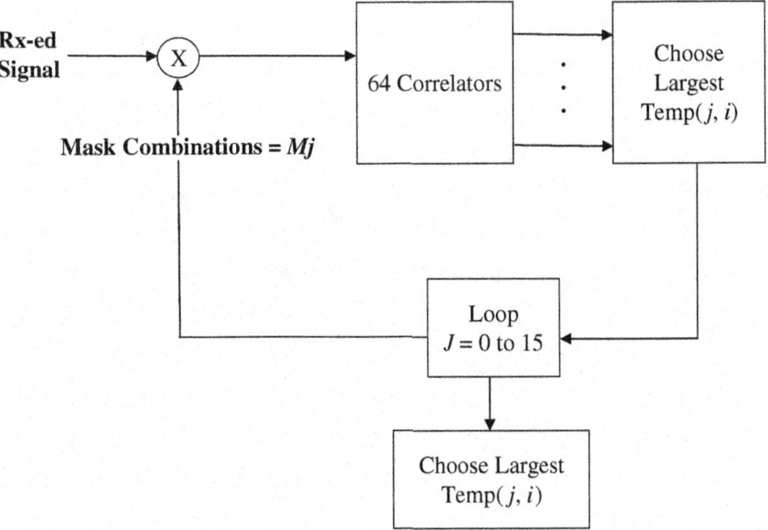

FIGURE 7.100 TFCI decoder block diagram.

```
For (k = 0 to 15)
{
    Premultiply the Rx signal by Mask combination # k
    For ( j = 0 to 63)
    {
        Correlate against each possibility
        Find Max energy
        M[k] = Max Energy
         I[k] = corresponding index
    }
}
For (k = 0 to 15)
    Find Max and save final info

Calculate index: TFCI Index = k*64 + corresponding index
```

FIGURE 7.101 TFCI Decoder Pseudocode Listing.

In Fig. 7.101, we present the pseudocode describing the TFCI decoding principle.

TFCI Decoder Performance Improvements. We have decided to separate the concept of performance improvement into two categories. The first being a reduction in implementation complexity, while the second category is the reduction of the required E_b/N_o operating point. We will discuss potential solutions for the first category and present a solution for the second category.

A complexity reduction method will aim to reduce the size or order of the FHT. In other words, we would like to reduce the number of parallel correlators to test against the orthogonal vectors. In fact, reduction by a factor of at least 2 is possible.

The link layer performance improvement can be accomplished as follows: We use soft decisions with erasure decoding. For typical cases, we will receive 30 bits out of the full 32-bit code word; the remaining 2 bits are inserted into the code word as 0, assuming bipolar signaling. A suggestion is to try all four possible combinations and choose the largest one. Here we can apply Chase's second algorithm as a start.

In Fig. 7.102, we plot the Word Error Rate (WER) of the above TFCI decoder, assuming both a single and dual receiver. Independent fading was assumed for the dual antenna receiver.

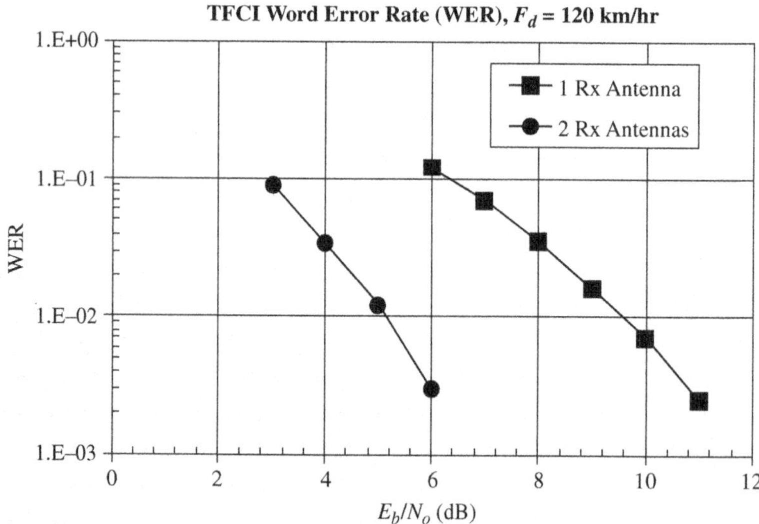

FIGURE 7.102 TFCI decoding performance.

We have concluded this section on RAKE finger signal processing that covered channel estimation, combining strategies, multipath performance, time tracking, power control, frequency compensation, and TFCI decoding. We have also provided various comments on implementation details and insight into some of the presented algorithms.

7.6.6 RACH/FACH Procedure

In this section, we describe the procedure of random access channel (RACH) transmission. The random access transmission is based on slotted ALOHA protocol with fast acquisition indication [16, 39]. The transmission consists of one or more RACH preambles of length 4096 chips and a PRACH message of length 10 or 20 msec. Users randomly transmit in time on boundaries of ASs of duration 5120 chips. The information on what ASs are available for random access transmission is given by the higher layers, which was extracted from decoding the system information on the BCH transport channel on the physical channel, P-CCPCH. The general access procedure is outlined in Fig. 7.103.

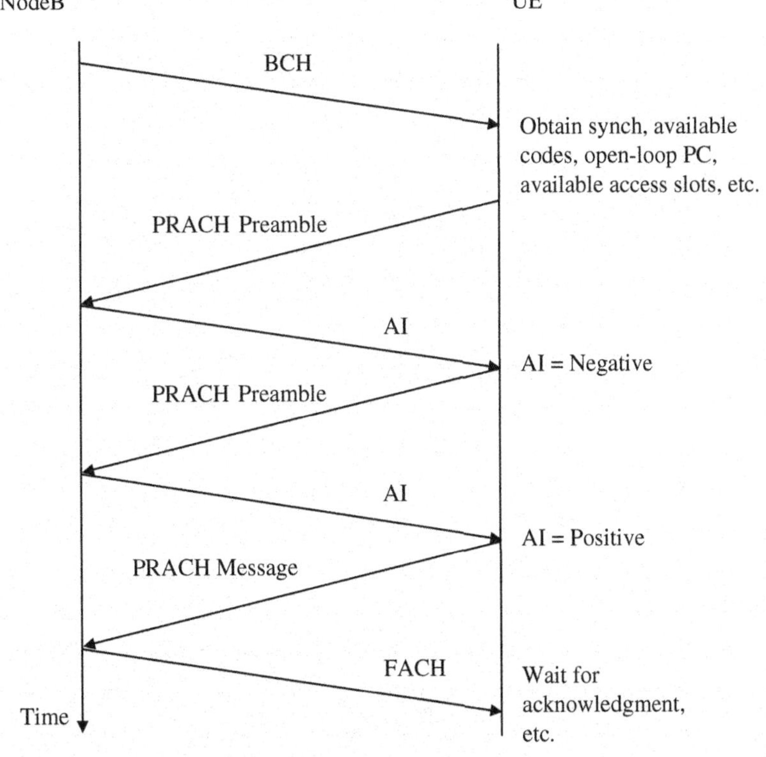

FIGURE 7.103 RACH and FACH procedure events.

The UE transmits a PRACH preamble and awaits an AI on the downlink AICH to indicate to the UE that the preamble was received properly. If the AI has a negative value, then the UE retransmits until either a maximum counter is reached or a positive AI is received. Once the awaited AI is received, the PRACH message is transmitted. As mentioned above, the RACH consists of two parts: preamble and message.

The preamble has the following features:

- Time duration = 4096 chips
- 16 possible signatures of length 16 (Hadamard code)
- Gold spread code
- 256 repetitions of the signatures
- Open-loop power control
- Special code construction
- Multiple simultaneous RACH attempts

The message has the following features:

- Duration = 10 or 20 msec
- HPSK spreading
- 4 possible data rates (15, 30, 60, and 120 kbps)
- No power control
- 8 pilot bits + 2 TFCI bits
- SF = 256 fixed.

The RACH preamble is a complex-valued sequence created from the preamble scrambling code and preamble signature.

$$C_{\mathrm{pre}}(k) = S_{\mathrm{pre}}(k) \cdot C_{\mathrm{sig}}(k) \cdot e^{j\left(\frac{\pi}{4}+\frac{\pi}{2}k\right)} \qquad (k = 0, \ldots, 4095) \tag{7.149}$$

There is a one-to-one relationship between the group of PRACH preamble scrambling codes in a cell and the primary scrambling code used in the cell.

The RACH access timing relationship is provided in Fig. 7.104 and taken from the 3GPP standard. The time slot and frame structure of the PRACH message is shown in Fig. 7.105.

RACH Access Performance. In this section, we will present the PRACH performance. We show the mean number of retransmissions required by the UE to successfully receive a FACH acknowledgment over the S-CCPCH. These results assumed the 64-kbps data rate for the RACH message. These results show that if the PRACH message E_b/N_o is below 6 dB, the performance significantly degrades (see Fig. 7.106).

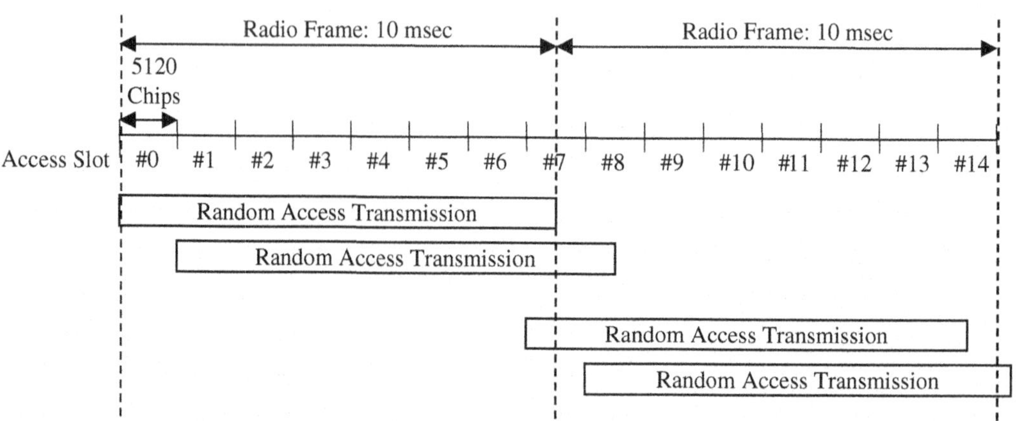

FIGURE 7.104 RACH random access time slots.

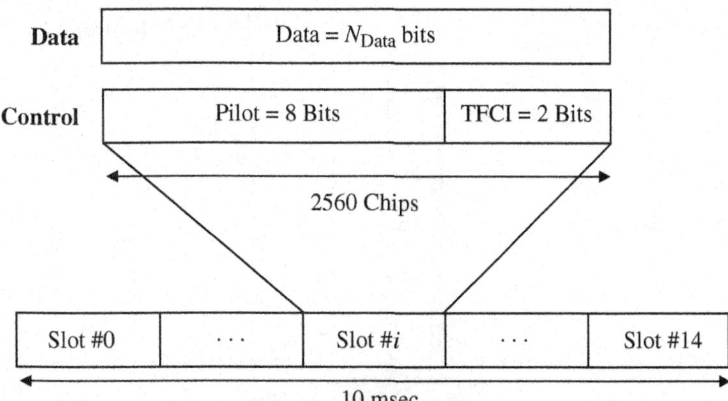

FIGURE 7.105 PRACH message time slot structure.

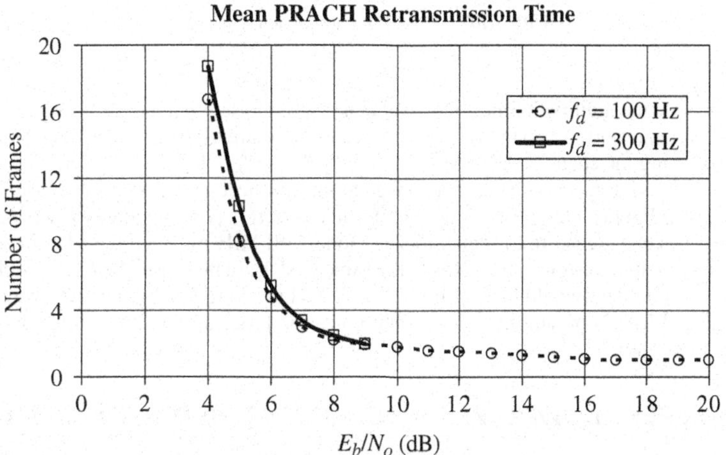

FIGURE 7.106 PRACH retransmission performance in a Rayleigh flat-fading channel.

A number of trials were conducted where for each E_b/N_o value, the number of retransmissions required to obtain a correct CRC was found was averaged.

We can also include a three-dimensional plot showing the PDF for each E_b/N_o value considered (see Fig. 7.107). Notice that for low E_b/N_o values, the PDF appears to be more flat than exponential.

It is expected that, as the E_b/N_o decreases, the PRACH message CRC would fail more often. This would place the UE into operating modes where higher transmit power is used to increase the probability of success. We point this out since this has an effect on the UE power consumption and thus decreases battery life. This can have a profound effect on talk time, especially when considering the location of the UE in the coverage area.

Some items to consider in the RACH preamble and message demodulation on the uplink will be discussed next. First, due to round-trip propagation time delay, the uplink PRACH preamble can be significantly delayed so that there is an overlap between adjacent valid ASs. Also, since the time duration between the end of the PRACH preamble and the beginning of the PRACH message is small, the

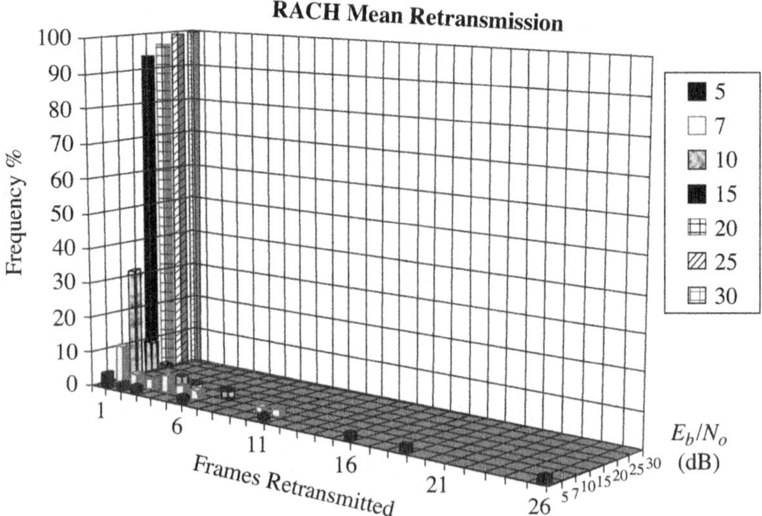

FIGURE 7.107 PRACH retransmission statistics.

same multipath times of arrival can be used to set up the RAKE fingers for demodulation of the message part. Even though the same multipaths are used, time-varying channel conditions require tracking the channel estimate.

Also, in order to not make the uplink access channel the bottle neck, multiple simultaneous PRACH attempts are possible. In this case, it is possible that two separate UEs randomly choose the same signature to transmit. The NodeB PRACH message RAKE can attempt to erroneously assign heterogeneous multipaths to the same RAKE. In this case, the CRC check would fail and no acknowledgment would be sent to the S-CCPCH carrying FACH. The UE would then return to transmitting PRACH preambles using a randomization procedure outlined in the 3GPP standard.

7.7 HIGH-SPEED DOWNLINK PACKET ACCESS (HSDPA) OVERVIEW

In this section, we will present an overview of this technique of providing packet services on the downlink. The important components/features of HSDPA are an increase in the downlink throughput and peak data rates, low-latency control, adaptive modulation and coding and physical layer retransmissions, fast channel adaptive scheduling, and dynamic resource sharing [23, 40–42].

The physical channels required to support HSDPA are shown in Fig. 7.108, with HS-DSCH being high-speed–downlink shared channel and HS-SCCH representing high-speed–shared control channel. In order to support the low-latency control, smaller frame sizes have been used for 2-msec duration. Moreover, the scheduling operation has been moved from the RNC to the NodeB to provide quick turnaround response times.

The basic HSDPA operation is as follows: The NodeB transmits packets of data at a high rate to the UE. The data rate highly depends on the UE capability, channel conditions, and NodeB capability. If the UE successfully receives them, then an ACK is transmitted and the NodeB continues to flush its data buffers. On the other hand, if a NACK is transmitted by the UE, then the NodeB will retransmit the packet with either the same information or additional information in order to help the UE successfully detect the packet of data.

The ACK/NACK information is carried in the uplink HS-DPCCH channel. In parallel to this loop, the UE is sending Channel Quality Indication (CQI) to the NodeB, giving the NodeB scheduler an

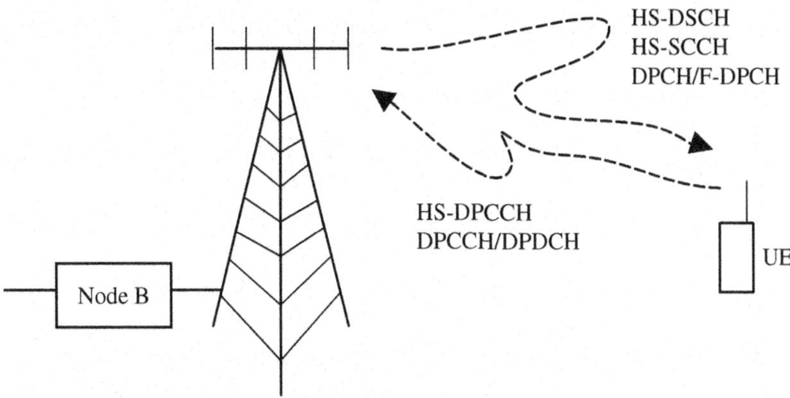

FIGURE 7.108 HSDPA physical channels.

indication of how good the channel is for that particular UE at that particular time instant. At this point, the NodeB can dynamically adjust its transmit resources to send data to the UE. We will define a resource consisting of a combination of the following: subframes, power allocation, modulation scheme, number of multicodes, and FEC rate.

HSDPA does not support handoffs due to a requirement of low latency, forcing the MAC functionality for the high-speed channels (MAC-hs) to reside in the NodeB. Hence cell reselection or WCDMA handoffs can cause a reduction in the user throughput.

7.7.1 Physical Channels

In this subsection, we will present the physical channels required for HSDPA operations.

High-Speed–Downlink Shared Channel (HS-DSCH). The HS-DSCH has a constant SF = 16, and can be scrambled by either the primary or secondary scrambling code. The time slot and subframe structure is given in Fig. 7.109. This HS-DSCH channel carries the packet data mentioned above.

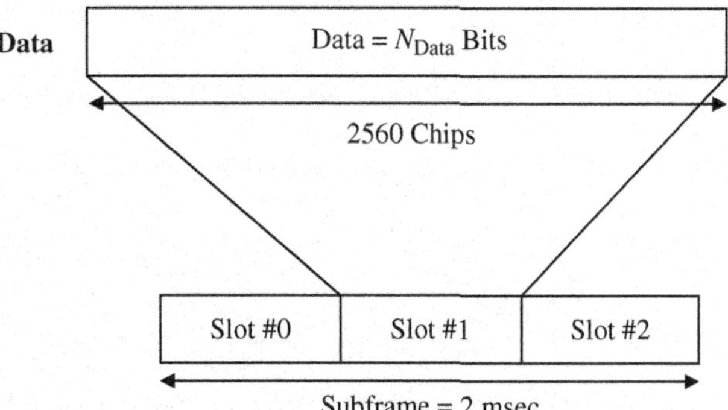

FIGURE 7.109 HS-DSCH time slot and subframe structure.

The HS-PDSCH in Fig. 7.109 corresponds to a single channelization code from the set of possible channelization codes reserved for the HS-DSCH transmission. Total number of possible codes = 15. HS-DSCH may use either QPSK or 16-QAM. In the later sections, in this chapter, we discuss higher-order modulation schemes (i.e., 64-QAM) that are added for certain deployment scenarios. Note the HS-PDSCH does not carry Layer 1 controls information; all relevant Layer 1 information is transmitted on the associated HS-SCCH. Moreover, while HSDPA services are provided, the uplink/downlink WCDMA link must be established to maintain a communication link.

High-Speed–Shared Control Channel (HS-SCCH). The SF is fixed at 128. The channelization codes used for HS-DSCH are allocated as follows: For X multicodes at offset Y, the codes are allocated contiguously.

$$C_{\mathrm{CH},16,Y} \cdots C_{\mathrm{CH},16,Y+X-1}$$

This information is signaled over the HS-SCCH. This channel can be scrambled by either the primary or a secondary code. The time slot and subframe structure is given in Fig. 7.110.

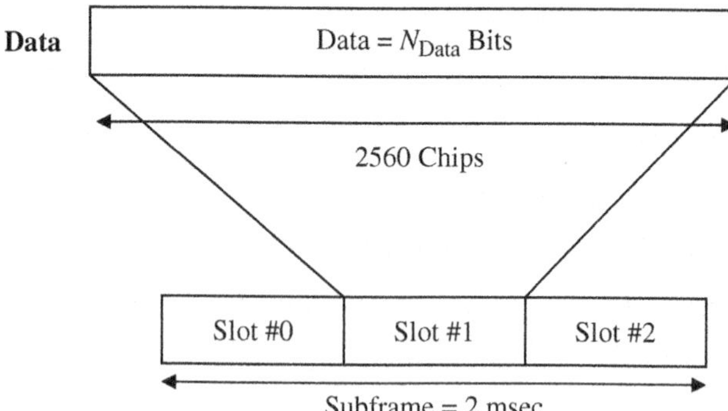

FIGURE 7.110 HS-SCCH time slot and subframe structure.

The information carried in the HS-SCCH physical channel is used to not only alert the specific UE the data is intended for, but also supply the UE receiver with some physical layer information, so it can prepare to demodulate the HS-DSCH channel.

The HS-SCCH will carry the downlink control signaling related to its associated HS-PDSCH transmission. Let us make a quick comment about the timing relationship between P-CCPCH, HS-DSCH, and HS-SCCH.

There are 5 subframes per single 10-msec radio frame. The P-CCPCH and HS-SCCH channels are frame aligned. The HS-DSCH overlaps the HS-SCCH by a single time slot (see Fig. 7.111). This

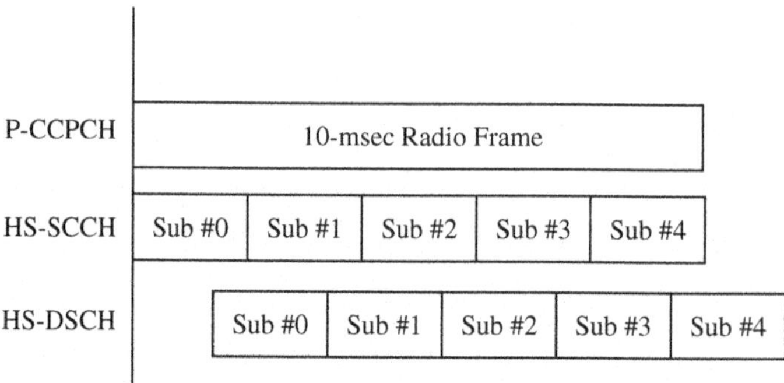

FIGURE 7.111 HS-DSCH and HS-SCCH subframe timing relationship.

allows the UE to use the first two slots of the HS-SCCH subframe to set up the HS-DSCH demodulator. The one-slot overlap reduces the round-trip latency as well as memory requirements of the UE.

HS-DPCCH Channel. The HS-DPCCH carries uplink signaling related to the downlink HS-DSCH transmission. This signaling consists of HARQ acknowledgment (HARQ-ACK) and Channel Quality Indication (CQI). The ACK/NACK is carried in the first time slot while the CQI is carried in the second and third time slots. The time slot and frame structure is given in Fig. 7.112.

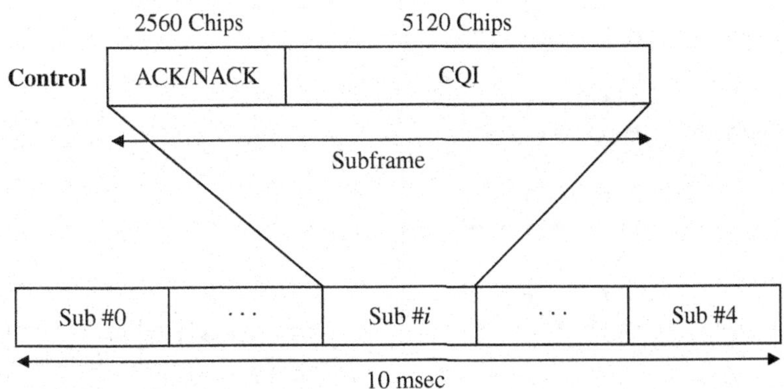

FIGURE 7.112 HS-DPCCH subframe structure.

The SF is constant at 256. HS-DPCCH can only exist together with an uplink DPCCH. The 3GPP standard provides details on mapping the downlink signal quality to uplink signaling, through the use of certain lookup tables. The HS-DPCCH time slot timing relationship with respect to the uplink DPCH time slot is not necessarily time aligned. This depends on the downlink timing offset given to the DPCH.

The uplink spreading and scrambling block diagram is shown in Fig. 7.113. The HS-DPCCH can be added in either the I- channel or the Q-channel, depending on the number of DPDCH channels. The exact conditions are clearly stated within the 3GPP standard.

7.7.2 Multicode Overview

The HS-PDSCH uses an SF = 16. Subtracting a code reserved for the common channels leaves us with 15 possible codes that can be transmitted by the NodeB. Depending on the UE capabilities as well as the channel conditions; this value varies. Hence multicodes are viewed as resources and are illustrated in Fig. 7.114.

In Fig. 7.114, we show $K + 1$ parallel data streams being spread with $K + 1$ consecutive channelization codes, followed by a single scrambling operation. The $K + 1$ data streams can be transmitted to either a single or multiple UE. In other words, we can have a scenario where UE #1 gets an allocation of 5 multicodes and UE #2 gets an allocation of 6 multicodes. The data traffic for both users is transmitted at the same time and can be separated with the knowledge of the channelization codes used.

Assuming a $(K + 1)$-multicode signal, an M-path channel, and a RAKE receiver with M fingers, we can draw the block diagram in Fig. 7.115. For each time of arrival, multicode despreading is performed.

Let us assume the receive signal is given as

$$r(t) = \sum_{l=1}^{M} \sum_{k=0}^{K} d_k(t - \tau_l) \cdot C_{\text{CH},j+k}(t - \tau_l) \cdot S_{cr}(t - \tau_l) \tag{7.150}$$

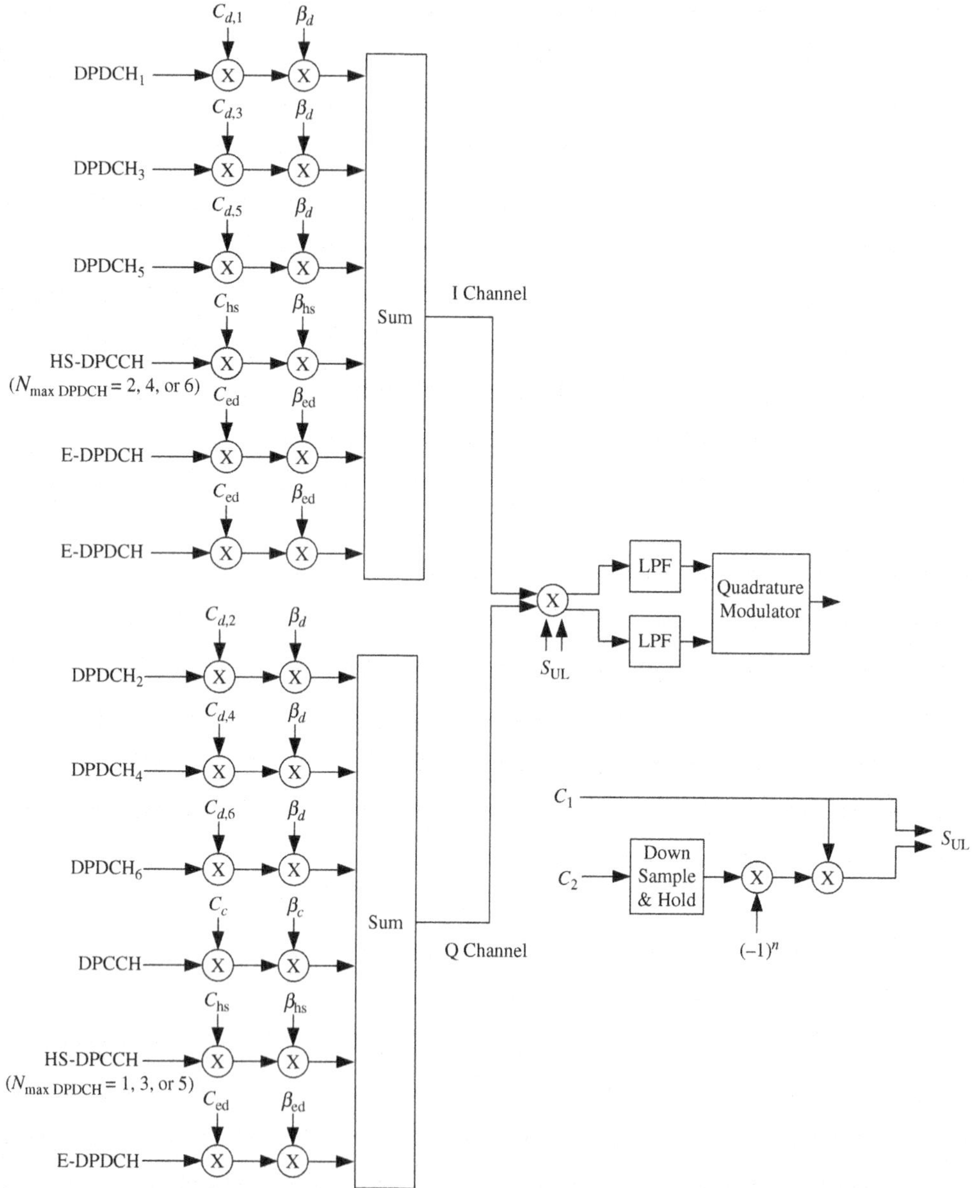

FIGURE 7.113 A simplified block diagram of the generation of uplink physical channels.

Then we simply show the first despreader output of the finger tracking the first arriving ray as

$$d_0(t) = \int_{SF} \left\{ \sum_{l=1}^{M} \sum_{k=0}^{K} d_k(t-\tau_l) \cdot C_{CH,j+k}(t-\tau_l) \cdot S_{cr}(t-\tau_l) \right\} \cdot S_{cr}^{*}(t-\tau_1) \cdot C_{CH,j}(t-\tau_1)\,dt \tag{7.151}$$

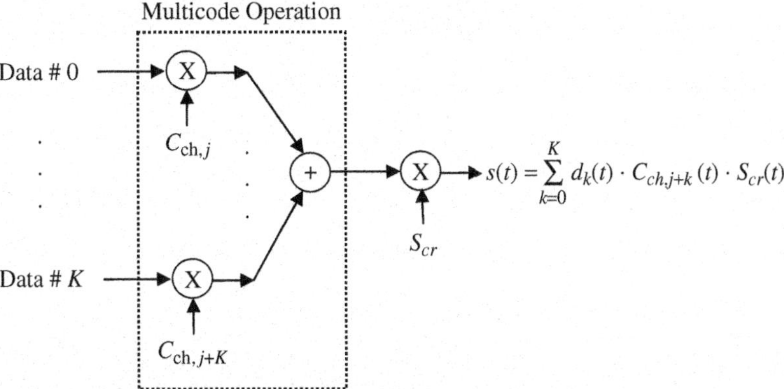

FIGURE 7.114 A multicode transmit example.

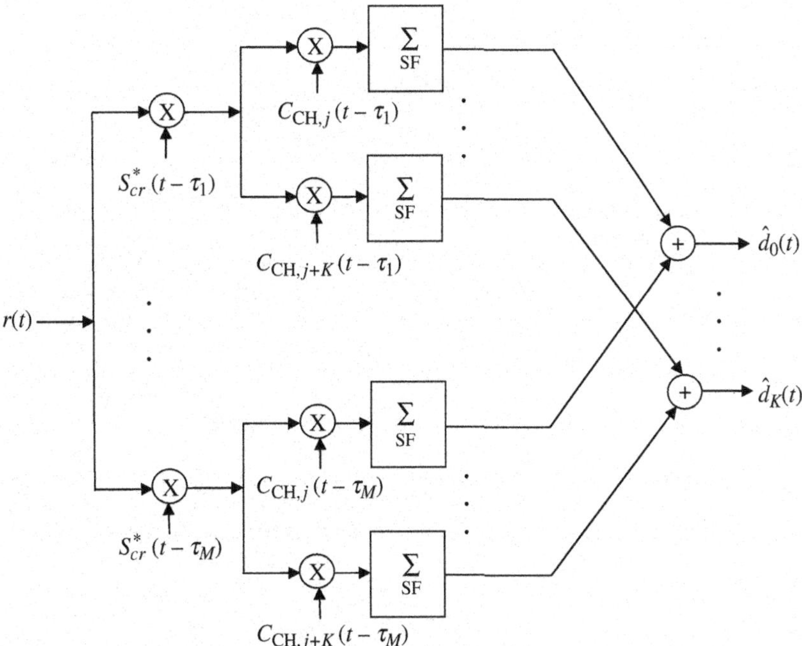

FIGURE 7.115 Block diagram of M-finger RAKE with $K + 1$ multicodes.

Let us separate the terms related to the first arriving multipath ray:

$$d_0(t) = \int_{\text{SF}} \left\{ \sum_{k=0}^{K} d_k(t-\tau_1) \cdot C_{\text{CH},j+k}(t-\tau_1) \cdot C_{\text{CH},j}(t-\tau_1) \cdot \left|S_{cr}(t-\tau_1)\right|^2 \right\} dt$$

$$+ \int_{\text{SF}} \left\{ \sum_{l=2}^{M} \sum_{k=0}^{K} d_k(t-\tau_l) \cdot C_{\text{CH},j+k}(t-\tau_l) \cdot C_{\text{CH},j}(t-\tau_1) \cdot S_{cr}(t-\tau_l) \cdot S_{cr}^*(t-\tau_1) \right\} dt \quad (7.152)$$

Next we extract the first multicode of the group.

$$d_0(t) = \int_{SF} d_0(t-\tau_1)\cdot\left|C_{CH,j}(t-\tau_1)\right|^2\cdot\left|S_{cr}(t-\tau_1)\right|^2 dt$$

$$+\int_{SF}\sum_{k=1}^{K} d_k(t-\tau_l)\cdot C_{CH,j+k}(t-\tau_1)\cdot C_{CH,j}(t-\tau_1)\cdot\left|S_{cr}(t-\tau_1)\right|^2 dt$$

$$+\int_{SF}\left\{\sum_{l=2}^{M}\sum_{k=0}^{K} d_k(t-\tau_l)\cdot C_{CH,j+k}(t-\tau_l)\cdot C_{CH,j}(t-\tau_1)\cdot S_{cr}(t-\tau_l)\cdot S^{*}_{cr}(t-\tau_1)\right\}dt \quad (7.153)$$

Hence, the finger output of the first arriving ray of the first multicode shows three terms. The first term shows the desired component; the second term shows the cross-correlation of the first multicode and the remaining codes. Under the assumption of orthogonality, this term equals zero. The third term represented the interference due to interpath interference (IPI). Recall IPI exists normally in a frequency-selective fading channel, which is exacerbated when multicodes are used.

7.7.3 Adaptive Modulation and Coding (AMC)

HSDPA uses link adaptation and physical channel retransmission in order to offer high data rates to the UE. The link adaptation uses feedback information from the UE about the present downlink channel conditions and potential throughput values to adjust the downlink resources to deliver packet data services. The turnaround or response time to the feedback information is shortened due to the smaller subframe duration (2 msec) used for the TTI. This can only be accomplished by moving the MAC functionality into the NodeB, instead of the legacy-based RNC location.

The downlink resources available to the NodeB are the number of subframes used in the downlink, the number of multicodes assigned to the UE, and lastly the modulation scheme used (QPSK, QAM). In an effort to improve throughput performance, physical channel retransmissions are supported when the UE cannot correctly decode the data packet sent to it. HSDPA allows for four transmissions before the data packet is removed from the NodeB transmit data buffer. In this case, the user would rely on the higher layers (i.e., RLC) to recover the data. A general block diagram of the HSDPA transmission is given in Fig. 7.116.

What this block diagram attempts to show is that the NodeB will use information from the UE, specifically the CQI and ACK/NACK responses, to determine the code rate and modulation scheme to be used for the next transmission. The CQI information will inform the NodeB what the best combination of modulation and code rate the UE can handle for that particular measurement interval is.

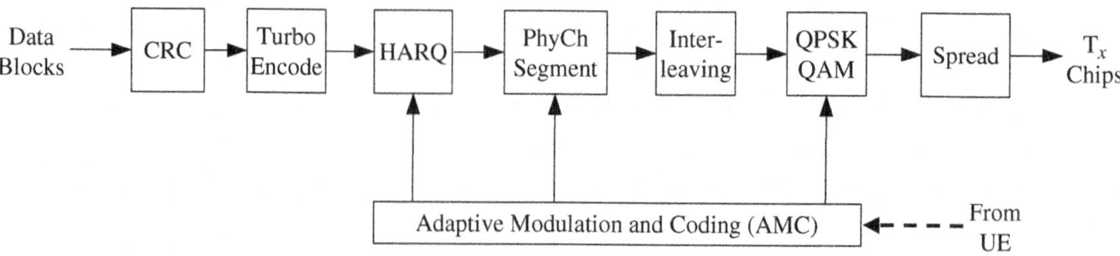

FIGURE 7.116 HSDPA block diagram.

The ACK/NACK information will inform the NodeB whether the last transmitted HS-DSCH packet was correctly/incorrectly received, respectively. The integrity of the packet data is determined by the CRC decoding. The NodeB will use this information to either retransmit the packet, so the UE can correctly receive it or adjust the HS-DSCH modulation and coding combination to deliver the data. The CQI mapping is clearly described in the 3GPP specifications. The retransmission is best described by HARQ.

There are two types of physical channel retransmissions: Chase/soft combining and Incremental Redundancy (IR). In the Chase combining case, identical bits are retransmitted. The UE stores each transmission in soft form and adds the presently received packets to the previously transmitted packet. This is done in order to improve decoding probability. In the IR case, varying parity and systematic bits are used in each retransmission in order to improve overall throughput performance. An example is given in Fig. 7.117 for illustrative purposes.

The HS-DSCH encoding procedure is given in Fig. 7.118. The transport block is appended with CRC bits and scrambled appropriately [43]. Turbo encoding with a rate = $^1/_3$ is used. The HARQ contains a virtual buffer used to vary the transmit information to the UE. This process will determine what additional systematic and parity bits get transmitted to the UE. Data is segmented into physical channels and later modulated by either QPSK or 16-QAM. Later in this chapter, we will provide throughput performance results comparing the various modulation schemes under different propagation conditions for multiple NodeB transmit power allocation scenarios. Note the HSDPA evolution discussions added higher-order modulation schemes, say 64-QAM, to the available set.

As discussed above, the HS-DSCH works in conjunction with the HS-SCCH. There is a time slot overlap between these two physical channels. The HS-SCCH is split up into two parts: The first part describes the number of multicodes and modulation schemes used, while the second part describes the redundancy version, HARQ process, transport channel block size, new data indicator, constellation order, UE identity, and channelization code set. The HS-SCCH encoding procedure is given in Fig. 7.119.

The HS-SCCH decoding performance is critical since this will affect the decoding of the HS-DSCH and possibly lower the downlink throughput. The 3GPP standard proposes various Fixed Reference Channels (FRCs) to evaluate the downlink receiver performance. The UE capability is currently placed into 16 categories (see Table 7.2).

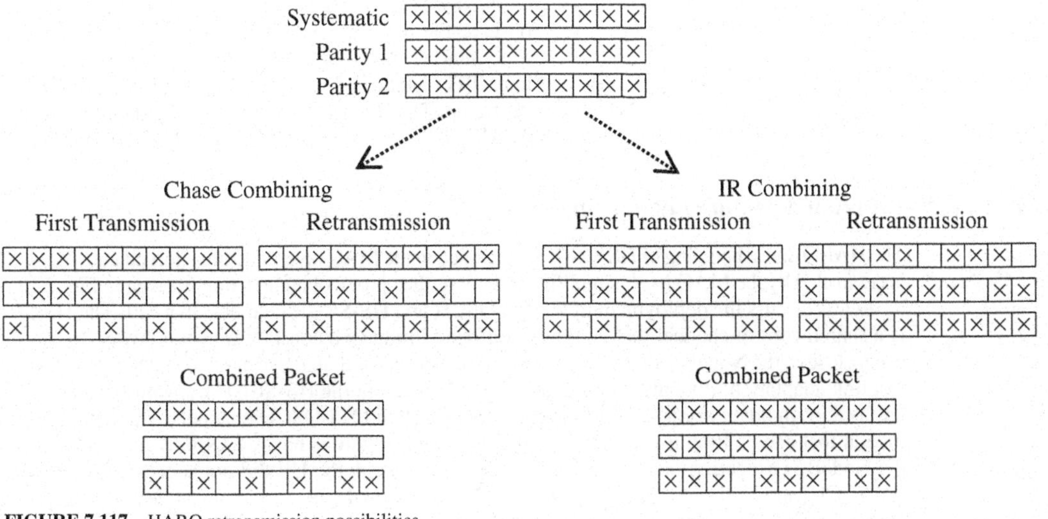

FIGURE 7.117 HARQ retransmission possibilities.

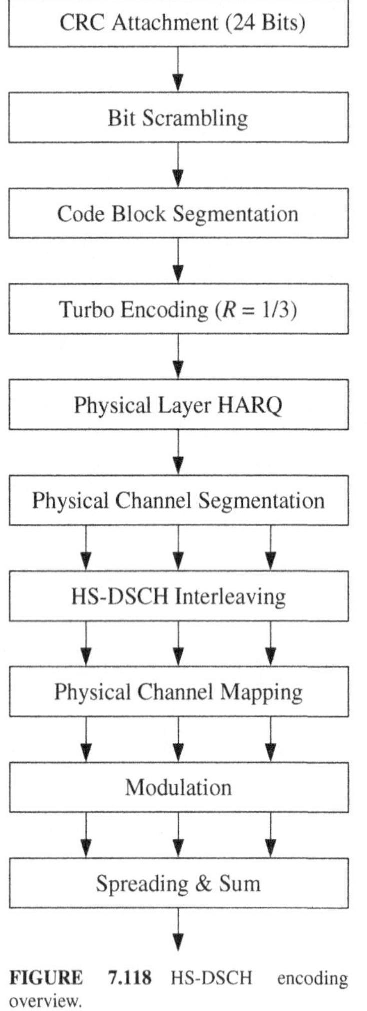

FIGURE 7.118 HS-DSCH encoding overview.

In Fig. 7.120 and Fig. 7.121, we present the H-Set 3 for Category 5/6 coding for QPSK and 16-QAM, respectively. The particular QPSK FRC example has a code rate equal to 3226/4800 and uses 5 multicodes to transmit the HS-DSCH user data.

The 16-QAM FRC example in Fig. 7.121 has a code rate equal to 4688/7680 and uses 4 multicodes to transmit the DSCH user data.

The timing relationship between the HSDPA-related physical channels and the other common and dedicated physical channels is provided in Fig. 7.122. What is important to note here is that the uplink HS-DPCCH has a certain timing relationship with the uplink DPCH. This timing relationship does not guarantee the respective time slots are time aligned.

Next we focus our attention to the HS-SCCH, HS-DSCH, and HS-DPCCH interactions. We have chosen to show the FRC H-Set 1 discussed above. This has an inter-TTI distance of 3. Also shown in Fig. 7.123 are two different processes interleaved within the retransmissions. The two time-interleaved processes are shown by dashed and gray bold boxes. For example, let's say that subframe number 0 contains the DSCH physical layer control information. The UE first decodes the HS-SCCH control information. Upon successful CRC decoding, the UE will attempt to demodulate the DSCH. The UE then has approximately 7.5 time slots to process the data and determine if an ACK or NACK should be transmitted on the uplink HS-DPCCH.

If the HS-DSCH demodulation was successful, then an ACK will be sent by the UE and the NodeB will send a new data indicator bit enabled along with new packet data in the next subframe. However, if a NACK was transmitted, then the new data bit indicator would be disabled and possibly a new redundancy version (HARQ) be applied.

Next we decreased the inter-TTI distance to 2, in order to observe the retransmission behavior. We have shown three time-interleaved processes in the figure in Fig. 7.124. An inter-TTI distance of 1 means that data is transmitted every subframe. This is used to generate more data throughput to the UE and is called H-Set 3 in the 3GPP standard.

7.7.4 NodeB HSDPA Architecture Enhancements

The HARQ and scheduling of HS-DSCH is included in the MAC layer. The transport channel HS-DSCH is controlled by the MAC-hs. A simplified radio interface protocol architecture is given in Fig. 7.125.

Let us discuss the deployment of the HS-DSCH and HS-SCCH channels in a cell. The HS-SCCH is a shared control channel that could be configured to have a constant transmit power. In this case, users close to the NodeB will have much better performance than those at the cell edge. Since we have fast link adaptation, it would be prudent to be able to adapt the HS-SCCH transmit power in a similar fashion. In other words, use TPC on the HS-SCCH and HS-DSCH channels. The feedback information that can be used is to count the number of times there is no ACK/NACK in the uplink when the NodeB is expecting one to be present. This implies that the UE did not successfully decode the HS-SCCH, and hence, its transmit power can be increased the next time the NodeB transmits to that UE. An ACK/NACK implies, although not strictly limited to, the HS-SCCH was decoded properly and the HS-DSCH demodulation failed due to poor channel conditions.

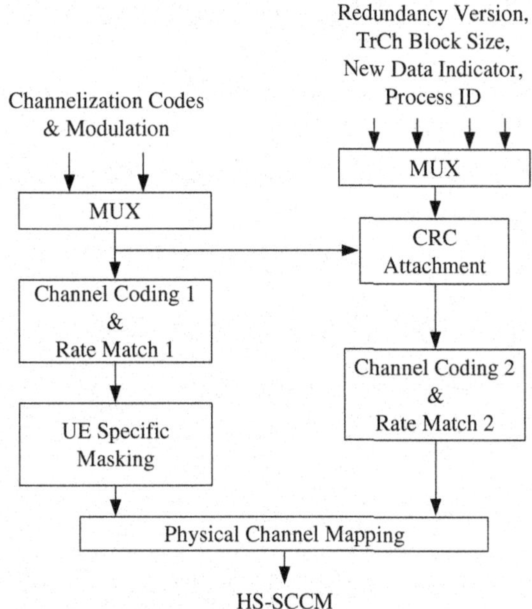

FIGURE 7.119 HS-SCCH encoding overview.

Lastly, the HS-SCCH transmit power can also be adjusted according to the CQI reported by the UE. When the CQI value is high, the channel conditions are good and the HS-SCCH can be decoded with reasonable rates. Moreover, based on the ACK/NACK signaling on the uplink, the transmit power of HS-DSCH can also be adjusted. All these power control suggestions are loosely discussed in the 3GPP specification and are left to the NodeB manufacturer's discretion [44].

Some authors proposed an inner– and outer–power control loop mechanism for the HS-DSCH and HS-SCCH channels, as shown in Fig. 7.126 [45].

TABLE 7.2 HSDPA Category Listing

HS-DSCH category	Modulation	H-Set	Max no. of HS-DSCH codes	Minimum inter-TTI interval	Data rate (QPSK/QAM), kbps
1	QPSK/16-QAM	1	5	3	534/777
2	QPSK/16-QAM	1	5	3	534/777
3	QPSK/16-QAM	2	5	2	801/1166
4	QPSK/16-QAM	2	5	2	801/1166
5	QPSK/16-QAM	3	5	1	1601/2332
6	QPSK/16-QAM	3	5	1	1601/2332
7	QPSK/16-QAM	3/6	10	1	3219/4689
8	QPSK/16-QAM	3/6	10	1	3219/4689
9	QPSK/16-QAM	NA	15	1	NA
10	QPSK/16-QAM	NA	15	1	NA
11	QPSK	4	5	2	534/NA
12	QPSK	5	5	1	801/NA
13	QPSK/16-QAM/64-QAM	3/6/8	15	1	13245
14	QPSK/16-QAM/64-QAM	3/6/8	15	1	13245
15	MIMO	9	15	1	4860/8650
16	MIMO	9	15	1	4860/8650

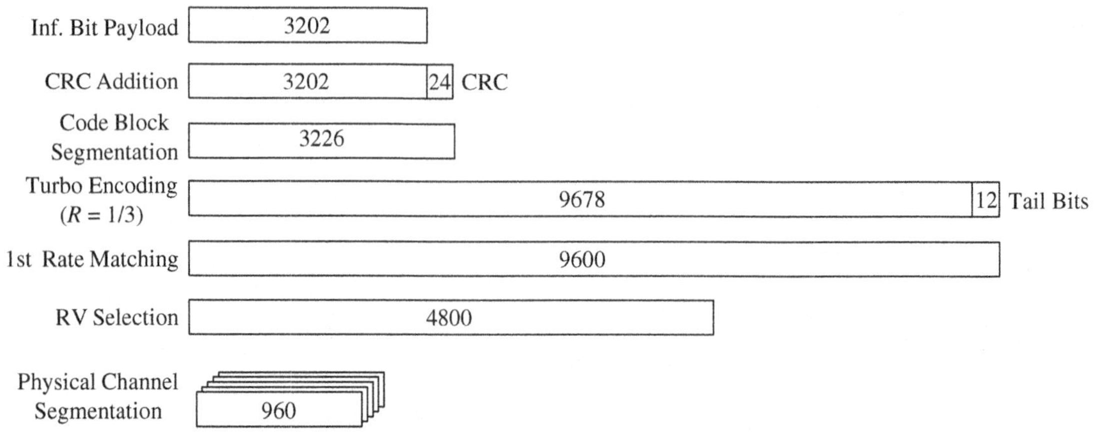

Inf. Bit Payload
3202
CRC Addition
3202
24
CRC
Code Block Segmentation
3226
Turbo Encoding (R = 1/3)
9678
12
Tail Bits
1st Rate Matching
9600
RV Selection
4800
Physical Channel Segmentation
960

FIGURE 7.120 Coding rate for FRC H-Set 3 (QPSK).

As stated earlier, if the HS-DSCH CRC fails, then a NACK will be sent on the uplink HS-DPCCH. Similarly, if the HS-DSCH CRC passes, then an ACK will be sent on the HS-DPCCH. Lastly, if the HS-SCCH decoding fails, then neither ACK nor NACK is transmitted on the HS-DPCCH. Since the NodeB may be expecting a response, it is certainly in a position to introduce time-varying power allocation schemes to improve overall system performance. The expected behavior is tabulated in Table 7.3.

The UE side has the 3GPP software architecture shown in Fig. 7.127, where we have simplified the block diagram for discussion purposes. The reader is encouraged to consult with the related 3GPP technical specifications for further information [46].

In order to provide some overview, the MAC functions have been split into three sections. The first is the MAC functionality for the dedicated channels (MAC-d), the second is the MAC functionality for the common and shared channels (MAC-c and MAC-sh). The third functionality is for the high speed channels (MAC-hs).

The NodeB side has the software architecture shown in Fig. 7.128, which has also been simplified for sake of discussion. At this point, it is necessary to show that MAC-hs functionality is performed in the NodeB while the other functions are performed in the RNCs.

Inf. Bit Payload
4664
CRC Addition
4664
24
CRC
Code Block Segmentation
4688
Turbo Encoding (R = 1/3)
14064
12
Tail Bits
1st Rate Matching
9600
RV Selection
7680
Physical Channel Segmentation
1920

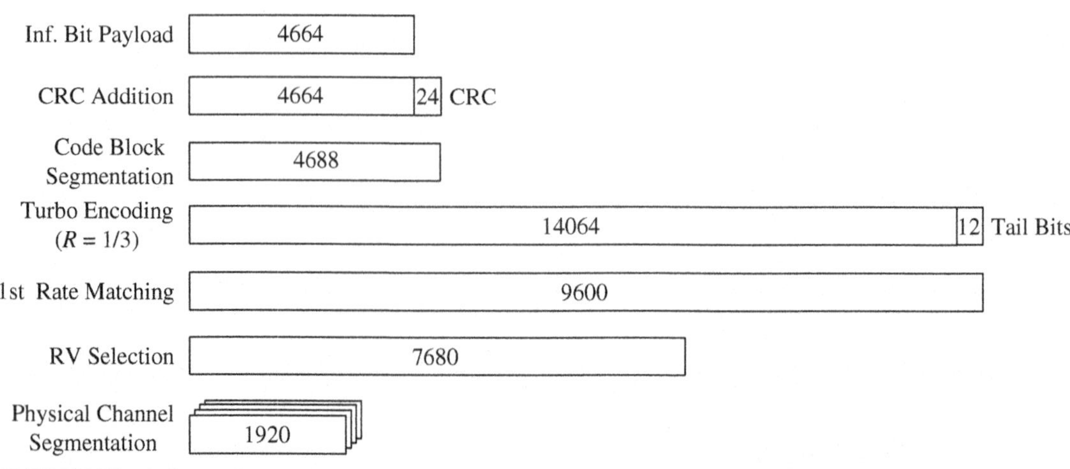

FIGURE 7.121 Coding rate for FRC H-Set 3 (16-QAM).

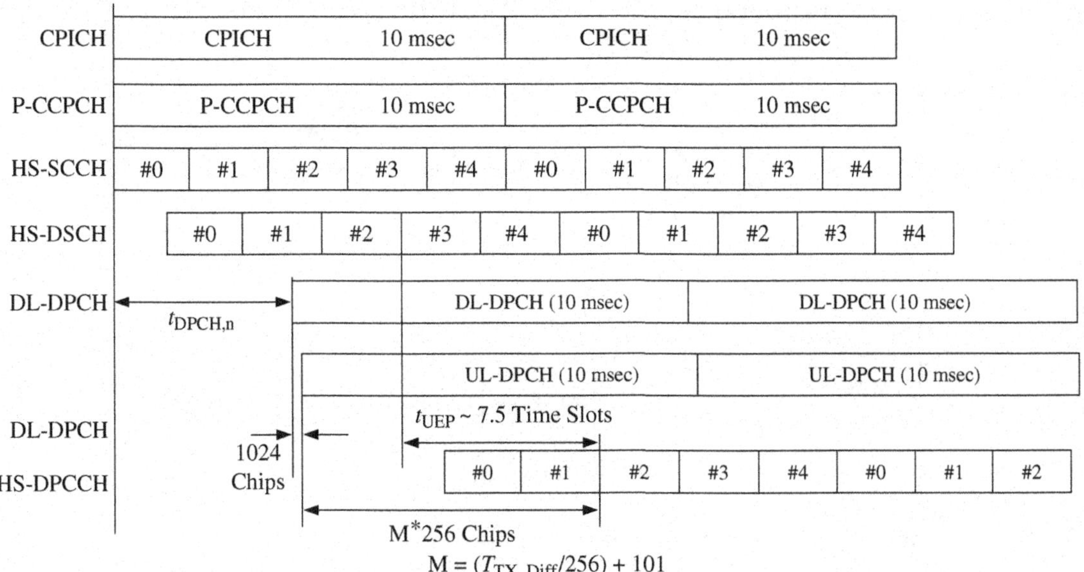

FIGURE 7.122 HSDPA timing relationships of associated physical channels.

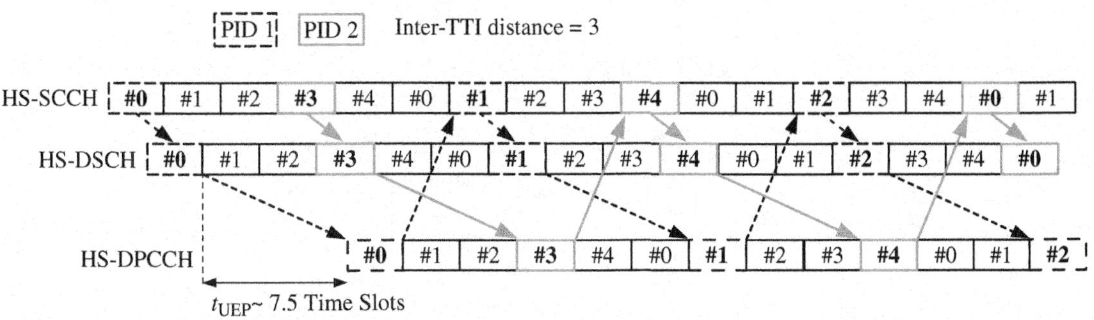

FIGURE 7.123 Inter-TTI distance of 3 timing relationship.

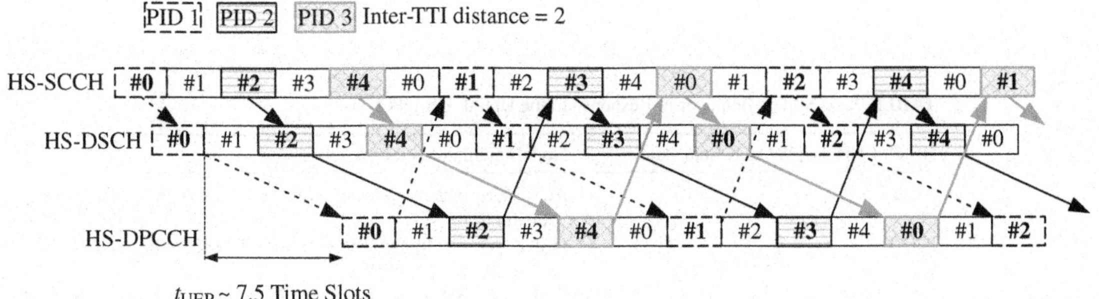

FIGURE 7.124 Inter-TTI distance of 2 timing relationship.

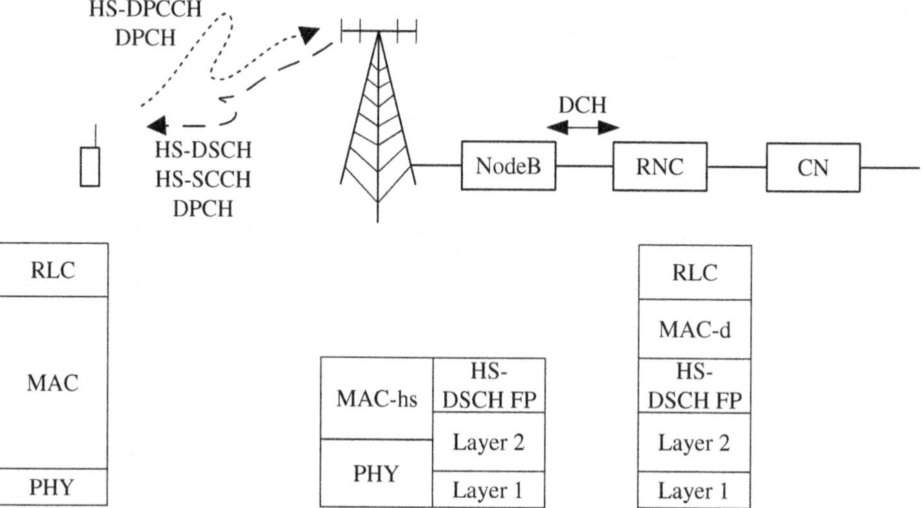

FIGURE 7.125 HS-DSCH radio interface protocol architecture.

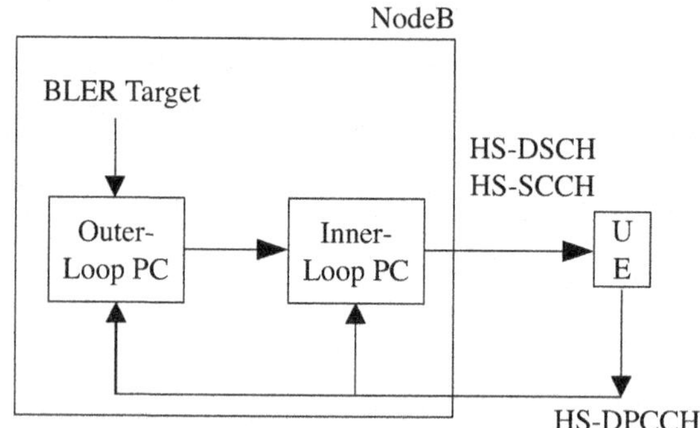

FIGURE 7.126 General power control mechanism for HSDPA-related channels.

TABLE 7.3 NodeB Behavior in Response to the Uplink HS-DPCCH

HS-DPCCH ACK/NACK field state	NodeB emulator behavior
ACK	ACK: new transmission using first redundancy and constellation version (RV)
NACK	NACK: retransmission using the next RV (up to the maximum permitted number or RVs)
DTX	DTX: retransmission using the RV previously transmitted to the same HARQ process

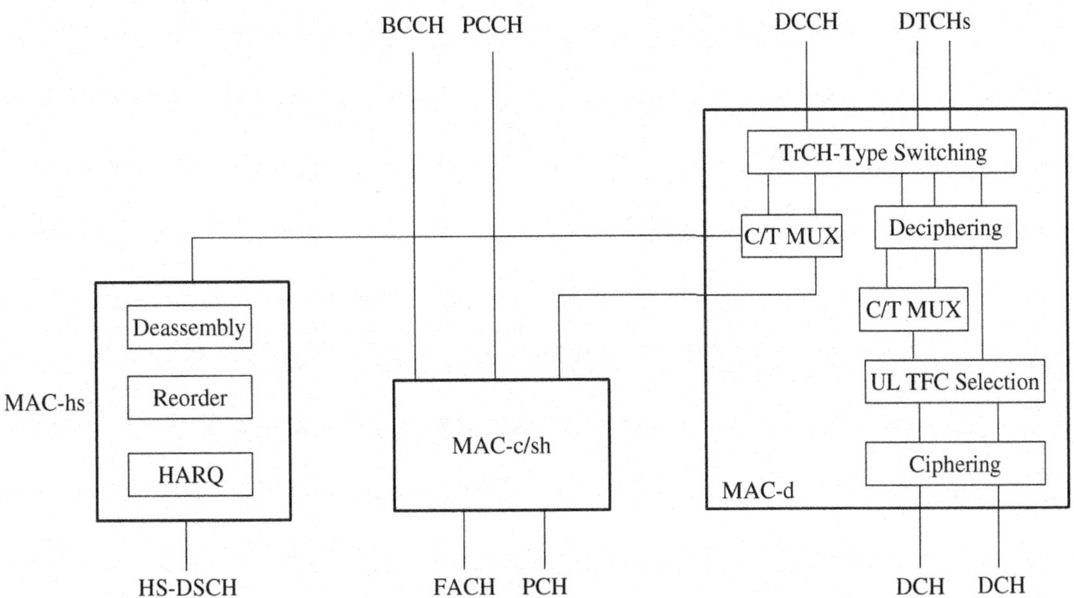

FIGURE 7.127 UE software structure.

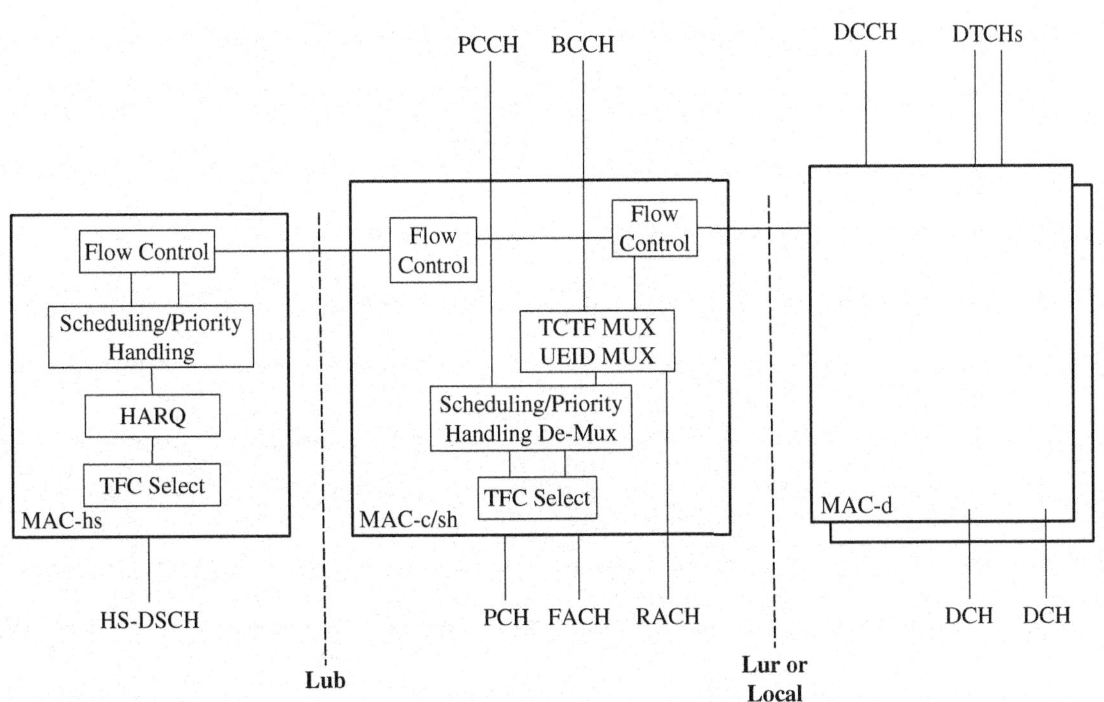

FIGURE 7.128 NodeB software architecture.

7.7.5 Throughput Performance Discussion

In this subsection, we will discuss throughput performance for a HSDPA receiver operating in various channel environments. Specifically, the 3GPP standards body has chosen to use the following four multipath channels to create the performance requirements for both the UE and the NodeB:

Pedestrian A, 3 km/hr (PA3)

Pedestrian B, 3 km/hr (PB3)

Vehicular A, 30 km/hr (VA30)

Vehicular A, 120 km/hr (VA120)

The performance results are highly dependent on the NodeB transmit power allocations: the common, shared, and dedicated channels. Here we provide an example of the downlink power allocation for the various channels.

Physical channel		E_c/I_{or}		Power allocated (%)
CPICH	=	−10 dB	=	10%
P-CCPCH	=	−12 dB	=	6.3%
SCH	=	−12 dB	=	6.3%
PICH	=	−15 dB	=	3.2%
DPCH	=	−16 dB	=	2.5%
HS-DSCH	=	Varies (−6 dB)	=	25.1%
HS-SCCH	=	Varies (−12 dB)	=	6.3%
OCNS	=	Varies (−3.3 dB)	=	46.6%
Total	=	**1**	=	**100%**

Note the other channel noise simulator (OCNS) was inserted here in order to model other users within the cell as well as to normalize the total transmit power to a constant value, say unity.

Performance curves are typically plotted against E_c/I_{or} or geometry factor (I_{or}/I_{oc}), depending on the message the reader is supposed to take away from the curves. We will provide throughput performance curves in the performance improvement section.

7.8 *HIGH-SPEED UPLINK PACKET ACCESS (HSUPA)*

In this section, we will review the HSUPA operations. Here the UE monitors the downlink channels and then decides what uplink channels should be used to transmit to achieve higher uplink throughput. With HSDPA, the link data rates are higher in the downlink than in the uplink. HSUPA attempts to equalize the uplink and downlink throughput to provide somewhat more symmetrical data rates. The term used in 3GPP is enhanced dedicated channel or E-DCH using short notation.

Similar techniques to HSDPA have been used by having fast uplink adaptation, scheduling, and multicodes transmission. The relevant operating channels are provided in the block diagram in Fig. 7.129. We have shown the UE communicating with two NodeBs, one declared as "serving" and the other as "nonserving." Note this distinction is important since certain channels and functionality are assumed for serving and nonserving cells, as will be discussed later.

Like HSDPA, HSUPA supports multicode operation. On the other hand, unlike HSDPA, HSUPA supports limited higher-order modulation on the uplink (BPSK and 16-QAM). This is simply due to the fact that more modulation levels lead to higher peak-to-average power ratios which require higher back-off in the transmit power amplifier device. Similar to the HSDPA HARQ procedure, the HSUPA procedure is such that the NodeB will combine each physical retransmission until the packet is received correctly or the maximum number of retransmissions has been reached. The UE will schedule such packet transmissions.

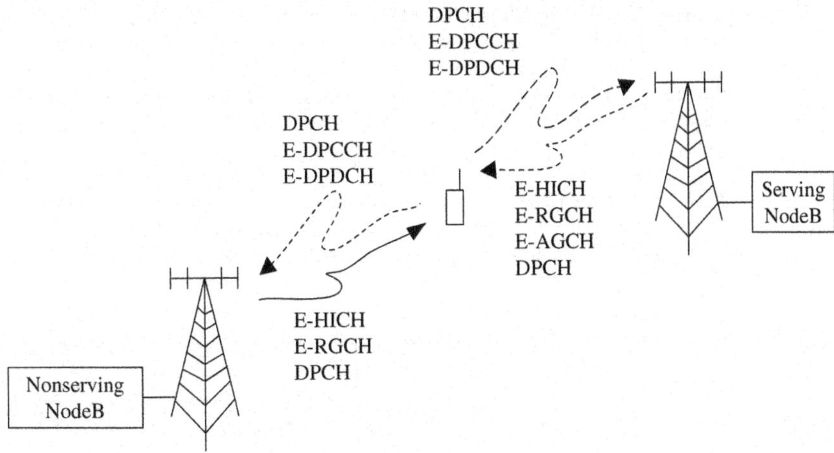

FIGURE 7.129 HSUPA service overview.

7.8.1 Physical Channels

In this section, we will discuss the HSUPA-relevant channels to support uplink packet access. A major difference is that HSUPA is not a shared channel but a dedicated channel. This implies it operates in soft handoff scenarios. The reason for this is as follows: On the downlink, the NodeB resources can be controlled and allocated to a single UE at a time (if needed), whereas the uplink resources (UE transmissions) cannot be shared and thus act very similar to the DPCH signals.

The uplink restriction is that when E-DCH is used, the maximum DCH data rate is 64 kbps. The E-DPDCH supports simultaneous transmission of two SF = 2 codes and two SF = 4 codes, which leads to a maximum physical layer bit rate of 5.76 Mbps (without higher-order modulation).

The E-DPDCH channel requires uplink DPCCH to be simultaneously transmitted to aid in CE, SIR estimation, and power control. Also E-DPDCH requires transmission of E-DPCCH to disclose the E-DPDCH format to the NodeB. The uplink timing relationship of the E-DCH and DPCH channels will be discussed later; for now, suffice it to say that they are time slot aligned.

E-DCH HARQ Indicator Channel (E-HICH). E-HICH has a fixed SF = 128 used to carry E-DCH HARQ ACK indicators. The indicator is transmitted using 3 or 12 consecutive slots which are used with E-DCH TTI set to 2 msec and 10 msec. Each slot contains 40 bits, which are set to an orthogonal sequence. The time slot and frame structure is given in Fig. 7.130.

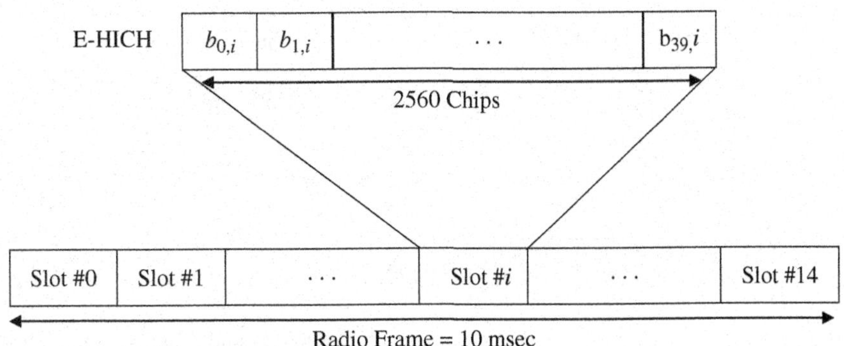

FIGURE 7.130 E-HICH time slot and frame structure.

The HARQ indicator is allowed to take on ACK or NACK values for radio link sets containing the serving E-DCH. For those cases when the radio link does not contain the serving E-DCH, the HARQ indicators can only take on ACK or DTX values. This latter case was created in order to reduce the downlink transmission power. Hence the UE will continue to retransmit until at least one cell responds with an ACK indicator. The relative timing of this physical channel will be discussed later in this section. No channel coding is applied to this channel.

E-DCH Relative Grant Channel (E-RGCH). The E-RGCH has a constant SF = 128 and is used to carry the uplink E-DCH relative grants that are transmitted using 3, 12, or 15 consecutive time slots. The time slot and frame structure is shown in Fig. 7.131.

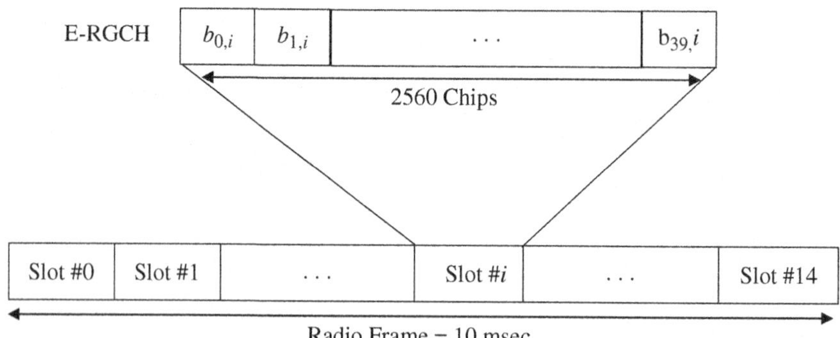

FIGURE 7.131 E-RGCH time slot and frame structure.

For Radio links in the serving E-DCH cell, use the 3 and 12 slot duration for E-DCH, TTI of 2, and 10 msec, respectively. The 15-slot duration is used for radio links not in the serving E-DCH cell. A 40-bit orthogonal sequence is transmitted in each time slot. The E-RGCH is used to signal to the UE a relative power up or down command to control the E-DPDCH transmission power. No channel coding is applied to this channel.

E-DCH Absolute Grant Channel (E-AGCH). The E-AGCH has a constant SF = 256 and is used to carry the E-DCH absolute grant. The time slot and frame structure is given in Fig. 7.132.

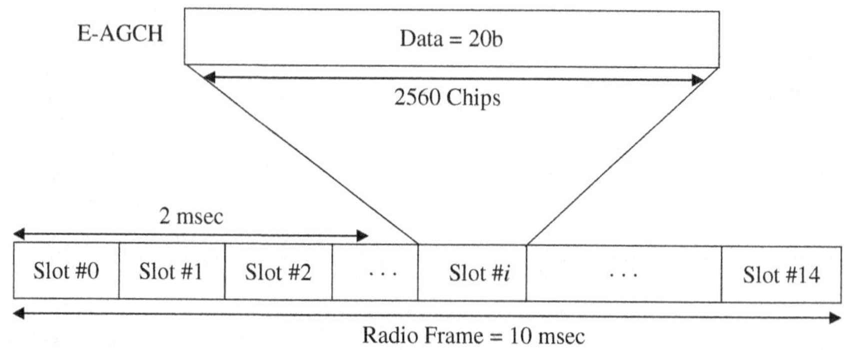

FIGURE 7.132 E-AGCH time slot and frame structure.

The absolute grant is used to tell the UE the maximum relative transmission power it is allowed to use. This relative power is E-DPDCH, with respect to DPCCH. This channel contains a 16-bit CRC and an $R = {}^1/_3$ convolutional code, and follows by rate matching. The resulting bit stream consists of 60 bits which are transmitted over 3 consecutive time slots.

E-DCH. The E-DCH consists of E-DPCCH and E-DPDCH channels. The E-DPDCH will carry the E-DCH transport channel, whereas the E-DPCCH will carry the control information. For the most part, E-DPDCH and E-DPCCH are transmitted simultaneously. E-DPCCH will not be transmitted unless DPCCH is also transmitted. The time slot and frame structure is provided in Fig. 7.133.

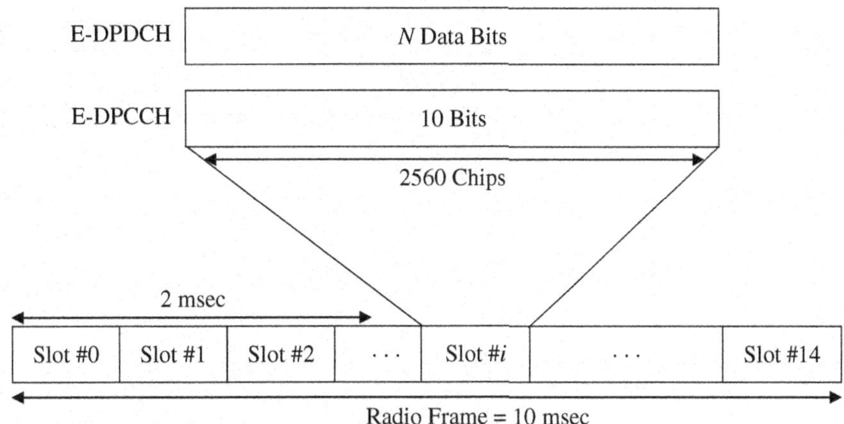

FIGURE 7.133 E-DCH time slot and frame structure.

The SF on E-DPDCH can vary from 2 up to 256, whereas the SF on E-DPCCH is constant at 256. The channelization code for E-DPCCH is always $C_{\text{CH},256,1}$. The HARQ process is similar to that presented earlier when discussing HSDPA. A rate $R = {}^1/_3$ turbo coding FEC is used.

7.8.2 Frame Timing Offsets

In this section, we will describe the frame timing relationships of the physical channels related to HSUPA services (see Fig. 7.134). Specifically, we provide the time offsets for E-HICH, E-RGCH, and E-AGCH with respect to P-CCPCH.

Channels E-RGCH, E-HICH, and E-AGCH assigned to a UE shall use the same scrambling code. The highest HSUPA data rate, without considering any higher-order modulation, is 5.76 Mbps, which is obtained by having 2 channels of SF = 4 and 2 channels of SF = 2. For an E-DCH TTI of 10 msec, the following equation applies:

$$\tau_{\text{E-HICH},n} = \tau_{\text{E-RGCH},n} = 5120 + 7680\left\lfloor \frac{(\tau_{\text{DPCH},n}/256) - 70}{30} \right\rfloor \tag{7.154}$$

and for E-DCH TTI of 2 msec, the following equation applies:

$$\tau_{\text{E-HICH},n} = \tau_{\text{E-RGCH},n} = 5120 + 7680\left\lfloor \frac{(\tau_{\text{DPCH},n}/256) + 50}{30} \right\rfloor \tag{7.155}$$

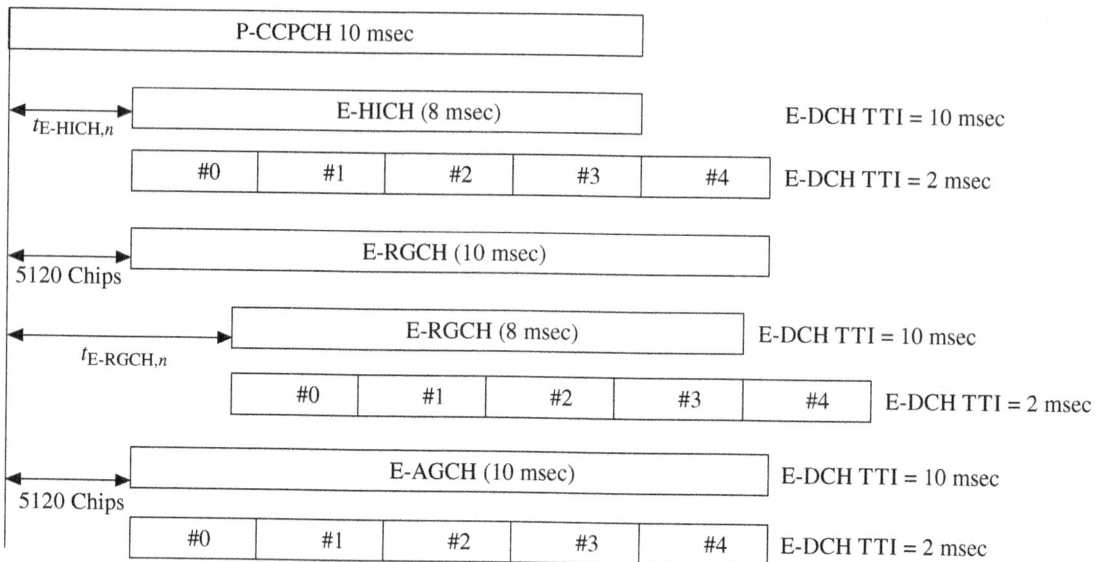

FIGURE 7.134 HSUPA timing relationship.

The possible combinations of the dedicated channels are listed in Table 7.4 (simultaneously configured). The DPCCH is always transmitted on the uplink, regardless of the service provided.

Generally speaking, the uplink DPCHs are created as in Fig. 7.135. The respective channels are spread and then summed prior to being scrambled.

The uplink consists of WCDMA-related physical channels such as DPCCH and DPDCH. It also contains HSDPA-related uplink signaling transmitted through the HS-DPCCH physical channel. Lastly, the HSUPA physical channels are E-DPCCH and E-DPDCH. We have previously drawn the uplink scrambling and spreading function in order to relate them to the HSUPA topic of interest (see Fig. 7.113).

7.8.3 NodeB Architecture Enhancements

As discussed above, in contrast to HSDPA, where only a single NodeB is communicating with various UE, HSUPA has a single UE communicating with multiple NodeBs. Therefore, HSUPA users can be involved in soft-handoff scenarios. A possible scenario can be best described with the help of Fig. 7.136.

We have the UE utilizing WCDMA + HSDPA + HSUPA services above. The HSDPA downlink channels will come from a single NodeB, whereas the HSUPA downlink channels will come from multiple NodeBs. This scenario covers a UE in a circuit-switched conversation while utilizing both the uplink and downlink packet data services.

As with HSDPA, a new MAC function was introduced to the NodeB to reduce round-trip latency. In Fig. 7.137, we show the functions in their respective positions in the WCDMA communications link. Specifically, we have drawn the UE, NodeB, and RNC interactions.

TABLE 7.4 HSDPA and HSUPA Configurations

Case	DPDCH	HS-DPCCH	E-DPDCH	E-DPCCH	Service
1	6	1	-	-	R99 + HSDPA
2	1	1	2	1	R99 + HSDPA + HSUPA
3	-	1	4	1	HSDPA + HSUPA

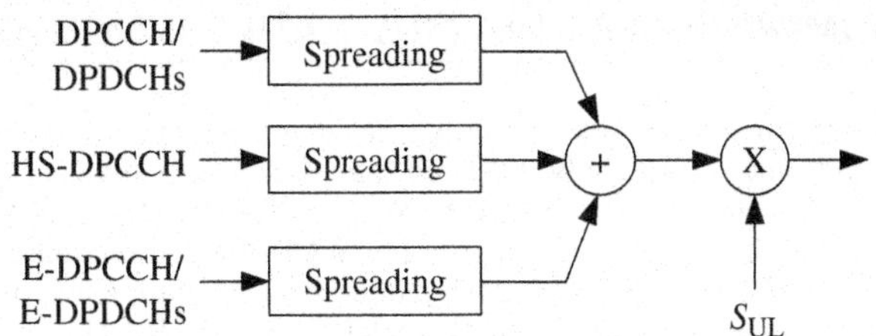

FIGURE 7.135 General block diagram of uplink dedicated channels.

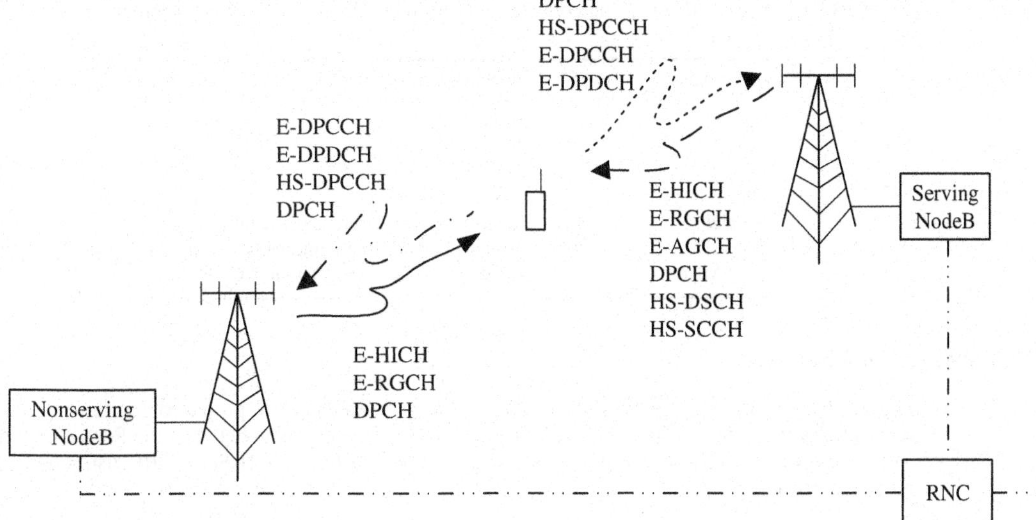

FIGURE 7.136 HSUPA and HSDPA possibilities.

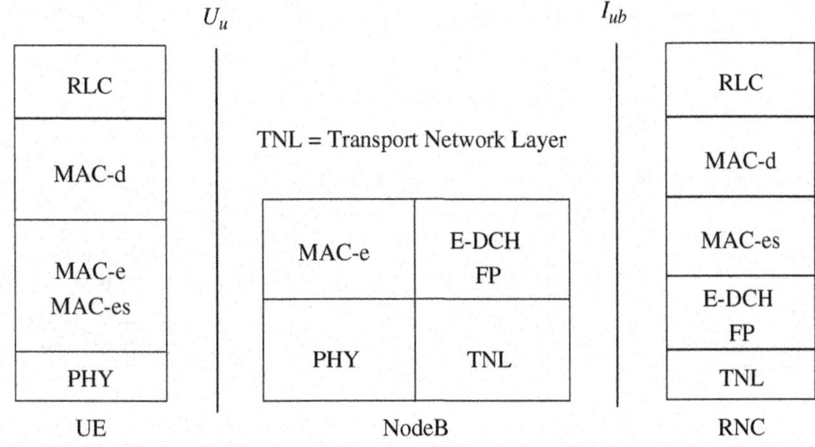

FIGURE 7.137 UE and NodeB overall software.

7.9 CAPACITY IMPROVEMENT IN A MULTIPATH ENVIRONMENT

Continuing efforts exist to improve the demodulation performance of the HS-DSCH (as well as support channels), especially at the cell edge. When a performance requirement is to be set, one needs to define a reference receiver in addition to the channel models, transport channel configuration, modulation scheme, and so forth. For WCDMA, the reference receiver is the well-known RAKE. However, for HSDPA setting the requirements became much more interesting. Naturally, the initial reference receiver for HS-DSCH was taken to be the RAKE. However, it is well known that as the SF decreases, the effects of ISI increases (as was shown earlier in this chapter) and so the RAKE receiver is clearly not the best choice [47–52].

In an effort to improve overall cell throughput and user capacity, other advanced reference receivers were investigated and used to establish performance requirements. We list the advanced receivers keeping the same naming convention as used within the 3GPP specification.

3GPP name		Reference receiver
Type 0	=	RAKE
Type 1	=	Diversity receiver (RAKE)
Type 2	=	Equalizer
Type 3	=	Diversity equalizer
Type 2i	=	Equalizer with interference awareness
Type 3i	=	Diversity equalizer with interference awareness
Type M	=	Multiple Input Multiple Output (MIMO)

7.9.1 Equalizer Requirements

For HS-DSCH, the Linear MMSE (LMMSE)-based equalizer was proposed for the reference receiver and is called a Type 2 receiver. The performance comparison between the RAKE and this Type 2 equalizer is given in Fig. 7.138 for the VA120 channel for H-Set 6 FRC using a single receive antenna.

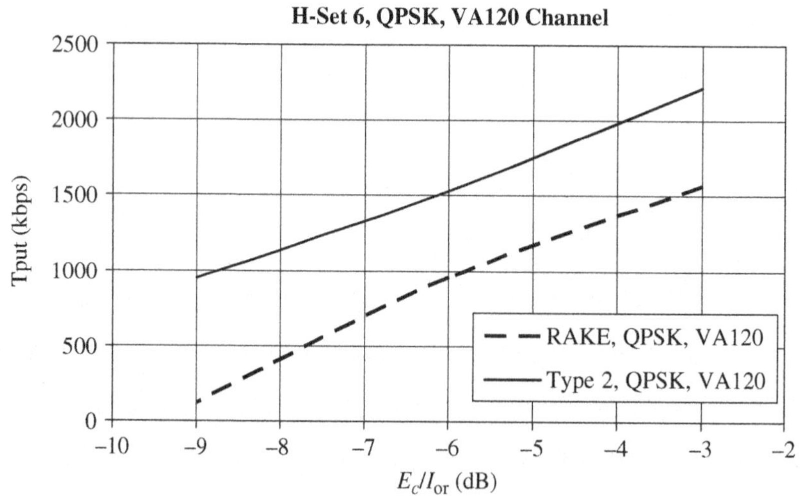

FIGURE 7.138 Equalizer and RAKE throughput performance comparison in a VA120 channel.

Next the modulation order was increased and the channel model changed to the PB3. These throughput simulation results are shown in Fig. 7.139.

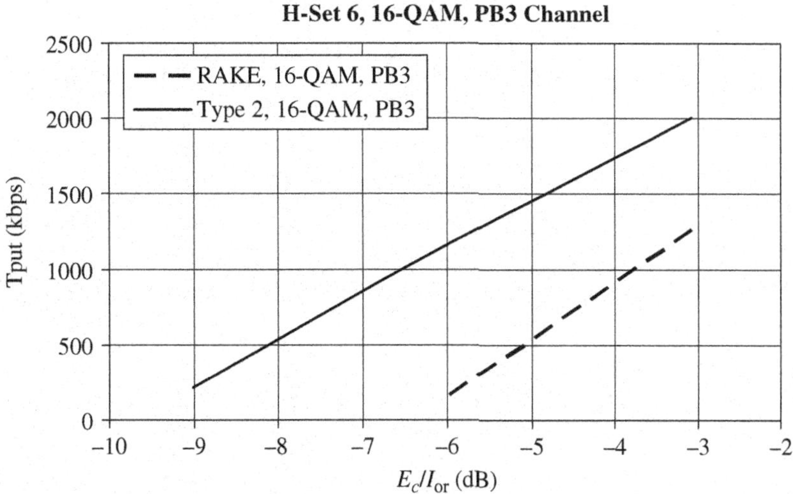

FIGURE 7.139 Equalizer and RAKE throughput performance comparison in a PB3 channel.

These results show the tremendous performance gain in using the LMMSE equalizer over the RAKE receiver. Hence additional performance requirements were defined for UE supporting this feature.

Next we attempt to show the benefits of using a spatial diversity receiver and performing a joint LMMSE equalizer across both receive antennas (Type 3) (see Fig. 7.140). Clearly the benefits of receive diversity are substantial.

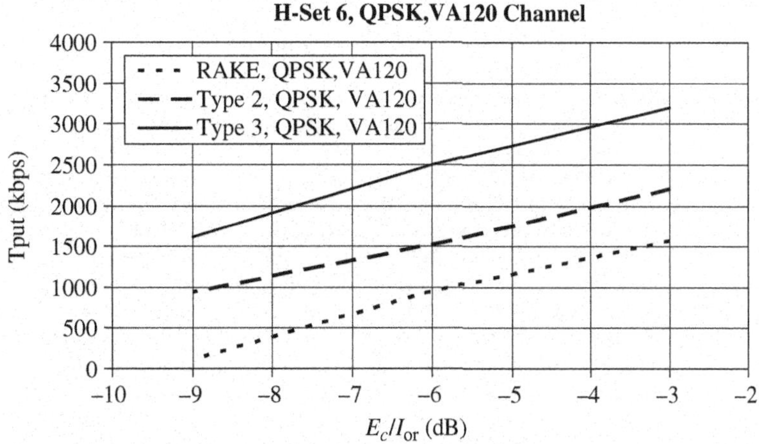

FIGURE 7.140 RAKE, Type 2, and Type 3 throughput performance results in a VA120 channel.

A point worthy of mention is that HS-DSCH doesn't operate in soft-handoff scenarios, and as such, this simplifies the equalizer structure (i.e., lower delay spread, etc.). We continued with the performance comparisons, except this time in the PB3 channel model. These simulation results are shown in Fig. 7.141.

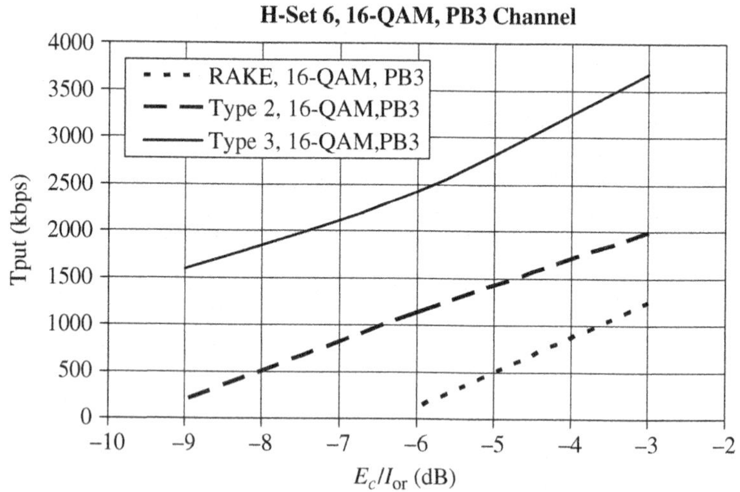

FIGURE 7.141 RAKE, Type 2, and Type 3 throughput performance results in a PB3 channel.

The implications of such improvements are polyfold. In one dimension, the UE can benefit from the additional data rate supplied, hence providing the user with faster download times and lower latencies. In another dimension, the network providers can use less transmit powers. This can allow for future services to grow, create a reduction in the out-of-cell interference, and potentially increase the cell radius, thus requiring less NodeBs to blanket a particular geographical area—not to mention possibly lowering the cost of network deployment. Also the network providers can translate this gain in throughput into an increase in capacity, that is, VOIP users. Quite frankly, these large gains eventually produce a better user experience, create larger service provider revenues, and offer a life extension to HSDPA-related service.

The last point to make in this section is that the LMMSE equalizer weights are based on a covariance matrix that consists of the desired signal's channel plus noise only. In other words, the interference terms have been ignored. Now when interference is added into the channel and the LMMSE equalizer is made aware of it, further performance improvement can be achieved. In fact, the performance improvement becomes highly dependent on the interference power profile (IPP). The Type 3i receiver creates a covariance matrix that includes the interfering channel information, thus offering performance improvement that is relative to the amount of interference present.

Simulation results comparing the Type 3 to the Type 3i are presented in Fig. 7.142, assuming perfect knowledge of the interfering channel matrix. Here we see, for large geometries, 16-QAM offers significant throughput advantages over QPSK. Also, as the geometry decreases, a crossover region exists such that QPSK is performing better than 16-QAM. Lastly, we can see that interference-aware receivers perform better at the lower geometries, where the UE is closer to the cell edge.

Similar observations can be made for the VA30 simulation results shown in Fig. 7.143. What these two performance figures have done was to show that further performance improvement can be obtained when the UE receiver is aware of interference. The gains appear to be more pronounced at the lower geometry values, which correspond to distances further away from the NodeB. In fact, this technique, in general, provides gain across the cell area, but more so at the cell edge. Regardless of

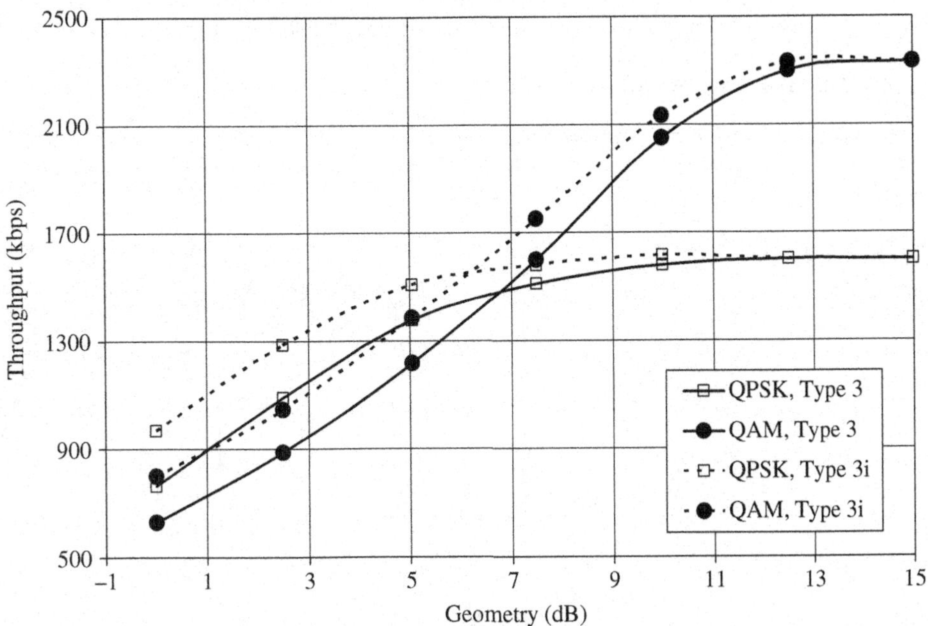

FIGURE 7.142 QPSK and 16-QAM performance comparison in a PB3 channel.

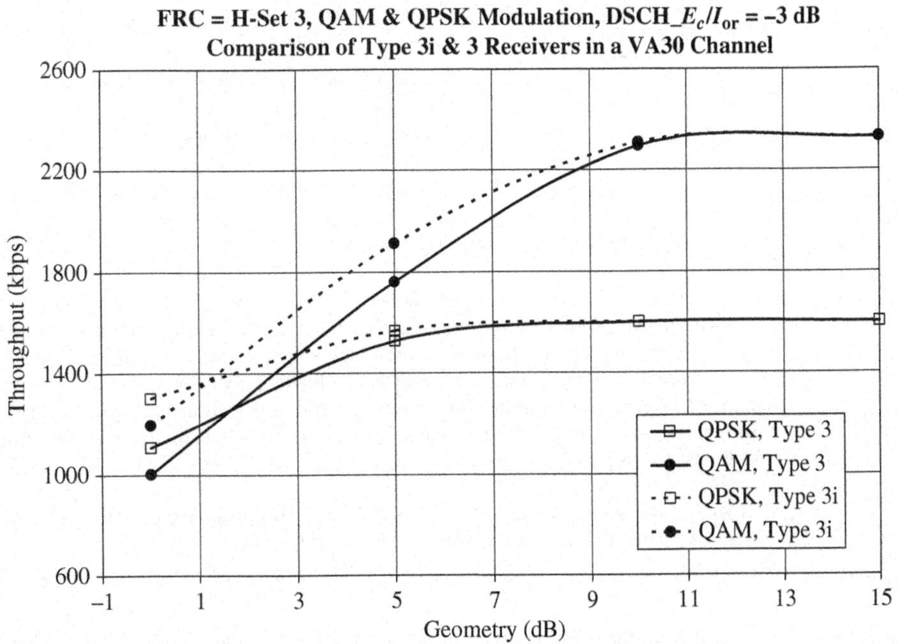

FIGURE 7.143 QPSK and 16-QAM performance comparison in a VA30 channel.

the advanced receiver technique chosen, the system or network benefits will only become visible when a significant number of the UE within the cell are using them. Having a single user operating with a type 3i receiver would negligibly impact the overall system. On the other hand, having all UE within the cell operating with a type 3i receiver would create sizable gains in the system.

7.9.2 Adaptive Antenna Arrays (AAA)

Another technique that can be used to improve performance is to increase the number of receive antennas and to optimally combine them in a per path basis. The block diagram is shown in Fig. 7.144 for a

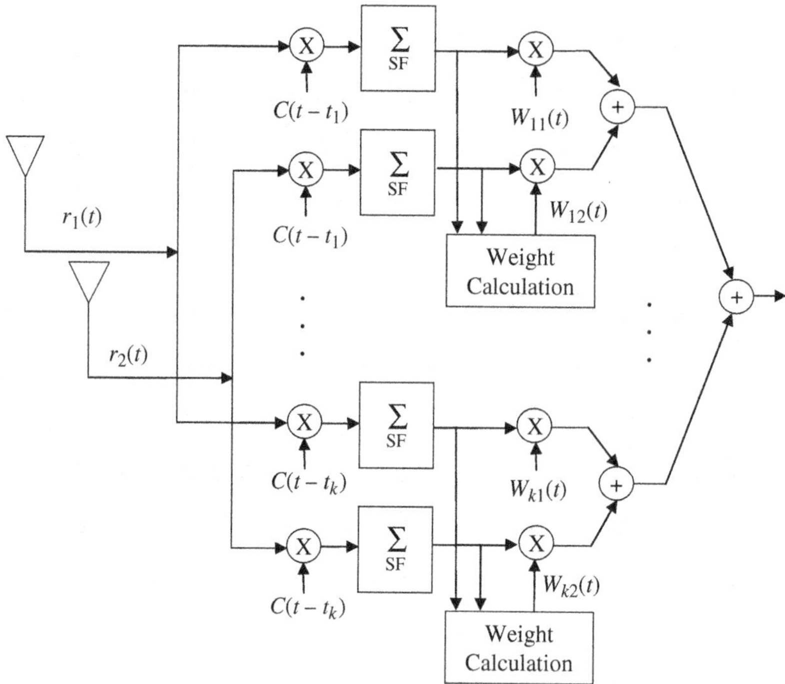

FIGURE 7.144 AAA receiver block diagram.

NodeB receiver, where we have chosen to show a dual receive diversity system capable of separately despreading *k* multipaths. Here the combining weights are jointly computed across antennas only [30].

Let us also mention that we could have shown a receiver where the combining weights were derived from the joint computation across antennas as well as multipaths. This would have resulted in a single, much larger matrix rather than the many small matrices shown above. Our point is to demonstrate many options exist.

MMSE Cost Function–based AAA receiver. The array weights that minimize the MSE were derived earlier and provided below for sake of reference.

$$\underline{w}_{\text{MMSE}} = R_{xx}^{-1} \cdot \underline{r}_{xd} \tag{7.156}$$

The BER performance comparing the AAA receiver versus the conventional MRC receiver is given in Fig. 7.145. A point we should mention is that we have assumed the NodeB receiver contains a data

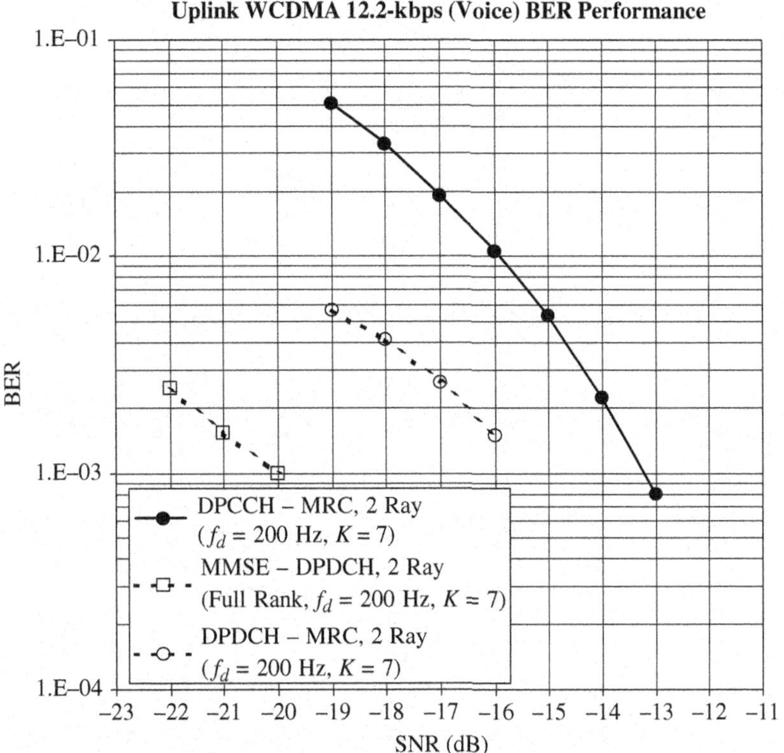

FIGURE 7.145 AAA receiver BER performance.

buffer storing approximately 10 msec of the received signal. This allows us to use noncausal signal processing to estimate the channel response and decode the TFCI bits to determine the DPDCH physical channel information. With these assumptions, we see that using the noncausal MMSE combing weights derived directly from the DPDCH channel offers significant BER performance improvement.

7.9.3 Link Budget Example

In this section we will present a link budget for both an Indoor and Outdoor scenario. This will concentrate on the specific location of the UE. We have always assumed the location of the NodeB to be outdoors thus far.

Indoors Case Study. The indoor link budget is provided in Table 7.5. We will briefly describe the parameters as well as the values used in the budget. For this indoor scenario we have included an outer wall penetration loss of 10 dB assuming the NodeB is located outside. We have considered a 3-dB noise rise due to other users occupying the same channel. Also a UE noise figure of 7 dB was chosen for the calculations [45, 53, 54]. Various implementation losses, such as fixed point implementation and power control rise, were lumped together. We have further assumed a handoff gain of 1 dB.

Outdoors Case Study. For this outdoor scenario, we have assumed the NodeB is located outside. We have assumed a noise rise of 3 dB due to other users and a handoff gain of 3 dB.

In comparing the above indoor and outdoor case studies, we purposely kept the E_b/I_o to be approximately 5 dB for both scenarios. This, coupled with the other parameters such as Handoff gain,

TABLE 7.5 Indoor Link Budget Example

a	Transmit Power	21	dBm	125 mWatts
b	Transmit Cable Loss	2	dB	
c	Transmit Antenna Gain	18	dBi	
d	Receiver Antenna Gain	0	dBi	
e	Receiver Cable Loss	0	dB	
f	Thermal Noise Density	−174	dBm/Hz	
g	Interference Density	−174	dBm/Hz	
h	Total Inter + Noise Density	−170.99	dBm/Hz	
i	Receiver Noise Figure	7	dB	
j	System Bandwidth	65.84331	dB	
k	Handoff Gain	1	dB	
l	Lognormal Fading Margin	7	dB	
m	Propagation Path Loss	134	dB	
n	Building Penetration Loss	10	dB	
o	Various Implementation Loss	3	dB	Implementation
p	**Rx Signal Power**	**−116**	dBm	a−b+c+d−e+k−l−m−n−o
q	$E_b/(I_o + N_o)$	4.988487	dB	
r	Processing Gain	21.0721	dB	
s	**Interference + Noise Floor**	**−98.1464**	dBm	h+i+j
t	$\mathbf{E_c/(I_o + N_o)}$	**−17.8536**	dB	p−s
u	**Sensitivity**	**−118.138**	dBm	h+i+q+v
v	User Data Rate (12.2 KBps)	40.8636	dB	

building penetration loss, and various implementation and system losses, forces the indoor scenario to support a 134-dB path loss versus a 143-dB path loss for the outdoor case. Note that these values are valid for the assumptions used to derive them; certainly, debates over the specific values can and should occur. However, it is not the intention Tables 7.5 and 7.6 to be all inclusive; rather they are for

TABLE 7.6 Outdoor Link Budget Example

a	Transmit Power	21	dBm	125 mWatts
b	Transmit Cable Loss	2	dB	
c	Transmit Antenna Gain	18	dBi	
d	Receiver Antenna Gain	0	dBi	
e	Receiver Cable Loss	0	dB	
f	Thermal Noise Density	−174	dBm/Hz	
g	Interference Density	−174	dBm/Hz	
h	Total Inter + Noise Density	−170.99	dBm/Hz	
i	Receiver Noise Figure	7	dB	
j	System Bandwidth	65.84331	dB	
k	Handoff Gain	3	dB	
l	Lognormal Fading Margin	7	dB	
m	Propagation Path Loss	143	dB	d = 1.5 km
n	Building Penetration Loss	0	dB	
o	Various Implementation Loss	6	dB	Implementation + Car Loss
p	**Rx Signal Power**	**−116**	dBm	a−b+c+d−e+k−l−m−n−o
q	$E_b/(I_o+N_o)$	4.988487	dB	
r	Processing Gain	21.0721	dB	
s	**Interference + Noise Floor**	**−98.1464**	dBm	h+i+j
t	$\mathbf{E_c/(I_o+N_o)}$	**−17.8536**	dB	p−s
u	**Sensitivity**	**−118.138**	dBm	h+i+q+v
v	User Data Rate (12.2 KBps)	40.8636	dB	

informative purposes, where the values can be changed depending on the deployment scenario, service providers, and UE manufacturers. One can quickly vary parameters such as path loss and noise rise to derive a first-order approximation to the cell's capacity, either number of users or normalized throughput (bps/Hz).

REFERENCES

[1] J. G. Proakis, *Digital Communications*, McGraw-Hill, 1989, New York.

[2] B. Sklar, *Digital Communications: Fundamental and Applications*, Prentice Hall, 1988, New Jersey.

[3] G. L. Turin, "Introduction to Spread-Spectrum Antimultipath Techniques and Their Application to Urban Digital Radio," *Proceedings of the IEEE,* Vol. 68, No. 3, March 1980, pp. 328–353.

[4] S. G. Glisic and B. Vucetic, *Spread Spectrum CDMA Systems for Wireless Communications*, Artech House, 1997, Massachusetts.

[5] A. Papoulis, *Signal Analysis*, McGraw-Hill, 1977, New York.

[6] IS-95 CDMA Spread Spectrum Digital Cellular Standard.

[7] G. E. Bottomley, T. Ottosson, and Y-P. E. Wang, "A Generalized RAKE Receiver for Interference Suppression," *IEEE Journal on Selected Areas in Communications,* Vol. 18, No. 8, Aug. 2000, pp. 1536–1545.

[8] J. Boccuzzi, S. U. Pillai, and J. H. Winters, "Adaptive Antenna Arrays Using Subspace Techniques in a Mobile Radio Environment with Flat Fading and CCI," *IEEE Vehicular Technology Conference,* 1999, pp. 50–54.

[9] X. Wu and A. M. Haimovich, "Adaptive Arrays for Increased Performance in Mobile Communications," in: *IEEE Personal, Indoor and Mobile Radio Communications Conference,* 1995, pp. 653–657.

[10] W. W. Peterson and E. J. Weldon, Jr., *Error-Correcting Codes*, The MIT Press, 1988, Massachusetts.

[11] J. S. Lee and L. E. Miller, *CDMA Systems Engineering Handbook*, Artech House, 1998, Massachusetts.

[12] D. V. Sarwate and M. B. Pursley, "Cross-correlation Properties of Pseudorandom and Related Sequences," *IEEE Proceedings,* Vol. 68, No. 5, May 1980, pp. 593–619.

[13] A. M. D. Turkmani and U. S. Goni, "Performance Evaluation of Maximal-Length, Gold and Kasami Codes as Spreading Sequences in CDMA Systems," *ICUPC,* 1993, pp. 970–974.

[14] www.3gpp.org

[15] 3GPP Technical Specification (TS) 43.051, "Radio Access Network: Overall Description—Stage 2," Release 7.

[16] 3GPP Technical Specification (TS) 25.213, "Spreading and Modulation," Release 7.

[17] 3GPP Technical Specification (TS) 25.401, "UTRAN Overall Description," Release 7.

[18] 3GPP Technical Specification (TS) 25.211, "Physical Channels and Mapping of Transport Channels on Physical Channels (FDD), Release 7.

[19] K. Laird, N. Whinnett, and S. Buljore, "A Peak-to-Average Power Reduction Method for Third Generation CDMA Reverse Links," in: *IEEE Vehicular Technology Conference,* 1999, pp. 551–555.

[20] N. Binucci, E. Hepsaydir, and E. Candy, "UE Receive Diversity as a Performance Enhancement to 3GPP Rel99, Rel5, and Rel6 Radio Features," in: *IEE International Conference on 3G Mobile Communication Technologies,* 2004, pp. 39–43.

[21] K. Higuchi, M. Sawahashi, and F. Adachi, "Fast Cell Search Algorithm in DS-CDMA Mobile Radio Using Long Spreading Codes," in: *IEEE Vehicular Technology Conference,* 1997, pp. 1430–1434.

[22] K. Higuchi, Y. Hanada, M. Sawahashi, and F. Adachi, "Experimental Evaluation of 3-Step Cell Search Method in W-CDMA Mobile Radio," *IEEE Vehicular Technology Conference,* 2000, pp. 303–307.

[23] F. Adachi, M. Sawahashi, and H. Suda, "Wideband DS-CDMA for Next Generation Mobile Communications Systems," *IEEE Communications Magazine,* Sept. 1998, pp. 56–69.

[24] J. Iinatti, "Comparison of Two Dwell Code Acquisition of DS Signal Using Different Threshold Setting Rules," in: *IEEE MILCOM Conference,* 1997, pp. 296–301.

[25] H. Hamada, M. Nakamura, T. Kubo, M. Minowa, Y. Oishi, "Performance Evaluation of The Path Search Process For the W-CDMA System," *IEEE Vehicular Technology Conference,* 1999, pp. 980–984.

[26] P. Lancaster and M. Tismenetsky, *The Theory of Matrices*, Academic Press, 1985, California.

[27] F. M. Gardner, "Interpolation in Digital Modems Part I: Fundamentals," *IEEE Transactions on Communications*, Vol. 41, No. 3, March 1993, pp. 501–507.

[28] H. Andoh, M. Sawahashi, and F. Adachi, "Channel Estimation Using Time Multiplexed Pilot Symbols for Coherent RAKE Combining for DS-CDMA Mobile Radio," in: *IEEE Personal, Indoor and Mobile Radio Communications Conference*, 1997, pp. 954–958.

[29] J. Boccuzzi and S. U. Pillai, US Patent #6,778,514, "Sub-Space Combining of Multi-Sensor Output Signals."

[30] J. Boccuzzi, "Adaptive Antenna Arrays for 3rd Generation Digital Cellular Systems," Ph.D. Dissertation, Polytechnic University of New York, 2004.

[31] M. B. Pursley, "Performance Evaluation for Phase-Coded Spread-Spectrum Multiple-Access Communication-Part I: System Analysis," *IEEE Transactions on Communications*, Vol. COM-25, No. 8, Aug. 1977, pp. 795–799.

[32] M. K. Simon, "Noncoherent Pseudonoise Code Tracking Performance of Spread Spectrum Receivers," *IEEE Transactions on Communications*, Vol. COM-25, No. 3, Mar. 1977, pp. 327–345.

[33] M. Sawahashi, F. Adachi, and H. Yamamoto, "Coherent Delay-Locked Code Tracking Loop Using Time-Multiplexed Pilot for DS-CDMA Mobile Radio," *IEICE Transactions on Communications*, Vol. E81-B, No. 7, July 1998, pp. 1426–1432.

[34] H. Kawai, H. Suda, and F. Adachi, "Outer-Loop Control of Target SIR for Fast Transmit Power Control in Turbo-coded W-CDMA Mobile Radio," *Electronics Letters*, April 1999, Vol. 35, No. 9, pp. 699–701.

[35] S. Niida, T. Suzuki, and Y. Takeuchi, "Experimental Results of Outer-Loop Transmission Power Control using Wideband CDMA for IMT-2000," *IEEE Vehicular Technology Conference*, 2000, pp. 775–779.

[36] C. S. Koo, S. H. Shin, R. A. DiFazio, D. Grieco, and A. Zeira, "Outer Loop Power Control Using Channel Adaptive Processing for 3G WCDMA," in: *IEEE Vehicular Technology Conference*, 2003, pp. 490–494.

[37] L. M. A. Jalloul, M. Kohlmann, and J. Medlock, "SIR Estimation and Closed Loop Power Control for 3G," *IEEE*, 2003, pp. 831–835.

[38] B. Kim, S. L. Kwon, B. J. Choi, H. G. Park, "Performance Analysis of TFCI Coding in Rayleigh Fading Channel for WDMA Systems," in: *IEEE TENCON Conference*, 2004, pp. 489–492.

[39] 3GPP Technical Specification (TS) 25.214, "Physical Layer Procedures (FDD)," Release 7.

[40] F. Adachi, K. Ohno, A. Higashi, T. Dohi, and Y. Okumura, "Coherent Multicode DS-CDMA Mobile Radio Access," *IEICE Transactions on Communications*, Vol. E79-B, No. 9, Sept. 1996, pp. 1316–1325.

[41] 3GPP Technical Specification (TS) 25.899, "High Speed Download Packet Access (HSDPA) Enhancements," Release 7.

[42] A. Baier et al. "Design Study for a CDMA-Based Third-Generation Mobile Radio System," *IEEE Journal on Selected Areas in Communications*, Vol. 12, No. 4, May 1994, pp. 733–743.

[43] 3GPP Technical Specification (TS) 25.212, "Multiplexing and Channel Coding," Release 7.

[44] R. Ratasuk, W. Xiao, A. Ghosh, N. Whinnett, and F. Wang, "Power Control of the High Speed Shared Control Channel," in: *IEEE Vehicular Technology Conference*, 2005, pp. 2449–2453.

[45] H. Holma and A. Toskala, *WCDMA for UMTS*, John Wiley & Sons, 2001, England.

[46] 3GPP Technical Specification (TS) 25.301, "Radio Interface Protocol Architecture," Release 7.

[47] M. Harteneck, M. Boloorian, S. Georgoulis, R. Tanner, "Practical Aspects of an HSDPA 14MBps Terminal," *Proceedings of 38th Asilomar Conference on Signals, Systems and Computers*, 2004, pp. 799–803.

[48] M. J. Heikkila and K. Majonen, "Increasing HSDPA Throughput by Employing Space-Time Equalization," in: *IEEE Personal, Indoor and Mobile Radio Communications Conference*, 2004, pp. 2328–2332.

[49] M. Harteneck and C. Luschi, "Practical Implementation Aspects of MMSE Equalization in a 3GPP HSDPA Terminal," in: *IEEE Vehicular Technology Conference*, 2004, pp. 445–449.

[50] T. Nihtila, J. Kurjenniemi, M. Lampinen, and T. Ristaniemi, "WCDMA HSDPA Network Performance with Receive Diversity and LMMSE Chip Equalization," *IEEE Personal, Indoor and Mobile Radio Communications Conference*, 2005, pp. 1245–1249.

[51] M. J. Heikkila, P. Komulainen, and J. Lilleberg, "Interference Suppression in CDMA Downlink through Adaptive Channel Equalization," in: *IEEE Vehicular Technology Conference*, 1999, pp. 978–982.

[52] Y. Iizuka, T. Nakamori, H. Ishii, S. Tanaka, S. Ogawa, and K. Ohno, "Field Experiment Results of User Throughput Performance in WCDMA HSDPA," *IEEE Personal, Indoor and Mobile Radio Communications Conference,* 2005, pp. 346–351.

[53] K. Sipila, M. Jasberg, J. L. Steffens, A. Wacker, "Soft Handover Gains in Fast Power Controlled WCDMA Uplink," in: *IEEE Vehicular Technology Conference,* 1999, pp. 1594–1598.

[54] S. Parkvall, E. Dahlman, P. Frenger, P. Beming, and M. Persson, "The High Speed Packet Data Evolution of WCDMA," in: *IEEE Personal, Indoor and Mobile Radio Communications Conference,* 2001, pp. G.27–G.31.

CHAPTER 8

COMPUTER SIMULATION ESTIMATION TECHNIQUES

Digital communication systems use some form of error rate as the figure of merit of how well the overall system is performing. This error rate can take on various forms such as bit error rate (BER), symbol error rate (SER), and frame error rate (FER). This chapter will present five error estimation techniques that can be used to evaluate system performance. We begin with the commonly used Monte Carlo (MC) technique, which essentially counts errors in the receiver. This technique makes no assumptions on the noise visible to the receiver, but can have a prohibitively long simulation run time when the BER of interest is extremely small. The estimation techniques that follow aim to reduce the long computer simulation run time.

The Conventional Importance Sampling (CIS) or Modified MC (MMC) techniques increase the likelihood that semirare events occur, thus reaching the expected error rate sooner, with our desired confidence level. This method was further enhanced with Improved Importance Sampling (IIS) which made those semirare events less rare. Next, the Tail Extrapolation (TE) technique, which makes certain assumptions regarding tail distributions, is used. Lastly, the Semi-Analytic (SA) technique, which combines both simulations and closed-form analytical solutions, is presented.

8.1 INTRODUCTION TO SIMULATION

Computer simulation has grown to become an integral part of the digital communication system design. Depending on the desired output, various degrees of system level abstractions exist that can be used in the performance investigations. A generic digital communication link is shown in Fig. 8.1.

The computer simulation is aware of what was transmitted. This information is delayed so that the received and transmitted bits (or symbols) are time aligned and bit-by-bit comparisons can be made. The diagram in Fig. 8.1 shows which points to view to aid in the calculation of the uncoded and coded BER. The uncoded error rate will provide a means to quantify the performance of the demodulation algorithms, while the coded error rate will be used to test the FEC technique. The noise and interference can be varied to control the SNR and SIR, respectively. Also the multipath channel is used to evaluate system performance in different delay spread environments, deployment scenarios, as well as different Doppler spreads.

Since we will be focusing our efforts on estimating the BER in the receiver, we wish to present the system of interest shown in Fig. 8.2 [1, 2].

Simply stated, the received signal $v(t)$ consists of the transmitted signal $s(t)$ perturbed by additive noise $n(t)$. For sake of discussion, we will assume binary signaling is used, and hence, we can describe hypothetical, probability density functions (PDFs) of the received signal corrupted by noise. These statistical functions are given in Fig. 8.3.

We have used $f_0(v)$ to represent the PDF of the received signal given a "0" that was transmitted. Similarly, $f_1(v)$ is the PDF of the received signal given a "1" was transmitted. A decision threshold labeled v_{th} is also shown to be used in the receiver to determine whether a 1 or a 0 was transmitted.

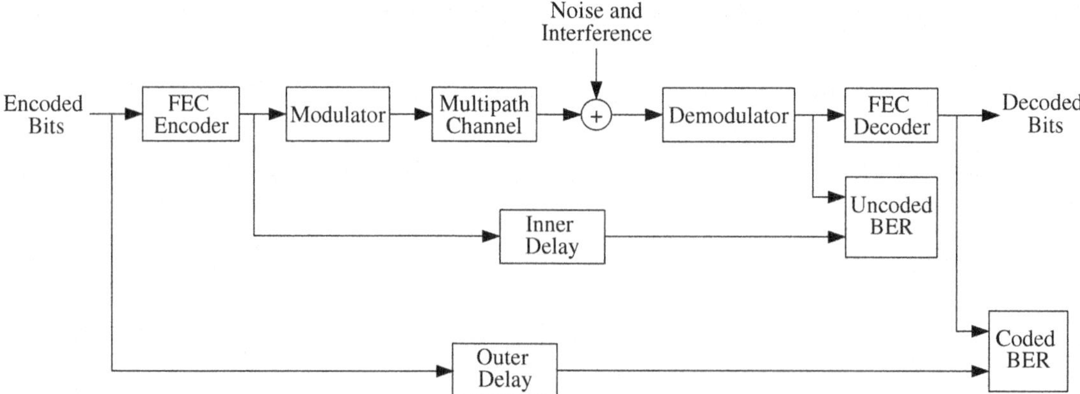

FIGURE 8.1 Typical digital communication simulation system.

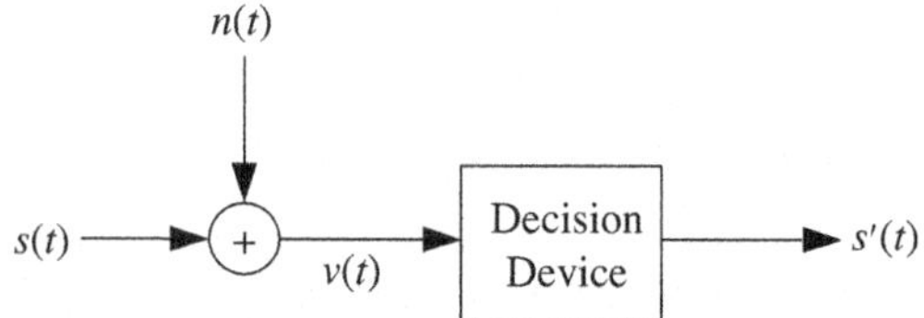

FIGURE 8.2 Error estimation decision device.

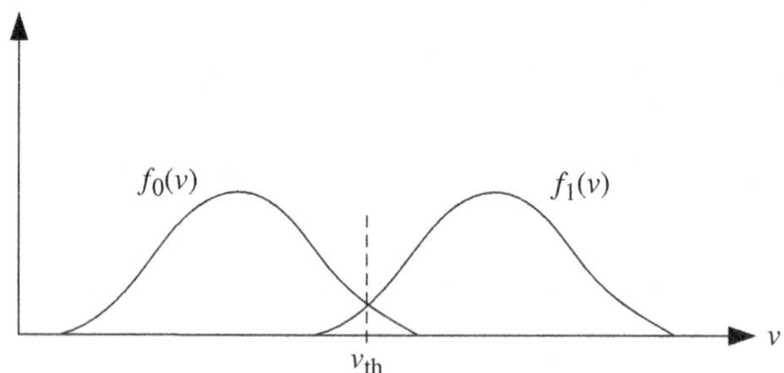

FIGURE 8.3 Received signal statistics.

8.1.1 Estimation Properties

In this section, we will present three properties of a "good" error rate estimator. If an estimator cannot produce accurate and consistent results, then caution must be exercised when using the questionable estimator, even though the error rate results converge quickly.

The three estimator properties of interest are

- Mean value
- Variance
- Confidence interval

We will attempt to preserve the notation used by the material in the reference section, whenever possible [3]. Hence we will consider the following estimator notation, where the estimate will be obtained by passing the variable Y through a function $G(x)$:

$$\hat{Q} = G(Y) \tag{8.1}$$

where we have used $\hat{Q}$ to represent the estimate of Q. Next we will present some general information surrounding the above-mentioned properties.

Mean. We say an estimator is unbiased if the following condition holds true:

$$E\{\hat{Q}\} = \int_{-\infty}^{\infty} G(Y) \cdot f_Y(y)\,dy \tag{8.2}$$

$$E\{\hat{Q}\} = Q \tag{8.3}$$

Variance. The variance of an estimator is calculated as follows:

$$\sigma^2(\hat{Q}) = E\{\hat{Q}^2\} - E^2\{\hat{Q}\} \tag{8.4}$$

$$\sigma^2(\hat{Q}) = \int_{-\infty}^{\infty} [G(Y)]^2 \cdot f_Y(y)\,dy - Q^2 \tag{8.5}$$

This property is a measure of dispersion of the estimate about its mean. Generally speaking, the smaller the variance is, the better the estimator. As the variance approaches zero, the number of observation samples, N, approaches infinity.

Confidence Interval. The confidence interval quantifies the measure of spread with an associated probability. In other words, it defines an interval (b_1, b_2) in which the true values of Q are within, with a specific probability. A 90% confidence interval corresponds to $\alpha = 0.1$

$$P(b_1 \leq Q \leq b_2) = 1 - \alpha \tag{8.6}$$

The addition of noise and interference in the communication system will cause errors to be made at the receiver. These errors can be mathematically represented as follows:

$$P(e|1) = \int_{-\infty}^{v_{th}} f_1(v)\,dv \qquad P(e|0) = \int_{v_{th}}^{\infty} f_0(v)\,dv \tag{8.7}$$

Here $P(e|1)$ is used to denote the probability of error, given a "1" was transmitted. Similarly, $P(e|0)$ is the probability of error, given a "0" was transmitted. The overall error probability P_e involves the weighted sum of the above conditional probabilities.

$$P_e = p(1) \cdot P(e|1) + p(0) \cdot P(e|0) \tag{8.8}$$

8.2 MONTE CARLO METHOD

In this section, we will present an error estimation technique known as the Monte Carlo method. Let us consider the probability of error, given a 0 was transmitted, and denote this value as P_0.

$$P_0 = \int_{v_{th}}^{\infty} f_0(v)\,dv \tag{8.9}$$

which we will rewrite as follows, after redefining the integration limits:

$$P_0 = \int_{-\infty}^{\infty} h_0(v) \cdot f_0(v)\, dv \tag{8.10}$$

where we have introduced the error detector $h_0(v)$, and it is defined as

$$h_0(v) = \begin{cases} 1, & v \geq v_{\text{th}} \\ 0, & v < v_{\text{th}} \end{cases} \tag{8.11}$$

We notice the error probability can then be represented in the following notation:

$$P_0 = E\{h_0(v)\} \tag{8.12}$$

This can be estimated by the sample mean, given N is the number of bits observed.

$$\hat{P}_0(v) = \frac{1}{N} \cdot \sum_{i=1}^{N} h_0(v_i) \tag{8.13}$$

The estimator $\hat{P}_0(v)$ is sometimes called an *error counter*. Assume n bits have been observed to be in error, out of the total N bits observed; then the error counter can be written as

$$\hat{P}_0 = \frac{n}{N} \tag{8.14}$$

As N approaches infinity, $\hat{P}_0$ will converge to the true value of P_0.

Let us take a moment to evaluate the estimator, given a finite value of N. Recall one such property of the estimator is the confidence interval and is given below for (y_+, y_-).

$$P(y_+ \leq p \leq y_-) = 1 - \alpha \tag{8.15}$$

Next we summarize the detailed steps carried out in [1] and [3]. It is known that $\hat{P}_0$ is binomially distributed and it converges to either a Poisson or Normal distribution, depending on the imposed constraints. A common practice is to use the Normal approximations due to insight and mathematical tractability. The end result is provided below, assuming the estimated error rate of $\hat{P}_0 = 10^{-k}$ and an estimation window size of $N = \eta \cdot 10^k$.

$$y_{\pm} = 10^{-k}\left\{1 + \frac{d_\alpha^2}{2\eta} \cdot \left[1 \pm \sqrt{\frac{4\eta}{d_\alpha^2} + 1}\right]\right\} \tag{8.16}$$

where d_α is chosen so that the following holds true:

$$\frac{1}{\sqrt{2\pi}} \int_{-d_\alpha}^{d_\alpha} e^{-t^2/2}\, dt = 1 - \alpha \tag{8.17}$$

The confidence interval is plotted in Fig. 8.4 for 90% and 99% values, assuming a hypothetical BER of 10^{-v}, where v has been set equal to 2.

What we can extract from this graph is, for a fixed 90% confidence interval and an observation length of 10^{v+2} bits, a variation of approximately $0.9\hat{P}_0$–$1.1\hat{P}_0$ is observed. In other words, giving the bit error event 100 opportunities to make itself public results in a small spread (approximately ±10%) around the expected BER value.

Let us consider using the sample mean estimate of the BER; then we have the following estimate, given e_i is the error at the ith observation.

$$\hat{P}_0 = \frac{1}{N} \cdot \sum_{i=1}^{N} e_i \tag{8.18}$$

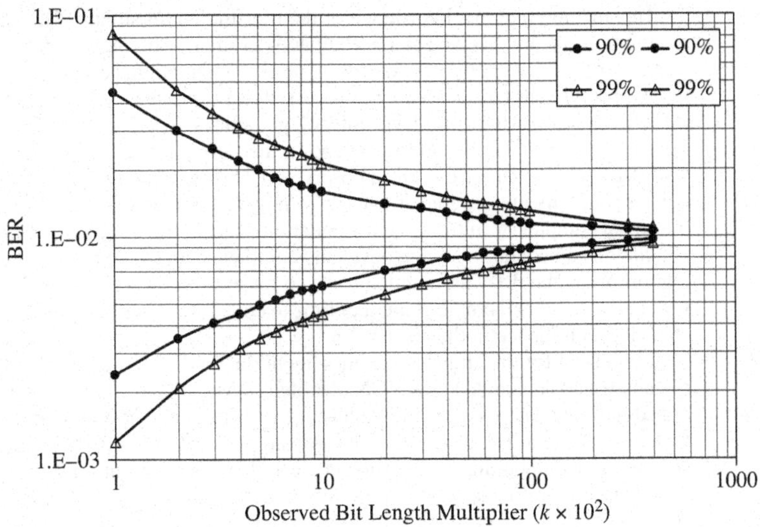

FIGURE 8.4 Confidence interval guideline for error estimation.

with the error signal expressed as

$$e_i = \begin{cases} 1, & \text{if error is present} \\ 0, & \text{otherwise} \end{cases} \tag{8.19}$$

The mean is $E\{\hat{P}_0\} = P_0$, which states the estimator is unbiased. We can also write down the variance of the estimator as [3]

$$\sigma^2(\hat{P}_0) = \frac{P_0(1 - P_0)}{N} \tag{8.20}$$

which can also be alternatively represented as

$$\sigma^2(\hat{P}_0) = \frac{1}{N} \cdot \int_{v_{\text{th}}}^{\infty} f_0(v) \cdot [1 - P_0]\, dv \tag{8.21}$$

Here we have assumed the errors are independent.

A system simulation using the MC method for BER estimation is given in Fig. 8.5.

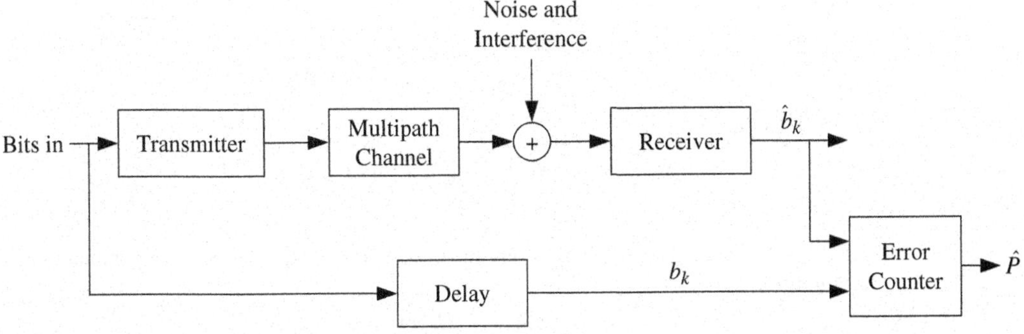

FIGURE 8.5 MC simulation system.

This method requires knowledge of the transmitted bit stream. This method makes no "a priori" assumptions of the received signal PDF and, as such, is the most general of all the estimation techniques. The price we must pay for this luxury is computational complexity. This is related to the number of observations (N) that must be processed to get an acceptable level of confidence in the resulting BER. Recall the earlier confidence interval guidelines, where a $10^{-\nu}$ BER for a 90% confidence interval and $N = 10^{\nu+3}$ produced low spread in the estimated values. For example, consider a target of 10^{-7} BER; this means we need to observe 10^{10} bits to obtain such statistical confidence. This is, in general, a highly undesirable drawback with this technique. An additional point worth mentioning is the actual complexity of the system simulation. Depending on the level of algorithmic abstraction, the simulation can contain very low level and complete details about the target implementation, receiver algorithms, technology assumptions, and so forth. If the system complexity is large, then the simulation can take a considerable amount of time to completely analyze the number of bits desired to obtain statistical confidence in the results.

The reason for this drawback comes about due to the inefficient use of the noise generated in the system. If $n(t)$ is modeled as AWGN, denoted as $N(0, \sigma^2)$, we see the majority of noise samples generated will be centered at a voltage level of 0. These noise levels do not cause errors. It is the large voltage levels that do cause errors; unfortunately, they rarely occur. It is for this reason that MC simulations are computationally inefficient since they must be run for a long period of time in order to give these rare events enough opportunity to occur [4–7].

8.3 MODIFIED MONTE CARLO OR IMPORTANCE SAMPLING METHOD

This next method involves a modification to the MC technique discussed above and is called the Modified MC (MMC) or Importance Sampling (IS) in the literature [8–11]. As discussed in the MC section, the error-producing noise voltages rarely occur at high SNR values due to the PDF chosen. Hence IS involves deliberate biasing of the noise statistics to artificially generate these errors or important events. At the end of reception, the BER estimate must be unbiased to remove these effects. Recall the following error event:

$$P_0 = \int_{-\infty}^{\infty} h_0(v) \cdot f_0(v)\, dv \tag{8.22}$$

This equation can be rewritten as follows to introduce the biased form:

$$P_0 = \int_{-\infty}^{\infty} h_0(v) \cdot \frac{f_0(v)}{f_0^*(v)} \cdot f_0^*(v)\, dv \tag{8.23}$$

where $f_0^*(v)$ is the biased PDF necessary to reduce the variance of the estimator, $\hat{P}_0$. We can define the IS weight as

$$w(v) = \frac{f_0(v)}{f_0^*(v)} \tag{8.24}$$

Then we have

$$P_0 = \int_{-\infty}^{\infty} h_0(v) \cdot w(v) \cdot f_0^*(v)\, dv \tag{8.25}$$

which is equivalent to the following, where the expectation operation has been assumed to be with respect to the biased PDF:

$$P_0 = E_*\{h_0(v) \cdot w(v)\} \tag{8.26}$$

The IS BER can be estimated by the sample mean estimator as follows:

$$\hat{P}_{o*} = \frac{1}{N_*} \cdot \sum_{i=1}^{N_*} h_0(v_i) \cdot w(v_i) \tag{8.27}$$

Notice we are removing the effects of biasing the noise statistics in the error counter itself by the multiplication of the IS weight.

Here are some properties of the IS BER estimator $\hat{P}_{o*}$.

Mean:

$$E\{\hat{P}_{0*}\} = P_0 \tag{8.28}$$

Variance:

$$\sigma^2(\hat{P}_{0*}) = \frac{1}{N_*} \cdot \int_{v_{\text{th}}}^{\infty} f_0(v) \cdot [w(v) - P_0]\, dv \tag{8.29}$$

Comparing these results to the MC given earlier can be done by using the ratio of the variance of the two estimators. The addition of the IS weight $w(v)$ allows us to reduce the variance of the IS estimator. Using Eq. 8.30, we can define the variance reduction factor by setting $N = N_*$

$$\frac{\sigma^2(\hat{P}_0)}{\sigma^2(\hat{P}_{0*})} = \frac{\frac{1}{N} \cdot \int_{v_{\text{th}}}^{\infty} f_0(v) \cdot [1 - P_0]\, dv}{\frac{1}{N_*} \cdot \int_{v_{\text{th}}}^{\infty} f_0(v) \cdot [w(v) - P_0]\, dv} \tag{8.30}$$

At this point, we can take two paths: The first involves equating the variances and enjoying the benefits of the smaller sample size $N_* < N$, while the second equates the sample sizes and benefits from the smaller variance $\sigma^2(\hat{P}_{0*}) < \sigma^2(\hat{P}_0)$. The first approach is typically chosen in order to reduce the computer simulation run time. However, it was our intention to inform the system designer of the available choices [12–15].

Let's consider the system in Fig. 8.6, where the goal is to derive the IS weight that is a function of the noise random variable.

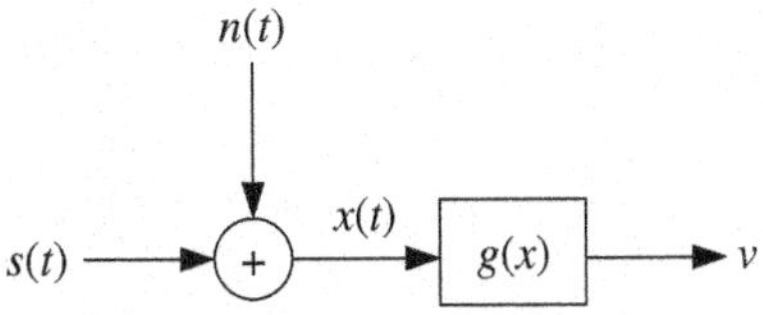

FIGURE 8.6 Simple error estimation receiver.

Let's define the following error estimator, with $v_i = g(x_i)$:

$$P_0 = \int_{-\infty}^{\infty} h_0(v) \cdot f_0(v) dv = \int_{-\infty}^{\infty} h_0[g(x)] \cdot f_X(x) dx \tag{8.31}$$

Employing the IS biasing technique discussed above, we arrive at

$$P_0 = \int_{-\infty}^{\infty} h_0[g(x)] \cdot \frac{f_X(x)}{f_X^*(x)} \cdot f_X^*(x)\, dx \tag{8.32}$$

$$P_0 = \int_{-\infty}^{\infty} h_0[g(x)] \cdot w(x) \cdot f_X^*(x)\, dx \tag{8.33}$$

which is equivalent to

$$P_0 = E_{X*}\{h_0[g(x)] \cdot w(x)\} \tag{8.34}$$

Using the sample mean estimator, the ensemble average can be replaced with the following equation:

$$\hat{P}_{0*} = \frac{1}{N_*} \cdot \sum_{i=1}^{N_*} h_0(v_i) \cdot w(x_i) \tag{8.35}$$

This involves biasing the $f_X(x)$ PDF; we prefer to bias the noise $f_N(n)$ PDF. Since $s(t)$ and $n(t)$ are assumed to be independent, this is possible.

Recall the decision output signal is

$$v_i = g(x_i) = g(s_i + n_i) \tag{8.36}$$

Let us start with the error estimator and use the joint PDF between the signal and noise components,

$$P_0 = \int_{-\infty}^{\infty} h_0[g(x)] \cdot f_X(x)dx = \int_{-\infty}^{\infty}\int_{-\infty}^{\infty} h_0[g(s+n)] \cdot f_{S,N}(s,n)\, ds\, dn \tag{8.37}$$

After invoking the independence assumption of the signal and noise, we have

$$P_0 = \int_{-\infty}^{\infty}\int_{-\infty}^{\infty} h_0[g(s+n)] \cdot f_S(s) \cdot f_N(n)\, ds\, dn \tag{8.38}$$

$$P_0 = \int_{-\infty}^{\infty} h_0[g(s+n)] \cdot f_N(n)\, dn \tag{8.39}$$

Using the IS biasing technique produces

$$\int_{-\infty}^{\infty} h_0\,[g(s+n)] \cdot \frac{f_N(n)}{f_N^*(n)} \cdot f_N^*(n)dn = \int_{-\infty}^{\infty} h_0[g(s+n)] \cdot w(n) \cdot f_N^*(n)\, dn \tag{8.40}$$

Lastly, using the sample mean estimator gives us the familiar form:

$$\hat{P}_{0*} = \frac{1}{N_*} \cdot \sum_{i=1}^{N_*} h_0(v_i) \cdot w(n_i) \tag{8.41}$$

The only difference is that now, the weighting is based on the noise samples rather than the received signal, as given earlier. Therefore, the noise PDF $f_N(n)$ can now be replaced with it biased version $f_N^*(n)$.

Thus far we have been able to mathematically show we can derive a BER estimator where the simulation noise PDF is biased, that is, $f_N^*(n)$ and the BER estimator remove this biasing through weighing of the noise samples. Let us slightly increase the complexity of the digital communication system simulation block diagram with the help of Fig. 8.7.

Next we discuss how to pick the biasing noise PDF, specifically $f_N^*(n)$. If $f_N(n)$ is Gaussian, then $f_N^*(n)$ is typically chosen also to be Gaussian, as shown in Fig. 8.8.

$$\begin{aligned} f_N(n) &= N(0, \sigma^2) \\ f_N^*(n) &= N(0, \sigma_*^2) \end{aligned} \tag{8.42}$$

Above we noticed the mean stays the same, but the variance is made larger.

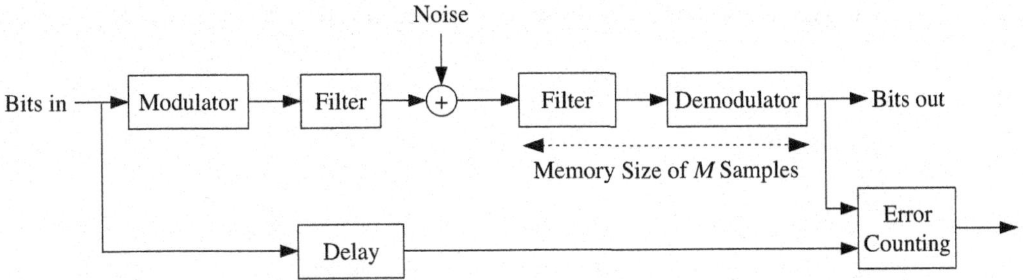

FIGURE 8.7 Modified digital communication simulation block diagram.

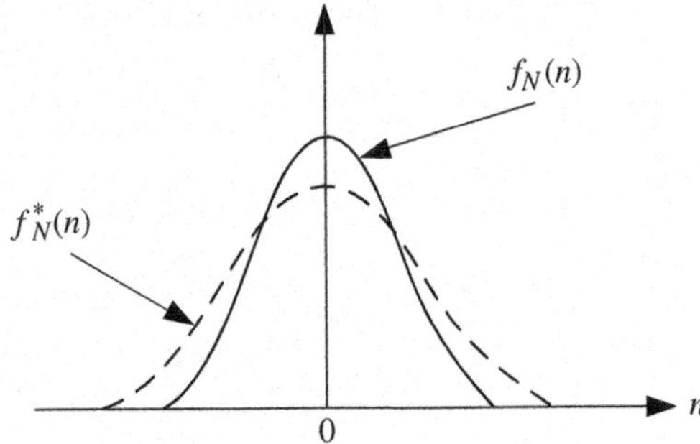

FIGURE 8.8 IS biasing.

Here we see that $\sigma_*^2 > \sigma^2$; thus the error events occur more frequently, allowing us to accurately count them. As a result of this, the simulation run time can be reduced significantly.

Consider the following weight function assuming Gaussian noise statistics:

$$w(n_k) = \frac{f_N(n_k)}{f_N^*(n_k)} = \frac{\dfrac{1}{\sigma\sqrt{2\pi}} \cdot e^{-n_k^2/2\sigma^2}}{\dfrac{1}{\sigma_*\sqrt{2\pi}} \cdot e^{-n_k^2/2\sigma_*^2}} \tag{8.43}$$

$$w(n_k) = \frac{\sigma_*}{\sigma} \cdot e^{-[1-\sigma^2/\sigma_*^2]n_k^2/2\sigma^2} \tag{8.44}$$

Typical literature notations use the following, with $0 < \alpha < 1$:

$$\frac{\sigma^2}{\sigma_*^2} = 1 - \alpha \tag{8.45}$$

Since we fix the IS parameter α prior to simulating, the weight function is a simple equation (assuming the Gaussian PDF).

In order to quantify the performance improvement, let's define the sample size reduction factor r as

$$r = \frac{N}{N_*} \approx \frac{\displaystyle\int_{v_{\text{th}}}^{\infty} f(v)\, dv}{\displaystyle\int_{v_{\text{th}}}^{\infty} f(v) \cdot w(v)\, dv} \tag{8.46}$$

Equivalently, it can be expressed using the noise PDF function, with $\sigma^2(\hat{P}_0) = \sigma^2(\hat{P}_{0*})$.

$$r = \frac{N}{N_*} \approx \frac{\displaystyle\int_{v_{\text{th}}}^{\infty} f_N(n)dn}{\displaystyle\int_{v_{\text{th}}}^{\infty} f_N(n) \cdot w(n)dn} \tag{8.47}$$

This variance-scaling technique can be visualized as follows: You specify a certain SNR; let's say SNR_1 is the desired point of interest.

$$\text{SNR}_1 = \frac{P_S}{\sigma^2} \tag{8.48}$$

The simulation SNR is actually lower due to the increased noise variance

$$\text{SNR}_2 = \frac{P_S}{\sigma_*^2} \tag{8.49}$$

This can be better understood with the BER performance curve in Fig. 8.9, assuming $\text{SNR}_2 < \text{SNR}_1$.

We need to consider a system with memory of length M. Let's define the system as follows: Y is to be determined by M independent samples.

$$Y = G(x_1, x_2, \ldots, x_M) \tag{8.50}$$

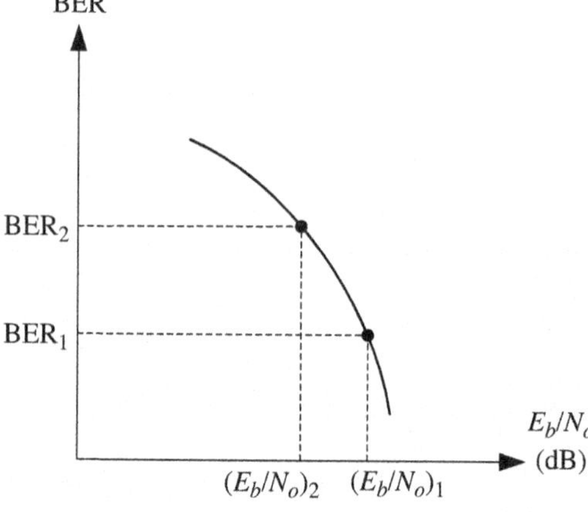

FIGURE 8.9 IS functional application.

Then the weight function becomes

$$w(n) = \frac{f_{NM}(n)}{f_{NM}^*(n)} \tag{8.51}$$

where $f_{NM}(n)$ is an M-dimensional PDF of the noise source. Assuming independent noise samples, we arrive at the product of the individual noise statistics.

$$w(n_i) = \prod_{j=1}^{M} \frac{f_N(n_{i+j})}{f_N^*(n_{i+j})} \tag{8.52}$$

The memory in the system has a negative impact on the system improvement. The reader can visit the applicable references in the end of this chapter to see various examples, in order to provide numerical results (see [3] and [8]). Some noteworthy observations are, as the memory increases, the improvement of IS over MC decreases. Also the optimum value of $1 - \alpha$ varies as the length of the memory changes.

Let us build a simple IS system and compare the error rate estimates. The noise samples are scaled to produce a larger variance and more frequent error events. The noise samples themselves and the biased variance value are inserted into the IS debiasing block which generated the weighting function to be used in the error estimator (see Fig. 8.10).

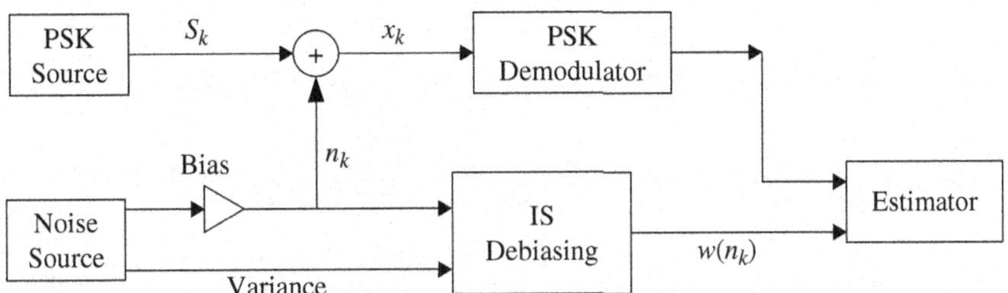

FIGURE 8.10 IS system block diagram.

In Fig. 8.11, we provide simulation results where we plot the SER versus the number of symbols observed. We can see that IS curve immediately converges to a smooth function, whereas the MC has a widely varying behavior. The curves are produced with an SNR = 13 dB and a bias value defined as

$$\frac{\sigma_*}{\sigma} = \sqrt{3.3} \tag{8.53}$$

As shown earlier in the MC section, using a sample size of 10^{k+1} to estimate a BER of 10^{-k} has a large confidence range of 0.45–2.2 of the target value (for 95% confidence interval). The MC estimate has not yet converged, as shown in Fig. 8.11, indicating that longer simulation runtimes are required.

8.4 IMPROVED IMPORTANCE SAMPLING METHOD

The IS method described above uses the variance-scaling technique to bias the noise PDF. This has also been called Conventional Importance Sampling (CIS) in the literature. An alternative IS

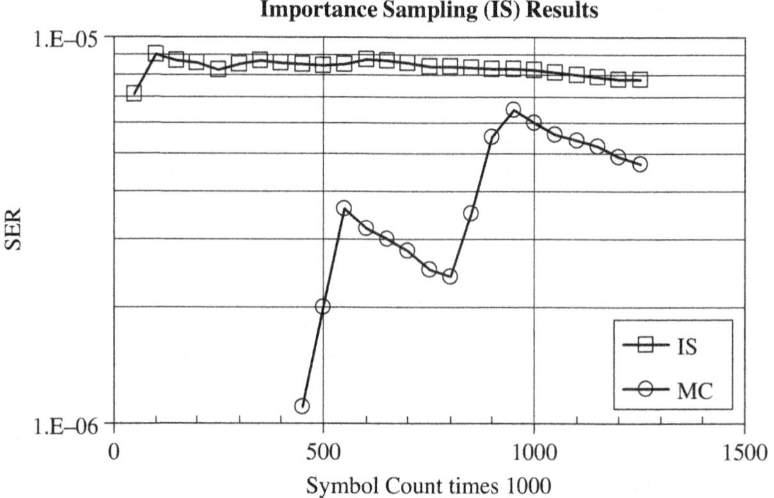

FIGURE 8.11 IS and MC convergence time comparison.

technique, called Improved Importance Sampling (IIS), is to vary the mean instead of the variance [16].

$$\begin{aligned} f_N(n) &= N(0, \sigma) \\ f_N^*(n) &= N(\mu, \sigma) \end{aligned} \tag{8.54}$$

This method gives further performance improvement, compared to IS. Recall IS performance depends on the length of the system memory, whereas IIS is less sensitive [1]. A major reason for this is the method in which the mean is varied or how the mean translation is accomplished. In IIS, the mean is varied in proportion to the system response. A more general approach is to not only vary the mean but also the variance.

In Fig. 8.12, we provide a block diagram where the mean of the noise is varied and the variance of the noise is kept at a constant value. The IIS debiasing accepts the noise samples, the mean biased value, and the noise variance. This information is used to create a IIS weights $w(n_k)$ to be used in the error estimator block.

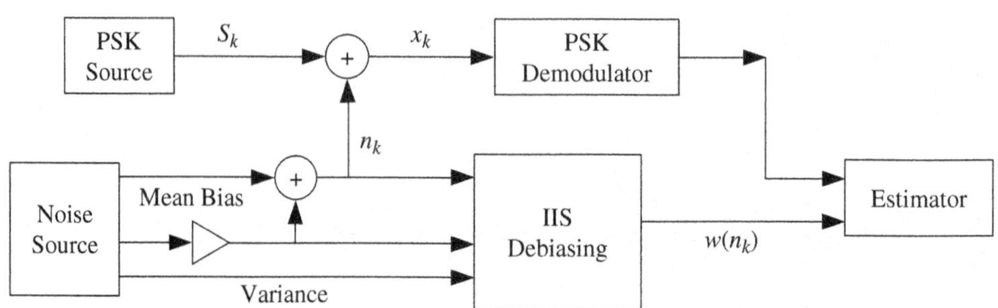

FIGURE 8.12 IIS system block diagram.

8.5 TAIL EXTRAPOLATION METHOD (TEM)

This technique is based on the assumption that the tail region of the PDF of the decision device input waveform is described by some member of the generalized exponential function (GEF) class defined below [17].

$$f_V(x) = \frac{v}{\sigma \cdot \Gamma\left(\frac{1}{v}\right)} \cdot e^{-\left(\frac{x}{\sigma}\right)^v} \qquad (0 \le x \le \infty) \tag{8.55}$$

where $\Gamma(v)$ is the gamma function, and σ and v are unknown.

This is sometimes written as (where μ = mean)

$$f_V(x) = \frac{v}{2\sqrt{2} \cdot \sigma \cdot \Gamma\left(\frac{1}{v}\right)} \cdot e^{[-|(x-\mu)/(\sigma \cdot \sqrt{2})|^v]} \tag{8.56}$$

The probability of error becomes expressed as follows, where the variable t is used to denote a pseudothreshold:

$$P_e = \int_{t+\mu}^{\infty} f_V(x)dx = \int_{\frac{t}{\sigma\sqrt{2}}}^{\infty} \frac{v}{2 \cdot \Gamma\left(\frac{1}{v}\right)} \cdot e^{-(y)^v} dy \tag{8.57}$$

where y is defined as

$$y = \frac{x - \mu}{\sigma\sqrt{2}} \tag{8.58}$$

For larger values of t, we approximate the above error probability as

$$P_e \cong e^{-\left(\frac{t}{\sigma\sqrt{2}}\right)^v \cdot [1-E(t)]} \tag{8.59}$$

$$P_e \cong e^{-\left(\frac{t}{\sigma\sqrt{2}}\right)^v} \quad \{E(t) \ll 1 \quad for\ t >> 1\} \tag{8.60}$$

$$\ln[P_e] \cong -\left(\frac{t}{\sigma\sqrt{2}}\right)^v \tag{8.61}$$

$$\ln\{-\ln[P_e]\} \cong v \cdot \ln\left[\frac{t}{\sigma\sqrt{2}}\right] \qquad (t > 0) \tag{8.62}$$

Hence the double log of P_e is asymptotically linear in log $t/\sigma\sqrt{2}$. Because of our positive-valued constraint ($t > 0$), we first translate the negative PDF above 0. Notice the value below is found to be proportional to the SNR.

$$\frac{t}{\sigma\sqrt{2}} \propto \sqrt{\text{SNR}} \tag{8.63}$$

Hence for each value of t (pseudothreshold), there exists an associated pseudo-BER value (Fig. 8.13).

Recall that we are assuming the tail part of the PDF can be represented by the GEF class. The example considered below focuses in the region near the threshold. In this region the true PDF can be approximated by an exponential.

This method involves a series of short MC simulation runs for various threshold levels, or since we have the PDF $f_V(x)$, we can simply calculate the BER, without knowledge of the transmit data.

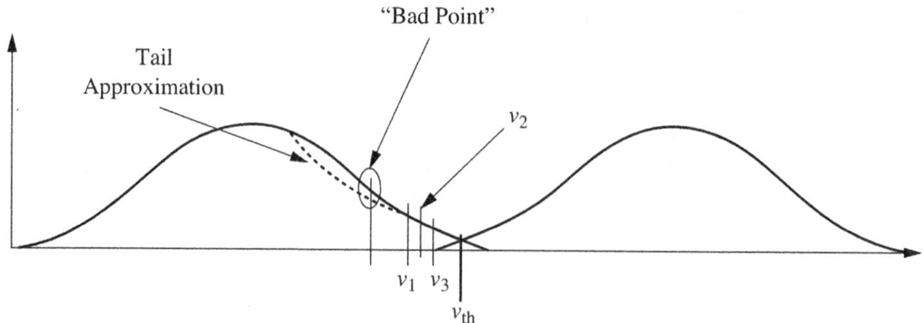

FIGURE 8.13 TEM emphasizing various pseudothresholds.

This can be accomplished by counting the number of times the received sample crosses the pseudothreshold; then divide this count by the total number of received bits to give a pseudo-BER. The following is a rule-of-thumb to be considered when using this estimation technique [17].

$$\mu - t_1 \geq 2 \cdot \alpha \tag{8.64}$$

The TEM procedure becomes running a series of short MC simulations using equally spaced thresholds. For each threshold, a BER value is obtained in order to plot the BER versus threshold results. Then draw a straight line intersecting the actual threshold value of interest. Now you can simply read off the BER values. The only drawback with this BER estimation technique is that you should run a few (e.g., 3) short MC simulations.

The basic idea behind the TEM is as follows: Choose the pseudothresholds, and obtain the corresponding pseudo-BER values. Plot the double log BER versus the log threshold, and draw a straight line between the points. Then extend or extrapolate the line to the threshold value of interest, and read off the BER. This is shown in Fig. 8.14.

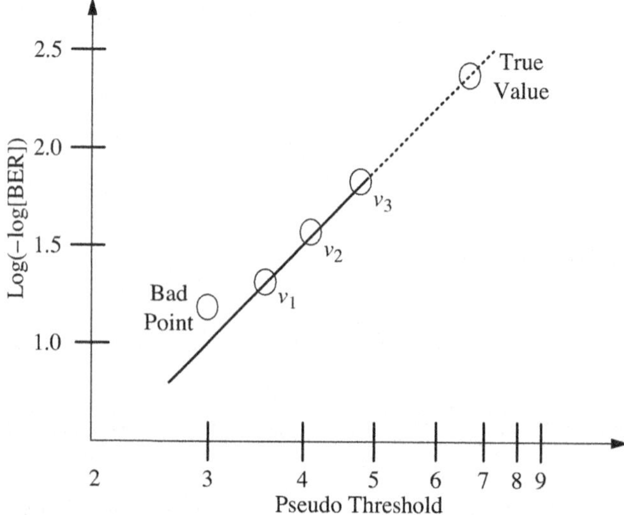

FIGURE 8.14 TEM pseudo-BER example.

In order to shorten the MC simulation run time, one would want to move the pseudothreshold as much as possible away from the true threshold in order to estimate a large error value in a short period of time. However, care must be exercised so as to not violate the exponential PDF assumption near the threshold region. In Fig. 8.13, we can see where the exponential PDF assumptions breaks down; it is labeled as a "Bad Point." Moreover, the impact of this "Bad Point" on the pseudo-BER is shown in Fig. 8.14.

8.6 SEMI-ANALYTIC METHOD

In all of the above-mentioned error estimation methods, we characterized the entire waveform (signal plus noise) at the input of the decision device by means of the received PDF. In the following method, the impact of the signal and noise are separated. The Semi-Analytic (SA) technique is also called Quasi-Analytical (QA) because it involves the mixture of simulation and analysis [1, 3, 18–20]. The BER is computed in analytical form, and the received samples are obtained from a noiseless simulation [18].

Here we separate the estimation technique into two parts: signal first, and then noise. All the effects of the noise are considered additive to the input of the decision device. Since we are separating the effects of the additive noise, we call this noise source an Equivalent Noise Source (ENS). In addition, we assume we know the PDF of the ENS to assist us in the BER calculation phase of the estimation technique.

Note the ENS statistics alone are not sufficient to compute the BER because we still need to take into account the system degradations. We let the simulation compute the effect of the system distortions (nonlinearity, group delay, etc.) in the absence of noise, and then we superimpose the noise (ENS) analytically onto the noiseless waveform.

Time saving is achieved with the SA BER estimation technique because we do not need to wait for rare errors to occur. We are calculating the error integral directly. In order to provide some insight, let us consider Fig. 8.15, where we show the transmitted and received waveforms. Here

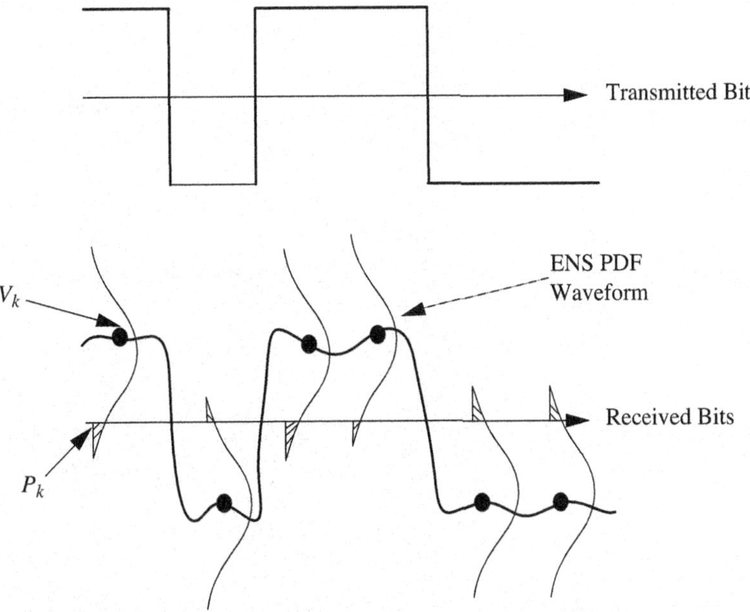

FIGURE 8.15 SA functional diagram.

the received waveforms include the system distortions (nonlinearity, ISI, etc.), but do not include the random noise.

In the diagram in Fig. 8.15, a short number of received bits are stored for processing. These received bits do not include the presence of noise, but do include other impairments. Each of the received bits has the PDF of the ENS essentially overlaid upon them. Each received bit produces an individual error probability P_k, which is evaluated mathematically. At the end of the observation window, the average probability of error is obtained by averaging each individual error probability P_k.

The received signal at the kth sampling instant is denoted as v_k, and the PDF of the ENS is denoted as $f_N(n)$. Here we can write the conditional error probability as follows (assuming the ENS has zero mean):

$$P_k = P(1|0) = P(\text{noise} > v_k | v_k < 0) = \int_{v_k}^{\infty} f_N(n)\,dn \tag{8.65}$$

$$P_k = P(0|1) = P(\text{noise} < v_k | v_k > 0) = \int_{-\infty}^{v_k} f_N(n)\,dn \tag{8.66}$$

Further insight into the SA reveals the expression of error probability P_k is an analytical form where the limit on the integrals v_k is obtained through simulation.

The average probability of error is obtained through (assuming successive decisions are independent)

$$P_e = \frac{1}{N} \cdot \sum_{k=1}^{N} P_k \tag{8.67}$$

where N is used to represent the length of the observed symbol sequence. Also the accurate assumption of the noise statistics can lead to a dramatic reduction in the entire simulation effort.

Next we need to determine the PDF of ENS; since this may or may not be Gaussian, we use a PDF from the family of GEF $f_V(n)$. Assume the decision device input signal consists of signal and noise components; then we can rewrite the previous expressions as

$$P(1|0) = \int_{0}^{\infty} \frac{v}{2\sqrt{2}\cdot\sigma\cdot\Gamma\left(\frac{1}{v}\right)} \cdot e^{-\left|\frac{n-v_k}{\sigma\sqrt{2}}\right|^{v}}\,dn \tag{8.68}$$

$$P(0|1) = \int_{-\infty}^{0} \frac{v}{2\sqrt{2}\cdot\sigma\cdot\Gamma\left(\frac{1}{v}\right)} \cdot e^{-\left|\frac{n-v_k}{\sigma\sqrt{2}}\right|^{v}}\,dn \tag{8.69}$$

If the received signal is rectified: $m_k = |v_k|$, then we have

$$P_k = \int_{0}^{\infty} \frac{v}{2\sqrt{2}\cdot\sigma\cdot\Gamma\left(\frac{1}{v}\right)} \cdot e^{-\left|\frac{n-m_k}{\sigma\sqrt{2}}\right|^{v}}\,dn \tag{8.70}$$

The received signal (v_k) amplitude fluctuations will generally be a function of a particular transmitted sequence. If the amplitude at any time instant is a function of a few adjacent symbols, then N can be made to be small. As this dependency (correlation) length increases, so must N. However, when compared to the observation length required for a target BER = 1E-8, which is at least $N = 10^{10}$ for the MC methods, the SA method will generally require a few thousand. This produces a significant amount of simulation run-time savings.

Typically, the ENS is Gaussian ($v = 2$), and for a linear receiver, the SA is the fastest and also exact in the BER estimation. Another assumption is that no errors occurred in the absence of noise; in other words, the received eye diagram is open. This assumption is important in order to not bias the final BER results incorrectly.

In Fig. 8.16, we present a flow chart of the SA procedure. For the Gaussian case, special care must be spent on the receiver noise bandwidth, since this defines the noise power. Also the received signal (v_k) power must be properly calculated. The combination of these two measurements constitutes the calibration phase.

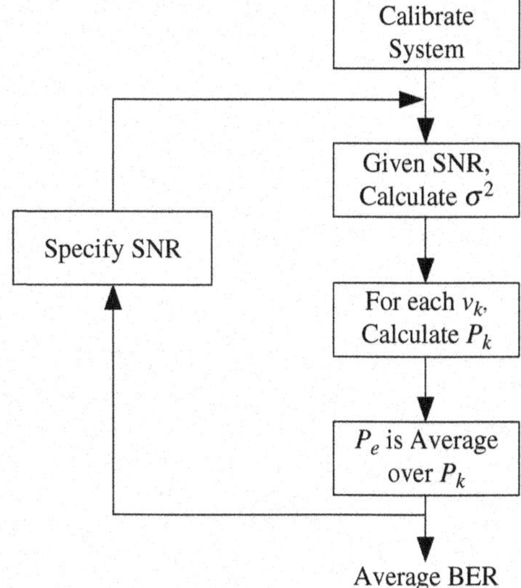

FIGURE 8.16 SA method flow chart description.

8.7 GENERAL DISCUSSION

We have briefly reviewed the MC, CIS, IIS, TE, and SA error estimation techniques. If the system designer's goal is to investigate transients or convergence effects, then the MC method is the preferred approach. If the goal is to evaluate specific algorithms for timing recovery, automatic frequency control, automatic gain control, and so forth, then the MC method will give more insight. If the communication system is relatively simple, then SA is the best approach to obtain fast and accurate BER results. The TE method also becomes useful in BER monitoring applications, where the transmitted data are unavailable. IS is very powerful, but we must consider the effects on system performance due to memory, since in some cases this memory can reduce the benefits.

In many applications, a combination of a few of the above-mentioned error estimation techniques is used to quantify the performance of a digital communication system. Hence the system designer should not preclude such a coordinated option.

REFERENCES

[1] M. C. Jeruchim, P. Balaban, and K. S. Shanmugan, *Simulation of Communication Systems*, Plenum Press, 1992, New York.

[2] *IEEE Journal on Selected Areas in Communications*, April 1993, all pages.

[3] M. C. Jeruchim, "Techniques for Estimating the Bit Error Rate in the Simulation of Digital Communications Systems," *IEEE Journal on Selected Areas in Communications*, Vol. 2, Jan. 1984, pp. 153–170.

[4] K. S. Shanmugam, "An Update on Software Packages for Simulation of Communication Systems (Links)," *IEEE Journal on Selected Areas in Communications*, Vol. 6, No. 1, Jan. 1988, pp. 5–12.

[5] M. C. Jeruchim, P. M. Hahn, K. P. Smyntek, and R. T. Ray, "An Experimental Investigation of Conventional and Efficient Importance Sampling," *IEEE Transactions on Communications*, Vol. 37, No. 6, June 1989, pp. 578–587.

[6] P. Balaban, "Statistical Evaluation of the Error Rate of the Fiberguide Repeater Using Importance Sampling," *Bell System Technical Journal*, Vol. 55, No. 6, July–Aug. 1976, pp. 745–766.

[7] D. Lu and K. Yao, "Improved Importance Sampling Technique for Efficient Simulation of Digital Communication System," *IEEE Journal on Selected Areas in Communications*, Vol. 6, No. 1, Jan. 1988, pp. 67–75.

[8] K. S. Shanmugam and P. Balaban, "A Modified Monte-Carlo Simulation Technique for the Evaluation of Error Rate in Digital Communication Systems," *IEEE Transactions on Communications*, Vol. COM-28, No. 11, Nov. 1980, pp. 1916–1924.

[9] M. K. Wyche, "Implementation of Importance Sampling Techniques Using the Signal Processing Worksystem (SPW)," in: *IEEE MILCOM*, 1992, pp. 701–706.

[10] P. M. Hahn and M. C. Jeruchim, "Developments in the Theory and Application of Importance Sampling," *IEEE Transactions on Communications*, Vol. COM-35, No. 7, July 1987, pp 706–714.

[11] J. Zou and V. K. Bhargava, "On the Optimum Biasing of Importance Sampling for Simulation of Communication Systems," in: *IEEE CCECE*, 1993, pp. 515–518.

[12] P. J. Smith, M. Shafi, and H. Gao, "Quick Simulation: A Review of Importance Sampling Techniques in Communications Systems," *IEEE Journal on Selected Areas in Communications*, Vol. 15, No. 4, May 1997, pp. 597–613.

[13] J. K. Townsend and K. S. Shanmugam, "On Improving the Computationally Efficiency of Digital Lightwave Link Simulation," *IEEE Transactions on Communications*, Vol. 38, No. 11, Nov. 1990, pp. 2040–2048.

[14] J. C. Chen, D. Lu, J. S. Sadowsky, and K. Yao, "On Importance Sampling in Digital Communications Part I: Fundamentals," *IEEE Journal on Selected Areas in Communications*, Vol. 11, No. 3, April 1993, pp. 289–299.

[15] J. C. Chen and J. S. Sadowsky, "On Importance Sampling in Digital Communications Part II: Trellis Coded Modulation," *IEEE Journal on Selected Areas in Communications*, Vol. 11, No. 3, April 1993, pp. 300–308.

[16] J. S. Stadler and S. Roy, "On the Use of Improved Importance Sampling (IIS) in the Simulation of Digital Communication Systems," in: *IEEE Globecom*, 1993, pp. 143–148.

[17] S. B. Weinstein, "Estimation of Small Probabilities by Linearization of the Tail of a Probability Distribution Function," *IEEE Transactions on Communications Technology*, Vol. COM-19, No. 6, Dec. 1971, pp. 1149–1155.

[18] W. H. Tranter and C. R. Ryan, "Simulation of Communication Systems Using Personal Computers," *IEEE Journal on Selected Areas in Communications*, Vol. 6, No. 1, Jan. 1988, pp. 13–23.

[19] N. Benvenuto, A. Salloum, and L. Tomba, "Performance of Digital DECT Radio Links Based on Semianalytical Methods," *IEEE Journal on Selected Areas in Communications*, Vol. 15, No. 4, May 1997, pp. 667–767.

[20] A. Morello and M. Pent, "Semianalytic BER Evaluation Method for FSK Demodulation with Discriminator and Post-Detection Filter," in: *IEEE ICC*, 1988, pp. 1698–1702.

CHAPTER 9
3G AND BEYOND DISCUSSION

In this chapter, we will discuss some evolutionary paths for the WCDMA 3G digital cellular system. We will begin by discussing the principles of Orthogonal Frequency Division Multiplexing (OFDM) since this has been chosen as the radio access technology of the Long-Term Evolution (LTE) for Universal Mobile Telecommunication System (UMTS). The Mobile TV delivery mechanism is presently being worked upon within the technical arena; the competition between dedicated single frequency networks versus shared frequency services will be introduced in this chapter. While on this topic, we present the Digital Video Broadcasting for handhelds (DVB-H) system that will be supplying Mobile TV to the end user, among other packet data services.

Since the Internet Protocol (IP) is increasingly becoming a mainstream interface to delivery platforms, we will present how 3G digital cellular systems address this increasing packet access need. The high-speed packet access (HSPA) evolution will be discussed as a means to achieving the aforementioned voice and data IP services. The evolution will consist of advanced receivers, higher-order modulation, broadcast and multicast services, packet connectivity, as well as other items. Lastly, we will briefly show the benefits of using multiple-input-multiple-output (MIMO) antenna systems to increase the system throughput and capacity of a wireless system. We hope to provide some insight into how the channel conditions affect the MIMO system capacity benefits.

9.1 INTRODUCTION

In recent years a variety of wireless systems have used OFDM to provide reliable communications. These systems range from satellite to terrestrial environments encompassing broadcasting such as Digital Audio Broadcasting (DAB) [1] and Digital Video Broadcasting (DVB) [2] as well as wireless LAN (IEEE802.11) [3] applications, to name a few. The most recent applications have been targeted to delivering IP-based data to mobile users. Such standards are WiMax (IEEE 802.16) [4] and the 3GPP LTE migration path of 3G cellular technology [5].

In OFDM multiple data symbols are simultaneously transmitted over the entire system bandwidth (BW). The transmission is divided into many narrowband subcarriers where each is orthogonal to each other. This composite signal comprises what is commonly called an OFDM symbol (see Fig 9.1). Since each subcarrier is narrow relative to the coherence BW of the channel, they will experience the flat fading phenomenon. As discussed in the earlier chapters, mobility brings forth the Doppler spread principle. This time varying behavior modulates the subcarriers dramatically degrading the orthogonality benefits; this is commonly called Inter-Carrier Interference (ICI) [6]. As would be expected, the degradation becomes more pronounced as the Doppler spread increases.

Consider a frequency in the 2 GHz band and a vehicle speed of 120 km/hr; this creates a maximum Doppler spread of approximately 220 Hz. Note some broadcasting systems begin to exhibit performance degradation at these Doppler spreads. Hence service on a High Speed Train (HST) can be temporarily interrupted if mitigation measures are not properly taken.

A particular point to make here is shifting to a lower frequency band; 700 MHz would reduce the maximum Doppler spread by nearly a factor of 3. This benefit coupled with better propagation

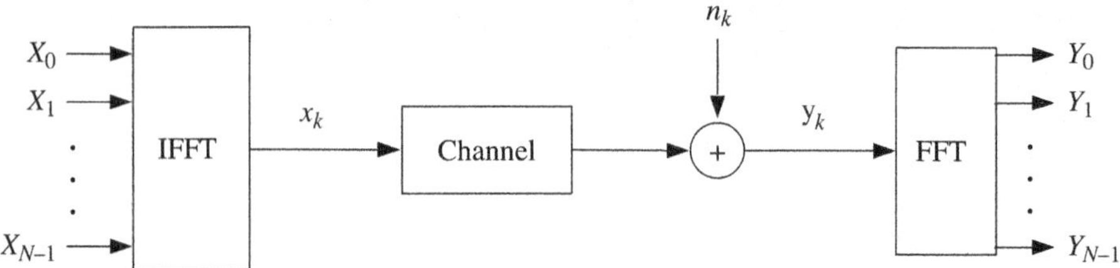

FIGURE 9.1 General OFDM communication system.

(i.e., path loss, shadowing, etc.) makes the lower frequency bands more attractive. A last note we wish to make in this area is lower frequency bands can mean less base stations required to blanket a broadcasting coverage area.

Recall the channel will have a time varying coherence BW which is caused by the delay spread or echoes present in the environment. These echoes have caused Inter-Symbol Interference (ISI) for other systems such as time division multiple access (TDMA) and Inter-Path Interference (IP) for code division multiple access (CDMA). In OFDM this interference is mitigated to a large degree, by the use of a cyclic prefix (CP) of length T_{CP}. This is accomplished by duplicating the last T_{CP} samples of the OFDM symbol and placing them in the front of the OFDM symbol. The length of the cyclic prefix is chosen to be larger than the expected maximum delay spread. When the delay spread is within the CP or guard interval, orthogonality can be preserved. This must be coupled with the assumption of perfect time and frequency synchronization in the receiver to fully extract the benefits (see Fig. 9.2) [7].

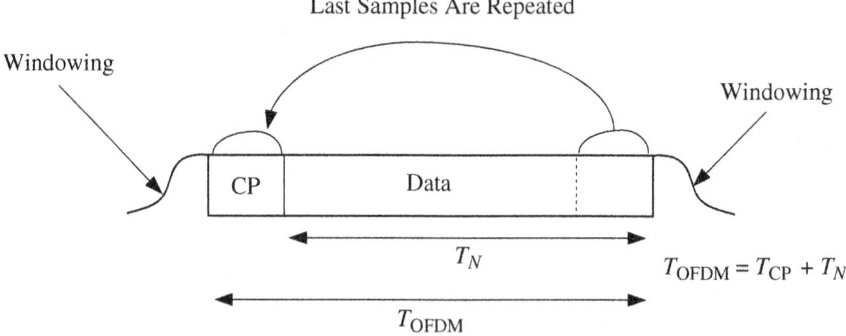

FIGURE 9.2 Cyclic prefix application in an OFDM symbol.

As noted above, each subcarrier will experience an amplitude and phase adjustment caused by the multipath channel. Hence mitigating frequency selective fading (FSF) channels can be as easy as equalizing the amplitude and phase of each subcarrier. The generation of the OFDM subcarriers is accomplished with the help of the inverse fast Fourier transform (IFFT). The separation of the OFDM subcarriers in the receiver is accomplished with the help of fast Fourier transform (FFT). Typically, the FFT size is chosen to be large enough so as to minimize the overhead of the CP. The CP is chosen to be large enough to avoid addition of ISI from the previous OFDM symbol as well as to effectively represent the channel in the frequency domain. Also the addition of the CP provides periodicity which allows the use of circular convolution operations in the receiver signal processing [8].

A last note to make here is that the literature defined a system using Forward Error Correction (FEC) with OFDM as COFDM, meaning coded OFDM. In order to handle the widely varying, harsh multipath environments, we have assumed FEC techniques are required. Having said this, no distinction is made to separate the terms.

9.1.1 General OFDM Transmission Discussion

In this subsection we will discuss the OFDM transmission or OFDM symbol generation. A baseband model of an OFDM transmitter is given in Fig. 9.3. The data bits enter an FEC encoder, followed by an interleaver before being modulated to create symbols, X_n. These symbols enter a serial-to-parallel (S/P) converter to support the IFFT operation. Next a cyclic prefix (CP) is inserted and spectral shaping is performed.

The output of the IFFT, x_n is given as ($n = 0, \ldots, N-1$)

$$x_n = \frac{1}{\sqrt{N}} \sum_{k=0}^{N-1} X_k \cdot e^{j\frac{2\pi nk}{N}} \tag{9.1}$$

where N is the number of subcarriers, X_k is the modulation output at subcarrier k.

Assume the OFDM symbol encounters an FSF channel that contains Q delay spread components, which can be written as

$$y_n = \sum_{t=0}^{Q-1} h_n(t) \cdot x_{n-t} + n_n \tag{9.2}$$

where $h_n(t)$ is the tth path of the FSF channel and n_n is the AWGN signal. Combining both of the above equations gives us the following.

$$y_n = \frac{1}{\sqrt{N}} \sum_{k=0}^{N-1} X_k \cdot \sum_{t=0}^{Q-1} h_n(t) \cdot e^{-j\frac{2\pi tk}{N}} \cdot e^{j\frac{2\pi nk}{N}} + n_n \tag{9.3}$$

The received OFDM symbol can be rewritten as

$$y_n = \frac{1}{\sqrt{N}} \sum_{k=0}^{N-1} X_k \cdot H_n(k) \cdot e^{j\frac{2\pi nk}{N}} + n_n \tag{9.4}$$

where $H_n(k)$ is the Fourier transform of the channel for subcarrier k at time instance n. Some key system design parameters are the size of the IFFT, number of useful data carrying subcarriers, the length of the cyclic prefix, and the symbol shaping [9].

The size of the FFT is important since this is the maximum number of possible subcarriers that can be used. Assume for the moment the transmission BW is fixed, then increasing the IFFT size will generate more subcarriers that are more closely spaced. This tends to amplify the ICI discussed above. Another important system design parameter is the length of the CP. On one hand, this should be as large as possible to accommodate large delay spreads which arise from large cell sizes or operation in a single frequency network (SFN). But power is wasted in this transmission. Hence this power concern leads us to making the CP as small as possible, but in environment with large delay spread, orthogonality will be degraded resulting in a degradation of system performance. Hence system performance trade-offs can be made.

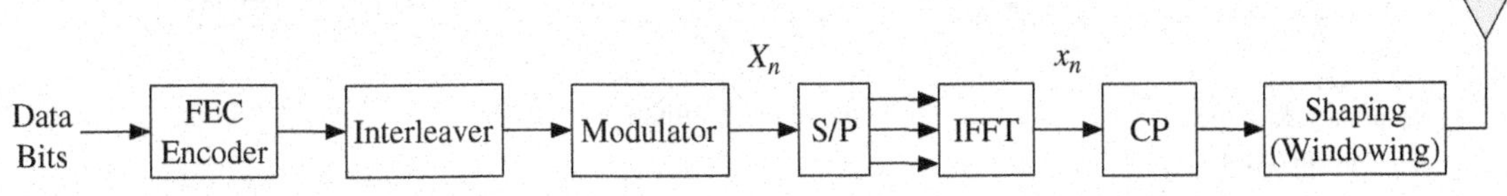

FIGURE 9.3 Baseband model of an OFDM transmitter using the cyclic prefix.

Transmission spectral side lobes are carefully monitored in a practical system in order to coexist with other systems. Windowing reduces the level of these side lobes and in turn generates less adjacent channel interference (ACI). The IFFT resembles that of using an ideal brick wall filter which generates large spectral side lobes. One windowing technique commonly used is a raised cosine response, others can be used; the reader is encouraged to peruse the references at the end of this chapter.

9.1.2 General OFDM Reception Discussion

In this subsection, we will discuss the OFDM receiving functions. Generally speaking, demodulation should be performed by a bank of matched filters (MFs), which are essentially matched to the OFDM symbol with the CP removed. Figure 9.4 represents a baseband model of the receiver [10].

In the above block diagram the data flow consists of removing the CP and performing the FFT to extract the subcarrier information. Next demodulation is performed (shown with equalization) to extract the bits/symbol from the received signal so that the FEC decoder can operate. In this case, we have also shown de-interleaving in order to help mitigate bursts of error due to the channel conditions.

Just as in any communication system certain parameters must be estimated in order to achieve the expected performance. For example, in the first block we assume that frequency and time synchronization was performed. Frequency offsets increase the ICI and thus degrade performance for all subcarriers. It is in the system designer's best interest to remove any residual frequency offset as early on in the receiver as possible. Similarly, time and frame synchronization is required for purposes of selecting a particular timing instance as well as determining where the beginning of the OFDM symbol is in order to remove the CP. These two estimators typically rely on pilot subcarriers to be used for reliable estimates. The demodulator also requires the channel estimates which are obtained from pilot symbols/subcarriers. Additionally, decision directed estimates can also be used depending on the reliability of the estimates [11].

After removal of the CP, the FFT produces (k = subcarrier number) the following signal.

$$Y_k = \frac{1}{\sqrt{N}} \sum_{n=0}^{N-1} y_n \cdot e^{-j\frac{2\pi nk}{N}} \tag{9.5}$$

We then wish to write the above equation in matrix notation and thus have the commonly used relationship.

$$\underline{Y} = H \cdot \underline{X} + \underline{\tilde{n}} \tag{9.6}$$

where the column vectors are defined as follows:

$$\underline{Y} = \begin{bmatrix} Y_0 \\ \vdots \\ Y_{N-1} \end{bmatrix}; \quad \underline{X} = \begin{bmatrix} X_0 \\ \vdots \\ X_{N-1} \end{bmatrix}; \quad \underline{\tilde{n}} = \begin{bmatrix} \tilde{n}_0 \\ \vdots \\ \tilde{n}_{N-1} \end{bmatrix} \tag{9.7}$$

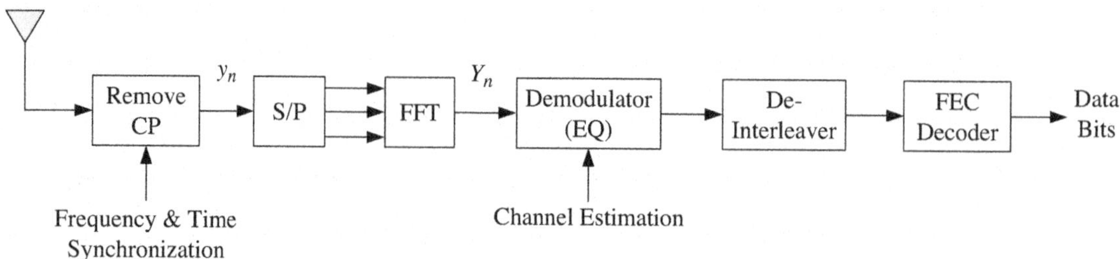

FIGURE 9.4 Baseband model of an OFDM receiver using the cyclic prefix.

And the channel matrix is given as a diagonal matrix when we assume the channel is time invariant.

$$H = \begin{bmatrix} \overline{H}_0 & 0 & \cdots & 0 \\ 0 & \overline{H}_1 & 0 & \vdots \\ \vdots & 0 & \ddots & 0 \\ 0 & \cdots & 0 & \overline{H}_{N-1} \end{bmatrix} \tag{9.8}$$

where the diagonal elements are defined with the assistance of the following equation:

$$\overline{H}_m = \frac{1}{N}\sum_{n=0}^{N-1} H_n(m) \tag{9.9}$$

The above matrix form allows us to view the OFDM system as containing a set of parallel Gaussian channels. In Fig. 9.5 we plot this parallel channel representation.

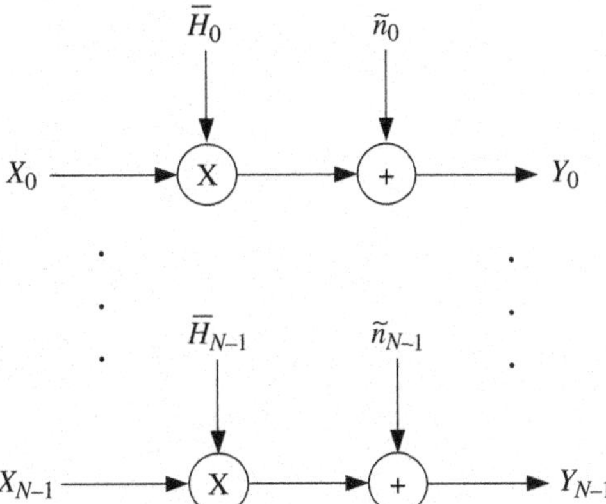

FIGURE 9.5 Parallel representation of an OFDM system.

At the end of this section we will revisit this derivation from the context of a circular matrix point of view; however, for the moment let us remove the assumption of a time invariant channel. The channel matrix no longer resembles a diagonal matrix and has off-diagonal elements that represent ICI caused by the Doppler spread of the channel. The new channel matrix can be written as

$$H = \begin{bmatrix} \overline{H}_0(0) & \cdots & \overline{H}_0(N-1) \\ \vdots & \ddots & \vdots \\ \overline{H}_{N-1}(0) & \cdots & \overline{H}_{N-1}(N-1) \end{bmatrix} \tag{9.10}$$

where the elements of the matrix are given by

$$\overline{H}_m(k) = \frac{1}{N}\sum_{n=0}^{N-1} H_n(k) \cdot e^{-j\frac{2\pi(m-k)n}{N}} \tag{9.11}$$

Typically a single tap equalizer provides reasonably good performance. However, as soon as the subcarrier orthogonality constraint is compromised, an MMSE-based FDE performs well [12]. The

MMSE solution has been presented a few times in this book (in various forms), the equalizer output is given as

$$\underline{z} = H^* \left[H \cdot H^* + \frac{\sigma_n^2}{\sigma_s^2} \cdot I_N \right]^{-1} \cdot \underline{Y} \tag{9.12}$$

Depending on the size of the FFT we notice that the matrix inversion operation can be prohibitively complex to perform in reality. Various solutions and approximations exist with minimal performance degradation [13, 14]. Also note assuming ideal time and frequency synchronization allows us to view OFDM system reception as a set of parallel channels.

As mentioned above, we wish to provide insight into how the use of the CP can allow us to represent the system with the help of a circulant matrix [15]. It is well known that the circulant matrix holds close ties to the discrete Fourier transform (DFT) operation. And it is a combination of all these reasons that we can replace the convolutional operators with multipliers in the frequency domain. Figure 9.6 shows a simple matrix formulation.

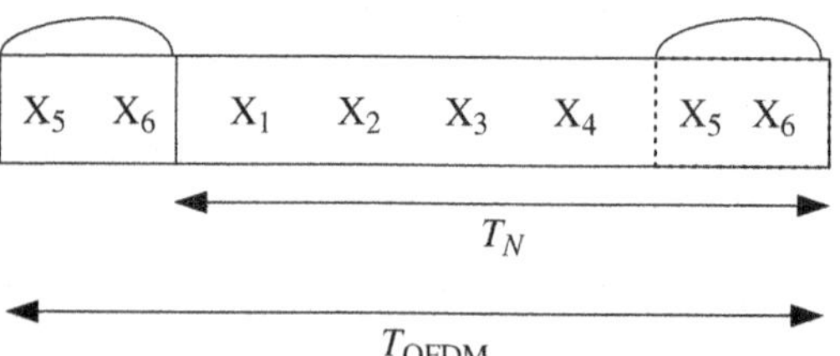

FIGURE 9.6 OFDM symbol example with cyclic prefix.

Let us consider the following OFDM symbol, consisting of six samples with a CP length of two samples.

In the absence of noise the received signal can be represented as follows, assuming a two-ray channel (h_1 and h_2) model is used.

$$\underline{y} = \begin{bmatrix} h_1 & 0 & 0 & 0 & 0 & 0 & 0 & 0 \\ h_2 & h_1 & 0 & 0 & 0 & 0 & 0 & 0 \\ 0 & h_2 & h_1 & 0 & 0 & 0 & 0 & 0 \\ 0 & 0 & h_2 & h_1 & 0 & 0 & 0 & 0 \\ 0 & 0 & 0 & h_2 & h_1 & 0 & 0 & 0 \\ 0 & 0 & 0 & 0 & h_2 & h_1 & 0 & 0 \\ 0 & 0 & 0 & 0 & 0 & h_2 & h_1 & 0 \\ 0 & 0 & 0 & 0 & 0 & 0 & h_2 & h_1 \end{bmatrix} \cdot \begin{bmatrix} x_5 \\ x_6 \\ x_1 \\ x_2 \\ x_3 \\ x_4 \\ x_5 \\ x_6 \end{bmatrix} \tag{9.13}$$

If we discard the CP part then the matrix notation is written as

$$\underline{y} = \widehat{H} \cdot \underline{x}$$

$$\underline{y} = \begin{bmatrix} h_1 & 0 & 0 & 0 & 0 & h_2 \\ h_2 & h_1 & 0 & 0 & 0 & 0 \\ 0 & h_2 & h_1 & 0 & 0 & 0 \\ 0 & 0 & h & h_1 & 0 & 0 \\ 0 & 0 & 0 & h_2 & h_1 & 0 \\ 0 & 0 & 0 & 0 & h_2 & h_1 \end{bmatrix} \cdot \begin{bmatrix} x_1 \\ x_2 \\ x_3 \\ x_4 \\ x_5 \\ x_6 \end{bmatrix} \tag{9.14}$$

One can clearly see that $\hat{H}$ is a circulant matrix, which is defined as follows: It is a matrix such that when a row is shifted to the right, the element that gets shifted out is inserted into the left side. This behavior extends from the last row to the first row. Specifically if one circularly shifts the last row, the first row would be visible. Next we mention some properties associated with the circulant matrix.

A circulant matrix can be diagonalized by a DFT, specifically the eigenvalues of the circular matrix are identical to the DFT. Recall that if a $N \times N$ matrix has N linearly independent eigenvectors then it can be diagonalized. The eigenvectors of a circulant matrix are linearly independent [16].

Let G be the $N \times N$ circulant matrix.

$$G = \begin{bmatrix} g(0) & g(1) & g(2) & \cdots & g(N-1) \\ g(N-1) & g(0) & g(1) & \cdots & g(N-2) \\ \vdots & & \ddots & & \vdots \\ g(1) & g(2) & \cdots & \cdots & g(0) \end{bmatrix} \tag{9.15}$$

The circulant matrix eigenvalues are given as follows:

$$\lambda_k = \sum_{p=0}^{N-1} g(p) \cdot e^{j\frac{2\pi}{N}kp} \tag{9.16}$$

And their corresponding eigenvectors are given as

$$\underline{w}_k = \begin{bmatrix} 1 \\ e^{j\frac{2\pi}{N}k} \\ \vdots \\ e^{j\frac{2\pi}{N}(N-1)k} \end{bmatrix} \tag{9.17}$$

We will collect these vectors to create a Fourier transform matrix given as

$$W = [\underline{w}_0, \underline{w}_1, \cdots \ \underline{w}_{N-1}] \quad (k = 0, 1, \ldots, N-1) \tag{9.18}$$

Each element of the Fourier transform matrix (kth row and pth column) is given as

$$w_{k,p} = e^{j\frac{2\pi}{N}kp} \tag{9.19}$$

We can show that the Fourier transform is Unitary ($WW^* = W^*W = I$) and as such we can write down the elements of the inverse of the Fourier transform matrix ($A = W^{-1}$) as

$$a_{k,p} = \frac{1}{N} \cdot e^{-j\frac{2\pi}{N}kp} \tag{9.20}$$

So we can continue with writing down the following representation of the received signal (this time with noise present), where circular convolution is used on the channel matrix.

$$\underline{y} = \hat{H} \cdot \underline{x} + \underline{n} \tag{9.21a}$$

We can then replace the channel matrix with its diagonalized equivalent.

$$\begin{aligned} \underline{y} &= W \cdot D \cdot W^* \cdot \underline{x} + \underline{n} \\ W^{-1} \cdot \underline{y} &= D \cdot W^* \cdot \underline{x} + W^{-1} \cdot \underline{n} \end{aligned} \tag{9.21b}$$

We can quickly see that the left-hand side of the above equation represents the Fourier transform of the received signal. While the right-hand side represents the Fourier transform of the data symbols, weighted by a diagonal matrix of eigenvalues (see Fig. 9.7).

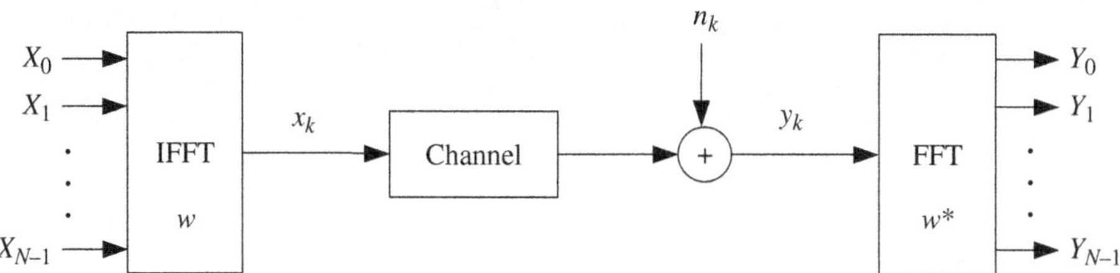

FIGURE 9.7 OFDM system emphasizing location of the transform matrices.

The above block diagram can be also used to write the equation for the received signal

$$\underline{Y} = D\underline{X} + \underline{\tilde{n}} \tag{9.22}$$

One last note to make regarding circular matrices, the product of circulant matrix, G, and a vector x can be commuted.

$$G\underline{x} = X\underline{g} \tag{9.23}$$

Alternatively written as follows where the G matrix is circulant with right circular shifts and the X matrix is circulant with left circular shifts.

$$\begin{bmatrix} g1 & g2 & g3 \\ g3 & g1 & g2 \\ g2 & g3 & g1 \end{bmatrix} \cdot \begin{bmatrix} x1 \\ x2 \\ x3 \end{bmatrix} = \begin{bmatrix} x1 & x2 & x3 \\ x2 & x3 & x1 \\ x3 & x1 & x2 \end{bmatrix} \cdot \begin{bmatrix} g1 \\ g2 \\ g3 \end{bmatrix} \tag{9.24}$$

9.1.3 3GPP Feasibility Study of OFDM

Similar constraints of 2G and 3G can be considered in to the development of an OFDM-based 3GPP digital cellular system [17]. One of the constraints is to have an uplink transmit signal with low peak-to-average-power ratio (PAPR) values, due to practical PA constraints. Here we refer to power not only in the context of transmit power, but also in the context of power consumption (current drain from the battery). Given this consideration single carrier FDMA (SC-FDMA) was chosen in the uplink physical layer for LTE.

Single-carrier systems have less PAPRs than traditional OFDM-based systems. This is because of the fact that in OFDM systems, many subcarriers are used instead of a single carrier. The reduction in the PAPR allows the transmitter to use more efficient PAs since less power back-off is required.

On the downlink the constraint is different. Here the base station transmitter must have other users as well as other control and pilot channels that will effectively increase the PAPR of the transmission anyway. Hence OFDMA is used on the downlink physical layer for LTE [18]. The uplink SC-FDMA system block diagram is provided in Fig. 9.8.

In single-carrier systems, the CP is added to enable the use of FDE methods. In conventional systems, as the system data rates increase, the complexity of the time domain equalizers exponentially increases. This increase in complexity can almost prohibit practical implementations. A solution to this complexity problem is to move the signal processing from the time domain to the frequency domain [19, 20]. Note that even in the frequency domain, applying the MMSE-based weights can be complex for a large number of subcarriers.

For this SC example, a CP is added in order to insert periodicity into the signal and thus the circular convolution solution can be used. This solution allows us to perform channel amplitude and phase compensation of the carrier with a single complex valued multiplication in the frequency domain. The reader is encouraged to review earlier work in this book where the topic of FDE was discussed.

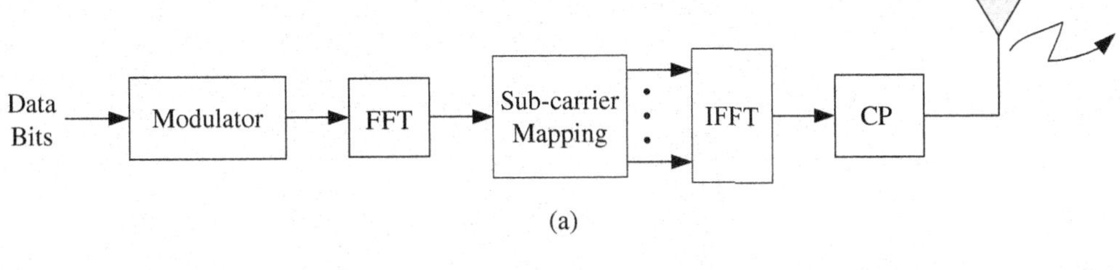

(a)

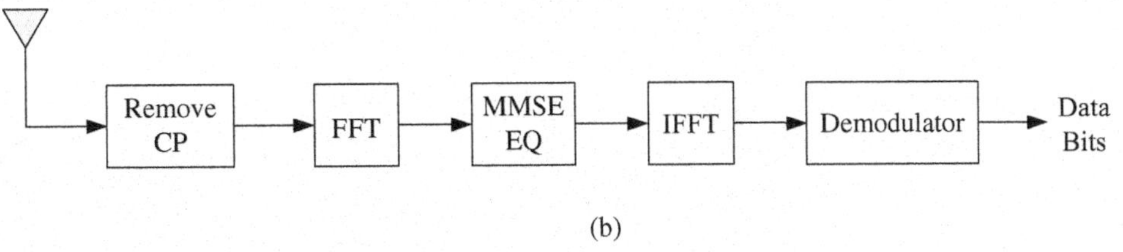

(b)

FIGURE 9.8 Single-carrier OFDM system: (a) Transmitter, (b) Receiver.

In order to keep 3G cellular competitive an evolutionary path has been identified that will not only increase user data rates, but also offer a variety of services. Some of the 3G LTE targets are given below:

- Data rates greater than 100 Mbps on the downlink and 50 Mbps on the uplink.
- Spectral efficiency improvement, measured in Bps/Hz/Cell
- Cell edge throughput improvement (coverage)
- Reduced system cost (operators, equipment vendors, and end users)
- Reduced complexity network (control plane, user plane, IP based)
- Reduced latency (call setup time, control plane, user plane)

A method used to provide an increase in data rates is through the use of multiple-input-multiple-output (MIMO) technology. Also variable BW allocation techniques can be used to address some of these performance targets. Specifically, 1.4 MHz, 1.6 MHz, 3 MHz, 3.2 MHz, 5 MHz, 10 MHz, 15 MHz, and 20 MHz spectrum BW have been identified. On the Downlink OFDM is used and on the uplink SC-FDMA is used. One comment is for sure; the LTE system evolution is surely not limited to the physical layer. The network and associated signaling must also change to accommodate the above-mentioned goals. This will be discussed in more detail in the HSPA evolution section that follows later in this chapter.

9.2 MULTIMEDIA AND MOBILE TV SERVICES

Several discussions pertaining to the delivery mechanism of multimedia services will be presented. They stem from dedicated systems such as DVB-H and MediaFLO to the evolution of existing shared systems such as Multimedia Broadcasting Multicast Services (MBMS) over the UMTS cellular system.

9.2.1 DVB-H Standard Overview

In this section, we will provide an overview of the Digital Video Broadcasting (DVB) system targeted to the handheld application, specifically called DVB-H.[1] This system builds upon the terrestrial version,

[1] Note there are also satellite and cable TV versions as well; these are denoted in the standards as DVB-S and DVB-C, respectively.

DVB-T, targeted for fixed and mobile terminals. DVB has the capability to deliver real-time services to a large group of terminals. Such services can be Mobile TV, live event broadcasting (such as sporting events), on demand video, and audio applications. When DVB-H is used with another mobile network such as GSM, CDMA2000, or WCDMA the application space can further increase. This rise in service options comes about from the fact that there is a reverse link to provide both feedback and data that can enable applications such as gaming. The reverse link allows the user to have a more interactive experience [21]. DVB-H delivers a downlink physical channel only. The additional features provided by DVB-H were aimed at accomplishing at least the following goals [22–24]:

- Reduce receiver power consumption
- Opportunity for seamless handoffs
- Improved system (S/N) performance
- Deployment trade-offs for service providers (cell size vs. mobility vs. data throughput)
- Backward compatibility to DVB-T
- Support of IP interface

The following additional features were added to support the above goals. They included time slicing, Multi Protocol Encapsulation with Forward Error Correction (MPE-FEC), a new FFT size, symbol (in-depth) interleaver, and necessary DVB-H signaling [25, 26].

Time slicing permits transmitting bursts of data to the handheld. This allows the receiver to be operating for a short period of time which in turn reduces the overall power consumption of the terminal. Below we present a figure expressing time slicing of three DVB-H terminals with different data rates and a single DVB-T terminal (see Fig. 9.9). Time slicing is mandatory in DVB-H. The time duration of data targeted for a user is called a burst. The time between bursts is defined as the OFF time. This is the time that the terminal doesn't need to be listening to the DVB-H downlink and hence can invoke power saving features.

One time-slicing burst includes one MPE-FEC frame; this creates a maximum size of approximately 2 Mb, generally representing 1–5 sec of the media content.

One can easily see that if the ratio of the burst time to the OFF time is low, then significant power saving is possible. In addition to reducing power consumption, time slicing can allow for seamless handovers. This can be achieved by using the OFF time to monitor/measure neighboring cells on other frequencies. In the absence of time slicing when a terminal would perform these neighbor cell measurements, an interruption in service would occur. The combination of time slicing and reasonable synchronization between DVB base stations can provide a seemingly nondisruptive reception of data when handing off from one cell to another. As presented earlier as long as the base station time differential plus delay spread is less than the CP the DVB-H system is operating properly.

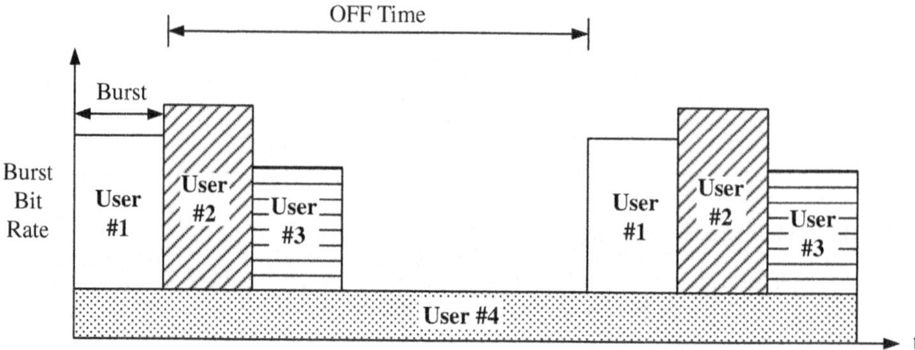

FIGURE 9.9 DVB-H and DVB-T transmission example.

Broadcasting of the "cell identifier" is mandatory in DVB-H and is used to help identify neighboring network with same services as the present one.

MPE-FEC adds another layer of error correction by adding additional parity information to the data to be transmitted. This is accomplished by using the Reed-Solomon RS(255,191) outer FEC code with error correction capability of $t = 32$ bytes. MPE-FEC is optional in DVB-H.

The DVB-H system is IP based hence the interface to higher layers is an IP interface. Since MPEG-2 transport streams are still the standard DVB interface, an adaptation protocol is needed. MPE is a protocol used to encapsulate multiple IP streams into the MPEG-2 DVB-T transport stream. For file download services Raptor FEC codes are applied in the application layer [27]. MPE-FEC is disabled when Raptor FEC is used.

The DVB-T supports 2K (2048) and 8K (8192) FFT sizes. DVB-H adds a 4K (4096) FFT size mode. This allows the network to have a trade-off of Doppler performance, mobility performance, cell size, and spectral efficiency. The 2K FFT system can tolerate the highest Doppler spread of the three possible values. However, it has a short guard interval preventing a large cell size deployment scenario. On the other hand, the 8K transmission mode has a large guard interval and suffers in performance in high Doppler spread cases. Lastly, a 5-MHz bandwidth has been added to the list of available BWs supported.

Symbol level interleaving has also been suggested to obtain better performance in channel conditions with larger bursts of error. The interleaving is applicable to the 2K and 4K transmissions modes.

Lastly, DVB-H signaling is required to indicate the various parameters associated with the above-mentioned additional features.

A block diagram of a network providing DVB-H services is shown in Fig. 9.10 where each cell is transmitting in a different frequency band (e.g., f_1, f_2, and f_3).

In Fig. 9.11 we provide an example protocol stack and software architecture of the DVB-H system. Note IP data can be sent over DVB-H through what is called IP data casting [26].

A few concerns should be addressed in such a broadcasting system. First deals with handoff measurements and execution. In the time slicing aspect this is a period of time denoted in the DVB-H standard as OFF time. However, note the receiver is not completely turned off. The receiver must decode the

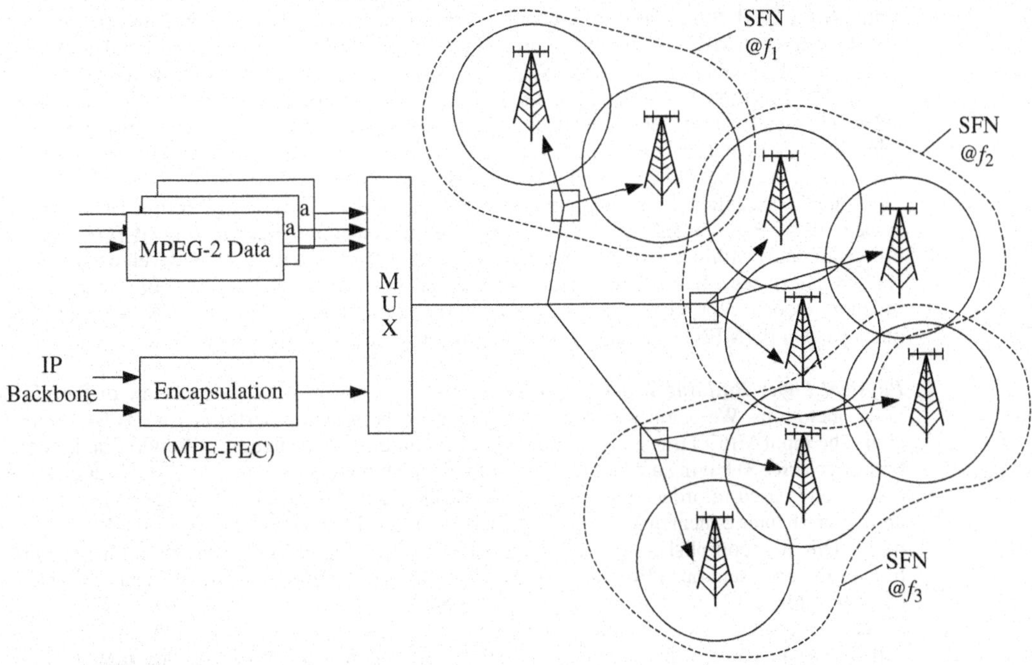

FIGURE 9.10 DVB-T and DVB-H network scenario.

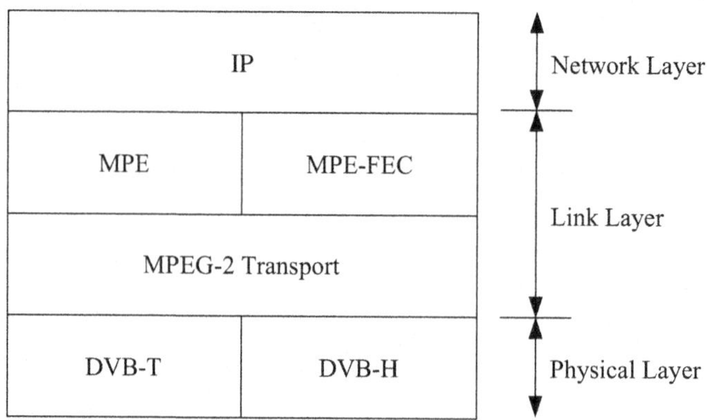

FIGURE 9.11 DVB-H protocol architecture.

data received over the DVB-H air interface and deliver the content to the specific media player. Certain sections of the terminal can be powered down for a reduction in power consumption, but the user must see a continuous data stream at the application end. Moreover, if neighboring cells need to be monitored to access the need for a handoff, then this power down benefit is lost during those periods of time. For this case, in addition to decoding the burst of data previously received the terminal must switch the frequency band to make certain measurements. Lastly, if a handoff must occur and the base stations are not reasonably time aligned then a disruption in service can occur. This disruption is somewhat dependent on the terminal capability (receiver architecture, size of memory, processor speed, and so on) and should be minimized in order to deliver a gratifying experience to the end user [28].

The second concern deals with channel switching. Here a user may want to view other channels and therefore the DVB-H may need to demodulate another time slice of data. In this case the terminal may need to wait for an MPE-FEC frame (of possibly different size than the existing one) and possibly compensate for other delays in the media. Needless to say this disruption to the user should be minimized.

Another concern deals with tune-in time for new linking terminals. Similar issues related to channel switching also apply here with any additional synchronization concerns due to the need to determine time-slice synchronization, and so on. In either case, it is always prudent to have a robust MPEG error concealment/handling algorithm to assist in these times.

Generally speaking, there are many issues affecting the overall system performance; below we list some of them: BW management, possible encryption, scheduling of services, DVB-H receiver capability variations within a coverage area, future IP-based applications, content development for small size display screens, base station capability, content server requirements, other FEC improvements, indoor/outdoor reception performance, coexistence with cellular systems, SFN and multiple frequency network (MFN) deployment, higher transmit power, and so on.

Physical Layer Building Blocks. In this section, we will provide an overview of the physical layer operations. We begin with a block diagram of the function performed in DVB-H transmission. The input MPEG transport stream (TS) is essentially scrambled by a Pseudo Random Binary Sequence (PRBS) having a generator polynomial equal to $g(x) = x^{15} + x^{14} + 1$. Concatenated error correction is used in order to get required performance to deliver the intended services. Recall in an earlier chapter concatenated coding performance results were presented for the RS + CC scenario. This is accomplished by having an outer Reed-Solomon FEC code and an inner convolutional FEC code separated by a convolutional interleaver operating on bytes of data having a depth of 12 (see Fig. 9.12).

In order to support variable bit rates the inner convolutional code is punctured. The convolutional code has a native rate of $R = 1/2$ with a constraint length of $K = 7$. This rate is punctured to allow $2/3$, $3/4$, $5/6$, and $7/8$ rates. A puncturing patterns table is provided in the DVB standard.

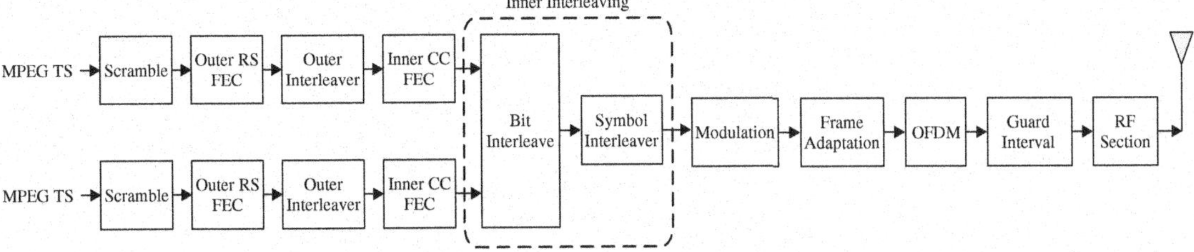

FIGURE 9.12 DVB-H physical layer transmission block diagram.

The Reed-Solomon code is a RS(204,188) with error correction capability of $t = 8$ bytes; this is shortened from the RS(255, 239) code by inserting 51 bytes of all zero value to the 188 input data.

The inner interleaver consists of a bit and symbol interleaving operation for the 2K and 8K transmission modes and has been extended to the 4K mode. As mentioned above there is an optional in-depth symbol interleaver for both the 2K and 4K modes in order to obtain better protection against burst of errors. Specifically, the in-depth interleaver doubles the protection for the 4K mode and quadruples the protection for the 2K mode.

The DVB system uses Orthogonal Frequency Division Multiplexing (OFDM) technology. Each subcarrier in the OFDM transmission is modulated by Quaternary Phase Shift Keying (QPSK), 16-QAM or 64-QAM. The DVB standard provides the technical details about the bit to symbol mapping for the various constellations supported.

The transmitted signal consists of OFDM frames. Each frame has a duration of 68 OFDM symbols. Each symbol is supported by a number of subcarriers, as shown in Table 9.1, for the different transmission modes.

TABLE 9.1 Transmission Mode Definitions

Transmission mode	Total no. of subcarriers	Useful data subcarriers	Continual pilot subcarriers	TPS subcarriers
2K	1,705	1,512	45	17
4K	3,409	3,024	89	34
8K	6,817	6,048	177	68

The OFDM signal consists of both pilot and data information. The pilots are extremely important in producing the required QoS performance. Namely, they should be used for time synchronization, frequency synchronization, channel estimation, as well as support other signal processing functions in the receiver. The pilot information is scattered across the subcarrier and across the OFDM frame. Note this pilot information is transmitted at a higher power level in order to aid in the aforementioned functionality.

It is important to now discuss the BWs that the DVB standard can operate in, namely 5 MHz, 6 MHz, 7 MHz, and 8 MHz. Hence for a given transmit BW, increasing the transmission mode will increase the number of subcarriers or decrease the intercarrier frequency spacing. As discussed earlier, this is detrimental in the presence of a large Doppler spread. Also as the BW decreases so does the maximum possible useful bit rate in the DVB system. The DVB standard identified the possible user data rates in the form of tables and will not be repeated here for sake of conserving space.

In this OFDM system a guard time or interval is inserted into the transmit signal which consists of a cyclic extension of the data. Each of the transmission modes has four guard interval time options. For sake of discussion below we present the guard interval durations for the 5-MHz bandwidth (see Table 9.2). The variety in the guard interval length allows for various deployment options.

TABLE 9.2 Guard Interval Duration for the BW = 5 MHz Option

Guard time ratio	8K mode	4K mode	2K mode
5/6	358.4 μsec	179.2 μsec	89.6 μsec
1/8	179.2 μsec	89.6 μsec	44.8 μsec
1/16	89.6 μsec	44.8 μsec	22.4 μsec
1/32	44.6 μsec	22.4 μsec	11.2 μsec

Signaling information about the transmission scheme parameters such as channel coder, modulation, frames number, etc., is sent by the Transmission Parameter Signaling (TPS) carriers. A TPS block consists of 68 bits. Each OFDM symbol in the frame contains 1 TPS bit. The TPS block is protected by a BCH code of (67, 53) with error correction capability $t = 2$ bits. Moreover, each TPC carrier uses DBPSK modulation where due to the differential encoding it is initialized in each TPS block.

The coexistence of the DVB system with the cellular systems will become more and more desirable from an end user perspective. In Fig. 9.13 we provide a simple example where the DVB-T and DVB-H share a frequency BW allocation. In addition, the terminal is also capable of receiving the cellular traffic as well; note this is a bidirectional link.

The interaction between the DVB-H and UMTS networks can take on a variety of forms. The collaborative block diagram presented in Fig. 9.13 shows the internetwork communications occurring at the IP level.

9.2.2 Multimedia Broadcasting Multicast Services

In this section, we will provide a description of the MBMS function within the UMTS standard. MBMS provides a means to deliver multimedia content to the UE. Capabilities exist to offer point-to-point (p-t-p) and point-to-multipoint (p-t-m) transmission modes [29, 30].

MBMS will use the MBMS notification indicator channel (MICH), MBMS p-t-m control channel (MCCH), MBMS p-t-m traffic channel (MTCH), and MBMS scheduling channel (MSCH). The MBMS p-t-p transmissions occur over the DTCH channels.

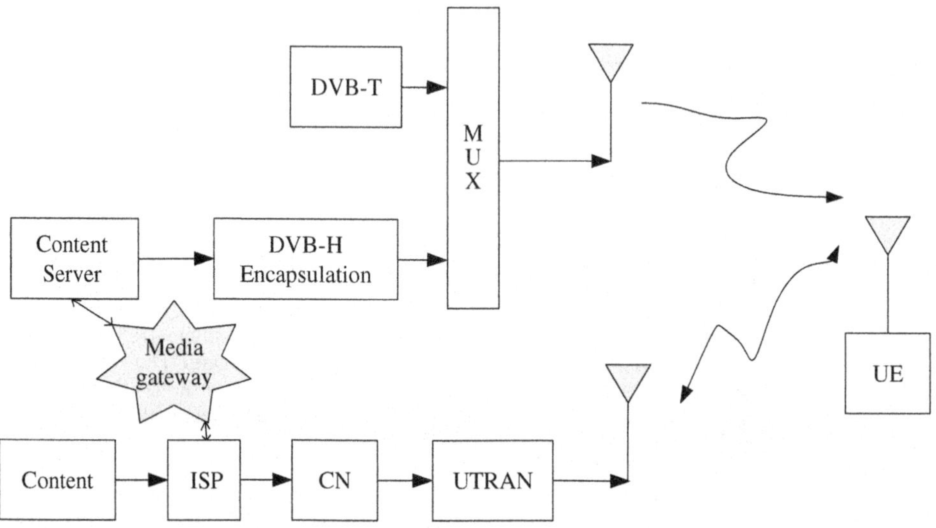

FIGURE 9.13 Mobile and broadcast operator collaboration.

The S-CCPCH will carry the MCCH channel. The MCCH information will be transmitted periodically using a reception interval. Changes to the MCCH can occur on the first MCCH transmission of the modification interval. UEs are notified about the changes by the MBMS notification indicators, MICH. The modification interval will be long enough so that UEs begin to access the MCCH at the beginning of the next modification interval. Once the MICH is detected, the UE begins to access the MCCH at the beginning of the next modification interval. The p-t-m data will be sent to the UE via the MTCH channel. The MSCH channel is used to convey the MBMS transmission schedule between the network and the UE. When the p-t-m service is active, the UE should periodically monitor the MCCH channel, which may be on a different S-CCPCH than the one carrying MTCH,

The channel mapping is provided in Fig. 9.14.

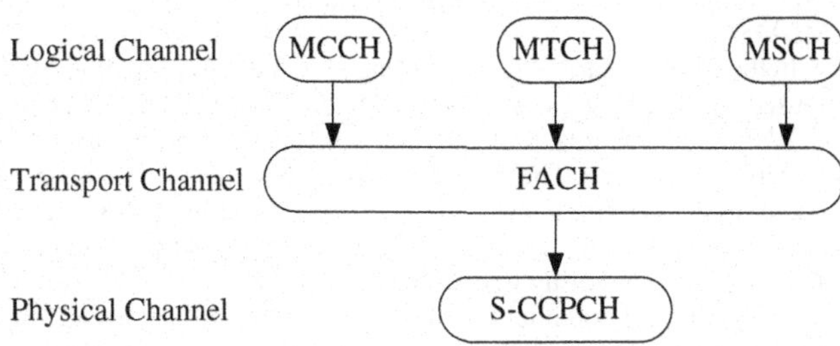

FIGURE 9.14 MBMS channel configurations.

For this particular MBMS service the UE capabilities will play a key role in not only supporting system performance, but also providing product differentiation.

System performance issues will be addressed by the TTI capable to be decoded, for example 40 msec or 80 msec or others. The larger TTI provides better interleaving capability and thus has improved performance. Since MBMS services are expected to handle mobility, then handoffs should be supported. The MBMS transmission allows for selective combining and soft combining between radio links. This is possible since the RLC Protocol Data Units (PDU) are numbered. As long as the NodeBs are reasonably synchronized and they have the same bit rate, selective combining can be used. We must also mention this assumes the RLC reordering capability of the UE is not exceeded. The 3GPP standard states that soft combining is possible given the delay between radio links is less than 1 TTI plus 1 time slot. Performance and memory requirement trade-offs should be conducted to determine acceptable QoS delivery to the end user.

Product differentiation issues will be addressed by simultaneous configurations. For example, the possibility to receive a DPCH/HS-DSCH and S-CCPCH carrying MTCH/MCCH simultaneously is a UE capability. The minimum MBMS bit rate and combining method are also under the control of the UE. This information is important and should be conveyed to the Radio Network Controller (RNC) in order to perform or potentially avoid additional signal processing. For example, if UEs can only support selective combining the RNC can resynch the cells to deliver reasonably synchronized MBMS transmissions to the UE.

In an effort to conserve UE power consumption and support simultaneous services, the UE will use DRX on the MTCH transmission based on scheduling information. In addition, for UE with enhanced capabilities (i.e., receive diversity, large memory) having intelligent signal processing can be utilized for further savings in power consumption. Since the UE will be operating in various states, there is a possibility that MBMS transmission will be interrupted due to the UE receiving a voice call, performing IRAT measurements, general handoff hazards, and so on. The 3GPP standard states that in general, "*data loss should be minimal.*"

There are many ways to address this requirement; one method is to often send critical system information over to the UE such as the neighboring cell information, and so on.

In MBMS services, application layer FEC is required for UEs; they are called Raptor codes [27]. As the 3GPP WCDMA system continues to evolve MBMS one method of improvement involves the UE being capable of simultaneously receiving MBMS and dedicated services on separate carriers. This will surely have an impact on not only the UE but also the network architecture and signal processing required to support such service. A particular application that would make use of MBMS is called Mobile TV. Deploying a successful service implies that MBMS will support a large amount of UEs. These resources should be efficiently used in order to not degrade other services within the same frequency carrier [31].

A potential scenario is to have the MBMS services delivered on a separate carrier frequency than the p-t-p transmissions. This allows UMTS operations to have flexibility to offer a variety of levels of service. The separate carrier frequency means the MBMS transmission from all the NodeB transmission will be on the same carrier frequency with some reasonable level of synchronization. This deployment strategy is called single frequency network (SFN) or MBMS single frequency network (MBMSFN).

In the SFN scenario signals coming from neighboring NodeBs will consist of valid multipaths to be used by either the GRAKE or equalizer receiver. The service area can be made as large as that of the RNC coverage area, thus providing the operator complete control over the content delivered over the deployment zone. The implications of such a deployment are profound. First the operator would need to either acquire a new frequency band to offer such services or it must take away some voice capacity from existing frequency bands. These decisions are made on a case-by-case situation requiring input from marketing trend analysis.

Neglecting the potential loss in voice capacity in such a spectrum deployment, the UE capabilities have increased. We mention this because the same RF section can be used for both the p-to-m as well as the voice/data services. This approach may make sense as an entry point into this Mobile TV market. With proper scheduling and an occasional loss of minimal data, the user satisfaction rate should be high.

Alternatively, demanding the UE to simultaneously monitor both the dedicated SFN and the Release 7 network has a tremendous impact on the UE complexity. A potential scenario is for the UE to join a MBMS session and periodically check the Release 7 network for incoming voice calls. If such an event occurs then the user must make a decision to either continue watching TV or answer the phone call. True simultaneous deployments would require a dual receiver in the UE. A potential UE capability would be for it to support the Type 1 diversity receiver when either service is being used and when a heterogeneous mix is required then a single receive antenna is used for each scenario. This seems to be the long-term solution for delivery of true customer satisfaction.

9.2.3 Competing Broadcast Technologies

In this section, we will compare two competing broadcast systems. Specifically we will compare DVB-H to the publicly available technical specifications of MediaFLO. It is not the intention of this section to "sway" the reader toward a particular direction. Rather we wish to have a place of reference where both standards can be compared and debated against each other.

MediaFLO and DVB-H Discussion. MediaFLO is a coined term representing Media Forward Link Only (FLO) [32]. It is a proprietary standard initially targeted for network scheduled delivery of media during peak and off-peak times. This media content can be saved inside the terminal to be viewed at a latter time. An intended spectrum usage in North America is the 700 MHz spectrum, this was channel 55 on UHF TV. Lower carrier frequencies have better propagation path loss and thus less base stations are needed to blanket a coverage area, when compared to much higher carrier frequencies.

Reference [33] provides some technical details on the physical layer. Both MediaFLO and DVB-H allow for various channel BWs. In Table 9.3 we provide a chart summary from the references herein.

At this time we would like to point out that various Mobile TV services have been and are planned to be deployed throughout the world. The table below provides a time snapshot of such global services available.

TABLE 9.3 Broadcasting System Comparison

System parameter	MediaFLO	DVB-H
FFT Size	4K	4K
Guard interval ratio	1/8	1/4, 1/8, 1/16, 1/32
Modulation	QPSK, 16-QAM	QPSK, 16-QAM, 64-QAM
Interactive capability	Yes w/cellular system	Yes w/cellular system
OFDM	Yes	Yes
Symbol duration	833.33 μsec	Variable
FEC	Turbo + Reed – Solomon	Convolutional + RS
Frequency bands	5 MHz, 6 MHz, 7 MHz, 8 MHz	5 MHz, 6 MHz, 7 MHz, 8 MHz
Power savings	TDM	Time slicing
Inbuilding services	Yes	Yes
BS separation	Variable	Variable
Channel switching times	~ few seconds	~ few seconds

USA

- DVB-H
- MediaFLO
- ATSC = Advanced Television System Committee

Japan

- ISDB-T = Integrated Services Digital Broadcasting—Terrestrial

Korea

- T-DMB = Terrestrial—Digital Multimedia Broadcasting

Europe and Asia

- DVB-H

China

- DTMB = Digital Terrestrial Multimedia Broadcasting.

9.2.4 Continuous Packet Connectivity

As packet data services expand, users will want to experience continuous connectivity. 3GPP has coined a term called Continuous Packet Connectivity (CPC) where users are essentially always connected with an occasional transmission. This technique removes the delays involved in connection establishment.

For large number of High Speed Downlink Packet Access (HSDPA) users having the associated dedicated physical data channel (DPDCH) and DPCCH, each user can quickly approach a code-limited scenario. Hence is one of the reasons F-DPCH is included in the Release 6 version. For this case the UE will monitor traffic on the HS-DSCH channel and uplink PC commands on the F-DPCH channel. Since users will have their PC command time domain multiplexed under the same scrambling code, downlink code usage can be dramatically reduced. On the uplink, supporting many E-DCH users creates an increase in noise (noise rise). In order to combat this rise in interference the UEs can either transmit with lower power or endorse some sort of DTX campaign (see Fig. 9.15). It is expected that there are common scenarios when not all the users will be transmitting all the time. This implies some users will transmit some of the time. In this case if we impose a restriction that the UE transmits packets only when it has information to transmit, then uplink noise rise can be significantly reduced [17]. When triggering the "always on" concept, the UE will be placed in the continuous packet connected mode. What this means is that the UE will be capable to enter and exit this mode depending on the particular service demand at that time instant.

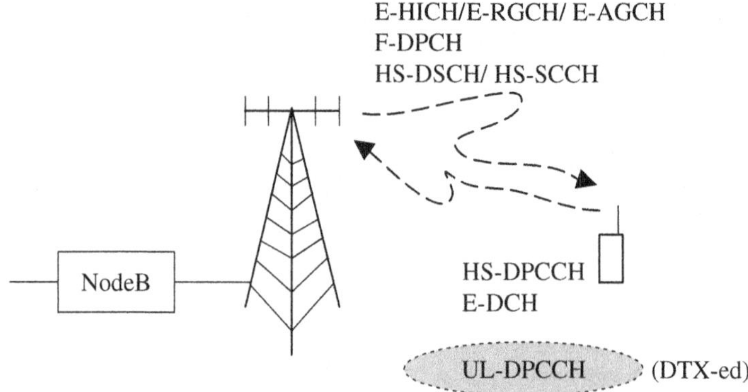

FIGURE 9.15 CPC general overview.

The concept of Continuous Packet Connectivity for HSUPA applies when there is no E-DCH or HS-DPCCH data to be transmitted, here we apply DTX on the UL-DPCCH. The best compromise is to maintain the connection and allow the UE to transmit only when required. However due to practical reasons having very long inactivity periods can be detrimental to the system especially if the NodeB would declare synchronization is lost. It would behoove the system to not terminate the connection and then re-establish it when data is ready to be transmitted. This approach would introduce unnecessary overhead and delays. Hence the CPC operational mode allows the uplink to be completely periodically DTX-ed in order to reduce the uplink noise rise, save on UE power consumption and reduce the interference to the other users who are actively transmitting. In order to further reduce the UE power consumption DRX operation is possible on the downlink related HSDPA and HSUPA channels. The system implications of such a service go beyond the CELL-DCH radio resource connected state and into the CELL-FACH, CELL-PCH and URA-PCH states as well, especially when HS-DSCH capability is present.

As shown in Fig. 9.15, where we described the physical channels involved in the CPC mode, it was required to define a new UL-DPCCH slot format. In this case only pilot and transmit power control (TPC) bits will be transmitted in the UL-DPCCH.

The UE is placed into CPC mode through signaling on the HS-SCCH and hence the UE begins transmitting the UL-DPCCH with a new time slot format described above. The new slot format should be configured by the RNC in order to potentially avoid blind slot format detection in the non-serving NodeBs.

In this case, it is important to clearly define the mechanisms to be used to detect the standard end of the total silence period. In order to help with the data channel re-activation, preambles are transmitted to help the NodeB with this transition. The 3GPP standard defines DTX cycle periods and various modes the UE must check in order to determine if the UL-DPCCH is allowed to be transmitted. For example, E-DCH and UL-DPCCH are always time slot aligned on the uplink whereas the HS-DPCCH may not be due to transmission time offsets. The reader should consult 3GPP TS25.214 for further details.

Besides the above-mentioned benefits of power consumption when the UE is in CPC mode, there are also some disadvantages to consider. Particularly, when the UE transmission is silenced the uplink TPC commands are not transmitted which has an overall effect on the closed loop power control behavior. While the UE is operating in CPC mode the NodeB should provide more information on the parameters rather than the RNC in order to reduce system latency.

We close this section by listing the related measurement that the system designer should pay considerable attention to:

- Handoff scenarios
- Power control behavior (HARQ support)

- NodeB resources
- Transitional behavior (Turning CPC On and Off, Preambles)
- Measurement support (Inter-RAT, Intra-RAT)
- BLER performance degradation cases, and so on

9.3 SOME SAMPLED 3G TERMINAL (UE) STATISTICS

In this section, we will provide some results of a statistical analysis performed on publicly available 3G UMTS terminals. Product differentiation can be achieved from a variety of ways. For starters talk time and standby times can be used in the comparison, but caution must be exercised here to normalize the results by the battery capacity. Display size and terminal weight are also key areas that deserve attention. As the UE's data rate capability is increased, the display size is also increased in order to provide a visual impact to the user's experience. The number of colors available in the display increased from the early arrivals using 4096 colors to the latest terminals using approximately 262K colors. Moreover, the number of displays has also doubled which is especially true for the famous "clamshell" form factor phones. Generally speaking, these features have improved year upon year from circa 2002 to the present.

Figure 9.16 presents select statistics for features such as weight, percentage of terminals having ~262K colors, and percentage having two displays.

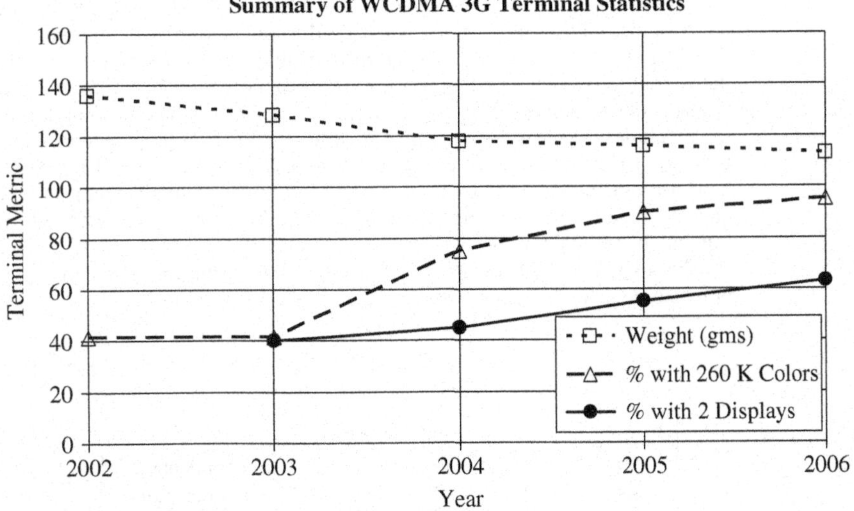

FIGURE 9.16 Select UMTS terminal statistics.

The weight trend clearly shows the downward behavior of making the phones smaller and lighter. The number of terminals having two displays is really applicable to those terminals with the clamshell form factor. We have presented the results based on the overall terminal population and decided not to present these results conditioned on the clamshell form factor. Also the number of colors used in the display has dramatically increased which helps in displaying pictures, videos, etc., and generally contributes to the viewing pleasure.

The number of phones that contain at least one camera has also increased as time progressed since the introduction of UMTS phones. Moreover, the quality and resolution of the cameras has increased to top a few megapixels (and growing). A last comment we wish to make in this section is that the size of the memories within the UEs has dramatically increased since their introduction. This size can conceptually be split into two regions, a public and private region. The private region is used for general terminal operations such as physical layer and upper layer support. The public region is closely tied to the features of the phone such as movie, picture, and audio storage. In addition to the storage demands, applications running on operating systems such as Symbian, Windows ME, Linux, and so on can benefit from the increase in memory capability.

9.3.1 Canonical/Classical UE Architecture Discussion

Next we provide a simple architecture so that we can discuss functionality partitioning and introduce the impacts of supporting simultaneous services with varying degrees of acceptable quality. The high level partitioning is provided in the block diagram shown in Fig. 9.17. We have decided to split the responsibility across four sections: spectral conversion (i.e., RF), communication processor, applications processor, and audio/video/data (multimedia) interface [34, 35].

Beneath each functional block we simply list a few items that deserve attention associated with that particular block. The spectral conversion block, labeled RF section in Fig. 9.17, is responsible to perform the frequency band switching, transmit and receive power adjustments, as well as other functions associated with spectral up and down conversion.

The communication processor (CP) overview is given in Fig. 9.18. The communication processor is responsible for maintaining the air interface link quality and basically delivering data "pipes" to the application processor or other devices.

The block contains the AFE interface such as ADC for the receiver and DAC for the transmitter. This also includes the RF controls such as AFC, AGC, operating frequency bands, and so on. A power management block is required in order to provide an overall low-power consumption design. We have inserted two programmable processors: microcontrol unit (MCU) and digital signal processor (DSP). We have shown both devices present; however their presence depends on the actual target application, UE capability, and data rates supported. In fact, nothing precludes the designer from inserting more devices as long as their cost, size, and power budgets are obviously met. For legacy reasons the audio codec (VSELP, AMR, and so on) functions have been inside the communication processor since terminals were originally voice centric. We have decided to leave this functionality within this section, but different functionality partitioning may be more desirable.

The application processor (AP) functional architecture is given in Fig. 9.19. The application processor essentially makes use of the data pipes delivered from the communication processor.

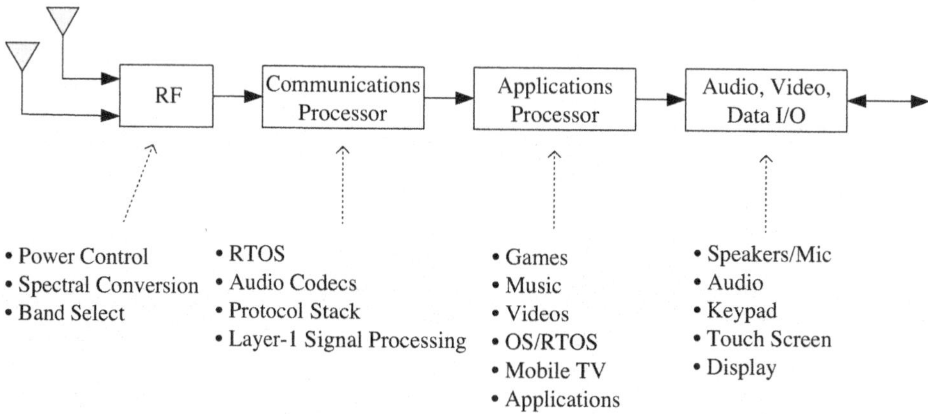

FIGURE 9.17 General receiver architectural functionality.

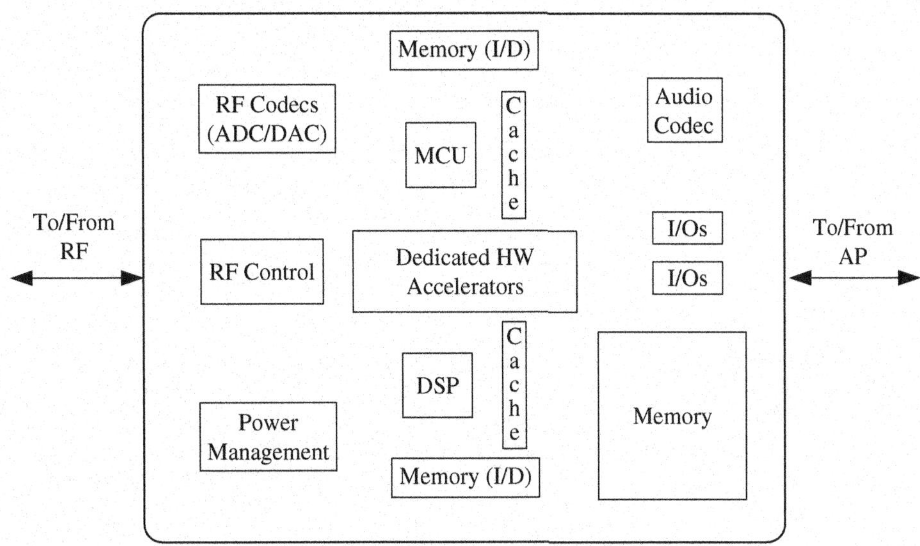

FIGURE 9.18 Communications processor functionality.

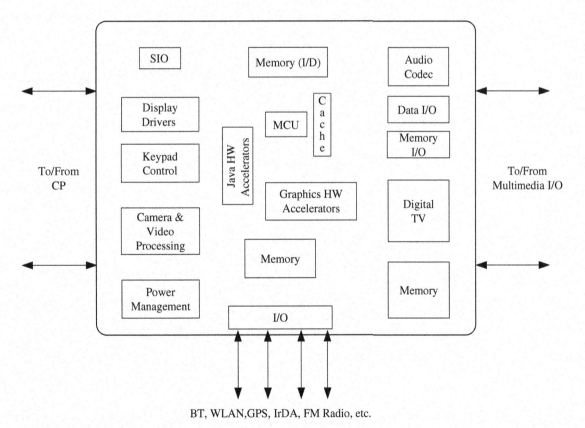

FIGURE 9.19 Applications processor functionality.

The block diagram includes the display(s) drivers and keypad control. As mentioned above, quite a few UE manufacturers produce terminals with two cameras. As time progresses the camera resolution, measured in megapixels, increases. The resolution comes at the expense of additional signal processing as well as memory required to store the new family photos, as well as the short videos.

We have decided to only show a single programmable processor (MCU) for sake of convenience, the same comment made earlier for the communication processor applies here as far as the number of required processors. The speed of the MCU is also increasing as well as the amount of memory (both public and private) available. There exist MCU cores that have or support HW-based accelerators. These accelerators should be used for the computationally extensive operations such as 3D gaming, and the like. Additional memory is needed to store and run the multiple applications, store the photos taken, keep music videos, and so on. As time progresses being able to record favorite Mobile TV programs will be a normal usage scenario.

Lastly, we inserted a variety of the I/O interfaces to handle potential other third-party vendors of Bluetooth (BT), Wireless LAN (WLAN), Global Positions System (GPS), IrDa and possibly, Mobile TV such as DVB-H and/or MediaFLO.

The complete AP + CP design should be stressed for a variety of combinations supporting simultaneous services such as **3G Voice + WLAN + BT** or **3G Voice + Gaming** or **3G Voice + Mobile TV**, and the like. We will present a simple overall system level block diagram that is used to demonstrate the proposed methodology and identify critical areas (defined as bottlenecks) in the overall design (see Fig. 9.20).

We have traced the data/control flows through the receiver for three cases. The first case is the traditional voice application which is outlined by the line with the long dashes. Generally the data samples flow through the CP, for example, the RAKE transport channel (TrCh) demultiplexing and then to the vocoder. In the WCDMA case this will consist of a single-code despreader with a high spreading factor. The second case is the simultaneous service scenario of both voice and data, which are outlined as the short- and long-dashed lines. In the WCDMA space this would consist of receiving not only the DPCH but also the HS-DSCH. Here a higher data throughput through the CP is required (possibly mandating advanced receiver architectures as previously discussed in the earlier chapters) and then entering a multimedia device. This application can be surfing the world wide web (WWW)

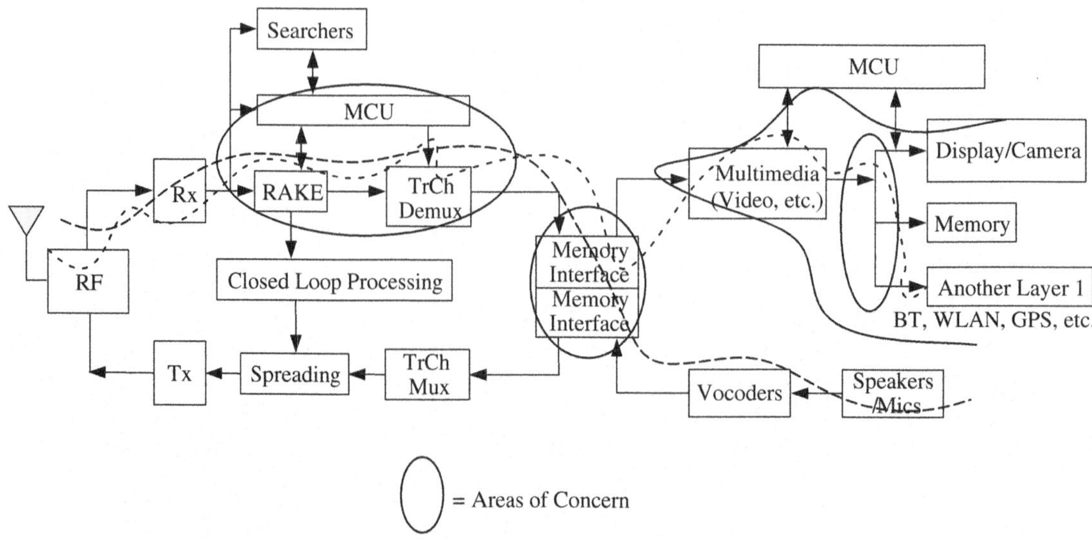

FIGURE 9.20 Overall transceiver block diagram emphasizing computational bottlenecks.

while maintaining a conversation. The last case simply involves using the UE as a multimedia device such as gaming, video, still picture camera and/or audio player (shown by the solid line) while using another Layer 1 interface (e.g., VoIP over WLAN). Bottlenecks that arise due to these options should be optimized for latency and power.

As shown above we have identified three areas of concern. First is maintaining the quality of service for the data pipes delivered from the CP to the AP. The second is the interface between the two processors which becomes more problematic as the data rates increase (which is especially true when MIMO technology is used to bring tens of Mbps to the terminal). The last area of concern is within the AP, dealing with other radio access technologies and high quality graphic, 3D gaming, etc.

9.4 HIGH-SPEED PACKET ACCESS EVOLUTION

In this section, we will discuss evolutionary paths to enhance the capabilities of HSPA systems [17, 36]. The optimal path is one that would provide a smoother transition toward supporting Long Term Evolution (LTE). Some general requirements of the evolution are higher data rates, increased system capacity, reduced latency, increased system coverage, reduced system operating costs, just to name a few. These requirements not only assume backward compatibility to legacy systems, but also apply to packet-based networks. In fact, IP multimedia subsystem (IMS) is a key component in the driving force to upgrade networks from circuit-switched services (CS) to packet-switched services (PS) domains.

In earlier chapters we have provided the current WCDMA network architecture and we would like to present a possible evolutionary path. The simplified architecture is provided in Fig. 9.21.

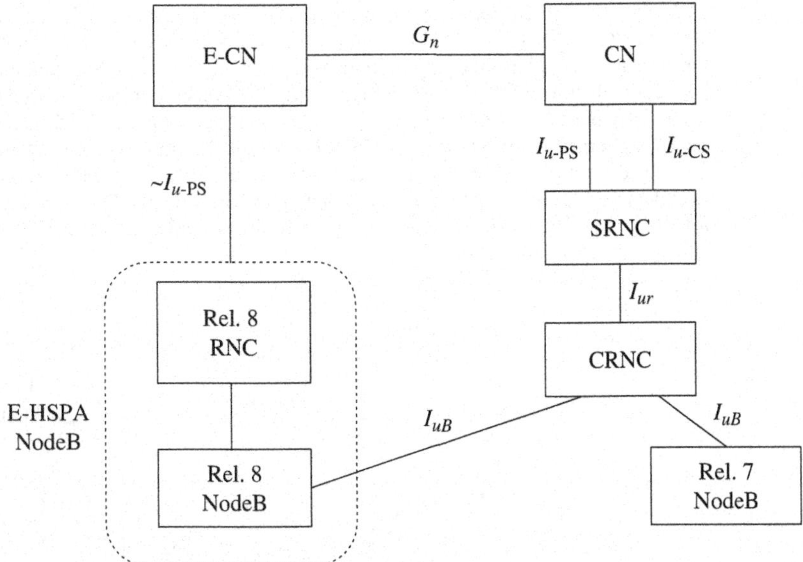

FIGURE 9.21 Possible evolutionary path for HSPA.

The basic premise behind the enhanced HSPA NodeB is that the less entities that actually "touch" the data and control before being sent through the CN the better off the system is in terms of reducing overall system latency. This statement has profound effect in that changes in signaling, functionality, and protocol stacks need to be seriously addressed. In Fig. 9.21 the functionality of the RNC and NodeB come closer together to reduce the overall system latency.

In an effort to address some of the other evolutionary requirements, such as increased data rates, additional signal processing functions are required. For example, HSDPA Release 6 and below assume use of QPSK and 16-QAM modulation schemes. Higher Order Modulation (HOM) scheme such as 64-QAM is used in the Release 7 version. Due to the increased number of modulation states, the most notable usage scenario would be in a large geometry setting with large NodeB power allocation in a slow moving, pedestrian-like environment. These comments have been added to stress operating conditions the UE should encounter in order to observe the expected performance improvement.

Similarly, HSUPA Release 6 and below assume utilizing BPSK uplink modulation scheme. The HOM solutions would also work in this case, for example, 16-QAM is used. A key outcome here is that the uplink PAPR would increase and thus force the UE to use more linear PAs or apply various power backoff strategies. This would also seem to point to usage cases with high geometries as a starting point. Note, generally speaking, using HOM tends to force tighter linearity requirements (i.e., smaller EVM targets) in both the transmit and receive end of the communication link. We should mention that uplink transmission of HOM also requires the NodeB receivers to move to more advanced receiver architectures to fully take advantage of the proposed benefits.

Continuing along the lines of increased data rates as well as increased capacity and increased coverage, we can mention the use of enhanced receivers. And under this umbrella we will include MIMO, receive diversity, and interference cancellation technologies. As far as MIMO is concerned, we will discuss this technique in more detail in a later section but suffice it to say that initial applications consider two transmit and two receive antenna scenarios. The complexity impacts of this technique are tremendous in that not only are two receivers needed, but also a more complicated baseband receiver. Moreover, since a maximum of two spatial data streams can be multiplexed into the spatial channel, two CQI estimators are required to drive the NodeB scheduler.

As discussed in earlier chapters, the usage of receive diversity with or without interference cancellation (IC) can lead to a few dB of improvement. We leave it to the system designer where to use this gain in order to have the largest impact. Currently, the two leading candidates are the dual receive diversity with and without IC capability. A simplified receiver block diagram of the signal processing blocks showing a reference receiver implementation is given in Fig. 9.22. The first block diagram uses the MMSE cost function in the creation of the array weights, but only the desired signal's channel matrix is estimated. In the second block diagram the desired and interference matrix is estimated to form the true MMSE-based weights. The spreading and despreading functions have been removed in order to generalize the operations and it is up to the system designer as to where the chips to symbol conversion should take place.

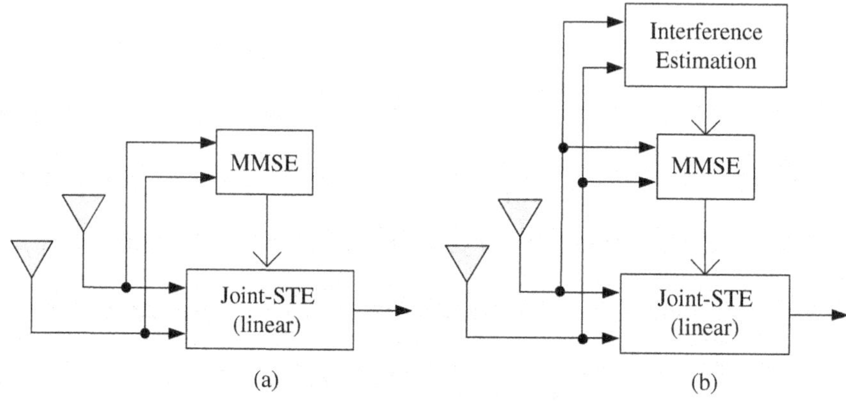

FIGURE 9.22 Dual receive diversity with and without IC capability.

Creating cost reduced, small size NodeBs for femto-cell[2] applications seems to be gaining some traction. In this case the usage scenario can be the femto-cell NodeB that will allow UE to act as your home cordless phone. Also with advancement in the NodeB and UE, the opportunity to also behave as a data service provider, such as WLAN exists.

Many possible problems can occur with this particular application the first deals with interference management. How and who will coordinate the femto-cell interference on the public network is a question deserving attention. Other potential issues deal with cost and capacity reduced versions of the public NodeB and the type of interface to the wireline side. The last issue we wish to address is that of handoff. This includes scenarios when the UE hands off between the private and public networks and vice versa. This is especially problematic when ping-ponging occurs.

What is important to mention is that there is significant marketing traction in this femto-cell or Home NodeB application. Below we provide potential benefits to both the operators and users [37].

Operators

- Increases coverage
- Increases capacity
- Enables quadruple play
- Competes with VoIP deployments
- Promotes more 3G traffic
- Increases revenues

Users

- Increases coverage
- Provides quadruple play options
- Multiple voice/data lines
- Reduces costs
- Handoff scenarios

As discussed above, moving to IMS coupled with reduced latency times makes VoIP in wireless more realistic. Here HS-DSCH will be used to deliver voice IP packets to the UE. The potential benefits in capacity can be huge depending on the vocoder used, channel conditions, users voice activity factor, header compression, congestion, receiver structures, and so on. A successful IP multimedia subsystem deployment should have at least the following characteristics:

- The circuit-switched voice service must be replaced with IP voice.
- VoIP quality should be as good as the circuit switched.
- Consistent delivery of performance.
- Interoperable IP services.
- Emergency call support.
- Capacity at least as good as circuit switched.
- Lower deployment costs.
- Flexible upgrading.
- Multimedia services (Voice/Video/Data).
- Low latency.

[2] We used the common terminology of femto-cell to represent a very small size cell. The cell size would be smaller than the much more familiar term pico-cell.

9.5 MIMO TECHNIQUES

In this section, we will present a multiple-input-multiple-output (MIMO) system. MIMO systems can be considered as extensions to the "Smart Antenna" techniques commonly deployed today. With a MIMO system a number of transmit antennas are used to transmit data streams to a number of receive antennas. Our goal is to jointly estimate and combine the spatial streams in their respective ways in order to improve overall system performance (i.e., BLER, Throughput, and so on). This can be viewed as having a system with joint transmit and receiver diversity gain [38–40].

In Fig. 9.23 we present a MIMO system using three transmit and three receive antennas. The spatial multiplexing of the data streams are highlighted by the solid and dashed lines representing the multipath.

What we can see is that using MIMO is alternatively represented as transmitting data over a matrix channel instead of a vector channel. In the case where scalar weights are used in the receiver, it assumes a flat fading channel is encountered. Otherwise creating a MIMO equalizer architecture can be considered an extension and will be discussed latter.

It has been shown by many authors that there exists a tremendous gain to be had when using MIMO technology. One approach to quantify these gains is through the use of information theory. The best way to approach this is to consider the scalar channel first. For this case the system capacity is given as [38]

$$C_1 = \log_2[1 + \text{SNR} \cdot |h|^2] \qquad \text{(Bps/Hz)} \tag{9.25}$$

where h is the channel response and SNR is the signal-to-noise ratio at the receiver antenna. Note that this type of channel is commonly called SISO, single-input-single-output. As we increase the number of receiver antennas the system capacity increases and thus we have the following assuming M receive antennas.

$$C_2 = \log_2\left[1 + \text{SNR} \cdot \sum_{i=1}^{M} |h_i|^2\right] \qquad \text{(Bps/Hz)} \tag{9.26}$$

Note this type of channel is commonly called SIMO, single-input-multiple-output. The performance gain shows up as an increase logarithmically in the average capacity. The flip side to this previous example is a system that uses N transmit antennas.

$$C_3 = \log_2\left[1 + \frac{\text{SNR}}{N} \cdot \sum_{i=1}^{N} |h_i|^2\right] \qquad \text{(Bps/Hz)} \tag{9.27}$$

Now this type of channel is commonly called MISO, multiple-input-single-output. Normalizing by N forces the total transmit power to be fixed.

Next we discuss the famous capacity equation, where N is the number of transmit antennas and M is the number of receive antennas.

$$C_4 = \log_2\left[\det\left(I_M + \frac{\text{SNR}}{N} \cdot H \cdot H^*\right)\right] \qquad \text{(Bps/Hz)} \tag{9.28}$$

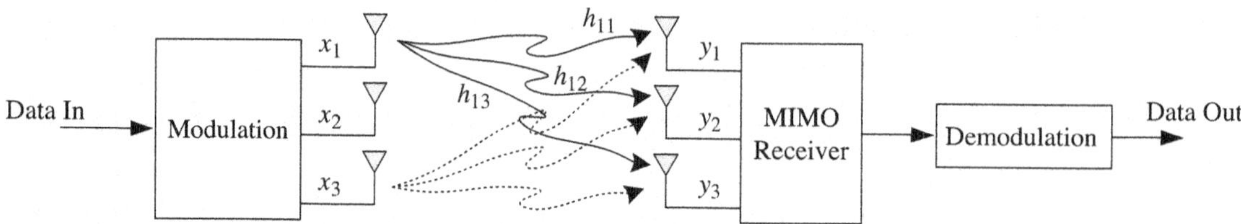

FIGURE 9.23 MIMO transmission example.

where H is an $M \times N$ channel matrix and I_m is the identity matrix of size $M \times M$. This can be rewritten below assuming $K = \min(M, N)$

$$C_4 = \sum_{j=1}^{K} \log_2\left[1 + \frac{\text{SNR}}{N} \cdot \lambda_j\right] \qquad \text{(Bps/Hz)} \tag{9.29}$$

where $\lambda_1, \lambda_2, \cdots, \lambda_K$ are the K nonzero eigenvalues of the matrix formed by $H \cdot H^*$ given the number of receive antennas is less than the number of transmit antennas. Note this matrix is a Wishart matrix. It is well known that the MIMO capacity grows linearly with K. We have provided some statistical properties of the eigenvalues in the earlier chapters when STE receivers were discussed. There we noticed the eigenvalues decreased rapidly. Applying that observation to the present topic, we notice the linear gain can also decrease rapidly on the distribution of λ_j's [41].

If we can take a temporary step backward to one earlier comment made about the channel resembling a matrix, then the following can be said. The rank of the MIMO channel is equal to the rank of the $M \times N$ channel matrix. Based on linear algebra, the rank is less than the number of transmit and receiver antennas. The significance of this rank is that it defines the number of independent streams that can be transmitted through the MIMO channel. As we will soon see MIMO transmission has similarities to what was presented earlier for OFDM regarding the transmission of parallel channels.

Let us provide the mathematical representation for a particular MIMO equalizer architecture where a linear equalizer is used on each receive antenna. Assuming a single transmit and receive antenna then one can write the received signal as

$$\begin{bmatrix} y(k) \\ \vdots \\ y_p(k) \\ \vdots \\ y(k-E+1) \\ \vdots \\ y_p(k-E+1) \end{bmatrix} = \begin{bmatrix} h(0) & \cdots & h(L-1) & \cdots & 0 \\ \vdots & & \vdots & & \\ h_p(0) & \cdots & h_p(L-1) & & \\ & & \ddots & & \\ \vdots & & h(0) & \cdots & h(L-1) \\ & & \vdots & & \vdots \\ 0 & \cdots & h_p(0) & \cdots & h_p(L-1) \end{bmatrix} \cdot \begin{bmatrix} x(k) \\ x(k-1) \\ \\ \vdots \\ \\ \\ x(k-E-L+2) \end{bmatrix} + \begin{bmatrix} n(k) \\ \\ \\ \vdots \\ \\ \\ n_p(k-E+1) \end{bmatrix} \tag{9.30}$$

where N is the number of transmit antennas, M is the number of receive antennas, P is the oversampling factor, L is the delay spread of the channel, and E is the equalizer duration. We can rewrite the above equation in the following vector notation:

$$\underline{y}(k) = \underline{H}(k) \cdot \underline{x}(k) + \underline{n}(k) \tag{9.31}$$

where $\underline{y}(k)$ is the received signal vector of size $PE \times 1$, $\underline{H}(k)$ is the channel matrix of size $PE \times (E + L - 1)$, $\underline{x}(k)$ is the transmitted signal vector of size $(E + L - 1) \times 1$, and $\underline{n}(k)$ is the noise vector of size $PE \times 1$.

Now if we consider N transmit and M receive antennas then we have the following:

$$\begin{bmatrix} \underline{y}_1(k) \\ \vdots \\ \underline{y}_M(k) \end{bmatrix} = \begin{bmatrix} H_{1,1} & \cdots & H_{1,N} \\ \vdots & \ddots & \vdots \\ H_{M,1} & \cdots & H_{M,N} \end{bmatrix} \cdot \begin{bmatrix} \underline{x}_1(k) \\ \vdots \\ \underline{x}_N(k) \end{bmatrix} + \begin{bmatrix} \underline{n}_1(k) \\ \vdots \\ \underline{n}_M(k) \end{bmatrix} \tag{9.32}$$

Which is rewritten as

$$\underline{\tilde{y}}(k) = \tilde{H}(k) \cdot \underline{\tilde{x}}(k) + \underline{\tilde{n}}(k) \tag{9.33}$$

Now the received signal vector is of size $PEM \times 1$, the channel matrix is of size $PEM \times N(E + L - 1)$, the transmitted signal vector is of size $N(E + L - 1) \times 1$, and the noise vector is of size

PEM × 1. The system designer would create a covariance matrix that needs inversion to be used in the MMSE weight calculation.

Generally speaking, the MIMO technique can be applied across any radio access technology (TDMA, CDMA, OFDMA). With respect to UMTS, MIMO is being used for HSDPA (specifically HS-DSCH physical channel) in order to increase the peak throughput data rates. Here data rates on the order of 20 Mbps can be expected when MIMO is used with HOM. In this application the architecture is a 2 × 2 MIMO, there can be a maximum of two spatial data streams, that is, primary and secondary. The NodeB HS-DSCH transmission builds upon the Closed Loop Transmit Diversity (CLTD) technique used in the previous releases (see Fig. 9.24).

The first two weights (e.g., w_1 and w_2) are actually signaled back to the NodeB whereas the last two weights can be derived from the first two. This reduces the uplink signaling required for the UE to adjust the NodeB transmit antenna weights. The two data streams will most likely have different modulation schemes operating on them along with differing FEC code rates. It is then expected that the UE would maintain the CQI for each of the data streams and communicate CQI values to the NodeB so relative adjustments can be made.

The HS-SCCH is also required for proper demodulation of the HS-DSCH channels. STTD is used for the HS-SCCH physical channels. Since more transmit and receive antennas are intended to be used then the assumption of independent fading across all antennas should be revisited to include more realism in the antenna modeling.

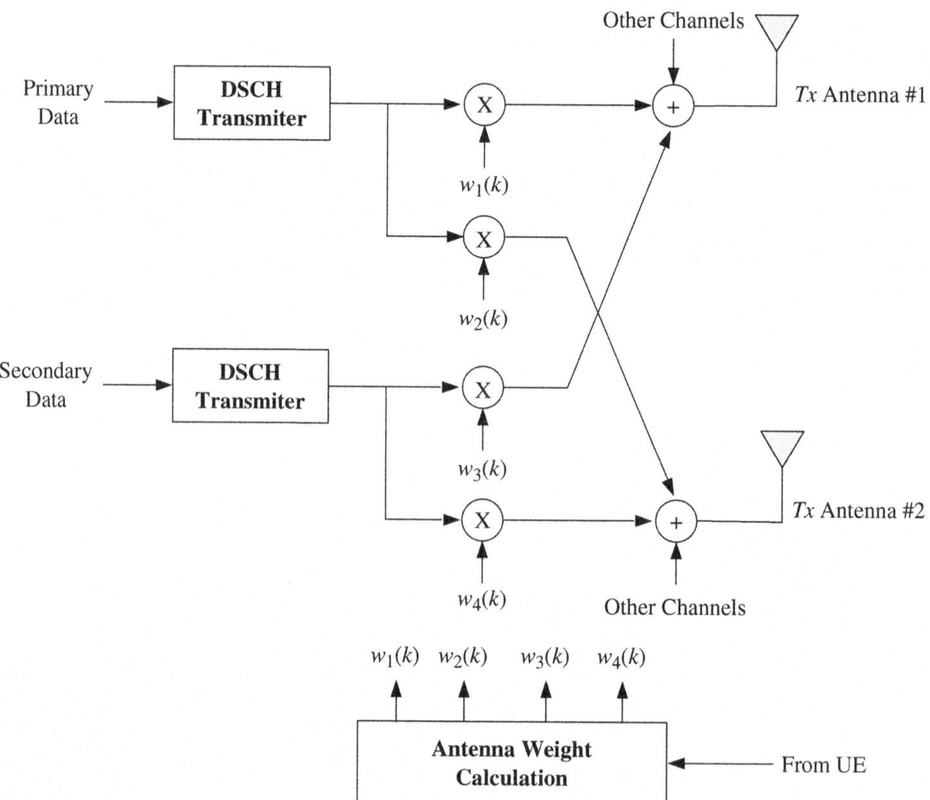

FIGURE 9.24 UMTS 2 × 2 MIMO example for the NodeB transmitter.

Receiver complexity and channel models are also other issues deserving special attention. Much work has been done on creating Spatial Channel Models (SCM) that would more accurately model the wireless communication channel. The SCM contains parameters such as angle of arrival (AOA), angle of departure (AOD), angular spread (AS) which are in addition to the already discussed parameters of path loss, delay spread, shadowing, Doppler spread, and so on.

For example, the AS with respect to the NodeB is discussed next. The antenna pattern, $A(\theta)$ is provided below assuming a three-sector deployment.

$$A(\theta) = -\min\left[12\left(\frac{\theta}{\theta_{3\,\mathrm{dB}}}\right), A_m\right] \qquad (-180° \le \theta \le 180°) \tag{9.34}$$

where θ is the angle between the direction of interest and the antenna boresight, $\theta_{3\ \mathrm{dB}}$ is the 3 dB beamwidth assumed to be 70°, and A_m is the maximum attenuation assumed to 20 dB.

The angular spread, σ, is referenced in the azimuth plane and this phenomenon is sometimes referred to as the Power Azimuth Spectrum (PAS). This typically is represented as a Laplacian distribution and is given below for sake of completion.

$$\mathrm{PAS} = C \cdot e^{\frac{-\sqrt{2}|\theta-\theta_D|}{\sigma_{\mathrm{RMS}}}} \cdot 10^{\frac{A(\theta)}{10}} \tag{9.35}$$

where θ_D is the angle of departure and we have also included the NodeB antenna gain in the equation. Angular spread values that are typically used for initial testing are on the order of 8° or so.

REFERENCES

[1] Digital Audio Broadcasting (DAB) Standard.

[2] European Telecommunications Standards Institute (ETSI), EN 300 744 v1.5.1: "Digital Video Broadcasting (DVB): Framing Structure, Channel Coding, and Modulation for Digital Terrestrial Television."

[3] Wireless LAN Standard, 802.11.

[4] WiMax 802.16 Standard.

[5] 3GPP, Technical Specification 26.201, "Long Term Evolution (LTE) Physical Layer – General Description," Release 8.

[6] R. Wang, J. G. Proakis, E. Masry, and J. R. Zeidler, "Performance Degradation of OFDM Systems Due to Doppler Spreading," *IEEE Transactions on Wireless Communications*, Vol. 5, No. 6, June 2006, pp. 1422–1432.

[7] B. Ai, Z. Yang, C. Pan, J. Ge Y. Wang, and Z. Lu, "On the Synchronization Techniques for Wireless OFDM Systems," *IEEE Transactions on Broadcasting*, Vol. 52, No. 2, June 2006, pp. 236–244.

[8] J. Liu and J. Li, "Packet Design and Signal Processing for OFDM-Based Mobile Broadband Wireless Communication Systems," *IEEE Transactions on Mobile Computing*, Vol. 5, No. 9, Sept. 2006, pp. 1133–1142.

[9] L. J. Cimini, Jr., "Analysis and Simulation of a Digital Mobile Channel Using Orthogonal Frequency Division Multiplexing," *IEEE Transactions on Communications*, Vol. COM-33, No. 7, July 1985, pp. 665–675.

[10] R. Kalbasi, S. Lu, and N. Al-Dhahir, "Receiver Design for Mobile OFDM with Application to DVB-H," in: *IEEE Vehicular Technology Conference*, 2006, pp. 1–5.

[11] R. Prasad, *OFDM for Wireless Communications Systems*, Artech House, Inc., 2004, Boston.

[12] M. Noh, Y. Lee, and H. Park, "Low Complexity LMMSE Channel Estimation for OFDM," *IEEE Proceedings on Communications*, Vol. 153, No. 5, Oct. 2006, pp. 645–650.

[13] O. Edfors, M. Sandell, J.-J. van de Beek, S. K. Wilson, and P. O. Borjesson, "OFDM Channel Estimation by Singular Value Decomposition," *IEEE Transactions on Communications*, Vol. 46, No. 7, July 1998, pp. 931–939.

[14] S. Kim and G. J. Pottie, "Robust OFDM in Fast Fading Channels," in: *IEEE Globecom*, 2003, pp. 1074–1078.

[15] P. Lancaster and M. Tismenetsky, *The Theory of Matrices*, Academic Press, 1985, California.

[16] B. R. Hunt, "A Matrix Theory Proof of the Discrete Convolution Theorem," *IEEE Transactions in Audio and Electroacoustics*, Vol. AU-19, No. 4, Dec. 1971, pp. 285–288.

[17] H. Holma and A. Toskala, *HSDPA/HSUPA for UMTS*, John Wiley & Sons, 2006, England.

[18] H. Ekstrom et al., "Technical Solutions for the 3G Long-Term Evolution," *IEEE Communications Magazine*, March 2006, pp. 38–45.

[19] D. Falconer, S. L. Ariyavisitakul, A. B. Seeyar, and B. Eidson, "Frequency Domain Equalization for Single Carrier Broadband Wireless Systems," *IEEE Communications Magazine*, April 2002, pp. 58–66.

[20] A. Czylwik, "Comparison Between Adaptive OFDM and Single Carrier Modulation with Frequency Domain Equalization," in: *IEEE Vehicular Technology Conference*, 1997, pp. 865–869.

[21] R. J. Crinon, D. Bhat, D. Catapano, G. Thomas, J. T. Van Loo, and G. Bang, "Data Broadcasting and Interactive Television," *IEEE Proceedings*, Vol. 94, No. 1, Jan. 2006, pp. 102–118.

[22] E. de Diego Balaguer, F. H. P. Fitzek, O. Olsen, and M. Gade, "Performance Evaluation of Power Strategies for DVB-H Services using Adaptive MPE-FEC Decoding," in: *IEEE PIMRC*, 2005, pp. 2221–2226.

[23] M. Kornfeld and U. Reimers, "DVB-H: The Emerging Standard for Mobile Data Communication," *EBU Technical Review*, No. 301, Jan. 2005, pp. 1–10.

[24] J. C. Kim and J. Y. Kim, "Single Frequency Network Design of DVB-H (Digital Video Broadcasting – Handheld) System," in: *ICACT*, 2006, pp. 1595–1598.

[25] G. Faria, J. A. Henriksson, E. Stare, and P. Talmola, "DVB-H: Digital Broadcast Services to Handheld Devices," *Proceedings of the IEEE*, Vol. 94, No. 1, Jan. 2006, pp. 194–209.

[26] European Telecommunications Standards Institute (ETSI), EN 302 304 v1.1.1, "Digital Video Broadcasting (DVB): Transmission System for Handheld Terminals (DVB-H)."

[27] www.digitalfountain.com, The Raptor Code Reference.

[28] M. Alard and R. Lassale, "Principles of Modulation and Channel Coding for Digital Broadcasting for Mobile Receivers," *EBU Technical Review*, No. 224, Aug. 1987, pp. 168–190.

[29] A. Boni, E. Launay, T. Mienville, and P. Stuckmann, "Multimedia Broadcast Multicast Service – Technology Overview and Service Aspects," in: *IEEE International Conference on 3G Mobile Communication Technologies*, 2004, pp. 634–638.

[30] S. Parkvall et al., "Evolving 3G Mobile Systems: Broadband and Broadcast Services in WCDMA," *IEEE Communications Magazine*, Feb. 2006, pp. 68–74.

[31] H. Holma and A. Toskala, *WCDMA for UMTS*, John Wiley & Sons, 2001, England.

[32] www.floforum.com, The MediaFLO web site.

[33] M. R. Chari et al. "FLO Physical Layer: An Overview," *IEEE Transactions on Broadcasting*, Vol. 53, No. 1, Mar. 2007, pp. 145–160.

[34] I. Held, O. Klein, A. Chen, C.-Y. Huang, and V. Ma, "Receiver Architecture and Performance of WLAN/Cellular Multi-Mode and Multi-Standard Mobile Terminals," in: *IEEE Vehicular Technology Conference*, 2004, pp. 2248–2253.

[35] R. Bagheri et al., "Software-Defined Radio Receiver: Dream to Reality," *IEEE Communications Magazine*, Vol. 44, Aug. 2006, pp. 111–118.

[36] F. Hartung, U. Horn, J. Huschke, M. Kampmann, T. Lohmar, and M. Lundevall, "Delivery of Broadcast Services in 3G Networks," *IEEE Transactions on Broadcasting*, Vol. 53, No. 1, Mar. 2007, pp. 188–199.

[37] H. Holma et al., "VOIP over HSDPA with 3GPP Release 7," in: *IEEE PIMRC*, 2006.

[38] D. Gesbert, M. Shafi, D. Shiu, P. J. Smith, and A. Naguib, "From Theory to Practice: An Overview of MIMO Space-Time Coded Wireless Systems," *IEEE Journal on Selected Areas in Communications*, Vol. 21, No. 3, April 2003, pp. 281–302.

[39] G. J. Foschini, "Layered Space-Time Architecture for Wireless Communication in a Fading Environment When Using Multi-Element Antennas," *Bell Labs Technical Journal*, 1996, pp. 41–59.

[40] A. Hottinen et al., "Industrial Embrace of Smart Antennas and MIMO," *IEEE Wireless Communications*, Aug. 2006, pp. 8–16.

[41] www.modern-wireless.com.

APPENDIX A
USEFUL FORMULAS

In this section, we will present several useful formulas typically used in digital communications system design. Characterizing the system performance typically involves evaluating integrals that contain the following forms [1–10].

The most popular integral is called the Gaussian Q function and can be written as

$$Q(x) = \int_x^\infty \frac{1}{2\pi} \cdot e^{\frac{-y^2}{2}} dy$$

It has the following properties:

$$\lim_{x \to \infty} Q(x) = 0$$

$$\lim_{x \to -\infty} Q(x) = 1$$

$$Q(0) = \frac{1}{2}$$

$$Q(-x) = 1 - Q(x)$$

Another representation is given as

$$Q(x) = \frac{1}{\pi} \int_0^{\pi/2} \exp\left[-\frac{x^2}{2 \cdot \sin^2\theta}\right] d\theta \qquad (x \geq 0)$$

This is related to the error function $\mathrm{erf}(x)$ and complementary error function $\mathrm{erfc}(x)$ as shown below. Recall $\mathrm{erf}(0) = 0$ and $\mathrm{erf}(\infty) = 1$.

$$Q(x) = \frac{1}{2} \cdot \mathrm{erfc}\left(\frac{x}{\sqrt{2}}\right) = \frac{1}{2} \cdot \left[1 - \mathrm{erf}\left(\frac{x}{\sqrt{2}}\right)\right] \qquad (x \geq 0)$$

$$\mathrm{erf}(x) = \frac{2}{\sqrt{\pi}} \cdot \int_0^x \exp(-t^2)\, dt$$

$$\mathrm{erfc}(x) = \frac{2}{\sqrt{\pi}} \cdot \int_x^\infty \exp(-t^2)\, dt = 1 - \mathrm{erf}(x)$$

Consider a normal variable y with mean m and variance σ^2; then the following relationship holds true:

$$P(y > z) = Q\left(\frac{z - m}{\sigma}\right)$$

A well-known upper bound is the Chernoff bound.

$$Q(x) \leq \frac{1}{2} \cdot \exp\left(-\frac{x^2}{2}\right)$$

In addition, we have the following lower and upper bounds:

$$Q(x) > \frac{1}{\sqrt{2\pi x}} \cdot \left(1 - \frac{1}{x^2}\right) \cdot \exp\left(\frac{-x^2}{2}\right)$$

$$Q(x) < \frac{1}{\sqrt{2\pi x}} \cdot \exp\left(\frac{-x^2}{2}\right)$$

Various approximations exist for the Q- function; below we list one to assist in obtaining numerical values. The reader should consult the references for other interesting expressions.

$$Q(y) \cong (ax + bx^2 + cx^3 + dx^4 + fx^5) \cdot \exp\left(\frac{-y^2}{2}\right) \qquad (y \geq 0)$$

along with the following definitions:

$$x = \frac{1}{1 + 0.231641888 \cdot y}$$

$$a = 0.127414796$$

$$b = -0.142248368$$

$$c = 0.7107068705$$

$$d = -0.7265760135$$

$$f = 0.5307027145$$

Often it is beneficial to have simple bounds in order to gain instant feedback and insight into performance. The Marcum Q-function is generally found in the analysis of communication systems. The generalized Marcum Q-function is defined by the following integral [1]:

$$Q_M(a,b) = \frac{1}{a^{M-1}} \cdot \int_b^\infty x^M \cdot \exp\left[-\left(\frac{x^2 + a^2}{2}\right)\right] \cdot I_{M-1}(ax)\,dx$$

with the modified Bessel function of the kth order expressed by the following integral:

$$I_k(x) = \frac{1}{2\pi} \cdot \int_{-\pi}^{\pi} \left(-j \cdot e^{-j\theta}\right)^k \cdot e^{-x\sin\theta}\,d\theta$$

The special case of the generalized Marcum Q-function is for the $M = 1$ case.

$$Q(a,b) = \int_b^\infty x \cdot \exp\left(-\frac{x^2 + a^2}{2}\right) \cdot I_0(ax)\,dx$$

with the modified zero-order Bessel function.

$$I_0(x) = \frac{1}{\pi} \cdot \int_0^{\pi} \exp(x \cdot \cos\theta)\, d\theta$$

There has been extensive work done to compare bounds for the Marcum Q-function. We provide a summary of the work presented in the reference section. We will first present the upper bound, assuming $(b \geq a)$ and using the following Bessel relationship $I_0(ax) \leq \exp(ax)$ in the region of $x \geq 0$ [2]:

$$Q(a,b) \leq \frac{I_0(ab)}{\exp(ab)} \cdot \left\{ \exp\left[-\frac{(b-a)^2}{2} \right] + a \cdot \sqrt{\frac{\pi}{2}} \cdot \operatorname{erfc}\left(\frac{b-a}{\sqrt{2}} \right) \right\}$$

$$Q(a,b) \leq \frac{b}{b-a} \cdot \exp\left[-\frac{(b-a)^2}{2} \right]$$

$$Q(a,b) \leq \exp\left(-\frac{a^2+b^2}{2} \right) \cdot I_0(ab) + a \cdot \sqrt{\frac{\pi}{8}} \cdot \operatorname{erfc}\left(\frac{b-a}{\sqrt{2}} \right)$$

$$Q(a,b) \leq \exp\left[-\frac{(b-a)^2}{2} \right]$$

Next we will present the lower bound, assuming $(b \geq a)$ and using the following Bessel relationship $I_0(x) \geq \frac{I_0(b)}{\exp(b)} \cdot \frac{\exp(x)}{x}$ in the region of $x \geq b$ [2]:

$$Q(a,b) \geq \frac{I_0(ab) \cdot b}{\exp(ab)} \cdot \sqrt{\frac{\pi}{2}} \cdot \operatorname{erfc}\left(\frac{b-a}{\sqrt{2}} \right)$$

$$Q(a,b) \geq \frac{b}{b+a} \cdot \exp\left[-\frac{(b+a)^2}{2} \right]$$

$$Q(a,b) \geq \exp\left(-\frac{a^2+b^2}{2} \right) \cdot I_0(ab)$$

$$Q(a,b) \geq \exp\left[-\frac{(b+a)^2}{2} \right]$$

For $b < a$, the following holds true:

$$Q(a,b) \geq 1 - \frac{1}{2} \cdot \left[\exp\left(-\frac{(b-a)^2}{2} \right) - \exp\left(-\frac{(b+a)^2}{2} \right) \right]$$

For large argument values, the following can be used for approximations:

$$I_0(x) \cong \frac{\exp(x)}{\sqrt{2\pi x}}$$

$$\operatorname{erfc}(x) \cong \frac{\exp(-x^2)}{\sqrt{\pi}x}$$

Some other useful variations to the above Marcum Q-function are given below.

$$Q(a,b) + Q(b,a) = 1 + \exp\left[-\frac{a^2 + b^2}{2}\right] \cdot I_0(ab)$$

$$Q_m(a,b) = Q(a,b) + \exp\left(-\frac{a^2 + b^2}{2}\right) \cdot \sum_{k=1}^{m-1} \left(\frac{b}{a}\right)^k I_k(ab) \qquad (m \geq 2)$$

$$1 - Q_m(a,b) = Q(b,a) - \exp\left(-\frac{a^2 + b^2}{2}\right) \cdot \sum_{k=0}^{m-1} \left(\frac{b}{a}\right)^k \cdot I_k(ab)$$

Next we will present various forms of the probability of bit error mathematical equations for some modulation schemes. The equations will consist of exact, approximate, and bounds in order to give the reader tools to quickly calculate the probability of error.

First we will provide a summary for the ***M*-ary PSK (MPSK)** modulation schemes. The exact form of the Symbol Error Probability for MPSK is given as

$$P_{\text{MPSK}} = \frac{1}{\pi} \cdot \int_0^{\pi - \pi/M} \exp\left[-\frac{\rho \cdot \sin^2\left(\frac{\pi}{M}\right)}{\sin^2(\theta)}\right] \cdot d\theta$$

where $\rho = \frac{E_b}{N_o} \cdot \log_2 M$.

A reasonable approximation exists for $M > 2$ in an AWGN channel given as

$$P_b \cong \frac{2}{\log_2 M} \cdot Q\left(\sqrt{\frac{2E_b \log_2 M}{N_o}} \cdot \sin\left[\frac{\pi}{M}\right]\right)$$

$$P_b \cong \frac{1}{\log_2 M} \cdot \left\{1 - \operatorname{erf}\left(\sqrt{\frac{E_b \log_2 M}{N_o}} \cdot \sin\left[\frac{\pi}{M}\right]\right)\right\}$$

Recall for QPSK, the error rate in an AWGN channel is given in the following forms:

$$P_b = Q\left(\sqrt{\frac{2E_b}{N_o}}\right) = \frac{1}{2}\operatorname{erfc}\left(\sqrt{\frac{E_b}{N_o}}\right) = \frac{1}{2} \cdot \left\{1 - \operatorname{erf}\left(\sqrt{\frac{E_b}{N_o}}\right)\right\}$$

Also the QPSK error rate in a flat Rayleigh fading channel is given as

$$P_b = \frac{1}{2} \cdot \left[1 - \sqrt{\frac{\frac{E_b}{N_o}}{\frac{E_b}{N_o} + 1}}\right]$$

Next we will provide a summary for the ***M*-ary DPSK (MDPSK)** modulation schemes. The Symbol Error Probability for MDPSK in an AWGN channel is given as

$$P_{\text{MDPSK}} = \frac{1}{\pi} \cdot \int_0^{\pi - \pi/M} \exp\left[-\frac{\rho \cdot \sin^2\left(\frac{\pi}{M}\right)}{1 + \cos\left(\frac{\pi}{M}\right) \cdot \cos(\theta)}\right] \cdot d\theta$$

Coherent detection of DPSK error rate in an AWGN channel is provided by the following:

$$P_b = \mathrm{erfc}\sqrt{\frac{E_b}{N_o}} \cdot \left[1 - \frac{1}{2} \cdot \mathrm{erfc}\left(\sqrt{\frac{E_b}{N_o}}\right)\right]$$

$$P_b = 2Q\left(\sqrt{\frac{2E_b}{N_o}}\right) \cdot \left[1 - Q\left(\sqrt{\frac{2E_b}{N_o}}\right)\right]$$

For coherent detection of DQPSK, it is given as

$$P_b = \mathrm{erfc}\left(\sqrt{\frac{E_b}{2N_o}}\right) - \mathrm{erfc}^2\left(\sqrt{\frac{E_b}{2N_o}}\right) + \frac{1}{2} \cdot \mathrm{erfc}^3\left(\sqrt{\frac{E_b}{2N_o}}\right) - \frac{1}{8} \cdot \mathrm{erfc}^4\left(\sqrt{\frac{E_b}{2N_o}}\right)$$

The general Bit Error Probability of coherent detection of MDPSK is expressed as

$$P_b = \frac{K}{\log_2 M} \cdot \mathrm{erfc}\left[\sqrt{\frac{E_b \log_2 M}{N_o}} \cdot \sin\left(\frac{\pi}{M}\right)\right] \cdot \left\{1 - \frac{1}{2} \cdot \mathrm{erfc}\left[\sqrt{\frac{E_b \log_2 M}{N_o}} \cdot \sin\left(\frac{\pi}{M}\right)\right]\right\}$$

where $K = 1$ for $M = 2$ and $K = 2$ for $M \geq 4$.

Noncoherent detection of DQPSK is given as

$$P_b = \frac{1}{2} \cdot I_o\left(\sqrt{\frac{2E_b}{N_o}}\right) \cdot e^{-\frac{2E_b}{N_o}} + e^{-\frac{2E_b}{N_o}} \cdot \sum_{k=1}^{\infty} (\sqrt{2} - 1)^k \cdot I_k\left(\sqrt{\frac{2E_b}{N_o}}\right)$$

Noncoherent detection of DPSK is given as

$$P_b = \frac{1}{2} \cdot e^{-\frac{E_b}{N_o}}$$

Noncoherent detection of MDPSK is given as

$$P_s \cong 2 \cdot Q\left(\sqrt{\frac{2E_b \log_2 M}{N_o}} \cdot \sin\left[\frac{\pi}{M\sqrt{2}}\right]\right)$$

The general noncoherent detection form is

$$P_b = Q(a\sqrt{\gamma}, b\sqrt{\gamma}) - \frac{1}{2} \cdot \exp\left(-\frac{(a^2 + b^2)\gamma}{2}\right) \cdot I_o(ab\gamma)$$

where for DPSK

$$a = 0 \qquad b = \sqrt{2}$$

and for DQPSK

$$a = \sqrt{2 - \sqrt{2}} \qquad b = \sqrt{2 + \sqrt{2}}$$

Next we will provide a summary for the ***M*-ary QAM (MQAM)** modulation schemes. The bit error rate for 16-QAM in an AWGN channel is given as

$$P_b = \frac{3}{8} \cdot \mathrm{erfc}\left(\sqrt{\frac{0.4E_b}{N_o}}\right) \cdot \left[1 - \frac{3}{8} \cdot \mathrm{erc}\left(\sqrt{\frac{0.4E_b}{N_o}}\right)\right]$$

Alternatively represented as

$$P_b = \frac{3}{4} \cdot Q\left(\sqrt{\frac{4E_b}{5N_o}}\right) + \frac{1}{2} \cdot Q\left(3\sqrt{\frac{4E_b}{5N_o}}\right) - \frac{1}{4} \cdot Q\left(5\sqrt{\frac{4E_b}{5N_o}}\right)$$

$$P_b = \frac{3}{8} \cdot \text{erfc}\left(\sqrt{\frac{0.4E_b}{N_o}}\right) - \frac{9}{64} \cdot \text{erfc}^2\left(\sqrt{\frac{0.4E_b}{N_o}}\right)$$

For square MQAM constellation with M = even number, we can use the following:

$$P_b = \frac{2}{\log_2 M} \cdot \left\{\frac{\sqrt{M} - 1}{\sqrt{M}}\right\} \cdot \left[1 - \text{erf}\left(\sqrt{\frac{E_b \log_2 M}{N_o}} \cdot \sqrt{\frac{3}{2(M - 1)}}\right)\right]$$

Also written as

$$P_b = \frac{4}{\log_2 M} \cdot \left\{\frac{\sqrt{M} - 1}{\sqrt{M}}\right\} \cdot Q\left(\sqrt{\frac{3E_b}{N_o} \cdot \frac{\log_2 M}{(M - 1)}}\right)$$

$$P_b = \frac{2}{\log_2 M} \cdot \left\{\frac{\sqrt{M} - 1}{\sqrt{M}}\right\} \cdot \text{erfc}\left(\sqrt{\frac{E_b \log_2 M}{2N_o(M - 1)}}\right) \cdot \left[1 - \left(\frac{\sqrt{M} - 1}{2\sqrt{M}}\right) \cdot \text{erfc}\left(\sqrt{\frac{3E_b \log_2 M}{2(M - 1)}}\right)\right]$$

For 64-QAM, we have the following:

$$P_b = \frac{4}{27} \cdot \text{erfc}\left(\sqrt{\frac{E_b}{7N_o}}\right) - \frac{49}{384} \cdot \text{erfc}^2\left(\sqrt{\frac{E_b}{7N_o}}\right)$$

For 256-QAM, we have the following:

$$P_b = \frac{15}{64} \cdot \text{erfc}\left(\sqrt{\frac{4E_b}{85N_o}}\right) - \frac{225}{2048} \cdot \text{erfc}^2\left(\sqrt{\frac{4E_b}{85N_o}}\right)$$

Next we will provide a summary for the ***M*-ary FSK (MFSK)** modulation schemes. Noncoherent detection of orthogonal signaling is given as

$$P_b \le \frac{M}{4} \cdot e^{-\frac{E_b \log_2 M}{2N_o}}$$

Here we have made use of the bit error rate and symbol error rate relationship

$$P_b = \frac{M}{2(M - 1)} \cdot P_s$$

which can also be represented as

$$P_s = \sum_{j=1}^{M-1} (-1)^{j+1} \cdot \frac{1}{j + 1} \cdot \frac{(M - 1)!}{j! \cdot (M - 1 - j)!} \cdot e^{-\frac{j}{j+1} \cdot \left(\frac{E_s}{N_o}\right)}$$

Noncoherent detection of BFSK is given as

$$P_b = \frac{1}{2} \cdot e^{-\frac{E_b}{2N_o}}$$

Coherent detection of orthogonal signaling is given as

$$P_b \leq \frac{M}{2} \cdot Q\left(\sqrt{\frac{E_b \log_2 M}{N_o}}\right)$$

$$P_b \leq \frac{M}{4} \cdot \left[1 - \operatorname{erf}\left(\sqrt{\frac{E_b \log_2 M}{2N_o}}\right)\right]$$

Coherent detection of BFSK is given as

$$P_b = Q\left(\sqrt{\frac{E_b}{N_o}}\right)$$

There has also been some work to provide a closed-form expression that takes into account the phase error in some practical coherent detection techniques. Below we list three commonly used expressions for the BER of BPSK, QPSK, and OQPSK modulation schemes, given a phase offset of ϕ.

$$P_{\text{BPSK}}(\phi) = Q\left(\sqrt{\frac{2E_b}{N_o}} \cdot \cos(\phi)\right)$$

$$P_{\text{QPSK}}(\phi) = \frac{1}{2} \cdot \left\{Q\left(\sqrt{\frac{2E_b}{N_o}} \cdot [\cos(\phi) + \sin(\phi)]\right) + Q\left(\sqrt{\frac{2E_b}{N_o}} \cdot [\cos(\phi) - \sin(\phi)]\right)\right\}$$

$$P_{\text{OQPSK}}(\phi) = \frac{1}{2} \cdot \left\{P_{\text{BPSK}}(\phi) + P_{\text{QPSK}}(\phi)\right\}$$

For phase references derived from PLL-based techniques, the phase-offset random variable can be represented as

$$f(\phi) = \frac{1}{2\pi \cdot I_o(\alpha)} \cdot \exp(\alpha \cos(\phi)) \qquad (-\pi \leq \phi \leq \pi)$$

where α is the SNR of the phase reference signal.

The probability of exactly m errors in a block of N bits is given as

$$\binom{N}{m} \cdot P_e^m \cdot (1 - P_e)^{N-m}$$

Then the probability of more than M errors within a block of N bits is given as

$$P(M, N) = \sum_{m=M+1}^{N} \binom{N}{m} \cdot P_e^m \cdot (1 - P_e)^{N-m}$$

$$P(M, N) = 1 - \sum_{m=0}^{M} \binom{N}{m} \cdot P_e^m \cdot (1 - P_e)^{N-m}$$

REFERENCES

[1] M. K. Simon, "A New Twist and the Marcum Q-Function and Its Application," *IEEE Communications Letters*, Vol. 2, No. 2, Feb. 1998, p. 39–41.

[2] G. Ferrari and G. E. Corazza, "Tight Bounds and Accurate Approximations for the BER of DQPSK Transmission from Novel Bounds on the Marcum Q-Function," in: *International Symposium on Information Theory and Its Applications*, Oct. 2004.

[3] G. E. Corazza and G. Ferrari, "New Bounds for the Marcum Q-Function," *IEEE Transactions on Information Theory*, Vol. 48, No. 11, Nov. 2002, pp. 3003–3008.

[4] M. K. Simon and M. S. Alouini, "Exponential-Type Bounds on the Generalized Marcum Q-Function with Application to Error Probability Analysis over Fading Channels," *IEEE Transactions on Communications*, Vol. 48, No. 3, March 2000, pp. 359–366.

[5] M. Chiani, "Integral Representation and Bounds For Marcum Q-function," *Electronic Letters*, March 1999, Vol. 35, No. 6, pp. 445–446.

[6] G. E. Corazza and G. Ferrari, "Tight Bounds and Accurate Approximations for DQPSK Transmission Bit Error Rate," *Electronic Letters*, Vol. 40, No. 20, Sept. 2004, pp.1284–1285.

[7] R. F. Pawula, "Generic Error Probabilities," *IEEE Transactions on Communications*, Vol. 47, No. 5, May 1999, pp. 697–702.

[8] M. K. Simon and M. S. Alouini, *Digital Communication over Fading Channels: A Unified Approach to Performance Analysis*, John Wiley & Sons, 2000, New York.

[9] W. H. Press, B. P. Flannery, S. A. Teukolsky, and W. T. Vetterling, *Numerical Recipes in C*, Cambridge University Press, 1990, New York.

[10] A. B. Carlson, *Communication Systems: An Introduction to Signals and Noise in Electrical Communication*, McGraw-Hill, 1986, New York.

APPENDIX B
TRIGONOMETRIC IDENTITIES

In this section, we present several useful trigonometric identities used in communication system design.

$$e^{\pm j\theta} = \cos(\theta) \pm j\sin(\theta)$$

$$\cos(A) = \frac{e^{jA} + e^{-jA}}{2}$$

$$\sin(A) = \frac{e^{jA} - e^{-jA}}{2j}$$

$$\cos(A) = \sin(A + 90°)$$

$$\sin(A) = \cos(A - 90°)$$

$$\cos^2(A) + \sin^2(A) = 1$$

$$\cos^2(A) - \sin^2(A) = \cos(2A)$$

$$2\cos^2(A) = 1 + \cos(2A)$$

$$2\sin^2(A) = 1 - \cos(2A)$$

$$\sin(2A) = 2\sin(A)\cos(A)$$

$$4\cos^3(A) = 3\cos(A) + \cos(3A)$$

$$4\sin^3(A) = 3\sin(A) - \sin(3A)$$

$$2\cos(A)\cos(B) = \cos(A + B) + \cos(A - B)$$

$$2\sin(A)\sin(B) = \cos(A - B) - \cos(A + B)$$

$$2\sin(A)\cos(B) = \sin(A - B) + \sin(A + B)$$

$$\sin(A)\sin(B) + \cos(A)\cos(B) = \cos(A - B)$$

$$\sin(A)\cos(B) + \cos(A)\sin(B) = \sin(A + B)$$

$$\tan\left[\frac{A + B}{2}\right] = \frac{\sin(A) + \sin(B)}{\cos(A) + \cos(B)}$$

$$\tan(A \pm B) = \frac{\tan(A) \pm \tan(B)}{1 \mp \tan(A)\tan(B)}$$

$$\sin(nA) = \text{Im}\{(\cos(A) + j\sin(A))^n\}$$

$$\cos(nA) = \text{Re}\{(\cos(A) + j\sin(A))^n\}$$

$$e = 2.718281828$$

$$\log_{10}e = 0.434294481$$

$$\ln 10 = 2.302585092$$

APPENDIX C
DEFINITE INTEGRALS

$$\int_{-\infty}^{\infty} e^{-x^2} dx = \sqrt{\pi}$$

$$\int_{-\infty}^{\infty} e^{\frac{-ax^2}{2}} dx = \sqrt{\frac{2\pi}{a}}$$

$$\int_{0}^{\infty} I_0(at) \cdot e^{-bt} dt = \frac{1}{\sqrt{b^2 - a^2}}$$

$$\int_{0}^{\infty} e^{-ax} \cdot I_o(2\sqrt{bx})\, dx = \frac{1}{a} \cdot e^{\frac{b}{a}}$$

$$\int_{0}^{\infty} x^2 \cdot \exp(-ax^2)\, dx = \frac{1}{4} \cdot \sqrt{\frac{\pi}{a^3}}$$

$$\int_{0}^{\infty} x \cdot \exp(-a^2x^2) \cdot \operatorname{erfc}(bx)\, dx = \frac{1}{2a^2} \cdot \left[1 - \frac{b}{\sqrt{a^2 + b^2}}\right]$$

$$\int_{0}^{\infty} e^{-ax} \cdot \cos(x) dx = \frac{a}{1 + a^2}$$

$$\int_{0}^{\infty} e^{-ax} \cdot \sin(x) dx = \frac{1}{1 + a^2}$$

$$\int_{0}^{\infty} e^{-a^2x^2} \cdot \cos(bx) dx = \frac{1}{2a}\sqrt{\pi} e^{-\left(\frac{b}{2a}\right)^2}$$

$$\int_{0}^{\infty} x^n e^{-ax} dx = \frac{n!}{a^{n+1}} \qquad (n \geq 1)$$

$$\int_{0}^{\infty} e^{-a^2x^2} dx = \frac{1}{2a}\sqrt{\pi}$$

$$\int_0^\infty x^2 e^{-x^2} dx = \frac{1}{4\sqrt{\pi}}$$

$$\int_0^z a^2 x e^{-ax} dx = 1 - (1 + az)e^{-az}$$

$$\int_{-\infty}^\infty x^2 e^{-bx^2} dx = \frac{1}{2}\frac{\sqrt{\pi}}{b\sqrt{b}}$$

$$\int_0^\infty \frac{\sin(x)}{x} dx = \frac{\pi}{2}$$

$$\int_0^\infty \frac{\tan(x)}{x} dx = \frac{\pi}{2}$$

$$\int_0^\infty \sin(x^2) dx = \frac{1}{2}\sqrt{\frac{\pi}{2}}$$

$$\int_0^\infty \cos(x^2) dx = \frac{1}{2}\sqrt{\frac{\pi}{2}}$$

$$\int_0^\infty \frac{\sin^2(x)}{x^2} dx = \frac{\pi}{2}$$

$$\int_0^\infty \frac{\cos(nx)}{1 + x^2} dx = \frac{\pi}{2} e^{-|n|}$$

$$\int_0^\infty \frac{\sin(\pi x)}{\pi x} dx = \frac{1}{2}$$

$$I_n(x) = \frac{1}{\pi} \cdot \int_0^\pi e^{x\cos(\theta)} \cdot \cos(n\theta) d\theta$$

$$I_n(x) = \sum_{k=0}^{\infty} \frac{\left(\frac{x}{2}\right)^{n+2k}}{k! \cdot \Gamma(n + k + 1)}$$

$$\int_0^\infty \exp(-xz) \cdot I_\nu(yz) dz = \frac{y^\nu}{\sqrt{x^2 - y^2}\,(x + \sqrt{x^2 - y^2})^\nu} \qquad (x > |y|)$$

$$\int_0^\infty \exp(-wz) \cdot Q(\sqrt{2pz}, \sqrt{2qz})\,dz = \frac{u + \sqrt{u^2 - v^2}}{\sqrt{u^2 - v^2}\,(u + \sqrt{u^2 - v^2} - 2p)} \qquad (q > p)$$

where $u = w + p + q$ and $v = 2\sqrt{pq}$.

$$\int_0^\infty x^{n-1} \cdot e^{-x}\,dx = \Gamma(n) \qquad (n > 0)$$

$$\Gamma(n + 1) = n \cdot \Gamma(n)$$

$$\Gamma\left(\frac{1}{2}\right) = \sqrt{\pi}$$

APPENDIX D
PROBABILITY FUNCTIONS

In this section, we list some of the well-known random-variable functions along with their mean value and variance.

Poisson

$$P_X(x) = e^{-K}\frac{K^x}{x!}$$

$$\bar{x} = K \qquad \sigma_x^2 = K$$

Binomial

$$P_X(x) = \binom{n}{x} \cdot a^x \cdot (1-a)^{n-x} \qquad (x = 0, 1, \ldots, n)$$

$$\bar{x} = na \qquad \sigma_x^2 = na(1-a)$$

Uniform

$$p_X(x) = \begin{cases} \dfrac{1}{b-a} & a < x < b \\ 0 & \text{otherwise} \end{cases}$$

$$\bar{x} = \frac{(a+b)}{2} \qquad \sigma_x^2 = \frac{(b-a)^2}{12}$$

Gaussian

$$p_X(x) = \frac{1}{\sigma\sqrt{2\pi}} e^{-\frac{(x-m)^2}{2\sigma^2}}$$

$$\bar{x} = m \qquad \sigma_x^2 = \sigma^2$$

Rayleigh

$$p_X(x) = \frac{x}{\sigma^2} \cdot e^{-\frac{x^2}{2\sigma^2}} \qquad (x \geq 0)$$

$$\bar{x} = \sigma\sqrt{\frac{\pi}{2}} \qquad \sigma_x^2 = \left(2 - \frac{\pi}{2}\right)\sigma^2$$

Exponential

$$p_X(x) = K \cdot e^{-Kx} \qquad (x \geq 0)$$

$$\bar{x} = \frac{1}{K} \qquad \sigma_x^2 = \frac{1}{K^2}$$

APPENDIX E
SERIES AND SUMMATIONS

$$\sin(x) = x - \frac{1}{3!}x^3 + \frac{1}{5!}x^5 - \frac{1}{7!}x^7 + \cdots$$

$$\cos(x) = 1 - \frac{1}{2!}x^2 + \frac{1}{4!}x^4 - \frac{1}{6!}x^6 + \cdots$$

$$\tan(x) = x + \frac{1}{3}x^3 + \frac{2}{15}x^5 - \frac{17}{315}x^7 + \cdots$$

$$(1 + x)^n = 1 + nx + \frac{n(n-1)}{2!}x^2 + \cdots$$

$$= 1 + \sum_{k=1}^{\infty} \frac{n(n-1)(n-2)\cdots(n-k+1)}{k!}x^k \qquad (x < 1)$$

$$e^x = 1 + x + \frac{1}{2!}x^2 + \frac{1}{3!}x^3 + \cdots$$

$$a^x = 1 + x\ln(a) + \frac{1}{2}[x\ln(a)]^2 + \cdots$$

$$\ln(1 + x) = x - \frac{1}{2}x^2 + \frac{1}{3}x^3 + \cdots$$

$$\frac{\sin(\pi x)}{\pi x} = 1 - \frac{1}{3!}(\pi x)^2 + \frac{1}{5!}(\pi x)^4 - \cdots$$

$$e = \sum_{k=0}^{\infty}\frac{1}{k!}$$

$$J_o(x) = \sum_{k=0}^{\infty}\frac{(-1)^k x^{2k}}{2^{2k}(k!)^2}$$

$$I_o(x) = \sum_{k=0}^{\infty}\frac{x^{2k}}{2^{2k}(k!)^2}$$

$$\sum_{k=1}^{K} k = \frac{K(K+1)}{2}$$

$$\sum_{k=1}^{K} k^2 = \frac{K(K+1)(2K+1)}{6}$$

$$\sum_{k=1}^{K} k^3 = \frac{K^2(K+1)^2}{4}$$

$$\sum_{k=0}^{K} x^k = \frac{x^{K+1} - 1}{x - 1}$$

$$\sum_{k=1}^{N} ax^{k-1} = \frac{a(1 - x^N)}{1 - x}$$

$$\sum_{k=1}^{\infty} ax^{k-1} = \frac{1}{1 - x}$$

Cauchy-Schwarz Inequality

$$\left(\sum_{k=1}^{N} a_k b_k\right)^2 \leq \left(\sum_{k=1}^{N} a_k^2\right) \cdot \left(\sum_{k=1}^{N} b_k^2\right)$$

Minkowski's Inequality

$$\left(\sum_{k=1}^{N} (a_k + b_k)^P\right)^{1/P} \leq \left(\sum_{k=1}^{N} a_k^P\right)^{1/P} + \left(\sum_{k=1}^{N} b_k^P\right)^{1/P}$$

Holder's Inequality

$$\sum_{k=1}^{N} a_k b_k \leq \left(\sum_{k=1}^{N} a_k^P\right)^{1/P} \cdot \left(\sum_{k=1}^{N} b_k^Q\right)^{1/Q} \quad \text{with} \left(\frac{1}{P} + \frac{1}{Q} = 1\right)$$

APPENDIX F
LINEAR ALGEBRA

Let A be an $n \times n$ matrix, then the following properties can be mentioned for determinants, where $\det[A] = |A|$.

$$|kA| = k^n|A|$$

$$|A^T| = |A|$$

$$|A||B| = |AB|$$

$$|A^{-1}| = \frac{1}{|A|}$$

$$(AB)^T = B^T A^T$$

$$(AB)^{-1} = B^{-1}A^{-1}$$

Let's consider the following relationship:

$$A = B - C$$

$$C^{-1} = (B - A)^{-1} = (I - B^{-1}A)^{-1}B^{-1} \cong (I + B^{-1}A)B^{-1}$$

$$C^{-1} \cong B^{-1} + B^{-1}AB^{-1} = B^{-1}(I + AB^{-1})$$

$$\det\begin{bmatrix} A & C \\ B & D \end{bmatrix} = \det(A) \cdot \det(D - CA^{-1}B)$$

$$\text{Trace}\,[A + B] = \text{Trace}\,[A] + \text{Trace}\,[B]$$

$$\text{Trace}\,[ABC] = \text{Trace}\,[BCA] + \text{Trace}\,[CAB]$$

$$\text{Trace}\,[UAU^*] = \text{Trace}\,[A] \qquad \text{if} \qquad U = \text{unitary matrix}$$

Inner product or dot product is given as

$$k = \underline{x}^* \cdot \underline{y}$$

Outer product is given as

$$A = \underline{x} \cdot \underline{y}^*$$

The inner and outer products are related as follows:

$$\underline{x}^* \cdot \underline{y} = \text{Trace}\,(\underline{x} \cdot \underline{y}^*)$$

Given a square matrix, A, we can write down the following relationships with λ = eigenvalue and $\underline{v}$ = eigenvector.

$$A \cdot \underline{v} = \lambda \cdot \underline{v}$$

$$\text{Trace }[A] = \sum_{i=0}^{M-1} \lambda_i$$

$$\det [A] = \prod_{i=0}^{M-1} \lambda_i$$

For Hermitian matrices, we provide the mathematical relationship between the arithmetic mean and the geometric mean.

$$\frac{1}{M} \cdot \sum_{i=0}^{M-1} \lambda_i \geq \left[\prod_{i=0}^{M-1} \lambda_i \right]^{1/M}$$

$$\frac{1}{M} \cdot \text{Trace }[A] \geq [\det [A]]^{1/M}$$

A square Hermitian matrix can be diagonalized as

$$A = U \cdot D \cdot U^*$$

where D = diagonal matrix, $\{\lambda_0, \lambda_1, \ldots, \lambda_{M-1}\}$, and U is a unitary matrix, $U^* \cdot U = I$ or $U^* = U^{-1}$.

We can define a Rayleigh quotient as

$$R(\underline{x}) = \frac{\underline{x}^* \cdot A \cdot \underline{x}}{\underline{x}^* \cdot \underline{x}}$$

The Frobenius or Euclidean norm is given as

$$\|A\|_F = \left(\sum_{i=0}^{M-1} \lambda_i \right)^{1/2}$$

INDEX

CPSIA information can be obtained
at www.ICGtesting.com
Printed in the USA

9 780071 489058